Modeling with Mathematics: A Bridge to Algebra II

Modeling with Mathematics: A Bridge to Algebra II

Teacher's Edition

Project Leadership

Jo Ann Wheeler
Region 4 ESC, Houston, TX

Solomon Garfunkel
COMAP Inc., Lexington, MA

David Eschberger
Region 4 ESC, Houston, TX

Roland Cheyney
COMAP Inc., Lexington, MA

Lead Authors

Gary Cosenza
Region 4 ESC, Houston, TX

Paul Gray
Region 4 ESC, Houston, TX

Julie Horn
Region 4 ESC, Houston, TX

Authors

Sharon Benson
Region 4 ESC, Houston, TX

Roland Cheyney
COMAP, Lexington, MA

David Eschberger
Region 4 ESC, Houston, TX

Jo Ann Wheeler
Region 4 ESC, Houston, TX

W. H. Freeman and Company
New York
www.whfreeman.com

Published and distributed by
W. H. Freeman and Company
41 Madison Avenue, New York NY 10010

ISBN 07167-6948-4 (EAN 9780716769484)

Printed in the United States of America
First Printing 2006

PUBLISHER: Craig Bleyer
ACQUISITIONS EDITOR: Cyndi WeissGoldner
MARKETING MANAGER: Mike Saltzman
SUPPLEMENTS AND NEW MEDIA EDITOR: Victoria Anderson
ASSOCIATE EDITOR: Brendan Cady
PRODUCTION MANAGER: George Ward
PRODUCTION DEPARTMENT: Daiva Kiliulis, Tim McLean, George Ward, Pauline Wright
PHOTO RESEARCHERS: COMAP Production
ILLUSTRATIONS: Lianne Dunn, George Ward
COVER AND TEXT DESIGN: Daiva Kiliulis
COVER IMAGES: Shutterstock
MANUFACTURING: RR Donnelley

Dear Teacher,

In *Principles & Standards for School Mathematics*, the National Council of Teachers of Mathematics (2000) states:

"In high school, students should build on their prior knowledge, learning more-varied and more-sophisticated problem-solving techniques. They should increase their abilities to visualize, describe, and analyze situations in mathematical terms. They need to learn to use a wide range of explicitly and recursively defined functions to model the world around them. Moreover, their understanding of the properties of those functions will give them insights in the phenomena being modeled." (p. 287)

In order to achieve this goal, the project team for *Modeling with Mathematics: A Bridge to Algebra II* uses functions-based mathematical modeling as a means of deepening high school students' mathematical understanding in this text.

"Modeling involves identifying and selecting relevant features of a real-world situation, representing those features symbolically, analyzing and reasoning about the model and the characteristics of the situation, and considering the accuracy and limitations of the model."

"High school students should study modeling in greater depth, generating or using data and exploring which kinds of functions best fit or model those data." (NCTM, 2000, p. 302)

To this end, engaging problems are presented with the intent that students will explore data and generate a function rule to best model the situation. Once the model is generated, students will apply their mathematical model to answer critical questions and solve meaningful, relevant problems related to the context. In doing so, students' understandings of algebraic and geometric concepts, procedures, and skills are strengthened.

Modeling with Mathematics: A Bridge to Algebra II is a series of lessons based on the 5E instructional model, which is an inquiry-based model used in science and mathematics instruction. Each chapter consists of a collection of 5E lessons. Each section within a chapter is one phase of the 5E lesson cycle. The Annotated Teacher's Edition contains notes to the teacher to provide guidance and flow within and between the 5E lesson cycles.

Modeling with Mathematics: A Bridge to Algebra II is designed to serve a course in mathematical modeling that falls in sequence after Algebra I and Geometry and before Algebra II. Students in this course will apply concepts typically covered in Algebra I and Geometry to solve meaningful real-world problems. In doing so, they will reinforce their algebra and geometry skills. For students that leave Algebra I with an unstable foundation in algebraic thinking, this reinforcement helps to solidify their mathematical understanding and empower them for success in Algebra II.

The remainder of the Annotated Teacher's Edition preface will explain key elements of the theoretical framework that is used in this textbook. A definition of mathematical modeling, descriptions of mathematical proficiency, descriptions and illuminations of the 5E instructional model, a few words concerning assessment, and a summary of key elements of cooperative learning will be discussed. These philosophies and instructional strategies were chosen because research shows that they have the most direct impact on student learning and student achievement.

We hope that you and your students enjoy your journey through mathematical modeling. As with other experiences, it is the passage rather than the destination that will reap the greater rewards.

Sincerely,

Jo Ann Wheeler
Region 4 ESC, Houston, TX

David Eschberger
Region 4 ESC, Houston, TX

Solomon Garfunkel
COMAP, Lexington, MA

Roland Cheyney
COMAP, Lexington, MA

Background and Philosophy for *Modeling with Mathematics: A Bridge to Algebra II*

What is mathematical modeling?

Mathematical modeling is the process by which students encounter a real-world situation, pose a question regarding the situation, and use mathematics to seek an answer to that question. Mathematical modeling involves identifying the relevant features of a real-world situation, representing those features symbolically, analyzing the model, and considering the accuracy and limitations of that model (NCTM, 2000).

The National Council of Teachers of Mathematics in their *Principles and Standards for School Mathematics* (2000), states, "one of the most powerful uses of mathematics is the mathematical modeling of phenomena" (p. 39). Further, high school students should be able to "develop models by drawing on their knowledge of many classes of functions—to decide, for instance, whether a situation would best be modeled with a linear function or a quadratic function" (p. 40).

In high school mathematics, students should use their middle school and Algebra I knowledge of the behaviors of different classes of functions (proportional, linear, quadratic, and exponential) in order to build and refine models for sets of data. Data can be gathered from any number of sources. Having students collect their own data makes the learning more concrete and interesting.

The analysis of the gathered data is the focal point of mathematical modeling. A graphical representation of the data empowers students to consider the physical features of the situation generating the data so that they can select an appropriate parent function for their model. Knowledge of transformations of functions helps students refine their model to achieve a "good fit." Technology such as graphing calculators or curve-fitting software, including regression models, allows students to quickly generate and test their models for suitability on the basis of the data itself and the physical parameters of the situation.

Engaging in mathematical modeling brings students full circle to the underpinnings of modern Western mathematical thought. Algebra, Geometry, and Calculus were all originally developed in order to explain or describe natural phenomena. Using a functions-based approach to mathematical modeling returns students to finding a purpose for doing mathematics. They begin to see mathematics as more than a series of steps to execute an algorithm; students see mathematics within the larger context of problem solving.

What is mathematical proficiency?

The National Research Council, with funding from the National Science Foundation and the U.S. Department of Education, has identified five strands that provide a framework for discussing the knowledge, skills, abilities, and beliefs that constitute mathematical proficiency. These proficiency strands are a result of careful analyses of relevant research on mathematics learning from pre-kindergarten through grade 8. These proficiencies express what it means for anyone to learn mathematics successfully and are identified in the research committee's report, *Adding It Up* (National Research Council, 2001), as well as in *Helping Children Learn Mathematics* (National Research Council, 2002), an independent review of *Adding It Up.*

The five interwoven and interdependent strands of mathematical proficiency are identified in *Helping Children Learn Mathematics* (p. 9). They are:

Conceptual Understanding

An integrated and functional grasp of mathematical ideas. "Comprehending mathematical concepts, operations, and relations—knowing what mathematical symbols, diagrams, and procedures mean." Students identify and combine the critical attribute(s) of a concept and make the correct relational inferences.

Computing (Procedural Fluency)

"Carrying out mathematical procedures, such as adding, subtracting, multiplying, and dividing numbers flexibly, accurately, efficiently, and appropriately."

Applying (Strategic Competency)

"Being able to formulate problems mathematically and to devise strategies for solving them using concepts and procedures appropriately." Students select an appropriate strategy to solve a problem and use appropriate representations to connect the procedure to the concept of the problem.

Reasoning (Adaptive Reasoning)

"Using logic to explain and justify a solution to a problem or to extend from something known to something not yet known." Students fully answer the question of "why" for strategy selection, explain procedures, and/or evaluate the reasonableness of a solution using age-appropriate mathematics vocabulary.

Engaging (Productive Disposition)

"Seeing mathematics as sensible, useful, and doable—if you work at it—and being willing to do the work."

Although these strands of proficiency are based on relevant research on mathematics learning from pre-kindergarten through grade 8, they can also be applied to high school mathematics. The National Council of Teachers of Mathematics (NCTM) states in *Principles and Standards for School Mathematics* (2000) that successful problem solving in grades 9 through 12 requires "knowledge of mathematical content, knowledge of problem-solving strategies, effective self-monitoring, and a productive disposition to pose and solve problems" (p. 341). The proficiency strands and the NCTM's principles and standards portray a more rounded approach to successful K–12 mathematics learning. *Helping Children Learn Mathematics* reinforces that "integrating all the strands of proficiency makes knowledge stronger, more durable, more adaptable, more useful, and more relevant" (p. 17).

What is the 5E model?

An effective mathematics lesson provides the most impact on student achievement when it ensures that students are actively engaged in learning as well as reflecting upon their learning to make sense of their activities. An effective lesson also provides opportunities to use, extend, and apply what is learned. The 5E Instructional Model developed and modified by Roger W. Bybee (Trowbridge & Bybee, 1996), past executive director of the National Research Council and Center for Science, Mathematics, and Engineering Education, provides such a model. Learning something new or understanding something familiar in greater depth involves making sense of both prior experiences and first-hand knowledge gained from new explorations. The lessons contained in *Modeling with Mathematics: A Bridge to Algebra II* are designed using the 5E Instructional Model.

The components of the 5E Instructional Model are:

1. **ENGAGE:** The instructor initiates this phase by asking well-chosen questions, by posing a problem to be solved, or by showing something intriguing. The activity should be designed to interest students in the problem and to make connections between past and present learning.

   Section 5.1 is the engagement portion of the first lesson in Chapter 5. The instructor uses a motion detector and a graphing calculator to collect data and create a distance vs. time graph. Questions are posed to the students about the rate and direction of the walkers with the intent of discovering what students already know about motion and trying to interest them in a further study of motion.

SECTION 5.1

Walking the Walk

In this section you will look at distance-versus-time graphs. For these graphs, you will consider time as the independent variable and distance as the dependent variable.

A motion detector can track a student walking toward or away from it. The motion detector along with a graphing calculator can be used to create a distance-versus-time graph of a student walk.

Imagine a student walking away from a motion detector.

1. As time increases, what happens to the student's distance from the motion detector?
2. Imagine a distance-versus-time graph of the walk. Would you expect the graph to be a straight line or a curved line? Why?
3. Would you expect the graph to increase or decrease from left to right? Why?

Use a motion detector and a graphing calculator to collect data and create a distance-versus-time graph for a classmate who walks away from the motion detector.

Sketch your graph.

4. How can you tell from the graph that the student was walking away from the motion detector?
5. Was the student walking at a constant rate? How can you tell from the graph?

Again use a motion detector and a graphing calculator to collect data and create a distance-versus-time graph for a second classmate who walks away from the motion detector.

Sketch your graph.

A sample graph is shown in **Figure 5.1.**

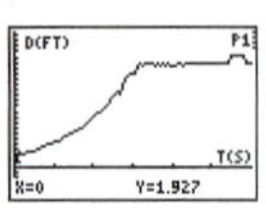

FIGURE 5.1.
A sample distance-versus-time graph.

258 Chapter 5 Motion: Quadratic Functions Section 5.1

6. How can you tell from the graph that the second student was walking away from the motion detector?
7. Was the second student walking at a constant rate? How can you tell from the graph?

Now you will analyze data taken from student walks in front of a motion detector. When you track the walk of another student, you can see how you would have to walk to produce the same distance-versus-time graph.

Cara walked in a straight line in front of a motion detector. The motion detector recorded her distance from it every 0.1 second for 6 seconds. Cara's distance-versus-time graph appears in **Figure 5.2.**

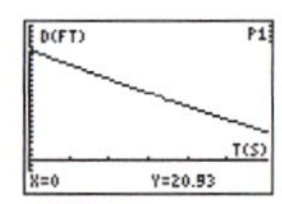

FIGURE 5.2.
Cara's distance-versus-time graph.

8. Based on Figure 5.2, did Cara walk toward or away from the motion detector? How can you tell from her graph?
9. Did she begin moving as soon as the motion detector began collecting data or did she pause first? How can you tell from her graph?
10. After she began moving, did she walk at a constant rate, or did she speed up or slow down? Explain how you decided.
11. What plan could you give to a walker so that he or she could walk a graph that is the same as Cara's?
12. Using your group's plans, have one group member walk in front of a motion detector. Is the graph of the walk close to the graph of Cara's walk?
13. How can you modify your plans so that the walk will more closely model Cara's walk?

Paul walked in a straight line in front of a motion detector. The motion detector recorded his distance from it every 0.1 second for 6 seconds. Paul's distance-versus-time graph appears in **Figure 5.3.**

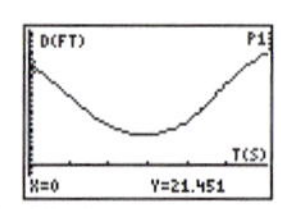

FIGURE 5.3.
Paul's distance-versus-time graph.

14. Based on Figure 5.3, did Paul walk toward or away from the motion detector? How can you tell from his graph?
15. After he began moving, did he walk at a constant rate, or did he speed up or slow down? Explain how you decided.

Section 5.1 Chapter 5 Motion: Quadratic Functions 259

2. **EXPLORE:** The exploration phase provides the opportunity for students to become directly involved with the key concepts of the lesson through guided exploration that requires them to probe, inquire, and question. As we learn, the puzzle pieces (ideas and concepts necessary to solve the problem) begin to fit together or have to be broken down and reconstructed several times. In this phase, instructors observe and listen to students as they interact with each other and the activity. Instructors provide probing questions to help students clarify their understanding of major concepts and redirect the questions when necessary.

   Section 2.2 is an example of the explore portion of a lesson. In this section students explore the relationship between the length of a person's head and height by measuring and collecting data.

SECTION 2.2

A Model for Estimating Height

One way to turn your guess of the dead peoples' heights into a scientific estimate is to explore the relationship between the length of certain bones and a person's height.

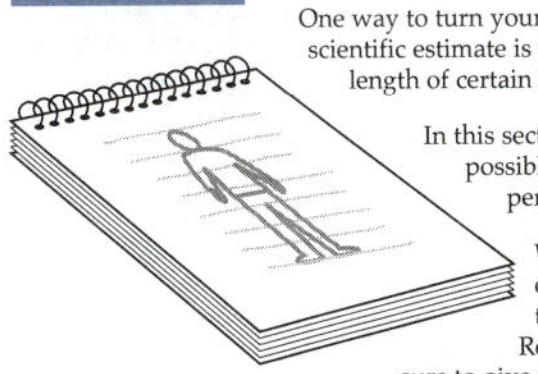

In this section you will gather some data to explore a possible relationship between head length and a person's height.

Within your group, measure the length of each person's head (from chin to the top of the head) and then measure their height. Record your data in a table like **Figure 2.3**. Be sure to give the units of measure you used at the top of the second and third columns.

Name	Head Length	Height

FIGURE 2.3. Group head-length and height data.

1. What patterns do you see in this data set?
2. We aren't sure yet that there is a relationship between head length and height, so why does it make sense to consider head length as the independent variable here?
3. What is a reasonable domain for this situation? Why?
4. What is a reasonable range for this situation? Why?
5. On a sheet of chart paper create a scatter plot of height versus head length. Sketch your graph.

74 Chapter 2 Bones: Linear Functions and Predictions Section 2.2

TREND LINES: LINES THAT APPROXIMATE DATA

The plot of height versus head length is fairly spread out, but it appears to have a linear form. This large amount of variability makes it hard to pick a line that you can use to make good predictions for heights. The following two methods will help you choose a line that approximates the pattern of the height versus head-length data from your class:

METHOD 1

1. Pick a point that appears to lie in the middle of the points displayed in your scatter plot. The point doesn't have to be a data point.
2. Using this point as one point on your line, adjust the slope of your line until you find a line that you think best describes the pattern of the data.
3. Find the equation of your line.

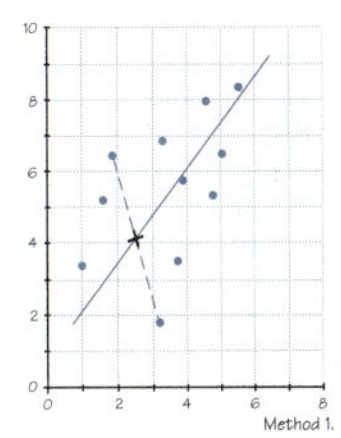

Method 1.

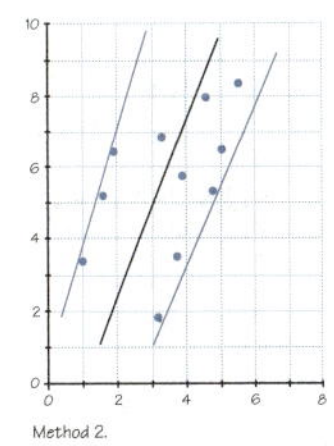

Method 2.

METHOD 2

1. Draw two lines in such a way that the points on your hand-drawn scatter plot are squeezed as tightly as possible between these lines. (The lines don't have to be parallel.)
2. Now draw one line halfway between the two lines that you have drawn.
3. Find the equation of this line.

6. Divide your group in half. Half of your group should use Method 1 and a meter stick to find a trend line for the graphed data. The other half should use Method 2 and two meter sticks to find a trend line. Draw both trend lines over the scatter plot and sketch your graph.
7. Write the equation of your trend line.

Section 2.2 Chapter 2 Bones: Linear Functions and Predictions 75

3. **EXPLAIN:** In the explanation phase, collaborative learning teams begin to logically sequence events and facts from the investigation and communicate these findings to each other and the instructor. The instructor, acting in a facilitation role, uses this phase to offer further explanation and provide additional meaning or information, such as formalizing correct terminology. Giving labels or correct terminology is far more meaningful and helpful in retention if it is done after the learner has had a direct experience. The explanation phase is used to record the learner's development and grasp of the key ideas and concepts of the lesson.

   For example, after students explore Hooke's law in Section 1.7, the instructor formalizes correct terminology by defining rate of change, slope, correlation, and slope-intercept form of an equation.

SECTION 1.6 This is Really a Stretch

In this section your teacher will conduct a class demonstration. Be sure to observe the demonstration carefully as you answer the following questions.

1. What do you observe when your teacher holds up the Slinky® with an empty film canister attached?
2. What observations can you make when your teacher holds up the Slinky with the attached film canister containing 5 pennies?
3. What observations can you make when your teacher holds up the Slinky with the attached film canister containing 10 pennies?
4. What observations can you make when your teacher holds up the Slinky with the attached film canister containing 15 pennies?
5. What patterns do you notice?
6. What are the input and output variables for this situation?
7. Do you think that the relationship between the amount of stretch of the Slinky and the number of pennies in the film canister is a proportional relationship? Why or why not?
8. Do you think that the amount of stretch of the Slinky and the number of pennies in the film canister are positively related? Why or why not?
9. What experiment could you do to further investigate the relationship between the stretch of a Slinky and the amount of weight attached to it?

SECTION 1.7 Hooke's Law

Robert Hooke (1635–1703) was an English scientist. He grew up during the Age of Enlightenment, a time of much scientific discovery. Hooke spent his life studying astronomy, geometry, mathematics, mechanics, and physics. He worked with such famous colleagues as Robert Boyle and Isaac Newton and served as the chief assistant to Christopher Wren, the famous architect who designed the rebuilt London after the Great Fire of 1666.

Today Hooke is known for his work with springs. In the 1600s, most people kept time with pendulum clocks. These clocks use a large pendulum that swings with a regular period to drive a wheel and turn the clock's hands. They work well on land. However, on a ship the large waves of the ocean disrupt the pendulum. Since determining a ship's longitude requires knowing the time in Greenwich, England when the sun is directly overhead, the ship's navigator needs an accurate way to keep time.

Hooke recognized this problem. In 1658, he began experiments using springs instead of a pendulum to control the wheel of a clock. By 1660, he made two significant improvements on spring clock design. While refining these improvements, he made many observations about the stretch of springs. One of these is a relationship between the amount of force applied to a spring and the length the spring stretches as a result. This relationship is known as Hooke's Law.

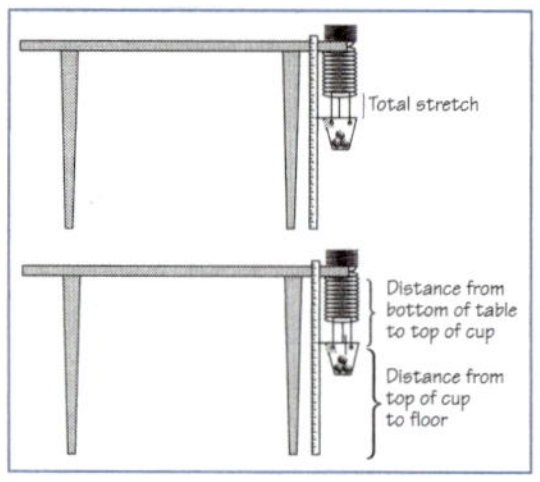

HOOKE'S LAW EXPERIMENT

Place paper clips through three holes in the sides of a cup and hang the cup from the bottom of a Slinky. Suspend the Slinky and cup from a desk to allow the cup to move up and down freely. Tape a bent paper clip to the top of the cup to use as a pointer. Tape a meter stick vertically to the table (see **Figure 1.15**).

FIGURE 1.15. Slinky set-up.

4. **ELABORATE:** The elaboration phase allows for students to extend and expand what they have learned in the first three phases and connect this knowledge with their prior learning in order to create understanding. It is critical that instructors verify students' understanding during this phase.

   In Section 6.14 students extend their knowledge of logarithmic functions to real-world situations such as Richter's formula relating the amount of energy released by an earthquake to its Richter magnitude.

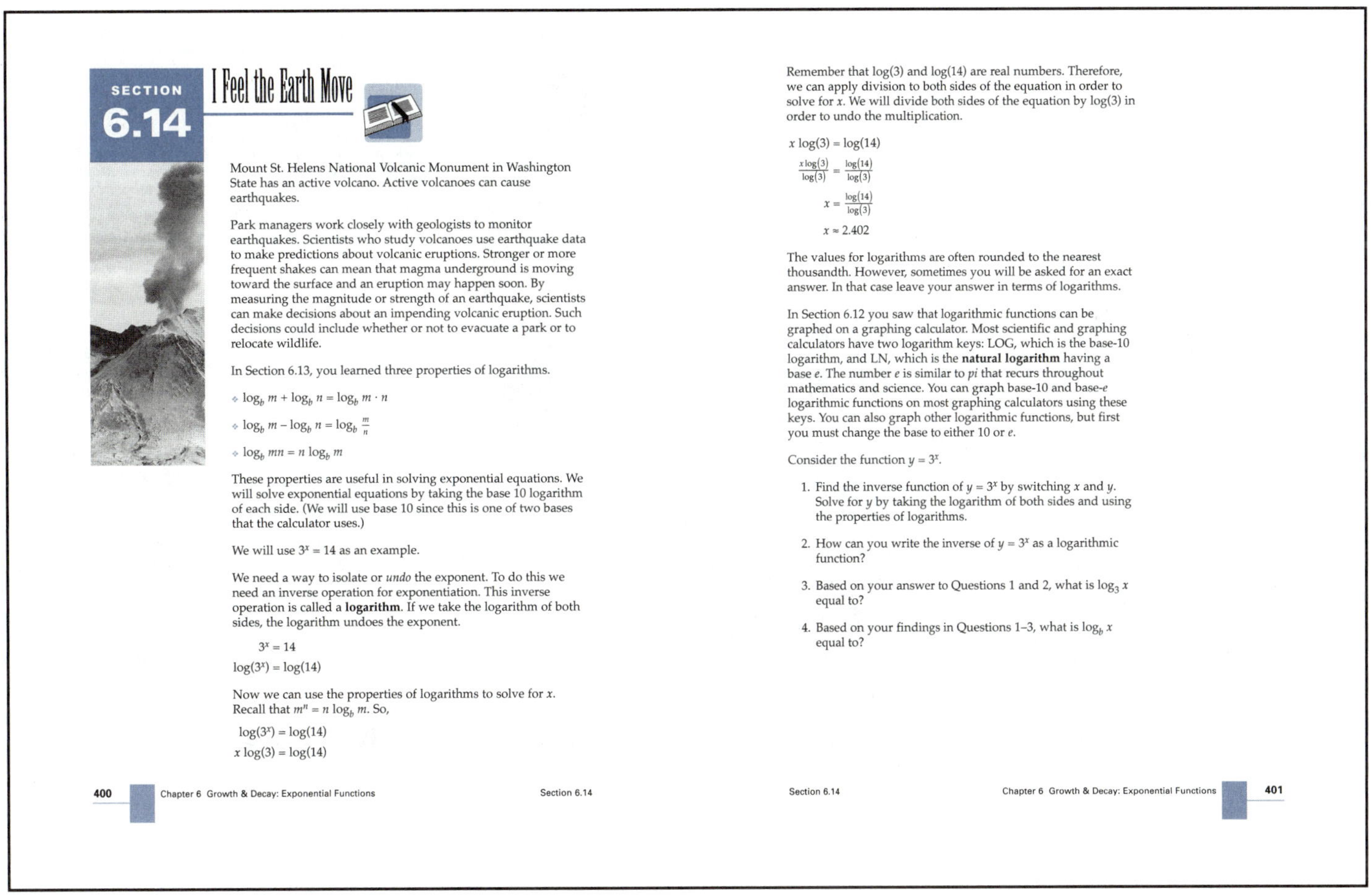

SECTION 6.14

I Feel the Earth Move

Mount St. Helens National Volcanic Monument in Washington State has an active volcano. Active volcanoes can cause earthquakes.

Park managers work closely with geologists to monitor earthquakes. Scientists who study volcanoes use earthquake data to make predictions about volcanic eruptions. Stronger or more frequent shakes can mean that magma underground is moving toward the surface and an eruption may happen soon. By measuring the magnitude or strength of an earthquake, scientists can make decisions about an impending volcanic eruption. Such decisions could include whether or not to evacuate a park or to relocate wildlife.

In Section 6.13, you learned three properties of logarithms.

- $\log_b m + \log_b n = \log_b m \cdot n$
- $\log_b m - \log_b n = \log_b \frac{m}{n}$
- $\log_b mn = n \log_b m$

These properties are useful in solving exponential equations. We will solve exponential equations by taking the base 10 logarithm of each side. (We will use base 10 since this is one of two bases that the calculator uses.)

We will use $3^x = 14$ as an example.

We need a way to isolate or *undo* the exponent. To do this we need an inverse operation for exponentiation. This inverse operation is called a **logarithm**. If we take the logarithm of both sides, the logarithm undoes the exponent.

$$3^x = 14$$
$$\log(3^x) = \log(14)$$

Now we can use the properties of logarithms to solve for x. Recall that $m^n = n \log_b m$. So,

$$\log(3^x) = \log(14)$$
$$x \log(3) = \log(14)$$

Remember that log(3) and log(14) are real numbers. Therefore, we can apply division to both sides of the equation in order to solve for x. We will divide both sides of the equation by log(3) in order to undo the multiplication.

$$x \log(3) = \log(14)$$
$$\frac{x \log(3)}{\log(3)} = \frac{\log(14)}{\log(3)}$$
$$x = \frac{\log(14)}{\log(3)}$$
$$x \approx 2.402$$

The values for logarithms are often rounded to the nearest thousandth. However, sometimes you will be asked for an exact answer. In that case leave your answer in terms of logarithms.

In Section 6.12 you saw that logarithmic functions can be graphed on a graphing calculator. Most scientific and graphing calculators have two logarithm keys: LOG, which is the base-10 logarithm, and LN, which is the **natural logarithm** having a base e. The number e is similar to *pi* that recurs throughout mathematics and science. You can graph base-10 and base-e logarithmic functions on most graphing calculators using these keys. You can also graph other logarithmic functions, but first you must change the base to either 10 or e.

Consider the function $y = 3^x$.

1. Find the inverse function of $y = 3^x$ by switching x and y. Solve for y by taking the logarithm of both sides and using the properties of logarithms.
2. How can you write the inverse of $y = 3^x$ as a logarithmic function?
3. Based on your answer to Questions 1 and 2, what is $\log_3 x$ equal to?
4. Based on your findings in Questions 1–3, what is $\log_b x$ equal to?

5. **EVALUATE:** Throughout the learning experience, the ongoing process of evaluation allows the instructor to determine whether the learner has reached the desired level of understanding of the key ideas and concepts. More formal evaluation can be conducted at this phase.

   The last section for each lesson is an open-middle performance assessment type of problem. For example, Section 8.5 asks students to solve a problem by deciding which fire truck can get to a fire first using the trigonometric concepts students learned in the previous sections.

SECTION 8.4 Assignment

1. What are the angle measures of a 3-4-5 right triangle?
2. A staircase, which takes up 12.5 feet of horizontal floor space, connects two floors that are 10 feet apart (**Figure 8.30**). What is the angle of elevation of the staircase?

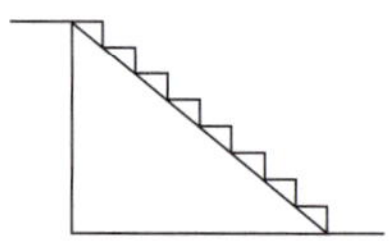

FIGURE 8.30. Staircase.

3. Go fly a kite! (Draw pictures with your work.)
   a) Shelia is standing 225 feet from a long line of trees 70 feet high. If she is holding the reel of string 4 feet off the ground, how far is it from the reel to the top of the trees?
   b) If Shelia lets out 300 feet of string so that the kite is now directly over a small building 195 feet away, what angle does the string make with the horizontal?
   c) Suppose Shelia moves away from the building so that her kite string makes a 40° angle with the horizontal. If Shelia still has 300 feet of string out, what is the ground distance between Shelia and the building?
4. A right triangle has a hypotenuse of 10 centimeters and one angle that measures 28°. What is the area of the triangle?

SECTION 8.5 The Two Towers

A distance of 2000 feet separates two towers in a state forest. One of the towers is directly west of the other. An observer in the West Tower notices a fire at an angle of 40° from the line of sight between the two towers. An observer in the East Tower notices the same fire due north from her tower (see **Figure 8.31**). Due to the terrain around the towers and through the forest, a fire truck from the West Tower travels at an average speed of 20 miles per hour. A fire truck from the East Tower travels at an average speed of 15 miles per hour. If both leave their towers at the same time, which one will get to the fire first? Justify your answer.

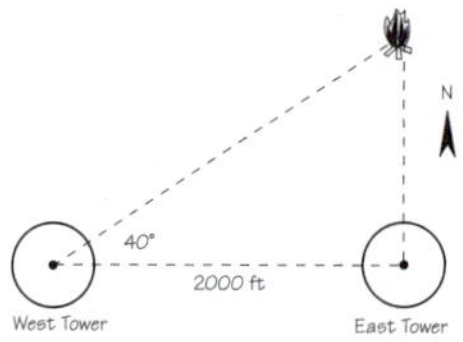

FIGURE 8.31. Two towers.

What is assessment?

Assessing student progress is a critical component of the curriculum, instruction, and assessment framework. Only through assessment can an accurate diagnosis be made of the students' strengths, weaknesses, and possible misconceptions.

There are key elements that should be considered when determining the assessment method to be used.

- Assessment should mirror good instruction.
- Assessment should happen continuously, and not obtrusively, as a part of instruction.
- Assessment should provide information to teachers, students, and parents about the levels of understanding that students are reaching.

There are two broad categories of assessment: formative assessment and summative assessment. Formative assessment is ongoing and serves as a source of feedback to improve teaching and learning. Summative assessment measures what students have learned at the end of a set of learning activities. Summative information should be used to plan instruction for the next major segment of instruction.

In *Modeling with Mathematics: A Bridge to Algebra II*, performance assessments are provided in each **Evaluate** phase to measure student progress. It is important to design a balanced assessment program that allows many routes to monitor student learning. The performance assessments are designed for formative use to provide checkpoints and define further instructional needs.

To provide meaningful summative assessment, each chapter contains two forms of selected-response assessment tests. The selected-response items are written at or above the "Apply" level of Bloom's taxonomy of cognitive rigor (Anderson & Krathwohl, 2001). They are designed to prompt students to apply what they have learned in the chapter, analyze new situations based on the concepts and skills covered in the chapter, and synthesize information in order to make and verify conjectures.

Error analysis patterns can be analyzed in order to diagnose students' procedural and conceptual misunderstandings. This information can be used by the teacher to prepare appropriate re-teaching strategies. In doing so, the teacher can save valuable class time by addressing only diagnosed student needs rather than wholesale re-teaching of the entire topic.

What is cooperative learning?

Cooperative learning encourages the exchange of mathematical ideas in a safe environment for the testing of conjectures. Because many of the lessons in *Modeling with Mathematics: A Bridge to Algebra II* call for cooperative learning, it is important for teachers to be familiar with the why and how of cooperative learning. **Simply placing students in a group does not promote a greater understanding of mathematical principles or the ability to communicate one's reasoning to others.** Cooperative learning looks like small groups of 2, 3, or 4 students working together to make sense of their own and each other's learning.

When compared to competitive or contest-based learning, where individuals are gauged against each other, Johnson, Johnson, and Holubec (1994) observed that purposeful cooperation results in:

1. Greater efforts to achieve
2. More positive relationships among students
3. Greater psychological health for students.

Basic Elements of Cooperative Learning:

- There must be a clear task and group goal.
- Students support and help each other to become successful.
- Each group member must be held accountable for contributing to the group's goals.
- Teamwork skills must be taught alongside academic skills.
- Groups must evaluate how well they are working together.

References:

Anderson, L. W., and Krathwohl, D. R., eds. (2001). *A taxonomy for learning, teaching, and assessing: A revision of Bloom's taxonomy of educational objectives.* New York: Addison Wesley Longman, Inc.

Johnson, D. W., Johnson, R. T., and Holubec, E. J. (1994). *The nuts and bolts of cooperative learning.* Edina, MN: Interaction Book Co.

National Council of Teachers of Mathematics. (2000). *Principles and standards for school mathematics.* Reston, VA: The National Council of Teachers of Mathematics, Inc.

National Research Council. (2002). *Helping children learn mathematics.* Mathematics Learning Study Committee, J. Kilpatrick and J. Swafford, Editors. Center for Education, Division of Behavioral and Social Sciences and Education. Washington, DC: National Academy Press.

National Research Council. (2001). *Adding it up: Helping children learn mathematics.* J. Kilpatrick and J. Swafford, and B. Findell (Eds.). Mathematics Learning Study Committee, Center for Education, Division of Behavioral and Social Sciences and Education. Washington, DC: National Academy Press.

Trowbridge, L. W., and Bybee, R. W. (1996). *Teaching secondary school science: Strategies for developing scientific literacy.* Englewood Cliffs, NJ: Merrill.

Program Components

- Student Text
- Annotated Teacher's Edition
- Teacher Resource CD-ROM, that includes
  Transparencies and Handouts
  Additional Activities
  Correlations to Regional and State Standards
  Correlations to State Mathematics Assessments
  Data Sets and TI-Link Files
  Assessments.

Modeling with Mathematics: A Bridge to Algebra II

Modeling with Mathematics: A Bridge to Algebra II

PROJECT LEADERSHIP

Jo Ann Wheeler
REGION 4 ESC, HOUSTON, TX

David Eschberger
REGION 4 ESC, HOUSTON, TX

Solomon Garfunkel
COMAP INC., LEXINGTON, MA

Roland Cheyney
COMAP INC., LEXINGTON, MA

LEAD AUTHORS

Gary Cosenza
REGION 4 ESC, HOUSTON, TX

Paul Gray
REGION 4 ESC, HOUSTON, TX

Julie Horn
REGION 4 ESC, HOUSTON, TX

AUTHORS

Sharon Benson
REGION 4 ESC, HOUSTON, TX

David Eschberger
REGION 4 ESC, HOUSTON, TX

Roland Cheyney
COMAP, LEXINGTON, MA

Jo Ann Wheeler
REGION 4 ESC, HOUSTON, TX

W. H. Freeman and Company
New York
www.whfreeman.com

Published and distributed by
W. H. Freeman and Company
41 Madison Avenue, New York NY 10010

Library of Congress Control Number: 2006921261

ISBN 07167-0780-2 (EAN 9780716707806)

Printed in the United States of America
First Printing 2006

Publisher: Craig Bleyer
Acquisitions Editor: Cyndi WeissGoldner
Marketing Manager: Mike Saltzman
Supplements and New Media Editor: Victoria Anderson
Associate Editor: Brendan Cady
Production Manager: George Ward
Production Department: Daiva Kiliulis, Tim McLean, George Ward, Pauline Wright
Photo Researchers: COMAP Production
Illustrations: Lianne Dunn, George Ward
Cover and Text Design: Daiva Kiliulis
Cover Images: Shutterstock
Manufacturing: RR Donnelley

Project Leaders:

Jo Ann Wheeler, Region 4 ESC, Houston, TX

David Eschberger, Region 4 ESC, Houston, TX

Solomon Garfunkel, COMAP, Lexington, MA

Roland Cheyney, COMAP, Lexington, MA

Lead Authors:

Gary Cosenza, Region 4 ESC, Houston, TX

Paul Gray, Region 4 ESC, Houston, TX

Julie Horn, Region 4 ESC, Houston, TX

Authors:

Sharon Benson, Region 4 ESC, Houston, TX

Roland Cheyney, COMAP, Lexington, MA

David Eschberger, Region 4 ESC, Houston, TX

Jo Ann Wheeler, Region 4 ESC, Houston, TX

Reviewer:

Sandra Nite, Texas A&M University, College Station, TX

Writers who contributed material used from COMAP's *Mathematical Models with Applications* and Region 4's *Mathematical Models with Applications/Algebra II Alignment*:

Marsha Davis, Eastern Connecticut State University, Williamantic, CT

Juliann Doris, Seguin High School, Seguin, TX

Anne Konz, Cypress-Fairbanks Independent School District, Cypress, TX

Jerry Lege, COMAP, Inc., Lexington, MA

Jennifer May, Independent Consultant

Paul Mlakar, St. Mark's School, Dallas, TX

Sandra Nite, Texas A&M University, College Station, TX

Richard Parr, Rice University, Houston, TX

Gary Simundza, Wentworth Institute of Technology, Boston, MA

Ann Worley, Spring Branch Independent School District, Houston, TX

Region 4 and COMAP would like to thank the following teachers and school districts for piloting *Modeling with Mathematics: A Bridge to Algebra II* and providing essential feedback:

Kim Armstrong
Lindale High School
Lindale Independent School District
Lindale, Texas

Toni Ericson
Hallsville High School
Hallsville Independent School District
Hallsville, Texas

Jason Hendricks
Madison High School
North East Independent School District
San Antonio, Texas

Sara Kamphaus
Summit High School
Mansfield Independent School District
Mansfield, Texas

Mary Ann Knight
Mansfield High School
Mansfield Independent School District
Mansfield, Texas

Connie Koehn
Elsik High School
Alief Independent School District
Houston, Texas

Garnette Lamm
Madison High School
North East Independent School District
San Antonio, Texas

Caroline Martin
Timberview High School
Mansfield Independent School District
Mansfield, Texas

Michael McCabe
Grapevine High School
Grapevine-Colleyville Independent School District
Grapevine, Texas

Robert McFarland
Aldine Senior High School
Aldine Independent School District
Houston, Texas

Jerry McHugh
MacArthur High School
North East Independent School District
San Antonio, Texas

Monica Merchant
Aldine Senior High School
Aldine Independent School District
Houston, Texas

Theresa Patton
Elsik High School
Alief Independent School District
Houston, Texas

Brenda Porter
Reagan High School
North East Independent School District
San Antonio, Texas

DeAnna Ramirez
Mansfield High School
Mansfield Independent School District
Mansfield, Texas

Clint Reynolds
Van High School
Van Independent School District
Van, Texas

Laura Salazar
Reagan High School
North East Independent School District
San Antonio, Texas

Michael Seibert
Grapevine High School
Grapevine-Colleyville Independent School District
Grapevine, Texas

Glenn T. Smith
Aldine Senior High School
Aldine Independent School District
Houston, Texas

Sheryl Smith
Taylor High School
Alief Independent School District
Houston, Texas

Diane Tobin
Aldine Senior High School
Aldine Independent School District
Houston, Texas

Maria Tolentino
Hastings High School
Alief Independent School District
Houston, Texas

Ron Van Raemdonck
Aldine Senior High School
Aldine Independent School District
Houston, Texas

Michael Williams
Gilmer High School
Gilmer Independent School District
Gilmer, Texas

Yangki Wojcik
Reagan High School
North East Independent School District
San Antonio, Texas

Aubrey Wright
Colleyville Heritage High School
Grapevine-Colleyville Independent School District
Colleyville, Texas

Randy Zelahy
Aldine Senior High School
Aldine Independent School District
Houston, Texas

Dear Student,

You have probably seen descriptions of how people use mathematics in their careers and lives in other math textbooks. The message of these little sidebars is simple, trust us … if you keep taking math courses you will have great career options and know more about our world. We agree!

But when we set out to make *Modeling with Mathematics: A Bridge to Algebra II* we asked a question: What if we do more? Could we not just tell you that mathematics is useful, but show you too? Why not create hands-on activities in science, business, design, and other fields and have you learn and use the mathematics in these real-world settings? This text uses a tool called mathematical modeling to explore these and other fields.

Mathematical modeling is used to clarify and solve a wide range of real-world problems. It is a tool used by engineers when they design a new style of MP3 player and by architects when they design a skyscraper. The best part about mathematical modeling is that it can simplify a tricky problem and help you focus on the things that really matter. The hard part about it is that it makes you read more, think more, and write more. We also think it makes learning math more interesting. The Introduction will tell you a little bit more about mathematical modeling.

In a few places in this text, you will see A Look Ahead section. These sections are a chance for you to see a little of the content you will do when you take Algebra II.

Some of you may not be confident about your math skills. Some of you may even feel that you can't take any more math courses. The goal of *Modeling with Mathematics: A Bridge to Algebra II* is to help you get back on track. It is a chance to look back at some of the math you may not have really understood and to look forward to a great experience in Algebra II.

Sincerely,

Jo Ann Wheeler
Region 4 ESC, Houston, TX

David Eschberger
Region 4 ESC, Houston, TX

Solomon Garfunkel
COMAP, Lexington, MA

Roland Cheyney
COMAP, Lexington, MA

Table of Contents

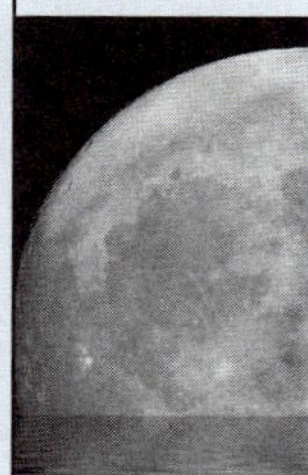

PREPARATION READING

The Modeling Process

INTRODUCTION

The process of starting with a situation or problem and gaining understanding about that situation through the use of mathematics is known as **mathematical modeling**.

Mathematical modeling attempts to describe real-world relationships in mathematical terms. Mathematical descriptions are useful because mathematics provides ways to obtain solutions. There are several ways to represent real-world relationships.

Mathematical modelers use a number of representations in their work. Among them are:

- verbal descriptions
- graphs, such as the graph of a straight line
- formulas, such as the formula for the area of a square
- tables, such as a table of values of length and width that give a rectangle with an area of 24 square inches
- drawings, such as a scale drawing of a room in a house
- diagrams, such as an arrow diagram or flow chart.

Length (l)	Width (w)
2	12
4	6
6	4
8	3
12	2

The mathematical modeling process can be described in steps.

Step 1. Identify the problem.

Read and ask questions about the situation. Identify a problem that you want to solve.

INTRODUCTION

Preparation Reading:
The Modeling Process

Description: The reading introduces students to the concepts of mathematical modeling through the context of modeling in science.

Materials: none

Teaching Ideas: Students should work in small groups (3 to 4 students). To help ensure that all students read the preparation reading, have students read silently then randomly select one student in each group to explain the passage to the other group members. Each student group should share their responses with the class. The text will later refer to information from the preparation reading.

Step 2. Simplify the situation.

Select the features of the situation that you think are most important. These are your assumptions that you will use to build a model. Note the features that you will ignore at first.

Step 3. Build the model and solve the problem.

Describe relationships among the parts of the problem in mathematical terms. Then find a mathematical solution to the problem. In this step you might do all or some of the following:

- define variables
- write equations
- draw shapes
- measure objects
- gather data and organize into tables
- make graphs.

Step 4. Evaluate and revise the model.

Go back to the original situation and see if the results of the mathematical model make sense. If so, use the model until new information becomes available or assumptions change. If not, reconsider the assumptions you made in step 2 and revise them.

You might go through a revision process several times before you have a good mathematical model. One helpful principle that guides all modelers is *keep it simple*. All models ignore something, and first models usually ignore several things.

A way of visualizing the mathematical modeling process is shown in **Figure 1.**

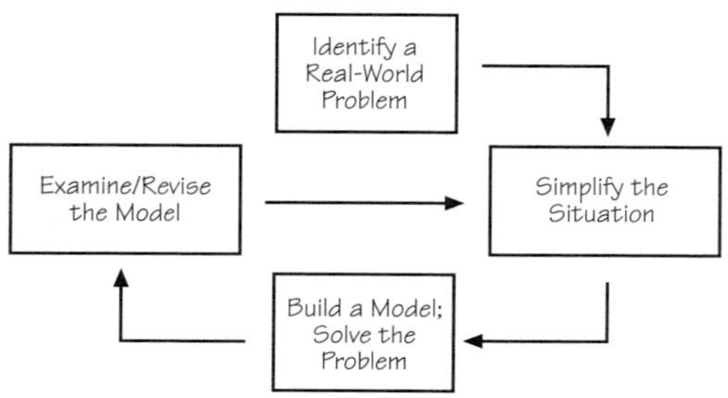

FIGURE 1. The modeling process.

Mathematical modeling is often used to describe the relationship between two or more real-world variables. (A variable is a quantity that changes.) Discovering and describing a relationship among variables allows modelers to predict values of one of the variables from values of the others.

Some examples of real-world situations where the modeling process might be used include:

- predicting the growth of a population based on variables such as the quantity of food available, disease rates, or birth rate
- predicting the profit for a company based on the costs of producing a product or the level of demand for the product
- predicting the effect on the economy of factors such as the unemployment rate or the value of exported goods
- predicting the effect of widening a road on traffic volume
- describing the relationship between the volume of blood flowing through an artery and the pulse rate of a person.

In this course you will use mathematics to create models that explore and explain a wide range of real-world issues. You will also use your imagination to create these models, then use mathematics to test them and see if they make sense. In order to create and test your models tools are needed, such as paper and pencil problem solving, using calculators, and doing research at your library and on the Internet. Most importantly you will need to find good ways to show your model to your teacher and to other students.

This may seem a very different way of learning math, but in this course you will see how mathematical modeling can help you succeed, both in school and in many careers. So give mathematical modeling a chance, you may find it changes how you feel about math.

CHAPTER 1

Science: Modeling with Direct and Inverse Variation

PREPARATION READING

Extra! Extra! Scientists Use Mathematical Modeling!

From ancient times, people have tried to make sense of the world around them. Nomadic Native Americans made observations about the behavior of the buffalo herds in the central plains of North America. These observations led them to make conjectures about the movement of the herd, which was essential to the tribe's survival. If they successfully followed the herd, then food was plentiful. If the herd left them behind, then they starved.

Some civilizations began to use numbers to describe relationships. In ancient Egypt, people began writing down numerals to keep count of things. They began to write simple equations to perform calculations.

These numeric observations prompted early scientists to want to learn more. They noticed that certain quantities are related, such as the length of day and the location of certain stars, and began to use mathematics to describe these relationships. These observations were the early stages of what we now call mathematical modeling.

In this chapter, you will explore some of the work of three notable scientists who used mathematics to describe relationships that they noticed in the natural world.

Archimedes (**Figure 1.1**) was a Greek scientist who lived in Syracuse, Sicily, during the third century BC. Archimedes wrote many books about geometry, mathematics, and physical science and was a good friend of King Hiero of Syracuse. Archimedes was instrumental in defending his home city against the Roman siege in 212 BC, before his death that same year.

FIGURE 1.1. *Archimedes.*

Archimedes is perhaps best known for what is today called Archimedes' Principle. This idea describes the forces that interact between a fluid, such as water, and an object that is submerged in that fluid.

CHAPTER 1

Science: Modeling with Direct and Inverse Variation

Preparation Reading:
Extra! Extra! Scientists Use Mathematical Modeling!

Description: Chapter 1 presents students with the foundation for a functions-based approach to mathematical modeling. The preparation reading is the students' first encounter in this course with using functions to build and using mathematical models to describe sets of data and the relationships between two or more variables. The reading presents a historical perspective of the uses of mathematical modeling in sciences. Each lesson in Chapter 1 features an important scientific principle that can be described using mathematical modeling: displacement of water, Hooke's Law, and Boyle's Law. Thus, students meet Archimedes, Robert Hooke, and Robert Boyle in the preparation reading.

Materials: none

Teaching Ideas: Students should work in small groups (3 to 4 students). To help insure that all students read the preparation reading, have students read silently then use a text-based reading sharing strategy to get students to explain what they have read. Using one of these strategies not only insures that students read the passage, but also allows the teacher to gauge the depth to which the students understand what they read.

One sharing strategy is to select at random one student in each group to explain the passage to the other group members. A variation of this strategy is to ask one student in the group to explain his or her understanding of the passage or to answer a focusing question that you pose. Then the other students in the group respond, one at a time for an equal amount of time to what they have heard their classmate say. Rounds continue until each student has had the opportunity to speak.

Another strategy is to jigsaw the reading into pertinent sections. Each student in the group reads his or her portion of the passage, then shares what they read with the other students in the group. For example, if students are working in groups of 3, each student can be assigned the portion of the passage relating to one scientist (Archimedes, Hooke, or Boyle). After each student reads his or her selection, he or she shares with the group a summary of what he or she has read.

A third strategy is to ask each student to identify 3 key ideas that he or she has read. Students should individually record these ideas on a sheet of paper or in their math journals. Then students should share their key ideas within their group, making a larger list of the ideas of the entire group. Each student group should share their combined list with the larger class while the teacher or a student volunteer records them on the overhead or on the chalkboard, generating a larger list of the key ideas that the class members feel are important in the reading.

While these strategies may take some class time, especially at first, it is important to remember that reading is essential for learning in any subject area; mathematics is no different. In order for students to become lifelong learners, they must learn to glean meaning from reading in a variety of contexts.

NOTES:

This chapter includes a number of experiments that require materials and setup time. While we feel students learn best by doing these activities, if you are unable to do the experiments, sample data sets are included on the Teacher Resource CD-ROM. If you choose this option, this data should be provided to the students.

Also if you would like to do additional experiments, such as collecting data for the Boyle's Law exploration in Section 1.14, the Teacher Resource CD-ROM has detailed instructions for doing this experiment.

FIGURE 1.2.
Robert Hooke.

Robert Hooke (**Figure 1.2**) was an English scientist who lived in London during the 1600s. This was an exciting time to be a scientist in England. Like most of Western Europe, England was enjoying the Age of Enlightenment, a time when scientists and philosophers explored new ideas about the world around them. After a long period of silence, they rediscovered the work of the ancient Greeks and Romans. As technology had become quite advanced for the day, scientists now had many new tools at their disposal, and they began to look for new ways to use them.

Robert Hooke began his work with springs as he searched for a way to build a better clock. Ships had been sailing from Europe to North America and Asia for over 200 years in Hooke's day. Yet, they had difficulty telling time in order to determine their longitude. European clocks at this time used a pendulum that swung back and forth, keeping steady time. However, on long sea voyages, waves disrupted the pendulum. So, Hooke began to explore the use of springs to make clocks tick. Along the way, he discovered a relationship between the length that a spring stretches and the amount of force required to make it stretch to that length. This relationship today is known as Hooke's Law.

FIGURE 1.3.
Robert Boyle.

Robert Boyle (**Figure 1.3**) was also an English scientist in London during the 1600s. In fact, Robert Hooke was one of his students. Boyle studied religion, chemistry, and physics. Like many of his colleagues, he wanted to take the mystery out of science, and he relied heavily on mathematics to do so.

Boyle was keenly interested in the properties of air, which were still largely misunderstood. Boyle's predecessors in Italy had found that air has weight and exerts a force on the ground called air pressure. Boyle learned through a series of experiments that not only did air have weight but also that fire cannot burn without air.

Boyle also studied the relationship between air pressure and volume. In 1662, he wrote in *Touching the Spring of the Air and Its Effects* about his experiments with a vacuum chamber to determine the relationship between pressure and volume of a gas. His findings are today known in chemistry as Boyle's Law.

Scientists continue to use mathematical modeling to describe natural events and to make predictions. Astrophysicists use mathematical modeling to chart the paths of stars and solar systems. Aeronautical engineers use mathematical modeling to build better airplanes and space vehicles. Social scientists use mathematical modeling to make predictions about populations and natural resource management.

Throughout this book, you will study how different people use mathematical modeling to make decisions about everyday life. You will also see how mathematics appears in unlikely places and is used by people from a variety of backgrounds.

In Chapter 1 you will explore some important ideas in science using mathematical modeling.

SECTION 1.1 Archimedes and the Crown

In the third century BC lived a famous mathematician and scientist named Archimedes. Archimedes is famous for many discoveries including an irrigation device, the law of the lever, the formula for the volume of a sphere, and possibly even the odometer. One of his discoveries is the Archimedes Principle, which describes the concept of buoyancy. Vitruvius, a Roman architect, tells a famous story about Archimedes and King Hiero of Syracuse.

King Hiero hired a craftsman to make a crown of gold. The king measured out the exact amount of gold for the craftsman to use. The craftsman later delivered to the king a beautiful crown. The crown weighed the same as the measured amount of gold. Rumors began floating around Syracuse that the crown was not made of pure gold. It was suggested that the craftsman had replaced part of the gold with an equal mass of silver. King Hiero asked Archimedes to prove or disprove the rumors without damaging the crown in any way. (Today this is called nondestructive testing.) As Archimedes began to sit down in a bath pondering the problem, he noticed the water level in the bath rising as he submerged more and more of his body into the bath. Realizing he had found a way to solve the problem he ran down the street, still naked, shouting, "Eureka!" ("I have found it!") The craftsman admitted he was guilty of stealing part of the gold and replacing it with silver.

1. What do you think Archimedes' solution might have been?
2. Gold has greater mass than silver. If a gold crown and a silver crown have the same weight, would they have the same volume? Why?
3. How can the volume of an object be determined by placing it in water?
4. Suppose your teacher places 15 pennies in a film canister and 20 pennies in another. Which canister has greater mass?

5. Which canister, the one with 15 pennies or the one with 20 pennies, would displace more water when submerged?
6. Does mass or volume cause displacement?

Your teacher will do a demonstration with film canisters, pennies, and a graduated cylinder. Answer the following questions as she/he does the demonstration.

7. What did you observe when your teacher placed the film canister into the graduated cylinder?
8. What did you observe when your teacher placed 20 pennies into the film canister?
9. Which canister, one with 15 pennies or one with 20 pennies, has greater mass?
10. Calculate the volume of a film canister.
11. What do you observe about the relationship between the calculated volume of the film canister and the amount of water displaced when the canister is submerged?

ENGAGE LESSON 1

Section 1.1: Archimedes and the Crown

Purpose: Students begin to work through the modeling process using the concepts of volume and displacement.

Description: Students are given several questions to answer, followed by a teacher-led demonstration. Students are asked to identify various factors that influence the situation.

Materials: large clear graduated cylinder (1 liter), water, film canisters with lids, pennies, small graduated cylinders (50 mL), digital balance, paper towels

Teaching Ideas: Students should work in pairs to answer the questions. At the end of this time, debrief by having each group share their answers to the questions. Be sure students realize that mathematical modeling is an important tool for a scientist to analyze his or her hypothesis.

Answers

1. Student answers will vary. Most students will not have a solution. After completing Section 1.3, explain the difference in displacement for silver and gold with equal masses.
2. No, because a gold crown would have to be somewhat smaller if gold is denser than silver.
3. You could measure the amount of water that is displaced when the object is submerged.
4. The canister with 20 pennies has greater mass.
5. They would displace the same amount of water.
6. Volume

At this point the teacher should do a demonstration. Use a clear graduated cylinder (1 liter) that is approximately two-thirds full of water. Have several film canisters available for your use and about 20 pennies. During the demonstration, try to keep the inside of the film canister and the pennies as dry as possible.

Demonstration:

- Fill the large graduated cylinder about $\frac{2}{3}$ full of water.
- Place 15 pennies in the empty film canister.
- Drop the empty film canister (with lid on) into the cylinder.

7. The canister sinks, and the water level rises.

- Place 5 more pennies into the canister, and then replace the lid and drop it back into the graduated cylinder.

8. The canister sinks and the water level rises about the same amount.

- Use the digital balance to find the mass of a canister with 15 pennies and the mass of a canister with 20 pennies.

9. The canister with 20 pennies has greater mass.
10. Give each pair of students an empty film canister and a ruler and have them approximate the volume in cubic centimeters of the canister by measuring.

If they need a clue, you might write the formula for the volume of a cylinder on the board.

Sample calculation of volume:

$V = B \cdot h$

$V = (\pi r^2)h$

$V = \pi \cdot (1.5\text{ cm})^2(5.1\text{ cm})$

$V = 36\text{ cm}^3$ (= 36 mL)

After students have calculated the volume, give each pair of students a small amount of water and a small graduated cylinder (50 mL) and have them approximate the volume of the canister using a different method.

Students should fill the film canister with water, replace the lid, and measure the volume by pouring the water into the small graduated cylinder. You might need to remind students that 1 cubic centimeter is equal to 1 milliliter. This method of approximating the volume will be slightly less than the measurement method since water will be inside the cylinder and the volume of the lid is neglected.

Sample data collected using this method was approximately 35 cubic centimeters.

11. They are approximately the same.

SECTION 1.2

Displacement Investigation

In this section you will explore displacement. You have a graduated cylinder filled with a given amount of water. You will record three things: the number of film canisters in the water, the total displacement of the canisters (the change in the volume of water), and the total volume contained in the cylinder.

CHECK THIS!
Displacement is the amount of water moved when an object is placed in the water.

1. Fill a large graduated cylinder to the 500 mL mark. In a table, record the amount of displacement when no film canisters have been placed in the cylinder. Also record the total volume contained in the cylinder.

2. Place a film canister with the minimum number of pennies to make it sink into the water in the large graduated cylinder (**Figure 1.5**). Read and record in your table (similar to the table in **Figure 1.4**) the volume of water displaced by 1 film canister and the total volume contained in the cylinder. Repeat this process adding 1 additional canister at a time. Continue until you submerge all 6 of your canisters.

Total Number of Submerged Canisters	Total Volume of Water Displaced (mL)	Total Volume Contained in the Graduated Cylinder (mL)

FIGURE 1.4.
Displacement data.

3. If you place 20 pennies in the film canister, does it displace more water than a film canister with 15 pennies?

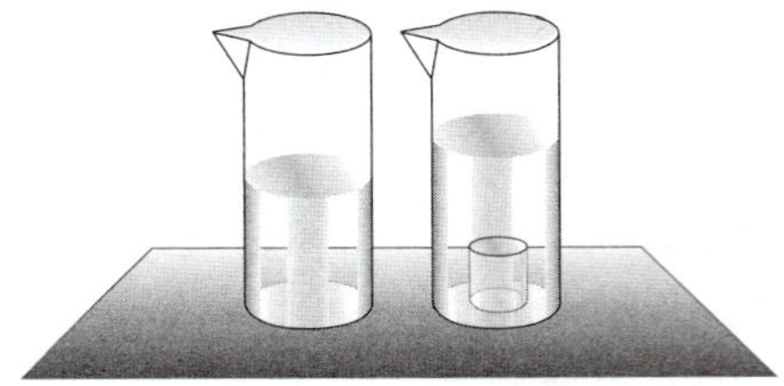

FIGURE 1.5.
Displacement illustration.

Your table has three columns. The first step of the mathematical modeling process is selecting a problem. In the rest of this section you will examine the relationship between displacement and the number of submerged canisters. You will examine the relationship between total volume and the number of submerged canisters in a later section.

4. Make a scatter plot of the pairs (displacement, number of submerged canisters) using the data in your table.

CHECK THIS!
A scatter plot is a graph of the ordered pairs representing a set of data.

5. Describe the pattern of the points that you observe in your scatter plot.

6. Predict the total volume of water displaced if you submerge 8 canisters.

7. Predict the number of canisters that you need to submerge in order to displace 500 mL.

EXPLORE LESSON 1

Section 1.2: Displacement Investigation

Purpose: In Section 1.1 students identified volume as the cause of displacement. In Section 1.2 students further explore the concept of displacement. They place film canisters weighted with pennies into a graduated cylinder and measure the volume of water displaced. They make a table and a scatter plot to represent the data. On completion, they are ready to formalize the concepts of proportional relationships and direct variation in Section 1.3.

Description: Students are given materials to collect data. They place the film canisters in the graduated cylinder containing water. As they add additional canisters they measure the total displacement in milliliters and the total volume contained in the cylinder. They record the data in a table and investigate the rate of change. Using their data, they make a scatter plot.

Materials per class: film canisters, pennies, large graduated cylinders, paper towels

Materials per group: large volume graduated cylinder (between 500 mL and 1000 mL), 6 film canisters, about 80 pennies, paper towels, water

Teaching Ideas: Students should work in small groups (3 to 4 students). Students may need instruction prior to the investigation on how to read a graduated cylinder. One student should be in charge of placing the canisters in the water. Another student should determine the total number of milliliters displaced by the canisters. A third student should record the data in a table. Use a pitcher of water to fill each group's graduated cylinder about halfway when students are ready to begin collecting data.

Answers

2. These are sample data.

Total Number of Submerged Canisters	Total Volume of Water Displaced (mL)	Total Volume Contained in the Graduated Cylinder (mL)
0	0	500
1	35	535
2	70	570
3	105	605
4	140	640
5	175	675
6	210	710

3. No, once a canister sinks, no matter what its weight, it displaces the same volume.

4. 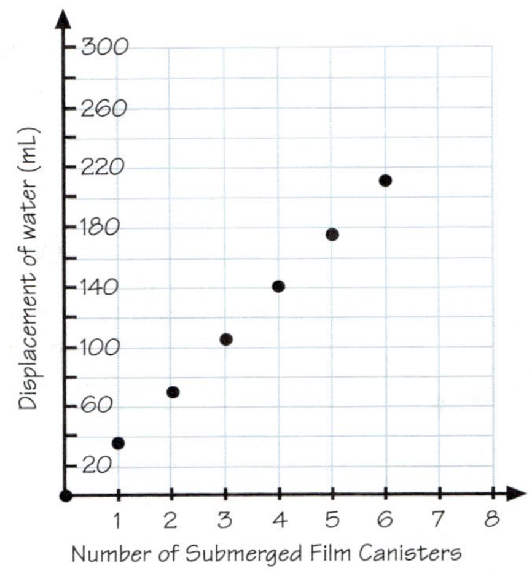

5. The points appear to lie on a line through (0, 0).

6. The total volume of water displaced would be approximately 280 mL. Students should be able to predict this by looking at a pattern in the table or in the scatter plot. You may want to allow one student group to test the prediction for the entire class.

7. Since each canister displaces about 35 mL, you would need to submerge approximately 14 or 15 canisters. You may want to allow one student group to test the prediction for the entire class.

SECTION 1.3

A Question of Variation

In the previous section, you collected and analyzed data regarding the relationship between volume of water displaced by canisters and the number of canisters submerged in water in a graduated cylinder.

You have already collected data and answered some questions about what you found. Now you will look at some important ideas about collecting data and building models based on that data.

The number of film canisters in the water and the total volume of water displaced are variables because their values change. A **variable** is a quantity that changes. Letters of the alphabet can be used to label variables. For example, c is a natural choice for the total number of film canisters in the water, and w is a good choice for the total volume of water displaced. The variable that you want to use to make predictions in your investigation is called the input or **independent** variable. The variable that you want to predict is called the output or **dependent** variable. Another way to think about this is that the dependent variable responds to changes in the independent variable or that changes in the independent variable explain changes in the dependent variable. In an experiment such as the one you just did, the independent variable is the one that you control.

When making a table of data in an experiment, it is customary to place the values of the independent variable in the first (left) column.

There are other variables in this situation, but in order to keep the modeling simple you will consider only two of them: the total number of canisters in the water (c) and the volume of water displaced (w). In modeling, variables have **units** that describe the kind of quantity the variable uses. For c, the units are numbers of canisters, for w, the units are milliliters

1. In the Displacement Investigation in Section 1.2, which variable is the independent variable? Why? Which variable is the dependent variable? Why?

2. The set of possible values of the independent variable is called the **domain of the situation**. What is a reasonable domain in this situation? Why?

3. The set of possible values of the dependent variable is called the **range of the situation**. What is a reasonable range for this situation? Why?

4. What patterns do you observe in your collected data?

5. One important aspect of mathematical modeling is the way the variables change in relation to each other. In the table where you recorded the number of submerged film canisters and the total displacement of the volume, find the value of the total displacement divided by the total number of film canisters for each row, $\frac{w}{c}$ (as in **Figure 1.6**).

Total Number of Submerged Canisters	Total Volume of Water Displaced (mL)	$\frac{w}{c} = \frac{\text{Total volume of water displaced}}{\text{Total number of film canisters}}$

FIGURE 1.6.
More displacement data.

6. A **proportional relationship** is one in which the ratio $\frac{\text{dependent variable}}{\text{independent variable}}$ is constant. The constant is called the **constant of proportionality**. In a proportional relationship, this constant is also a **rate of change**. A **rate of change** is a ratio that compares the change in the output variable to the change in the input variable. What do you observe in your table?

7. Real-world data are seldom perfect. Is the constant of proportionality the same or does it vary in your table?

EXPLAIN LESSON 1

Section 1.3: A Question of Variation

Purpose: The purpose of this section is to formalize the definitions of independent and dependent variables, domain and range, proportional relationship, positive relationship, and to demonstrate for students how to use the scatter plot feature of the graphing calculator.

Description: This section is designed to be a teacher-led discussion of the concepts in the lesson. Use the questions below to facilitate the class discussion.

Materials: graphing calculator, overhead graphing calculator, a container such as a Mini M&M's® cylinder or a film canister, sand

Teaching Ideas: Debrief students after Section 1.2 by having them share their answers to the questions. As they share, be sure to fill in any missing concept, process, or skill that could have been used to answer the questions. Work through the questions in this section with the entire class. Once again, be sure all necessary concepts, processes, or skills are addressed.

Answers

1. The independent variable is the total number of canisters, and the dependent variable is the total volume of water displaced. This is because the volume changes in response to changes in the number of canisters, which we are controlling.

2. Answers may vary. Sample answer: A reasonable domain would be whole numbers $0 \leq c < 15$ if the graduated cylinder has a maximum volume of 1000 mL and you started with 500 mL, because adding 15 canisters would cause the water to overflow.

3. Answers may vary. Sample answer: A reasonable range would be $0 \text{ mL} \leq w \leq 500 \text{ mL}$ if the graduated cylinder has a maximum volume of 1000 mL and started with 500 mL of water in the cylinder. Adding 15 canisters would displace almost 525 mL of water and cause the water to overflow.

4. Answers may vary. Sample answers: The total volume of water displaced is increasing by approximately 35 or 36 mL every time a canister is added. The total volume of water displaced is approximately thirty-five times the total number of containers in the water.

5.

Total Number of Submerged Canisters	Total Volume of Water Displaced (mL)	$\frac{w}{c} = \frac{\text{Total volume of water displaced}}{\text{Total number of film canisters}}$
0	0	undefined
1	35	35
2	70	35
3	105	35
4	140	35
5	175	35
6	210	35

6. Answers will vary. Guide students through a discussion of the dependent variable divided by the independent variable. This value, w/c, turns out to be a constant 35.

7. The constant of proportionality is constant.

One way to visualize a relationship between two variables is with a graph of enough data to get a sense of the pattern they form. In mathematics, a pair of perpendicular **axes** (number lines) creates a **coordinate plane**, in which a pair of numbers locates a point. The two numbers, a value of the independent variable and the corresponding value of the dependent variable, are called the **coordinates** of the point. In a graph, values of the independent variable are located along the horizontal axis (traditionally called the x-axis), and values of the dependent variable are located along the vertical axis (traditionally called the y-axis).

A **scatter plot** (**Figure 1.7**) is a graph of pairs (x, y) where the x-values are the independent values from your data and the y-values are the corresponding dependent values from your data. You created a scatter plot with the number of submerged film canisters on the x-axis and the total displacement of water on the y-axis.

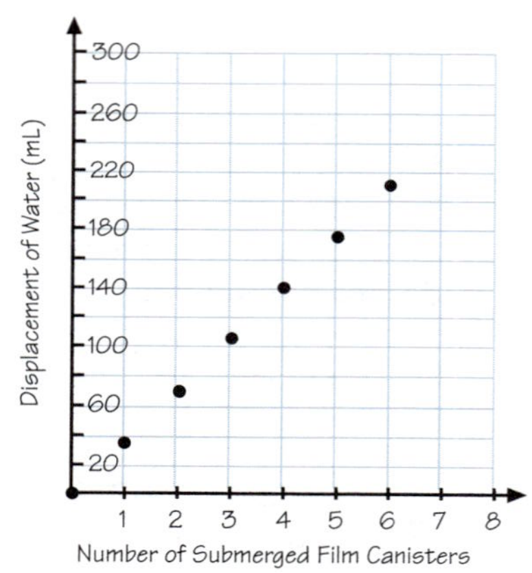

FIGURE 1.7.
A scatter plot.

This graph is called **discrete** because it shows distinct (separated) points. It does not make sense to connect the points since you cannot have a part of a canister. Since you cannot have a fraction of a canister, the domain is restricted to whole numbers. However, to use a scatter plot to make predictions, it is convenient to connect the points with a straight line, which creates a type of **continuous** graph. However, doing so changes the domain shown in the graph from *'some'* values to *'all'* values. Even though it makes mathematical sense, drawing a continuous graph can cause confusion when working with real-world situations.

8. Does it make sense to ask how much water 1.5 film canisters displace? Why or why not?

In a proportional relationship:

- The continuous graph is a straight line that passes through the origin;
- The ratio $\frac{\text{dependent variable}}{\text{independent variable}}$ is constant (or nearly so) for all data.

A proportional relationship can be described with an equation of the form $y = kx$, which is called a **direct variation function**. When a function is used to describe a relationship among variables, it is sometimes said that the function models the relationship.

9. Do your data demonstrate a proportional relationship between displacement and number of canisters? How do you know?

10. Describe how the constant of proportionality is displayed in the graph.

11. Based on your table or graph, what is the volume of 1 film canister in cubic centimeters?

Two quantities have a **positive relationship** (sometimes called a positive correlation) when the dependent variable increases as the independent variable increases.

CHECK THIS!
A relationship is called negative when one quantity decreases as the other increases.

12. Is there a positive relationship between the total displacement of the water and the number of canisters submerged? Why or why not?

13. Use a graphing calculator to make a scatter plot of your data from the Displacement Investigation in Section 1.2. Your teacher will give you instructions on how to do this.

Answers (continued)

8. No, because we are not physically able to submerge 1.5 film canisters in the real world.
9. The data appear to lie on a line through (0, 0), the origin. The ratio $\frac{y}{x}$ is close to 35 for all of the data. Moreover, the equation $w = 35 \cdot c$ fits the data.
10. The number of canisters is on the horizontal axis, and the volume of water displaced is on the vertical axis. For every one canister increase or decrease, 35 mL of water are displaced. So for every canister moved right or left, 35 mL must be moved up or down to get to the next point on the graph.
11. Each canister has a volume of approximately 35 cm^3.
12. Yes. The displacement increases as the number of submerged canisters increases.
13.

Press	Result
STAT, 1, ▲ to top of list to be cleared, CLEAR ▼ Repeat to clear all lists.	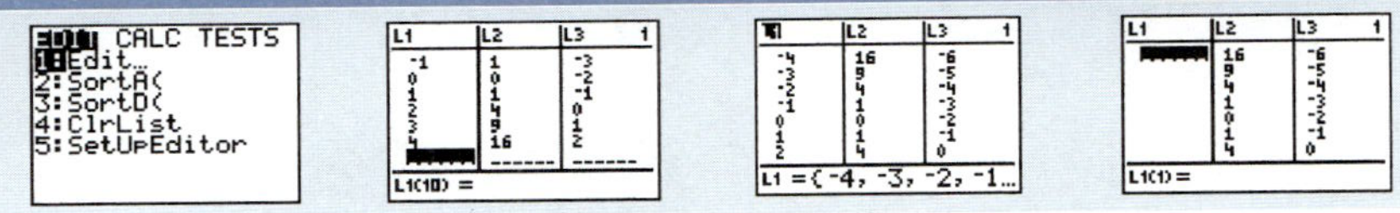
◄ to L1, type the data for the figure in L1 and L2, 2nd Y=, 1 ◄ ENTER ▼ ► ENTER ▼ 2nd 1 ▼ 2nd 2 ▼ ENTER	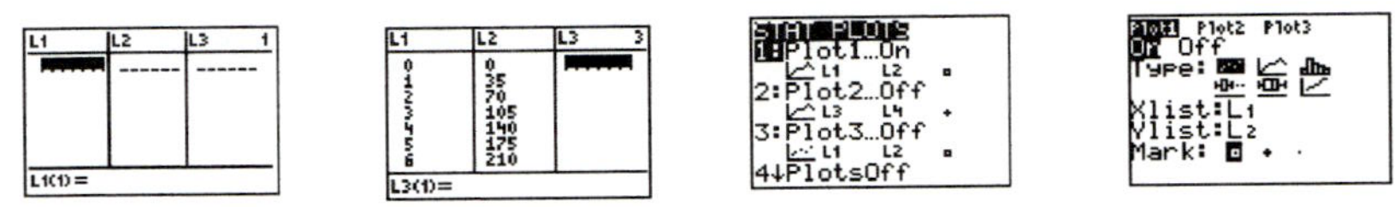
Set an appropriate window. WINDOW, 0, ▼, 8, ▼, 1, ▼, 0, ▼, 240, ▼, 20, ▼, 1	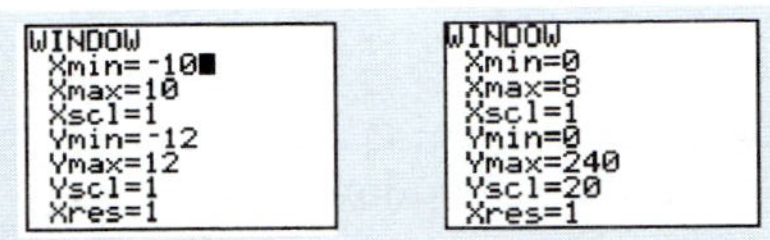
Clear the function plotter and graph. Y= CLEAR GRAPH	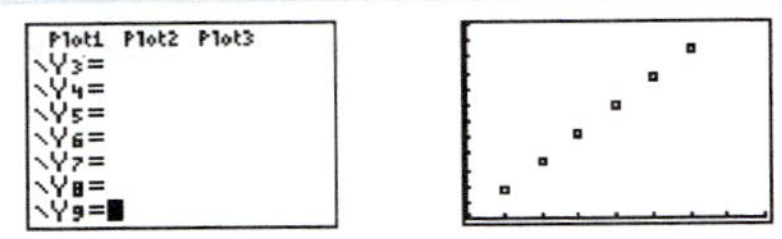
Enter the equation of the line into the function plotter and graph. Y=, 35, X GRAPH	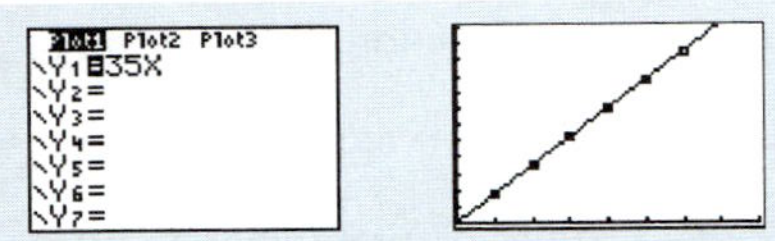

NOTES: If students are interested after completing the Displacement Investigation, you can explain how Archimedes knew that the crown was not made of 100% gold.

You should guide them through these calculations. Since the density of gold is 19.3 g/cm^3 and the density of silver is 10.6 g/cm^3, gold is almost twice as dense as silver. As an example for students,

100 grams of gold has a volume of:

$$\frac{100\text{ g}}{1} \cdot \frac{1\text{ cm}^3}{19.3\text{ g}} = 5.2\text{ cm}^3.$$

100 grams of silver has a volume of:

$$\frac{100\text{ g}}{1} \cdot \frac{1\text{ cm}^3}{10.6\text{ g}} = 9.4\text{ cm}^3.$$

To demonstrate for students how mass, density, and volume are related; fill a film canister with sand. Fill a Mini M&M's cylinder (or a similar cylinder such as the one glucose tablets come in) with water. (Mini M&M's come in a cylinder that is about twice as tall as a film canister.) Weigh both cylinders making sure their masses are the same. If necessary, you may need to remove a small amount of water or sand from either one of the cylinders.

- How are the two cylinders (film and M&M's) alike, and how are they different?

Students should point out that the cylinders weigh the same amount but have different volumes. They are filled with different substances.

- Fill two identical graduated cylinders with the same amount of water. Place the M&M's cylinder into one graduated cylinder and the film canister into the other. What do you notice?

Students should notice that the cylinder with the larger volume displaces a larger volume of water.

- Use your observations to explain how Archimedes knew that the crown contained silver.

Give students time to talk over their answers with a partner. Have partners volunteer to explain their thinking. Since equal masses of gold and silver have different volumes, a crown made with silver would have a greater volume than a crown of the same mass made of pure gold. The water level would rise higher when the crown is made of silver than it would when it is made of gold.

SECTION 1.3 Assignment

For their science project, Judy and Jose plant a kudzu vine seed in order to investigate how quickly it grows. Every four days after the seed sprouts, they measure and record the height of their plant. Their results are shown in **Figure 1.8.**

Day Number	Height (cm)
0	0
4	3
8	6
12	9
16	12
20	15

FIGURE 1.8. Kudzu data.

1. For this situation, what is the dependent variable and what is the independent variable?
2. State a reasonable domain and range for this situation.
3. Make a scatter plot for these data.
4. Is there a proportional relationship between height and day number? Why or why not?
5. What is the constant of proportionality for this situation?
6. Which kind of graph, continuous or discrete, is more appropriate for this situation? Why?
7. What is an appropriate direct variation function to model the relationship between height and day number?

Alexia and Alex collected data by placing marbles in a graduated cylinder partially filled with water and measuring the volume of displaced water. Their data are shown in the scatter plot in **Figure 1.9.**

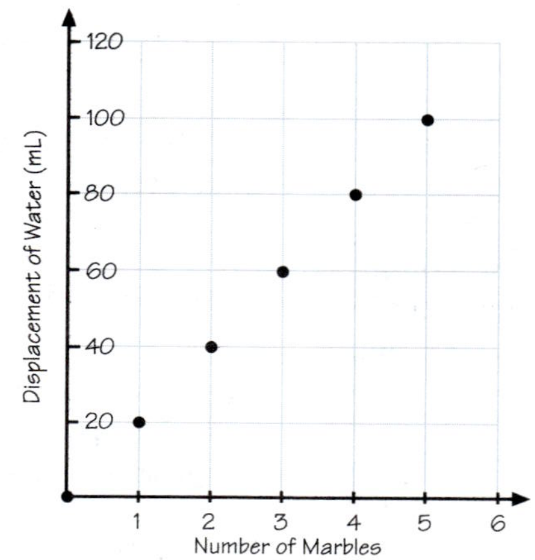

FIGURE 1.9. Marble data.

8. Make a table for the data displayed in the scatter plot.
9. For this situation, what is the dependent variable and what is the independent variable?
10. State a reasonable domain and range for this situation.
11. Is this a proportional relationship? Why or why not?
12. What is the constant of proportionality for this situation?
13. Which kind of graph, continuous or discrete, is more appropriate for this situation?
14. What is an appropriate direct variation function to model the relationship between displacement and number of marbles?
15. Use the volume of 1 marble to find the approximate radius of a marble.

CHECK THIS!

A marble is spherical in shape. The volume of a sphere is $V = \frac{4}{3}\pi r^3$.

Section 1.3 Assignment

Answers

1. The independent variable is the day number. The dependent variable is the height.
2. The domain is $0 \le d \le 20$. The range is $0 \le h \le 15$.
3. Sample answer:

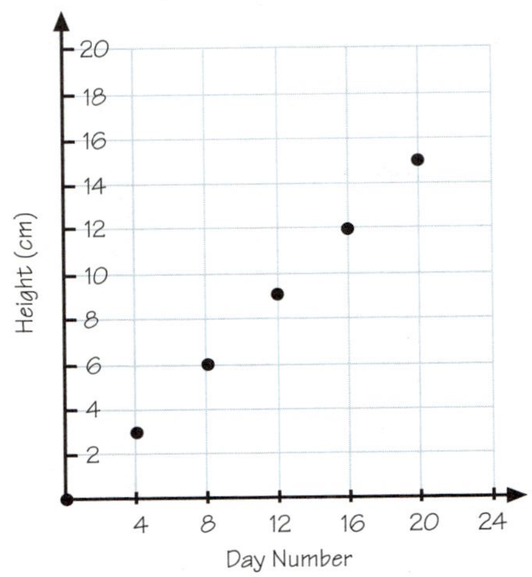

4. Yes. For each data pair, the ratio of height to day number is a constant 0.75. The line also contains the origin (0, 0).
5. The constant of proportionality is 0.75.
6. A continuous graph is more appropriate because a plant grows continuously. It does not grow 3 cm instantaneously.
7. $h = 0.75d$
8.

Number of Marbles	Displacement of Water (mL)
0	0
1	20
2	40
3	60
4	80
5	100

9. The independent variable is the number of marbles, and the dependent variable is the total displacement of water.
10. The domain is {0, 1, 2, 3, 4, 5}. The range is $0 \le d \le 100$.
11. Yes, it is a proportional relationship. For each data point, the ratio of displacement to marbles is a constant 20. A line drawn through the data would contain the origin (0, 0).
12. The constant of proportionality is 20.
13. A discrete graph is more appropriate because you cannot place fractions of marbles into the graduated cylinder.
14. $d = 20$ m
15. $V = \frac{4}{3}\pi r^3$

    $20 = \frac{4}{3}\pi r^3$

    $\frac{3}{4} \cdot \frac{1}{\pi} \cdot 20 = r^3$

    $4.775 \approx r^3$

    $\sqrt[3]{4.775} \approx r$

    $1.684 \text{ cm} \approx r$

SECTION 1.4

Displacement Investigation Round 2

In Section 1.2 you collected data and recorded the data in a table. You made a scatter plot of the relationship between displacement and number of submerged canisters, and found that the data demonstrated a proportional relationship. The direct variation model that you made involves two variables: the number of submerged film canisters (c) and the amount of water displaced (w).

If you change the variables you examine in a situation, your model may change.

In Section 1.2 you also recorded the total volume of the graduated cylinder. For this next investigation, continue to use c for the number of submerged canisters, your independent variable. Use v for the total volume contained in the graduated cylinder, your dependent variable.

1. Make a table showing the number of submerged canisters and the total volume of the graduated cylinder.
2. Make a scatter plot of the data in your table.
3. What are the domain and range for this situation?
4. When the independent variable has a value of 0, what is the value of the dependent variable?
5. Describe the relationship displayed in your scatter plot.
6. Do the variables appear to be positively related?
7. Do you think there is a proportional relationship between the total volume of the graduated cylinder and the number of submerged canisters?
8. What is the value of the dependent variable when the independent variable is 0?
9. What does your answer to the Question 8 value mean in this situation?

10. What do you predict the total volume contained in the cylinder will be if you submerge 7 film canisters?
11. How many canisters do you need to submerge for the total volume to be 900 mL?
12. When the points in a scatter plot lie along a straight line, the relationship between the two variables is called **linear**. Not all linear relationships are proportional. Do you think the relationship between total volume and number of canisters is linear?

ELABORATE LESSON 1

Section 1.4: Displacement Investigation Round 2

Purpose: In Section 1.3 students examined a proportional relationship. In this section they make a table and a scatter plot for a relationship that is not proportional.

Description: In this section students apply and extend their knowledge. They work with data they collected in Section 1.2. They compare their model to the earlier model. Students then use the model to predict the total volume contained in a graduated cylinder when a number of containers have been submerged in the water.

Materials: graphing calculators

Teaching Ideas: Have students solve problems using multiple representations (graph or table). They should be able to create a scatter plot and make predictions based on their model. They should be able to decide if a relationship is proportional.

Answers

1. These are sample data.

Total Number of Submerged Canisters	Total Volume Contained in the Graduated Cylinder (mL)
0	500
1	535
2	570
3	605
4	640
5	675
6	710

2.

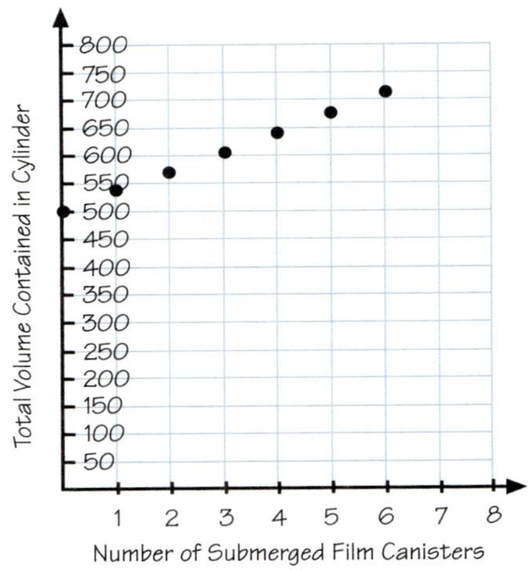

3. The domain is {0, 1, 2, 3, 4, 5, 6}. The range is $500 \le V \le 750$, depending on the students' data.

4. When $c = 0$, $V = 500$.

5. The data appear to lie on a line. The line starts at (0, 500). As the number of submerged canisters increases, the total volume increases.

6. Yes, as the number of canisters increases, the total volume increases.

7. No. By investigating the data in the table, students should notice that $\frac{V}{c}$ is not a constant.

   A line through these data does not contain the point (0, 0).

Total Number of Submerged Canisters	Total Volume Contained in the Graduated Cylinder (mL)	$\frac{V}{c}$
0	500	*undefined*
1	535	535
2	570	285
3	605	202
4	640	160
5	675	135
6	710	118

8. 500

9. It means that when you have 0 film canisters the volume is initially 500 mL.

10. The total volume in the cylinder would be approximately 745 mL according to the sample data. Student answers may vary.

11. Answers may vary. By looking at the table and the scatter plot, students might conjecture 11 or 12 film canisters.

12. Yes. The points in the scatter plot appear to lie along a straight line.

SECTION 1.4

Assignment

Tell whether the relationship in each situation is proportional or not. Justify your answer.

1. Traveling on interstate highway I-10 with data shown in **Figure 1.10.**

Time Elapsed (hr)	Distance Traveled (mi)
0	0
3	180
6	360
9	540

FIGURE 1.10. Highway travel.

2. Fare in a taxicab: it costs $2.00 to get in the cab and an additional $0.20 for each $\frac{1}{4}$ mile traveled.

   (Hint: Make a table!)

3. Temperature conversions shown in **Figure 1.11.**

Celsius Scale (°C)	Fahrenheit Scale (°F)
5	41
10	50
15	59
20	68

FIGURE 1.11. Temperature conversions.

4. Money spent on many cans of coffee when the sale price is $1.89/can.

A bank savings plan book keeps track of the amount of money in the account at the end of each month. The data are recorded in **Figure 1.12.**

Number of Months	Account Balance
0	$220
1	$250
2	$280
3	$310
4	$340

FIGURE 1.12. Savings account.

5. Explain how the balance can be something other than $0 after 0 months.
6. Make a scatter plot of the savings account data.
7. What are the domain and range for this situation?
8. When the independent variable has a value of 0, what is the value of the dependent variable?
9. Describe the type of relationship displayed in your scatter plot.
10. Do the variables appear to be positively related? Why or why not?
11. Is the relationship proportional? Why or why not?
12. What does the pair (0, 220) mean in this situation?
13. Is this set of data better represented by a continuous or discrete graph?
14. If the person deposits money into his savings account each month at the same rate, how much money is in the account after 10 months?
15. How many months does it take for this person to save $1000?
16. Make a scatter plot on the graphing calculator of the savings account data.

Section 1.4 Assignment

Answers

1. Proportional
2. Not proportional. Sample table:

Number of Miles Traveled	Cost
0	$2.00
$\frac{1}{4}$	$2.20
$\frac{2}{4}$	$2.40
$\frac{3}{4}$	$2.60
1	$2.80

3. Not proportional
4. Proportional
5. The person who opened the account deposited $220 to open the account.
6. Sample answer.

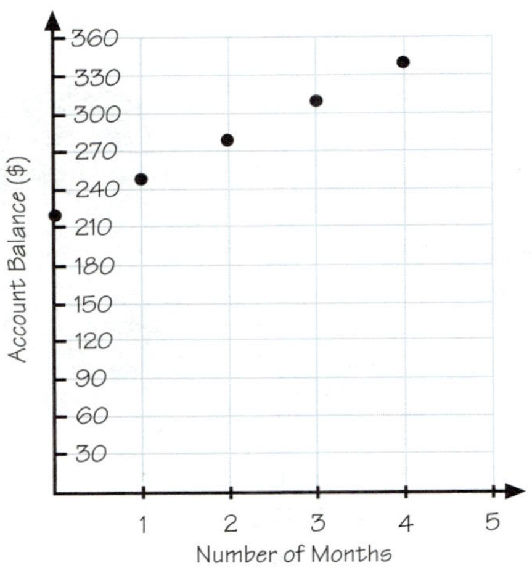

7. The domain is {0, 1, 2, 3, 4}, and the range is {220, 250, 280, 310, 340}.
8. The dependent variable has a value of 220 when the independent variable has a value of 0.
9. The graph appears linear and includes the point (0, 220). It is not a proportional relationship, but it appears to be a linear relationship.
10. Yes, the variables appear to be positively related. As the number of months increases, the account balance increases.
11. No, the relationship is not proportional. The ratio of account balance to number of months does not remain constant.
12. (0, 220) means the person deposited $220 to start the account.
13. This set of data is better represented by a discrete graph. The account holder probably does not deposit money on a continuous basis. He deposits $30 each month.
14. $520. Students may determine this by using either the table or the graph.
15. 26 months. Students may determine this by using either the table or the graph.
16.

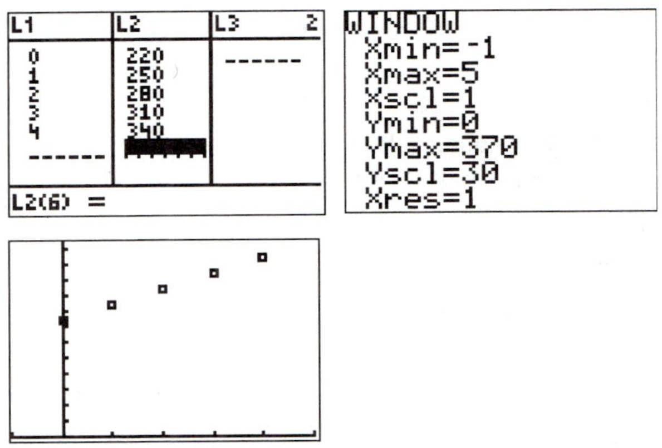

SECTION 1.5

Mystery Metals

Tom and Tara's teacher gives them one of the metals shown in **Figure 1.13.**

Metal	Density $\left(\frac{g}{cm^3}\right)$
Aluminum	2.70
Zinc	7.13
Silver	10.49
Mercury	13.55
Gold	19.32

FIGURE 1.13.
Mystery metals.

Tom and Tara collect the following data (**Figure 1.14**) in the science lab on their metal. They use a graduated cylinder to find the volume and a balance to find the mass.

Volume (cm^3)	Mass (g)
10	135
20	270
30	410
50	680
100	1350

FIGURE 1.14.
Metal data.

1. Use three different methods (table, graph, and equation) to examine the relationship between the mass of the metal and its volume. Then use the results of your investigation to determine the volume of 500 grams of the metal. Justify your answer.

2. Which metal do you think Tom and Tara were given? Why?

SUMMARY

In these sections you learned that proportional relationships have certain characteristics.

1. Continuous graphs of proportional relationships are linear and contain the point (0, 0).

2. The ratio $\frac{y}{x}$ is a constant in a proportional relationship and is also called the **constant of proportionality.**

3. Proportional relationships can be described by a function of the form $y = kx$, which is called a **direct variation function**.

You also saw that there are relationships that are linear, but not proportional.

EVALUATE LESSON 1

Section 1.5: Mystery Metals

Answers

1. Sample answers:

Volume (cm^3)	Mass (g)	$\frac{\text{Mass}}{\text{Volume}}$
10	135	13.50
20	270	13.50
30	410	13.67
50	680	13.60
100	1350	13.50
?	500	13.55

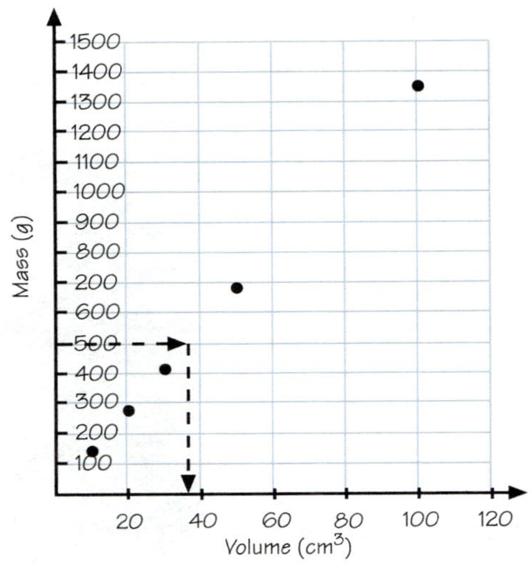

Equation using direct variation:

mass = 13.55 · volume

or, $m = 13.55 \cdot v$

$500 = 13.55 \cdot v$

$v = 37\ \text{cm}^3$

2. Mercury, since the constant of proportionality, or density, is $\frac{1355\ \text{g}}{100\ \text{cm}^3}$ or $\frac{13.55\ \text{g}}{1\ \text{cm}^3}$.

SECTION 1.6

This is Really a Stretch

In this section your teacher will conduct a class demonstration. Be sure to observe the demonstration carefully as you answer the following questions.

1. What do you observe when your teacher holds up the Slinky® with an empty film canister attached?
2. What observations can you make when your teacher holds up the Slinky with the attached film canister containing 5 pennies?
3. What observations can you make when your teacher holds up the Slinky with the attached film canister containing 10 pennies?
4. What observations can you make when your teacher holds up the Slinky with the attached film canister containing 15 pennies?
5. What patterns do you notice?
6. What are the input and output variables for this situation?
7. Do you think that the relationship between the amount of stretch of the Slinky and the number of pennies in the film canister is a proportional relationship? Why or why not?
8. Do you think that the amount of stretch of the Slinky and the number of pennies in the film canister are positively related? Why or why not?
9. What experiment could you do to further investigate the relationship between the stretch of a Slinky and the amount of weight attached to it?

SECTION 1.7

Hooke's Law

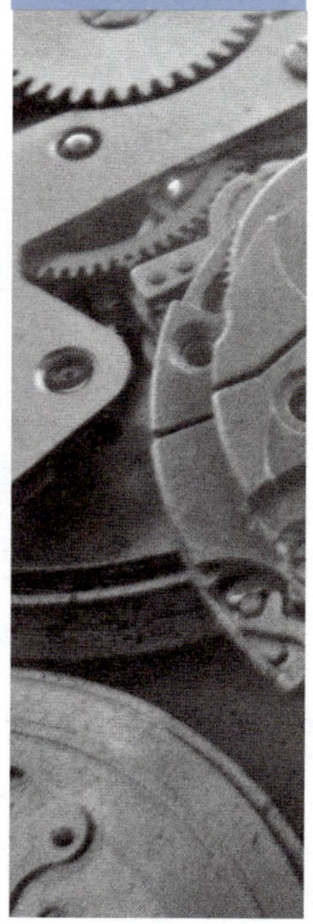

Robert Hooke (1635–1703) was an English scientist. He grew up during the Age of Enlightenment, a time of much scientific discovery. Hooke spent his life studying astronomy, geometry, mathematics, mechanics, and physics. He worked with such famous colleagues as Robert Boyle and Isaac Newton and served as the chief assistant to Christopher Wren, the famous architect who designed the rebuilt London after the Great Fire of 1666.

Today Hooke is known for his work with springs. In the 1600s, most people kept time with pendulum clocks. These clocks use a large pendulum that swings with a regular period to drive a wheel and turn the clock's hands. They work well on land. However, on a ship the large waves of the ocean disrupt the pendulum. Since determining a ship's longitude requires knowing the time in Greenwich, England when the sun is directly overhead, the ship's navigator needs an accurate way to keep time.

Hooke recognized this problem. In 1658, he began experiments using springs instead of a pendulum to control the wheel of a clock. By 1660, he made two significant improvements on spring clock design. While refining these improvements, he made many observations about the stretch of springs. One of these is a relationship between the amount of force applied to a spring and the length the spring stretches as a result. This relationship is known as Hooke's Law.

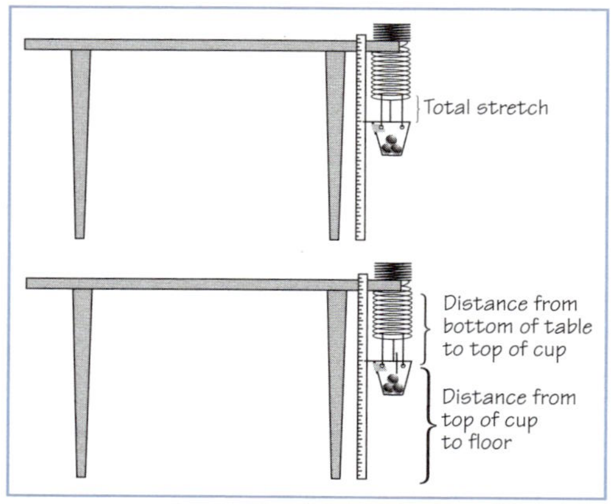

FIGURE 1.15. Slinky set-up.

HOOKE'S LAW EXPERIMENT

Place paper clips through three holes in the sides of a cup and hang the cup from the bottom of a Slinky. Suspend the Slinky and cup from a desk to allow the cup to move up and down freely. Tape a bent paper clip to the top of the cup to use as a pointer. Tape a meter stick vertically to the table (see **Figure 1.15**).

ENGAGE LESSON 2

Section 1.6: This is Really a Stretch

Purpose: In Sections 1.1 through 1.5, students investigated proportional relationships that can be modeled with direct variation functions and non-proportional relationships (relationships of the form $y = mx + b$ where b is not 0). In this lesson, Sections 1.6 through 1.10, students explore non-proportional relationships that can be modeled with linear functions. In Section 1.6 students will use characteristics of functions to describe a situation that they will explore further in Section 1.7.

Description: This section is a whole-class demonstration. The teacher will use a Slinky® with a film canister basket (constructed prior to class). Begin with an empty film canister and ask students to make observations. Ask students to continue making and recording observations as the teacher adds weight to the basket. These observations allow students to identify characteristics of the functional relationship between weight and the stretch of the Slinky; they will use these observations in Section 1.7 to gather and analyze data.

Materials: film canister, awl or ice pick, 2 jumbo paper clips, 15 pennies, a Slinky

Teaching Ideas: The teacher should demonstrate for the entire class. Prior to class, construct a basket using a film canister and a paper clip.

Step 1: Use an ice pick or awl CAREFULLY to punch small holes near the top of the film canister on opposite sides.

Step 2: Take a paper clip (string could be used in a pinch, but paper clips work better) and unfold it to make a straight line. Bend about a quarter-inch of each end to be perpendicular to the unfolded clip, facing the same direction, creating the handle of your "basket."

Step 3: Using this paper clip handle, insert each of the bent ends into the holes in your film canister.

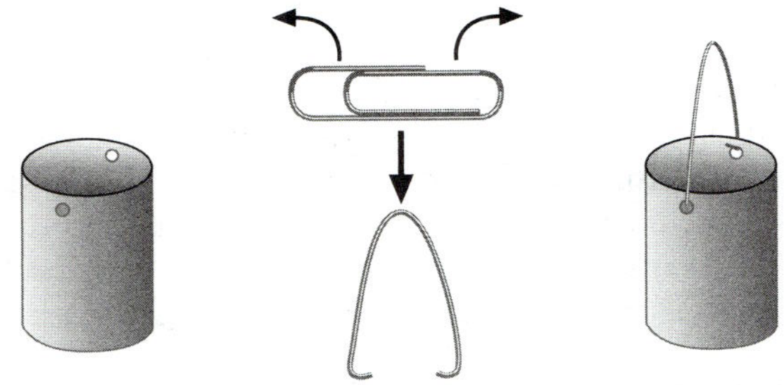

Using another paper clip, attach a film canister to a Slinky. Grasp the Slinky so that 10 coils of the Slinky are below your hand. Hold the Slinky with the attached film canister in front of the class. Prompt students to make observations. Place 5 pennies in the film canister then hold the Slinky again so that 10 coils of the Slinky are below your hand. Prompt students to make observations. Continue this procedure using 10 then 15 pennies. Give students a limited amount of time to answer the questions posed in the textbook. Debrief the students after this activity by discussing the answers to the questions. This discussion should create a desire to investigate further the relationship between weight and the stretch of the Slinky.

Answers

1. Sample answer: The Slinky is stretched. There are 10 coils of the Slinky showing.
2. Sample answer: The Slinky is stretched more than it was when no pennies were in the canister. There are still 10 coils of the Slinky showing.
3. Sample answer: The Slinky is stretched more than it was when 5 pennies were in the canister. There are still 10 coils of the Slinky showing.
4. Sample answer: The Slinky is stretched more than it was when 10 pennies were in the canister. There are still 10 coils of the Slinky showing.
5. Sample answer: It appears that stretch of the Slinky increases by the same amount each time 5 additional pennies are added to the canister. The number of coils of the Slinky showing stays the same.
6. The input is the weight or number of pennies placed in the canister. The output is the distance the Slinky stretches.
7. Answers can vary. Some students may say yes because they think the ratio of stretch to number of pennies is constant. Other students may say no because they think the amount of stretch with zero pennies in the basket is not zero. Students should agree that the relationship is probably linear because it appears that the stretch increases by the same amount each time a penny is added to the basket.
8. Sample answer: Yes, it appears that stretch of the Slinky increases as the number of pennies in the canister increases.

9. Sample answer: Attach a cup to a Slinky and measure the length of the stretched Slinky. Add one object at a time to the cup and measure the length of the stretched Slinky after each addition. Record the data and look for relationships

NOTE: This discussion should lead to the Hooke's Law investigation in the next section.

EXPLORE

Section 1.7: Hooke's Law.

Purpose: In Section 1.6 students used prior knowledge of proportional and linear relationships to discuss characteristics of a functional relationship. In Section 1.7 students collect data that can be used to describe Hooke's Law. In Section 1.8 terms such as linear relationships, direct variation, positive correlation, slope, rate of change, y-intercept, and vertical shift are connected to Hooke's Law.

Description: In this activity, students make a spring (Slinky) with a basket then add marbles to the basket. They measure the stretch of the Slinky each time one marble is added, and record the data in a table. They identify independent and dependent variables, and domain and range. They make a scatter plot of the stretch of the Slinky versus the number of marbles and describe the graph. They also make a scatter plot of the distance from the top of the table versus the number of marbles added and describe the graph. Students use the table and a graph to identify a proportional relationship.

Materials: graphing calculator for each student; each group will need 1 meter stick, 1 metal Slinky, 1 plastic cup, 5 paper clips, and about 10 marbles (or other uniform objects) NOTE: 1 Slinky can be cut in half or thirds using a good pair of metal scissors or tin snips, but be sure to file the sharp metal ends.

Teaching Ideas: Students should work in small groups (3 to 4 students). Demonstrate the experiment setup and clarify the measurements the students must make. The Slinky and cup are then suspended from a tabletop. The Slinky must be allowed to stretch freely as objects are added to the cup one at a time.

Prior to class, punch 3 holes in the cups as shown below. Students will use the holes to attach paper clips to the cup. Rubber bands may be substituted for the Slinky, but instead of marbles, more (or heavier) uniform objects would be needed.

In this experiment, measure the stretch that is explained by the marbles, but not the stretch that is explained by the cup or the weight of the Slinky. Place a mark or a piece of tape on the table or the meter stick at the level of the top of the empty cup and measure the stretch from this mark.

1. You will be adding one object at a time to the cup and measuring the total stretch of the Slinky as the objects are added. What is the independent variable? What is the dependent variable? What are a reasonable domain and range for this situation? Why?
2. You will also be measuring the distance from the bottom of the table to the top of the cup. What is the independent variable? What is the dependent variable? What are a reasonable domain and range for this situation? Why?
3. The third measurement that you will take is the distance from the top of the cup to the floor. What is the independent variable? What is the dependent variable? What are a reasonable domain and range for this situation? Why? Record and keep these data. You will need them in Section 1.9.
4. Sketch what you predict a scatter plot of total stretch versus number of objects placed in the cup would look like. Explain why you made the scatter plot the way you did.

CHECK THIS!

In scatter plots, the term "versus" (or the abbreviation "vs.") describes the position of the variables. The first variable (the dependent variable) is associated with the vertical axis and the second variable (the independent variable) with the horizontal axis.

5. Sketch what you predict a scatter plot of the distance from the bottom of the table to the top of the cup versus the number of objects placed in the cup would look like. Explain why you made the scatter plot the way you did.
6. Add objects to the cup one at a time. After each object is added, record the total stretch of the Slinky, the distance from the bottom of the table to the top of the cup, and the distance from the top of the cup to the floor (see **Figure 1.16**).

Number of Objects in the Cup	Total Stretch of the Slinky® (cm)	Distance from the Bottom of the Table to the Top of the Cup (cm)	Distance from the Top of the Cup to the Floor (cm)
0	0		
1			
2			
3			
4			
5			

FIGURE 1.16.
Slinky data.

7. Is the relationship between the total stretch and the number of objects placed in the cup a proportional relationship? Use your table to justify your answer.
8. Use the data in your table to create a scatter plot of total stretch versus number of objects placed in the cup. Record your window. Sketch and describe your graph.
9. Do you think that your scatter plot shows a positive relationship between the total stretch and the number of objects placed in the cup? Why?

CHECK THIS!

Recall that the rate of change is a ratio that compares change in the dependent variable to change in the independent variable.

10. What characteristics of your graph verify that the relationships between total stretch and number of objects placed in the cup is or is not a proportional relationship?

Mathematicians use Δ, the Greek letter delta, for the change in a variable. Thus, Δs means "the change in the variable named s," or simply "the change in s." For example, between $s = 3$ and $s = 5$, $\Delta s = 2$.

You can calculate a rate of change between any two pairs of values in your table. To do so, divide the change in the dependent variable by the change in the independent variable. That is, find the ratio $\frac{\Delta s}{\Delta c}$, where s is the total stretch in the spring and c is the number of objects in the cup.

Answers

1. The independent variable is the number of objects added to the cup. The dependent variable is the length of the stretch. The domain is the number of objects added to the cup, so whole numbers from 0 to 10 would be reasonable. The range is the number of centimeters that the Slinky stretches. Since an average tabletop is about 75 centimeters above the floor, real numbers from 0 to 60 would be reasonable for the range.

2. The independent variable is the number of objects added to the cup. The dependent variable is the number of centimeters from the bottom of the table to the top of the cup. The domain is the number of objects added to the cup, so whole numbers from 0 to 10 would be reasonable. The range is the number of centimeters from the bottom of the table to the top of the cup. Since an average tabletop is about 75 centimeters above the floor, real numbers from 0 to 70 would be reasonable for the range.

3. NOTE: Be sure students keep these data.

   The independent variable is the number of objects added to the cup. The dependent variable is the number of centimeters from the top of the cup to the floor. The domain is the number of objects added to the cup, so whole numbers from 0 to 10 would be reasonable. The range is the number of centimeters from the top of the cup to the floor. Since an average tabletop is about 75 centimeters above the floor, real numbers from 0 to 70 would be reasonable for the range.

4. Sample answer: As the number of objects increases, the length of the Slinky should increase at a constant rate.

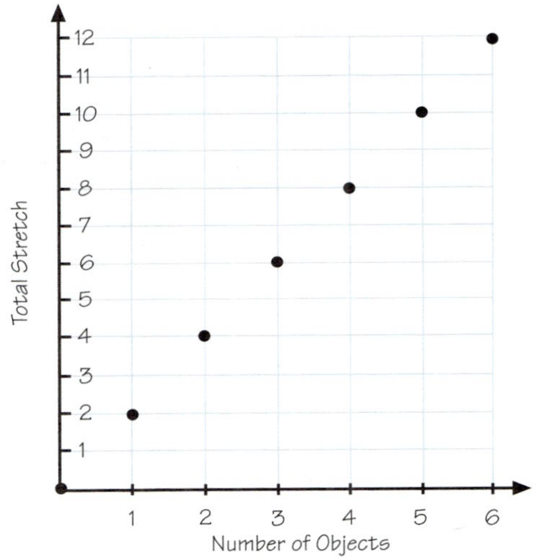

5. Sample answer: The distance from the bottom of the tabletop to the top of the cup will start at some number other than 0 when the cup is empty and increase at a constant rate as objects are added.

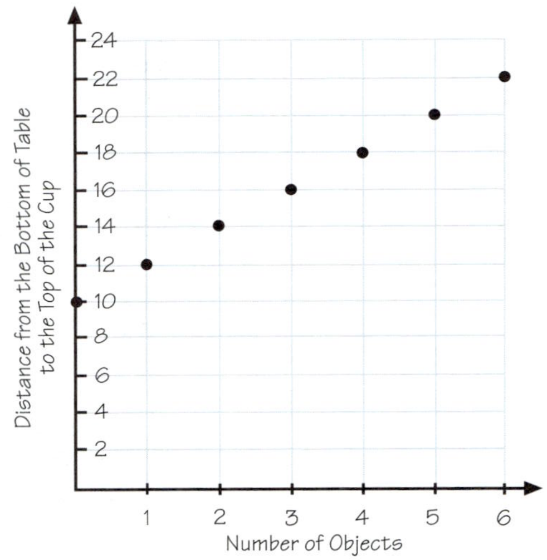

6. NOTE: The total stretch starts at 0 when the cup is empty and increases as objects are added. The distance from the bottom of the tabletop to the top of the cup, and the distance from the top of the cup to the floor starts at some number other than 0 when the cup is empty and increases or decreases as objects are added.

   Sample data: NOTE: Answers that follow are based on these data.

Number of Objects	Total Stretch of the Slinky (cm)	Distance from the Bottom of the Table to the Top of the Cup (cm)	Distance from the Top of the Cup to the Floor (cm)
0	0	17.2	30.0
1	1.9	19.1	28.1
2	4.0	21.2	26.0
3	5.3	22.5	24.6
4	8.4	25.6	21.5
5	10.4	27.6	19.5

7. Sample answer: The relationship is approximately proportional. In each case, the ratio of the total stretch to the number of objects placed in the cup is approximately 2. Therefore 2 is the constant of proportionality.

8. Sample answer: The graph appears to be linear.

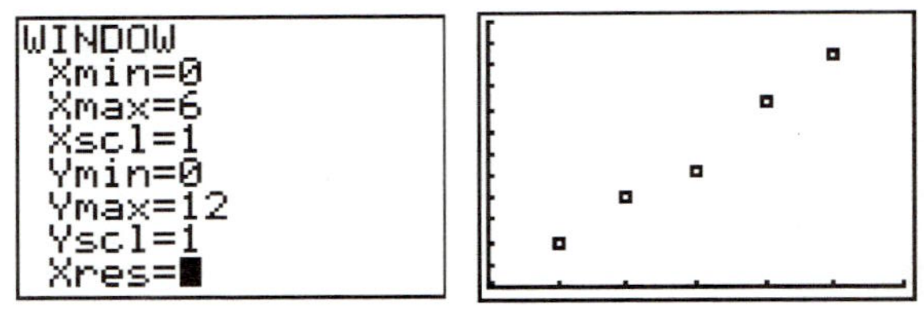

9. The graph shows a positive relationship because the dependent variable increases when the independent variable increases.

10. Sample answer: This is a proportional relationship since the data appear to be linear and include the point (0, 0).

11. Find the rate of change between two pairs of values in your table. How does the rate compare to the constant of proportionality? What does the rate mean in this situation?

CHECK THIS!

A proportional relationship can be modeled with a direct variation function. The function's equation is in the form $y = kx$ where k, the constant rate of change, is called the constant of proportionality.

12. What is an appropriate function rule for a direct variation function that models the relationship between total stretch and the number of objects placed in the cup?

A **function rule** is a symbolic equation that describes a relationship between two variables. For example, if each value of the independent variable is triple the corresponding value of the dependent variable, then a function rule is $y = 3x$.

13. Graph your function rule over your scatter plot. Sketch your results. Does the graph show that your function models the relationship well?

14. Use your function rule to determine the total stretch of the Slinky when 12 objects are added to the cup.

15. How many objects should it take to make the Slinky stretch a total of 32 cm?

16. Is the relationship between the distance from the bottom of the table to the top of the cup to the number of objects placed in the cup a proportional relationship? What characteristics in your table justify your answer?

17. Use the data in your table to create a scatter plot of the distance from the bottom of the table to the top of the cup versus the number of objects placed in the cup. Record your window. Sketch and describe your graph.

18. What characteristics of your graph show that the relationship between the distance from the bottom of the table and the top of the cup versus the number of objects placed in the cup is or is not a proportional relationship?

19. Find the rate of change between two pairs of values in your table. Do the same for another two pairs. What is the meaning of the rate in this situation?

20. How does the rate of change for the relationship between the distance from the bottom of the table to the top of the cup and the number of objects placed in the cup compare to the rate of change for the relationship between total stretch and the number of objects placed in the cup? How are the graphs related?

Answers (continued)

11. The rate of change is 2 for every two pairs. The rate is the same as the constant of proportionality. The rate means that for every object added to the cup, the Slinky stretches an additional 2 centimeters.

12. $y = 2x$

13. This function appears to be a good model.

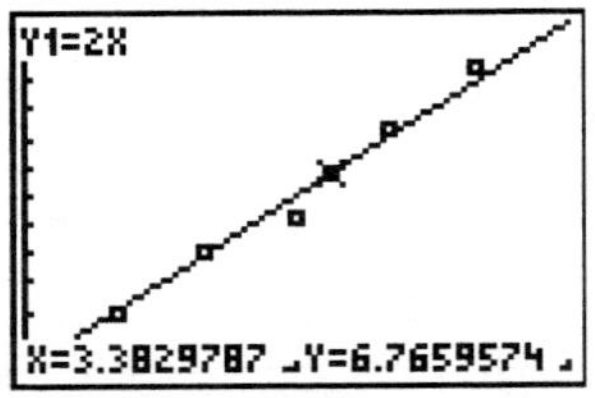

14. When 12 objects are added to the cup, the Slinky should stretch a total of 24 centimeters.

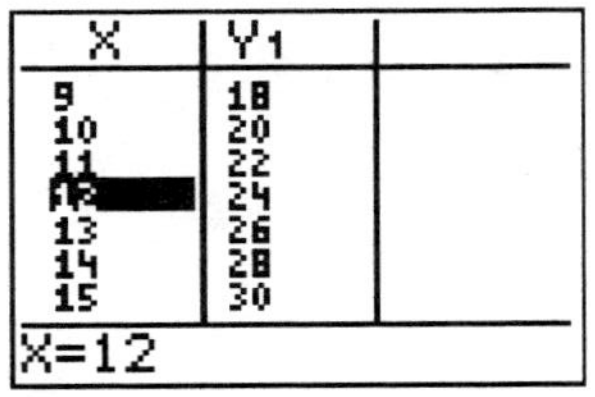

15. To make the Slinky stretch a total of 32 centimeters, 16 objects would have to be added to the cup.

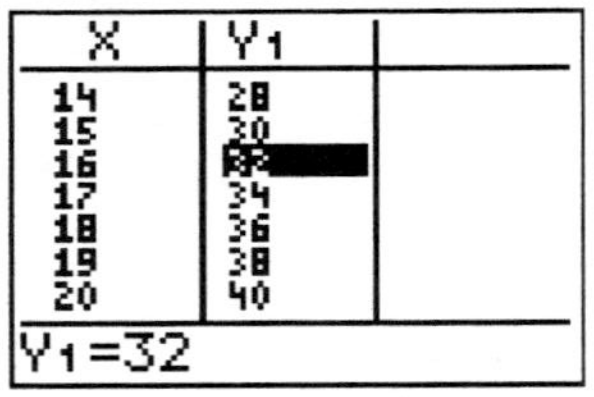

16. Sample answer: The ratio of the distance from the bottom of the table to the top of the cup to the number of objects placed in the cup is not a constant ratio. Therefore the relationship is not a proportional relationship.

17. Sample answer: The graph appears to be linear but does not pass through (0, 0).

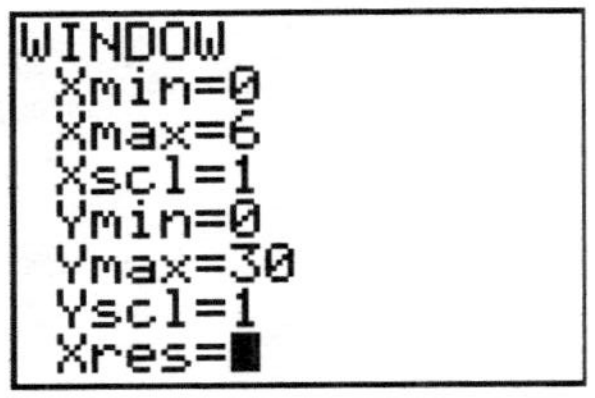

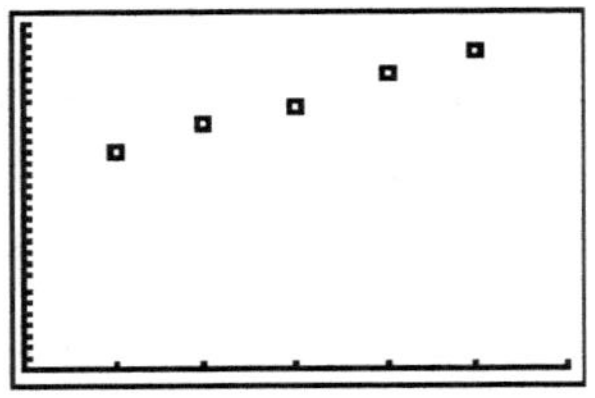

18. Sample answer: Although the graph appears to be linear, it does not pass through (0, 0).

19. The rate is 2. This means that for every object added to the cup, the Slinky stretches an additional 2 centimeters.

20. Sample answer: Both relationships have a rate of change of 2 centimeters per object. The graphs of the relationships appear to be 2 parallel lines.

SECTION 1.7 Assignment

For Questions 1 and 2, decide whether the relationship between the two variables is or is not a proportional relationship. Explain how you made your decision.

1.

x	0	0.2	0.6	2.5	6.1
y	0	1.2	3.6	15.0	36.6

2.

t	0	1.6	2.4	3.0
d	0	3.2	4.2	5.1

For Questions 3 and 4, decide whether the function rule describes a direct variation function. Explain how you made your decision.

3. $x \cdot y = 12$

4. $y = 0.3875x$

For Questions 5 and 6, decide whether the graph depicts a proportional relationship between the two variables. Explain how you made your decision.

5. The graph shown in **Figure 1.17**

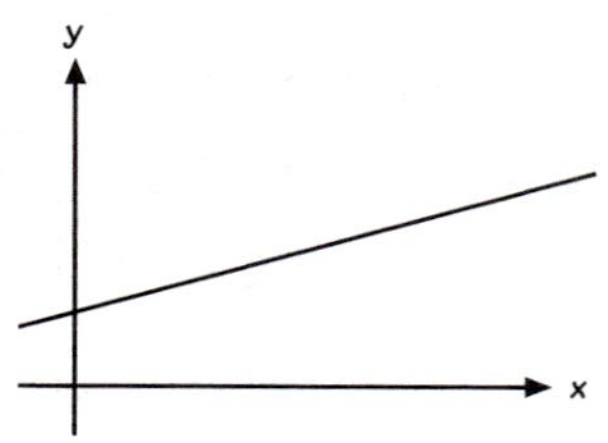

FIGURE 1.17.
A graph.

6. The graph shown in **Figure 1.18**

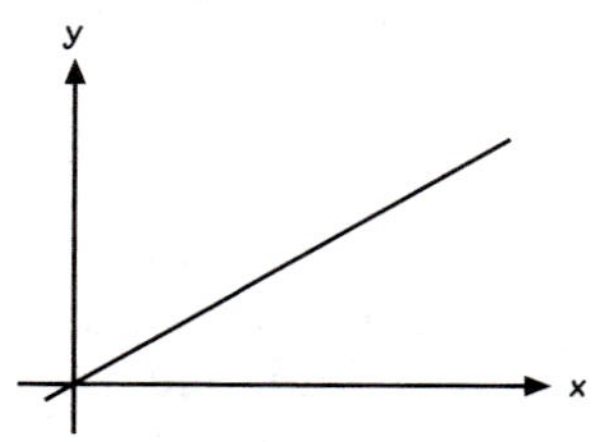

FIGURE 1.18.
A graph.

For Questions 7 and 8, let the relationship between x and y be a proportional relationship, and let $y = 12$ when $x = 5$.

7. What is the constant of proportionality?

8. If $x = 11$, what is the value of y?

Use the following information for Questions 9–17. Marbles were placed on a scale and weighed. The collected data are shown in **Figure 1.19.**

Number of Marbles (m)	Mass in Ounces (w)
1	1.5
2	3.0
4	6.0
8	12.0
10	15.0

FIGURE 1.19.
Mass of marbles.

9. What is the independent variable in this situation? Why?

10. What is the dependent variable in this situation? Why?

11. What are a reasonable domain and range for this situation?

12. Create a scatter plot of mass in ounces versus number of marbles.

13. Is this a proportional relationship? How do you know?

14. Find the rate of change between two pairs in the table. Do the same for two other pairs. What do you notice?

15. What is a function rule that models this relationship?

16. Predict the weight of 27 marbles.

17. Predict the number of marbles that have a mass of 54 ounces.

Section 1.7 Assignment

Answers

1. Yes, compare the ratio of $\frac{y}{x}$ for each pair.

   $\frac{1.2}{0.2} = 6 \quad \frac{3.6}{0.6} = 6 \quad \frac{15.0}{2.5} = 6 \quad \frac{36.6}{6.1} = 6$

   Since the ratios are all equal and the table includes the point (0, 0), the relationship between x and y is a proportional relationship.

2. No, compare the ratio of $\frac{d}{t}$ for each pair. Note, however, that since you cannot divide by 0, you cannot compute the ratio for the first pair.

   $\frac{3.2}{1.6} = 2 \quad \frac{4.2}{2.4} = 1.75 \quad \frac{5.1}{3.0} = 1.7$

   Since the ratios are not all equal, the relationship between x and y is not a proportional relationship.

3. No, solve the equation for y in terms of x:

   $xy = 12$

   $\frac{xy}{x} = \frac{12}{x}$

   $y = \frac{12}{x}$

   A direct variation function must be of the form $y = kx$, and the given equation is not. So this function is not a direct variation.

4. Yes, this function is of the form $y = kx$, so it is a direct variation.

5. No, the graph is a line, but it doesn't go through the origin, so it is not the graph of a proportional relationship.

6. Yes, the graph is linear and passes through the origin, so it could be the graph of a proportional relationship.

7. Since the relationship is a direct variation, the constant of proportionality is equal to the ratio of $\frac{y}{x}$ for any data pair. Determine the ratio of $\frac{y}{x}$ for the given pair:

   $\frac{y}{x} = \frac{12}{5} = 2.4$

8. Since the constant of proportionality is 2.4, substitute 2.4 into the generic form of a direct variation equation, $y = kx$:

   $y = 2.4x$

   Now substitute $x = 11$ into this equation and simplify.

   $y = 2.4(11) = 26.4$

9. The number of marbles on the scale is the independent variable because it explains the mass and it is the variable that is being controlled.

10. The mass of the marbles is the dependent variable because that is what results from the number of marbles on the scale.

11. The number of marbles must be a whole number, so a reasonable domain is all whole numbers. If there is a limited number of marbles available, there will be a maximum value. According to the table, the domain is whole numbers between 1 and 10.

    It appears that the mass in ounces could be a decimal number, but mass must be positive; therefore, a reasonable range is a set of positive real numbers. Again, if there is a limited number of marbles available, there is a limit to the total mass of the marbles. According to the data table, the range of the marbles' mass is real numbers between 1.5 and 15.0.

12. Sample answer:

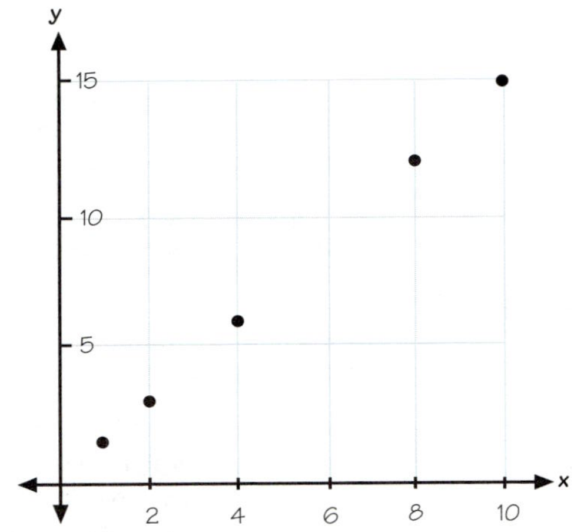

13. Yes, compare the ratios of the mass in ounces to the number of marbles for each pair:

    $\frac{1.5}{1} = 1.5 \quad \frac{3.0}{2} = 1.5 \quad \frac{6.0}{4} = 1.5$

    $\frac{12.0}{8} = 1.5 \quad \frac{15.0}{10} = 1.5$

    Since all of the ratios are equal, the relationship is a proportional relationship.

14. The rate of change is equal to the constant of proportionality, 1.5.

15. $w = 1.5m$

16. $w = 1.5(27) = 40.5$ ounces

17. $54 = 1.5m$, so $m = \frac{54}{1.5} = 36$ marbles

SECTION 1.8

Variation on a Theme

In the previous section, you collected and analyzed data for the relationship between the amount of stretch in a Slinky and the number of marbles in a plastic cup.

You have already collected data and answered some questions about what you found. Now you will look at some important ideas about data collection and building models based on those data.

1. How did you decide which is the independent and which is the dependent variable for this situation?
2. How did you decide on a reasonable domain and range for this situation?
3. Why did you sketch your predicted scatter plots the way you did?

4. How did you use your table to determine if the relationship between total stretch and the number of objects in the cup is a proportional relationship? What happens when you divide w by n?

NEGATIVE RELATIONSHIP

A negative relationship (correlation) is one in which as one variable increases, the other decreases.

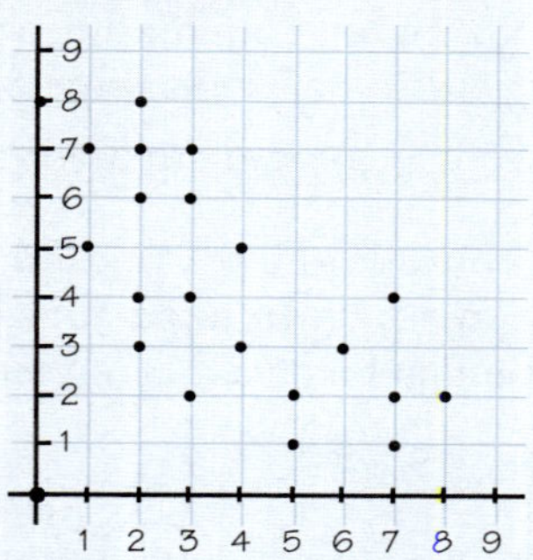

NO CORRELATION

If no trend is apparent between two variables, then we say there is no correlation between the variables.

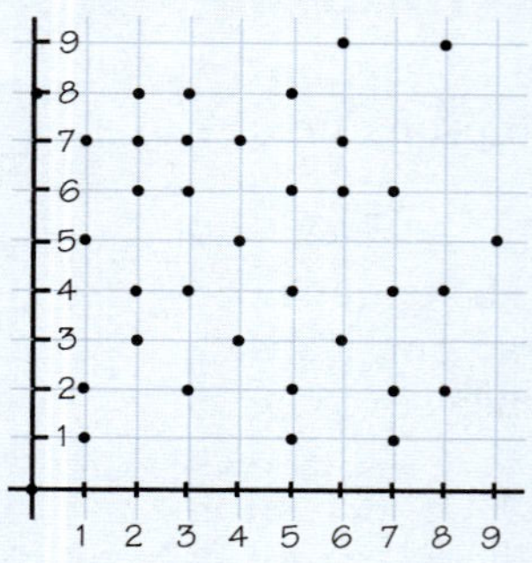

5. Why do you think that there is a positive relationship between the total stretch of the Slinky and the number of objects in the cup?
6. What do you think is meant by a negative relationship?
7. What do you think is meant by no relationship?
8. How did you determine rates of change in your table?

To find the rate of change $\frac{\Delta w}{\Delta n}$ for any two pairs of values in the table, subtract the values of the dependent variable and divide by the corresponding difference in the values of the independent variable.

9. Find the rate of change for the pairs (2 objects, 4 cm) and (5 objects, 10 cm).
10. Explain how the rate of change for pairs of values in the table appears in the graph of this relationship.
11. In a proportional relationship, how can you use the rate of change and the values of the independent variable to find the values of the dependent variable?
12. Write a function rule to model the relationship between the total stretch of the Slinky, w, and the total number of objects in the container, n.

For a linear function including direct variation functions, the rate of change is also the slope of the line. In terms of x and y, the formula for the slope of the line through (x_1, y_1), (x_2, y_2) is $m = \frac{\Delta y}{\Delta x} = \frac{y_2 - y_1}{x_2 - x_1}$. In Question 12, you used your table to write a direct variation function rule, $w = 2n$, to relate the variables. In terms of x and y, the equation of the line is $y = 2x$.

13. Show how you can use the slope formula to determine the slope of the line that relates the total amount of stretch to the number of objects in the cup.

Another way to find the equation of a line through (x_1, y_1) and (x_2, y_2) is to use the **point-slope formula**: $y - y_1 = m(x - x_1)$.

14. Show how you can use the point-slope formula to determine the equation of the line that relates the total amount of stretch to the number of objects in the cup.

EXPLAIN LESSON 2

Section 1.8: Variation on a Theme

Purpose: Section 1.8 formalizes the mathematical terminology and characteristics of direct variation.

Description: This section is designed to be a teacher-led discussion of the concepts in the lesson. Use the questions below to facilitate the class discussion.

Materials: graphing calculator

Teaching Ideas: Debrief students after Section 1.7 by having them share their answers to the questions. As they share, be sure to fill in any missing concept, process, or skill that could have been used to answer the questions. Then work through the questions in this section with the entire class. Once again, be sure all necessary concepts, processes, or skills are addressed.

In this section our data was modified so that the concepts involved could be clearly communicated to the students. Students' actual data will most likely be messier than the data used in the explanation. The point is that students understand the concepts and that you reinforce that in modeling real-world situations. Often we must look at trends because data collection does not always yield perfect data. This would be a good time to talk about sources of error. When students collect data, what experimental conditions create irregular measurements? How precise are the measurements being taken? What degree of rounding are students using?

Answers

1. Answers may vary. Since the number of objects in the cup explains the stretch, it is the independent variable.

2. The stretch of the Slinky (the range) will never be less than zero or more than the distance to the floor. The number of objects (the domain) cannot be less than zero or greater than the number of objects necessary to fill the cup. Students should understand that by observing the collected data and the reasonableness of possible values in the situation, they can determine a domain and range.

3. Sample answer: For total stretch versus the number of objects: As the number of objects increases, the length of the Slinky should increase at a constant rate so the graph should be linear and increase from left to right.

   For the distance from the bottom of the table to the top of the cup versus the number of objects placed in the cup: The distance from the bottom of the table to the top of the cup begins at some value other than 0 when the cup is empty and increases at a constant rate as objects are added, so the graph has a *y*-intercept other than 0, is linear, and increases from left to right.

4. Answers will vary. Guide students through a discussion of the dependent variable divided by the independent variable. This value turns out to be a constant. This constant *k* is the constant of proportionality. A proportional relationship includes (0, 0) and has a constant rate of change. Make sure that students see that the rate of change and the constant of proportionality are the same value.

n, Number of Objects in the Cup	*w*, Total Stretch of the Slinky (cm)	
0	0	
1	2	$\frac{w}{n} = \frac{2\text{ cm}}{1\text{ object}} = 2\ \frac{\text{cm}}{\text{object}}$
2	4	$\frac{w}{n} = \frac{4\text{ cm}}{2\text{ objects}} = 2\ \frac{\text{cm}}{\text{object}}$
3	6	$\frac{w}{n} = \frac{6\text{ cm}}{3\text{ objects}} = 2\ \frac{\text{cm}}{\text{object}}$
4	8	$\frac{w}{n} = \frac{8\text{ cm}}{4\text{ objects}} = 2\ \frac{\text{cm}}{\text{object}}$
5	10	$\frac{w}{n} = \frac{10\text{ cm}}{5\text{ objects}} = 2\ \frac{\text{cm}}{\text{object}}$

5. This is a positive relationship because as the number of objects increases, the stretch of the Slinky increases.

6. Sample answer: As one variable increases, the other decreases.

7. Sample answer: There is no pattern in the relationship of the variables.

8. Answers will vary. Guide students through a discussion of the change in the dependent and independent variables. Ask students to calculate the rate $\frac{w}{n}$ for several pairs of pairs in their tables.

Change in the Total Number of Objects in the Cup or Δn	*n*, Number of Objects in the Cup	*w*, Total Stretch of the Slinky (cm)	Change in the Total Stretch of the Slinky or Δw
	0	0	
+ 1 object	1	2	+ 2 cm
+ 1 object	2	4	+ 2 cm
+ 1 object	3	6	+ 2 cm
+ 1 object	4	8	+ 2 cm
+ 1 object	5	10	+ 2 cm

   The rate is constant at $\frac{2\text{ cm}}{1\text{ object}}$.

9. $\frac{w}{n} = \frac{10\text{ cm} - 4\text{ cm}}{5\text{ objects} - 2\text{ objects}} = \frac{6\text{ cm}}{3\text{ objects}} = \frac{2\text{ cm}}{1\text{ object}}$

Answers (continued)

10. The rate of change describes how to get from one point to another on the scatter plot. The numerator Δw of the rate $\frac{\Delta w}{\Delta n}$ is the vertical distance you move to find the next point on the scatter plot, while the denominator Δn is the horizontal distance you move to find the next point on the scatter plot.

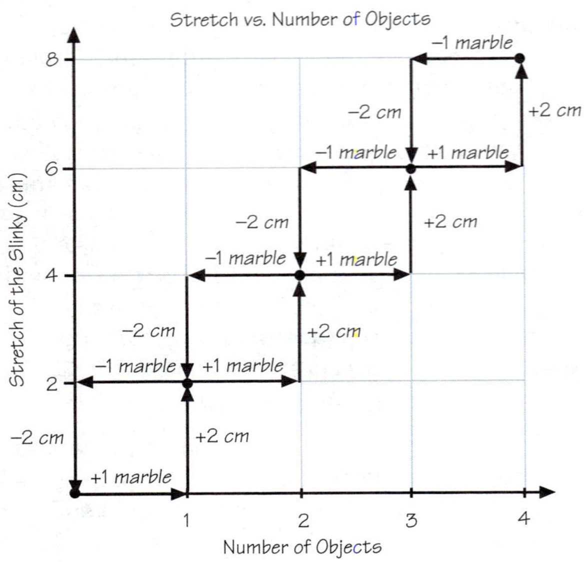

11. The value of the dependent variable is the rate times the value of the independent variable.

    For example, $\frac{2\text{ cm}}{1\text{ object}} \times 4$ objects = 8 cm.

12. $w = \frac{2\text{ cm}}{1\text{ object}} \times n$ objects or $w = 2n$

13. Select any 2 points (x_1, y_1), (x_2, y_2) say, (1, 2) and (5, 10). Substitute into the slope formula and simplify. $m = \frac{\Delta y}{\Delta x} = \frac{y_2 - y_1}{x_2 - x_1} = \frac{10-2}{5-1} = \frac{8}{4} = \frac{2}{1}$

14. Select any point, then substitute it and the slope of the line into the point-slope formula. If you pick (5, 10), substitute and simplify.

    $y - y_1 = m(x - x_1)$

    $y - 10 = 2(x - 5)$

    $y - 10 = 2x - 10$

    $y - 10 + 10 = 2x - 10 + 10$

    $y = 2x$

15. How can you predict the total stretch of the Slinky when 12 objects are added to the cup?

16. How can you predict the number of objects it takes to stretch the Slinky 32 cm?

Now, consider the situation where you compared the number of objects in the cup with the distance from the bottom of the table to the top of the cup. Data that one group gathered are in **Figure 1.20.** Use these data to answer the remaining questions.

Number of Objects in the Cup (n)	Distance (cm) from the Bottom of the Table to the Top of the Cup (d)
0	17
1	19
2	21
3	23
4	25
5	27

FIGURE 1.20.
More Slinky data.

17. How can you decide if this relationship is or is not proportional?

18. How can you find the slope of a function that models this relationship?

19. The ***y*-intercept** of a function's graph is the point where the graph crosses the y-axis. For the function that models this relationship, where do you find this point's coordinates in the table?

20. What is the meaning of the y-intercept in this situation?

Another representation for a linear function is called **slope-intercept form**: $y = mx + b$. This general form of a linear function expresses the dependent variable, y, in terms of the independent variable, x, the slope of the line (m), and the y-intercept of the line whose coordinates are $(0, b)$.

The slope-intercept form of a linear function is useful when you know the rate of change, or slope, and the starting point, or y-intercept, for a relationship that can be described with a linear function.

21. Use your answers to Questions 18 and 19 to write a function rule that can be used to find the distance between the bottom of the table and the top of the cup for n marbles.

CHECK THIS!

Recall that the slope of the line through (x_1, y_1), (x_2, y_2) is

$$m = \frac{\Delta y}{\Delta x} = \frac{y_2 - y_1}{x_2 - x_1}.$$

22. How does the function relating the distance from the table to the cup and the number of objects in the cup (Question 21) compare to the function relating the total stretch of the Slinky and the number of objects in the cup (Question 14)?

Answers (continued)

15. Sample answer: Students may use a table, substitution, or the graph to solve this problem.

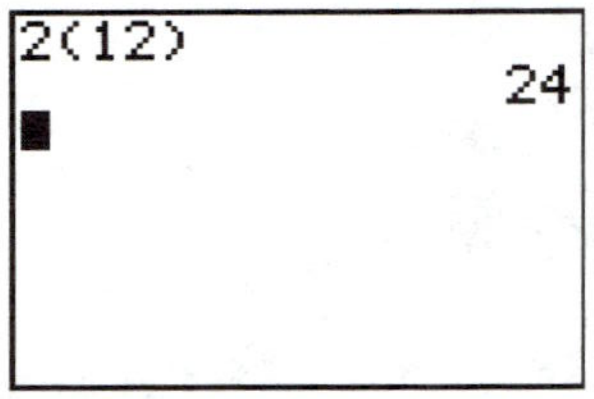

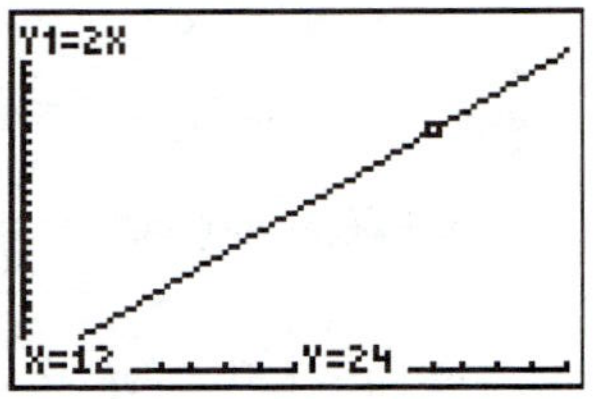

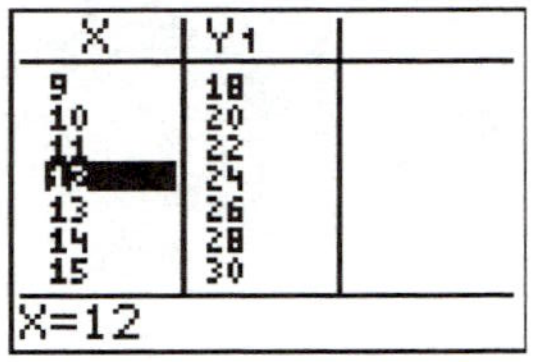

16. Sample answer: Students may use a table, solve an equation, or use the graph to solve this problem.

$y = 2x$

$32 = 2x$

$\frac{32}{2} = \frac{2x}{2}$

$16 = x$

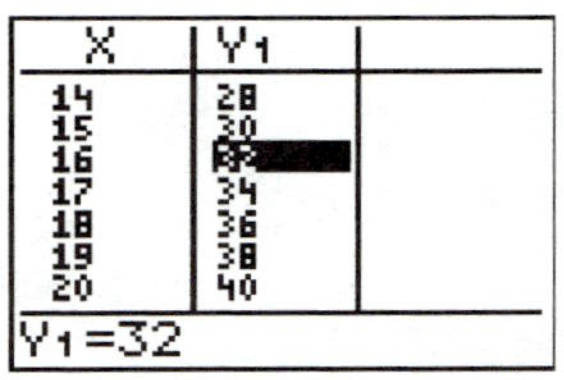

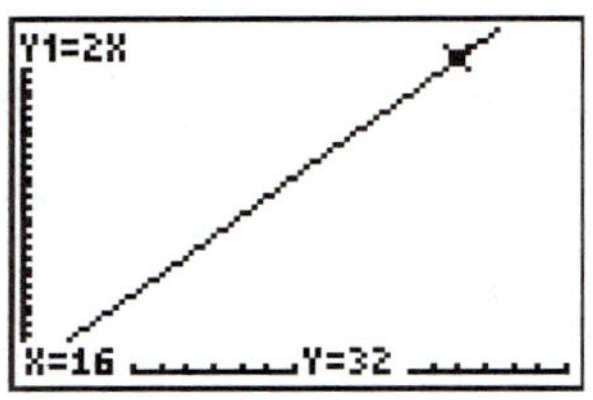

17. Answers may vary. Sample answer: Compare the ratio of the distance to the number of objects (d to n).

$\frac{19}{1} = 19 \quad \frac{21}{2} = 11.5 \quad \frac{23}{3} = 7\frac{2}{3} \quad \frac{25}{4} = 6.25 \quad \frac{27}{5} = 5.4$

Since the ratios are not equal, the relationship is not proportional.

18. Answers may vary. Students may have used a table similar to this one:

Change in the Total Number of Objects in the Cup or Δn	Number of Objects in the Cup (n)	Distance (in cm) from the Table to the Cup (d)	Change in Total Stretch of the Slinky or Δd
	0	17	
+1 object	1	19	+2 cm
+1 object	2	21	+2 cm
+1 object	3	23	+2 cm
+1 object	4	25	+2 cm
+1 object	5	27	+2 cm

The slope is constant at $\frac{2 \text{ cm}}{1 \text{ object}}$.

19. The y-intercept is (0, 17). This is the value of the dependent variable, d, when the independent variable, n, is equal to 0.

20. The point (0, 17) represents the fact that when there are 0 marbles inside the cup, the top of the cup is located 17 centimeters below the bottom of the table.

21. $y = 2x + 17$. We know from Question 18 that the slope of the linear function is +2. We know from Question 19 that the y-intercept is (0, 17). Thus, $m = 2$ and $b = 17$.

22. Answers may vary. Sample answer: The functions have the same slope but different y-intercepts. The slope represents the change in the length of the Slinky, which is the same as the change in the distance between the table and the cup for each marble added. When there are no marbles in the cup, there is no stretch of the Slinky. However, the top of the cup is 17 centimeters below the bottom of the table. So the functions have different values when x is 0 but increase at the same rate. The graphs of these two functions are parallel lines.

SECTION 1.8 Assignment

Predicting values from a proportional relationship can be done using a **proportion** (two fractions that are equal to each other), by writing an equation that looks like this:

$$\frac{12}{5} = \frac{y}{11}.$$

In this proportion 12 and 5 are corresponding values of y and x. The y-value that corresponds to the x-value 11 is not known.

1. Why must the two fractions be equal?

One algebraic method to solve a proportion involves "cross-multiplying" in the manner shown in **Figure 1.21**, and setting the two products equal to each other.

FIGURE 1.21.
Cross-multiplying.

2. In this case, what are the two products formed?

3. What equation do you get when you set the products equal to each other?

4. How do you solve that equation for y?

Another way to solve the equation $\frac{12}{5} = \frac{y}{11}$ is to multiply both sides of the equation by 11.

5. Why does that one step *solve* the equation?

6. In order to find the answer using this method, what arithmetic steps must you do? How does this compare to the arithmetic you did in Questions 2 through 4?

Solve each of the following proportions:

7. $\frac{4}{x} = \frac{15}{23}$

8. $\frac{7}{6} = \frac{y}{25}$

9. x and y form a proportional relationship. When $x = 13$, $y = 5$. Find x, when $y = 13$.

Write a direct variation function rule to model each of the following situations. Be sure to list the constant of proportionality and its units. Then use the rule to solve the problem.

10. On a map scale, 1 cm represents a distance of 25 miles. Two cities on the map are 3.2 cm apart. What actual distance separates the two cities?

11. On a typing test, Jane typed 98 words correctly in 2 minutes. Assuming she can type for longer periods of time without losing her concentration, how many minutes would it take her to type an essay that contains 343 words?

12. Hanging 3 washers on a spring stretches it a total of 4.5 cm. If 13 washers are placed on it instead, how far will the spring stretch? (Assume that the spring can be stretched several feet without problems.)

A local theme park charges $15 for admission. The amount of revenue generated by admission sales can be found using the function rule $R = 15n$, where R is the total revenue and n is the number of admissions sold.

13. Complete the table in **Figure 1.22**, and then use that information to answer Questions 14–15.

n	2	20		1000	
R			3000		30,000

FIGURE 1.22.
Theme park admissions.

14. For each of the five pairs in the table, calculate the ratio $\frac{R}{n}$. What do you notice?

15. Find $\frac{\Delta R}{\Delta n}$ (the rate of change of the revenue compared to the number of tickets sold) between the first and second entries of the table. What units does your answer have?

In Questions 16–18, find the slope of the line.

16. The line whose equation is $y = \frac{4}{3}x + 12$

17. The line that goes through the points $P(4, 9)$ and $Q(7, 21)$

18. The line that goes through the points $P(12, 5)$ and $Q(2, 8)$

EXPLAIN LESSON 2

Section 1.8 Assignment

Answers

1. They are equal because the ratio y/x is constant in a proportional relationship, and both sides of the proportion form a ratio comparing a value of y to a value of x.
2. The products are $11(12) = 132$, and $5(y) = 5y$.
3. $5y = 132$
4. Divide both sides by 5.
5. Multiplying both sides by 11 eliminates the denominator of the fraction on the right side (i.e., "undivides by 11") that contains the variable, leaving y by itself.
6. Multiply 11 by 12, then divide by 5. These are the same arithmetic steps in both methods!
7. $x \approx 6.13$
8. $y \approx 29.17$
9. Since x and y form a proportional relationship, $\frac{y}{x}$ is a constant ratio. So we can set up the proportion $\frac{5}{13} = \frac{13}{x}$ and solve.

   $\frac{5}{13} = \frac{13}{x}$

   $5x = 13(13)$

   $5x = 169$

   $\frac{5x}{5} = \frac{169}{5}$

   $x = 33.8$

10. $\frac{25 \text{ miles}}{1 \text{ cm}} = 25$ miles per centimeter, so $y = 25x$ where y represents the distance in miles and x represents the distance in centimeters. Substitute $x = 3.2$ and simplify.

    $y = 25(3.2) = 80$ miles

11. $\frac{98 \text{ words}}{2 \text{ minutes}}$ words per minute, so $y = 49x$, where y represents the number of words and x represents the number of minutes. Substitute $y = 343$ and solve for x.

    $343 = 49x$

    $\frac{343}{49} = \frac{49x}{49}$

    $x = 7$ minutes

12. $\frac{4.5 \text{ cm}}{3 \text{ washers}} = 1.5$ centimeters per washer, so $y = 1.5x$ where y represents the stretch in centimeters and x represents the number of washers. Substitute $x = 13$ and simplify.

    $y = 1.5(13) = 19.5$ cm

13.

n	2	20	200	1,000	2,000
R	30	300	3000	15,000	30,000

14. $\frac{30}{2} = 15$ $\frac{300}{20} = 15$ $\frac{3000}{200} = 15$

    $\frac{15,000}{1000} = 15$ $\frac{30,000}{2000} = 15$

    All of the ratios are equal.

15. $\Delta n = 20 - 2 = 18$, and $\Delta R = 300 - 30 = 270$, so the rate of change is calculated as $\frac{\Delta R}{\Delta n} = \frac{270}{18} = 15$ dollars per ticket.
16. The value for m in the equation is $\frac{4}{3}$, so the slope is $\frac{4}{3}$.
17. $m = \frac{\Delta y}{\Delta x} = \frac{y_2 - y_1}{x_2 - x_1} = \frac{21-9}{7-4} = \frac{12}{3} = 4$
18. $m = \frac{\Delta y}{\Delta x} = \frac{y_2 - y_1}{x_2 - x_1} = \frac{8-5}{2-12} = \frac{3}{-10} = -\frac{3}{10}$

SECTION 1.9 This Floors Me

Recall that in Section 1.7 you took three measurements during the Hooke's Law experiment. You measured the total stretch of the Slinky, the distance from the bottom of table to the top of cup, and the distance from the top of the cup to the floor (**Figure 1.23**). In this section you will investigate the relationship between the number of objects placed in the cup and the distance from the top of the cup to the floor.

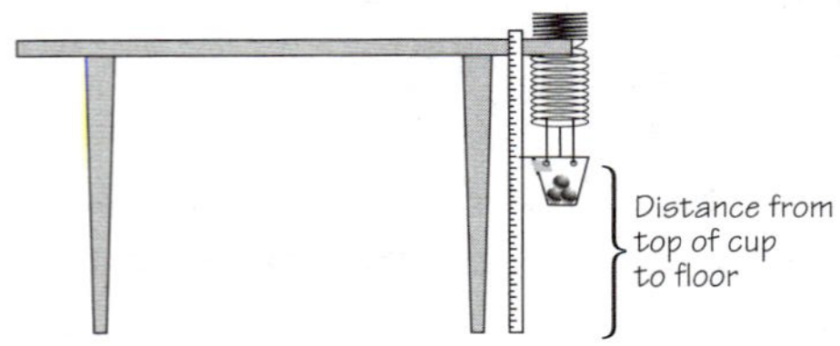

FIGURE 1.23. Slinky stretch.

Copy your data from Section 1.7 into a table like **Figure 1.24.**

Number of Objects in the Cup	Distance from the Top of the Cup to the Floor (cm)
0	
1	
2	
3	
4	
5	

FIGURE 1.24. Distance from cup to floor data.

1. In the relationship between the distance from the top of the cup to the floor and the number of objects in the cup, what is the independent variable? What is the dependent variable? What are a reasonable domain and range for this situation? Why?

2. Is the relationship between the distance from the top of the cup to the floor and the number of objects in the cup a proportional relationship? Use your table to justify your answer.

3. Suppose the starting position (empty cup) is 10 cm. How would this change your table?

4. Sketch what you predict a scatter plot of the distance from the top of the cup to the floor versus the number of objects in the cup would look like. Explain why you made the scatter plot the way you did.

5. Use the data from your table to create a scatter plot of the distance from the top of the cup to the floor versus the number of objects in the cup. Record your window, sketch and describe your graph.

6. Does your scatter plot show a positive, negative, or no correlation between the distance from the top of the cup to the floor and the number of objects in the cup? How do you know?

7. What characteristics of your graph verify that distance from the top of the cup to the floor versus the number of objects in the cup is or is not a proportional relationship?

8. Suppose the top of your cup is 10 cm higher. How would this change your graph?

CHECK THIS!

Recall that the rate of change compares the change in the value of the dependent variable to change in the value of the independent variable.

CHECK THIS!

Recall that the rate of change is also the slope of the linear function that models a linear relationship.

9. Find the rate of change between successive pairs of values in your table. To estimate the slope of a linear model for this relationship, find the average of these rates of change. What is the meaning of the slope in this situation?

CHECK THIS!

Recall that the y-intercept is the point where the graph of a function intersects the y-axis, which happens when $x = 0$. The y-intercept's coordinates are $(0, y)$.

ELABORATE LESSON 2

Section 1.9: This Floors Me

Purpose: In Sections 1.7 and 1.8, students formalized different methods to write linear functions to describe data from proportional and non-proportional situations. In Section 1.9 students investigate data yielding a negative relationship and formalize the idea of negative correlation. This section prepares students to demonstrate their understanding of developing a function to model a relationship in Section 1.10.

Description: In this section students apply and extend their knowledge. Having worked with linear data whose model passes through the origin and with linear data whose model has a non-zero *y*-intercept, students now work with linear data with a non-zero *y*-intercept that has a decreasing rate of change, or negative slope. Students do not need to collect new data; they can use data they collected in Section 1.7 but have not yet analyzed.

Materials: graphing calculator, data from Section 1.7

Teaching Ideas: Students should work in groups of 3 or 4. Encourage students to communicate with each other and assist each other as necessary. Remind students of the different ways to develop function models that they used in Section 1.8.

Answers

Sample data: Answers that follow are based on these data.

Number of Objects in the Cup	Distance from the Top of the Cup to the Floor (cm)
0	30.0
1	28.1
2	26.0
3	24.6
4	21.5
5	19.5

1. The independent variable is the number of objects added to the cup. The dependent variable is the number of centimeters from the top of the cup to the floor. The domain is the number of objects added to the cup; therefore, whole numbers from 0 to 10 would be reasonable for the domain. The range is the number of centimeters from the top of the cup to the floor. Since an average tabletop is about 75 centimeters above the floor, real numbers from 0 to 70 would be reasonable for the range.
2. Sample answer: The ratio of the distance from the top of the cup to the floor to the number of objects in the cup is not a constant ratio. Therefore there is no constant of proportionality, and the relationship is not proportional.
3. Each value in the second column of the table would increase by 10.
4. Sample answer: As the number of objects increases, the distance from the top of the cup to the floor should decrease at a constant rate.

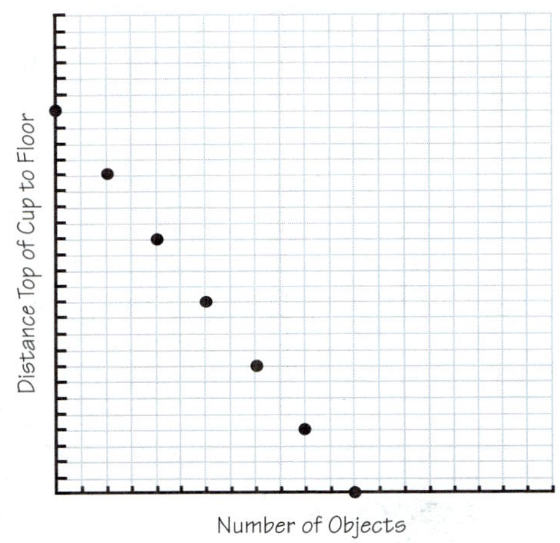

5. Sample answer: The graph decreases from left to right and appears to be linear.

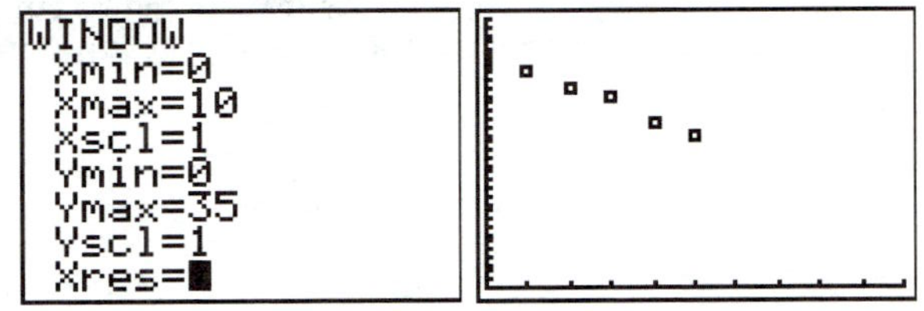

6. The graph shows a negative correlation because the dependent variable decreases when the independent variable increases.
7. Sample answer: Although the graph appears to be linear, it does not pass through (0, 0); therefore, the relationship is not proportional.
8. Sample answer: The graph would shift vertically a distance of 10.
9. 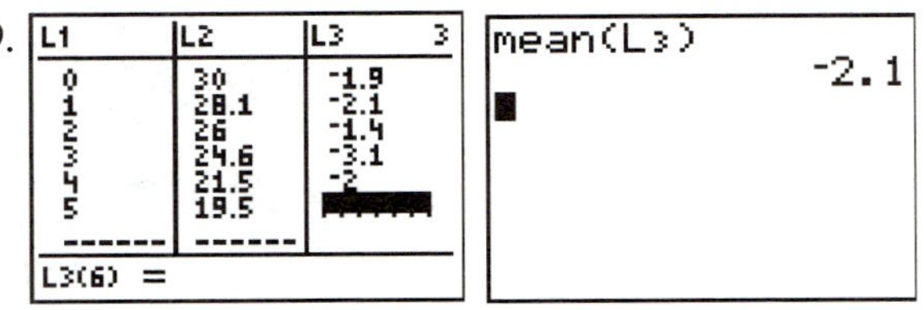

Since the mean of the changes is –2.1, the slope is –2.1. This means that for every object added to the cup, the top of the cup will get about 2.1 centimeters closer to the floor.

10. What is the y-intercept of the graph of a linear function that models the relationship between the distance from the top of the cup to the floor and the number of objects in the cup? What is the meaning of the y-intercept in this situation?

CHECK THIS!

Recall that the slope-intercept form of a linear equation is $y = mx + b$, where m is the slope of the line and b is the y-intercept of the line.

11. Write the slope-intercept form of a function rule that models the relationship between the distance from the top of the cup to the floor and the number of objects in the cup.

12. Suppose the top of your cup is 10 cm higher. Write the new function rule.

13. Graph your function rule on your scatter plot. Sketch your results. Do you think your function models the data well?

14. Predict the distance from the top of the cup to the floor when 9 objects are added to the cup.

15. Predict the smallest number of objects to place in the cup to make the bottom of the cup touch the floor.

SECTION 1.9 Assignment

In Questions 1–3, an equation in slope-intercept form is provided. Use the equation to fill in the table. Then calculate Δx, Δy, and the ratio $\frac{\Delta y}{\Delta x}$ for successive pairs. (In these questions, x and y have no units.)

1. $y = 3x - 8$

x	y
4	
8	
12	
16	

2. $y = 1.2x + 15$

x	y
5	
7	
9	
11	

3. $y = 4.1x - 12$

x	y
13	
19	

Use the tables from Questions 1–3 to answer Questions 4–6.

4. In a table, do the values of the input variable (x) <u>have</u> to increase by a constant amount?

5. What is the smallest number of ordered pairs needed to calculate the ratio $\frac{\Delta y}{\Delta x}$?

6. What is the relationship between the rate of change ratio $\frac{\Delta y}{\Delta x}$ and the equation given?

Answers (continued)

10. The y-intercept is (0, 30). This means that when there are no marbles in the cup, the top of the cup is 30 centimeters above the floor.

11. $y = -2.1x + 30$

12. $y = -2.1x + 40$

13. This function appears to be a good model.

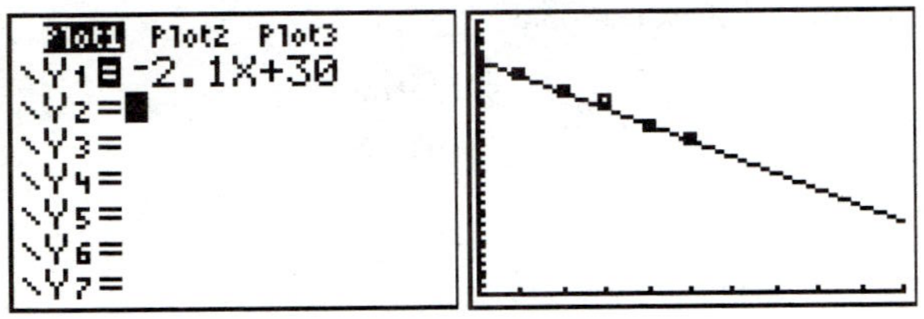

14. When 9 objects are added to the cup, the top of the cup is 11.1 centimeters above the floor.

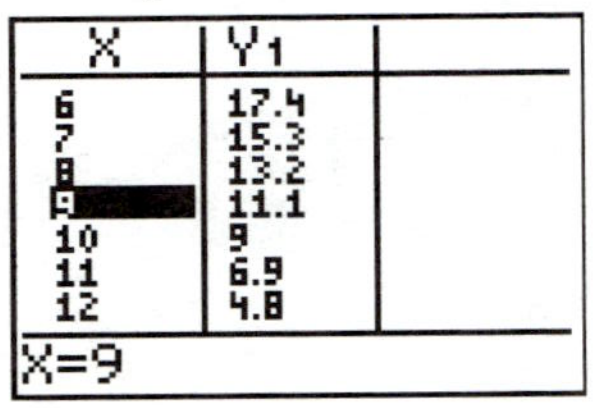

15. Assume that the height of the cup is 10 centimeters. When the 10th object is added to the cup, the top of the cup should be 9 centimeters above the floor. When the top of the cup is at this distance above the floor, the bottom of the cup is 1 centimeter below the floor. Since it is not possible to go below the floor, when the 10th object is added the cup will hit the floor.

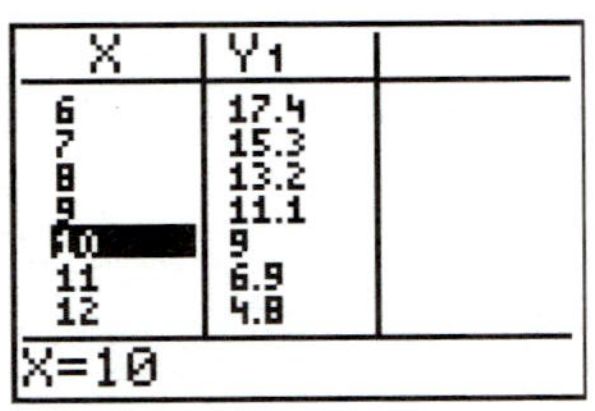

Section 1.9 Assignment

1. $\Delta x = 4$; $\Delta y = 12$; $\frac{\Delta y}{\Delta x} = \frac{12}{4} = 3$

x	y
4	4
8	16
12	28
16	40

2. $\Delta x = 2$; $\Delta y = 2.4$; $\frac{\Delta y}{\Delta x} = \frac{2.4}{2} = 1.2$

x	y
5	21.0
7	23.4
9	25.8
11	28.2

3. $\Delta x = 6$; $\Delta y = 24.6$; $\frac{\Delta y}{\Delta x} = \frac{24.6}{6} = 4.1$

x	y
13	41.3
19	65.9

4. No, but it's much easier when they do, since for linear equations that would make the output variable also go up by a constant amount.

5. You need two ordered pairs so there can be some kind of change that takes place in going from the first pair to the second one.

6. The rate of change ratio is the value for m (the slope) in the equation.

7. The following functions were graphed with a computer drawing utility: $y = 1.2x + 5$, $y = 0.8x + 5$, $y = 2.3x - 5$, $y = -0.5x + 10$, and $y = -1.4x + 10$. The graphs are shown in **Figure 1.26.**

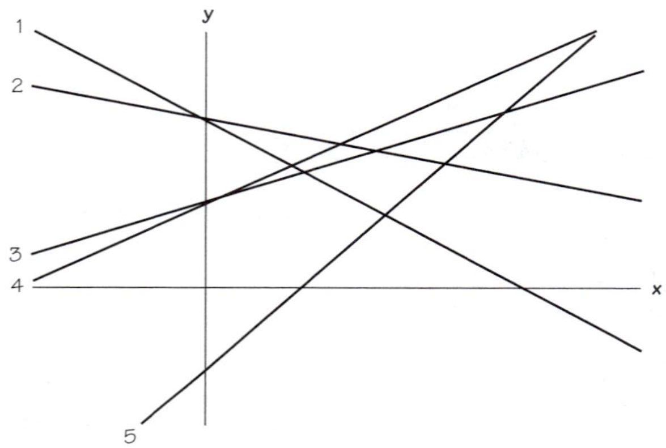

FIGURE 1.26.
Graph of 5 functions.

The order in which the lines were drawn is unknown, and the axes don't show a scale. Explain how you can identify the graph of each equation.

For Questions 8–11, find the equation of the line that goes through the two points given. Express your answer in slope-intercept form.

8. $P(-4, 12)$ and $Q(2, 3)$

9. $P(3, 8)$ and $Q(7, 15)$

10. $P(5, 12)$ and $Q(9, 14)$

11. $P(4, 14)$ and $Q(8, 4)$

Recall that the equation for a line in point-slope form looks like this: $y - k = m(x - h)$.

12. Graphs of direct variation functions always go through the origin (0, 0). What does $y = mx$ look like if written in point-slope form?

13. If a function is written in slope-intercept form, like $y = 2x + 5$, you already know one point on its graph, the y-intercept. What are the coordinates of that point? What does the point-slope form of that same equation look like?

14. Suppose that Garth Brooks can sell out the Cotton Bowl, which has 25,704 seats, when the ticket price is set at \$50, but a crowd of 16,000 is estimated when the admission is raised to \$80. Assume that ticket sales drop off in a linear pattern as the price increases. Find a function rule to model the relationship between the ticket sales and ticket price.

Use the following information for Questions 15–17.

A concert promoter keeps track of advance ticket sales. Those sales can be used to estimate the attendance at the performance and to help determine how much help is needed in planning for the event. The data shown in **Figure 1.27** compares ticket sales 4 weeks in advance of several concerts with the actual attendance.

Advance Ticket Sales (*A*)	Actual Attendance (*n*)
5628	13,581
7043	15,902
8912	19,873
9117	21,683
9741	22,705

FIGURE 1.27.
Ticket sales.

15. Use the grid in **Figure 1.28** to make a scatter plot that shows the relationship between the actual attendance and the advance ticket sales.

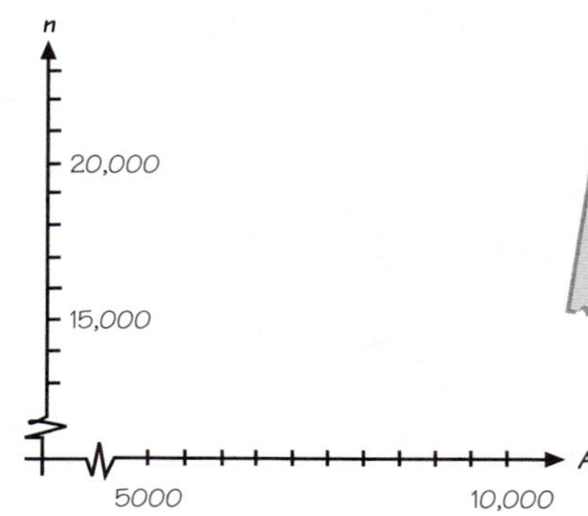

FIGURE 1.28.
Grid for ticket sales data.

16. When the pattern in a scatter plot is nearly linear, but not perfectly so, you can estimate a linear model that you can use to make fairly accurate predictions. Draw a line on the scatter plot you made in Question 15 that you think captures the trend of the data fairly well. Then find an equation for the line. Explain how you found the equation.

17. If a performance features a popular act, would you expect the data point to be above the line or below it? Explain.

Answers (continued)

7. Graphs 1 and 2 have the highest y-intercepts, and both have negative slopes. Graph 1 is steeper, so it has the equation $y = -1.4x + 10$, and graph 2 has the equation $y = -0.5x + 10$. Graphs 3 and 4 also have positive y-intercepts but clearly below the other two. Graph 3 is not as steep as graph 4 so it has the equation $y = 0.8x + 5$, and graph 4 has the equation $y = 1.2x + 5$. Graph 5 has the steepest slope and is the only one with a negative y-intercept, so its equation is $y = 2.3x - 5$.

8. $m = \frac{\Delta y}{\Delta x} = \frac{y_2 - y_1}{x_2 - x_1} = \frac{3-12}{2-(-4)} = \frac{-9}{6} = -\frac{3}{2}$

   Using slope-intercept form with the point (2, 3),

   $y = mx + b$

   $3 = \left(-\frac{3}{2}\right)(2) + b$

   $3 = -3 + b$

   $6 = b$

   Therefore, the equation of the line is:

   $y = -\frac{3}{2}x + 6.$

9. $m = \frac{\Delta y}{\Delta x} = \frac{y_2 - y_1}{x_2 - x_1} = \frac{15-8}{7-3} = \frac{7}{4}$

   Using slope-intercept form with the point (3, 8),

   $y = mx + b$

   $8 = \left(\frac{7}{4}\right)(3) + b$

   $8 = \frac{21}{4} + b$

   $\frac{32}{4} - \frac{21}{4} = \frac{21}{4} - \frac{21}{4} + b$

   $\frac{11}{4} = b$

   Therefore, the equation of the line is: $y = \frac{7}{4}x + \frac{11}{4}$.

10. $m = \frac{\Delta y}{\Delta x} = \frac{y_2 - y_1}{x_2 - x_1} = \frac{14-12}{9-5} = \frac{2}{4} = \frac{1}{2}$

    Using point-slope form with the point (5, 12),

    $y - y_1 = m(x - x_1)$

    $y - 12 = \frac{1}{2}(x - 5)$

    $y - 12 = \frac{1}{2}x - \frac{5}{2}$

    $y = \frac{1}{2}x - \frac{5}{2} + 12$

    $y = \frac{1}{2}x + \frac{19}{2}$

11. $m = \frac{\Delta y}{\Delta x} = \frac{y_2 - y_1}{x_2 - x_1} = \frac{14-4}{4-8} = \frac{10}{-4} = -\frac{5}{2}$

    Using point-slope form with the point (8, 4),

    $y - y_1 = m(x - x_1)$

    $y - 4 = -\frac{5}{2}(x - 8)$

    $y - 4 = -\frac{5}{2}x + 20$

    $y = -\frac{5}{2}x + 20 + 4$

    $y = -\frac{5}{2}x + 24$

12. $y - 0 = m(x - 0)$

13. For this example, the equation is $y - 5 = 2(x - 0)$ in point-slope form. In general, the y-intercept has coordinates $(0, b)$. So the point-slope form is $y - b = m(x - 0)$.

14. The two data points on which this model is based are (50; 25,704) and (80; 16,000). The slope of the line is

    $m = \frac{\Delta y}{\Delta x} = \frac{y_2 - y_1}{x_2 - x_1} = \frac{16,000 - 25,704}{80 - 50} \approx -323.5.$

    The point-slope form equation is $y - 25,704 = -323.5(x - 50)$, and in slope-intercept form that equation becomes $y = -323.5x + 41,879$.

15. Sample answer:

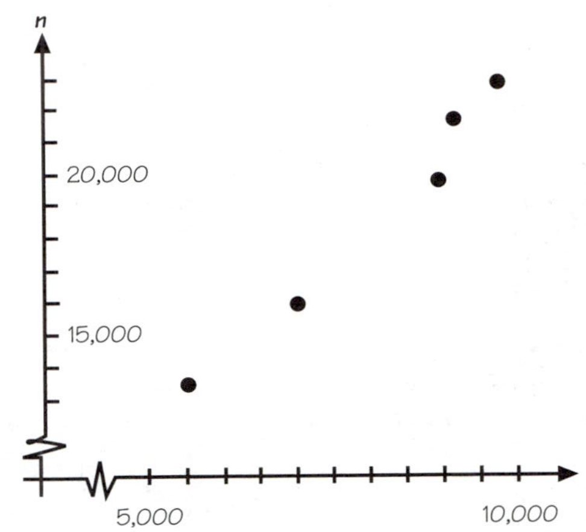

16. Sample answer:

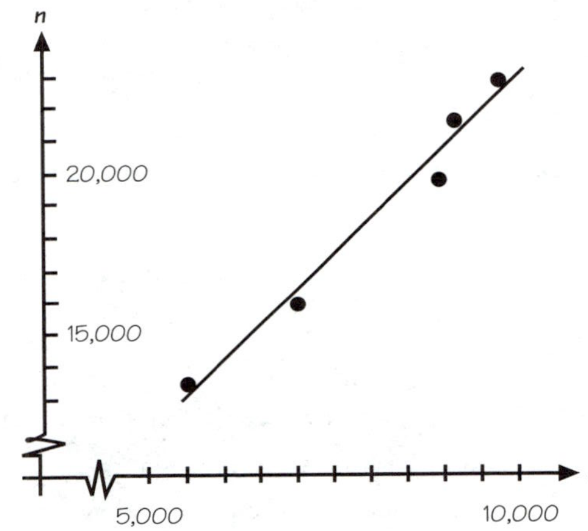

Estimate the coordinates of two points on the line to be (6000; 13,600) and (10,000; 22,700). The slope is $m = \frac{\Delta y}{\Delta x} = \frac{y_2 - y_1}{x_2 - x_1} = \frac{22,700 - 13,6000}{10,000 - 6000} = 2.275.$

The point-slope form is either $n - 13,600 = 2.275(A - 6,000)$ or $n - 22,700 = 2.275(A - 10,000)$. The slope-intercept form is $n = 2.275 \cdot A - 50$.

17. Answers will vary. Popular groups sell out quickly, so attendance is maximized; however, advanced ticket sales are also greater. If the data represent sold-out shows at various venues, then the data point should be below the line.

SECTION 1.10 Distance vs. Ramp Height

After investigating the relationship between the number of objects in a cup and the stretch of a Slinky, the students in the Math Club decided to roll a tennis ball down a ramp and measure how far it travels. To see if there is a relationship between the distance that a tennis ball rolls and the steepness of the ramp, they stacked books beneath one side of the ramp (**Figure 1.29**). As they changed the number of books, they rolled the ball and measured the distance of the roll.

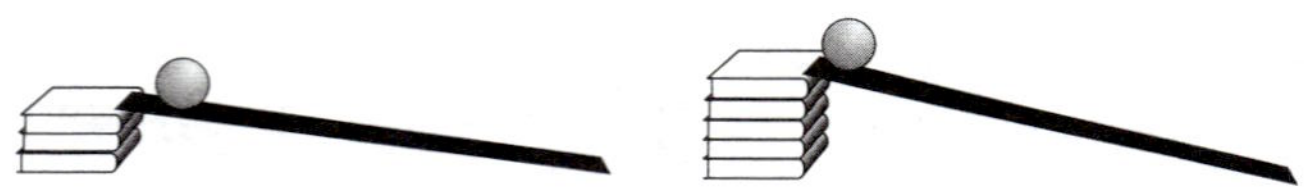

FIGURE 1.29.
Changing ramp height.

The data from their experiment are in **Figure 1.30.**

Ramp Height in Books	Distance Traveled from the End of the Ramp (in)
1	27
2	53
3	84
4	110
5	129
6	160
7	180

FIGURE 1.30.
Ramp height data.

1. Is there a proportional relationship between the distance traveled and the number of books? Justify your answer.

2. Use three different methods (table, graph, and function rule) to predict the distance a tennis ball travels if the ramp is 10 books high.

SUMMARY

In this lesson you learned about different methods to get the value of m in $y = mx$, function rules that model proportional relationships.

- Find the constant of proportionality by dividing each y-value by its corresponding x-value.
- Find the rate of change.

You also learned about methods to write symbolic rules for functions that model non-proportional, linear relationships.

- From a table, find the rate of change (slope) and the y-intercept. Then use the slope-intercept formula, $y = mx + b$.
- If you know the coordinates of two points, (x_1, y_1), (x_2, y_2), use the slope formula, $m = \frac{\Delta y}{\Delta x} = \frac{y_2 - y_1}{x_2 - x_1}$, to find the slope. Then use the point-slope formula, $y - y_1 = m(x - x_1)$.

Two variables can have a positive relationship, a negative relationship, or no relationship at all.

Positive Relationship:

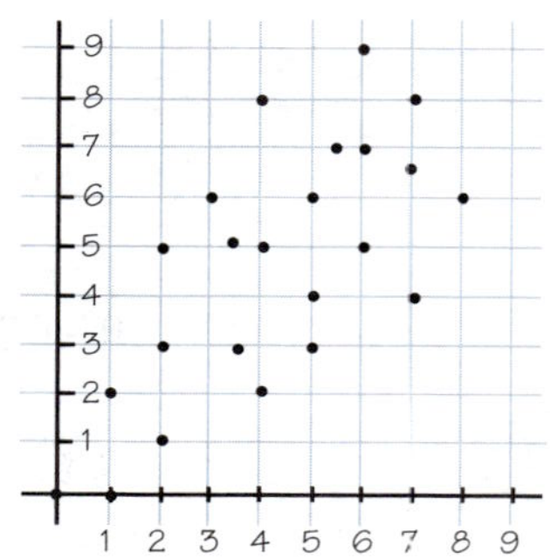

If one variable increases as the other variable increases, then there is a **positive relationship (correlation)** between the two variables.

Negative Relationship:

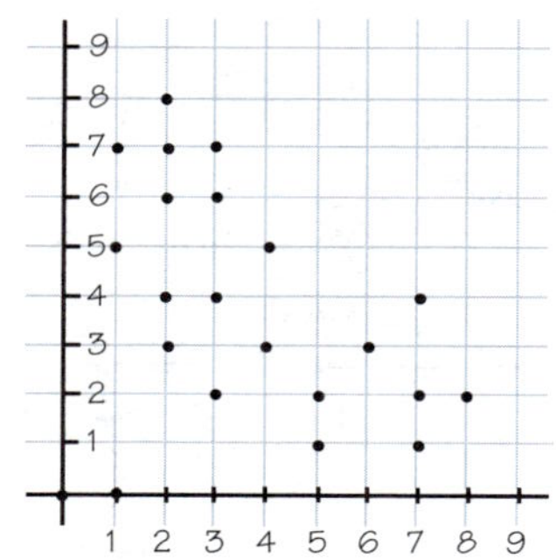

If one variable decreases as the other variable increases, then there is a **negative relationship (correlation)** between the two variables.

No Relationship:

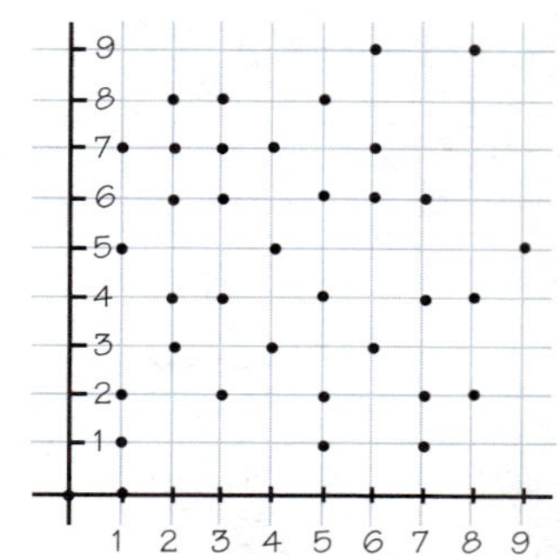

If there is no apparent trend in the scatter plot, then there is **no relationship (correlation)** between the two variables.

EVALUATE LESSON 2

Section 1.10: Distance versus Ramp Height

Purpose: In Section 1.10 students do an experiment that demonstrates direct variation. Students then create a direct variation function that describes the experiment.

Description: Students determine and implement a method to solve the problem. This problem is an open-middle performance assessment. There are several methods that may be used to arrive at the answer. In their justification, students should demonstrate that they have mastered the concepts presented in this lesson.

Materials: graphing calculators

Teaching Ideas: This task may be assigned as group work or individual work.

Answers

1. Sample answer: With 0 books, it is reasonable to say that the ball would not travel at all. The $\frac{y}{x}$ values are approximately equal.

2. Sample answer: Create a scatter plot of distance versus the number of books.

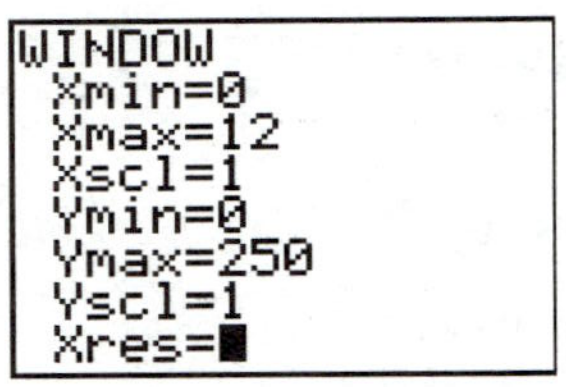

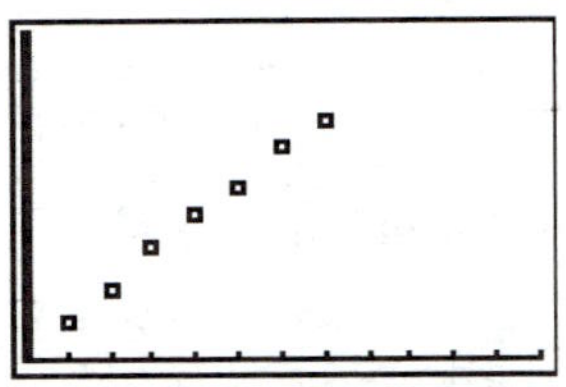

The scatter plot appears to be linear, and a line through the points would probably pass through the origin. The graph indicates that the relationship may be proportional. It seems reasonable that the distance traveled increases by the same amount for each additional book. It also seems reasonable that the distance traveled is 0 if no books are in the stack.

Use the data table to determine a constant of proportionality.

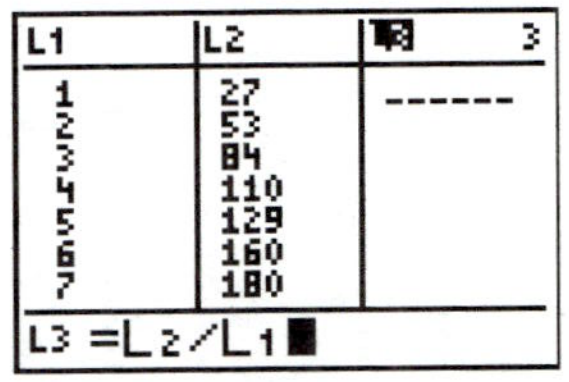

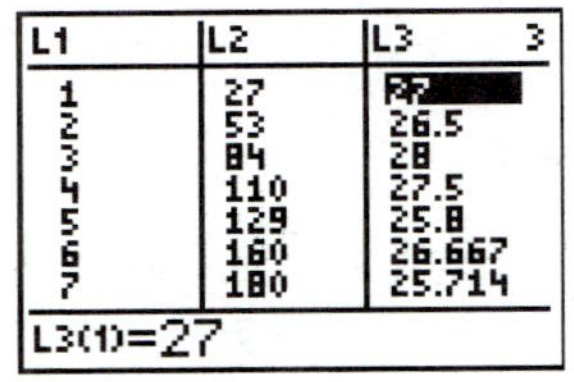

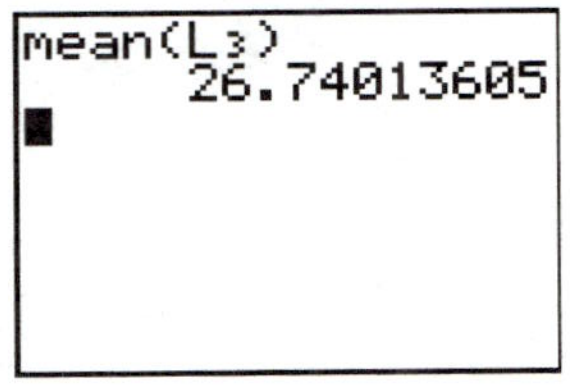

First find the ratio $\frac{y}{x}$ for each row of the table, then find the mean of these ratios. The mean is 26.7. A direct variation function is $y = 26.7x$. Test by graphing the equation over the scatter plot. Then use the graph, table, or home screen to determine that the tennis ball should travel 267 inches when the stack of books is 10 high.

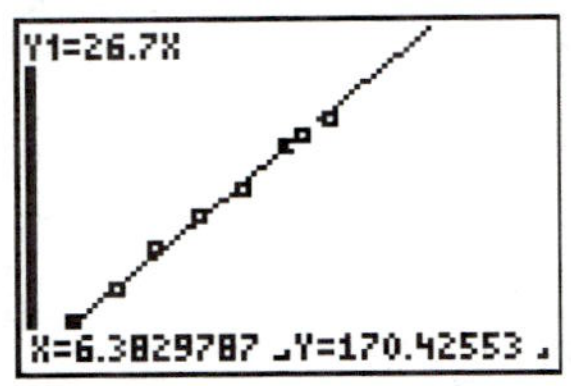

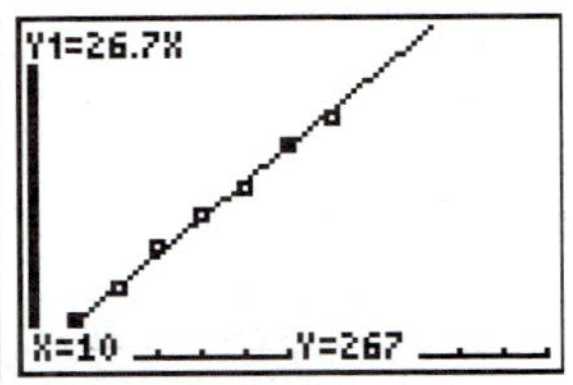

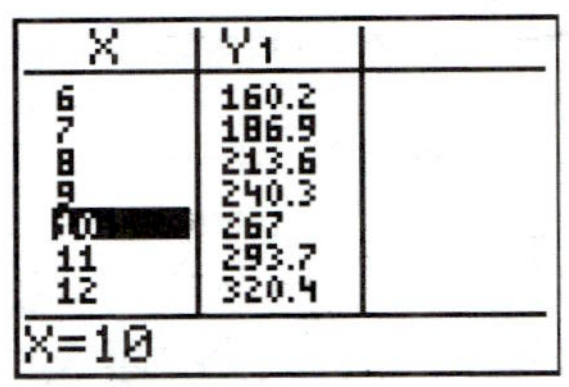

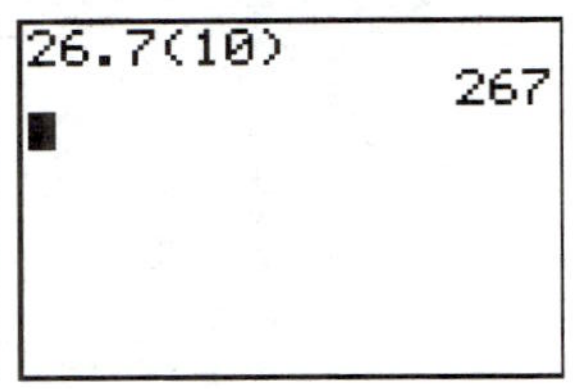

SECTION 1.11 Apparent Size

In this section you will mark the apparent size of a meter stick that your teacher has attached to a wall of the room. You will mark the apparent size on a pencil that you hold a short distance from your eyes.

1. Hold a pencil a few inches from your eyes. Align one end of the pencil with the top of the stick. Use your thumb to mark the bottom of the stick. Move the pencil farther from your eyes and repeat. What do you notice?
2. What are the variables in this situation?
3. Which is the independent variable? Why?
4. How are these variables related?
5. Again, mark the apparent size of the meter stick by holding the pencil a few inches from your eyes. After you have done so, keep the distance between your eyes and the pencil the same. This time move farther away from the meter stick and mark the apparent size. What do you notice?
6. What are the variables in this second situation?
7. Which is the independent variable? Why?
8. How are the variables related?
9. What experiment could you do to investigate the relationship between an object's apparent size and the distance that an observer is from the object?

SECTION 1.12 The Farther You Go

In this section you will conduct an experiment or use data provided by your teacher to further examine a relationship that you discussed in Section 1.11. The variables are the distance that an observer stands from a meter stick taped to a wall and the apparent length of the meter stick.

To do the experiment, begin by taping a meter stick vertically to one of your room's shorter walls or use a meter stick that your teacher taped to the wall in advance. The middle of the ruler should be about eye level.

Use a tape measure to measure a distance of 8 feet from the wall or use an 8-foot mark made by your teacher. Stand with your toes on the mark and hold a ruler vertically at arm's length. Align the top of the ruler with the meter stick. Use your thumb to mark the place on the ruler that is aligned with the bottom of meter stick (see **Figure 1.31**). Record this apparent length of the meter stick.

FIGURE 1.31.
Apparent length of a meter stick.

1. Continue taking measurements, moving an additional 2 feet from the meter stick each time. Record the measurements in a table like the one in **Figure 1.32.**

Distance from Meter Stick (ft)	Apparent Length of Stick (in)
8	
10	
12	
14	
16	
18	
20	
22	
24	
26	
28	
30	

FIGURE 1.32.
Meter stick data.

ENGAGE LESSON 3

Section 1.11: Apparent Size

Purpose: In previous lessons students investigated and defined aspects of proportional relationships, direct variation functions, and linear relationships. In Section 1.11 students explore an inverse variation relationship: the relationship between the apparent size of an object and the distance from an observer to the object. This section sets the stage for students to learn more formal methods to describe and use inverse variation in Sections 1.12 through 1.15.

Description: In this section of the lesson, students answer questions about key aspects of an inverse relationship.

Materials: Transparency 1.1, meter stick

Teaching Ideas: This section is designed to be a teacher-led, whole-class discussion. Students can record their answers to the questions. The discussion sets the stage for data collection in Section 1.12. Before beginning the discussion, tape a meter stick vertically to the wall in the front of the classroom.

Answers

1. Sample answer: When you move the pencil farther from your eyes, the apparent height of the stick gets larger.
2. The variables are the distance the pencil is from your eyes and the apparent size of the stick.
3. The independent variable is the distance the pencil is from your face, because it explains the apparent height of the stick. It is the variable that we are controlling.
4. The variables are positively related because when one increases so does the other.
5. The apparent size gets smaller.
6. The variables are the distance from the meter stick and its apparent size.
7. The independent variable is the distance from the meter stick because it explains the apparent size of the stick, and it is the variable that we are controlling.
8. The variables have a negative relationship because as one increases, the other decreases.
9. Sample answer: Use a ruler instead of a pencil to take measurements of the apparent size as you move from relatively close to the object to farther away from it. Keep the distance between your eyes and the ruler constant.

NOTE: This discussion should lead naturally to the apparent size investigation in the next section.

EXPLORE LESSON 3

Section 1.12: The Farther You Go

Purpose: In Section 1.11 students discussed a situation that leads to an inverse variation relationship. In Section 1.12 students gather data and develop an inverse variation function to describe the data that they gather. This section prepares students for a more formal explanation of inverse variation and its applications in Section 1.13

Description: In this section students perform a simple apparent size experiment. They use a ruler held at arm's length to measure the apparent length of a meter stick taped to a wall. They take several measurements at increasing distances from the wall. Then students look for a numerical relationship between the apparent size of the meter stick and the distance from the wall. They develop an inverse variation function to describe the relationship and use it to make predictions.

Materials: (For each group) meter stick, tape measure, ruler, masking tape; each student needs a graphing calculator

Teaching Ideas: Students should work in small groups (3 to 4 students). Three is ideal: one student to measure the distance from the wall, another to measure apparent height, and a third to record the measurements taken by the other two.

You can reduce the equipment needs by preparing your room in advance. Tape several meter sticks to one of the room's short walls at intervals of a few feet. Use a tape measure and masking tape to mark off distances in 2-foot increments from about 8 to about 30 feet from each meter stick. Write the distance on each piece of tape. With this setup, each group needs only a ruler. If you have more groups than meter stick setups, it is reasonable to let groups share setups since data collection should take each group a few minutes.

You can let students choose units for their apparent size measurements or tell them to use a particular unit. If inches are used, measuring to the nearest quarter inch is sufficient precision.

Answers

1. Sample answer:

Distance from Meter Stick (ft)	Apparent Length of Stick (in)
8	9.25
10	7.75
12	6.5
14	5.75
16	5.0
18	4.5
20	4.0
22	3.5
24	3.25
26	3.0
28	2.75
30	2.5

2. Enter the data points from the table into a graphing calculator and make a scatter plot. Record your viewing window and sketch the graph.

3. What happens to apparent length as distance from the meter stick increases?

4. Can the distance from the meter stick equal 0? Why or why not?

5. Do you think a decreasing linear function would be a good model for the relationship between apparent length and distance from the meter stick? Why?

6. Add a column to your table for the products of the apparent length and the distance from the meter stick. Calculate the product of each pair. What do you notice?

7. How can you use your observation in Question 6 to write a function rule for the relationship between apparent size and distance from the meter stick?

8. Use your equation to predict the distance you should stand from the meter stick for its apparent size to be 2 inches.

9. Change your viewing window to include negative values of x and y. (For example, if your scatter plot had Xmin = 0 and Xmax = 35, change Xmin to –35.) How does the graph change? Do you see more or fewer data points?

10. Enter your function from Question 7 into your calculator and graph it over your scatter plot. Use the same window you used in Question 9. Sketch your graph.

11. The domain and range of the mathematical function you just graphed are different from the domain and range of the situation shown in your table. Explain the differences.

SECTION 1.12 Assignment

For each table of values determine an appropriate viewing window, create a scatter plot, write a function rule, and graph your function rule over your scatter plot.

1.

x	y
–3	–2
–2	–3
–1	–6
1	6
2	3
3	2

2.

x	y
–3	–3
–2	–4.5
–1	–9
1	9
2	4.5
3	3

3.

x	y
–3	$-\frac{2}{3}$
–2	–1
–1	–2
1	2
2	1
3	$\frac{2}{3}$

4.

x	y
–3	–4
–2	–6
–1	–12
1	12
2	6
3	4

5. A group of senior citizens is planning a trip from Houston to Seattle. The cost to charter a private airplane is $10,000. The plane holds a maximum of 80 passengers.

Number of Passengers	Cost per Passenger
10	
20	
30	
40	
50	
60	
70	
80	

a) Fill in the table for the cost per passenger.

b) Which is the independent variable? Why?

c) Which is the dependent variable? Why?

d) What is a reasonable domain for this situation?

Answers (continued)

2. Sample answer:

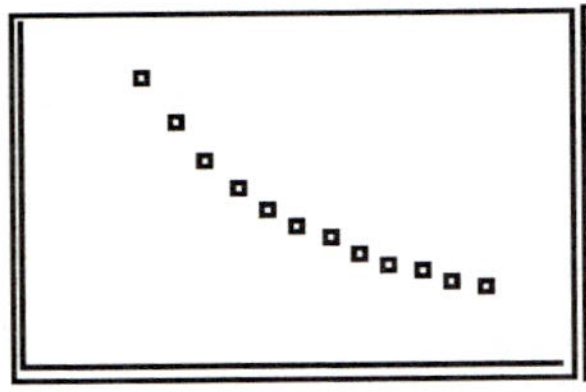

```
WINDOW
 Xmin=0
 Xmax=35
 Xscl=0
 Ymin=0
 Ymax=11
 Yscl=0
 Xres=1█
```

3. The distance decreases. The farther you are away from something, the smaller it looks.

4. No, because you cannot stand 0 feet from the meter stick; the arm with which you are holding the ruler would have to go through the wall.

5. Sample answer: The graph does not appear linear. You can also see from the table that the rates of change are not constant.

6.

Distance from Meter Stick (ft)	Apparent Length of Stick (in)	Product
8	9.25	74.0
10	7.75	77.5
12	6.5	78.0
14	5.75	80.5
16	5.0	80.0
18	4.5	81.0
20	4.0	80.0
22	3.5	77.0
24	3.25	78.0
26	3.0	78.0
28	2.75	77.0
30	2.5	75.0

The products are nearly equal.

7. Sample answer: To estimate the constant product, you can average the products in the table. In the sample data, the average is 78. Thus, the equation for the sample data is

$xy = 78$, or $y = \frac{78}{x}$.

8. For the sample equation given in the answer to Question 7, you should stand 39 feet from the meter stick.

9. You can now see all four quadrants, but the same points are shown.

```
WINDOW
 Xmin=-35
 Xmax=35
 Xscl=0
 Ymin=-11
 Ymax=11
 Yscl=0
 Xres=1█
```

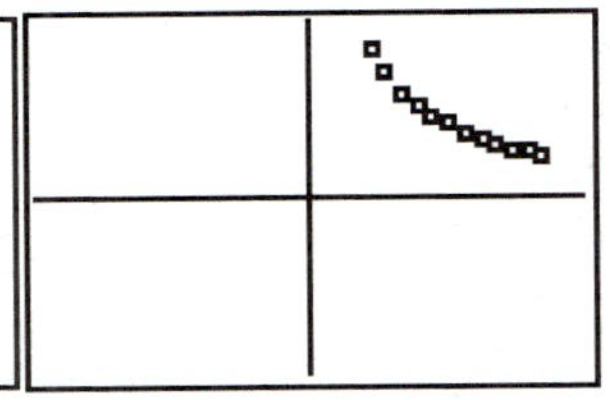

10.

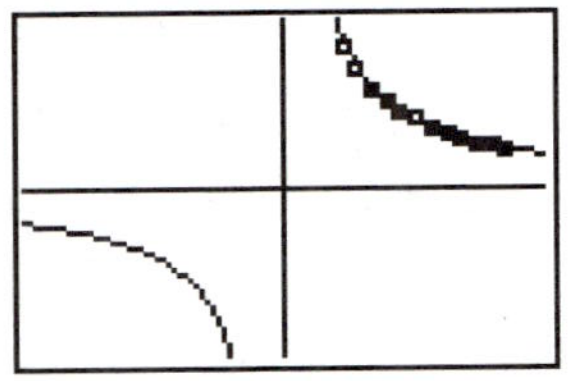

11. The data points only appear in the first quadrant, but the function's graph appears in the first and third quadrants. In the first quadrant, values for distance from the stick (x-values) and apparent length (y-values) are positive, and these values make sense in this situation. However, in the third quadrant, values for distance from the stick (x-values) and apparent length (y-values) are negative, which do not make sense in this situation since either the distance from the meter stick nor its apparent size can be negative. From the graph, it appears that the domain and range of the mathematical function are all real numbers except 0, which makes sense in the function rule because you cannot divide by 0.

Section 1.12 Assignment

Answers

1. Sample answer:

```
WINDOW
 Xmin=-5
 Xmax=5
 Xscl=1
 Ymin=-10
 Ymax=10
 Yscl=1
 Xres=█
```

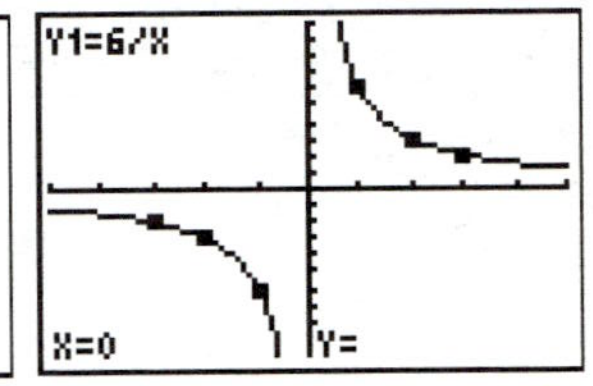

2. Sample answer:

```
WINDOW
 Xmin=-5
 Xmax=5
 Xscl=1
 Ymin=-12
 Ymax=12
 Yscl=1
 Xres=█
```

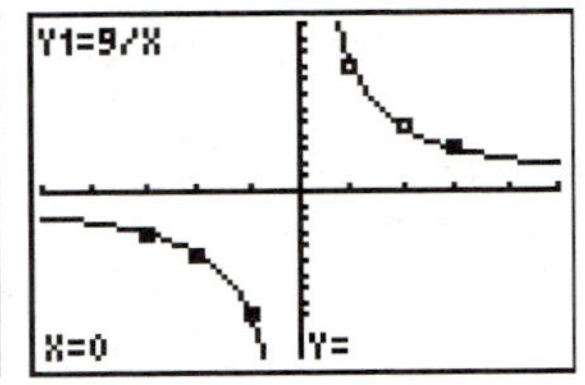

3. Sample answer:

```
WINDOW
 Xmin=-5
 Xmax=5
 Xscl=1
 Ymin=-5
 Ymax=5
 Yscl=1
 Xres=█
```

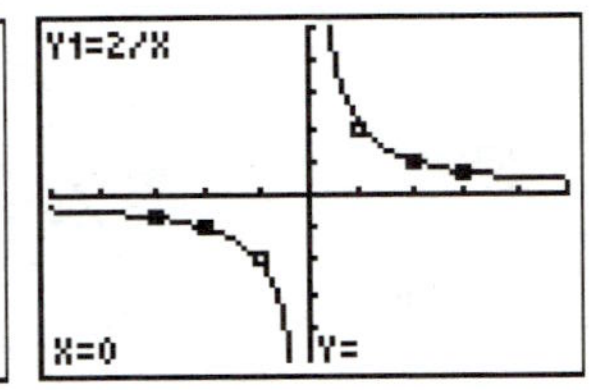

4. Sample answer:

```
WINDOW
 Xmin=█5
 Xmax=5
 Xscl=1
 Ymin=-13
 Ymax=13
 Yscl=1
 Xres=1
```

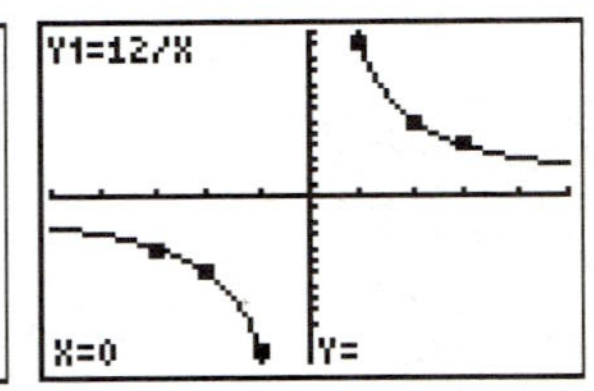

5. a)

Number of Passengers	Cost per Passenger
10	1000.00
20	500.00
30	333.33
40	250.00
50	200.00
60	166.67
70	142.86
80	125.00

b) The number of passengers is the independent variable because it explains the cost per passenger.

c) The cost per person is the dependent variable because it depends on the number of passengers.

d) In this situation, the domain is the number of passengers. The plane can hold at most 80 passengers, so a reasonable domain is whole numbers between 0 and 80.

e) What is a reasonable range for this situation?

f) Indicate an appropriate viewing window, create a scatter plot, write a function rule, and graph your function rule over your scatter plot.

g) If 52 people buy tickets on the flight, what is the cost per person?

6. In some situations, it is not clear which variable is independent. For example, consider a rectangle that has a fixed area of 20 square centimeters. If the length is 10 centimeters, then the width must be 2 centimeters. But neither the length nor the width is truly independent of the other. So you can choose either to be the independent variable.

a) Use x and y to write a function rule that describes the relationship between the length and the width in this 20 square centimeter rectangle.

b) What are a reasonable domain and range for this situation?

c) Graph your function rule on a graphing calculator and record the viewing window you used.

d) If the length of the rectangle is 13 cm, what is the width?

SECTION 1.13 One Goes Up, One Goes Down

In the last section you investigated the relationship between the distance from an object and its apparent size. You found a decreasing non-linear relationship.

1. What function rule describes the relationship between the distance from a meter stick and its apparent size?
2. What function rule describes the relationship between the length and width of a rectangle that has an area of 20 square units?
3. What do these function rules have in common? How do they differ?
4. Graph the functions on separate axes. Sketch your graphs.
5. What do the graphs have in common? How do they differ?
6. As the independent variable increases, what happens to the dependent variable in each situation? Does this change happen at a constant rate?
7. Think back to the apparent size situation in Section 1.12. The table in **Figure 1.33** contains some sample data that one student group collected.

Distance from Meter Stick (x)	Apparent Length of Stick (y), in cm	xy
8	24.5	
10	20.0	
12	16.3	
14	13.5	
16	12.3	
18	11.3	

FIGURE 1.33. *Sample data.*

For each pair, multiply the apparent length of the stick by the distance from it. Record your results in a table like Figure 1.33.

Answers (continued)

e) In this situation, the range is the cost per passenger. The range can be any decimal amount rounded to the nearest hundredth between 125 and 10,000.

f) 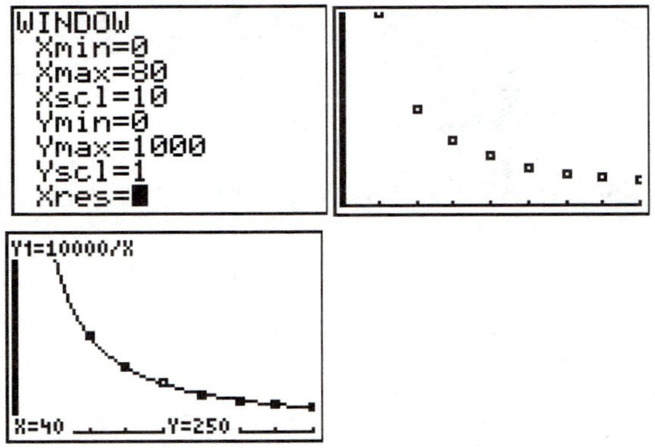

g) \$192.31 per person

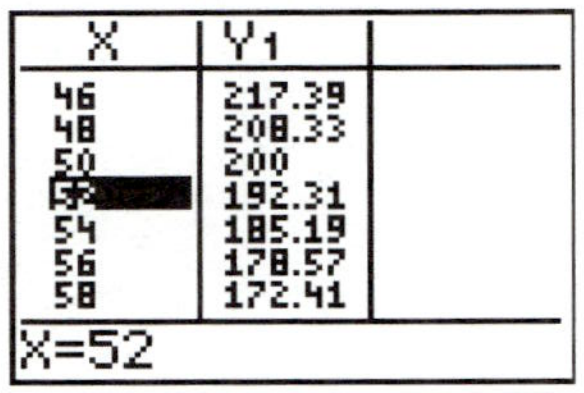

6. a) $y = \frac{20}{x}$

b) Neither dimension can be negative; both must be less than 20. Therefore, all real numbers greater than 0 but less than 20 form the domain as well as the range.

c) Sample answers:

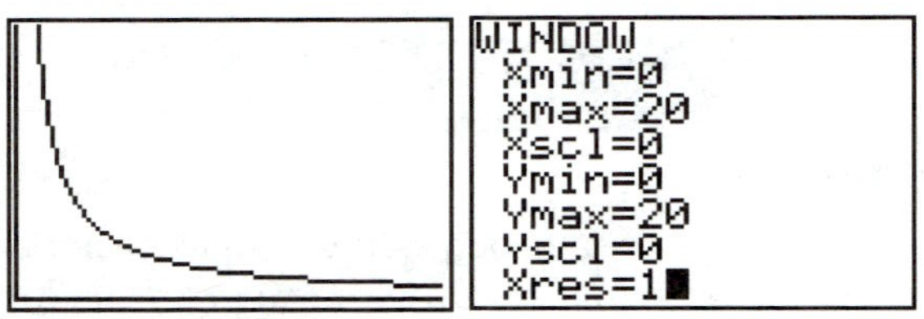

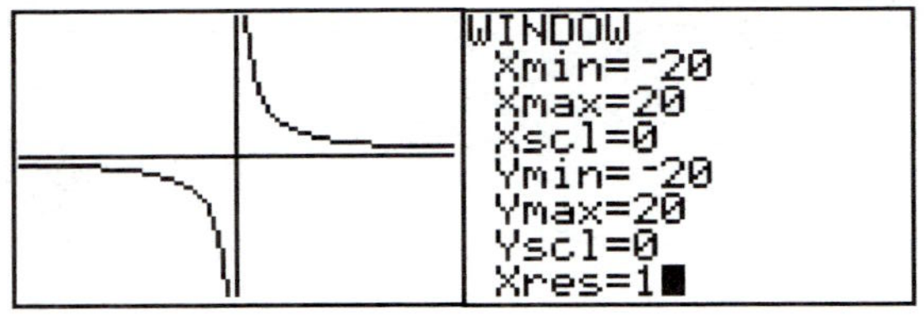

d) $\frac{20}{13}$ cm, or about 1.54 cm

EXPLAIN LESSON 3

Section 1.13: One Goes Up, One Goes Down

Purpose: In Section 1.12 students explored an inverse variation relationship where increasing the distance from an object resulted in a decrease in its apparent size. In this section students formalize the concept of an inverse variation relationship. In Section 1.14 students will apply the concept of inverse variation to Boyle's Law, a physical science relationship between the volume of a gas and its pressure.

Description: In this section students formalize the concept of inverse variation in light of their experiences in Sections 1.11 and 1.12. They develop the general form of the inverse variation equation, $y = \frac{k}{x}$.

Materials: graphing calculators

Teaching Ideas: Debrief students after Section 1.12 by having them share their answers to the questions. As they share, be sure to fill in any missing concept, process, or skill that could have been used to answer the questions. Then work through the questions in this section with the entire class. Once again be sure all necessary concepts, processes, or skills are addressed. In this section vocabulary terms such as "inverse variation" will be formalized.

Answers

1. $y = \frac{78}{x}$
2. $y = \frac{20}{x}$
3. Both are of the form $y = \frac{k}{x}$. However, the constant in the numerator is different for each situation.
4. 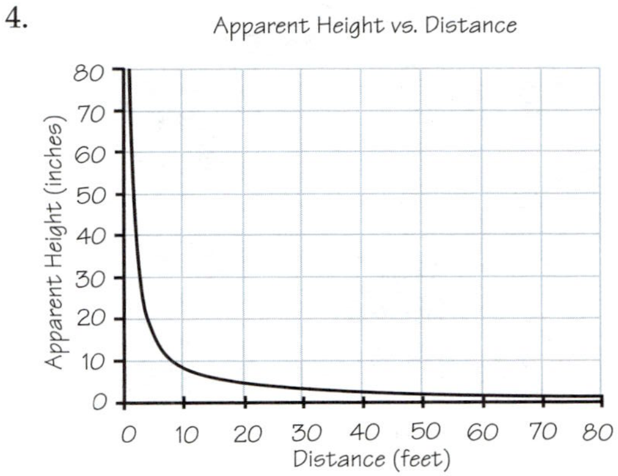

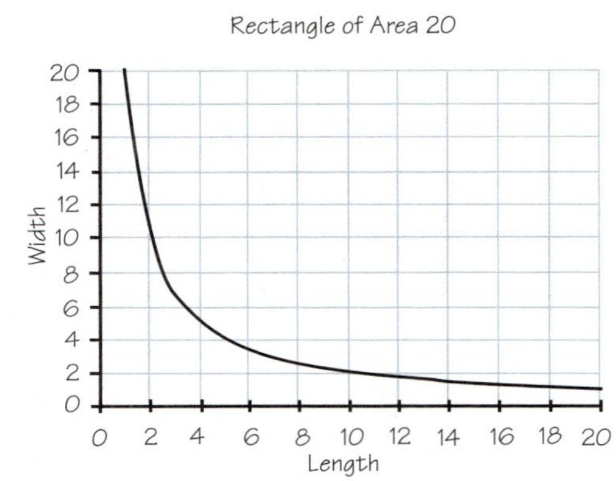

5. Answers may vary. Sample answer: Both graphs are nonlinear and on both graphs as x increases y decreases. However, the coordinates of points on the two graphs are not the same.

6. In each situation, as the independent variable increases the dependent variable decreases but not at a constant rate. At first the decrease happens quickly, but as the x-values get larger, the decrease in y-values becomes smaller.

7. Sample answer:

Distance from Meter Stick (x)	Apparent Length of Stick (y), in cm	xy
8	24.5	196.0
10	20.0	200.0
12	16.3	195.6
14	13.5	189.0
16	12.3	196.8
18	11.3	203.4

8. What is the average of the products of the distance from the meter stick and its apparent length?

9. In a proportional relationship, $y = kx$, or $k = \frac{y}{x}$. This means that $\frac{y}{x}$ is a constant value. Add a fourth column to your table and use it to find $\frac{y}{x}$ for each of your data points. Is this ratio constant? Are these ratios close to each other?

10. Is the relationship between distance from the meter stick and apparent size a proportional relationship? How do you know?

11. Use your graphing calculator to make a scatter plot of apparent size versus distance from the stick. Sketch your graph and identify your window.

12. If $xy = 196.8$, solve this equation for y in terms of x. Graph your equation on this scatter plot. Based on your graph, how well do you think this function models the relationship between apparent size and distance from the meter stick? Explain.

SUMMARY

In some relationships, the product of the values of the independent variable and the corresponding values of the dependent variable is a constant. In other words, xy equals (or is close to) a constant, k. These relationships are called **inverse variation** or **indirect variation**. In these situations, the variables are said to be **inversely proportional**.

The general form of an inverse variation function rule is $y = \frac{k}{x}$. The parent function for inverse variation functions is $y = \frac{k}{x}$, and the graphs of this family of functions have two branches (see **Figures 1.34 and 1.35**).

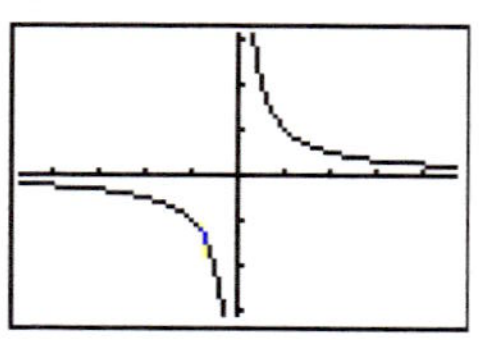

FIGURE 1.34. Graph of two branches of inverse functions.

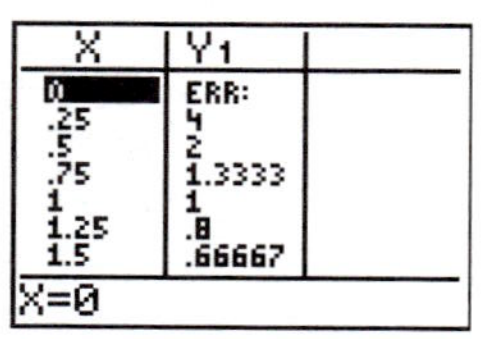

X	Y1
0	ERR:
.25	4
.5	2
.75	1.3333
1	1
1.25	.8
1.5	.66667

X=0

FIGURE 1.35. Table of family of inverse functions.

When k is positive, one branch of the graph is in the first quadrant where small x-values yield large y-values and large x-values yield small y-values. This branch is commonly used to model inverse variation situations in the real world, such as the relationship between the apparent size of an object and distance from the object (see **Figure 1.36**).

Direct Variation	Inverse Variation
$\frac{y}{x}$ is a constant value, k	xy is a constant value, k
$y = kx$	$y = \frac{k}{x}$
x and y are positively related	x and y are negatively related
Linear graph	Non-linear graph

FIGURE 1.36. Traits of direct and inverse variation.

Answers (continued)

8. 196.8

9. Sample answer:

Distance from Meter Stick (x)	Apparent Length of Stick (y), in cm	xy	$\frac{y}{x}$
8	24.5	196.0	3.063
10	20.0	200.0	2.000
12	16.3	195.6	1.358
14	13.5	189.0	0.964
16	12.3	196.8	0.769
18	11.3	203.4	0.628

The ratio of y to x is not constant. The ratios get smaller as distance from the meter stick increases. This make sense because the farther you are from something, the smaller is its apparent size.

10. No, the ratios of $\frac{y}{x}$ are not constant for each ordered pair.

11. 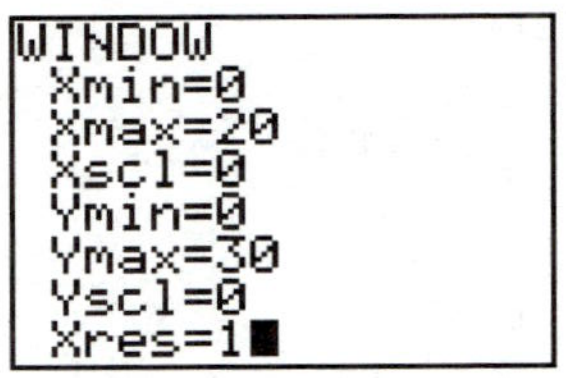

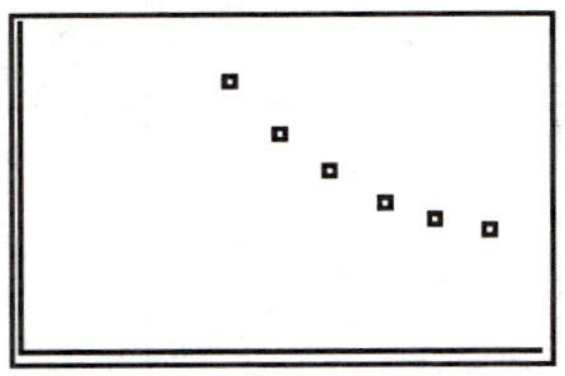

12. $y = \frac{196.8}{x}$

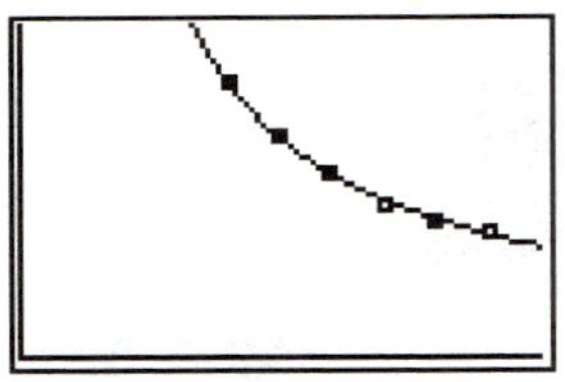

Answers may vary. Based on these data, the graph of $y = \frac{196.8}{x}$ seems to model the relationship quite well.

SECTION 1.13 Assignment

For each inverse function, complete the table and graph the function using the ordered pairs from the table.

1. $y = \frac{1}{x}$

x	y
−4	
−2	
−1	
−0.5	
−0.25	
0.25	
0.5	
1	
2	
4	

2. $y = \frac{2}{x}$

x	y
−4	
−2	
−1	
−0.5	
−0.25	
0.25	
0.5	
1	
2	
4	

3. $y = \frac{5}{x}$

x	y
−4	
−2	
−1	
−0.5	
−0.25	
0.25	
0.5	
1	
2	
4	

4. $y = \frac{-1}{x}$

x	y
−4	
−2	
−1	
−0.5	
−0.25	
0.25	
0.5	
1	
2	
4	

5. Compare the graphs in Questions 1, 2, and 3. In what ways are they alike, and in what ways do they differ? How does the equation of each support your answers?

6. Compare the tables in Questions 1, 2, and 3. In what ways are they alike, and in what ways do they differ? How does the equation of each support your answers?

7. How do the graph and table in Question 4 differ from those in Questions 1, 2, and 3? How does the equation of each support your answers?

8. For each table, is the relationship an inverse variation, a direct variation, or neither?

a)

x	y
3	1
6	2
9	3
12	4
15	5

b)

x	y
2	12
7	32
11	48
13	56
14	60

c)

x	y
3	8
4	6
6	4
8	3
12	2

d)

x	y
1	5
2	$\frac{5}{2}$
3	$\frac{5}{3}$
4	$\frac{5}{4}$
5	1

9. Describe a method for determining if a direct variation function is a good model for a relationship in a table.

10. Describe a method for determining if an inverse variation function is a good model for a relationship in a table.

Section 1.13 Assignment

Answers

1.

x	y
−4	−0.25
−2	−0.5
−1	−1
−0.5	−2
−0.25	−4
0.25	4
0.5	2
1	1
2	0.5
4	0.25

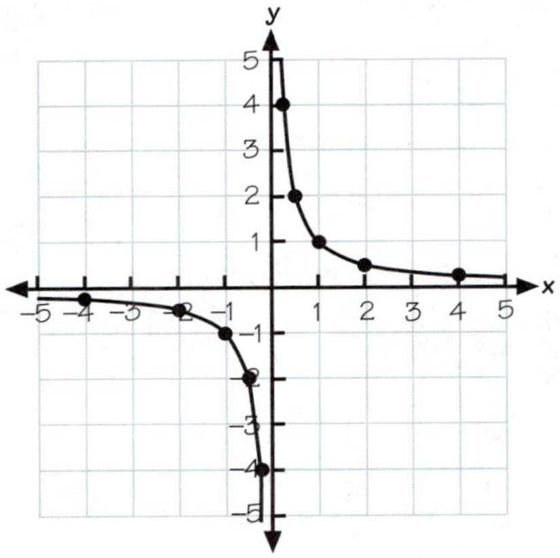

2.

x	y
−4	−0.5
−2	−1
−1	−2
−0.5	−4
−0.25	−8
0.25	8
0.5	4
1	2
2	1
4	0.5

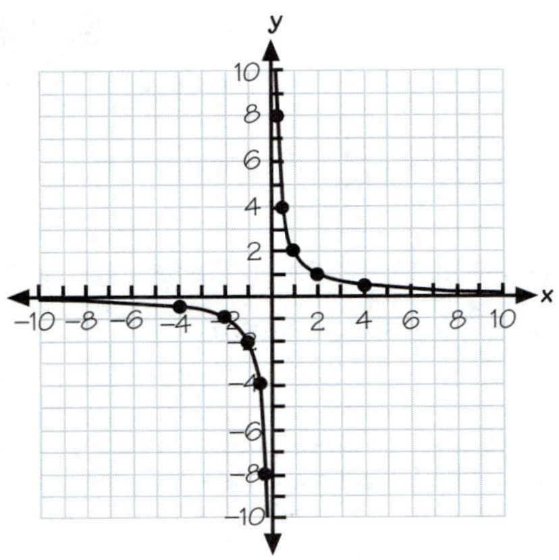

3.

x	y
−4	−5/4
−2	−5/2
−1	−5
−0.5	−10
−0.25	−20
0.25	20
0.5	10
1	5
2	5/2
4	5/4

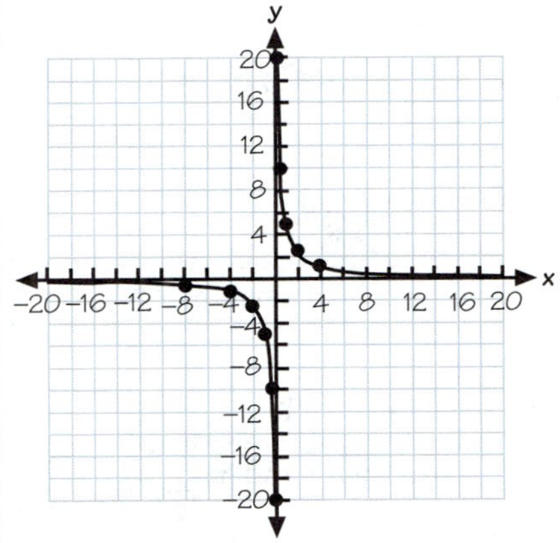

4.

x	y
−4	0.25
−2	0.5
−1	1
−0.5	2
−0.25	4
0.25	−4
0.5	−2
1	−1
2	−0.5
4	−0.25

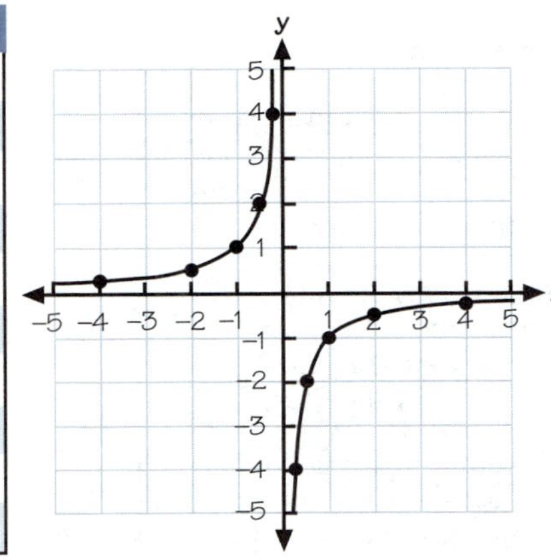

5. Sample answer: The graphs are all in quadrants 1 and 3. Although the graphs have similar shapes, they pass through different points. The constant, k, is positive meaning that x and y will have the same sign for all points (x, y).

6. Sample answer: Positive x's yield positive y's; negative x's yield negative y's. A given x-value does not yield the same y-value in every table. However, the domains and ranges of the mathematical functions are the same: all real numbers except 0.

7. Sample answer: Graph is in quadrants 2 and 4; positive x's yield negative y's and vice versa; equation has negative constant.

8. a) direct

   b) neither

   c) inverse

   d) inverse

9. Find the value for y/x for each pair. This value should be constant.

10. Find the value for xy for each pair. This value should be constant.

SECTION 1.14 Boyle's Law

In 1662, Robert Boyle and an assistant found a way to increase the pressure on a quantity of air and measure both the volume of the air and the pressure exerted on it. They did these things by pouring mercury into a large U-shaped tube containing air. As more mercury was added, the air compressed. Boyle measured the volume of the air by measuring its height in the tube. He measured the pressure on the air by measuring the height of the mercury in the tube.

Some of Boyle's original data are shown in **Figure 1.36.**

Volume (height of air in inches)	Pressure (height of mercury in inches)
12	117.5625
16	87.875
20	70.6875
24	58.8125
28	50.3125
32	44.1875
36	39.3125
40	35.3125
44	31.9375
48	29.125

FIGURE 1.36. Boyle's data.

1. Enter the data from the table into your graphing calculator and make a scatter plot. Sketch your graph and record your viewing window.
2. As volume increases, what happens to pressure?
3. Based on the scatter plot, do you think a direct variation function is a good model for the relationship between volume and pressure? Why?
4. Does the scatter plot appear to represent an inverse variation relationship? How do you know?

5. Add a third column to the table and label it product. Calculate and record the product for each pair.
6. Find the average product of volume and pressure.
7. Use the mean (average) product as the constant of variation and write an inverse variation function rule to model the relationship between pressure and volume for a gas.
8. Graph this function on your scatter plot. Based on the graph, does the function rule appear to model the relationship well? Explain.

Graphing a function rule on a scatter plot is one way to assess how well a function models a relationship. Another way is to compare numerical values. For example, you could use your rule to predict the pressure for a volume value of 12 and compare the prediction to the actual pressure that Boyle measured, 117.5625.

9. Use your function rule to calculate a predicted value for each volume value in your table. Compare the actual pressure values to the predicted ones. How well do you think the function models this relationship?

In the seventeenth century, most Europeans did not understand the nature of air. In 1644, Evangelista Torricelli, an Italian scientist and colleague of Galileo Galilei, correctly identified the concept of air pressure and built the first mercury barometer to measure it.

Robert Boyle, a prominent English chemist, set up a series of experiments with his research assistant, Robert Hooke (whom you met in Section 1.6), to further investigate properties of air. Among other things, Boyle and Hooke proved in the 1660s that air is necessary for sound, fire, and life.

The proof that fire needs air to exist and that sound cannot travel in a vacuum were radical at the time. Many scientists and philosophers insisted there was no such thing as a vacuum, or absence of air. Thus, the work of Boyle and Hooke was groundbreaking. Boyle's Law continues to be used as one of the fundamental principles of branches of science such as fluid dynamics and meteorology.

In 1662, Boyle published an appendix to his 1660 textbook *New Experiments Physio-Mechanicall, Touching the Spring of the Air and its Effects.* In this appendix, Boyle wrote about the mathematical relationship between the volume of air and its pressure.

ELABORATE LESSON 3

Section 1.14: Boyle's Law

Purpose: In Section 1.13 students examined inverse variation relationships and distinguished those relationships from direct variation (proportional) relationships. In Section 1.14 students apply inverse variation to Boyle's Law, a scientific principle stating that the volume of a gas is inversely proportional to the pressure of the gas at a constant temperature. This section prepares students to build a model of an inverse variation function and use that model to make predictions in Section 1.15.

Description: In this section students use graphing calculators to examine some of Robert Boyle's original data. Then they apply principles of inverse variation in order to develop Boyle's Law, $P = \frac{k}{V}$.

Materials: graphing calculators

Teaching Ideas: If equipment is available, students can collect their own data. See the Instructor's Manual for additional information.

Answers

1. 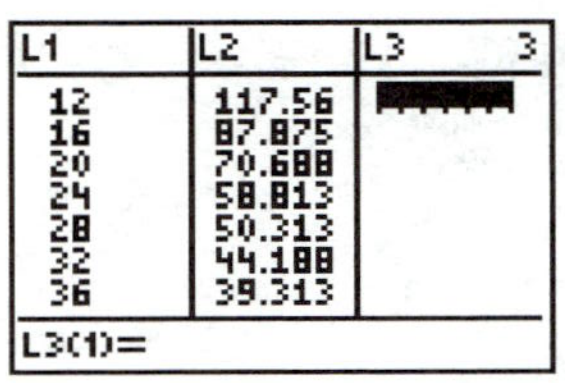

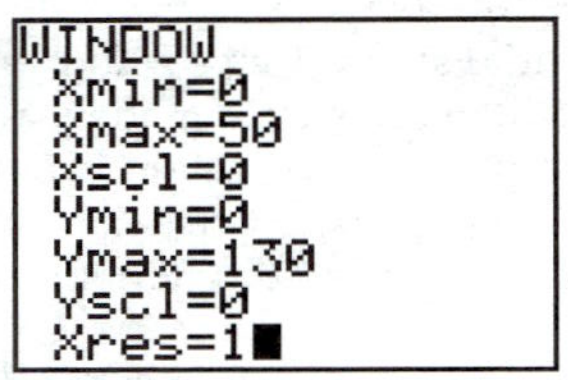

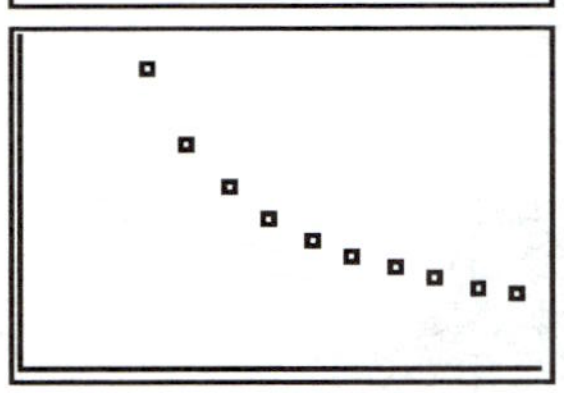

2. As volume increases, pressure decreases.
3. No, the pattern is not linear and therefore the relationship is not proportional. Remember that direct variation functions are used to model proportional relationships, and they have graphs that are linear and pass through the origin.
4. Answers may vary. Sample answer: The scatter plot appears to show an inverse variation relationship. The pressure decreases quickly at first but decreases more slowly as volume increases.
5. Sample answer:

Volume (height of air in inches)	Pressure (height of mercury in inches)	Product
12	117.5625	1410.8
16	87.8750	1406.0
20	70.6875	1413.8
24	58.8125	1411.5
28	50.3125	1408.8
32	44.1875	1414.0
36	39.3125	1415.3
40	35.3125	1412.5
44	31.9375	1405.3
48	29.1250	1398.0

6. The mean product is 1409.575.

Enter L3 = L1 * L2 into the List 3 header of the graphing calculator then press ENTER.

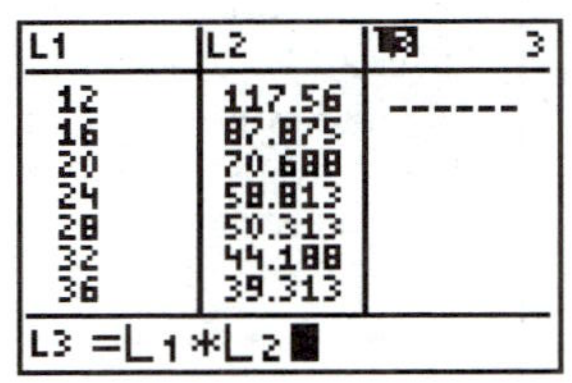

Return to the home screen. From the LIST menu (2nd-STAT), arrow over to the MATH options. Select option 3, mean.

Type L3 (2nd-3), then press ENTER.

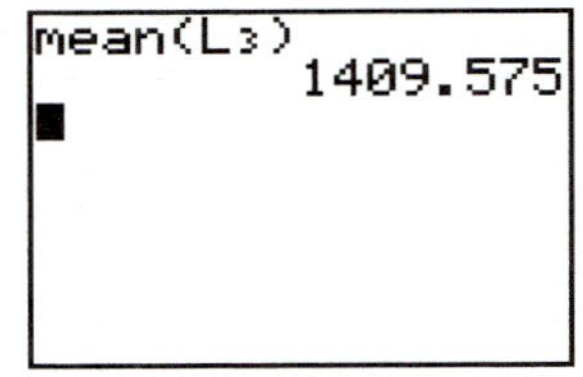

7. $P = \frac{1409.575}{V}$
8. 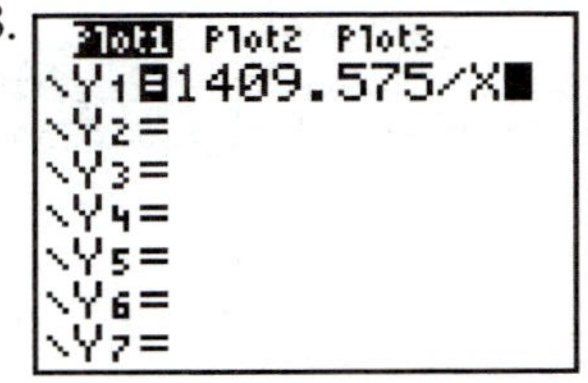

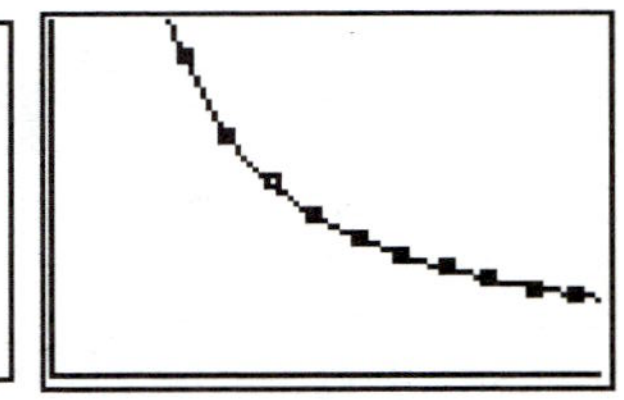

The function rule seems to model the relationship quite well because the graph appears to pass very close to each data point.

9. These predictions were made with $P = \frac{1409.575}{V}$ and are rounded to 4 decimal places.

Volume (height of air in inches)	Pressure (height of mercury in inches)	Predicted Pressure
12	117.5625	117.4646
16	87.875	83.0984
20	70.6875	70.4788
24	58.8125	58.7323
28	50.3125	50.3420
32	44.1875	44.0492
36	39.3125	39.1549
40	35.3125	35.2394
44	31.9375	32.0358
48	29.125	29.3661

Sample answer: Most of the actual and predicted values are quite close. The largest discrepancy is for a volume value of 16.

10. Based on your work in this section and Boyle's ideas about the relationship between the volume of a gas and its pressure, make a conjecture about a function rule that could model this relationship. How can this function rule be generalized to model the pressure and volume in any data set? Explain your answer.

11. Boyle's Law is one example of inverse variation. Can you think of some others?

SUMMARY

Inverse variation can be used to describe many relationships in the natural world. In 1662, Robert Boyle established an inverse variation relationship between the pressure and volume of a gas when it is kept at a constant temperature. This relationship is known as Boyle's Law and is represented mathematically by the equation $P = \frac{k}{V}$, where P is the pressure, V is the volume, and k is the constant of variation.

SECTION 1.14 Assignment

1. A group of students decide to conduct an experiment to verify Boyle's law. They use a CBL and a pressure sensor attachment to vary the volume and record the resulting pressure. Their data are in **Figure 1.38.**

Volume (cubic cm)	6	8	10	12	14	16	18
Pressure (atmospheres)	3.94	3.00	2.41	2.00	1.67	1.44	1.26

FIGURE 1.38. Student data.

a) Which is the independent variable? Why?

b) Which is the dependent variable? Why?

c) What is a reasonable domain for this situation?

d) What is a reasonable range for this situation?

e) Decide an appropriate viewing window, create a scatter plot, write a function rule, and graph this function rule on your scatter plot.

f) Predict the pressure that results from a volume of 2 cubic centimeters.

g) Predict the volume when the pressure is 1 atmosphere.

2. In a recent experiment, Javier used a 9-volt battery to test a circuit. He set various current levels and measured the resulting resistance. The data are in **Figure 1.39.**

Current (*I*)	Resistance (*R*)
2	4.5
5	1.8
6	1.5
10	0.9
12	0.75
15	0.6
20	0.45

FIGURE 1.39. Javier's data.

Answers (continued)

10. Answers will vary. Sample answer: The data suggest an inverse variation relationship between the pressure and the volume of air. The equation $P = \frac{1409.575}{V}$ models the relationship in this case. In general, an equation of the form $P = \frac{k}{V}$ can be used to describe the relationship between pressure and volume for any gas.

11. Possible examples:

    - The point value of each question on a test varies inversely with the number of questions on the test.
    - The quantity of an item that you can buy for a fixed total amount varies inversely with the unit price of the item.

Section 1.14 Assignment

Answers

1. a) The volume is the independent variable because the students control it.

   b) The pressure is the dependent variable. It responds to volume changes made by the students.

   c) A reasonable domain is numbers between 0 and 20.

   d) A reasonable range is numbers between 0 and 4.

   e)

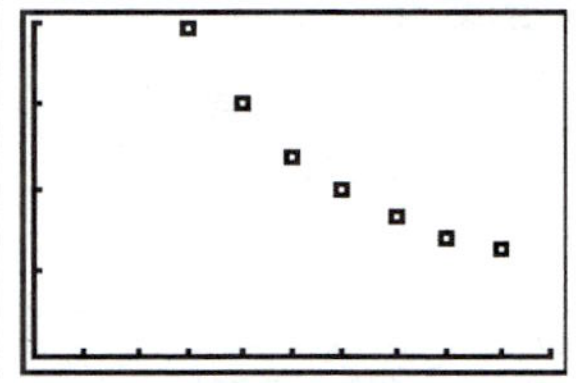

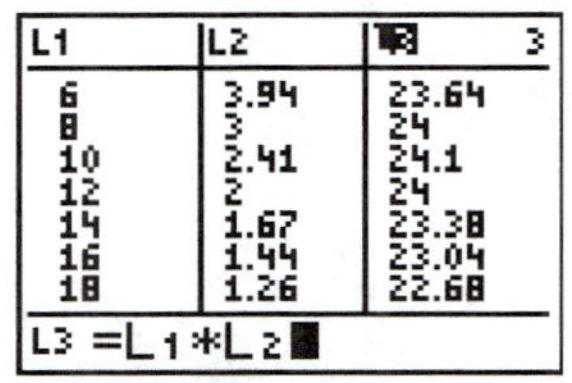

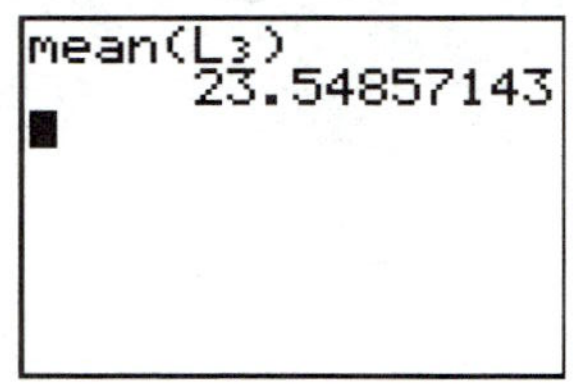

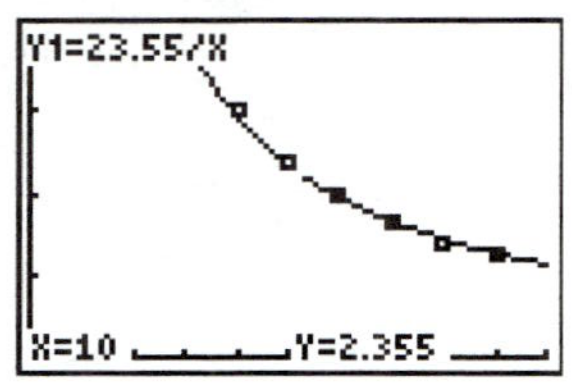

   f) 11.775 atmospheres

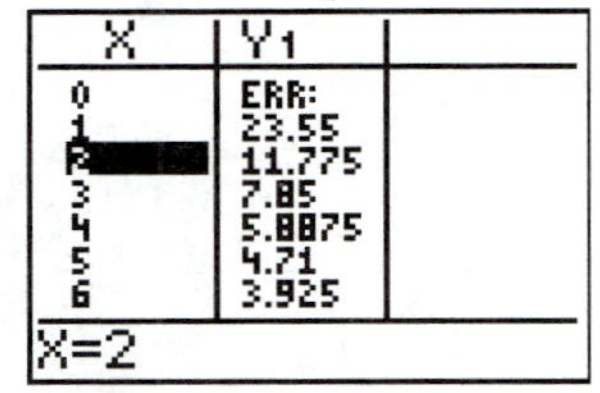

   g) 23.5 cc

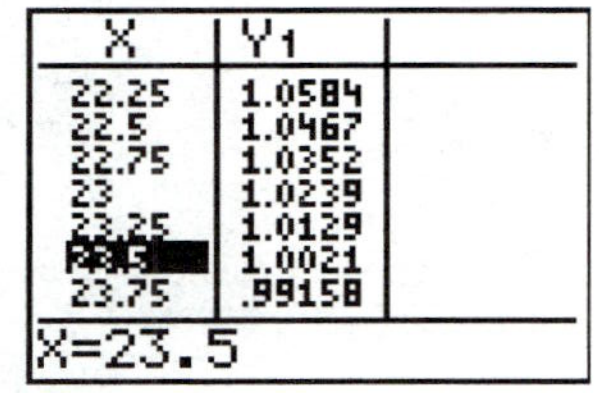

a) Which is the independent variable? Why?

b) Which is the dependent variable? Why?

c) What is a reasonable domain for this situation?

d) What is a reasonable range for this situation?

e) Decide an appropriate viewing window, create a scatter plot, write a function rule, and graph your function rule on your scatter plot.

f) If the current is 7, predict the resistance.

g) Predict the current when the resistance is 1.75.

SECTION 1.15 Using Inverse Variation

1. Pete's Construction Company has hired a team of engineers to study the company's policy for buying cement. Pete's foreman always orders 108 cubic feet of cement to pour a 12-foot wide driveway. The engineers conducted an experiment. They collected data by changing the length of a driveway, while keeping the width constant at 12 feet, and measuring the depth of cement after pouring 108 cubic feet of cement. Their data are in the table in **Figure 1.40.**

Width (feet)	Length (feet)	Depth (feet)	Volume of Cement (cubic feet)
12	1	9.00	108
12	2	4.50	108
12	3	3.00	108
12	4	2.25	108
12	5	1.80	108

FIGURE 1.40.
Pete's Construction data.

City code requires that driveways be 4 inches deep. What is the longest 12-foot wide driveway that can be constructed using the foreman's order? Justify your answer.

Answers (continued)

2. a) The current is the independent variable because it explains the resistance measurements. The current was under Javier's control.

   b) The resistance is the dependent variable because it responds to current changes.

   c) A reasonable domain is numbers between 0 and 20.

   d) A reasonable range is numbers between 0 and 5.

   e) 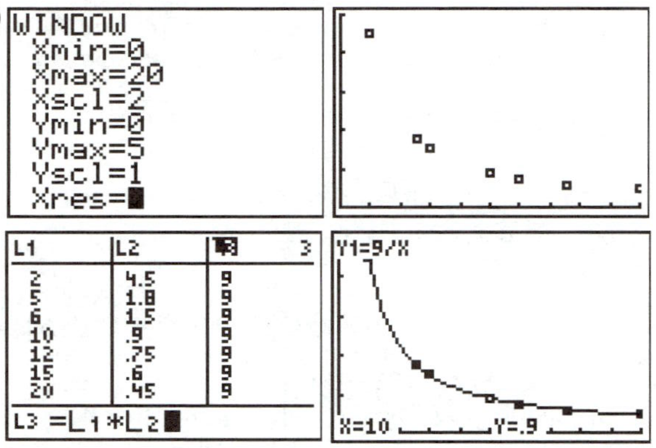

   Since the products are all 9, a function rule is $y = \frac{9}{x}$.

   f) 1.2857

   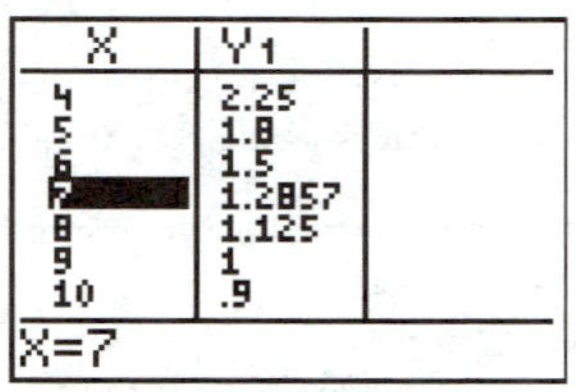

   g) About 5.14

   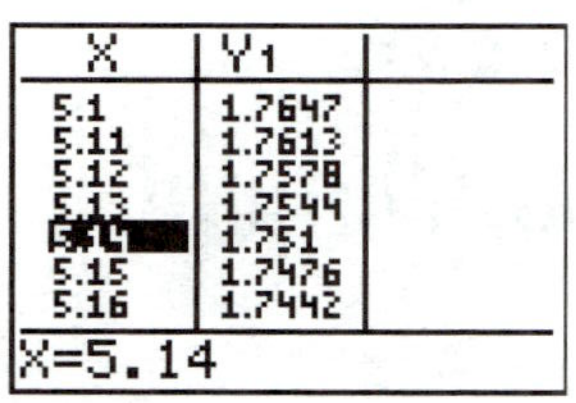

EVALUATE LESSON 3

Section 1.15: Using Inverse Variation

Purpose: In Sections 1.11 through 1.14, students discovered multiple ways to solve problems regarding inverse variation. In this section students have the opportunity to demonstrate their knowledge by applying those concepts and skills to solve a problem in a new context. This section prepares students to delve deeper into mathematical modeling in Chapter 2.

Description: Facing a problem in a new context, students determine and implement a method to solve it. This is an open-middle performance assessment task; in other words, there is only one correct answer but there are a variety of valid methods to obtain it. In their justification, which is the most important part of a performance assessment task, students should be able to communicate their knowledge of mathematics and demonstrate that they have mastered the concepts presented in Sections 1.11 through 1.14.

Materials: graphing calculator

Teaching Ideas: This task may be assigned as group work or individual work. The mathematics performance assessment rubric should be used to evaluate students' work.

Answer

1. Sample answer: Examine the relationship between the length and depth of the driveway.

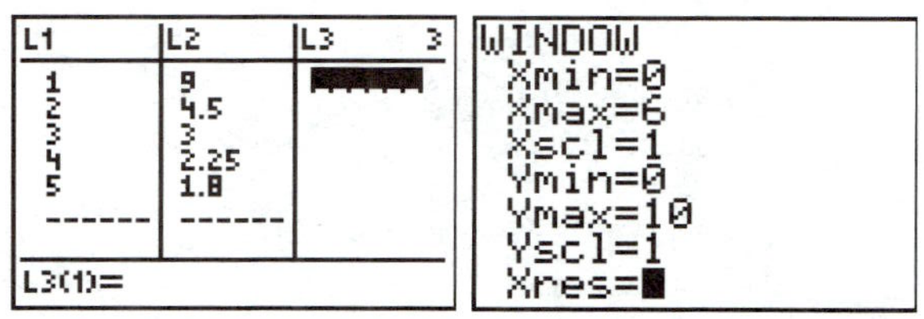

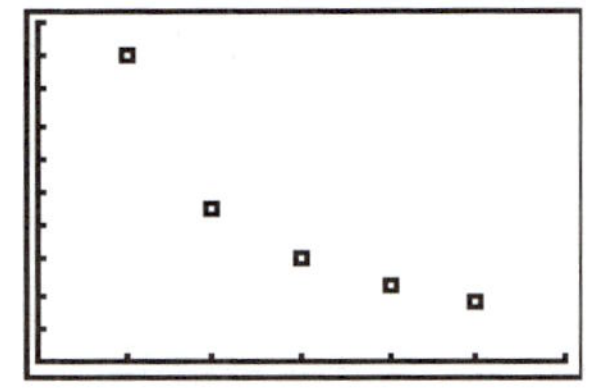

Find the product of length and depth. The function rule is $xy = 9$ or $y = \frac{9}{x}$. Verify the function rule by graphing over the scatter plot then use the table.

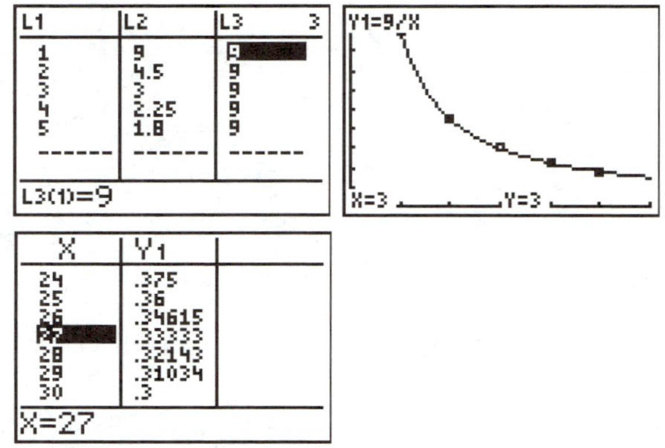

A 12-foot-wide driveway that is 0.333 feet (4 inches) deep should be 27 feet long in order to use all 108 cubic feet of cement. So a 27-foot-long driveway that is 12 feet wide and 4 inches deep can be constructed with 108 cubic feet of cement.

Modeling Project Scope It Out

When you look through a telescope, the amount you can see depends on several variables.

In this project, you will investigate the relationship between the size of a telescope's viewing circle and two other variables.

You will need several paper towel rolls, masking tape, a meter stick, a tape measure, and scissors.

Make a simple telescope by cutting a paper towel roll from one end to the other so that another paper towel tube can fit inside of it. Do the same with a third paper towel roll. Put the uncut tube inside one of the cut tubes. Then put these two inside the other cut tube. You can now vary the length of your scope from one tube to nearly three.

Next, tape a meter stick to a wall horizontally. Back off and look at the stick through your scope. If you align the left edge of your viewing circle with the left end of the meter stick, you can estimate the viewing circle's diameter. Now change the length of the scope. The diameter of the viewing circle should change.

There are several variables in this situation:

- The diameter of the viewing circle;
- The length of the scope;
- The distance you are standing from the meter stick.

In mathematical modeling, it can be difficult to analyze the relationship between several variables at once. So mathematical modelers use a simple trick: they keep all but two of the variables constant and examine just those two. To do this, they make changes in one variable and record the effect on the other. After they have figured out how these two variables are related, they do something similar with a different pair of variables.

So here is an approach to this situation:

- Keep the distance from the meter stick constant and examine the relationship between viewing circle diameter and telescope length;
- Keep telescope length constant and examine the relationship between viewing circle diameter and distance from the meter stick.

Conduct an investigation into these two relationships. Use the scope you have made to gather data. Prepare a report on your findings.

Practice Problems

1. A concert promoter is considering setting the ticket price at \$30. **Figure 1.41** shows a table, function rule, and graph for revenue (R) based on number of tickets sold (n).

n	R
10	\$300
20	\$600
30	\$900
40	\$1200
50	\$1500

$$R = 30n$$

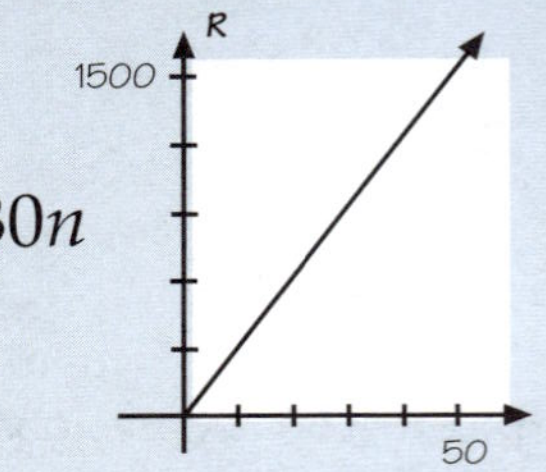

FIGURE 1.41.
Data for Question 1.

Explain how to tell if the relationship between R and n is a proportional relationship from the

a) table of values.

b) function rule.

c) graph.

2. For each linear relationship, find the slope (m). That is, find the constant rate of change $\frac{\Delta y}{\Delta x}$.

a) The table in **Figure 1.42**

x	y
5	23
9	29
13	35
17	41
21	47

FIGURE 1.42.
Data for Question 2a.

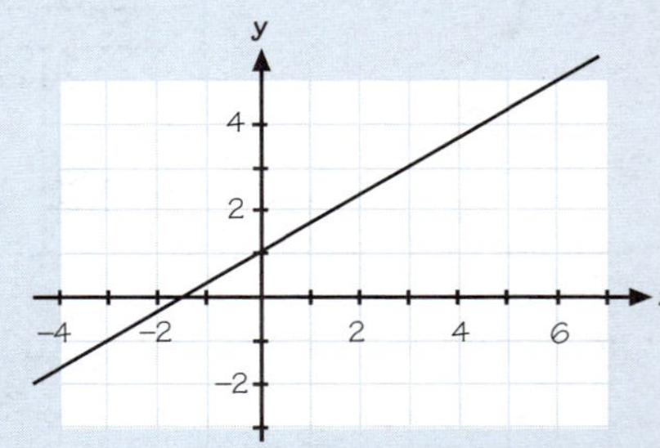

FIGURE 1.43.
Graph for Question 2c.

b) The line going through the points $P(5, 13)$ and $Q(10, 10)$

c) The line shown in **Figure 1.43**

Practice Problems

Answers

1. a) The ratio R/n is a constant value of 30, which is the ticket price.

   b) The equation has only one operation, in which you multiply by the variation constant.

   c) The graph is a straight line going through the origin.

2. a) For any two consecutive pairs in the table, $\Delta y = 6$ and $\Delta x = 4$. The slope or rate of change is $m = \frac{\Delta y}{\Delta x} = \frac{6}{4} = \frac{3}{2} = 1.5$.

   b) Using the slope formula, $m = \frac{\Delta y}{\Delta x} = \frac{y_2 - y_1}{x_2 - x_1} = \frac{10 - 13}{10 - 5} = \frac{-3}{5}$.

   c) The line goes through the points (0, 1) and (3, 3). The change in the horizontal direction Δx is 3, and the change in the vertical direction Δy is 2. So the slope of the line is $\frac{2}{3}$.

3. Solve each of the following equations for the indicated variable:

   a) $\frac{5}{12} = \frac{x}{18}$, for x.

   b) $4.20n + 67000 = 76030$, for n.

   c) $C = 2.45 \cdot n + 108000$, for n.

4. Two long-distance telephone carriers have quoted costs shown in **Figure 1.44:**

Company	Monthly Service Charge	Rate (per min)
NCO	$5.00	$0.20
Jog	$3.00	$0.25

FIGURE 1.44. Comparison of phone company rates.

Write equations describing the total cost associated with each company plan for one month. Use t as the variable to represent the amount of time (number of minutes) spent on the phone.

5. Sam gathered data on the advertising costs and ticket sales from concerts at the Cotton Bowl. The data are shown in **Figure 1.45.** Sam is interested in the relationship between advertising costs and ticket sales.

Advertising Costs	Ticket Sales
$10,000	12,478
$15,000	16,395
$20,000	19,882
$25,000	22,043
$30,000	25,512
$40,000	25,704
$50,000	25,704

FIGURE 1.45. Advertising costs vs. ticket sales.

   a) The pattern in the table shows increasing ticket sales as advertising costs go up. However, when the large sums are spent, there isn't an increase in ticket sales. What could explain this?

   b) On a copy of the grid in **Figure 1.46**, make a scatter plot to show how the ticket sales are related to advertising costs.

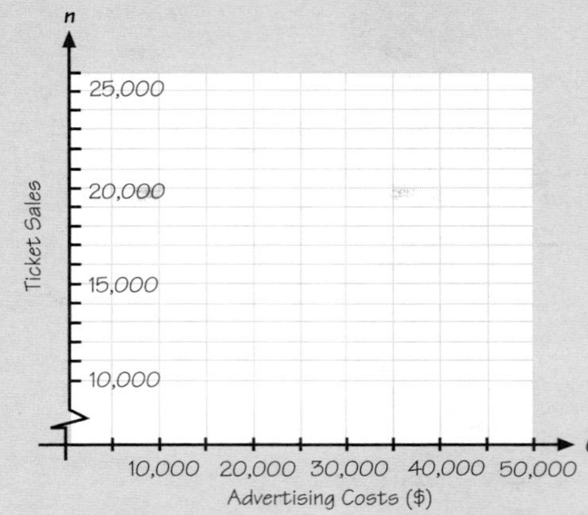

FIGURE 1.46. Ticket sales and advertising costs.

   c) The pattern for the first five data points is fairly linear. Draw a line on the scatter plot that you think captures this linear trend. Find the line's equation.

6. Data collected from ticket sales at the Cotton Bowl are shown in **Figure 1.47.**

Price (p)	Ticket Sales (n)
$10	25,000
$20	24,000
$30	22,000
$40	19,000
$50	16,000
$60	14,000
$70	10,500

FIGURE 1.47. Price vs. ticket sales.

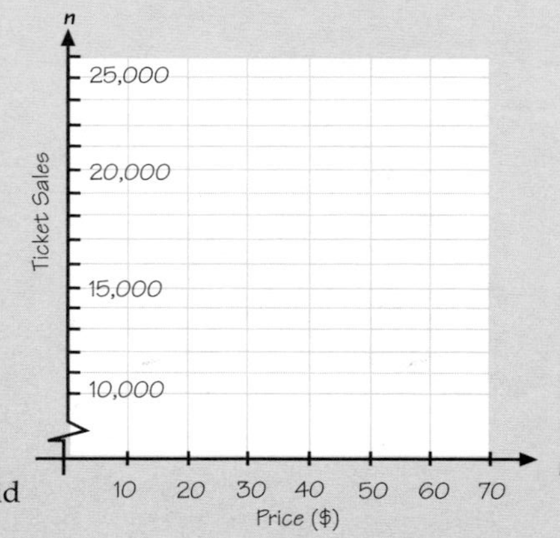

FIGURE 1.48. A blank grid plot comparing price and ticket sales.

   a) Make a scatter plot on a copy of the grid in **Figure 1.48.**

   b) Draw a line on the graph that you think captures the trend. Find the line's equation.

   c) On a graphing calculator, put the ticket price information in list L1 and the ticket sales information in list L2. Adjust the window settings to match the scatter plot you drew in a. What window settings did you use?

Answers (continued)

3. a) Cross-multiplying, $12 \cdot x = 90$ so $x = 7.5$. Alternatively, 18 is 1.5 times bigger than 12 so $x = (1.5)(5) = 7.5$. Or, simply multiply both sides by 18 and evaluate: $5 \cdot 18/12 = 7.5 = x$.

   b) Subtract 67,000 then divide by 4.20. The solution is $n = 2150$.

   c) Subtract 108,000 then divide by 2.45. Therefore $n = (C - 108{,}000)/2.45$, or in slope-intercept form $n \approx 0.408C - 44{,}081.6$.

4. NCO: $C = 0.20t + 5.00$; Jog: $C = 0.25t + 3.00$

5. a) There is a limited seating capacity. According to the data table, spending more than \$30,000 in advertising results in a sellout.

   b) Sample answer:

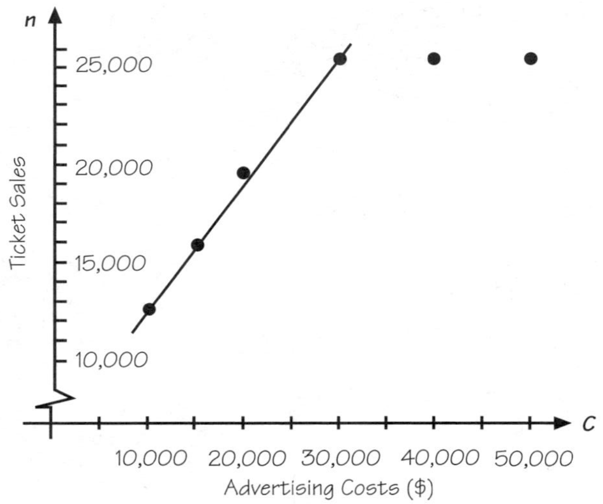

   c) See the answer to b for the graph of the line. Estimate the coordinates of two points on the line: (10,000; 12,500) and (30,000; 25,500). The slope is $(25{,}500 - 12{,}500)/(30{,}000 - 10{,}000) = 0.65$. The point-slope equation would be either $n - 12{,}500 = 0.65(C - 10{,}000)$ or $n - 25{,}500 = 0.65(C - 30{,}000)$. The slope-intercept equation is $n = 0.65 \cdot C + 6000$.

6. a) Sample answer:

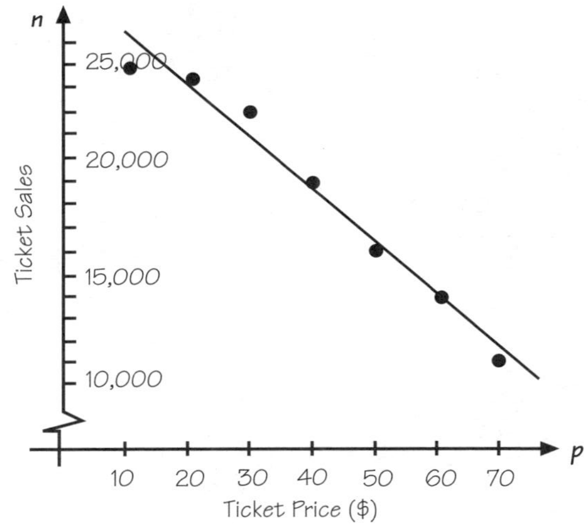

   b) See the answer to a for the line's graph. Student answers may vary slightly. Estimate the coordinates of two points on the line: (20; 24,000) and (60; 14,000). The slope is $(14{,}000 - 24{,}000)/(60 - 20) = -250$. The point-slope equation would be either $n - 24{,}000 = -250(p - 20)$ or $n - 14{,}000 = -250(p - 60)$. The slope-intercept equation for the model is $n = -250 \cdot p + 29{,}000$.

   c) Answers will vary. Appropriate WINDOW settings might be Xmin = –1, Xmax = 75, Xscl = 10, Ymin = 8000, Ymax = 26,000, Yscl = 1000.

7. Suppose that a concert will have a sell-out attendance of 20,000 when the admission is $25, but the attendance drops 2000 for every $5 added to the ticket price.

   a) What is a function rule that models the relationship between attendance and admission price?

   b) What is a reasonable domain for this situation?

8. Direct variation functions have graphs that are straight lines, but not all functions with straight-line graphs are direct variations.

   a) Compare slope-intercept and point-slope forms of function rules for lines in general with function rules for direct variations.

   b) Compare the graphs of direction variations with the graphs of linear functions in general.

9. Data on various attendance projections for a concert are in **Figure 1.49:**

Ticket Price (p)	$10	$20	$30	$40	$50
Ticket Sales (n)	25,000	23,500	22,250	21,100	19,900

FIGURE 1.49. Projection of the ticket sales by ticket price.

   a) Draw a scatter plot on a copy of the grid in **Figure 1.50.**

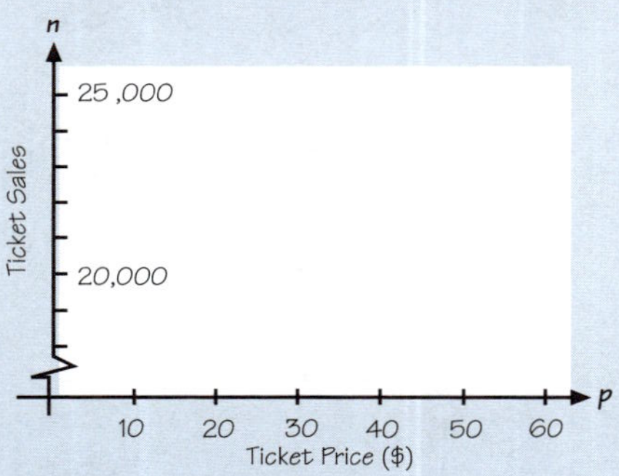

FIGURE 1.50. Grid for Question 9a.

   b) Draw a line on your scatter plot that you think captures the trend of the data. Find the line's equation.

For each table in Questions 10 through 13, determine the constant of variation. Then match the table with a graph (a–d) and an equation (e–h) below.

10.

X	Y1
.5	32
1	16
2	8
4	4
8	2
16	1

$k =$ _____

graph ____

equation ____

11.

X	Y1
-16	-.25
-8	-.5
-4	-1
-2	-2
-1	-4
-.5	-8

$k =$ _____

graph ____

equation ____

12.

X	Y1
-16	-.5
-8	-1
-4	-2
-2	-4
-1	-8
-.5	-16

$k =$ _____

graph ____

equation ____

13.

X	Y1
-8	.5
-4	1
-2	2
-1	4
-.5	8
-.25	16

$k =$ _____

graph ____

equation ____

a)
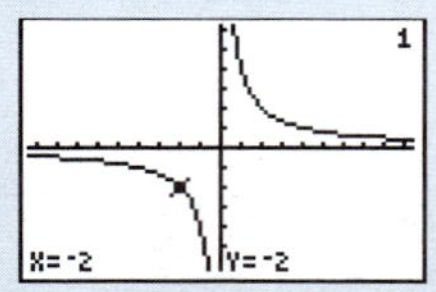

b)
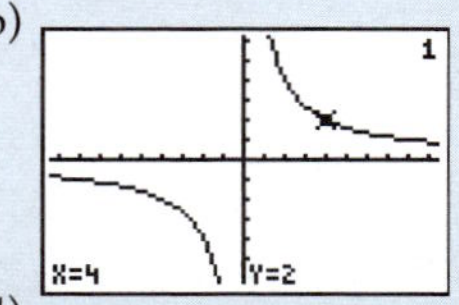

c)
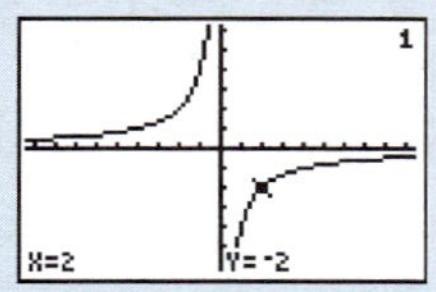

d)
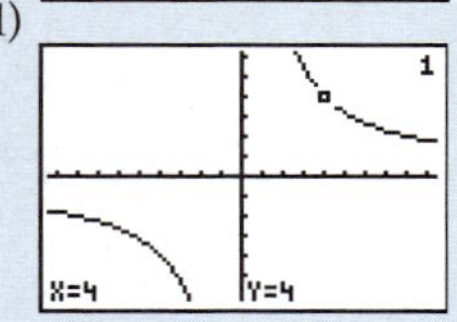

e) $y = \frac{4}{x}$

f) $y = \frac{-4}{x}$

g) $y = \frac{16}{x}$

h) $y = \frac{8}{x}$

Answers (continued)

7. a) In point-slope form, the equation modeling the demand drop-off is $n - 20{,}000 = -400(p - 25)$. In slope-intercept form the equation is $n = -400p + 30{,}000$.

   b) The domain is ticket prices of $25 or higher, but only up to $75.

8. a) The direct variation: $y = mx$. The slope-intercept form: $y = mx + b$. The point-slope form: $y - y_1 = m(x - x_1)$.

   b) Direct variations graph as lines that go through the origin. Lines in general go through a point on the y-axis, but it does not have to be the origin.

9. a) Sample answer:

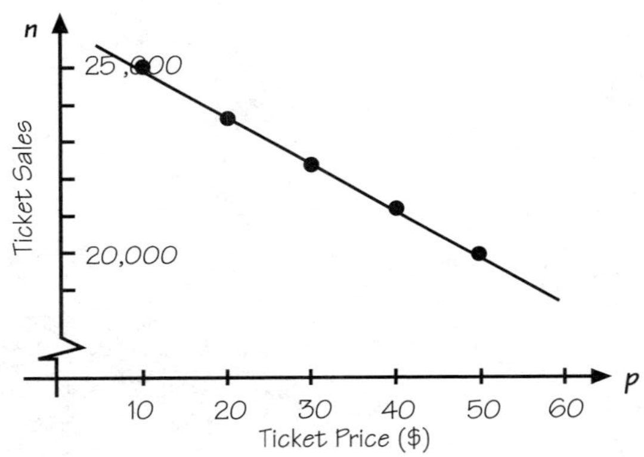

   b) Answers may vary slightly. Using the data points (10; 25,000) and (50; 20,000),
   $m = \frac{25{,}000 - 20{,}000}{10 - 50} = \frac{5000}{-40} = -125.$

   Using the point-slope form of the equation, $n - 25{,}000 = -125(p - 10)$.

   Therefore, an equation is $n = -125p + 26{,}250$. (A rough sketch of its graph is shown on the graph in the answer to a.)

10. $k = 16$; d; g

11. $k = 4$; a; e

12. $k = 8$; b; h

13. $k = -4$; c; f

For Questions 14 and 15, fill in the missing values in the table and write the function rule.

14.

X	Y1
	ERROR
.25	
.5	48
1	
2	12
	6
	4

15.

X	Y1
-10	-10
-5	
	-100
0	
1	
5	20
	10

16. Students studying strengths of materials performed an experiment. They cut Styrofoam into beams of different lengths, all with the same width and thickness. Then they fastened each beam between two desktops and gradually placed one-ounce weights on it until it broke. Their data are shown in **Figure 1.51.**

Length of Beams (inches)	Breaking Weight (ounces)
4	28
5	22
6	19
7	16
8	14
9	13

FIGURE 1.51.
Data on length of beams vs. breaking weight.

a) Examine the relationship between beam length and breaking weight. What kind of relationship do you think it is?

b) Which is the independent variable in this situation? Why?

c) What function rule would you use to model this relationship? Explain.

17. The students in Question 16 conducted another experiment. This time they kept the length and thickness of the beams constant, but varied the width. Their data are shown in **Figure 1.52.**

Width of Beams (inches)	Breaking Weight (ounces)
0.5	6
1.0	13
1.5	19
2.0	22
2.5	29
3.0	37

FIGURE 1.52.
More data on width of beams vs. breaking weight.

Investigate the relationship between breaking weight and beam width. Give a function rule that models the relationship.

Glossary

KEY CONCEPTS

Axes: Number lines drawn at right angles used to locate points. Axes is the plural of axis.

Boyle's Law: A law that states that the pressure of a gas is inversely related to the volume of the gas.

Constant of proportionality: The constant ratio of y/x for (x, y) values of a direct variation function.

Continuous graph: A graph with the data points connected. For example, the relationship between the amount of hamburger purchased and the amount paid at a supermarket has a continuous graph. A shopper can order any amount of hamburger and pays more as the amount of hamburger increases.

Coordinate plane: A system for finding the location of points (x, y) in two dimensions, using axes with scales.

Coordinates: A pair of numbers that describes the location of a point in a coordinate plane. The first number is the x-coordinate, the second number is the y-coordinate.

Δ: The Greek capital letter, delta, used to show "the change in." The notation Δy means the change in the y variable.

Dependent variable: The output variable. In modeling, the dependent variable is the one you want to predict.

Direct variation function: A function with an equation of the form $y = kx$. Direct variation functions are used to describe proportional relationships.

Discrete graph: A graph with the data points unconnected. For example, the relationship between the number of oranges purchased at a supermarket and the price paid has a discrete graph. You cannot order 4.5 oranges, so connecting lines between 4 and 5 does not make sense.

Domain: The set of values that the input variable in a mathematical function can have.

Domain of the situation (or model): The set of reasonable values of the independent variable for a given situation.

Function: A set of ordered pairs of real numbers in which no two pairs with the same first coordinates have different second coordinates.

Function rule: A symbolic equation that describes a relationship between two variables.

Hooke's Law: A law that states that the force applied to a spring is directly proportional to the stretch of the spring.

Independent variable: The input variable. In modeling, the input variable is the one that is used to make predictions.

Inverse variation or indirect variation: A relationship in which one variable decreases as the other increases and the product of corresponding values of the variables is a constant. In other words, xy equals a constant, k.

Inversely proportional: Variables in an inverse variation relationship are said to be inversely proportional. That is, as one variable increases the other decreases, and the product of corresponding values of the variables is constant.

Linear relationship: When the points in a scatter plot lie along a straight line, the relationship between the two variables is called linear.

Mass: In science, mass is the amount of matter in an object. It is common to use weight and mass as the same.

Negative relationship: Two variables have a negative relationship (sometimes called a negative correlation) when the dependent variable decreases as the independent variable increases.

Answers (continued)

14.

X	Y1
0	ERROR
.25	96
.5	48
1	24
2	12
4	6
6	4

$y = \frac{24}{x}$

15.

X	Y1
-10	-10
-5	-20
-1	-100
0	ERROR
1	100
5	20
10	10

$y = \frac{100}{x}$

16. a) Sample answer:

A scatter plot shows a decreasing relationship that is not linear:

```
WINDOW
 Xmin=0
 Xmax=10
 Xscl=0
 Ymin=0
 Ymax=30
 Yscl=0
 Xres=1
```

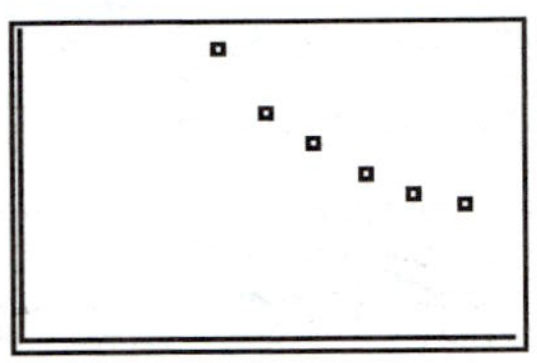

The products of the pairs are nearly constant:

L1	L2	L3
4	28	112
5	22	110
6	19	114
7	16	112
8	14	112
9	13	117

L3(7) =

Therefore, the relationship appears to be an inverse variation.

b) Beam length, because it explains the breaking weight. Beam length is the variable being controlled.

c) Sample answer: The average of the products is 112.83. Therefore, a function rule is $y = \frac{112.83}{x}$ where y is the breaking weight and x is the length of the beam. When graphed on the scatter plot, this function appears to model the relationship well:

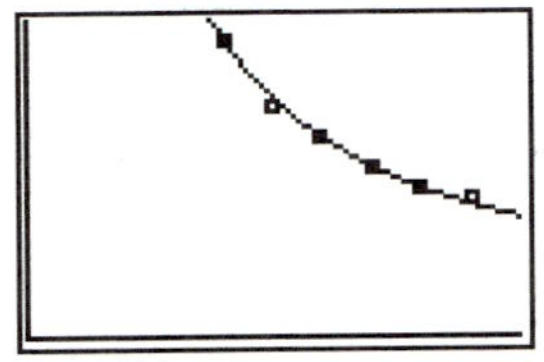

17. Sample answer:

```
WINDOW
 Xmin=0
 Xmax=4
 Xscl=0
 Ymin=0
 Ymax=40
 Yscl=0
 Xres=1
```

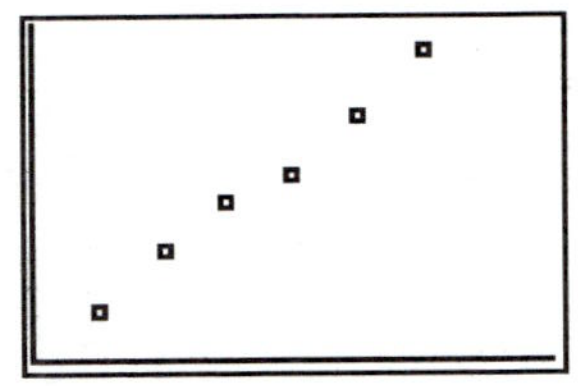

A scatter plot shows an increasing relationship that appears to be linear. If the points were connected with a line, the line would come close to the origin. Moreover, the ratios are nearly constant:

L1	L2	L3
.5	6	12
1	13	13
1.5	19	12.667
2	22	11
2.5	29	11.6
3	37	12.333

L3(7) =

Therefore, the relationship is probably a direct variation. The average of the ratios is 12.1, so a function that models the relationship is $y = 12.1x$, where x is the beam width and y is the breaking weight. This function appears to capture the trend of these data well:

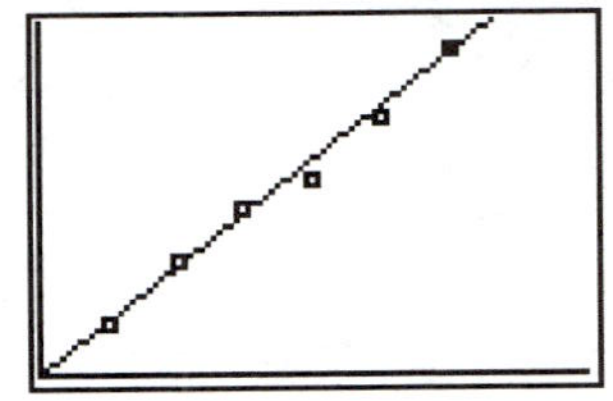

Point-slope formula: A method for finding the equation of a line. The formula is $y - y_1 = m(x - x_1)$.

Positive relationship: Two quantities have a positive relationship (sometimes called a positive correlation) when the dependent variable increases as the independent variable increases.

Proportional (relationship): A relationship is proportional if its continuous graph is a straight line that passes through the origin and the ratio $\frac{y}{x}$ is constant (or nearly so) for all data.

Proportions: Two fractions that are equal to each other.

Range: The values that the output variable in a mathematical function can have.

Range of the situation (or model): The set of reasonable values of the dependent variable for a given situation.

Rate of change: A ratio that compares the change in the output variable to the change in the input variable

Scatter plot: A graph of pairs (x, y) where the x-values are the independent values from your data and the y-values are the corresponding dependent values from your data.

Slope: A measure of the steepness of a line: the ratio of the vertical change to the horizontal change between two points on a line, or "rise over run."

Slope-intercept form: A form represented by the equation $y = mx + b$. This general form of a linear function expresses the dependent variable, y, in terms of the independent variable, x, the slope of the line (m), and the y-intercept of the line whose coordinates are $(0, b)$.

Units: Units describe the kind of quantity or information that a variable uses. Inches, centimeters, and milliliters are all units.

Variable: A quantity that changes.

Versus (vs.): When used in the phrase y versus x, it describes a scatter plot in which y is the dependent variable (vertical axis) and x is the independent variable (horizontal axis).

Volume: Generally, refers to the size of an object. That is, volume is a measure of the amount of space that a three-dimensional object occupies.

***y*-intercept:** The point at which a graph crosses the y-axis. The y-intercept is the term b in the slope-intercept form for the equation of a line.

CHAPTER 2

Bones: Linear Functions and Predictions

PREPARATION READING

The Disappearance of Amelia Earhart

Amelia Earhart is one of the most famous pilots of all time. She was the first woman to fly solo across the Atlantic Ocean and later across the Pacific. On June 1, 1937 she set off on a flight around the world along with her navigator. On July 2, their plane vanished between New Guinea and Howland Island. The U.S. Navy searched but did not find a trace of them. To this day their fate is unknown.

A group trying to solve this mystery heard about bones that were found on a Pacific island in 1940. At the time a Dr. D. W. Hoodless examined the bones. He concluded that they were from a male about 5 feet 5 inches tall. Some of his measurements are given in **Figure 2.1**.

Bones	Length (cm)
Humerus	32.4
Tibia	37.2
Radius	24.5

FIGURE 2.1. Bones analyzed by Dr. Hoodless.

Statements in his report raised doubt about his knowledge of the human skeleton. Lacking the bones, Dr. Karen Burns and Dr. Richard Jantz studied the measurements left by Dr. Hoodless. Using data from a data bank at the University of Tennessee they built models to predict height, gender, and ethnic background. They concluded that the bones were from a white female about 5 feet 7 inches tall, which fits Earhart. Could the bones have been hers? This question can only be answered if the bones are recovered.

In this chapter you will be asked to think like a forensic anthropologist. Given measurements of a set of bones, you will investigate the clues about a dead person. You will also collect data and build models to predict height and gender.

CHECK THIS!

A forensic anthropologist is a scientist who uses information about the human body and its bones to try to identify a dead person based on bones and teeth. This is one of many kinds of work that a forensic anthropologist does.

CHAPTER 2

Bones: Linear Functions and Predictions

Preparation Reading:
The Disappearance of Amelia Earhart

Description: The reading introduces the major context of the chapter, prediction of height from the lengths of long bones. The story focuses on whether bones found on the island of Nikumaroro were Amelia Earhart's. The only clues are a few measurements.

Materials: none

Teaching Ideas: Students should work in small groups (3 to 4 students). To help insure that all students read the preparation reading, have students read silently then randomly select one student in each group to explain it to the other group members. Ideas from the reading are referred to several times in the text.

SECTION 2.1

Mysterious Findings

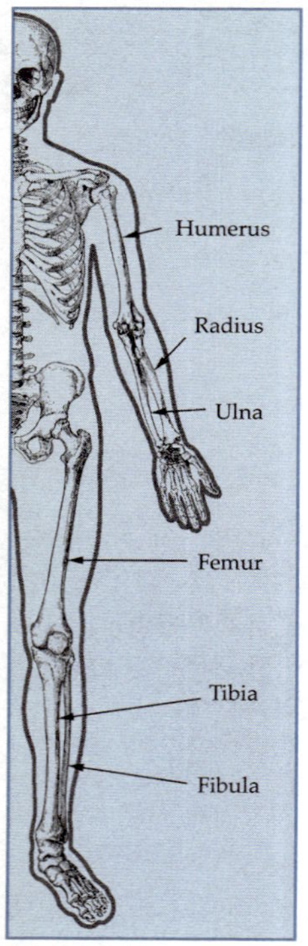

A human skeleton.

Archaeologists and forensic scientists often use mathematical models to help investigate human remains found at historical sites and crime scenes. In this section you will explore models that describe the relationship between the length of a person's head and the person's height.

From time to time, bones are found in rugged areas. A hiker in Arizona's Superstition Mountains found a skull, eight long bones, and many bone fragments. He called the local police who sent a team to investigate. The team recorded facts about the bones, including their size. **Figure 2.2** shows some data that are similar to what the team recorded.

Bone Type	Number Found	Length (cm)
Femur	3	41.5, 41.4, 50.8
Tibia	1	41.6
Ulna	2	22.9, 29.0
Radius	1	21.6
Humerus	1	35.6
Complete Skull (including jaw)	1	23.0
Fragments	More than 10	From 3.0 to 5.0 cm

FIGURE 2.2. Measurements of bones found at Superstition Mountains.

Use all the facts you have so far to answer these questions.

1. Based on the data in the table, what is the precision of the measurements?

2. Study the data in Figure 2.2. The team reports that these bones belong to at least two people. How do they know?

3. Which bones do you think belong to the same person? What assumptions did you make to get your answer? Explain your answer. (To make it easier to classify the bones, refer to the dead people as Skeleton 1, Skeleton 2, and so on.) You will need the answers to this question later on.

4. Do the lengths of the bones make you think they belonged to a male or a female? What did you use to get your answer? (Remember, Dr. Hoodless concluded that the bones found on the island belonged to a male, but Burns and Jantz disagreed with his findings.)

5. Do you think the dead people were children or adults? Defend your answer.

6. Guess the heights of the dead people. How good do you think your guesses are?

ENGAGE LESSON 1

Section 2.1: Mysterious Findings

Purpose: Students review the concept of precision and make simple predictions from a collection of bones. They will test these predictions later using a variety of models.

Description: Students are given a description of a group of bones. They begin to sort out the bones and work to answer six questions about the people to whom these bones belonged. They discover that the skull is particularly difficult to classify. To classify the skull, they will need to know the relationship between head length and height and femur length and height. One way to find these relationships is to collect data and fit a model to the data. In the next phase of this lesson students will explore the relationship between head length and height.

Materials: either Transparency 2.1 or Handout 2.1 of Figure 2.2 "Measurements of bones found at Superstition Mountains"

Teaching Ideas: Students should work in small groups (3 to 4 students). Give students a limited amount of time to work on the questions in their groups (around 15–20 minutes, perhaps longer if students continue to be engaged with the problem). At the end of this time, debrief the activity by discussing the students' answers to Questions 1–6. It is not important that students arrive at correct answers. What is important is the reasoning they use to arrive at their answers. For example, some students may decide that the 41.6-cm tibia belongs to the person with a 41.4-cm femur. They may have arrived at this conclusion based on the diagram of the skeleton. (In that diagram, the femur and the tibia appear to be roughly the same length.) Other students may argue that the femur is the longest bone in the body, and thus, the 41.6-cm tibia must belong to the person with the 50.8-cm femur. Let students discuss these issues themselves. Don't give them the answer.

If students struggle with an answer to Question 6, encourage them to use their general knowledge of people's heights. For instance, no one would guess that the person was 20 feet tall because none of us knows anyone who is 20 feet tall. At the very least, students should be able to come up with a range of reasonable heights and select values that lie somewhere in the middle of this range for their estimates.

CHECK THIS!

You may need to remind students that, in this chapter, precision refers to the units (centimeters, meters, etc.) and the place value of the units we use to measure something.

Answers

1. Nearest tenth of a centimeter. Be sure students understand that consistent precision is important when recording data.

2. There are three femurs. Two of the femur bones are very close in length (41.4 cm and 41.5 cm) and probably belong to the same person. So, these three bones most likely belong to two people. It is possible that the bones actually belong to more than two people.

3. Sample answer: Assume that there are only two people and that Skeleton 1 is the shorter of the two.

   Skeleton 1: femurs—41.4, 41.5; tibia—41.6; ulna—22.9; and radius—21.6.

   Skeleton 2: femur—50.8; ulna— 29.0; humerus—35.6.

   The skull could belong to either Skeleton 1 or Skeleton 2.

   Since two femurs—41.4 cm and 41.5 cm—are close in length, they most likely belong to Skeleton 1. From the diagram of the skeleton, it appears reasonable to assume that a person's tibia should be fairly close in length to the femur, so the 41.6-cm tibia belongs to Skeleton 1. The person with the shorter leg bones probably also has the shorter arm bones. That means that the 22.9-cm tibia belongs to Skeleton 1. The length of a person's radius should be close to the length of the ulna, and hence the 21.6-cm radius belongs to Skeleton 1.

   Skeleton 2, the taller of the two people, has a 50.8-cm femur and a 29.0-cm ulna. A person's humerus should be just a bit longer than the ulna (based on the diagram of the skeleton). So, the 35.6-cm humerus belongs to Skeleton 2.

NOTE: Student answers may vary. What is important is the reasoning behind their classifications. For example, some students may know that the femur is the longest bone in the body. This knowledge would lead them to conclude that the 41.6-cm tibia belongs to Skeleton 2.

4. Sample answer: There is very little evidence to say whether the bones belong to a male or female. Because the femur bones of one of the decedents are considerably shorter than the other, it's possible that Skeleton 1 (shorter femur) is female and Skeleton 2 is male. However, it could just as easily be the case that the decedents are two males, one short and one tall, or a child and an adult.

5. Sample answer #1: Most likely, the decedents were adults. What would young children be doing out in the Superstition Mountains?

   Sample answer #2: When I measured my own forearm to get an estimate of how long an adult's ulna might be, it measured about 26.0 cm. (The actual length of my ulna would be less than this measurement.) So, Skeleton 2 is probably an adult. Skeleton 1 might be a child.

6. Sample answer: Assuming that Skeleton 1 is female and Skeleton 2 is male, my guesses are that Skeleton 1 is 5 feet 5 inches tall and Skeleton 2 is 5 feet 10 inches tall. These guesses were based on my estimates of average heights for women and men, respectively. These are very rough estimates but they should be within a foot of the actual heights.

SECTION 2.2

A Model for Estimating Height

One way to turn your guess of the dead peoples' heights into a scientific estimate is to explore the relationship between the length of certain bones and a person's height.

In this section you will gather some data to explore a possible relationship between head length and a person's height.

Within your group, measure the length of each person's head (from chin to the top of the head) and then measure their height. Record your data in a table like **Figure 2.3**. Be sure to give the units of measure you used at the top of the second and third columns.

Name	Head Length	Height

FIGURE 2.3. *Group head-length and height data.*

1. What patterns do you see in this data set?
2. We aren't sure yet that there is a relationship between head length and height, so why does it make sense to consider head length as the independent variable here?
3. What is a reasonable domain for this situation? Why?
4. What is a reasonable range for this situation? Why?
5. On a sheet of chart paper create a scatter plot of height versus head length. Sketch your graph.

TREND LINES: LINES THAT APPROXIMATE DATA

The plot of height versus head length is fairly spread out, but it appears to have a linear form. This large amount of variability makes it hard to pick a line that you can use to make good predictions for heights. The following two methods will help you choose a line that approximates the pattern of the height versus head-length data from your class:

METHOD 1

1. Pick a point that appears to lie in the middle of the points displayed in your scatter plot. The point doesn't have to be a data point.
2. Using this point as one point on your line, adjust the slope of your line until you find a line that you think best describes the pattern of the data.
3. Find the equation of your line.

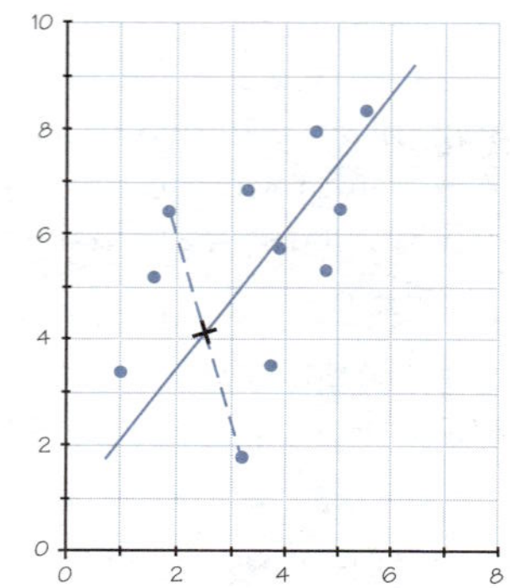

Method 1.

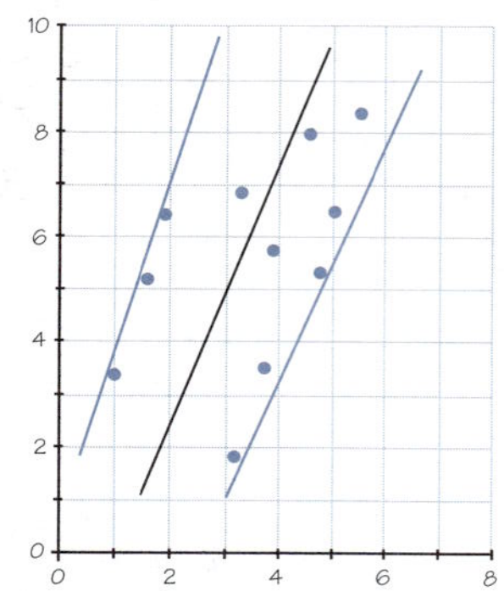

Method 2.

METHOD 2

1. Draw two lines in such a way that the points on your hand-drawn scatter plot are squeezed as tightly as possible between these lines. (The lines don't have to be parallel.)
2. Now draw one line halfway between the two lines that you have drawn.
3. Find the equation of this line.

6. Divide your group in half. Half of your group should use Method 1 and a meter stick to find a trend line for the graphed data. The other half should use Method 2 and two meter sticks to find a trend line. Draw both trend lines over the scatter plot and sketch your graph.
7. Write the equation of your trend line.

EXPLORE LESSON 1

Section 2.2: A Model for Estimating Height

Purpose: In Section 2.1 students were given the lengths of several bones, and used their prior knowledge and reasoning to classify bones and then guess the height of the dead people. In Section 2.2 students explore to see if there is a relationship between head length and height. They also compare different methods of finding trend lines and identify the method that yields a function rule that appears to better model the data. In Section 2.3 methods of finding trend lines continue to be examined so that the rule that better models the data can be used to make predictions.

Description: Students measure the head lengths and heights of classmates, then use the data to examine relationships between height and head length. First they use chart paper and a meter stick to create a scatter plot of the data and find a function rule (trend line) that models the data. Next students use a measure of central tendency to calculate a constant of proportionality (multiplier) and use it to write a function rule. Using a graphing calculator, students graph their function rule over a scatter plot. If the rule does not appear to be a good fit to the data the students adjust the multiplier to get a better fit. Students then compare graphs of the two function rules to see which appears to better model the data. Students use their models to predict the height of a person given his/her head length and to predict the head length of a person given his/her height.

Materials: graphing calculator, meter sticks, chart paper, markers, rulers

Teaching Ideas: Students should work in small groups (3 to 4 students). Stack two meter sticks end to end and tape them to the wall so that students can measure their heights. (If you have additional meter sticks, you may want to set up more stations for measuring height.) Distribute rulers to the groups at the beginning of the activity. Students need them to measure head length. They may want to use two rulers—one to place on top of the head for a level and one to measure the length. They can make their own tables similar to the one in Figure 2.3 or you can distribute a copy of the table to each group. Groups need to get results from at least one other group so that they have sufficient data to answer the questions.

Answers:

Sample answer:

Name	Head Length (cm)	Height (cm)
Jan	23.0	168.0
Horace	25.0	178.0
Betty	24.0	166.5
John	25.0	184.0
Joy	23.5	170.5
Lora	22.0	159.0

1. Answers may vary. Sample answer: The people with the greater head lengths are the taller people and the people with the smaller head lengths are the shortest people.
2. Answers may vary. Sample answer: In the context of this situation scientists will be using the length of human skulls to predict the height of decedents. So in a way we can say that head length will predict height.
3. Answers may vary. Sample answer: A reasonable domain would be 20 cm $< x <$ 27 cm because in our sample the students' head lengths ranged from 22 cm to 25 cm.
4. Answers may vary. Sample answer: A reasonable range would be 150 cm $< x <$ 190 cm because in our sample the students' heights ranged from 159 cm to 184 cm.
5. Graphs may vary. Two sample graphs are shown.

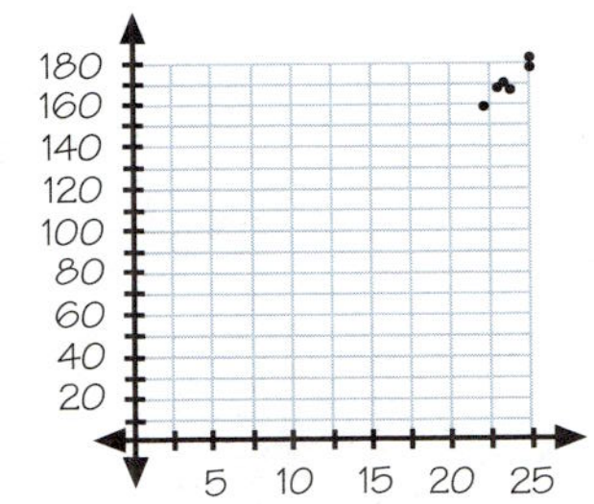

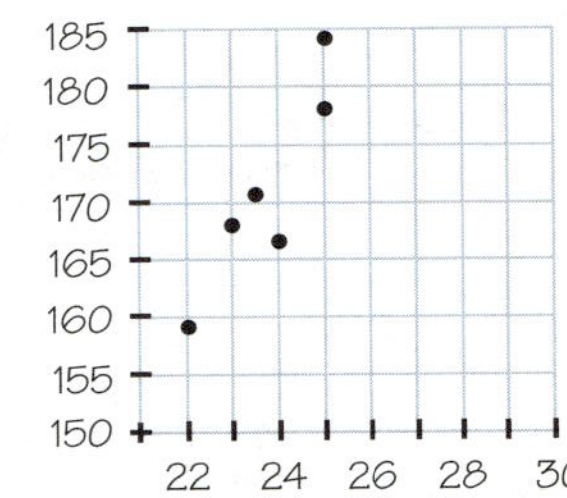

6. Trend lines may vary. A sample trend line is shown.

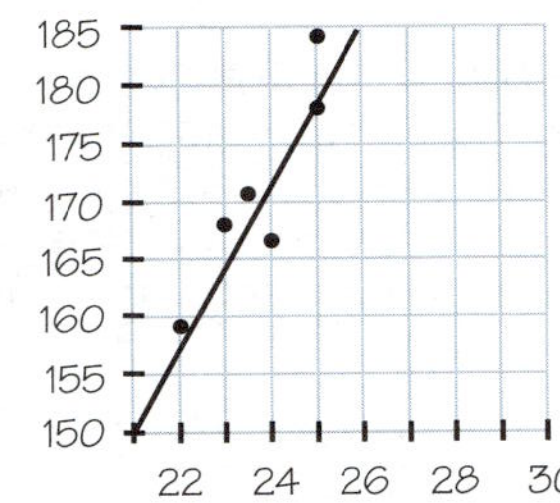

7. Answers may vary. The equation for the trend line above is $y = 7.1x$.

8. Find the ratio of head length to height by dividing each person's height by his or her head length. Record your results in a table like **Figure 2.4.**

Name	$\frac{\text{Height}}{\text{Head Length}}$

FIGURE 2.4. Ratio of head length to height.

Now you have a ratio of head length to height for each member of the class. You can start using these data to predict a person's height.

You need to find the best constant of proportionality (multiplier) that can be used to predict the height of the whole class.

One way to find the constant of proportionality is to use a measure of central tendency. You may have seen three before: mean, median, and mode. Let's review these and decide which one would be useful in making a prediction.

CHECK THIS!

The mean is the average value. Here's how to calculate the mean:

$$\text{mean} = \frac{\text{sum of all observations}}{\text{number of all observations}}.$$

9. What is the mean of the ratios of the head length to height?

CHECK THIS!

The median is the middle value. Here's how to calculate the median:

1. Arrange the data in order from smallest to largest.
2. If the number of data is odd, the median is the middle number. To find it, add 1 to the number of data then divide by 2. Start at the bottom of the ordered data and count up that number of observations.
3. If the number of data is even, the median is the average of the middle two numbers. To find it divide the number of data by 2. Again, start at the bottom and count up that number of observations. Add this number to the next number in the data set. Divide the sum by 2.

10. What is the median of the ratios of the head length to height?

CHECK THIS!

The mode is the most frequent data point in a data set.

11. What is the mode of the data set of ratios of the head length to height?

The question you are trying to answer is which measure of central tendency is the best constant of proportionality (multiplier).

In general, use the mode when you want to know what is the most common, as in the most common shoe size among American men. Use the mean when no values are much larger or much smaller than the rest of the data. Use median when a few relatively large or small values make the mean larger or smaller than most of the data.

Since all of the measures of central tendency are close in this case, it does not really matter which one we use. So, we will use the mean as the constant of proportionality (multiplier).

12. Use the constant of proportionality (multiplier) to write a function rule that could be used to predict a person's height given their head length.

13. Use a graphing calculator to create a scatter plot of height versus head length. Sketch your scatter plot and record your window.

Answers (continued)

8.

Name	$\frac{\text{Height}}{\text{Head Length}}$
Jan	7.3
Horace	7.1
Betty	6.9
John	7.4
Joy	7.3
Lora	7.2

9. The mean of the ratios is 7.2.

10. The median of the ratios is 7.3.

11. The mode of the ratios is 7.3.

12. $y = 7.2x$

13.

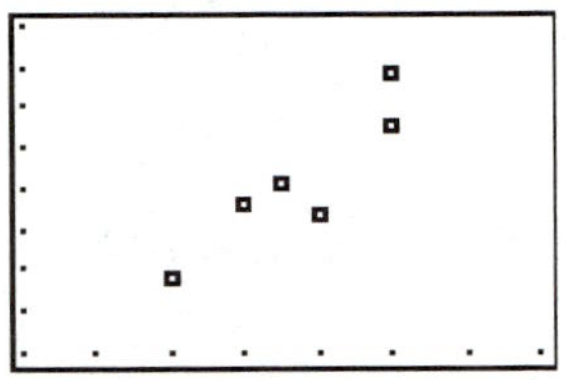

x-min: 20
x-max: 27
y-min: 150
y-max: 190

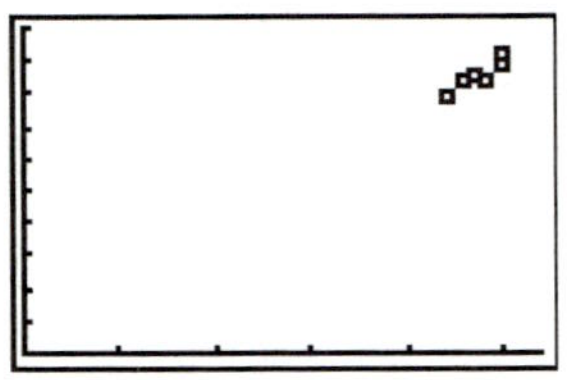

x-min: 0
x-max: 27
y-min: 0
y-max: 190

14. Graph your function rule on the scatter plot. If needed adjust the equation. Write your new equation and sketch your graph.
15. On the same scatter plot, graph the trend line that you found using the meter stick.
16. How does the trend line found using the meter stick compare to the function rule found with the constant of proportionality?
17. Does the relationship of height and head length appear to be a proportional relationship? Why or why not? Give at least two reasons to support your answer.

CHECK THIS!

Recall that the slope of a line = $\frac{\Delta y}{\Delta x}$.

In words, the slope is the ratio of the change in the output variable *y* for a given change in the input variable *x*.

18. What is the equation of a line that you think fits the data well? What is the slope of this line?
19. What meaning does the slope have in this context?
20. Using words, describe the relationship between the length of a person's head and the height of the person.
21. Predict the height of a person whose head length is 25 cm.
22. Predict the head length of a person whose height is 175 cm.

SECTION 2.2 Assignment

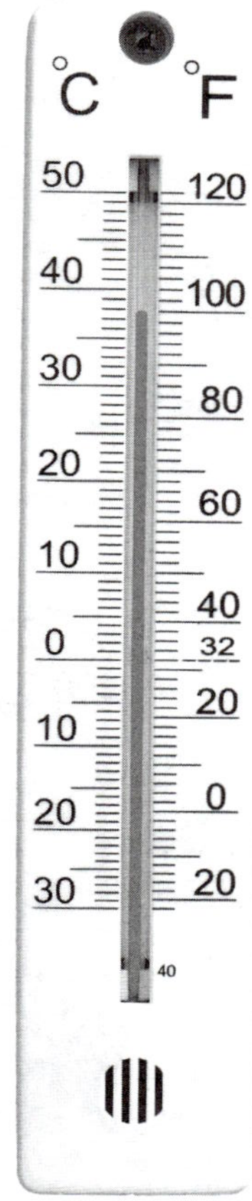

1. A class has the following heights (in centimeters):

   168, 178, 161, 155, 156, 155, 171, 188, 166, 161, 178, 180, 155, 167, 155, 157, 160, 158, 152, 152, 159, 155, 172, 188

   a) What is the mean height for this group of students?

   b) What is the median height?

   c) What is the mode of the heights?

2. The line plot in **Figure 2.5** shows the average daily temperature in a city for each day during the month of May. Each dot represents one day when the average daily temperature was that number of degrees Fahrenheit. For example, there were three days when the average daily temperature was 50°F.

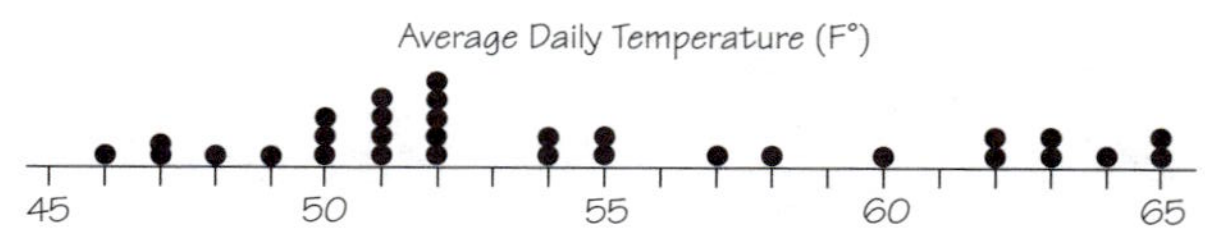

FIGURE 2.5. Line plot.

   a) What is the average (mean) daily temperature for the month of May? Explain how you found your answer.

   b) What is the median daily temperature? Explain how you found your answer.

   c) What is the mode daily temperature? Explain how you found your answer.

Answers (continued)

14.

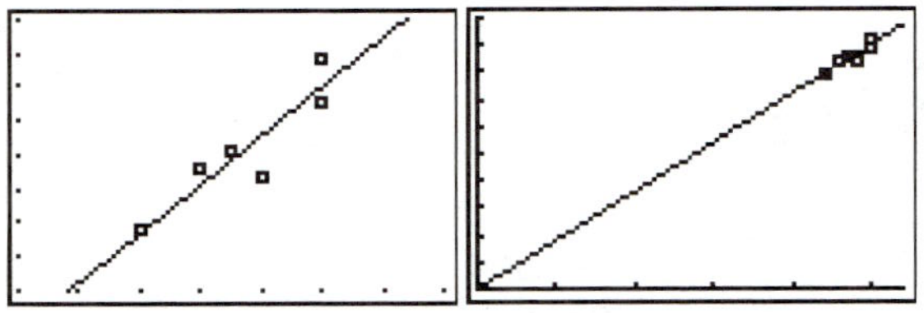

15.

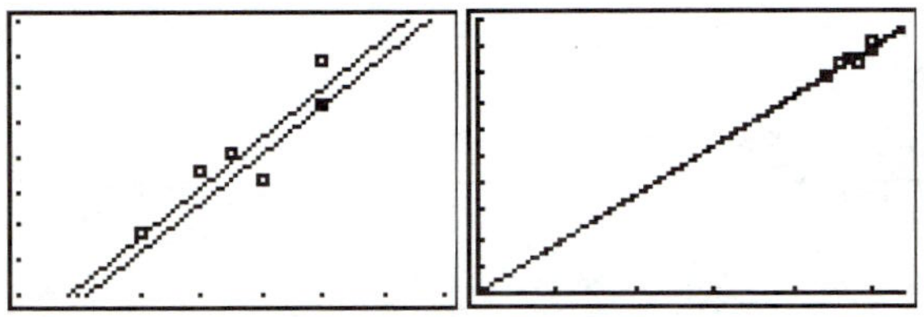

16. Both lines are very similar. When the lines are examined using the window [x-min: 20, x-max: 27, y-min: 150, y-max: 190] the line $y = 7.2x$ appears to model the data more closely.

17. It is a proportional relationship because, (1) the graph is linear and passes through the origin; (2) the ratio of y/x is close to constant for all of the data; and (3) the form of the equation to fit the data is $y = kx$, a direct variation function.

18. $y = 7.2x$; the slope of the line is 7.2.

19. Students should realize that the slope is the same as the constant of proportionality, $k = 7.2$, which means that 1 centimeter of head length corresponds to 7.2 centimeters of height.

20. A person's height is 7.2 times the head length.

21.

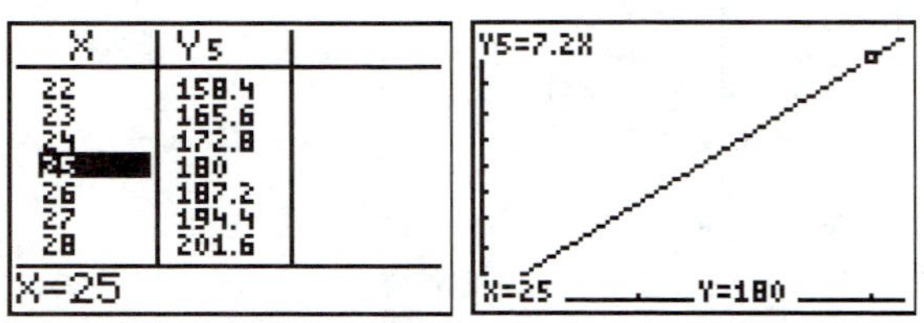

$y = 7.2x$
$y = 7.2(25 \text{ cm})$
$y = 180 \text{ cm}$

22.

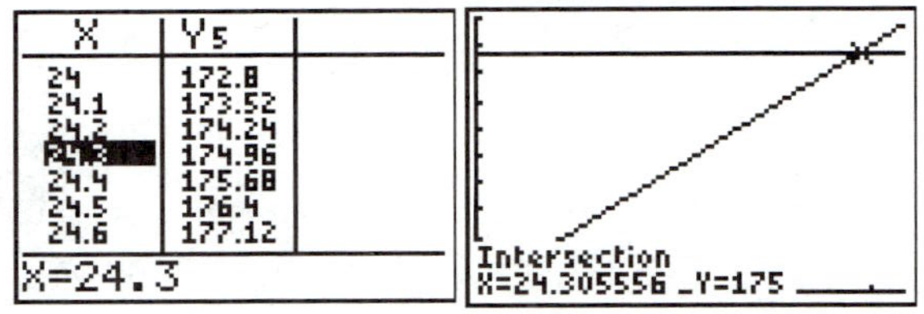

$175 \text{ cm} = 7.2x$

$\frac{175 \text{ cm}}{7.2} = x$

$24.3 \text{ cm} = x$

Section 2.2 Assignment

Answers

1. a) 164.5 cm

   b) 160.5 cm

   c) 155 cm

2. a) 54.5°F. One possible explanation is that students will sum the 31 temperatures, then divide by 31 days.

   b) 52°F. One possible explanation is that students will list the temperatures in order from greatest to least, then select the 16th temperature (since 31 is an odd number, 15 temperatures will be lower than the 16th temperature, and 15 temperatures will be higher than the 16th temperature).

   c) 52°F. One possible explanation is that students will count the frequency of the temperatures and find that on 5 days during May the temperature was 52°F. This temperature occurred more frequently than any other temperature during the month. A second possible explanation utilizes the power of a line plot; students may observe that the stack of dots at 52°F is taller than the other stacks of dots, indicating the most frequently occurring average daily temperature.

Use **Figure 2.6**, which shows the height of students in Ms. Jaczinko's tenth-gradeclass, for Questions 3–19.

Female		Male	
Name	Height (cm)	Name	Height (cm)
Alicia	157	Ahmed	173
Bia	166	Brian	177
Christi	164	Jesús	174
Chantalle	164	Davis	192
Coral-Mae	161	José	172
Juanita	164	Kelvin	180
Ji-Hyun	167	Lenny	174
Kim	162	Luis	175
Kristen	175	Mike	185
Maria	166	Pedro	185
Tianna	172	Sang	178
Teresa	176		

FIGURE 2.6. Height data from Ms. Jaczinko's class.

3. What can you say about the precision of these measurements?

4. Use the table in **Figure 2.7** to find the mean, median, and mode for the male students, the female students, and all students in one class.

	Mean	Median	Mode
Male students			
Female students			
All students			

FIGURE 2.7. Measures of central tendency for Ms. Jaczinko's class.

5. Look over the data in Figure 2.6. What is the range in heights from the shortest student to the tallest?

6. A student joined Ms. Jaczinko's class. Is it reasonable to predict that the tenth-grader would be between 164 cm and 180 cm tall? Explain your answer.

7. What is the range of girls' heights from shortest to tallest?

8. If you know the new student in Ms. Jaczinko's class is a girl, what would you say about this person's height based on these data?

9. A girl who is 196 cm tall joins Ms. Jaczinko's class. How does this new girl's height compare with the other heights in the class?

10. What effect does this new height have on the mean, median, and mode of the girls' height data?

One way to predict a new student's height is to take the **mean** of all the heights.

11. Use the mean of the students' heights as a prediction of a new student's height. Record the mean student height.

12. If the new student is as short as the shortest student in the class, how far off is the prediction in Question 11?

13. If the new student is as tall as the tallest student in the class, how far off is the prediction in Question 11?

14. Do you think that the mean is a good prediction? Explain. Can you make a better one?

15. A new student, Melissa, joins the class. Does the new student, a girl, change your prediction? Find the mean for girls' heights. Use this average to predict Melissa's height.

16. If Melissa is as short as the shortest girl in Ms. Jaczinko's class, how far off is your prediction? What if Melissa is as tall as the tallest girl?

17. Do you think this is a better prediction than the one made in Question 11? Explain.

18. If the new student turns out to be Martin (a boy), not Melissa, can you choose a way to predict Martin's height? Give your prediction and your method.

19. If the new student's height is between the shortest boy and the tallest boy, what is the largest possible error for your prediction?

Answers (continued)

3. The data are recorded to the nearest centimeter.

4.

	Mean	Median	Mode
Male students	178.6	177	174 or 185
Female students	166.2	165	164
All students	172.1	173	164

5. The heights of the students vary from 157 cm to 192 cm; a total difference of 35 cm.

6. Sample answer: There are only three students in Ms. Jaczinko's class with heights below 164 cm and three above 180 cm. The heights of 17 of the 23 students fall within the 164- to 180-cm range. So, if heights of students in Ms. Jaczinko's class are representative of tenth grade students in general, chances are pretty good that this prediction will be correct. Even if this prediction is wrong, it will be off by at most 12 cm (provided tenth grade students, in general, have heights similar to those of the students in Ms. Jaczinko's class).

7. The heights of the girls vary from 157 cm to 176 cm; a total difference of 19 cm.

8. Sample answer: Her height was between 157 cm and 176 cm.

9. This girl is considerably taller than the other students in the class.

10. The new mean is 168.5, an increase of 2.3 cm; the new median is 166, an increase of 1 cm; the new mode is 164, which is unchanged.

11. 172.1 cm

12. If the new student is 157 cm, then the prediction will be off by 15.1 cm.

13. If the new student is 192 cm, then the prediction is off by 19.9 cm.

14. Sample answer #1: The prediction in Question 11 is somewhere near the middle of the height data. If the unknown student's height lies somewhere in the middle of the heights of students in Class A, this prediction is close to the actual height of the student. If the student's height is as small as the shortest student in Class A or as tall as the tallest student, this prediction is off by at most 19.9 cm. So this prediction seems reasonable.

    Sample answer #2: Perhaps the median height, 173 cm, might be better.

    Sample answer #3: Maybe we should use 174.5 cm, midway between the shortest and tallest student's heights. Then our prediction would be off by no more than 17.5 cm.

15. 166.2 cm

16. If Melissa is 157 cm, this prediction is off by 9.2 cm; if she is 176 cm, this prediction is off by 9.8 cm.

17. The prediction from Question 11, 172.1 cm, is probably too high if you know that the student is a girl. (Only three girls are at least 172 cm tall.) If the new student's height is in the middle of the girls' heights for students in Class A, then using the prediction of 166.2 cm is off by no more than 9.8 cm. On the other hand, the prediction of 172.1 cm may be off by as much as 15.1 cm.

18. Sample answers:

    Possible predictions: Average of boys' heights—approximately 178.6 cm; median of boys' heights—177 cm. Some students may just eyeball what seems to be a typical boy's height.

19. This answer depends on the answer to Question 18.

    Sample answer: If the prediction is 178.6 cm, then it could be off by at most 13.4 cm.

SECTION 2.3 Height vs. Bone Lengths

In the last section, you collected and analyzed data on the relationship between a person's height and the length of their head from the bottom of the chin to the top of the skull. This relationship is important because it allows anthropologists who find skull bones to estimate the height of the person. Forensic scientists, such as those on crime-solving television shows, can also use this relationship to help find whose bones they were.

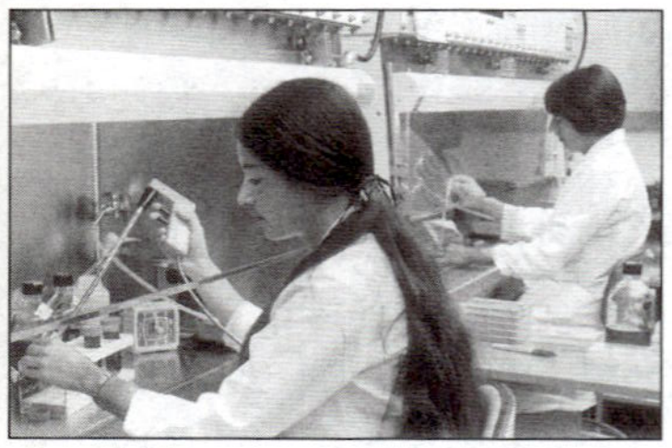

You have already collected data and answered questions about what you found. Now, let's look at some important ideas about data collection and building linear models based on those data.

1. When you measured the head length of your group members, what method did you use? How did you know that your method would yield good data?
2. How did you decide on a reasonable domain and range here?
3. How did you place your trend line?
4. How did you find the equation of your trend line?
5. When is it best to use mean as a measure of central tendency?
6. When is it best to use median as a measure of central tendency?
7. When is it best to use mode as a measure of central tendency?
8. What are the features of a direct variation function?
9. What effect does changing the value of the multiplier have on the graph of a direct variation function?

10. How might you use your model to make a rough prediction of the height of the person whose skull length is recorded in Figure 2.2?

    a) Recall that the skull measured 23.0 cm in length. Predict the height of the person in centimeters (cm). Describe the process you used to make your prediction.

    b) Does your prediction result in a reasonable height for a person? Explain.

11. How might you use your model to make a rough prediction of the head length of a person whose height was 175 centimeters?

    a) Predict the head length of the person in centimeters. Describe the process you used in making your prediction.

    b) Does your prediction result in a reasonable head length for a person? Explain.

12. What information do you think might give better estimates of the heights of the dead people whose bone lengths are recorded in Figure 2.2? How or where might you get this information?

MODELS OF THE FORM $y = mx$

- The graph of a function in the $y = mx$ family is a line that passes through the origin.
- The steepness and incline or decline of each of these lines is controlled by the value m, which is the slope of the line.
- These are also called direct variation functions, which are used to model proportional relationships.

EXPLAIN LESSON 1

Section 2.3: Height vs. Bone Lengths

Purpose: Students use a linear model to predict heights.

Description: This section is designed to be a teacher-led discussion of the concepts in Section 2.2. Use the questions below to facilitate the class discussion.

Materials: none

Teaching Ideas: Debrief students on Section 2.2 by having them share their answers to the questions. As they share, be sure to fill in any missing concept, process, or skill that could have been used to answer the questions. Then work through the questions in this section with the entire class. Once again be sure all necessary concepts, processes, or skills are addressed.

Answers

1. Answers may vary. During this discussion be sure that students came to a group decision on methods for measuring the heights of students and the lengths of their heads. They should have tested their methods. They may have had two students use different methods to measure the height of the same student. Both height measurements should have been roughly the same. Students should have agreed on a degree of precision for measurements. Be sure they understand precision and that it is important to ensure quality data.
2. Students should understand that by observing the collected data and the reasonableness of possible values for the situation they are able to find the domain and range.
3. Answers may vary. Be sure students understand how to find a trend line and that the line is an approximate model for the data. Revisit the two methods discussed in Section 2.2.
4. Answers may vary. Students should understand that there are several methods that may be used to find the equation. Any two points could be selected and used to find the slope of the line, slope could be counted on the grid or rate of change could be found from a table of values.
5. Use mean when the data set is compact and there are no very large or very small values as compared to the rest of the data set.
6. Use median when there are very large or very small values that do not match the rest of the data set.
7. Use mode when you want to know what is the most common, as in what shoe size is the most common among American men?
8. The features of a direct variation include: (1) the graph is linear and will pass through the origin; (2) the ratio of y/x is close to constant for all of the data; and (3) the form of the equation to fit the data is $y = kx$, a direct variation equation.
9. The multiplier controls the steepness of the line; the larger the multiplier the steeper the slope of the line.
10. Students should be able to use a graph, a table, and symbolic manipulation to solve this type of problem.

    a) 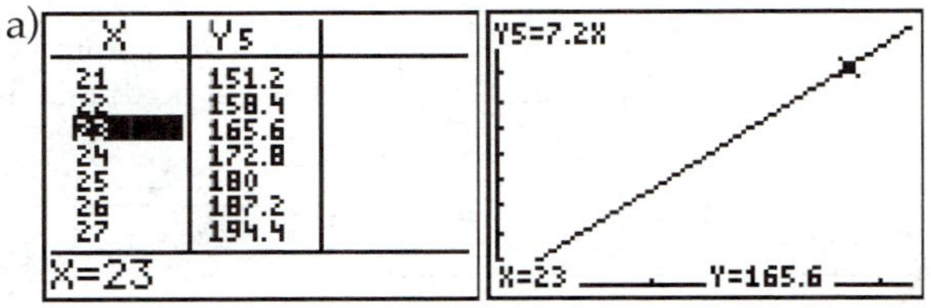

    $y = 7.2x$
    $y = 7.2\ (23\text{ cm})$
    $y = 165.6\text{ cm}$

    b) Answers may vary. Students should be able to justify their answer.
11. Students should be able to use a graph, a table, and symbolic manipulation to solve this type of problem.

    a) 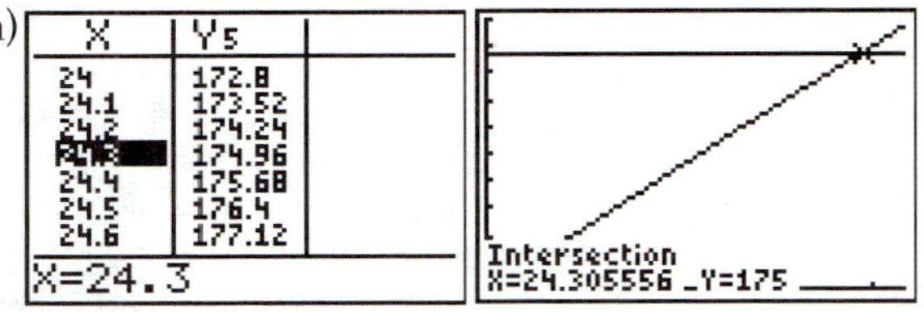

    $175\text{ cm} = 7.2x$

    $\dfrac{175\text{ cm}}{7.2} = x$

    $24.3\text{ cm} = x$

    b) Answers may vary. Students should be able to justify their answer.
12. It would be helpful to know the relationship between bone lengths and heights. Perhaps we could use the data from the Forensic Data Bank to make our own models.

    Another possibility is from an artist's handbook. Since these handbooks contain ratios of head length to height, they might also contain ratios of leg length or arm length to height.

MODELS OF THE FORM $y = mx + b$

- The graph of a function in the $y = mx + b$ family is a line that intersects the y-axis at $(0, b)$.
- The steepness and incline or decline of each of these lines is controlled by the value of m, which is the slope of the line.
- The vertical shift of the graph is controlled by the value of b, which is the y-intercept of the line.

SUMMARY

In Sections 2.1, 2.2, and 2.3 you explored the relationship between the length of certain bones and a person's height. You then used specific functions to predict a person's height from these bones.

- You also gathered data and applied a measure of central tendency—mean, median, or mode—to the data to find a multiplier (constant of proportionality) and a formula to predict height based on the length of the person's head.
- Since these models give rough estimates of a person's height, you will need to do more investigating to find a better model that gives better predictions. One way to improve your models is to collect data and use the data to test your model. Analyzing data forms the basis of the next section.

SECTION 2.3 Assignment

1. Which graph(s) in **Figure 2.8** show(s) a trend line that appears to model the given data well? Explain your reasoning.

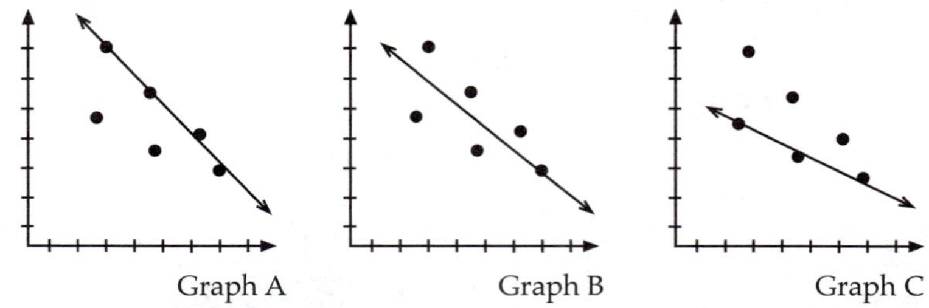

FIGURE 2.8. Three graphs of trend lines.

2. Hernando kept a record of his salon sales and tips for one afternoon as shown in **Figure 2.9.** Plot the data on a copy of the grid in Figure 2.9, number and label the graph appropriately. Use a straightedge to draw a trend line. Use this graph to answer the questions that follow.

Sale Amount ($)	Tip Amount ($)
10	2
15	3
20	2
25	4
30	4
35	6
40	5
45	7

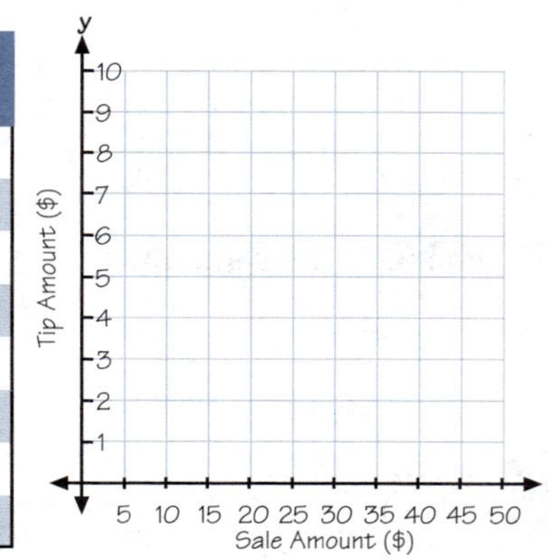

FIGURE 2.9. A table of Hernando's tips with a grid to plot the data.

a) Use what you know about slope and y-intercept to write an equation for your trend line.

Section 2.3 Assignment

Answers

1. Although the line in Graph A hits more points, a best-fit line is a line that takes into account the distance of each point from the line. Looking at Graph B, we can see it is the best choice because the line is in the center of the negative and positive values. So the negative vertical distances offset the positive vertical distances, and the sum of these appears close to zero.

2. a) The students should use their knowledge of slope and y-intercept to estimate a trend line. Students should then adjust the slope and/or y-intercept until they find a line that best models the data. While answers may vary, a sample equation for this data is $y = 0.14x + 0.39$.

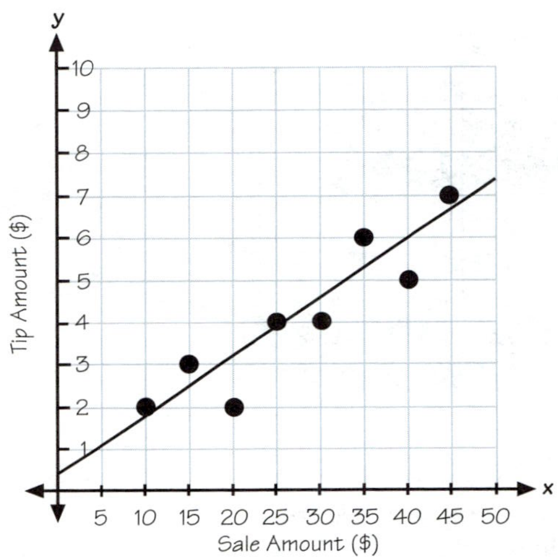

b) What does the slope mean here? Does this value make sense in this context?

c) What does the y-intercept mean? Does this value make sense in this context?

d) Predict Hernando's tip if his salon sale is $55. Justify your answer.

3. Nedra buys beads in bulk, by the pound, for her jewelry business. She has kept a record of her bead purchases (which are small) for the last 2 months. Use the data in **Figure 2.10** to graph a trend line and to answer the questions that follow.

Pounds of Beads Purchased	Total Amount Paid
1	1
1.25	1.50
2	2
2.25	2.5
2.75	3
3	3.5
3.75	4

FIGURE 2.10. Nedra's data.

a) Use your knowledge of slope and y-intercept to write an equation for your trend line in slope-intercept form.

b) What does the slope mean here? Does this value make sense in this context?

c) What does the y-intercept mean? Does this value make sense in this context?

d) How much would you expect Nedra to pay for 5 lbs of beads? Justify your answer.

4. Ten students' semester grades from Mr. Johnson's last year's class are given in **Figure 2.11.**

First Semester Grade (x)	55	62	64	71	75	78	82	85	95	97
Second Semester Grade (y)	50	54	64	65	60	70	77	72	88	88

FIGURE 2.11. Grades for 10 students.

a) Use these data to make a scatter plot. Find an appropriate trend line.

b) What does the slope mean here? Does this value make sense in this context?

c) What does the y-intercept mean? Does this value make sense in this context?

d) If you made a 72 the first semester, what is your predicted score for the second semester?

5. Psychologists are trying to find a relationship between test anxiety and success on a test. Examine the data in **Figure 2.12**, and answer the questions that follow.

X	0	7	14	14	15	20	20	21	23
Y	77	50	59	48	51	52	46	51	43

x = score on a test anxiety evaluation.
y = exam grade.

FIGURE 2.12. Test anxiety data.

a) Looking at the overall trend in the table, as x-values increase do the y-values increase or decrease?

b) Use these data to create a scatter plot. Find the equation of a good trend line.

c) What does the slope mean here? Does this value make sense in this context?

d) If a person scores an 18 on the anxiety evaluation, what is the predicted exam grade?

e) What is the expected anxiety score if the exam grade is a 60?

Answers (continued)

b) 0.14 is the average tips to sales ratio or the average percentage of tip. Ask students what the recommended tip percentage is. In this situation, Hernando receives an average tip of 14%, which is close to the standard American tip of 15%.

c) Hernando makes 39 cents on $0.00 sales. The y-intercept in this case is not a valid real-world value.

d) Around $8. Students should explain how they could find that answer by using their graph, table, and equation.

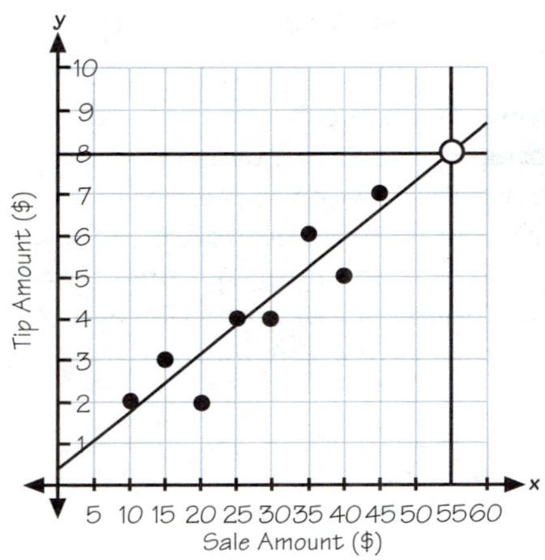

x	y
0	0.39
5	1.09
10	1.79
15	2.49
20	3.19
25	3.89
30	4.59
35	5.29
40	5.99
45	6.69
50	7.39
55	8.09
60	8.79
65	9.49
70	10.19

$y = 0.14x + 0.39$
$y = 0.14x(55) + 0.39$
$y = 8.09$

3. a) The students should use their knowledge of slope and y-intercept to estimate a trend line. Students should then adjust the slope and/or y-intercept until they find a line that best models the data. A sample equation for this data is $y = 1.10x - 0.02$.

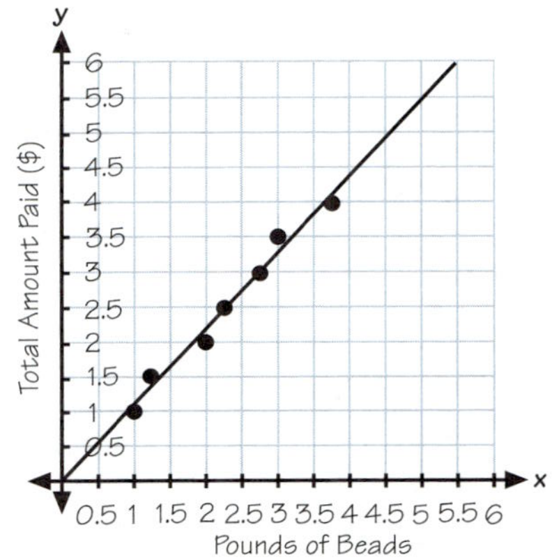

b) Slope is a rate of change. In this situation, it is the average price per pound Nedra paid for her purchases, or about $1.10. It is a reasonable price for one pound of beads.

c) About $0.00 for 0 lbs of beads, which makes sense in the context of the problem.

d) Around $5.50. Students should explain how to find that answer using their graph, table, and equation.

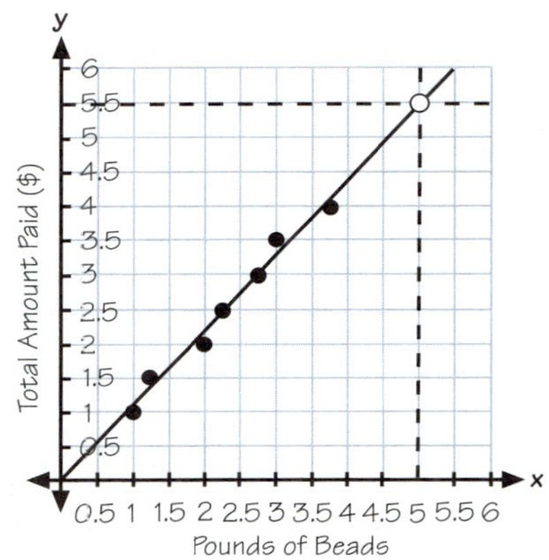

x	y
0.0	−0.02
0.5	0.53
1.0	1.08
1.5	1.63
2.0	2.18
2.5	2.73
3.0	3.28
3.5	3.83
4.0	4.38
4.5	4.93
5.0	5.48
5.5	6.03
6.0	6.58
6.5	7.13
7.0	7.68

$y = 1.10x - 0.02$
$y = 1.10(5) - 0.02$
$y = 5.48$

4. a)

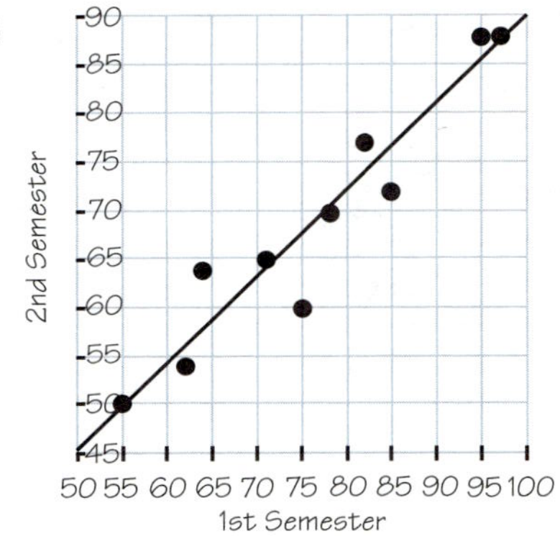

b) The slope represents the change in second semester scores compared to first semester scores. In this situation, the second semester scores are expected to be about 89% of the first semester scores. Depending on the difficulty of the course, this could be a reasonable model.

c) The y-intercept represents the second semester score if the first semester score was zero. In this situation, the second semester score would be 0.91, which does not make sense. However, if students calculated a trend line with a y-intercept of 0, then a student who makes a 0 score for the first semester would also make a 0 score for the second semester, which is reasonable.

d) $y = 0.89(72) + 0.91 = 65$

Answers (continued)

5. a) decrease

   b) $y = -1.06x + 68.8$

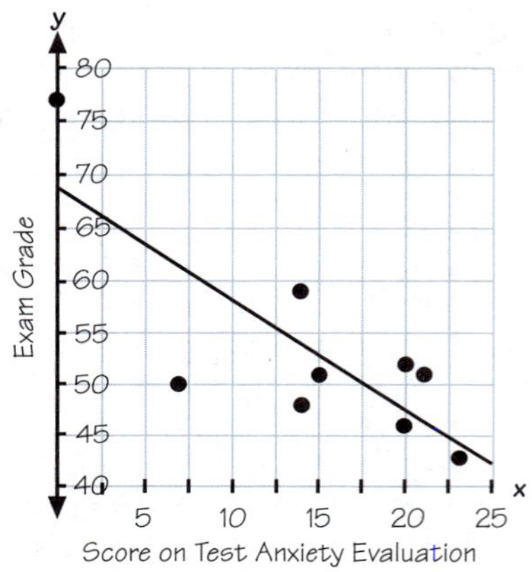

   c) The slope represents the change in exam grade compared to a score on a test anxiety evaluation. In this situation, the exam grade should drop 1.06 points for every point scored on the test anxiety evaluation. This makes sense in this context, since students who are more anxious about exams would score lower on the exam.

   d) $y = -1.06(18) + 68.8 = 49.72 \approx 50$

   e)
$$\begin{aligned} 60 &= -1.06x + 68.8 \\ 60 - 68.8 &= -1.06x + 68.8 - 68.8 \\ -8.8 &= -1.06x \\ \frac{-8.8}{-1.06} &= \frac{-1.06x}{-1.06} \\ 8.30 &\approx x \end{aligned}$$

SECTION 2.4

I Think We Have Your Thighbones

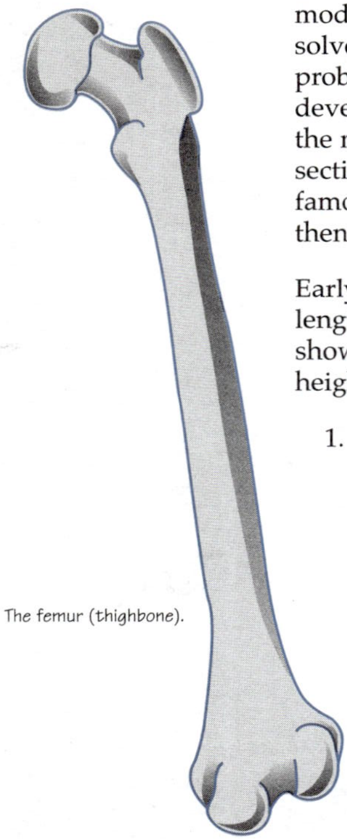

The femur (thighbone).

In Section 2.2 you completed a process known as mathematical modeling, which is similar to using the Scientific Method, to solve problems. The first step in this process is to identify a problem that you need to solve. Collecting and analyzing data to develop a mathematical model and create a hypothesis is often the next step. The model is then used to make predictions. In this section you will analyze data that may have been collected by a famous doctor, develop a mathematical model of the data, and then use that model to make predictions.

Early models (from the late 1800s) that predicted height from the length of long bones were based on ratios. (See Figure 2.2 that shows some of the long bones.) For example, the ratio of the height H and femur length F is given by the formula $H/F = 3.72$.

1. Based on this model, estimate the heights of the people whose femur lengths are given in **Figure 2.13.**

Bone Type	Number Found	Length (cm)
Femur	3	41.5, 41.4, 50.8
Tibia	1	41.6
Ulna	2	22.9, 29.0
Radius	1	21.6
Humerus	1	35.6
Complete Skull (including jaw)	1	23.0
Fragments	More than 10	From 3.0 to 5.0 cm

FIGURE 2.13. Measurements of bones found at Superstition Mountains.

2. For most adults, femurs range in size from 38 cm to 55 cm. According to this model, what is the range of heights of most adults? Are these reasonable heights for adults? (Recall that 2.54 cm = 1 in.)

3. What function rule would you enter into your graphing calculator in order to graph this model for the relationship between height and femur length?

The function rule that you wrote for Question 3 should be a member of the $y = mx$ family, as were the first models used to predict height from the lengths of long bones.

Dr. Mildred Trotter

Dr. Mildred Trotter (1899–1991), a physical anthropologist, was well known for her work in the area of height prediction based on the lengths of long bones.

During World War II, the armed services sometimes had problems identifying the remains of dead soldiers. Dr. Trotter was asked to help. She decided to measure each person's height and the length of each person's femur. Using these data she refined earlier models by adding a constant, thereby producing models of the form $y = mx + b$.

Data similar to the data she obtained are shown in **Figure 2.14.**

Femur Length (cm)	Height (cm)
45	165
53	190
46	168
38	150
54	188
52	185
39	155
51	186
41	157
42	160
47	178
40	160
55	195
50	175
43	165
48	175
44	166
49	178

FIGURE 2.14. Adult height versus femur length.

4. Why does it make sense to consider femur length as the independent variable?

5. What is a reasonable domain for this situation? Why?

6. What is a reasonable range for this situation? Why?

7. Use a graphing calculator to make a scatter plot of height versus femur length. Sketch your scatter plot and record your window.

8. Recall that Dr. Trotter's model is of the form $y = mx + b$. Find a function rule for the trend line for these data. So that you can compare your function rule to Dr. Trotter's model, be sure your rule is of the form $y = mx + b$.

9. Graph your function rule on the scatter plot. If necessary adjust the equation. Write your adjusted equation, sketch your graph, and record your window.

Dr. Trotter models the relationship between the height measurements and femur measurements with the equation $H = 2.38F + 61.41$ where height, H, and femur length, F, are in centimeters.

10. Is your function rule similar to Dr. Trotter's model? How do you know?

ELABORATE LESSON 1

Section 2.4: I Think We Have Your Thighbones

Purpose: Students use the actual algebraic equations used by an anthropologist to predict height from specific bones.

Description: In this section students apply and extend their knowledge. Students work with data similar to the data collected by Dr. Mildred Trotter to develop a model in the form $y = mx + b$. They compare their model to an earlier model and to Dr. Trotter's model. Students then use the model to predict people's heights based on femur lengths. They use the same model to solve for femur length in terms of height.

Materials: graphing calculator

Teaching Ideas: Have students solve problems using multiple representations (graph, table, symbolic manipulation). They should be able to create scatter plots, find function rules for trend lines, and apply the function rule to make predictions.

BACKGROUND ON DR. TROTTER

Dr. Mildred Trotter had a long distinguished career as a physical anthropologist that included working as a special consultant to the U.S. government during World War II. Her task during the war involved the identification of skeletal remains of servicemen. At the time, she realized that bone size and proportions vary based on age, gender, and racial/ethnic background. Forensic scientists and law enforcement agencies are still using Dr. Trotter's formulas for estimating stature based on the lengths of long bones.

[Conroy, Glenn, et. al. 1992.]

Answers

1. The answer below is based on the assumption that the bones in Figure 2.13 were from two people.

   Sample answer:

   Height of Skeleton 1: Using the average femur length $(41.5 + 41.4)/2 = 41.45$, $H = 3.72(41.45) = 154.194$ cm (about 5 ft 1 in).

   Height of Skeleton 2: $H = 3.72(50.8) = 188.976$ cm (about 6 ft 2 in).

2. 141.36 cm (about 4 ft 8 in) to 204.6 cm (about 6 ft 9 in)

   Sample answer: Most adults are taller than 4 ft 8 in and shorter than 6 ft 9 in. Certainly there are adults whose heights are at the extremes of this range (and a few who are outside of this range).

3. $y = 3.72x$

4. Answers may vary. Sample answer: In the context of this situation scientists will be using the length of human femurs to estimate the height of dead people. So in a way we can say that femur length will predict height.

5. Answers may vary. Sample answer: A reasonable domain would be 35 cm $< x <$ 60 cm because in our data the femur lengths range from 38 cm to 55 cm.

6. Answers may vary. Sample answer: A reasonable range would be 150 cm $< x <$ 200 cm because in our data heights range from 150 cm to 195 cm.

7. 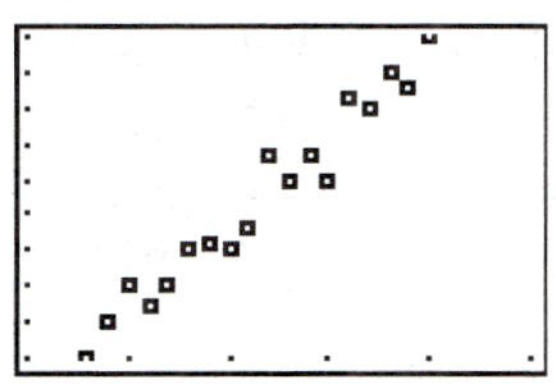

   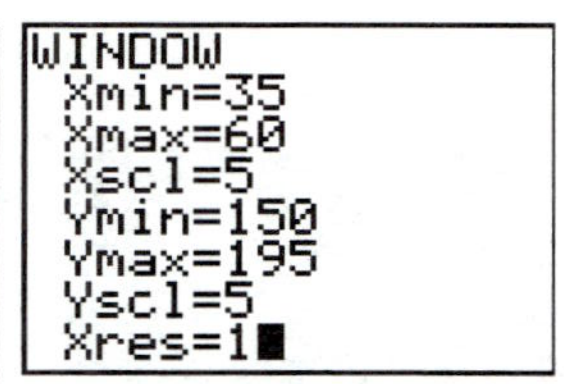

   Describe your window: x-min: 35
   x-max: 60
   y-min: 150
   y-max: 195

8. $y = 2.6x + 50$

9. Answers may vary. In the example below, $y = 3x$ was graphed first. The slope of the line appeared to be close to the scatter plot. To move the line vertically a y-intercept of 40 was added. By continuing to adjust the slope and y-intercept we arrived at the equation $y = 2.6x + 50$.

   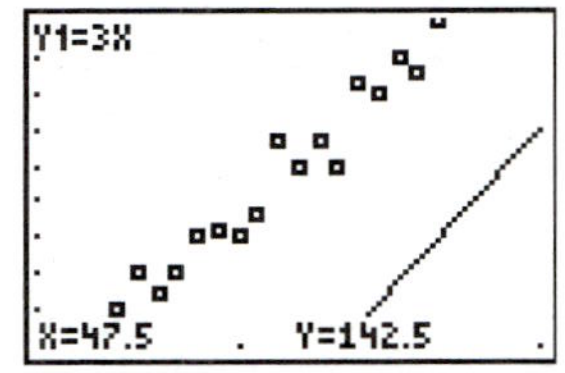

   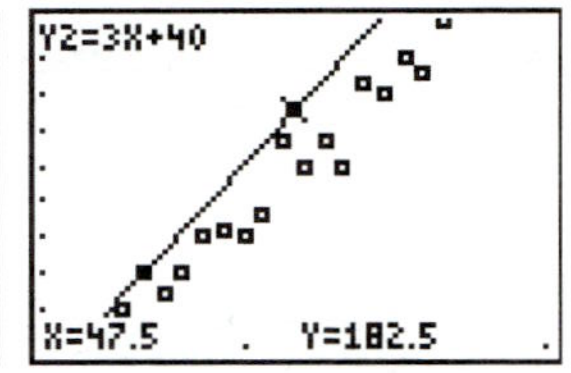

   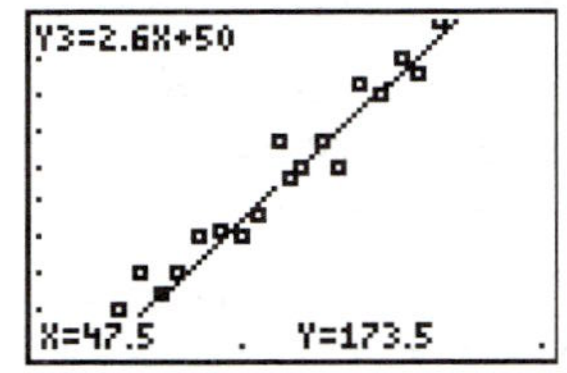

10. Answers may vary. Yes. Using the equation $y = 2.6x + 50$ as our rule we can make a comparison to Dr. Trotter's rule using the table feature of a graphing calculator. The values obtained using Dr. Trotter's rule are similar to the values obtained using our rule.

    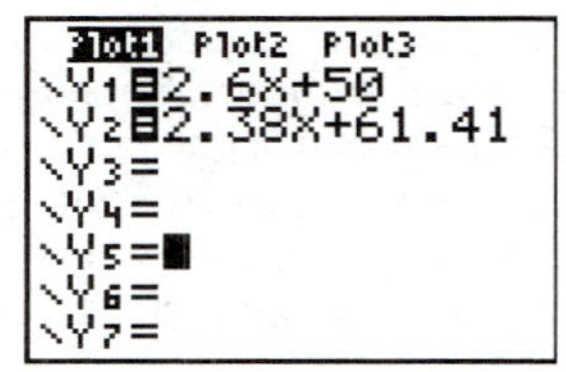

    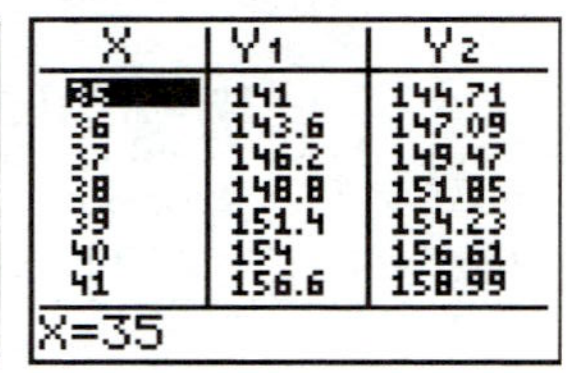

    Students may also compare the graphs of their rule to Dr. Trotter's rule. In our sample the graphs of the functions yield two lines that have very little visible difference when they are graphed over the range of the data.

    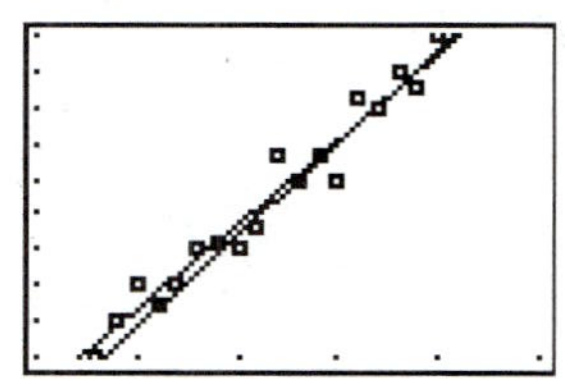

11. How does the graph of the function rule for the early model ($y = 3.72x$) compare to the graph of the function rule for Dr. Trotter's model?

12. Would the early model or Dr. Trotter's model be the best way to predict a person's height given their femur length? Why?

13. Predict the height of an adult whose femur length is 42.5 centimeters.

14. If an adult is 170 centimeters tall, what is the approximate length of his or her femur?

SUMMARY

To build and test a model, you often need to collect data. In this section you did an analysis of collected data and made predictions. You also compared different models that are used to predict height. Most models used in this lesson are of the $y = mx$ family. Sometimes a model contains a non-zero constant such as those in the $y = mx + b$ family. These models often are used due to the variability of real-world data.

SECTION 2.4 Assignment

1. Jackson's femur measures 39 cm and his brother's measures 40 cm. Based on Dr. Trotter's first formula, $H = 2.38F + 61.41$ where H is the person's height (cm) and F is the length of the femur (cm), predict the difference in the brothers' heights.

2. Femurs that belong to two women are found. One femur is one centimeter longer than the other. Predict the difference in their heights. Explain how you were able to get your answer even though the lengths of the two women's femurs were not given. In addition, tell how you could get your answer from Dr. Trotter's first formula.

3. If a man is 178 cm (about 5 ft 10 in) tall:

   a) Explain how you could use your graph to estimate the length of his femur. What is your estimate?

   b) Write a set of algebraic steps to solve Dr. Trotter's first formula, $H = 2.38F + 61.41$, for F. (A medical doctor might use such an equation to check that the length of a person's femur is normal for a person of that height.)

   c) Use your equation in b to predict the length of a man's femur if the man is 178 cm tall. Compare your answer to the femur length you estimated in a.

4. Another of Dr. Trotter's equations predicts height, H, from a known tibia length, T (this is called the second formula): $H = 2.52T + 78.62$ where H and T are measured in cm.

   a) The length of the tibia listed in Figure 2.13 is 41.6 cm. Use Dr. Trotter's second formula to predict the person's height. Is your answer a reasonable height for a person?

   b) Use your predicted height from 4a and your formula from 3b to predict the length of the femur of a person with the 41.6-cm tibia. Which femur length from Figure 2.13 is closest to your prediction?

Answers (continued)

11. We can make a comparison to Dr. Trotter's rule using the table and graphing features of a graphing calculator. The values obtained using the early rule are only similar to the values obtained using Dr. Trotter's rule for people whose femur lengths are toward the center of the data. This is demonstrated in the table and the graph below. Dr. Trotter's rule is more accurate for the general population, and the earlier rule is more accurate for people of average height and less accurate for short or tall people.

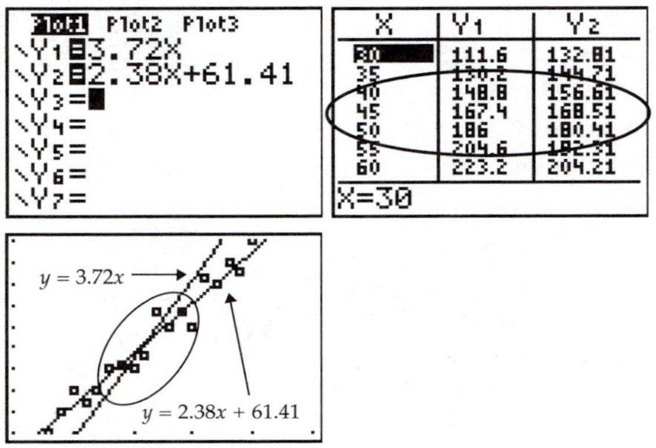

12. Answers will vary (see #11). Dr. Trotter's rule is more accurate for the general population, and the earlier rule is most accurate for people of average height and less accurate for short or tall people.

13. Answers will vary. Sample answer is based on Dr. Trotter's rule.

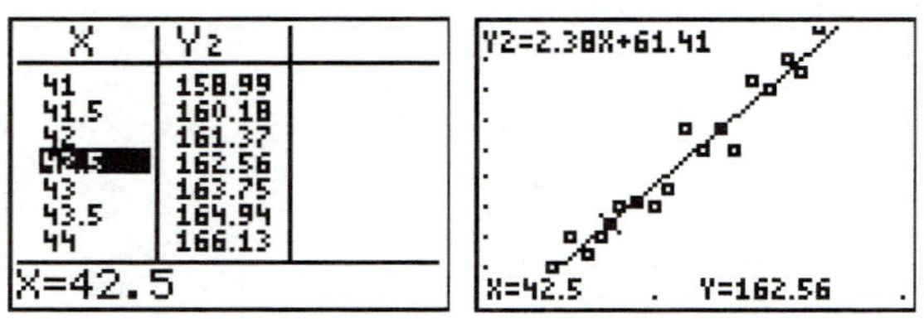

$y = 2.38x + 61.41$

$y = 2.38(42.5) + 61.41$

$y = 162.56$ cm

14.

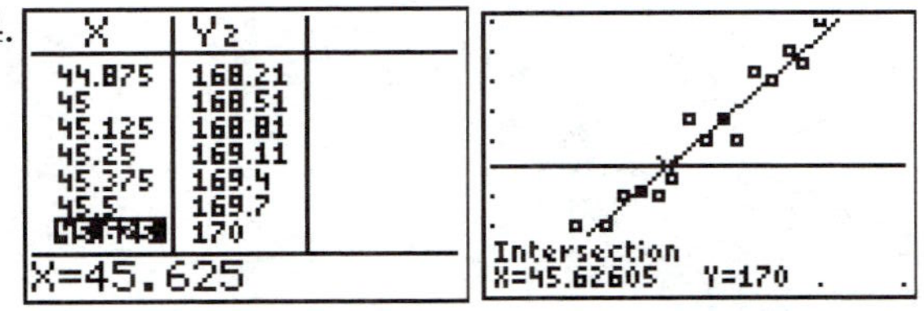

$$170 = 2.38x + 61.41$$
$$170 = 2.38x + 61.41$$
$$170 - 61.41 = 2.38x + 61.41 - 61.41$$
$$\frac{108.59}{2.38} = \frac{2.38}{2.38}x$$
$$45.63 \text{ cm} = x$$

Section 2.4 Assignment

Answers

1. Jackson: $H = 2.38(39 \text{ cm}) + 61.41 = 154.23$ cm

   Brother: $H = 2.38(40 \text{ cm}) + 61.41 = 156.61$ cm

   Difference = 156.61 cm – 154.23 cm = 2.38 cm

2. Each time you increase the length of the femur by 1 cm, the height changes by 2.38 cm. This difference has the same value as the multiplier of F, the slope of the linear function.

3. a) Sample answer: Draw a horizontal line at approximately $H = 178$ cm. Find the F-coordinate that corresponds to the point where the horizontal line and the graph of Dr. Trotter's equation intersect. That value should be around 49 cm.

   b)
   $$H = 2.38F + 61.41$$
   $$H - 61.41 = 2.38F + 61.41 - 61.41$$
   $$H - 61.41 = 2.38F$$
   $$\frac{H - 61.41}{2.38} = \frac{2.38F}{2.38}$$
   $$\frac{H - 61.41}{2.38} = F$$

   c) $F = \frac{178 - 61.41}{2.38} \approx 49$ cm. This answer is about the same as the estimate in a.

4. a) $H = (2.52)(41.6) + 78.62 \approx 183.5$ cm. This person is approximately 6 ft tall. This is a reasonable height for a person.

   b) $F = (183.5 - 61.41)/2.38 \approx 51.3$ cm. The femur length of 50.8 cm is closest.

c) Write a set of algebraic steps to solve Dr. Trotter's second formula, $H = 2.52T + 78.62$, for T. (A medical doctor might use such an equation to check that the length of a person's tibia is normal for a person of that height.)

5. In a third formula, Dr. Trotter used both the tibia and the femur to predict height:
$H = 1.30(F + T) + 63.29$. (All measurements are in cm.)

a) A group of students measure the femur and tibia of a skeleton. They find that the femur is 42 cm long and the tibia is 43 cm long. Predict the height of the person using Dr. Trotter's third formula, $H = 1.30(F + T) + 63.29$.

b) Compare the prediction in a with the predicted height using Dr. Trotter's equation, $H = 2.38F + 61.41$.

c) Compare the predictions in a and b with the predicted height using Dr. Trotter's second formula, $H = 2.52T + 78.62$.

d) You should have found a clear difference between your predictions in a–c. If a man is 175 cm tall (about 5 ft 9 in), based on Dr. Trotter's equations in b and c, would you expect his tibia or his femur to be longer and by how much?

SECTION 2.5 Height vs. Forearm Length

Let's apply what you have learned thus far about building and using linear models to a new situation.

The table in **Figure 2.15** shows the heights and forearm lengths of students in Ms. Jaczinko's class.

Female			Male		
Name	Forearm Length (cm)	Height (cm)	Name	Forearm Length (cm)	Height (cm)
Alicia	24	157	Ahmed	26.5	173
Bia	24.5	166	Brian	27	177
Christi	27	164	Jesús	27	174
Chantalle	24	164	Davis	31	192
Coral-Mae	23	161	José	28	172
Juanita	27.5	164	Kelvin	29	180
Ji-Hyun	27	167	Lenny	27	174
Kim	26	162	Luis	28	175
Kristen	26	175	Mike	32	185
Maria	28.5	166	Pedro	30	185
Tianna	26.5	172	Sang	30	178
Teresa	25.5	176			

FIGURE 2.15. Heights and forearm data from Ms. Jaczinko's class.

1. Two new students will be added to the class next Monday. Each new student will get a made-to-fit class shirt. Ms. Jaczinko finds the sleeve lengths of shirts she orders by doubling the student's forearm length. Student A is 160 centimeters tall. Student B is 182 centimeters tall. Based on the class data, what sleeve lengths should Ms. Jaczinko order for the two students? Should she order only boys' shirts, only girls' shirts, or one of each? Justify your answer.

Answers (continued)

c) $$H = 2.52T + 78.62$$
$$H - 78.62 = 2.52T + 78.62 - 78.62$$
$$H - 78.62 = 2.52T$$
$$\frac{H - 78.62}{2.52} = \frac{2.52T}{2.52}$$
$$\frac{H - 78.62}{2.52} = T$$

5. a) The predicted height is 173.8 cm.

   b) The predicted height is $(2.38)(42) + 61.41 \approx$ 161.4 cm. This is 12.4 cm (or about 5 in) shorter than the prediction in a.

   c) The predicted height is $(2.52)(43) + 78.62 \approx$ 187 cm. This is 13.2 cm larger than the prediction in a and 25.6 cm more than the prediction in b.

   d) Using the equation from 3b:
   $F = (175 - 61.41)/2.38 \approx 47.7$ cm.

   Using the equation from 4c:
   $T = (175 - 78.62)/2.52 \approx 38.2$ cm.

   Based on these results, for a person 175 cm tall, you would expect the femur to be approximately 9.5 cm longer than the tibia.

EVALUATE LESSON 1

Section 2.5: Height vs. Forearm Length

Purpose: Students practice using a trend line to make decisions for a group of students.

Description: Students find and implement a method to solve the problem. This is an open middle performance assessment. There are several methods that may be used to arrive at the answer. In their justification students should demonstrate that they have mastered the concepts presented in this lesson.

Materials: none

Teaching Ideas: This task may be assigned as group work or individual work.

Answers

1. Sample solution: Create a scatter plot of height vs. forearm length.

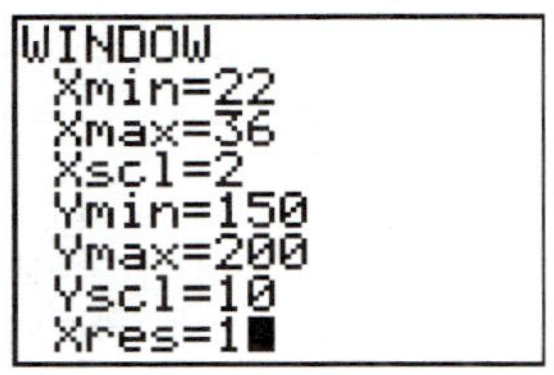

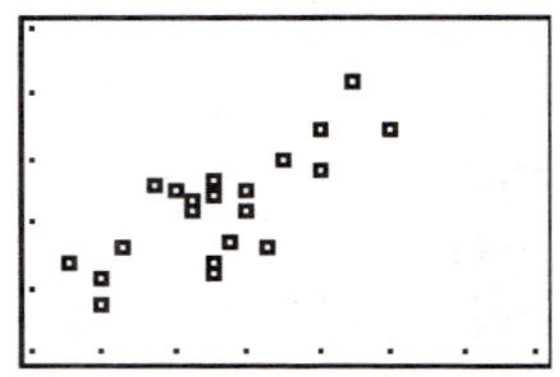

Find a trend line then use the table feature of the calculator to estimate each student's forearm length.

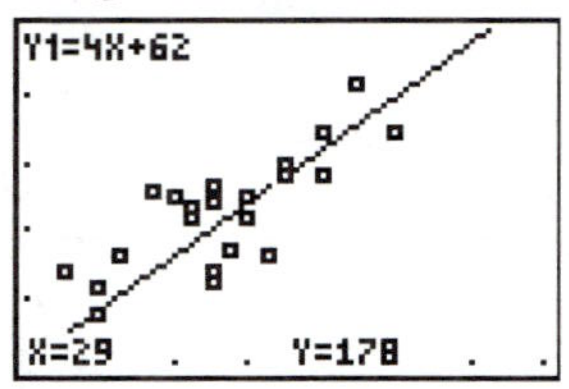

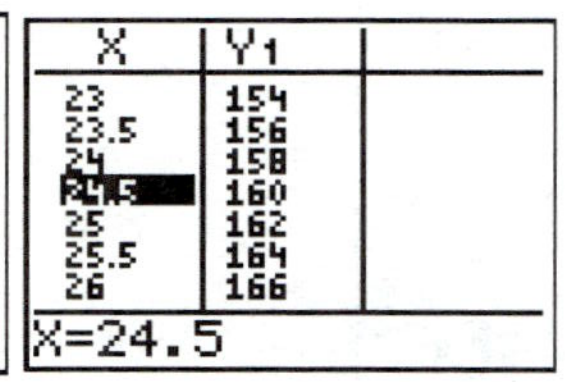

Student A

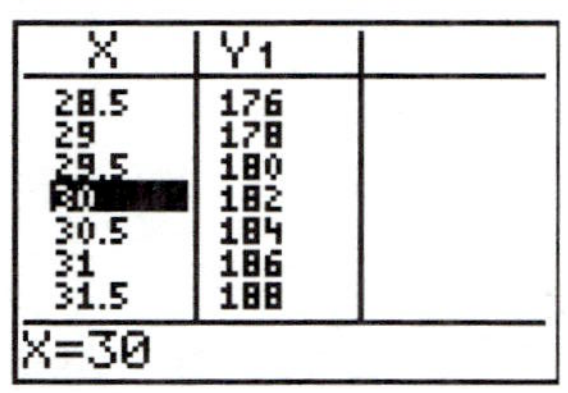

Student B

Student A's forearm length should be about 24.5 cm. Since this length is between 23 cm and 28 cm, Student A is most likely a girl whose sleeve length is 2(24.5 cm) = 49 cm.

Student B's forearm length should be about 30 cm. Since this length is between 26.5 cm and 32 cm, Student B is most likely a boy whose sleeve length is 2(30 cm) = 60 cm.

Ms. Jaczinko should order one girl's shirt with a sleeve length of 49 centimeters and one boy's shirt with a sleeve length of 60 centimeters.

SECTION 2.6

Manatees and Powerboats

On the coast of Florida lives the manatee, a large and friendly marine mammal. However, the gentle Florida manatee does not live an easy life. It is one of the most endangered marine mammals in the United States. One major threat to the manatee is the large number of them killed each year by powerboats.

Should the Florida Department of Environmental Protection limit the number of registered boats in order to protect the manatee population? Before they decide, they will have to present a convincing argument to the public.

In a later section you must make a good case for whether or not to restrict powerboats. These are the main steps to present a good case to the authorities:

- Find a model that describes the relationship between manatee deaths and powerboat registrations.
- Show that the model does a good job of describing the data.
- Use the model to make your prediction.

Figure 2.16 contains data on the number of powerboats registered in Florida (in thousands) and the number of manatees killed.

1. Does there appear to be a relationship between powerboat registrations and manatee deaths? Explain why.

2. To make a recommendation about the number of powerboat registrations, you must find a model that represents the relationship between manatee deaths and powerboat registrations. To investigate this relationship, choose which variable—powerboat registrations or manatee deaths—should be the independent variable and which should be the dependent variable. Explain why.

Year	Powerboat Registrations (in thousands)	Manatees Killed
1977	447	13
1978	460	21
1979	481	24
1980	498	16
1981	513	24
1982	512	20
1983	526	15
1984	559	34
1985	585	33
1986	614	33
1987	645	39
1988	675	43
1989	711	50
1990	719	47

FIGURE 2.16. Powerboat registrations and manatee deaths in Florida, 1977–1990.

ENGAGE LESSON 2

Section 2.6: Manatees and Powerboats

Purpose: Students have previously developed trend lines for sets of data. This section engages them in the relationship between the number of manatee deaths and the number of powerboat registrations in Florida. In Section 2.7 they will find a trend line as they have before to get them ready to find the line of best fit using the least-squares regression method with the graphing calculator in Section 2.8.

Description: This section is designed to interest students in the relationship between the number of manatees killed each year in Florida and the number of powerboat registrations. In a later section students will present a case for whether or not to restrict powerboat registrations based on a least-squares linear regression model.

Materials: Transparency 2.2

Teaching Ideas: After students have read and examined the information in the table (Transparency 2.2), encourage class discussion in order to facilitate students' answers to the two questions in the text.

Answers

1. By examining the table, students may suggest that there appears to be a cause and effect relationship. They appear to be positively related; the number of manatee deaths appears to increase as the number of powerboat registrations increase.
2. Powerboat registrations should be the independent variable and manatee deaths the dependent variable. Increased manatee deaths clearly do not increase powerboat registrations. While we aren't sure yet whether an increase in powerboat registrations increases manatee deaths, it is a theory we can test. To make a recommendation for limiting the number of powerboat registrations the student should be able to convince the Florida Department of Environmental Protection that the number of manatee deaths is related to the number of powerboat registrations.

SECTION 2.6 Assignment

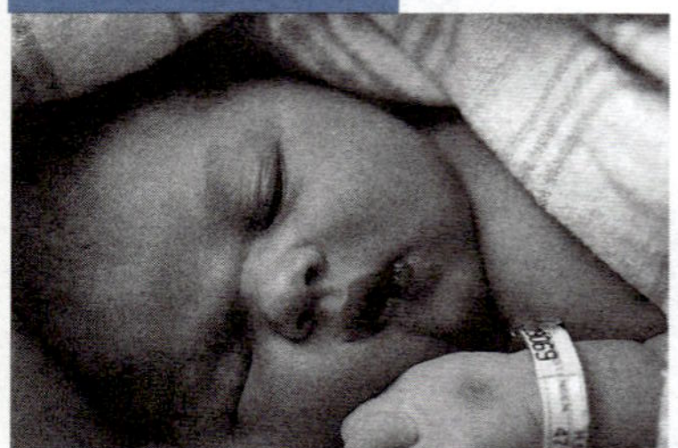

1. The table in **Figure 2.17** lists the weights at birth for two groups of babies. The first group is babies whose mothers never smoked. The second group is babies whose mothers smoked at least ten cigarettes per day. From this small data set, does it appear that smoking affects a baby's birth weight? Explain your answer.

FIGURE 2.17. Babies' birth weights.

Never Smoked	6.3	7.3	8.2	7.1	7.8	9.7	6.1	9.6	7.4	7.8	9.4	7.6
Smoked Ten or More Cigarettes per Day	6.3	6.4	4.2	9.4	7.1	5.9	6.8	8.2	7.8	5.9	5.4	6.3

2. Two groups of high school students were asked how much they typically spend on a date. The first group includes 12 students who did not exercise. Students in the second group exercised at least twice a week. **Figure 2.18** shows the results.

FIGURE 2.18. Cost of a date (in dollars).

Does Not Exercise	10	5	20	4	20	20	15	0	8	40	8	15
Does Exercise	15	15	15	5	10	5	5	6	30	25	30	60

a) Find the mean, median, and mode for the Does Not Exercise group.

b) Find the mean, median, and mode for the Does Exercise group.

c) Based on this information, which of the two groups of students spends more on a date?

d) Make a scatter plot of the Does Not Exercise group vs. the Does Exercise group.

e) Is it valid to claim that there is a relationship between the exercise and non-exercise groups? Could you use the typical amount spent by someone in the Does Exercise group to predict how much a person in the Does Not Exercise group would spend? Explain.

CHECK THIS!

If the relationship between two variables is strong and has linear form, then the points in a scatter plot fall very close to a line. For weaker relationships, the data are more scattered.

Section 2.6 Assignment

Answers

1. Never Smoked: Average weight—approximately 7.9 pounds.

   Smoked: Average weight—approximately 6.6 pounds.

   Answers will vary. A sample answer might say that in this data set, mothers who smoked had, on average, babies that weighed less than mothers who did not smoke. The mothers of the four babies who weighed less than 6 pounds all smoked. None of the babies whose mothers did not smoke weighed less than 6 pounds. It is a good point to reinforce to students that making strong statements about relationships between two variables is tricky, and requires a series of procedures in selecting and analyzing data.

2. a) The mean is $13.75. The median is $12.50. The mode is $20.00.

   b) The mean is $18.42. The median is $15.00. The modes are $15.00 and $5.00.

   c) The Does Exercise group spends more on a date based on both the mean and the median. By inspecting the table, indications are that the Does Exercise group spends more. The calculation of the mean and median confirms this.

   d)

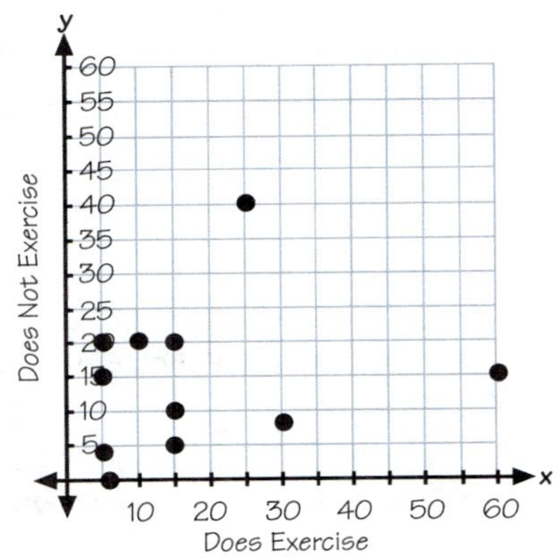

   e) The scatter plot doesn't make any sense in this situation. The spending levels for the Does Exercise group and the Does Not Exercise group are not paired. In other words, the x- and y-coordinates were matched up only by the order in the table. They don't belong together in any meaningful way.

SECTION 2.7

A Scatter Plot of Manatees Killed vs. Registrations

The last section established a problem: More manatees (an endangered species) are being killed as the number of powerboat registrations in the state of Florida rises. In this section, you will use the data comparing the number of manatee deaths and the number of powerboat registrations in order to build a mathematical model. You will use this model to make predictions in preparation for your report to the Florida Department of Environmental Protection on this issue.

1. Make a scatter plot of manatee deaths versus powerboat registrations on a sheet of graph or chart paper, and then make a scatter plot on your graphing calculator.
2. Are the two variables positively or negatively related? What does this mean?

CHECK THIS!

- If one variable increases as the other variable increases, then the two variables are positively related. This is sometimes called positive correlation.
- If one variable increases as the other variable decreases, then the two variables are negatively related. This is sometimes called negative correlation.

3. With your group, find an equation of a trend line that approximates the pattern of the data in your scatter plot. Draw the trend line on your scatter plot.
4. Use your model to predict the number of manatee deaths when 500,000 powerboats are registered. Does your prediction make sense?
5. Did all groups have the same prediction? Why or why not?

SECTION 2.7

Assignment

In this assignment you will make scatter plots of some data sets, and in some cases fit a line to the scatter plot and make predictions. Pay particular attention to the features of the relationship between the variables.

Do the points in the scatter plot appear to be scattered close to the straight line? Then the scatter plot has linear form and it makes sense to describe it with a linear equation (a member of the $y = mx + b$ family).

Does the pattern made by the points move upward as you look from left to right? If so, the two variables are positively related (as one variable increases the other tends to increase). If the pattern drifts downward, then the two variables are negatively related (as one variable increases the other tends to decrease).

1. Linda heats her house with natural gas. She wonders how her gas consumption is related to the weather. The table in **Figure 2.19** shows the average outside temperature (in degrees Fahrenheit) each winter month and the average amount of natural gas Linda used (in hundreds of cubic feet) each day that month.

Month	Sep	Oct	Nov	Dec	Jan	Feb	Mar	Apr	May
Outdoor Temperature F	48	46	38	29	26	28	49	57	65
Gas Used Per Day (x100 ft^3)	5.1	4.9	6.0	8.9	8.8	8.5	4.4	2.5	1.1

FIGURE 2.19. Gas usage and temperature data.

a) Make a scatter plot of these data. Which is the independent variable and which is the dependent variable? How did you decide?

b) Describe the features of the relationship between outside temperature and natural gas consumption. Why does the relationship have this direction?

c) Draw a trend line that you think best describes the pattern of these data. What is the equation of your line?

EXPLORE LESSON 2

Section 2.7: A Scatter Plot of Manatees Killed vs. Registrations

Purpose: The purpose of this section is for students to explore the relationship between the number of manatees killed and the number of powerboat registrations in Florida and develop a trend line as they have in previous sections. They will be prepared to find the least-squares line in Section 2.8.

Description: The students make a scatter plot to describe the relationship between manatees killed and powerboat registrations, find an equation of a trend line, and make a prediction based on the trend line.

Materials: graph or chart paper, graphing calculator

Teaching Ideas: Place students in small groups of 2 to 3 for this portion of the lesson. Each group should make a scatter plot of the data on either chart paper or graph paper. Guiding questions to ask might include:

- What are the domain and range for the data?

  A reasonable domain to use is [400, 750].

  A reasonable range to use is [0, 60].
- How do you decide what values to use on the x- and y-axis?

  Use the reasonable domain and range and make the x-axis have a scale of 50 and the y-axis a scale of 10.

Continue to work through the questions in the text with students as they plot their trend line on their scatter plot. Screenshots from the TI-83 are shown for teachers to assist students in the use of the calculator. Use the manual for your calculator if you are using a different model.

Answers

1.

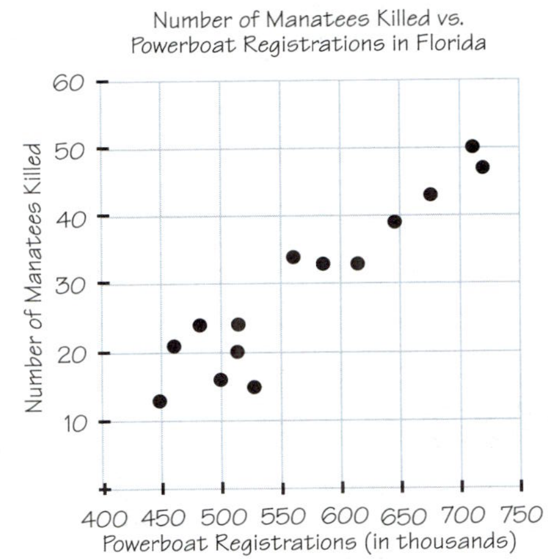

Enter the number of powerboat registrations into L1 and the number of manatees killed into L2 on a graphing calculator. Then use STAT PLOT to investigate the scatter plot.

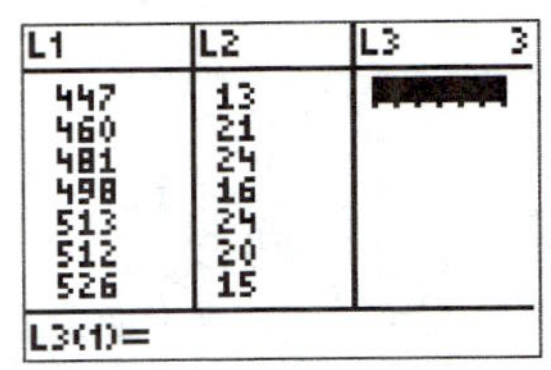

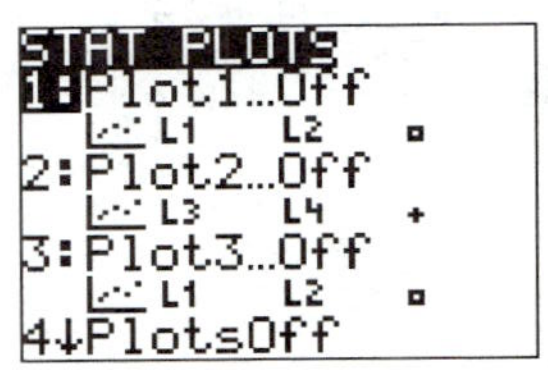

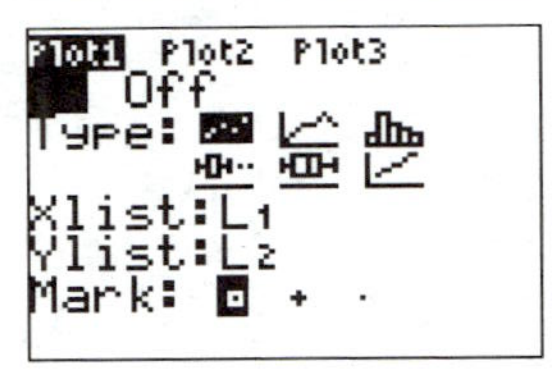

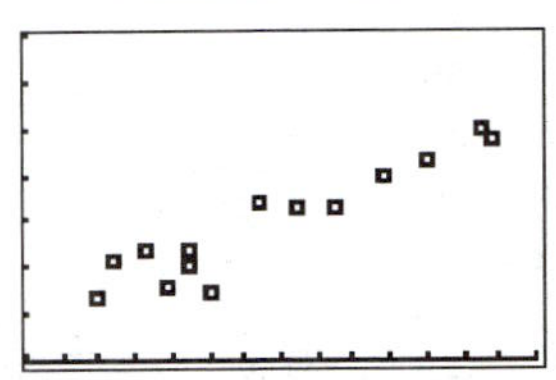

2. They are positively related. As the number of powerboat registrations goes up, the number of manatee deaths also rises.

3. Students may try to guess and check on the calculator. From the scatter plot they may notice that for every 100 (thousand) additional powerboat registrations, approximately 15 additional manatees are killed. They may assume that if there are no powerboat registrations, then no manatees will be killed. Their first guess might be $y = 15/100x$. However, if they plot this function, the graph will not be very close to the points.

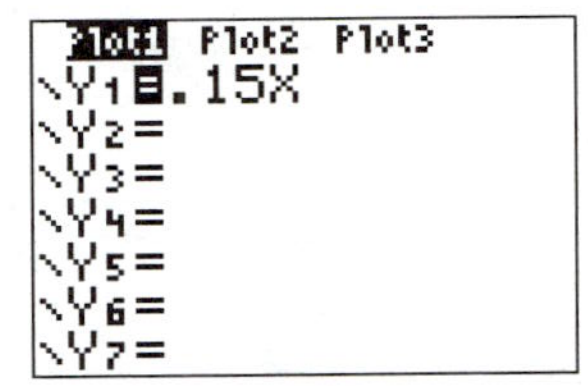

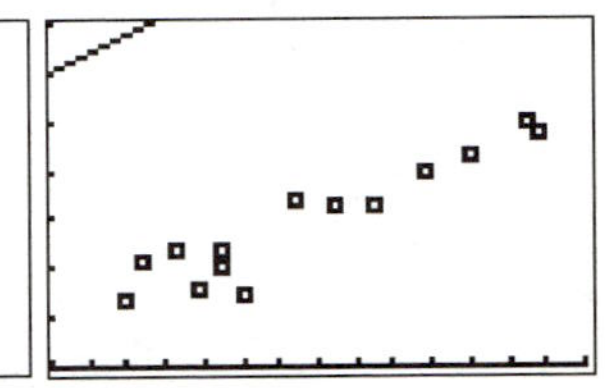

They might try to shift the line down by subtracting a constant, such as 56.

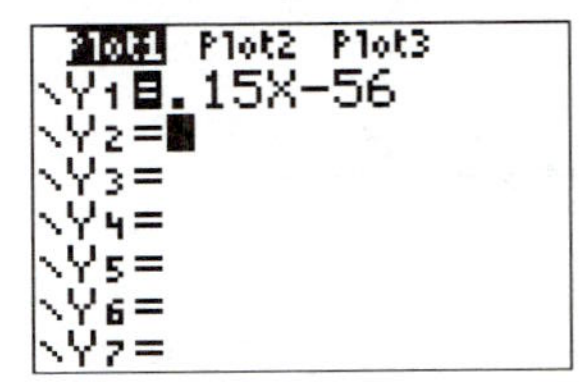

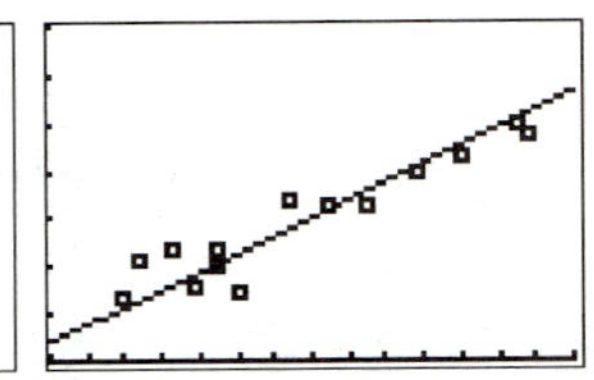

Students may also choose two points that appear to be in the middle of the cluster of the data points, but far apart, and find an equation of a line that goes through those two points.

They may also decide on the slope they think looks the best and pick a good point and use the point-slope formula.

They may also find the average number of manatees killed and the average number of powerboat registrations to use as an (x, y) point and then eyeball a value for the slope. They may then use the point-slope formula.

Answers (continued)

4. Answers may vary. Based on the model above, the number of manatees killed when there are 500,000 powerboat registrations is 0.15(500) – 56, or 19.

   Have each group of students explain the model they chose, how they decided on the model, and what their prediction is for the number of manatees killed when the powerboat registrations reach 500,000.

5. Answers will vary for different student groups depending on the model they chose.

Section 2.7 Assignment Answers

1. a) The independent variable is temperature and the dependent variable is the gas used per day. Generally people want to predict the amount of fuel they use. Generally temperature is one of the key factors in finding fuel usage.

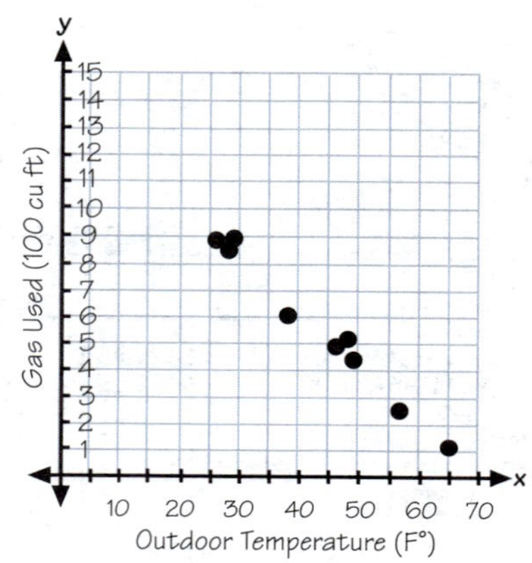

   b) There is very little scatter around a straight line. The direction of the line is downwards, so the variables have a negative relationship. An increase in the mean monthly temperature produces a decrease in the monthly gas consumption.

   c) Answers may vary. Using the points (26, 8.8) and (65, 1.1): $y = -0.20x + 14$. These points were selected because the line passed through them or came close to passing through them and they were the farthest apart on the graph.

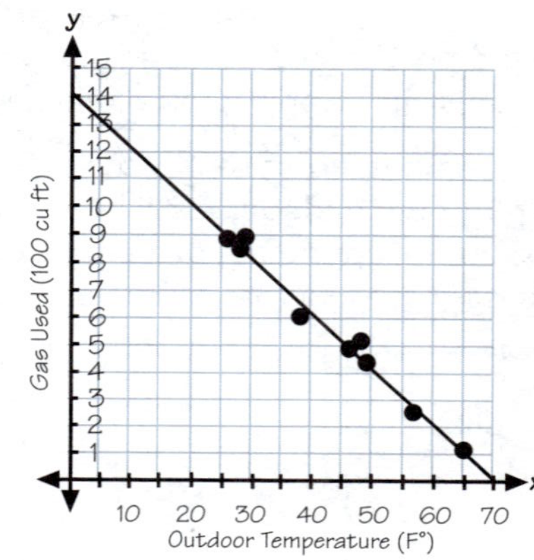

d) Use your equation from c to predict the gas used during a month when the average temperature is 60°F.

2. The 11 members of a college women's golf team play a practice round, then the next day play a round in competition on the same course. Their scores appear in **Figure 2.20.** (A golf score is the number of strokes required to complete the course, so low scores are better.)

Player	1	2	3	4	5	6	7	8	9	10	11
Practice	89	90	87	95	86	81	105	83	88	91	79
Competition	94	85	89	89	81	76	89	87	91	88	80

FIGURE 2.20. Golf scores.

a) Make a scatter plot of competition score versus the practice score.

b) Describe the relationship between practice and competition scores. Is there a positive or negative relationship? Explain why you would expect the scores to have a relationship like the one you observe.

OUTLIER

A datum, or a single data point, is called an outlier if it is well outside the general pattern of the rest of the data. More generally, an outlier appears to be outside the general pattern of the remaining data.

c) One point falls clearly outside the overall pattern. Circle this point in your plot. A good golfer can have a bad round, or a weak golfer can have a good round. Can you tell from the given data whether the unusual point is produced by a good player or a poor player? What other data would you need to distinguish between the two possibilities?

d) You might expect a player to have about the same score on two rounds played on the same course. Draw on your graph the line that represents the same score on both days. Does this line fit the data well when you ignore the unusual point? If you don't like this line, draw a line that you would prefer to use to predict the competition score from the practice score.

e) Another golf team member shot a 95 in practice. Predict her score in competition.

3. **Figure 2.21** contains two scatter plots that might show the relationship between human height and ulna length (forearm-bone length). In both displays, a line describing the pattern of the data has been added.

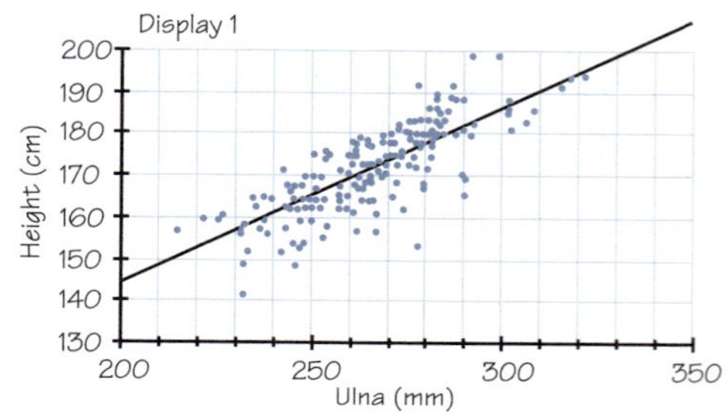

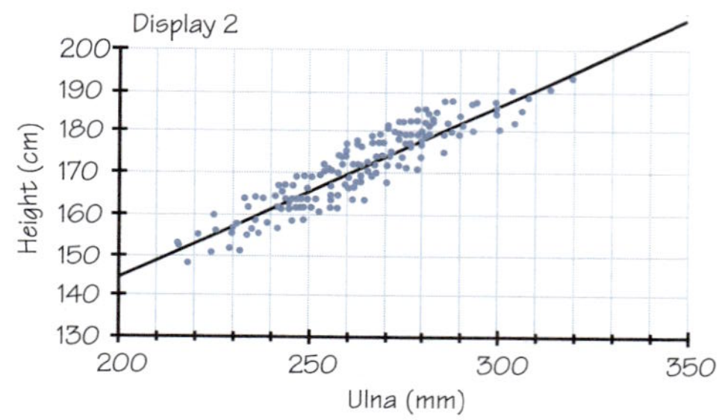

FIGURE 2.21. Two scatter plots showing height versus ulna length.

a) Both scatter plots have a linear form. Which of the two shows a stronger linear relationship? Explain.

b) A scientist used the model $y = 60.2 + 4.2x$ (where x is ulna length in cm and y is height in cm) to predict the height of a dead person whose ulna length of 29.0 cm is shown in Figure 2.21. What did the scientist predict as his or her height?

c) How close is the scientist's prediction if Display 1 shows the real data? How close is the scientist's prediction if Display 2 shows the real data? Which display better supports the scientist's prediction?

d) Describe the connection between a scatter plot and the predictions you can make using a trend line.

Answers (continued)

d) $y = -0.20x + 14$
$y = -0.20(60) + 14$
$y = 200$ cu ft of gas

2. a)

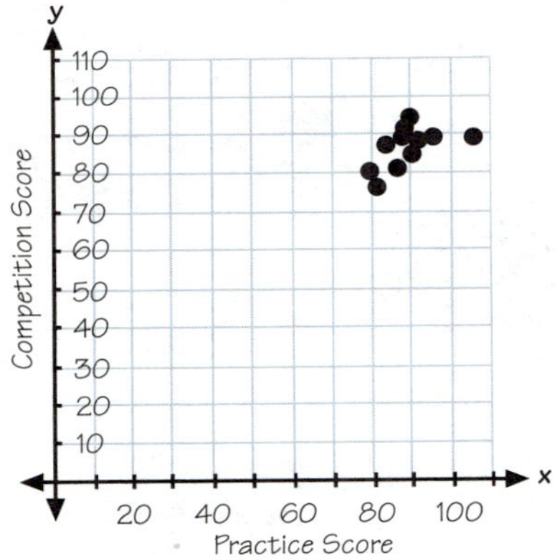

b) The points are loosely scattered around a straight line with a positive association between the variables. This indicates that the better golfer should have lower scores for both practice and competition.

c)

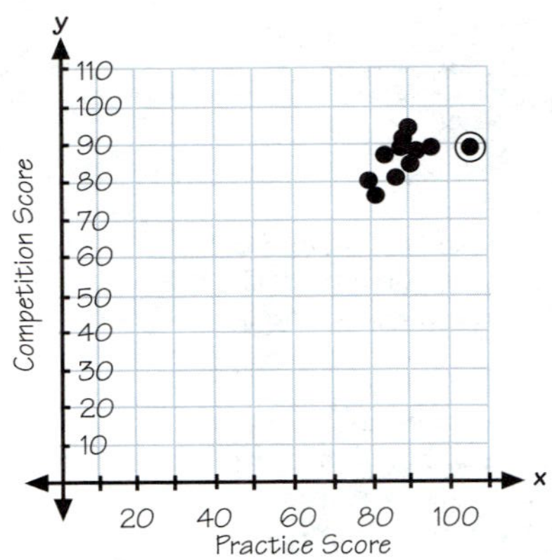

It is not possible to tell if this player is a good player who had a bad practice round or a weaker player who had an unusually good competition. You would need more data on this player to decide.

d) The graph below shows the data with the unusual value omitted and the line $y = x$ superimposed. It appears to do a reasonable job in describing these data.

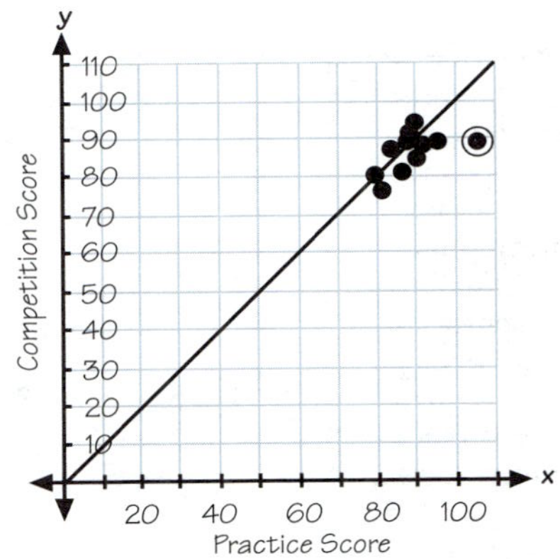

e) Prediction: 95 in competition round.

3. a) Display 2: The data in Display 2 appear to lie in a narrower band about the line than the data in Display 1.

b) $y = 60.2 + 4.2(29.0 \text{ cm})$
$y = 181.2$ cm

x	$60 + 4.2x$
24	160.8
25	165.0
26	169.2
27	173.4
28	177.6
29	181.8
30	186.0
31	190.2
32	194.4
33	198.6
34	202.8
35	207.0
36	211.2
37	215.4

c) Sample Answer:

Display 1: Draw a vertical line at $x = 29$ and find the y-values corresponding to x-values around 29 cm. Based on the graph, the actual height might be as low as 165 cm or as high as 195 cm.

Display 2: The actual height might be as low as 175 cm or as high as 185 cm.

The prediction is closer if the real data are from Display 2.

d) When the data in a scatter plot form a narrow band about the linear model, the predictions are **closer** than when the data form a wide band about the linear model. You can make **better** predictions when you are dealing with a strong relationship than with a weaker one.

SECTION 2.8

Residuals and Least-Squares Regression Lines

Previously, you learned methods for finding a trend line that describes a pattern of data. You used the equations of the functions you found to make predictions. In this section you will learn about a method used to find a line that approximates the pattern of data.

Whenever you make predictions based on data, you want to check the accuracy of your predictions. One way to evaluate your predictions is to calculate what are called the residual errors.

CHECK THIS!

Residual error: When a prediction is made from data, the difference between the actual value and the predicted value is called the residual error.

Residual error = Actual value − Predicted value.

You can represent the residuals graphically by drawing a vertical line from each point on the scatter plot to the line that represents your model as shown in **Figure 2.22.**

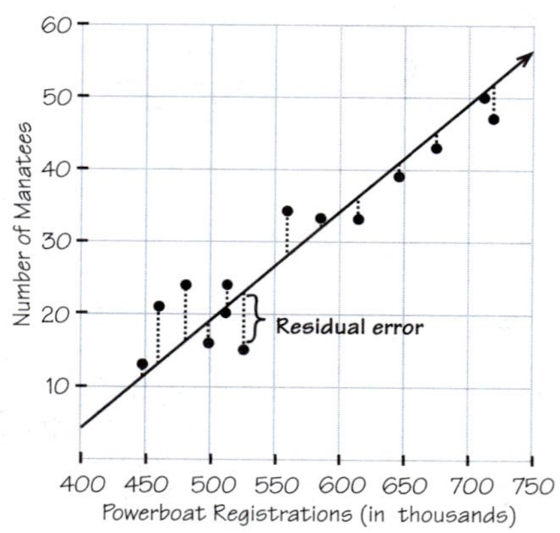

FIGURE 2.22. Residuals for the plight of the manatee.

A good line should meet a few tests. One test is that the sum of the residual errors should be close to 0. The sum of the residual errors is often found with a calculator or computer software. At times, the graph allows you to see that the sum of the residuals is fairly close to 0, but often it is easier to use a table, such as **Figure 2.23,** where you can total the residuals.

Registrations (in thousands)	Manatees Killed	Predicted Number of Manatees Killed ($y = 0.15x - 56$)	Residual Error
447	13	11	2
460	21	13	7
481	24	16	8
498	16	19	−3
513	24	21	3
512	20	21	−1
526	15	23	−8
559	34	28	6
585	33	32	1
614	33	36	−3
645	39	41	−2
675	43	45	−2
711	50	51	−1
719	47	52	−5
Total			2

FIGURE 2.23. Residual table for manatee deaths.

As you can see in Figure 2.23, the sum of the residual errors can be close to zero even if most of the residual errors are not. This happens because some of the residuals errors are positive and others are negative. To give more emphasis to values that deviate from the overall pattern, the residual errors are squared and then totaled. The squares of the residuals are shown in **Figure 2.24**. The goal is then to find a line that makes the sum of the squared residual errors as small as possible.

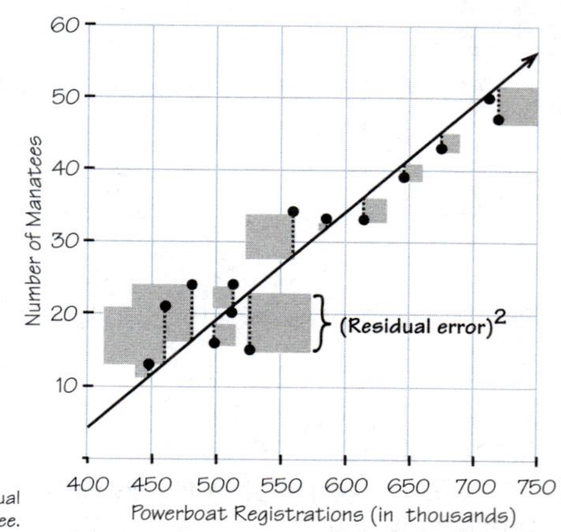

FIGURE 2.24. Squared residual errors for the plight of the manatee.

EXPLAIN LESSON 2

Section 2.8: Residuals and Least-Squares Regression Lines

Purpose: In previous sections students found the equation of a trend line that models linear data. The purpose of this section is to informally present the mathematics behind why we use the least-squares regression line found by a graphing calculator to fit a model for linear data that has been collected. (If you would like to delve more deeply into the sum of the residual errors, residual plots, and the sum of the squares, there is an additional lesson in the supplementary material.) This section will prepare students for Section 2.9 when we continue to discuss the least-squares regression line and how it does not always predict accurately outside the range of the independent variable.

Description: Students find a trend line to describe the relationship between the number of manatees killed per year and the number of powerboat registrations. They use the sum of the residual errors to assess whether their model is adequate to describe the data. They should also understand that the line of best fit is defined as the line where the sum of the squares of the residual errors is a minimum. Students also use the linear regression feature of the graphing calculator to find the equation of the least-squares line.

Materials: graphing calculator, Transparencies 2.3, 2.4, and 2.5

Teaching Ideas: This section is designed to be a whole class discussion with guided explanation.

To try and find the "line of best fit" by trial and error might take you a while. **Figure 2.25** shows some examples of equations that you might investigate as a trend line and the sum of squares for each line.

Equation	Sum of Squares
$y = 0.14x - 56$	745
$y = 0.13x - 50$	670
$y = 0.13x - 45$	228
$y = 0.12x - 45$	783
$y = 0.12x - 40$	246
$y = 0.125x - 41$	223

FIGURE 2.25. Sum of the squared residual errors.

Your graphing calculator has built-in commands that calculate the equation of the **least-squares regression line**. Statisticians also call this line the **line of best fit**. To find the model that fits the data the best, they look for several criteria:

- The pattern of data in the scatter plot is randomly scattered close to the line.
- The sum of the residual errors is close to 0.
- The sum of the squared residual errors is as small as possible.

The general name for finding an equation that fits the data is **regression**. Since you are looking for a linear relationship, the technique used in fitting the "best" line is called **linear regression**.

1. Use your calculator (or spreadsheet) to calculate the values of the slope and y-intercept of the least-squares line. Write its equation.
2. What do the slope and y-intercept of the linear regression line tell you about boats and manatees?
3. Plot both the data and the least-squares line so that you can see both graphs in the same window. Does the least-squares line appear to fit the data well? Support your answer.

SUMMARY

After you collect data on two variables and display the data in a scatter plot, you may describe a relationship between the two variables by finding a line of best fit. Statisticians often use the least-squares regression line as the best model linear relationship. Criteria for a good regression line include the following:

- The line appears to go through the middle of the points in the scatter plot.
- The sum of the residual errors is close to 0.
- The sum of the squared residual errors is as small as possible.

Graphing calculators (and spreadsheets) can calculate the least-squares regression line for you. However, you should check your linear regression model to see if it fits the criteria above. If it does not fit, you should look for a different model to describe your data.

Answers

1. Demonstrate for students how they can find the line of best fit with a graphing calculator. You may need to make sure that the students have the correct data in L1 and L2 and the correct window before proceeding.

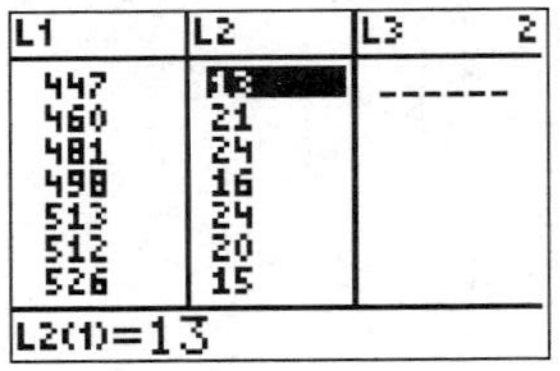

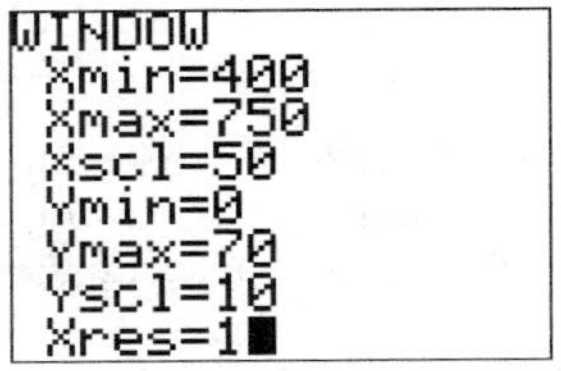

Press the STAT key, arrow over to CALC, arrow down to 4:LinReg(ax+b), indicate L1 and L2 as the locations of the two lists of data you wish to use:

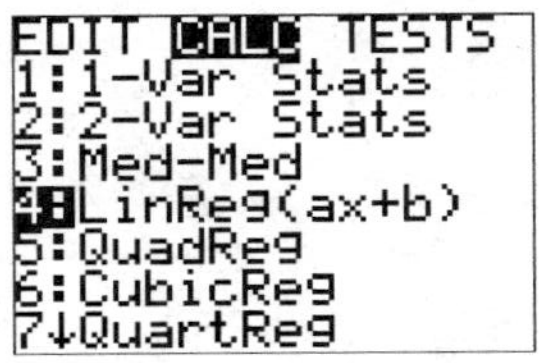

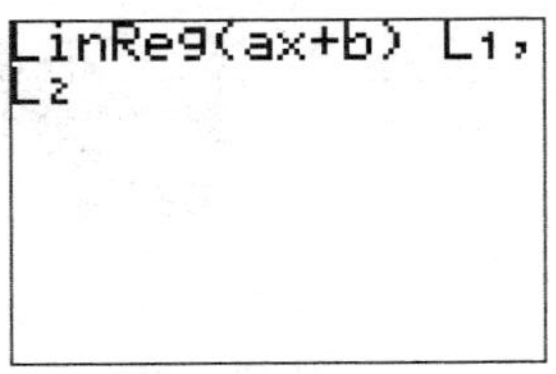

Press ENTER again and the line of best fit will be displayed:

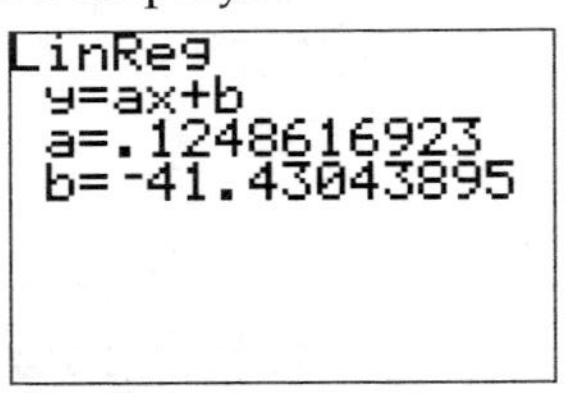

The slope is about 0.125 and the y-intercept is about –41.4. The least-squares equation is $y = 0.125x - 41.4$. The values for slope and y-intercept have been rounded. Some calculators give the value of Pearson's correlation coefficient, r, as part of the output of a linear regression. This is true of the TI-83 unless you specifically turn off the feature. If students ask, you may want to tell them that r is a measure of the strength and direction of a linear relationship. If the value of r is positive, the regression line (least-squares line) will have a positive slope and if the value of r is negative, it will have a negative slope. The value of r is always between –1 and 1. When $r = \pm 1$, the data lie exactly on a line. As the value of r moves closer to 0, the strength of the linear relationship weakens and the data are more spread out about the line.

2. The slope tells you that each time you increase the number of powerboats by one unit (1000 boats), you can expect 0.125 additional manatee deaths. The value of the y-intercept is meaningless in this context. The y-intercept would tell you the number of manatees killed by powerboats when no powerboats were registered. When statisticians find a linear regression equation they limit their predictions to the range of the independent variable. So, in this case the y-intercept occurs when the value of x is not between 447 and 719.

3. Students should graph the least-squares line along with the scatter plot:

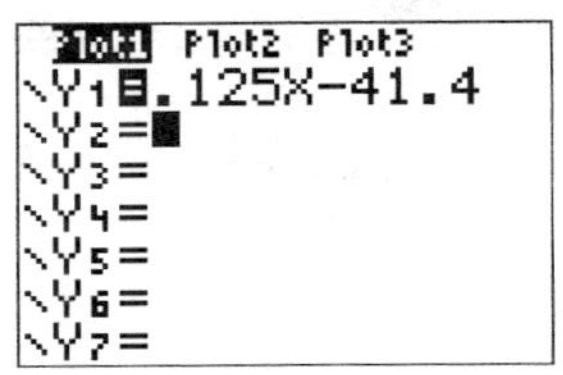

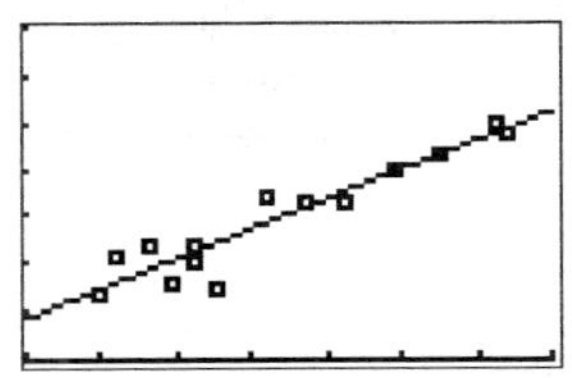

In the plot there appear to be about as many points above the line as below the line. Most of the points are fairly close to the line. So, the line appears to fit the data fairly well.

Students can also paste the equation into Y= without rounding by following these steps:

After calculating linear regression, go to the Y= screen, press VARS, 5: Statistics, ENTER.

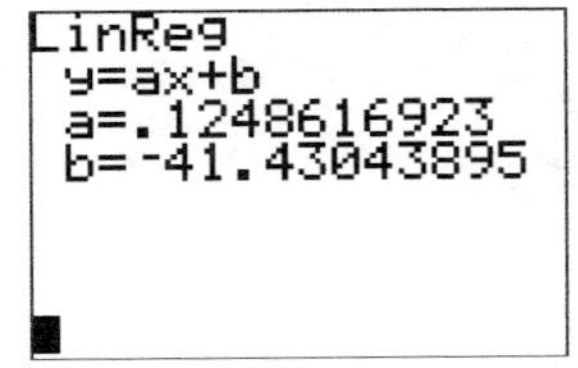

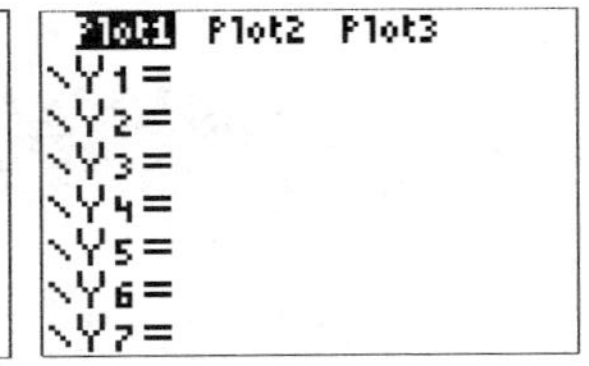

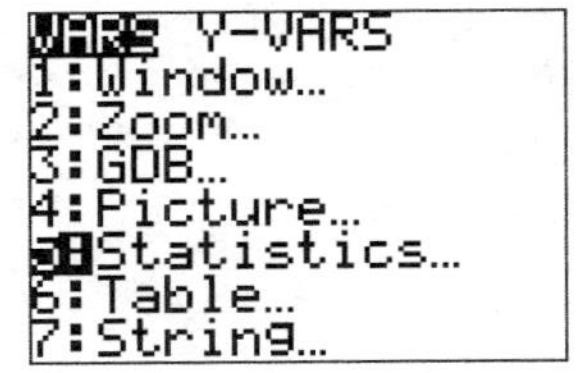

Arrow over to EQ, highlight RegEQ, press ENTER. The equation will be pasted into Y1=.

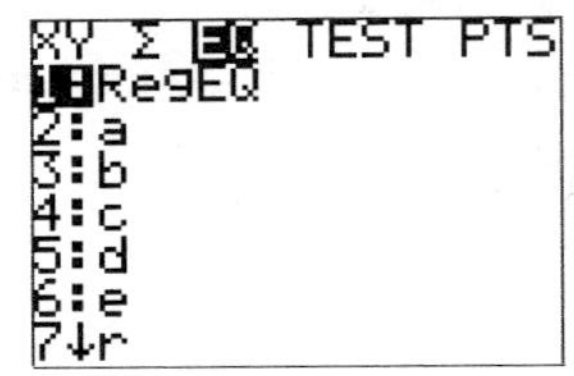

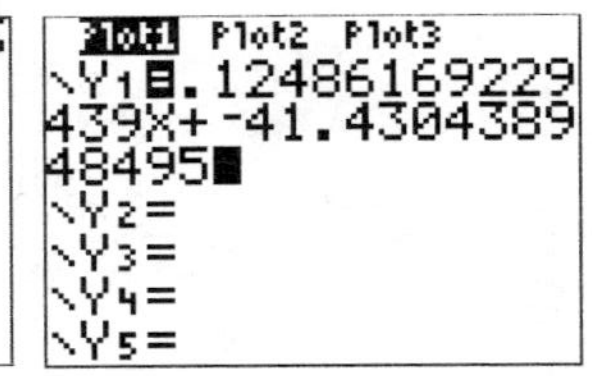

Some students may be familiar with a second method of calculating the linear regression equation and pasting it into Y1 all in one step:

Press STAT, CALC, 4:LinReg(ax+b), and then enter L1, L2, Y1. The calculator will find the line of best fit for L1 and L2 and paste the equation into Y1.

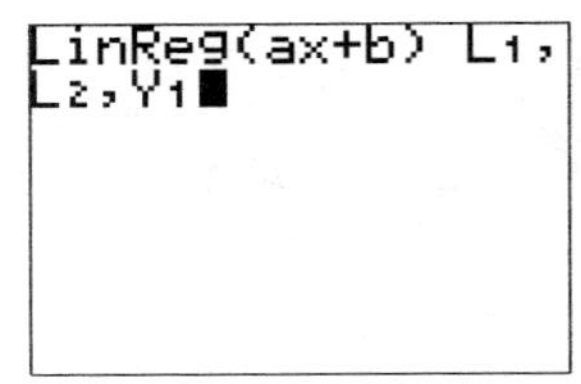

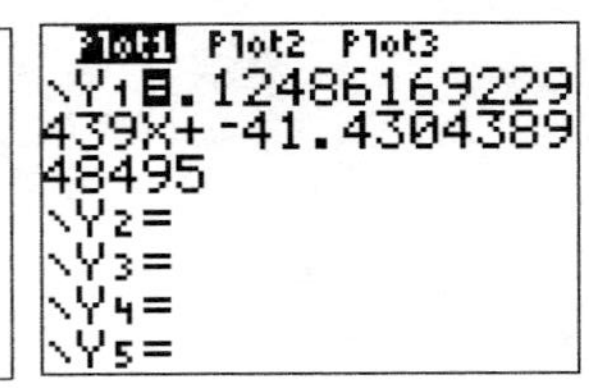

SECTION 2.8

Assignment

1. On graph paper, make a scatter plot of Ms. Jaczinko's class data (see **Figure 2.26**). Let forearm length be the independent variable and height be the dependent variable. Remember to label each axis and use an appropriate scale. To distinguish the boys from the girls, use two colors (or different shapes), one to represent the boys' data and the other to show the girls'.

FIGURE 2.26. Height and forearm data from Ms. Jaczinko's class.

Female			Male		
Name	Forearm Length (cm)	Height (cm)	Name	Forearm Length (cm)	Height (cm)
Alicia	24	157	Ahmed	26.5	173
Bia	24.5	166	Brian	27	177
Christi	27	164	Jesús	27	174
Chantalle	24	164	Davis	31	192
Coral-Mae	23	161	José	28	172
Juanita	27.5	164	Kelvin	29	180
Ji-Hyun	27	167	Lenny	27	174
Kim	26	162	Luis	28	175
Kristen	26	175	Mike	32	185
Maria	28.5	166	Pedro	30	185
Tianna	26.5	172	Sang	30	178
Teresa	25.5	176			

2. Find a linear regression model to predict a girl's height from her forearm length. Find a second linear regression model to predict a boy's height from his forearm length.

3. Use your scatter plot and least-squares model to make these predictions.

   a) If a girl the same age as the students in Ms. Jaczinko's class has a forearm that measures between 25 cm and 27 cm, what would you predict for her height?

   b) Predict the height of a tenth-grade boy with a 28.5-cm forearm. Explain your answer.

   c) What would be the forearm length of a student who is 163 cm tall? Explain your answer.

The 16% Model

Next you will use the data from Ms. Jaczinko's class to examine and refine a model that represents the relationship between height and forearm length.

4. Archaeologists study ancient human life. They use general rules of proportions. So an archaeologist might use the rule of thumb that the forearm of a typical female teenager is 16% of her height.

   a) Write this rule as an equation that predicts height, y, from forearm length, x. We call this equation the "16% model." Test a few pairs. Does this model make sense?

   b) Sketch a graph of the 16% model on the same set of axes as your scatter plot.

   c) Compare the 16% model with the data from Ms. Jaczinko's class. Is the model true for all the people in the class? Justify your answer based on your graph.

5. Use the 16% model to make the following predictions.

   a) Predict the height of a student whose forearm is 27 cm. Use the data from Ms. Jaczinko's class to check your prediction.

   b) Predict the height of a student whose forearm is 33 cm.

   c) The forearm lengths of two students differ by 1 cm. Use the 16% model to predict how much their heights differ. What if their forearm lengths differed by 2 cm? Justify your answers.

Section 2.8 Assignment

Answers

1. Student graphs should be similar to the one below. Circles represent boys, and squares girls.

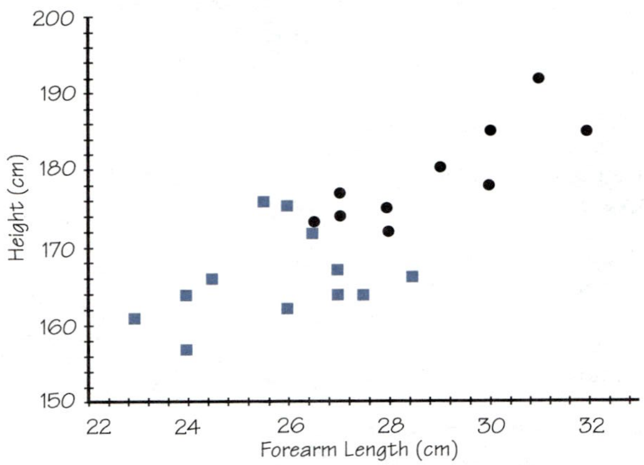

2. Using least-squares regression:
   Girl's height: $y = 1.057x + 138.9$
   Boy's height: $y = 2.863x + 96.5$

3. a) Sample answer: Using the least-squares model from Question 2, her height should be between 165 cm and 167 cm.

   b) Sample answer: Using the least squares model from Question 2, his height should be about 178 cm.

   c) Student answers will vary.

   Sample answer: Use the least-squares models from Question 2. If the student is a girl, then her forearm length will be 22.8 cm. However, if the student is a boy, then his forearm length will be 23.2 cm.

Girl's Model:

$$\begin{aligned} y &= 1.057x + 138.9 \\ 163 &= 1.057x + 138.9 \\ 163 - 138.9 &= 1.057x + 138.9 - 138.9 \\ 24.1 &= 1.057x \\ \frac{24.1}{1.057} &= \frac{1.057x}{1.057} \\ 22.8 &\approx x \end{aligned}$$

Boy's Model:

$$\begin{aligned} y &= 2.863x + 96.5 \\ 163 &= 2.863x + 96.5 \\ 163 - 96.5 &= 2.863x + 96.5 - 96.5 \\ 66.5 &= 2.863x \\ \frac{66.5}{2.863} &= \frac{2.863x}{2.863} \\ 23.2 &\approx x \end{aligned}$$

4. a) The 16% model: $x = 0.16y$. Solving for y in terms of x gives $y = x/0.16$.

   b) Student graphs should be similar to the one below.

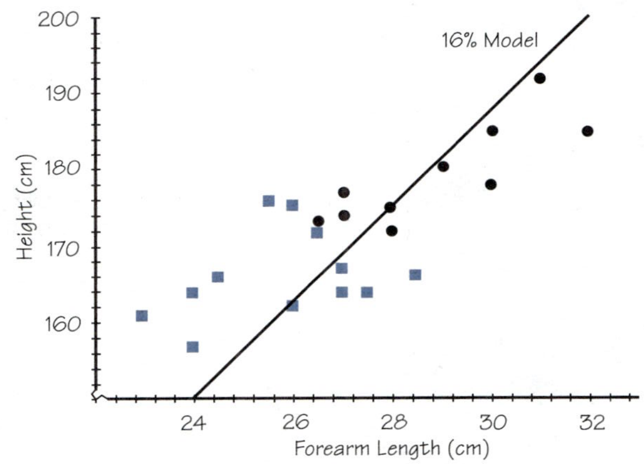

   c) No, this relationship is not true for all students in Ms. Jaczinko's class. If it were true, all the points would be on the line representing the 16% model.

5. a) According to the 16% model, a student with a forearm length of 27 cm would be 168.75 cm tall. If this student were female, the 16% model overestimates height. If this student were male, the 16% model underestimates height.

   b) According to the 16% model, a student with a forearm length of 33 cm would be 206.25 cm tall.

   c) According to the 16% model, a forearm length of 1 cm corresponds to a height of 6.25 cm (since $1/0.16 = 6.25$). Thus, if there is an additional 1 cm of forearm length, then the student will have an additional 6.25 cm of height. Two students whose forearm lengths differ by 1 cm will have heights that differ by 6.25 cm, and two students whose forearm lengths differ by 2 cm will have heights that differ by 12.5 cm.

SECTION 2.9

What is Your Recommendation?

Now that you have a model for the relationship between manatee deaths and powerboat registrations, you are ready to make your recommendation to the Florida Department of Environmental Protection.

1. Present to the class, using your scatter plot and the equation of the least-squares regression line, a convincing case for whether or not to restrict powerboat registrations. Your argument should include the three main steps listed in Section 2.6:
   - Find a model that describes the relationship between manatee deaths and powerboat registrations.
   - Show that the model does a good job of describing the data.
   - Use the model to make your prediction.
2. Say you want to reduce the number of manatee deaths to about 30 per year. How many powerboat registrations should there be? Explain how you can use your model and algebra to answer this question. What number of powerboat registrations do you recommend?
3. Say, instead, that you recommend 700,000 powerboat registrations (slightly below the number of registrations in 1989). Predict the number of manatees that would be killed each year, on average, if this proposal were adopted. Use your scatter plot to see how good your prediction is.
4. Fill in the blank: Every time the number of boat registrations is raised by 50,000, one can predict that, on average, an additional ________ manatees would be killed by powerboats each year. Justify your response.
5. It is possible that the Florida Department of Environmental Protection did not follow your recommendations back in 1991. In 2002, powerboats killed 95 manatees, and there were 923,000 powerboat registrations. Even though the least-squares model should only be used to predict the number of manatee deaths when the number of powerboat

registrations is between 447,000 and 719,000, how well does the least squares model predict the number of manatees killed in 2002?

6. The year 2002 had more manatees killed than the model predicts. **Figure 2.27** shows data for the years 1996 through 2003. Add these data to the previous data.

Year	Powerboat Registrations (in thousands)	Actual Number of Manatees Killed
1996	751	60
1997	797	54
1998	806	66
1999	805	82
2000	841	78
2001	903	81
2002	923	95
2003	978	73

FIGURE 2.27. Data on powerboats and manatees from 1996–2003.

*These data are from the Florida Fish and Wildlife Research Institute and the United States Coast Guard.

a) What is an equation of a new line of best fit?

b) Graph both the old and new lines of best fit on the calculator scatter plot.

c) What does the new slope tell you?

d) What is the predicted number of manatee deaths with the new model?

e) Is the new prediction closer to the actual number than the old model?

In this section, you used the least-squares model that you calculated in Section 2.8 to make predictions. In the process, you were given additional data to test how well your model predicted future events. You observed that a model that fits one data set may or may not fit an expanded data set. A line that fits a set of data well cannot always be trusted to predict outside the range of the data.

ELABORATE LESSON 2

Section 2.9: What is Your Recommendation?

Purpose: The purpose of this section is to give students an additional opportunity to use the graphing calculator to find a linear regression model with additional data and to investigate the idea of statistical predictions being limited to the range of the independent variable.

Description: Students will present to the class a recommendation for the Florida Department of Environmental Protection officials based on their model. They will extend and expand the relationship between the number of manatees killed and the number of powerboat registrations with data for additional years. They will investigate how accurately the least-squares model predicted the number of manatees killed outside the range of the independent variable. Using the graphing calculator students will calculate a new linear regression model using the original and additional data.

Materials: graphing calculator

Teaching Ideas: Have each student group make a brief presentation of their recommendation. Guiding questions to ask various groups might include:

- What model did you use?
- Is your recommendation based on your model?
- Do you think that powerboat owners will go along with your recommendation?
- How will your recommendation affect the manatees?
- Should the number of powerboat registrations be restricted? Why or why not?
- Do you have any additional recommendations to make to the Florida Department of Environmental Protection?

If students are comfortable with using the graphing calculator, you may want to allow them to complete Questions 2 through 6 individually. If they are still not comfortable, complete the questions as a class.

Answers

1. Answers will vary. Students should back up their answers with predictions based on their regression and scatter plot.

2. Sample answer: $x = 571.2$ units or about 571,200 registrations.

   Solve: $30 = 0.125x - 41.4$ for x.

   Or, students can look in the TABLE on the calculator and find 30 in the Y1 list.

   The number of powerboat registrations from the least-squares regression model is between 569,000 and 576,000 since you would round the manatee values to whole numbers. (There is no fraction of a manatee.)

   Or, they may use the TRACE key on their graph.

X	Y1
570	29.741
571	29.866
572	29.99
573	30.115
574	30.24
575	30.365
576	30.49

X=573

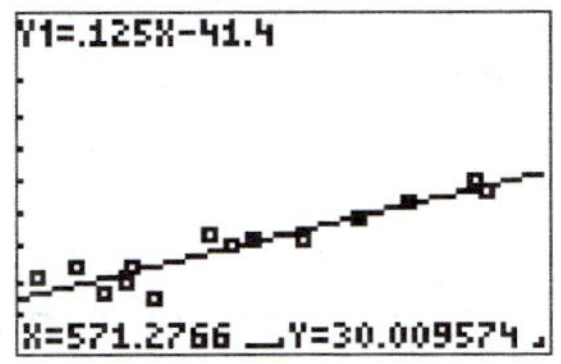

3. Using substitution, $y = 0.125(700) - 41.4 = 46.1$ or approximately 46 manatees would be killed by powerboats. Or, students could investigate using the TABLE in the graphing calculator. Or, they may use TRACE on the graph.

X	Y1
696	45.473
697	45.598
698	45.723
699	45.848
700	45.973
701	46.098
702	46.222

X=700

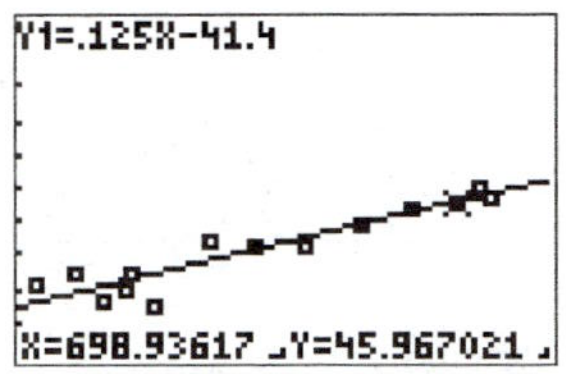

Again, approximately 46 manatees would be killed. The answers are slightly different since $y = 0.125x - 41.4$ has been rounded.

Since $x = 700$ is between observations for $x = 675$ and $x = 711$, we can base our answer on values (675, 43) and (711, 50). The actual numbers of manatees killed might be as small as 43 or as large as 50, even though we have predicted 46 deaths.

4. Because the units for the powerboats are in terms of 1000 boats, a 50,000 increase in the number of powerboats corresponds to an increase of 50 units for the independent variable. Using our model, we would predict an increase of $(0.125)(50) = 6.25$ deaths. Based on this calculation you could respond to the Florida Department of Environmental Protection as follows:

   Every time the limit is raised by 50,000 registrations, we predict that, on average, an additional 6 manatees would be killed by powerboats each year.

5. Students may look up this value in the TABLE of the calculator. Set the TBLSET as shown, and then look in the TABLE.

X	Y1
923	73.817
924	73.942
925	74.067
926	74.191
927	74.316
928	74.441
929	74.566

X=923

Based on the least-squares line we would predict that powerboats would kill 74 manatees when there are 923,000 powerboat registrations.

Or, substitute and calculate: $y = 0.125\,(923) - 41.4 = 73.975 \approx 74$ manatees.

Since the actual number of manatees killed in 2002 was 95 and the predicted number is 74, the prediction is 21 manatees killed fewer than the actual number killed. The prediction is not very accurate outside the range of the independent variable.

Answers (continued)

6. a) Students should add these new data points to List 1 and List 2 in the calculator. Then find the line of best fit. The new line of best fit with the additional data points is $y = 0.146x - 52.4$.

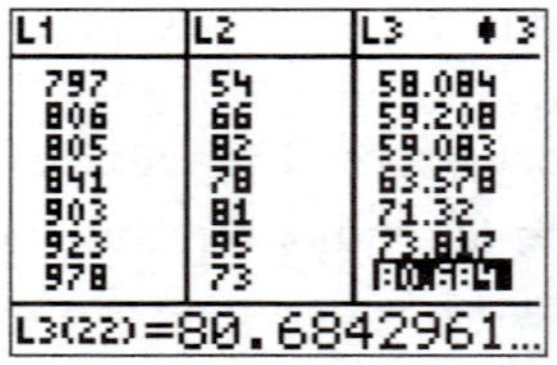

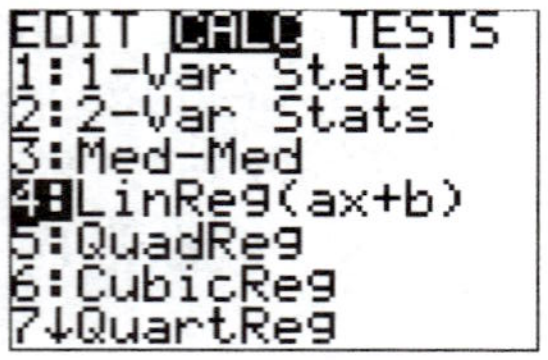

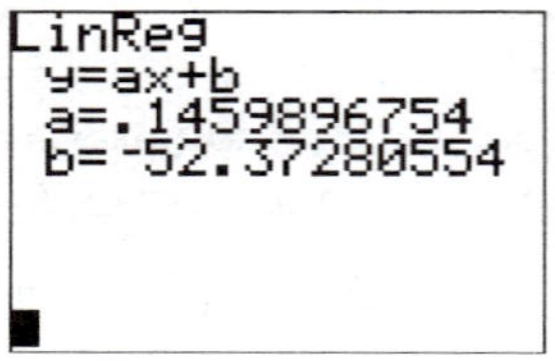

b)

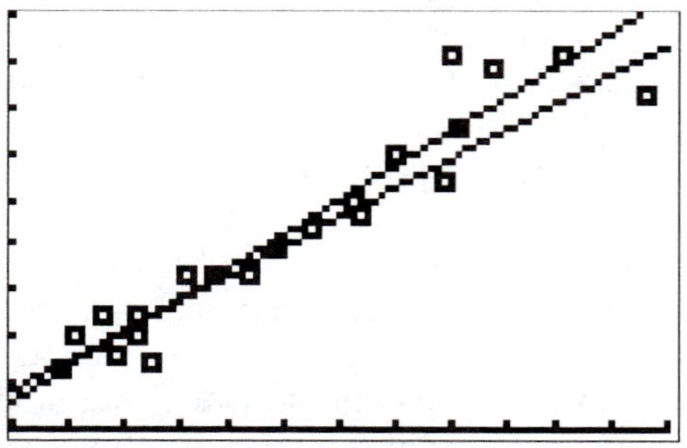

c) The new slope of 0.146 tells you that you can expect 0.146 additional manatee deaths each time the number of powerboats is increased by one unit (1000 boats). So, not only has the actual number of manatees killed by powerboats increased, but the rate of change (slope) in the number of manatees killed versus powerboat registrations has also increased. The more data you collect, the better the linear regression model will be.

d) If students investigate the two models in their calculators, they can set the TBLSET to Ask and view both predictions at the same time.

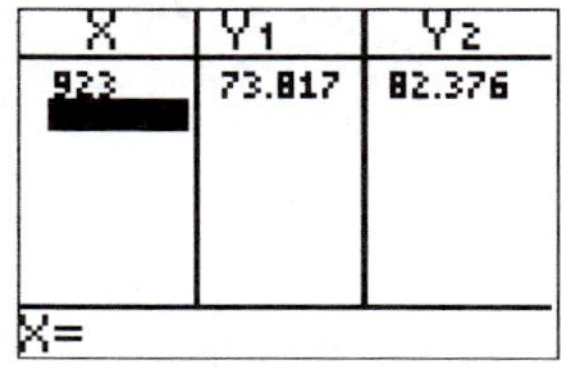

e) The new model predicts 83 manatee deaths. This is still not equal to the actual number of 95, but it is more in line with adjacent years. In 2001, there were 81 deaths, and in 2003, there were 73 deaths. (Note of interest: The number of manatee deaths due to powerboats decreased in 2004 to 69. Some environmentalists theorize that this drop was due to the four hurricanes that hit Florida during 2004, consequently, powerboat enthusiasts were not able to go boating as often during the 2004 season.)

Hence, often when you gather additional data, you may need to refine your mathematical model. The more data points that you can collect, the more accurate your model will be.

Not every data set can be modeled well with a linear regression equation. There are many other types of relationships between variables. We will explore some of these relationships in the following chapters.

SECTION 2.9 Assignment

How would you choose which of two models makes better predictions? Let's explore some data to help you decide. The data in **Figure 2.28** were collected from a ninth grade class. The asterisks (*) indicate missing values.

FIGURE 2.28. Data on height, stride length, and forearm length.

Height (cm)	Stride Length (cm)	Forearm Length (cm)
166.0	58.2	28.5
164.5	55.9	27.2
175.0	59.1	28.6
184.0	68.9	30.5
161.0	72.5	26.5
164.0	*	28.2
171.0	*	29.0

1. a) Make a scatter plot of height versus stride length. Make a second scatter plot of height versus forearm length. (For both plots, height should be on the vertical axis. Use the same scale on the vertical axis for each plot.) Which of the two scatter plots shows the stronger relationship? How can you tell?

   b) Find the least-squares line that represents each of these relationships.

   c) Use the equations of your two least-squares lines to predict the height of a person whose stride is 73 cm and whose forearm length is 27 cm. Which of the two estimates is more reliable? Explain.

   d) State, in your own words, how you would choose, in general, between two different models that predict the same quantity.

Section 2.9 Assignment

Answers

1. a) The graph of height vs. forearm length shows a more linear relationship since the two longest stride lengths were made by the shortest and tallest persons. Also, the points appear to be closer to a line.

Graphing Window

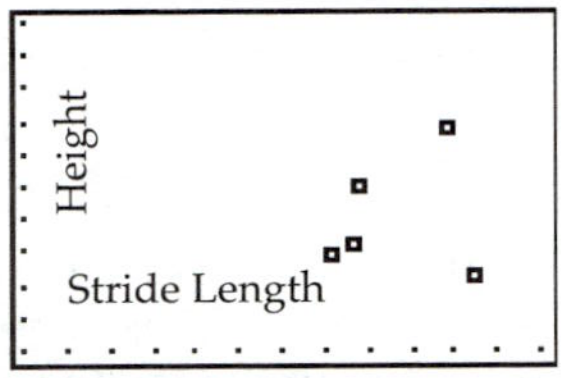

Height vs. Stride Length

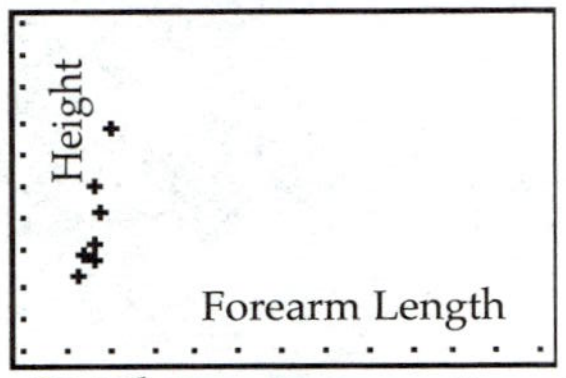

Height vs. Forearm Length

b) Height = 160 + 0.168 (Stride Length)
Height = 10.9 + 5.59 (Forearm Length)

c) Sample answer:

Height-forearm model:
Height = 10.9 + 5.59(27) ≈ 161.8 cm

Height-stride model:
Height = 160 + 0.167(73) ≈ 172.2

The first estimate is more reliable, since the data appears more concentrated along the least-squares line.

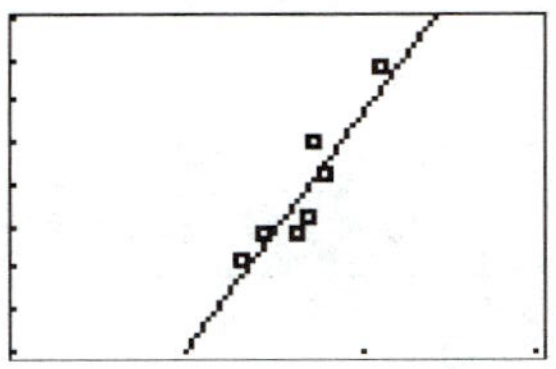

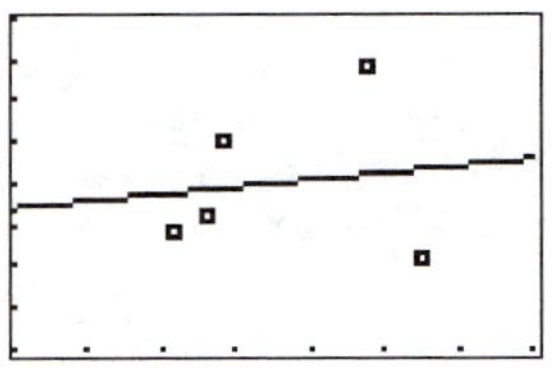

d) Select the model for which the data appear most concentrated along the least-squares line.

The Missing Manatee

FIGURE 2.29.
The missing manatee.

A school's mascot is stolen (see **Figure 2.29**). The thief has left some clues: a plain blue sweater and a set of footprints under a window. The footprints appear to have been made by a boy's sneaker.

The distance between the footprints, from the back of the heel on the first footprint to the back of the heel on the second, reveals that the thief's stride is approximately 58 cm. The length of the thief's forearm can be estimated by measuring the sweater from the center of a worn spot on the elbow to the turn where the cuff meets the sleeve. The thief's forearm is between 26 cm and 27 cm.

School officials suspect that the thief is a student from a rival high school. **Figure 2.30** contains data from a 1st period class about students' genders, heights, stride lengths, and forearm lengths. Use these data to answer the following questions.

2. Find the mean, median, and mode for the whole class and then find the mean, median, and mode for the boys and the girls.

   a) Is there a difference between the mean stride length for a boy and the mean stride length for a girl?

3. The footprint was made by a boy's sneaker. However, sometimes girls wear boys' sneakers. So you'll need to decide whether or not the thief was male. Use what you know about the thief's stride length and forearm length.

   a) Do the class stride length data tend to confirm that the footprints belonged to a boy and not a girl? Use the data you gathered to support your answer.

   b) Do the forearm data tend to confirm that the sweater belonged to a boy and not a girl? Use the data you gathered to support your answer.

Next you will predict the height of the thief. You have two possible independent variables you can use for your prediction: stride length and forearm length. You will try to decide which is better. Divide the work in Questions 4 and 5 among the members of your group.

Name	Gender	Height (in cm)	Stride Length (in cm)	Forearm Length (in cm)
Scott	Male	166.0	58.25	28.5
John	Male	178.0	68.5	29.0
Matt	Male	171.0	58.5	27.2
Will	Male	165.0	50.125	28.0
Michael	Male	177.5	58.75	31.3
Jeffrey	Male	166.0	62.875	28.3
Even	Male	175.5	59.125	28.6
Brad	Male	171.0	67.75	31.5
Lonnie	Male	184.0	68.875	30.5
William	Male	184.5	66.25	30.8
Robert	Male	183.5	79.5	30.5
Karim	Male	172.0	70.5	30.3
Meredith	Female	164.5	55.875	24.2
Lee	Female	166.0	52.375	27.3
Pilar	Female	168.0	55.375	28.0
Ansley	Female	178.5	59.75	29.1
Julie	Female	166.0	48.375	27.9
Becton	Female	159.0	57.125	28.0
Elizabeth	Female	166.0	64.0	27.4
Shannon	Female	154.5	57.75	25.8
Janie	Female	161.0	63.5	27.0
Jeris	Female	177.0	69.75	30.1
Kat	Female	161.0	72.5	26.5
Blaie	Female	164.0	75.25	28.2
Frances	Female	174.0	58.5	28.4
Eliz	Female	164.0	59.75	26.8
Baily	Female	168.0	55.25	26.4

FIGURE 2.30. Height, stride length, and forearm length for students in 1st period class.

4. a) Use your data to find models for the relationship between height and forearm length. First find a model based on the entire data set. Then use the girls' data to find a second model, and finally use the boys' data to find a third model.

   b) Compare the three models. Is there much difference? Explain.

Answers (continued)

2.

	Mean	Median	Mode (round to nearest cm)
Class	62	59.75	59
Boys	64	64.56	59
Girls	60.34	58.5	55, 60, 64

a) The mean stride length for all students in this class is 62 cm. There is one stride length of 79.5 cm that tends to inflate the mean of the class somewhat. The average stride length for girls is 60.34 cm or a bit more than 60 cm; for boys it is approximately 64 cm.

3. a) Sample answer: There is a cluster of students whose stride lengths were around 58 cm. Some were boys and some girls. The stride length does not appear to provide helpful information about the gender of the thief.

b) Sample answer: Four people from this class had forearms with measurements between 26 and 27 cm. All of these students were girls. So, the forearm data tends to indicate that the thief was a girl!

4. a) Class Data: $y = 75.5 + 3.325x$

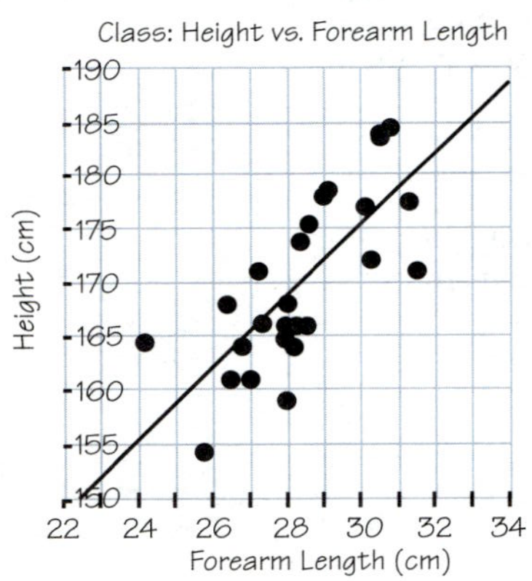

Girls' Data: $y = 84.4 + 2.980x$

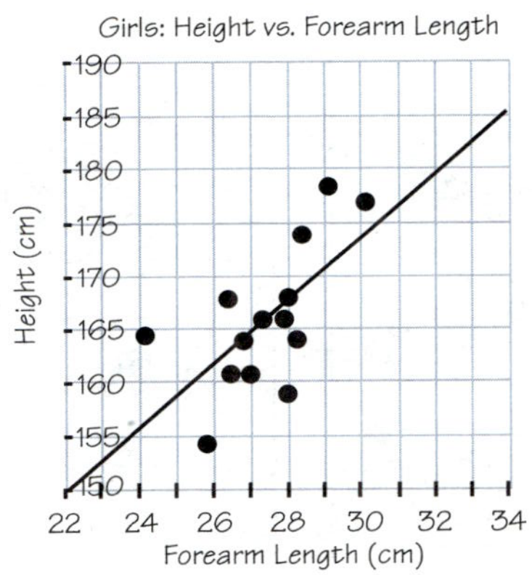

Boys' Data: $y = 87.9 + 2.933x$

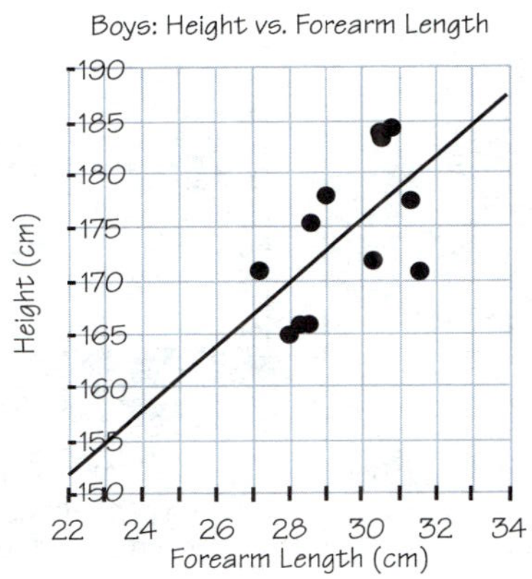

b) The slopes for the boys' data and girls' data are fairly close. However, the boys' line is shifted vertically from the girls' line. It appears, according to these two models, that boys are, on average, 2.8 cm taller than the girls with similar forearm lengths.

c) Which model gives a better prediction if, in fact, the thief were male? Justify your answer.

d) Which model gives a better prediction if, in fact, the thief were female? Justify your answer.

5. Repeat Question 4 for the relationship between height and stride length.

6. Select the best model for the job of predicting the height of the thief. Support your choice.

7. Use this model to predict the height of the thief. Explain your answer.

8. **Figure 2.31** shows the winners of the Indianapolis 500 Auto Race and their average speed every 4 years from 1912 to 1972. No race was run in 1944 due to World War II.

Year	Number of Years Since 1912	Winner	Winner's Average Race Speed (mph)
1912	0	Joe Dawson	79
1916	4	Dario Resta	84
1920	8	Gaston Cheverolet	89
1924	12	L. L. Corum & J. Boyer	98
1928	16	Louis Meyer	99
1932	20	Fred Frame	104
1936	24	Louis Meyer	109
1940	28	Wilbur Shaw	114
1944	32	—	—
1948	36	Mauri Rose	120
1952	40	Troy Ruttman	129
1956	44	Pat Flaherty	128
1960	48	Jim Rathmann	139
1964	52	A. J. Foyt	147
1968	56	Bobby Unser	153
1972	60	Mark Donohue	163

FIGURE 2.31. Indianapolis 500 Auto Race results.

Finding a linear regression equation is easier if you use the number of years since 1912 as the independent variable. This also gives meaning to the y-intercept.

a) Make a scatter plot of the data.

b) Find the equation of a linear regression model.

c) What do the slope and y-intercept in this model show?

d) The speed of the winner in 1957 was 136 mph. How well does the linear regression model predict this value?

e) Even though you know that 200 mph is outside the range of the data, you would like to use your model to predict when a driver will reach a race speed of 200 mph. Based on the linear regression model, when would you expect a driver to average 200 mph in the Indianapolis 500 race?

f) The actual race speed of the winner, Buddy Rice, in 2004 was 139 mph. Use your line of best fit to predict the speed in 2004. Is the least-squares line a good predictor? Why or why not?

Answers (continued)

c) Sample answer: If you are fairly certain that the thief is male, use the boys' model.

d) Sample answer: If you are fairly certain that the thief is female, use the girls' model.

5. Class Model: $y = 145 + 0.399x$

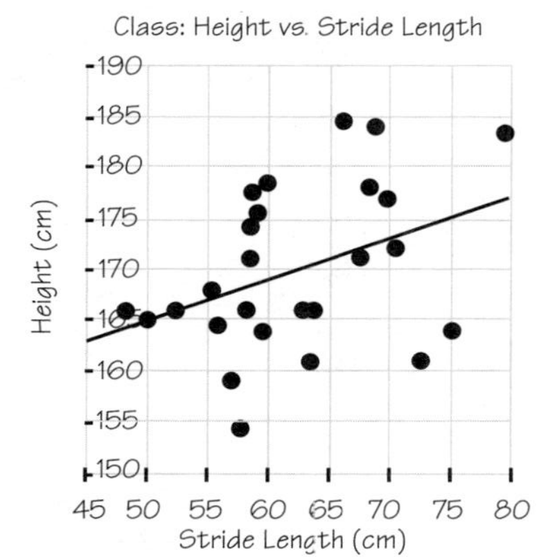

Girls' Model: $y = 165 + 0.019x$

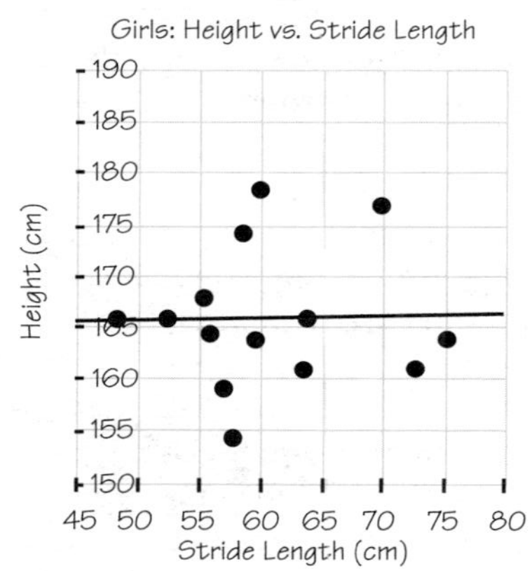

Boys' Model: $y = 137 + 0.590x$

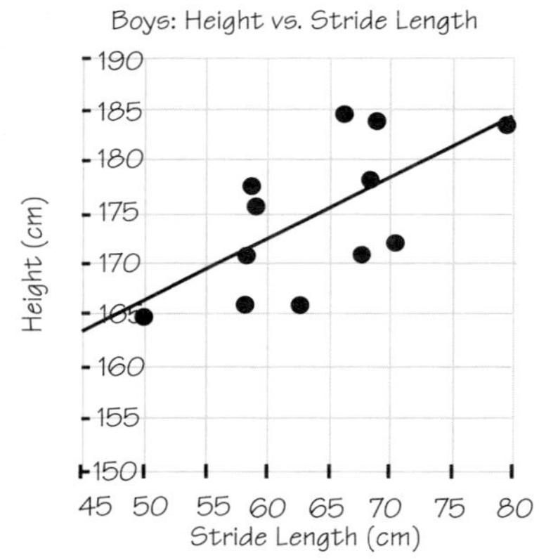

The slopes for these three models are very different. Because the slope of the girls' model is so small, it offers very little help in predicting height from stride length. Because you have decided that it is likely that the thief is a girl, stop here and use the girls' model from Question 4.

6. Sample answer:

Girls' model based on forearm length:
$y = 84.4 + 2.980x$

7. Students should predict that the thief is <u>between 162 and 165</u> cm tall. But the thief might be as short as <u>155 cm</u> or as tall as <u>168 cm</u>.

They can reach this conclusion by looking at heights corresponding to girls with forearm lengths near 26 or 27 cm.

8. a)

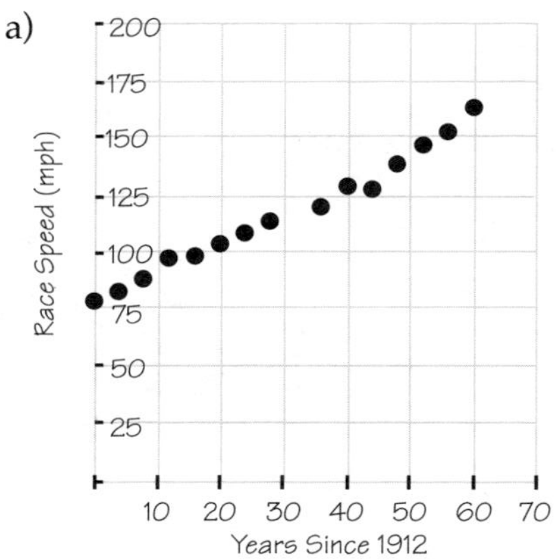

b) The linear regression model for this data is $y = 1.298x + 78.231$.

c) Every year the speed increases on the average 1.298 miles per hour. Every 4 years this would amount to approximately a 5 mph increase. The y-intercept indicates that at year 0, the speed is predicted to be approximately 78.231 mph.

d) First we need to find out how many years it has been since 1912. 1957 – 1912 = 45.

Then substitute 45 into our model. 1.298(45) + 78.231 = 136.641, or approximately 137 mph. The prediction was off by only 1 mile per hour. The year 1957 is within the data range, and it is a good predictor in this case.

e) Using algebra: $200 = 1.298x + 78.231$, $x \approx 94$ years since 1912 or 2006. Students may also use the table or the graph for the linear regression equation to make the same prediction.

Answers (continued)

f) Using algebra:
$y = 1.298(2004 - 1912) + 78.231 = 197.647$, or approximately 198 mph. No, the least-squares line is not a good predictor outside the data range in this case. This is why you should be cautious with predictions. The prediction is off by 59 mph. In 1973, because of two accidents at the Indianapolis 500 more safety measures were adopted to reduce the possibility of fire, speeds were reduced, and fuel capacity was cut; consequently, cars were not able to continue the increase in speed at the same rate. Other factors that might affect the race speed would be weather, including both rain and wind, changes in fuel specifications, or changes in vehicle specifications, such as changes in carburetor design.

SECTION 2.10

Line Fitting with Domain Restrictions

Let's take what we have learned about building linear models thus far, and apply it to a new situation.

Marietta has been working as a waitress for 7 weeks. Since she earns an hourly wage plus tips, there is no set amount she makes per hour. However, after examining her 7 pay stubs she notices an apparent trend. **Figure 2.32** shows how many hours she worked and how much money she makes per week.

Time (T) (in hours)	8	15	23	21	16	20	22
Money (M) (in dollars)	53	101	155	131	124	138	157

FIGURE 2.32. Marietta's data table.

1. Label the axes appropriately and draw a scatter plot of money (M) versus time (T).
2. Use a straightedge to approximate a trend line for the data. Write an equation for your trend line.
3. Use a graphing calculator to make a scatter plot, and then find the line of best fit using linear regression.
4. How are your trend line and the calculator's line of best fit alike and how are they different?
5. What do the slope and y-intercept of the line of best fit mean in this context?
6. Marietta's boss tells her that she must work at least 10 hours every week in order to keep her job. Furthermore, state law says that a restaurant employee cannot work more than 60 hours during any week. Considering these restrictions, sketch a more appropriate trend line. Do not re-plot the data. Number and label your axes appropriately.
7. Use the second graph to predict how much Marietta will make if she works for 30 hours. Explain your answer.
8. Use the second graph to predict how many hours Marietta needs to work to make $75. Explain your answer.

Modeling Project Who Am I?

You are an anthropologist that finds bones of two or more individuals. The table in **Figure 2.33** contains the information about those bones.

Skeleton 1 (possibly female)	Skeleton 2 (taller of the two)	Uncertain
Femur: 413, 414	Femur: 508	Skull: 230
Ulna: 228	Ulna: 290	Humerus: 357
		Radius: 215
		Tibia: 416

FIGURE 2.33. Classification of bones, measurements in millimeters.

Figure 2.34, on the following page, contains actual data from the Forensic Anthropology Data Bank (FDB) at the University of Tennessee. The FDB contains metric, nonmetric, demographic, and other kinds of data on skeletons from all over the United States. These individuals most likely are unidentified bodies that went to forensic anthropologists for analysis and identification.

Use the data in Figure 2.34 to answer the following items. Present your findings in a formal report. All of your conclusions must be supported by statistical analysis of the data in Figure 2.34.

1. Determine two models to predict people's height from the lengths of various long bones in their arms and legs. Explain which of these models you would prefer to use and why.
2. Based on these data, do you agree that Skeleton 1 is female? Do the data provide any information that would help you decide whether Skeleton 2 is male or female?
3. Find models for relationships between pairs of long bones that would help you decide whether the bones in the uncertain column belong to Skeleton 1 or Skeleton 2. (Or is there strong evidence that one of these bones belongs to a third person?)
4. Predict the heights of Skeleton 1 and Skeleton 2. Explain why you chose the model that you used to make your predictions.

EVALUATE LESSON 2

Section 2.10: Line Fitting with Domain Restrictions

Purpose: The purpose of this section is to evaluate students' understanding of a line of best fit and to see if they are able to apply the concept to a real-world problem with a restricted domain.

Description: This section allows students the opportunity to demonstrate their knowledge about finding a line of best fit and how the line is related to a restricted domain in a problem involving earnings versus time.

Materials: graphing calculator

Teaching Ideas: Students may complete these questions individually in order for you to evaluate their understanding of the concept of finding the line of best fit.

Answers

1. Domain of data: [8, 23]
   Range of data: [53, 157]

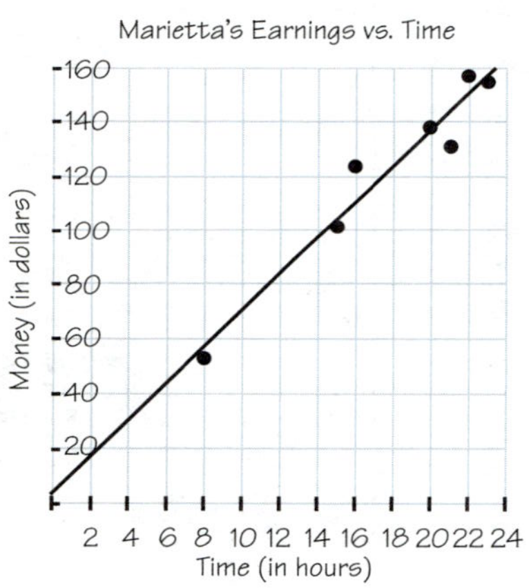

2. Answers will vary. Students may assume that if she works zero hours, she makes no money and write a direct variation equation such as $M = 7T$, or $M = 6.68T$.
3. The linear regression equation from using the calculator is $y = 6.67x + 3.52$.
4. They should have slopes and y-intercepts that are similar.
5. The slope represents a rate of change of approximately $6.67 for every additional hour worked. The y-intercept is meaningless in this situation since she does not earn any money when she works 0 hours, and $x = 0$ is not within the range of the data.
6. Domain: [10, 60]
   Range: [$70.27, $404.00]

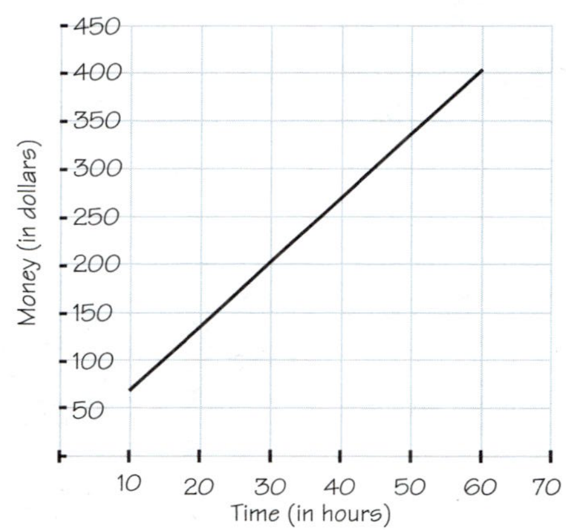

Make sure the students understand that they must restrict the domain to [10, 60]. The new range should be something close to [$70, $404], based on the linear regression equation. It is not crucial that the students have perfect graphs for this exercise, just reasonable ones.

7. From the graph, she would make about $200. From the linear regression equation, she would make about $203.62.
8. From the graph, she needs to work about 10 hours. From the linear regression equation, she needs to work about 10.7 hours.

1	168	307	240	258	448	384	368
1	178	336	247	261	463	404	390
1	161	294	213	227	413	335	322
1	155	324	262	279	465	395	375
1	165	314	243	258	432	364	364
1	168	303	223	244	441	355	342
1	165	311	231	254	436	362	360
1	173	312	248	266	483	405	401
1	165	322	229	246	448	368	352
1	163	298	221	245	443	355	361
1	153	280	218	234	410	345	344
1	165	294	220	235	448	354	353
1	170	311	235	253	440	360	347
1	160	316	214	226	437	356	348
1	159	292	223	233	419	346	336
1	163	315	228	251	438	356	347
1	165	303	237	249	451	356	348
1	165	308	234	248	439	348	344
1	165	315	227	240	448	363	353
1	175	316	244	260	473	390	374
1	180	333	256	278	475	391	381
1	168	321	230	248	450	365	362
1	163	299	219	236	435	357	339
1	165	304	246	264	467	392	383
1	160	309	236	248	432	364	358
1	158	319	246	268	442	371	364
1	165	325	242	250	448	378	365
1	170	335	248	263	474	400	382
1	182	334	254	273	514	420	407

1	165	307	230	248	452	363	355
1	163	297	240	260	435	356	356
1	143	282	216	233	398	334	318
1	154	297	228	248	423	344	334
1	171	342	272	290	485	418	407
1	162	303	237	262	433	367	364
1	150	308	220	247	383	352	341
1	157	288	201	215	429	363	350
1	158	314	239	263	432	371	358
1	162	306	250	268	444	355	352
1	159	310	238	255	449	362	352
2	169	337	254	273	460	396	385
2	153	296	223	243	407	337	338
2	175	339	256	271	470	390	381
2	179	343	242	263	464	378	371
2	179	352	253	269	484	407	397
2	198	354	263	292	508	417	412
2	173	327	256	276	463	383	387
2	180	357	268	278	494	401	390
2	178	344	254	269	464	371	366
2	175	339	245	272	456	374	366
2	177	343	250	266	483	361	365
2	180	353	260	281	490	420	415
2	170	303	235	249	435	366	361
2	191	364	263	278	511	430	417
2	188	349	269	288	498	427	423
2	179	323	256	276	486	398	400
2	180	350	263	280	480	419	418
2	181	350	263	282	488	391	381

2	178	337	272	272	475	393	390
2	172	344	255	281	470	400	393
2	188	360	269	283	510	422	416
2	189	347	272	283	547	432	445
2	177	330	246	262	462	386	370
2	166	322	242	258	442	373	374
2	186	332	267	283	478	391	388
2	177	322	245	265	457	397	395
2	176	332	259	274	458	382	378
2	180	323	251	275	448	390	387
2	173	335	253	273	497	404	389
2	175	330	253	274	470	384	382
2	169	313	252	265	472	391	385
2	175	336	256	274	464	388	377
2	181	390	284	303	521	440	435
2	193	356	297	318	522	451	433
2	182	362	275	293	499	424	405
2	169	322	249	266	426	366	356
2	180	337	265	281	482	412	399
2	185	363	286	302	520	429	420
2	180	355	274	292	490	422	424
2	170	378	272	291	512	404	390
2	180	370	278	292	523	429	420
2	175	333	260	273	484	398	386
2	168	342	262	280	484	404	385
2	170	347	269	291	476	396	393
2	166	315	240	260	456	377	362
2	185	363	295	309	524	446	427
2	191	382	299	316	537	479	466

FIGURE 2.34. Data from Forensic Anthropology Data Bank (FDB).

Key to Data (in order from left to right): sex (1 = female, 2 = male), height (cm), humerus (mm), radius (mm), ulna (mm), femur (mm), tibia (mm), fibula (mm)

Practice Problems

1. For each of the descriptions that follow, find an equation of the line.

   a) The line through the point (2, 3) that has slope $\frac{1}{2}$.

   b) The line that has y-intercept 3 and slope $-\frac{1}{2}$.

   c) The line passing through points (3, 2) and (2, 5).

   d) The line has y-intercept 5 and is parallel to the line $y = 7x - 1$.

2. On December 18, 1994, three amateur spelunkers (cave explorers) found a cave in France, now known as Chauvet Cave, that had ancient paintings. In later explorations, human footprints were found. According to prehistorian Michel-Alain Garcia, the footprints belonged to a boy who was about 4.5 feet tall.

   a) Scientists have used simple models to predict height from footprints since the mid-1800s. One model, still in use today, predicts height by dividing the maximum foot length by 0.15. Write an equation for this model.

   b) The model in a can be expressed as a member of the $y = mx$ family. What is the value of m?

   c) Use this model to predict the maximum length of the footprints found in Chauvet Cave. Give your answer in centimeters. (Remember 2.54 cm = 1 in.)

   d) Suppose that another footprint was found and it measured 1 cm more than the one discussed by Garcia. By how much would you increase your estimate for height? (Be careful of the units you use to report your answer.)

3. Anthropologists have refined early models for estimating a person's height from the length of a footprint. One revision suggests different models depending on whether a footprint is from the right foot or from the left. Here are two models designed for use with adult footprints.

Practice Problems

Answers

1. a) $y - 3 = \frac{1}{2}(x - 2)$

   An equivalent equation is $y = \frac{1}{2}x + 2$.

   b) $y = -\frac{1}{2}x + 3$

   c) The slope is $(5 - 2)/(2 - 3) = -3$. Using the first point, $y - 2 = -3(x - 3)$, which is equivalent to $y = -3x + 11$.

   d) $y = 7x + 5$

2. a) $H = L/0.15$

   b) $m = 1/0.15 \approx 6.67$

   c) $4.5 = L/0.15$; $L = 0.675$ ft, approximately 8 inches, or approximately 20.6 cm.

   d) If L is increased by 1 cm, H will increase by about 6.67 cm.

For right foot: $H = 3.641L + 72.92$
For left foot: $H = 4.229L + 56.49$
(where all measurements are in cm)

a) Suppose a footprint measures 22 cm. Predict the person's height, first by assuming the print is from the right foot and then by assuming it is from the left foot.

b) Use your calculator to graph each function. Based on your graphs, what footprint length gives the same height prediction from both models? How can you determine this value using algebra? Try it and check to see if you get the same result.

c) Both the right-foot and left-foot models are from the $y = mx + b$ family. For each of the models, interpret the meaning of m in the height-footprint context. What does b mean in this context?

4. a) Solve the right-foot equation in Question 3 for L in terms of H. Then do the same for the left-foot equation.

b) Suppose a person is 153 cm tall. Use your equations from a to predict the lengths of the person's right and left footprints. How much longer is the larger foot than the shorter foot?

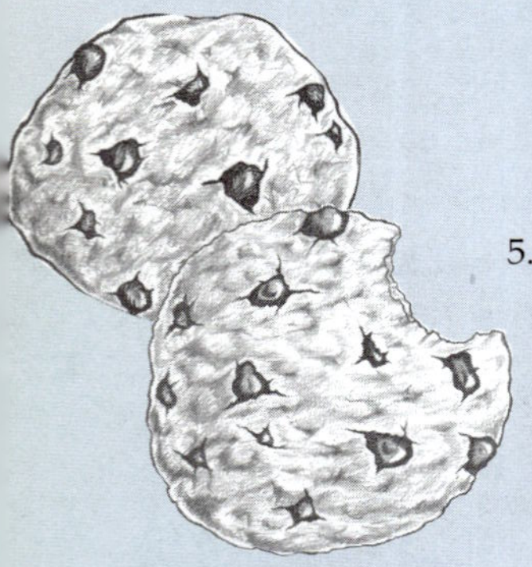

5. Regular Chips Ahoy® chocolate chip cookies boast "1000 chips in every bag." You can also buy reduced fat Chips Ahoy cookies. Do you think both types of Chips Ahoy cookies contain roughly the same number of chips? To find out, a statistics class opened bags of regular Chips Ahoy and reduced fat Chips Ahoy cookies, randomly selected 15 cookies from each bag, and counted the number of chips in each cookie. Their data appear in **Figure 2.35**.

Reduced Fat	13	15	14	12	15	17	13	10	15	18	19	18	20	21	16
Regular	18	20	17	22	20	16	21	18	16	19	27	22	19	16	24

FIGURE 2.35. Chip counts in reduced fat and regular Chips Ahoy cookies.

a) From the data in Figure 2.35, make two dot plots using the same scaling on each. Place one dot plot directly above the other.

b) Suppose you select a cookie from the regular Chips Ahoy bag. Predict the number of chips in the cookie. How far off might your prediction be?

c) Suppose instead, you select a cookie from the reduced fat bag. Predict the number of chips in the cookie. How far off might your prediction be?

d) Does it appear that the number of chips is changed in order to produce a lower fat product? Explain.

e) George wants to make a scatter plot of the row 1 data in Figure 2.35 versus the row 2 data. Do you think his scatter plot would reveal useful information about these two types of cookies? Explain.

6. A newspaper article in the *Worcester Sunday Telegram* reported that scores on intelligence tests are going up at a rate of 3 IQ points per decade.

a) In 1932 the average IQ test score for Americans was 100 points. Use the information given in the article to write a function rule that predicts the average IQ score for years after 1932. Let $x = 0$ represent 1932 to simplify your model.

b) **Figure 2.36** provides data on the average IQ scores of Americans. Make a scatter plot of these data. (Let $x = 0$ represent 1932.)

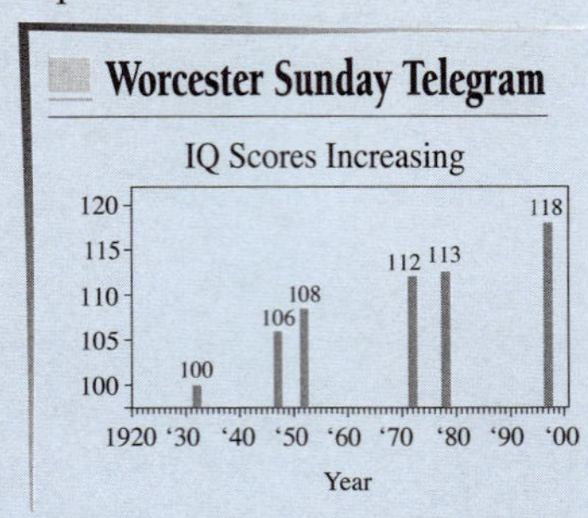

FIGURE 2.36. Average IQ scores from 1932 to 1997.

c) Based on these data, do IQ scores and years have a positive or negative relationship? Explain.

d) Fit a least-squares line to these data. What is the equation for this line? Add this line to your scatter plot in b.

e) Based on the least-squares equation from d, was the newspaper correct in reporting that IQ scores are rising at a rate of 3 points per decade? Explain.

Answers (continued)

3. a) Height prediction from right foot: 153.022 cm or about 153 cm.

   Height prediction from left foot: 149.528 or about 150 cm.

   b) The two graphs intersect at approximately (27.9, 174.7). If the footprint is approximately 27.9 cm, both models will predict heights of approximately 174.7 cm. You could determine this value algebraically by setting the right sides of the two equations equal to each other, $3.641L + 72.92 = 4.229L + 56.49$, and then solving for L:

   $(4.229 - 3.641)L = 72.92 - 56.49$
   $0.588L = 16.43$
   $L \approx 27.94$

   c) Right-foot model: If you change the length of footprint by 1 cm, then predicted height will change by 3.641 cm.

   Left-foot model: If you change the length of footprint by 1 cm, then predicted height will change by 4.229 cm.

   The value of b gives the intercept of the vertical axis. From the right-foot model, you would interpret the value $b = 72.92$ to be the height of a person whose footprint measures 0 cm. This is nonsense! The value of b has no meaning in this context.

4. a) Right-foot equation: $L = (H - 72.92)/3.641$.

   Left-foot equation: $L = (H - 56.49)/4.229$.

   b) For right foot: $153 = 3.641L + 72.92$; $L \approx 21.99$ cm

   For left foot: $175 = 4.229L + 56.49$; $L = 22.82$ cm

   The difference in foot length is 0.83 cm or around 8 mm.

5. a)

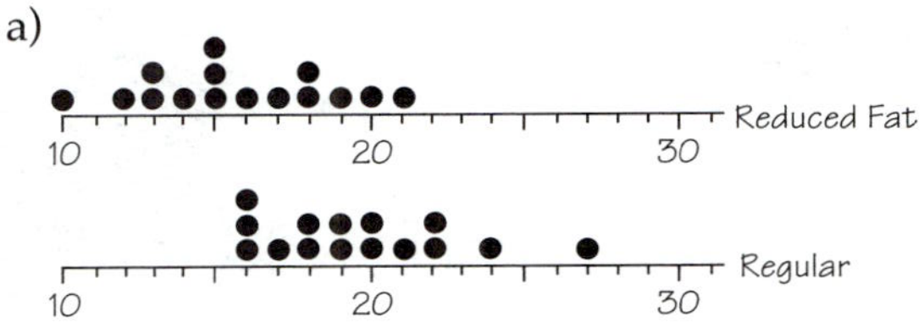

   b) The average is 19.7 chips. However, the actual count could be as low as 16 or as high as 27. So the prediction could be off by as many as 7 chips.

   c) The average number of chips is 15.7. The actual number of chips could be as low as 10 or as high as 21, so we could be off by as many as 6 chips.

   d) The dots corresponding to the reduced fat cookies appear shifted to the left from the dots corresponding to the regular cookies. In addition, the average number of chips for the reduced fat cookies is 4 chips lower than the average for the regular cookies. It appears that the number of chips is reduced in order to help reduce the fat content of the cookies.

   e) There is no natural pairing between the cookies listed in each row. The scatter plot is not appropriate in this situation.

6. a) The article reported that IQ scores were increasing by 3 points per decade, that's an average of 0.3 points per year. Let x be the years since 1932 and y be the average IQ score. Then $y = 0.3x + 100$.

   b) See answer to d.

   c) The two variables have a positive relationship. As the value for x increases, the IQ scores increase.

   d)

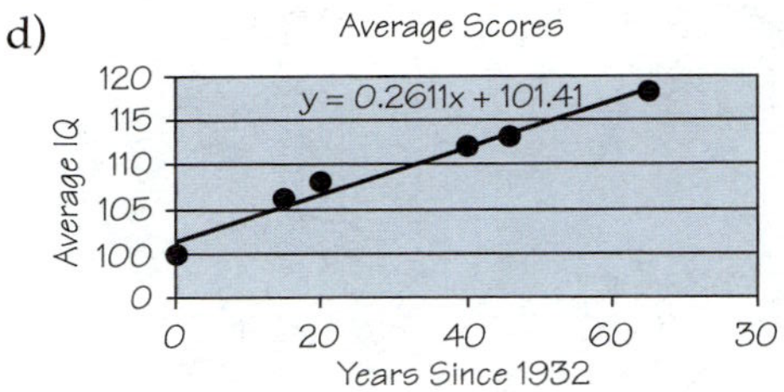

   e) Sample answer:

   The slope of the least-squares equation is approximately 0.26. That means that IQ scores are increasing by 0.26 points per year or by 2.6 points per decade. The newspaper reporter rounded up to 3 points.

f) A genius is a person with an IQ over 140. If the rate of increase continues, in what year will the average IQ be at the level of genius? Do you believe this? Explain.

The Boston Marathon is the world's oldest and best-known marathon. The first Boston Marathon was held in 1897. Fifteen men participated. Over 17,000 people entered the 104th Boston Marathon held April 17, 2000. Questions 7–8 are based on the first-place times in **Figure 2.37**.

Men's	
Marathon Number	Time (hr: min: sec)
31	2:40:22
36	2:33:36
41	2:33:20
46	2:26:51
51	2:25:39
56	2:31:53
61	2:20:5
66	2:23:48
71	2:15:45
76	2:15:30
81	2:14:46
86	2:8:51
91	2:11:50
96	2:8:14

Women's	
Marathon Number	Time (hr: min: sec)
82	2:34:28
83	2:26:46
84	2:29:33
85	2:22:42
86	2:29:28
87	2:34:6
88	2:24:55
89	2:25:21
90	2:24:30
91	2:24:33
92	2:25:24
93	2:24:18
94	2:23:43
95	2:25:27
96	2:21:45
97	2:25:11
98	2:27:13
99	2:58:0
100	2:40:10

FIGURE 2.37. Men's and women's Boston Marathon times.

7. The men's data in Figure 2.37 contain first-place times for selected marathons from 1927, marathon 31, to 1992, marathon 96.

   a) Make a scatter plot of the men's times versus the marathon number. (What unit did you use for time?)

   b) Fit a least-squares line to the data in your scatter plot. What is the equation? Add a graph of the least-squares line to your scatter plot.

   c) Interpret the slope of your linear model in the context of the Boston Marathon.

   d) Use your model to predict the time of the first-place finisher for the 104th Boston Marathon. The actual winning time was 2:09:47. How close is your prediction?

8. Figure 2.37 shows women's times for marathons 82–100, held in years 1980–1996.

   a) Fit a least-square's line to these data. Interpret the slope in the context of the marathon. Does this seem reasonable? Explain.

   b) Make a scatter plot of the women's data. Add the least-squares line to your plot. Does the least-squares equation do a good job of describing the women's data?

   c) Remove the times corresponding to the 99th and 100th marathons. Refit the least-squares line to the remaining data. What affect did the removal of these points have on the slope?

Answers (continued)

f) Solve $140 = 0.2611x + 101.41$; $x \approx 148$. If this model continues to hold, then the average IQ would be over 140 in the year 2080.

Sample answer:

This model is based on only six data points. There is no way to tell that the upward pattern will continue 83 years beyond the time when the last data point was recorded. I don't think that Americans will be that much smarter in 2080.

7. a) Sample answer: Times were converted to minutes.

See answer to b for the scatter plot.

b) Answer is based on time measured in minutes. Students' equations will differ if they measured time in seconds or hours.

Least-squares equation: $y = -0.465x + 171.7$, where x is the marathon number and y is the time in minutes.

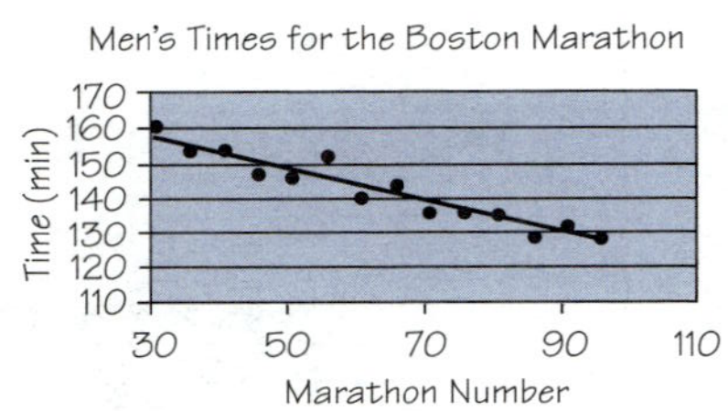

c) The slope is –0.465. Times are decreasing by 0.465 minutes per year or about 28 seconds per year.

d) Predicted time = $-0.465(104) + 171.7 = 123.34$ minutes. The winner of the 104th marathon finished in 2:09:47, which is approximately 129.78. The prediction was off by 6.44 minutes or 6 minutes 26 seconds.

8. a) Least-squares equation: $y = 0.389x + 113.42$. Times are increasing by about 0.389 minutes per year or about 23 seconds per year.

Sample answer: This does not seem reasonable. If you look at the data, times tended to go down, but in the last three races they went back up again.

b) The least-squares line does not do a good job in describing these data. In fact, the data do not appear to have a linear form. If times corresponding to the 99th and 100th race were removed the pattern would appear more linear.

c) The least-squares line is $y = -0.3783x + 180.49$. The slope is now negative, indicating that marathon times are decreasing.

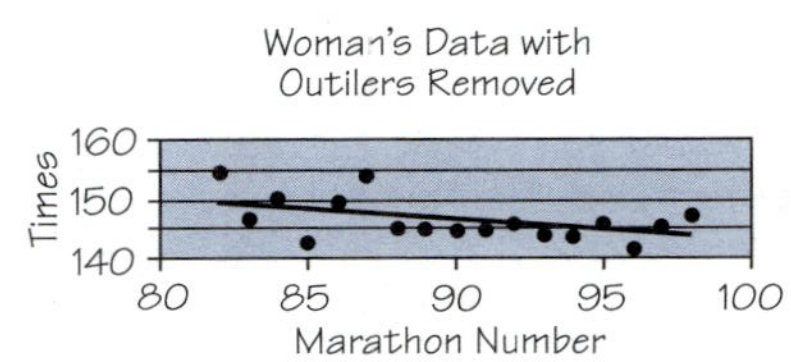

Men's		Women's	
Marathon Number	Time (hr: min: sec)	Marathon Number	Time (hr: min: sec)
82	2:40:10	82	3:48:51
83	2:26:57	83	3:52:53
84	2:38:59	84	3:27:56
85	1:55:0	85	2:49:4
86	2:0:41	86	2:38:41
87	1:51:31	87	2:12:43
88	1:47:10	88	2:27:7
89	2:5:20	89	2:26:51
90	1:45:34	90	2:5:26
91	1:43:25	91	2:9:28
92	1:55:42	92	2:19:55
93	1:43:19	93	2:10:44
94	1:36:4	94	1:50:6
95	1:29:53	95	1:43:17
96	1:30:44	96	1:42:42
97	1:26:28	97	1:36:52
98	1:22:17	98	1:34:50
99	1:25:59	99	1:40:41
100	1:30:11	100	1:52:54

FIGURE 2.38. Men's and women's wheelchair times in the Boston Marathon.

9. a) Make scatter plots for the men's and women's wheelchair times in **Figure 2.38**. So that you can compare the men's and women's data, use the same scale on the axes of both plots.

   b) Is there a positive or negative relationship between marathon number and times for the men's data? What about for the women's data? Interpret what this means in the context of the marathon.

   c) Comment on the form of the data. Does the pattern of the data appear linear or nonlinear? Explain.

The data in **Figure 2.39** provide information on the heights of people as children and again as adults. Use the data in this figure to answer Questions 10–13.

Girl's Height at 1.5	Girl's Adult Height	Boy's Height at 2	Boy's Adult Height
78.0	157.0	89.0	178.0
79.4	158.4	89.9	177.1
80.4	161.4	90.3	179.6
81.3	164.7	90.8	181.8
81.3	160.4	90.9	184.0
82.1	163.7	91.0	180.5
83.2	164.4	91.1	182.0
83.2	170.2	91.2	183.1
83.9	170.5	91.4	180.1
84.9	166.5	91.5	185.1
86.2	171.3	92.9	182.0
87.9	170.7	93.3	186.3
88.2	179.7	94.7	187.4
89.4	176.9	95.4	187.9
90.1	176.9	96.1	189.4

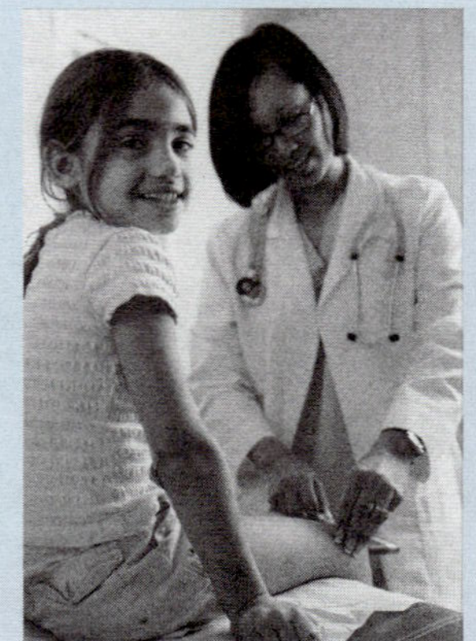

FIGURE 2.39. Height data on children and adults.

10. Suppose one of the people from this study plans to visit your school.

    a) If you find out the visitor is a woman, predict her height. How did you decide on your prediction?

    b) What if the visitor is a man?

11. Create a display that compares the men's heights with the women's heights. Write a description of the information your display conveys.

12. Suppose you want to predict how tall a 1 1/2-year-old girl will be when she reaches adulthood.

    a) Which is the independent variable and which is the dependent variable?

    b) Make a scatter plot of women's adult heights versus their heights when they were 1 1/2 years old.

Answers (continued)

9. a)

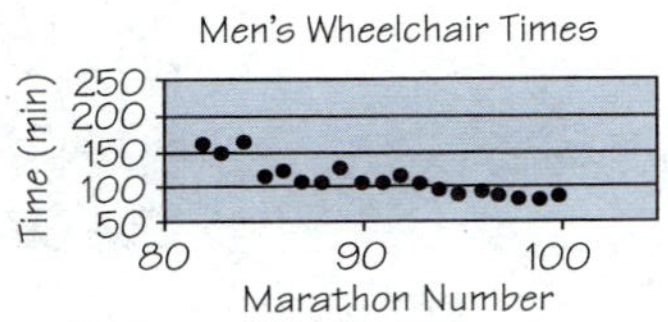

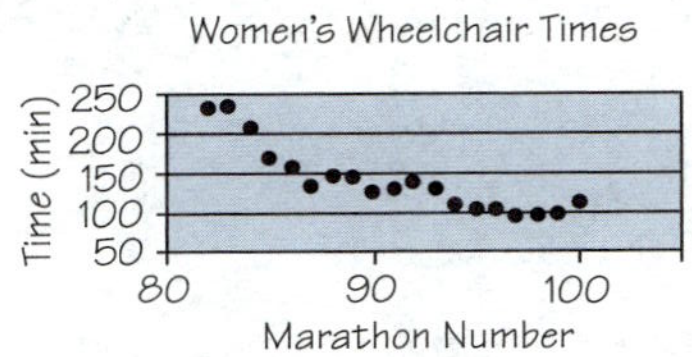

b) In both cases, the relationship between marathon number and times appears to be negative. That means the times tend to go down from race to race.

c) Sample answer: The pattern of the men's data appears fairly linear. The pattern for the women's data appears to bend upward.

10. a) Sample answer: The mean height for the women was approximately 167.5 cm. Since there do not appear to be any outliers in this data set, the mean appears to be a good prediction.

b) Sample answer: The mean height for men is approximately 183 cm.

11.

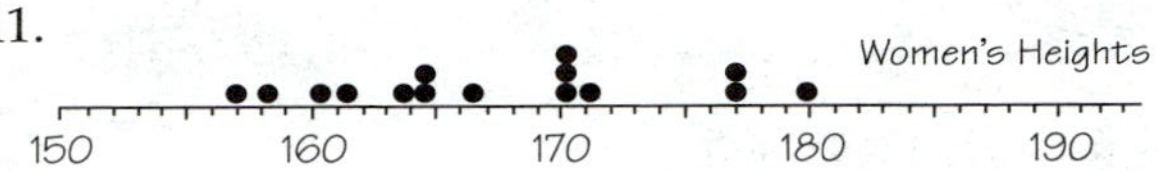

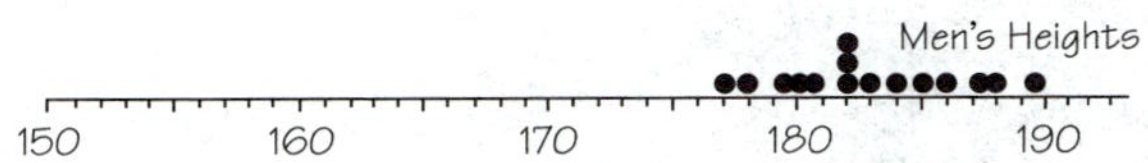

Although some women are taller than some men, it is clear from the comparative dot plots that women tend to be shorter than men. The women's heights ranged from around 157 cm to 180 cm, and the men's from around 177 cm to 189 cm. In fact, only three of the men were shorter than the tallest woman.

12. a) The explanatory variable is girl's height at $1\frac{1}{2}$ years old. The response variable is girl's adult height.

b) See answer in d.

c) Would you describe the relationship between women's heights and girls' heights as linear or nonlinear? Does this relationship appear to be positive or negative?

d) Fit a least-squares line to the data. Write its equation, and sketch its graph on your scatter plot.

e) Use your equation to predict the adult height of a 1 1/2-year-old girl who is 82.5 cm tall.

13. a) Determine the least-squares line for predicting men's heights from their heights when they were 2 years old.

b) If two 2-year-old boys differ in height by 1 cm, predict how much their heights will differ when they are adults.

c) What if their heights as 2-year-olds differ by 2 cm?

d) Does the y-intercept of the least-squares line have any meaning in this context? Explain.

e) Does the slope of the least-squares line have any meaning in this context? Explain.

14. The grades in **Figure 2.40** are from a high school statistics class.

a) What is the average grade on the midterm exam? On the final exam?

b) Make a scatter plot of these data.

c) Does the pattern in your scatter plot appear to have linear form? Does there appear to be a positive or negative relationship between students' midterm grades and students' final grades?

d) Fit a least-squares line to the data.

Name	Midterm exam grade	Final exam grade
	50	54
	53	52
	57	59
	62	70
	62	68
	68	70
	69	68
	73	79
	73	82
	74	83
	74	80
	75	82
	78	88
	79	89
	79	90
	81	88
	81	89
	81	92
	82	91
	85	94
	87	98
	100	99
	100	100

FIGURE 2.40.
Midterm and final exam grades.

e) Add a graph of your model from d to your scatter plot. Do you think your line does a good job in describing these data? Explain.

f) One student missed the final exam and had to take it at a later date. She got 77 on the midterm. Predict her grade on the final exam.

Answers (continued)

c) The pattern in the scatter plot appears to have a linear form. The points appear scattered about the line. The relationship between women's heights and girls' heights is positive. Taller $1\frac{1}{2}$-year-old girls tend to become taller women.

d) Women's height = 20.7 + 1.75(child's height).

Scatter plot of women's heights versus girls' heights with least-squares line is shown below.

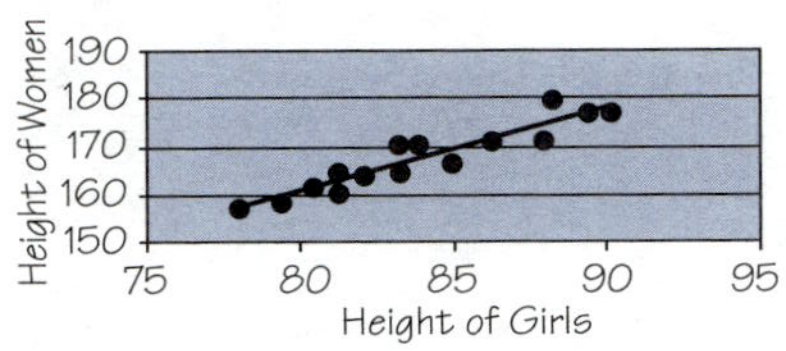

e) $20.7 + 1.75(82.5) \approx 165.1$ cm.

13. a) Man's height = 36.7 + 1.59(child's height).

b) Their heights will differ by approximately 1.59 cm.

c) Then their adult heights will differ by approximately 3.18 cm.

d) No. The y-intercept says that a 2-year old who is 0-cm tall will grow up to be 36.7 cm tall. However, since a 2-year old will never be 0-cm tall, this prediction is nonsense.

e) Yes. The slope tells you the rate of change between adults' heights per one-unit change in 2-year olds' heights.

14. a) The average grade on the midterm is approximately 73 and on the final is approximately 79.

b)

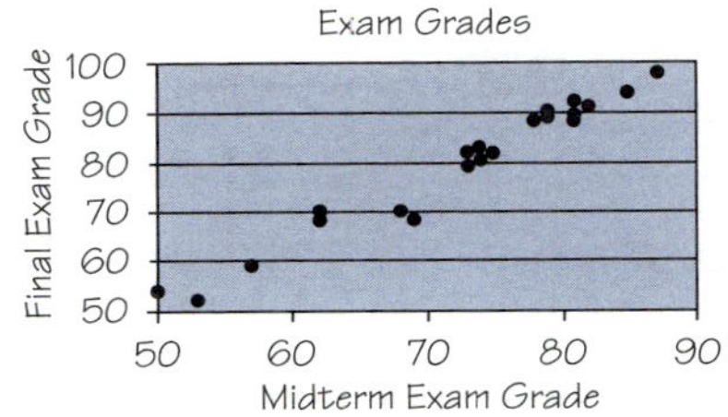

c) The data appears to have linear form. There is a positive relationship between students' midterm grades and their final grades.

d)

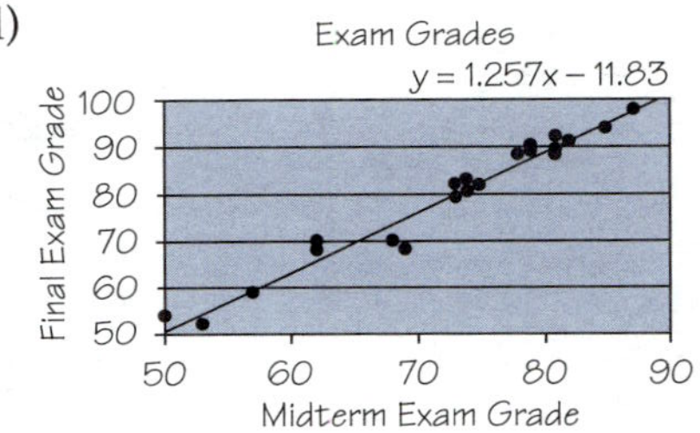

e) See answer to d. The linear model appears to fit the data well.

f) Predicted grade: $1.257(77) - 11.83 \approx 85$.

Glossary

KEY CONCEPTS

Average: To find the average of a data set, sum the data and divide by the number of data in the set.

Dot plot: Display in which dots are placed above a number line to represent the values of a single variable. (Also called line plot.)

Least-squares line: The line with the smallest sum of squared residual errors.

Linear equation: An equation relating two variables, x and y, that can be put in the form $y = mx + b$.

Linear form: The form of a scatter plot for which it is possible to draw a line that describes the general trend of the data.

Linear regression: The process of fitting a line to data using the least-squares criterion.

Mean: To compute the mean, sum the data and divide by the number of data. Sometimes mean and average are used interchangeably.

Median: The middle value in a set of data.

Mode: The value in a set of data that appears most often.

Nonlinear form: The form of a scatter plot for which the general pattern of the data is not well described by a straight line.

Outlier: In a collection of data, an individual data point that falls outside the general pattern of the other data.

Residual error: Actual value of the dependent variable minus the predicted value.

Strong linear relationship: A scatter plot of the data that lies in a narrow band.

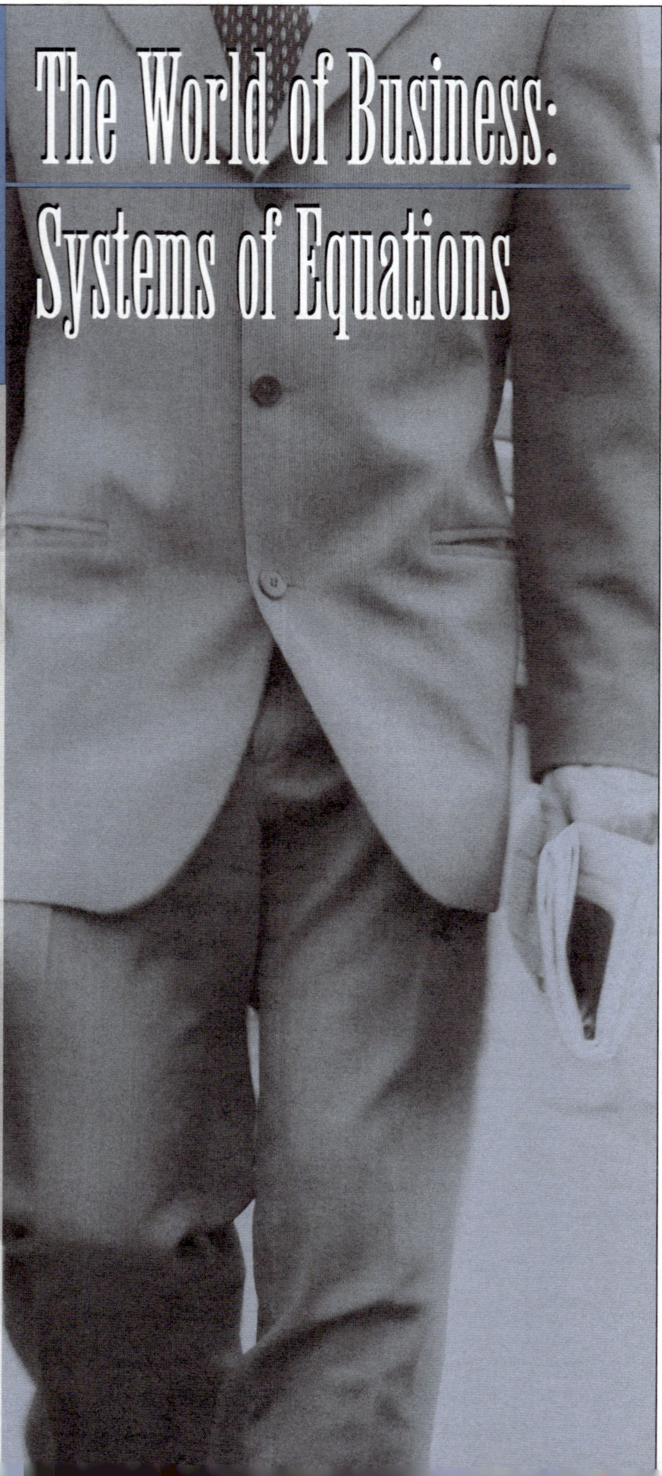

3 CHAPTER

The World of Business: Systems of Equations

PREPARATION READING

Running a Business

Four years ago, Warren Brown left his job as a lawyer to bake cakes. It all started when he decided to become an expert baker. After work, he tried his hand cake making. Then he had dessert parties for his friends. His friends loved his cakes. This made him want to see if he could support himself as a baker. He rented a small kitchen and found customers mostly by word of mouth. His business went so well that he moved into a storefront that he called Cake Love. At first he used his credit cards to get the business started. Then he got a $125,000 loan from the Small Business Administration. Now, Cake Love sells around 40 cakes per day at about $55 each. Cake Love was often very crowded, so Warren opened the Love Cafe across the street. The café serves sandwiches and soups as well as cake.

Founder/Owner: Warren Brown
Company: Cake Love and Love Café
Location: Washington, D.C.
Industry: Retail baked goods

Warren plans to open more stores this year. Even though managing takes most of his time, he still bakes most mornings. He also works on new types of cakes. His business is now so successful that he has the same salary he used to make as a lawyer.

People start their own businesses for lots of reasons. They know that if they succeed the rewards will be great. Also this gives them a way to make more money than if they worked for someone else, and they can be the person in charge.

Of course there is no guarantee that all businesses will be successful. In fact, for every successful business, there are many that fail. Often a business fails not because it makes a poor product but because of poor financial decisions.

In this chapter you will examine a start-up business that makes and sells athletic wear. You will have the opportunity to create mathematical models that can be used to make good financial decisions.

CHAPTER 3

The World of Business: Systems of Equations

Preparation Reading:
Running a Business

Description: The preparation reading introduces the major context of the chapter, running a business and making financial decisions based on mathematical models. Students read about Warren Brown, a successful businessperson. They should realize that for a business to be successful a businessperson must make several decisions. These decisions set the stage for the financial modeling that follows in this chapter.

Materials: none

Teaching Ideas: Students should work in small groups (3 to 4 students). To help ensure that all students read the preparation reading, have them read silently, then randomly select one student in each group to explain the passage to the other group members. After the reading, students should want to know more about running a business and how to make financial decisions.

SECTION 3.1 Do You Have Issues?

In this section you will assume the role of a businessperson like Warren Brown. Imagine that last year you started making baseball caps in your house. You gave them to a few close friends as birthday gifts. Your friends liked the caps and offered to pay you to make caps for a few of their friends. Soon you were getting hundreds of requests from people wanting to buy caps. Seeing this response as a great opportunity, you decide to start a business. There is not enough room in your house or time for you to make all these caps on your own. What issues, both practical and financial, do you need to consider before you start your business and agree to produce the caps?

Talk with your group members and make a list of issues that need to be considered before starting the business.

SECTION 3.2 Comparing Pay Structures

One important issue a businessperson must consider is employee pay. After doing some research you find that there are two companies in your city that make caps. Hats, Inc., pays its employees $1000.00 a month plus $5.00 for every hat they complete in the month. Caps, Inc., pays its employees $500.00 a month plus $10.00 for every hat they complete in the month.

In the next activity you will compare these pay structures and see if one will fit your company.

1. Use the description of the pay structure from Hats, Inc., to complete a table like **Figure 3.1.** Then use the process column to develop an algebraic rule for this pay structure.

Number of Caps Completed During the Month (c)	Process Column	Pay (p)

FIGURE 3.1. Hats, Inc., pay structure.

2. What algebraic rule models the pay structure for Hats, Inc.?

3. Enter your rule into your graphing calculator. Set the table to begin at 0 and increase by 10. Compare the table generated by the calculator to the table you created for Hats, Inc. Do the tables have the same values? What does this tell you about your algebraic rule?

   If the tables do not have the same values, adjust them to match.

ENGAGE LESSON 1

Section 3.1: Do You Have Issues?

Purpose: In previous chapters students explored functional relationships and used linear functions to model data. In Section 3.1 students will discuss issues that should be considered when starting a business. In the next phase of this lesson, students will explore the use of systems of linear equations as models to make a decision about one of these issues, employee salaries.

Description: Students assume the role of a businessperson starting a company that manufactures athletic apparel. They are asked to brainstorm issues that should be considered when starting the business. They discover that several decisions must be made and these decisions can make or break the company. One tool a businessperson can use to help make the best decisions is mathematical modeling. In the next section students will explore the use of systems of linear equations to make a decision about employee salaries.

Materials: none

Teaching Ideas: Students should work in small groups (3 to 4 students). Give students a limited amount of time to brainstorm issues that should be considered when starting the business. At the end of this time, debrief the students by having each group share their list of issues. Be sure students realize that mathematical modeling is an important tool for a businessperson to use when making financial decisions regarding these issues.

Sample answers: How much to charge per cap? How much should each employee be paid? Should employees be paid by the hour? If so, what hourly rate? Should employees receive a fixed amount plus an amount for each cap or each part of a cap constructed? How many caps need to be sold to break even? When does the company start to see a profit?

NOTE: If no group brings up employee salaries, be sure to bring up the issue of salaries for discussion. Comparison of employee pay structures is the basis for the exploration in the next phase of this lesson.

EXPLORE LESSON 1

Section 3.2: Comparing Pay Structures

Purpose: In Section 3.1 students identified several issues that a businessperson must consider when starting a business. In Section 3.2 students will model two different employee pay scenarios and use a table and then a graph to compare them. Students will then be ready to solve systems of equations algebraically, using linear combination and substitution in Section 3.3.

Description: Again assuming the role of the businessperson, students explore different models for employee pay. They will create tables and graphs representing two different pay structures. They will interpret the tables and graphs in order to make an informed decision about the issue of employee pay.

Materials: graphing calculator

Teaching Ideas: Students should work in small groups (3 to 4 students). This section is designed for students to make connections among the paper tables, the calculator tables, and the graph of a system of equations. As students are working through this activity, circulate to be sure they are making connections among the different representations. Also monitor calculator use and assist as necessary.

Answers

1. Sample answers in the table.

Number of Caps Completed During the Month (c)	Process Column	Pay (p)
0	1000 + 5(0)	1000
10	1000 + 5(10)	1050
20	1000 + 5(20)	1100
30	1000 + 5(30)	1150
40	1000 + 5(40)	1200
50	1000 + 5(50)	1250
100	1000 + 5(100)	1500
150	1000 + 5(150)	1750
200	1000 + 5(200)	2000
c	1000 + 5(c)	

2. $1000 + 5(c) = p$

3. Students should realize that if the table on paper matches the table generated when entering the function rule into the calculator, then their function rule is a good model for this situation.

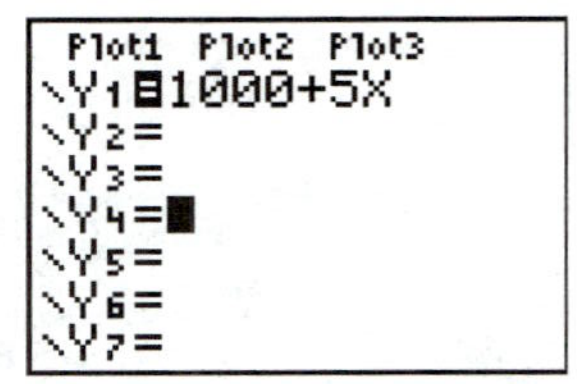

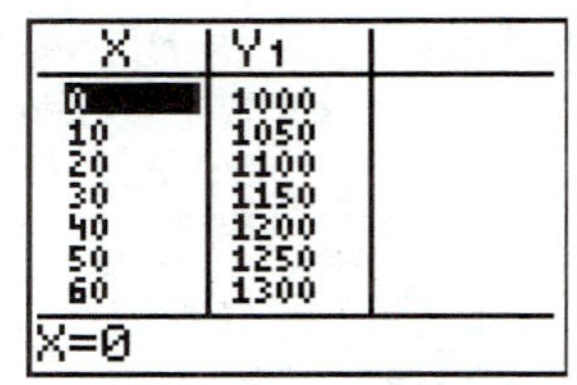

4. Use the description of the pay structure from Caps, Inc., to complete a table like **Figure 3.2.** Then use the process column to develop an algebraic rule for this pay structure.

FIGURE 3.2. Caps, Inc., pay structure.

Number of Caps Completed During the Month (c)	Process Column	Pay (p)

5. What algebraic rule models the pay structure for Caps, Inc.?

6. Enter your rule into your graphing calculator. Set the table to begin at 0 and increase by 10. Compare the table generated by the calculator to the table you created for Caps, Inc. Do the tables have the same values? What does this tell you about your algebraic rule?

7. Look at the table values for both companies. What patterns do you observe?

8. If an employee makes 97 hats in one month, how much will she earn under each pay structure?

9. If an employee makes 103 hats in one month, how much will she earn under each pay structure?

SOLVE THE SYSTEM OF EQUATIONS USING A TABLE

SYSTEM OF EQUATIONS

A *system of equations* is a set of two or more equations that we use at the same time. A solution to a system of equations is a set of points that are solutions to each of the individual equations. For example

$3x - 2y = -1$

$x + y = 3$

is a system of equations.

The pair (1, 2) is a solution to this system because $3(1) - 2(2) = -1$ and $1 + 2 = 3$ are *both* true.

10. Use the table feature on your calculator to solve the system of equations. In other words, for what number of hats will two employees, who are on different pay structures, make the same amount of money?

11. As a businessperson, you want to keep the cost of doing business as low as possible. So, which pay structure is best for you? Why?

12. If you were an employee, which pay structure would you prefer? Why?

13. If it takes somewhere between one and three hours for an employee to complete one hat, what is a reasonable domain for this situation? Why?

14. What is a reasonable range for this situation? Why?

15. Use your calculator. Find an appropriate viewing window then graph both functions on the same grid. Sketch your graph and record your window.

SOLVE THE SYSTEM OF EQUATIONS GRAPHICALLY

16. What are the coordinates of the point where the lines intersect? What does this point tell us?

17. If you sell hats for $20 each, which pay structure would you select? Why?

Answers (continued)

4. Sample answers in the table.

Number of Caps Completed During the Month (*c*)	Process Column	Pay (*p*)
0	500 + 10(0)	500
10	500 + 10(10)	600
20	500 + 10(20)	700
30	500 + 10(30)	800
40	500 + 10(40)	900
50	500 + 10(50)	1000
100	500 + 10(100)	1500
150	500 + 10(150)	2000
200	500 + 10(200)	2500
c	500 + 10(*c*)	

5. $500 + 10(c) = p$

6. Students should realize that if the table on paper matches the table generated when entering the function rule into the calculator, then their function rule is a good model for this situation.

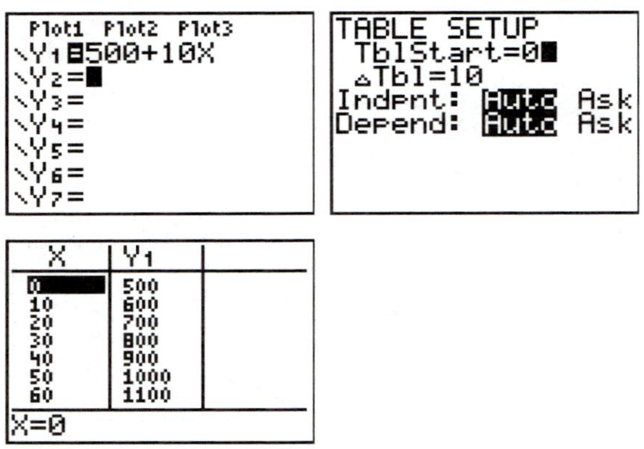

7. Answers may vary. Sample answer: If they make under 100 hats, employees at Caps, Inc. make less money per month, but if they make more than 100 hats in a month then they make more than employees at Hats, Inc.

8. At Hats, Inc., the employee would receive \$1485.00. At Caps, Inc., the employee would receive \$1470.00.

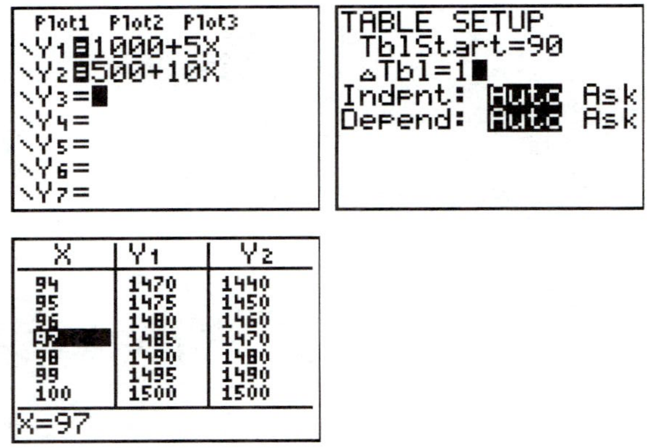

9. At Hats, Inc., the employee would receive \$1515.00. At Caps, Inc., the employee would receive \$1530.00.

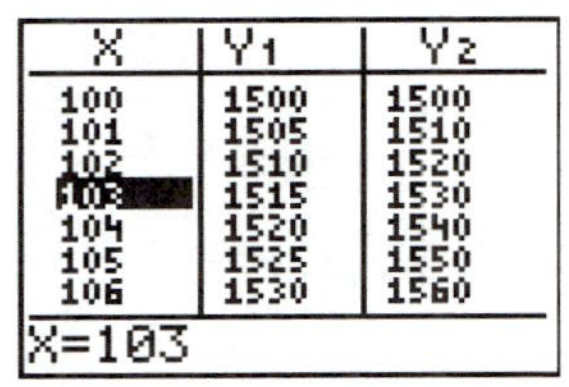

10. The table shows that the solution to this system of equations is (100, 1500). This ordered pair of numbers means that if an employee completes 100 hats in a month he will make \$1500 on either pay structure.

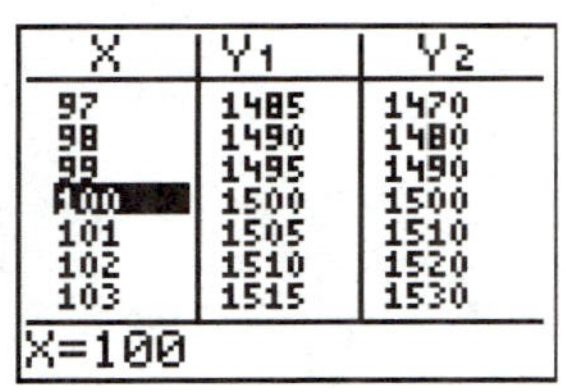

11. Answers may vary. Sample answer: It depends. If I expect my employees to make less than 100 hats each month, I would use Caps, Inc.'s structure. If I expect my employees to make more than 100 hats each month, I would use Hats, Inc.'s structure.

12. Answers may vary. Sample answer: I am a very fast worker. I like Caps, Inc. because I think I can make more than 100 hats in a month.

13. Answers may vary. Sample answer: The domain is the number of hats completed by an employee. A reasonable domain would be $0 < x < 160$. An employee who completes a hat every 3 hours will make about 53 hats in a month, and an employee who completes a hat every hour will make about 160 hats in a month. It is possible to complete no hats so zero is reasonable.

14. Answers may vary. Sample answer: The range is the amount of money each employee will be paid at the end of a month. A reasonable range would be $0 < x < 2100$. Under the two plans, the least an employee will make in one month is \$500.00. The most an employee can make, based on the information above, is \$2100. An employee who completes no hats may be terminated, so zero pay is reasonable.

15. Graphs may vary. A sample window and graph is shown.

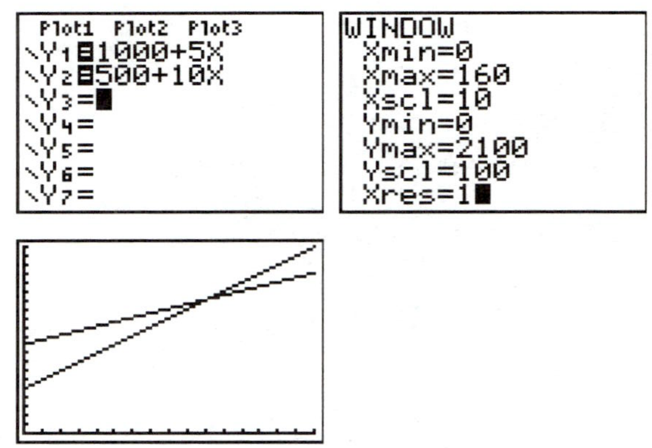

Answers (continued)

16. The lines intersect at the point (100, 1500). This tells us that under each pay structure, if an employee makes 100 hats, they will earn $1500.00.

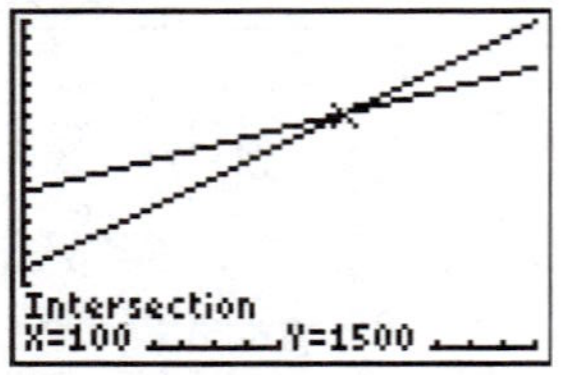

17. Selling hats at $20 each will take 50 hats to equal an employee's pay using the Caps, Inc. structure. It will take about 67 hats to equal an employee's pay using the Hats, Inc. structure. If most employees will make less that 100 hats per month, the Caps, Inc. structure would yield more profit. If most employees will make more than 100 hats per month, the Hats, Inc. structure will yield a greater profit.

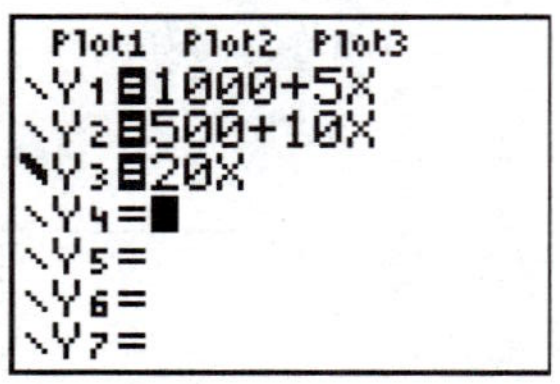

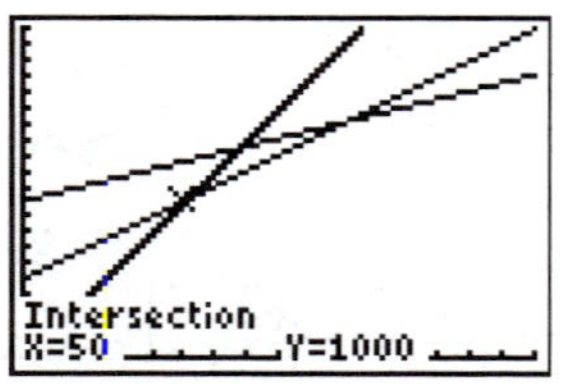

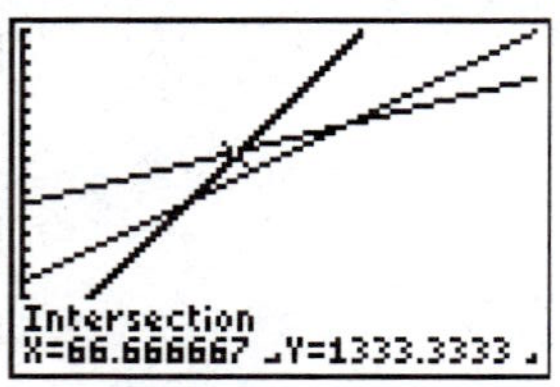

18. If you sell hats for $10 each, which pay structure would you choose? Why?

19. Your research also shows that the average time it takes an employee to finish one hat is two hours. If every employee works forty hours a week and a month has four complete weeks, which pay structure would you choose? Why?

SECTION 3.2 Assignment

Dolly Wagner plans to open a T-shirt shop in Galveston, Texas. She plans to sell each of her T-shirts for the same low price of $15.00. Use a table like **Figure 3.3** to find her income from T-shirt sales. Then use the process column to write an algebraic rule to represent her total income (y) in terms of the number of T-shirts sold (x).

Number of T-Shirts Sold (x)	Process Column	Total Income (y)
10		
20		
30		
50		
100		
x		

FIGURE 3.3. Income from T-shirt sales.

1. Write the algebraic rule that models the amount of income earned as a function of the number of T-shirts sold.
2. How much income will Dolly earn for selling 200 T-shirts?
3. How many T-shirts would she have to sell to earn $1050?

As a businesswoman, Dolly knows that it takes money to make money. She needs $525 to get her business started. She also knows that each T-shirt costs $8 to produce. Use a table like **Figure 3.4** to find the cost of producing the T-shirts. Then use the process column to write an algebraic rule to represent her total cost (y) in terms of the number of T-shirts produced (x).

Number of T-Shirts Produced	Process Column	Total Cost
10		
20		
30		
50		
100		
x		

FIGURE 3.4. Cost of producing T-shirts.

Answers (continued)

18. Selling hats at $10 each will take 200 hats to equal an employee's pay using the Hats, Inc. structure. Since the graph of the Caps, Inc. structure $500 + 10x$ and the graph of the sales function $10x$ each have a slope of 10, the lines will never intersect. Therefore, it will be impossible to equal an employee's pay using the Caps, Inc. structure. The Hats, Inc. structure should be selected since there will never be a profit using the Caps, Inc. pay structure.

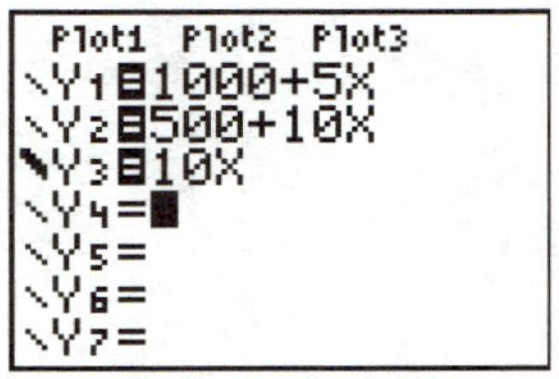

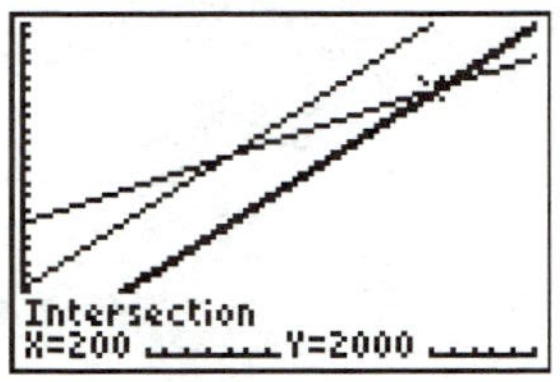

19. A mean time of 2 hours to complete 1 cap would indicate that the average employee should complete about 80 caps per month. Using the table for the Hats, Inc. structure, employee pay for 1 month will be $1400. For the Caps, Inc. structure, employee pay for 1 month will be $1300. Selecting the Caps, Inc. structure is the best decision based on our information.

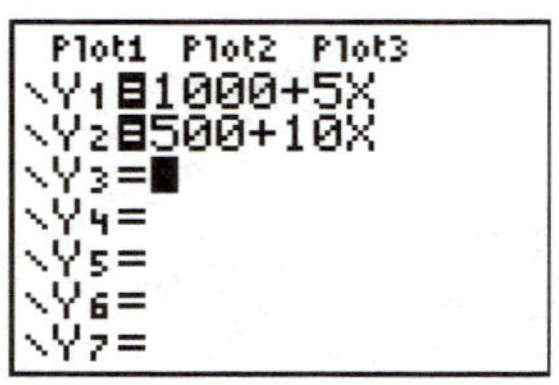

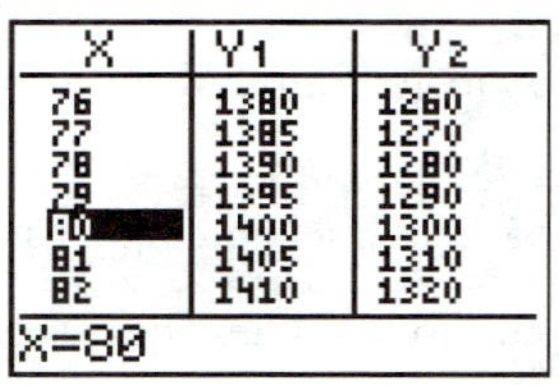

Section 3.2 Assignment

Answers

Number of T-Shirts Sold (x)	Process Column	Total Income (y)
10	15(10)	150
20	15(20)	300
30	15(30)	450
50	15(50)	750
100	15(100)	1500
x	15(x)	

1. $y = 15x$
2. $3000.00
3. 70 T-shirts

Number of T-Shirts Produced	Process Column	Total Cost
10	525 + 8(10)	605
20	525 + 8(20)	685
30	525 + 8(30)	765
50	525 + 8(50)	925
100	525 + 8(100)	1325
x	525 + 8(x)	

4. Write the algebraic rule that models the cost of producing T-shirts as a function of the number of T-shirts produced.

5. What is the total cost of producing 60 T-shirts?

6. Has Dolly made a profit yet if she has only produced 60 T-shirts and sold them all? Why or why not?

7. Use your graphing calculator to determine an appropriate viewing window, then graph both the income and the cost functions on the same grid. Sketch your graph and record your window.

CHECK THIS!

Remember here an ordered pair is made up of the variables x and y, or the input and output variables. We write the ordered pair as (x, y).

8. What are the coordinates of the point of intersection of the two lines? Write this point as an ordered pair.

9. Use words to describe the meaning of this point in terms of Dolly's business.

SECTION 3.3 Do the Combo!

In the last section you made decisions about employee pay structures by solving systems of equations using tables and graphs. Solving systems of equations is important because it helps make decisions when there are several things to think about.

Let's take another look at the pay structure equations for Hats, Inc., and Caps, Inc.

1. In the algebraic rule for Hats, Inc., $1000 + 5c = p$, what does 1000 represent? What does 5 represent?

2. In the algebraic rule for Caps, Inc., $500 + 10c = p$, what does 500 represent? What does 10 represent?

3. How did you use your table to solve (find the number of hats and the pay where both pay structures are the same) the system of equations $1000 + 5c = p$ *and* $500 + 10c = p$?

4. How did you decide on a reasonable domain and range for this situation?

5. How did you use your graph to solve (determine the number of hats and the pay where both pay structures are the same) the system of equations $1000 + 5c = p$ and $500 + 10c = p$?

There are two other methods to consider when solving systems of equations: linear combination and substitution.

LINEAR COMBINATION METHOD

A system of equations can be solved by the **linear combination** method. This method uses the following steps to eliminate or *cancel out* one of the variables by adding the equations together.

Write the equations one above the other.

$$500 + 10h = p$$
$$1000 + 5h = p$$

Answers (continued)

4. $y = 525 + 8x$
5. $1005.00
6. No, she has not made a profit. The cost to produce 60 T-shirts is $1005.00; however, her income would be only $900 from selling the 60 shirts.
7. 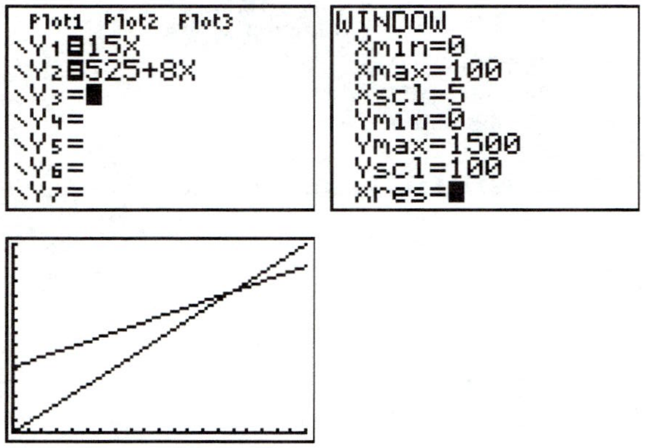

8. The intersection of the two lines is the ordered pair (75, 1125).

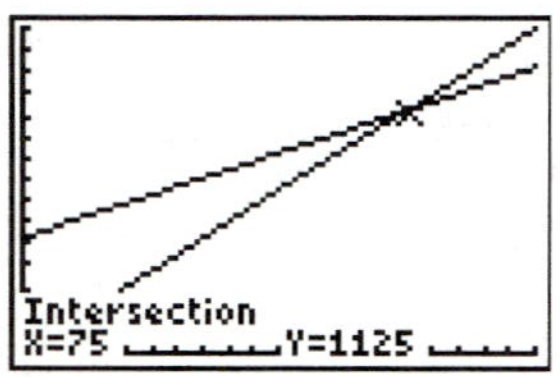

9. When Dolly sells 75 T-shirts, her cost and income are both $1125. This is her break-even point. If she sells less than 75 T-shirts, she will lose money. If she sells more than 75 T-shirts, she will make a profit.

EXPLAIN LESSON 1

Section 3.3: Do the Combo!

Purpose: In Section 3.2 students modeled two different employee pay scenarios and used tables and graphs to compare the pay structures. In general they solved a system of equations using tables and graphs. In Section 3.3 solving systems of equations using tables and graphs will be formalized. Students will also solve systems of equations algebraically using linear combination and substitution. Students will then be ready to use all of these methods as they make further business decisions in Section 3.4.

Description: This section is designed to be a teacher-led discussion of the concepts in the lesson. Use the questions to facilitate class discussion.

Materials: graphing calculators

Teaching Ideas: Debrief students after Section 3.2 by having them share their answers to the questions. As they share, be sure to fill in any missing concept, process, or skill that could have been used to answer the questions. Then work through the questions in this section with the entire class. Once again be sure all necessary concepts, processes, or skills are addressed.

Answers

1. 1000 represents the fixed amount of money ($1000.00) an employee receives each month. 5 represents the additional money ($5.00) an employee would receive for each completed hat.
2. 500 represents the fixed amount of money ($500.00) an employee receives each month. 10 represents the additional money ($10.00) an employee receives for each completed hat.
3. Answers may vary. Students should be able to explain what each column on the table represents and why the solution occurs where it does on the table.
4. Answers may vary. Students should indicate that they established the domain by making a reasonable estimate of the amount of hats one employee could complete in one month. They should have established the range by making a reasonable estimate of the total amount of money one employee could earn in one month.
5. Answers may vary. Students should have found the coordinates of the point of intersection of the graphs of the function rules.

Decide what to multiply by, so that when the equations are added one variable is eliminated.

$$500 + 10h = p$$
$$-2[1000 + 5h = p]$$

Multiply every term in the equation by that number.

$$500 + 10h = p$$
$$-2000 + (-10h) = -2p$$

Add the equations and solve for the remaining variable.

$$-1500 = -p$$
$$1500 = p$$

Substitute that value into one of the original equations and solve.

$$500 + 10h = p$$
$$500 + 10h = 1500$$
$$-500 + 500 + 10h = 1500 - 500$$
$$10h = 1000$$
$$\frac{10h}{10} = \frac{1000}{10}$$
$$h = 100$$

Use words to describe the meaning of your answer.

The solution to this system of equations is (100, 1500). This means if an employee completes 100 hats in a month, he or she will make $1500 on either pay structure.

SUBSTITUTION METHOD

A system of equations can be solved by the substitution method. This method uses the following steps to isolate one of the variables in one equation and then substitute into the other equation.

Isolate one variable. (Note: In the pay structure system of equations, p is already by itself on one side of the equal sign.)

$$500 + 10h = p$$

Substitute the expression that is equivalent to that variable into the other equation.

$$1000 + 5h = 500 + 10h$$

Solve the equation.

$$1000 + 5h = 500 + 10h$$
$$1000 + 5h - 5h = 500 + 10h - 5h$$
$$1000 = 500 + 5h$$
$$-500 + 1000 = 500 + 5h - 500$$
$$500 = 5h$$
$$\frac{500}{5} = \frac{5h}{5}$$
$$100 = h$$

Substitute that value into one of the original equations and solve.

$$1000 + 5h = p$$
$$1000 + 5(100) = p$$
$$1500 = p$$

Use words to describe the meaning of your answer.

The solution to this system of equations is (100, 1500). This means if an employee completes 100 hats in a month he or she will make $1500 on either pay structure.

After you have made decisions and hired workers, one of your workers comes to you with a question. She has a box containing 144 silver stars. Her job is to complete each hat by attaching 2 stars to every red hat and 6 stars to every green hat. She must complete 50 hats and use all of the stars. She wants to know how many of each color hat she will need to complete.

6. Write a system of equations to determine the solution to this problem.
7. Solve the system using a table.
8. Solve the system graphically.
9. Solve the system using linear combination.
10. Solve the system using substitution.

Answers (continued)

6. Work through this problem with the whole class. Be sure to define the variables before writing the equations.

   Let r = the number of red hats

   Let g = the number of green hats

   The number of red hats plus the number of green hats is 50. So one equation is $r + g = 50$.

   Each red hat gets 2 stars and each green hat gets 6 stars. Together they get a total of 144 stars. So the second equation is $2r + 6g = 144$.

7. First solve each equation for r in terms of g.

$$r + g = 50 \qquad r + g - g = 50 - g \qquad r = 50 - g$$

$$2r + 6g = 144 \qquad 2r + 6g - 6g = 144 - 6g \qquad 2r = 144 - 6g \qquad \frac{2r}{2} = \frac{144}{2} - \frac{6g}{2} \qquad r = 72 - 3g$$

   Enter each equation into a graphing calculator then use the table.

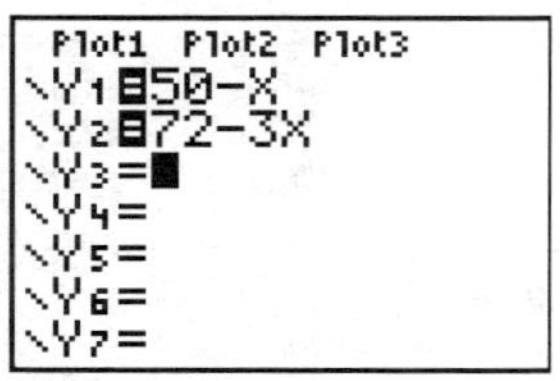

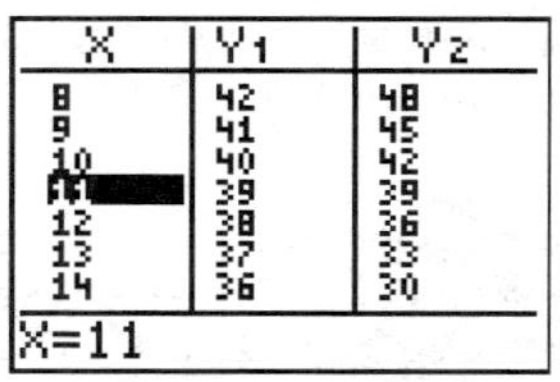

   When x is 11, y is 39 for both equations. The solution is (11, 39). Since x represents g and y represents r, she will need to complete 11 green hats and 39 red hats.

8. Enter each equation into a graphing calculator, set an appropriate window, graph the functions, then find the coordinates of the point of intersection.

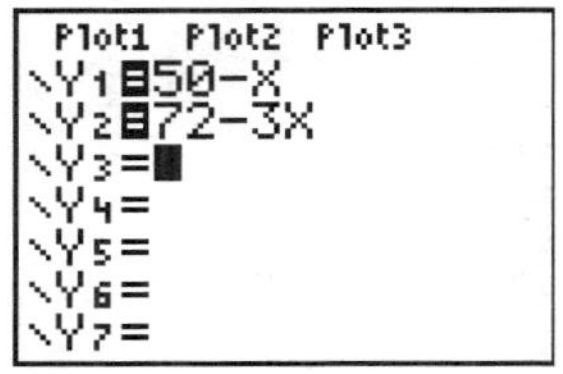

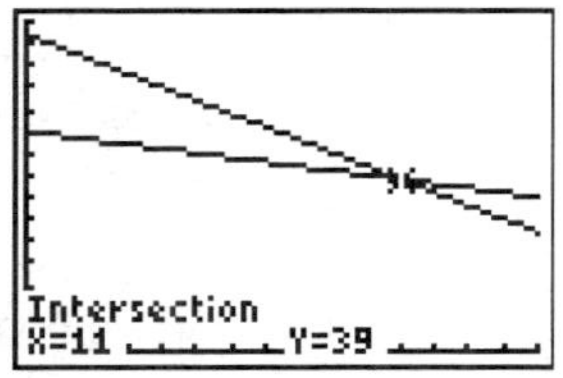

   When x is 11, y is 39 for both equations. The solution is (11, 39). Since x represents g and y represents r, she will need to complete 11 green hats and 39 red hats.

9. $\begin{aligned} r + g &= 50 \\ 2r + 6g &= 144 \end{aligned} \rightarrow \begin{aligned} -2[r + g &= 50] \\ 2r + 6g &= 144 \end{aligned} \rightarrow$

$$\begin{aligned} -2r + -2g &= -100 \\ 2r + 6g &= 144 \end{aligned}$$

$$4g = 44 \qquad \frac{4g}{4} = \frac{44}{4} \qquad g = 11 \rightarrow \quad r + g = 50 \qquad r + 11 = 50 \qquad r + 11 - 11 = 50 - 11 \qquad r = 39$$

   $g = 11$, and $r = 39$. The solution is (11, 39). She will need to complete 11 green hats and 39 red hats.

10. $r + g = 50 \qquad r + g - g = 50 - g \qquad r = 50 - g \rightarrow$

$$2r + 6g = 144 \qquad 2(50 - g) + 6g = 144 \qquad 100 - 2g + 6g = 144 \qquad 100 + 4g = 144 \qquad -100 + 100 + 4g = 144 - 100 \qquad 4g = 44 \qquad g = 11 \rightarrow$$

$$r + g = 50 \qquad r + 11 = 50 \qquad r + 11 - 11 = 50 - 11 \qquad r = 39$$

   $g = 11$, and $r = 39$. The solution is (11, 39). She will need to complete 11 green hats and 39 red hats.

SECTION 3.3

Assignment

For each of these problems, define the variables and write a system of two equations. Solve the system using a table, a graph, and an algebraic method (linear combination or substitution). Then use words to explain the meaning of your solution in the context of the problem.

1. Jed's market has a total of 26 employees. Each full-time employee makes \$275/week, and each part-time employee makes \$140/week. Jed pays a total of \$5935 in wages each week. How many full-time and part-time employees does he have?

2. Tony has \$100 in the bank and is spending it at a rate of \$3 each day. Kai has \$20 in the bank and is adding to it by saving \$5 each day. How many days will it take for the two boys to have the same amount of money?

3. A wildlife manager has rabbits and pheasants in a cage. All together there are 35 heads and 94 feet in the cage. Every rabbit has 4 feet and every pheasant has 2 feet. How many rabbits and how many pheasants are in the cage?

4. A carnival booth has small stuffed bears and large stuffed bears that it uses for prizes. Each small bear is worth \$2.50 and each large bear is worth \$5. If the booth contains a total of 200 bears with a total value of \$625, how many bears are there of each size?

5. An order of MP3 players and CD players totals \$4003 without any taxes or other charges. The cost of each MP3 player is \$194.50 and the cost of each CD player is \$159.50. The shipment contains a total of 24 players, how many of each is in the order?

SECTION 3.4

Buyer Beware!

Now that you've explored how to pay your employees, you have to decide where to buy the materials for your hats.

There are several options. First, you can buy the materials from Cloth City, a retail store in your town where you have been buying them up to this point. Since it is local, Cloth City must charge 8.25% sales tax. Another option is ordering your materials from Discount Fabrics, a wholesaler in another state. By doing so, you avoid paying sales tax. However, the wholesaler charges a processing fee of \$13 per order, and 5% of the cost of materials for shipping and handling.

You must decide which option is the most cost-effective way to purchase your materials.

1. Define your variables.

2. Write a system of equations in terms of x and y.

3. Solve the system using a table, a graph, and an algebraic method of your choice.

4. What is the intersection point, or solution to the system? What does it mean in this context?

5. Which option is the most cost-effective way to buy your materials?

6. While shopping on the Internet, you discover Textile City, a textile factory in another state that charges a \$37 processing fee, but only 1% of the cost of materials for shipping and handling. In light of this new option, where should you buy your materials?

   To answer this question, write an equation in terms of x and y for the Textile City offer.

7. Use a graph, a table, and the algebraic method of your choice to compare Cloth City with Textile City. Between these two, who has the better offer?

Section 3.3 Assignment

Answers

1. Let f = the number of full-time employees
   Let p = the number of part-time employees

$$f + p = 26$$
$$275f + 140p = 5935$$

```
Plot1 Plot2 Plot3
\Y1■26-X
\Y2■(5935/140)-(275/140)X
\Y3=■
\Y4=
\Y5=
\Y6=
```

X	Y1	Y2
13	13	16.857
14	12	14.893
15	11	12.929
16	10	10.964
17	9	9
18	8	7.0357
19	7	5.0714

X=17

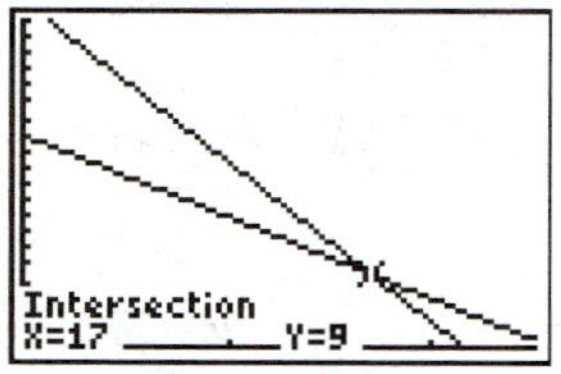

$$f + p = 26$$
$$-f + f + p = 26 - f$$
$$p = 26 - f$$

$$275f + 140p = 5935$$
$$275f + 140(26 - f) = 5935$$
$$275f + 3640 - 140f = 5935$$
$$135f + 3640 = 5935$$
$$135f = 2295$$
$$f = 17$$

$$f + p = 26$$
$$17 + p = 26$$
$$p = 9$$

The solution is (17, 9). There are 17 full-time employees and 9 part-time employees.

2. Let d = the number of days
   Let m = the amount of money after d days

Tony: $m = 100 - 3d$
Kai: $m = 20 + 5d$

```
Plot1 Plot2 Plot3
\Y1■100-3X
\Y2■20+5X
\Y3=■
\Y4=
\Y5=
\Y6=
\Y7=
```

X	Y1	Y2
7	79	55
8	76	60
9	73	65
10	70	70
11	67	75
12	64	80
13	61	85

X=10

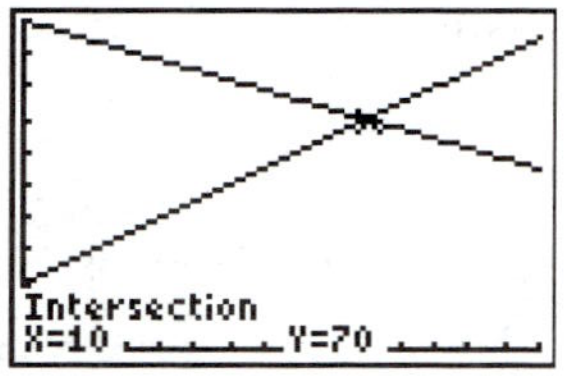

$$m = 100 - 3d$$
$$m = 20 + 5d$$

$$100 - 3d = 20 + 5d$$
$$80 = 8d$$
$$10 = d$$

$$m = 100 - 3(10)$$
$$m = 100 - 30$$
$$m = 70$$

The solution is (10, 70). On the 10th day both boys will have the same amount of money ($70.00).

3. Let r = the number of rabbits
   Let p = the number of pheasants

$$r + p = 35$$
$$4r + 2p = 94$$

```
Plot1 Plot2 Plot3
\Y1■35-X
\Y2■-2X+47
\Y3=■
\Y4=
\Y5=
\Y6=
\Y7=
```

X	Y1	Y2
9	26	29
10	25	27
11	24	25
12	23	23
13	22	21
14	21	19
15	20	17

X=12

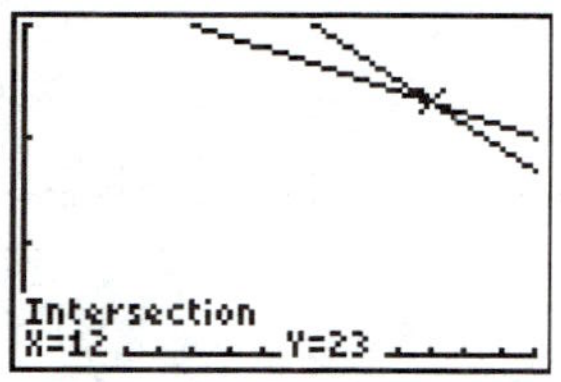

$$r + p = 35$$
$$p = 35 - r$$

$$4r + 2p = 94$$
$$4r + 2(35 - r) = 94$$
$$4r + 70 - 2r = 94$$
$$2r + 70 = 94$$
$$2r = 24$$
$$r = 12$$

$$r + p = 35$$
$$p = 35 - r$$
$$p = 35 - 12$$
$$p = 23$$

The solution is (12, 23). There are 12 rabbits and 23 pheasants in the cage.

4. Let s = the number of small bears
   Let l = the number of large bears

$$s + l = 200$$
$$2.50s + 5.00l = 625$$

```
Plot1 Plot2 Plot3
\Y1■200-X
\Y2■-.5X+125
\Y3=■
\Y4=
\Y5=
\Y6=
\Y7=
```

X	Y1	Y2
147	53	51.5
148	52	51
149	51	50.5
150	50	50
151	49	49.5
152	48	49
153	47	48.5

X=150

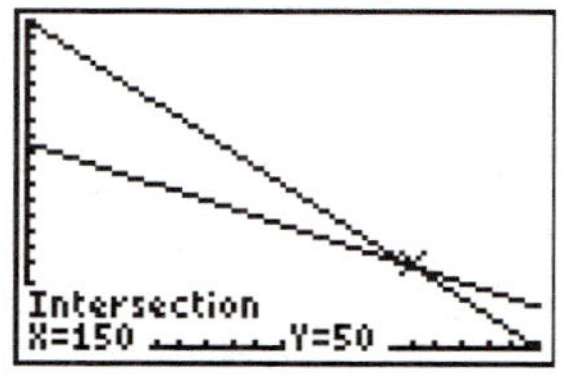

$$s + l = 200$$
$$l = 200 - s$$

$$2.50s + 5.00(200 - s) = 625$$
$$2.50s + 1000 - 5s = 625$$
$$-2.5s = -375$$
$$s = 150$$

$$s + l = 200$$
$$l = 200 - s$$
$$l = 200 - 150$$
$$l = 50$$

The solution is (150, 50). There are 150 small bears and 50 large bears.

Answers (continued)

5. Let m = the number of MP3 players
   Let c = the number of CD players

$$m + c = 24$$
$$194.50m + 159.50c = 4003.00$$

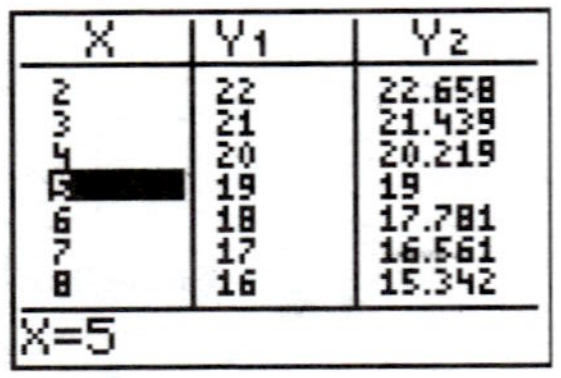

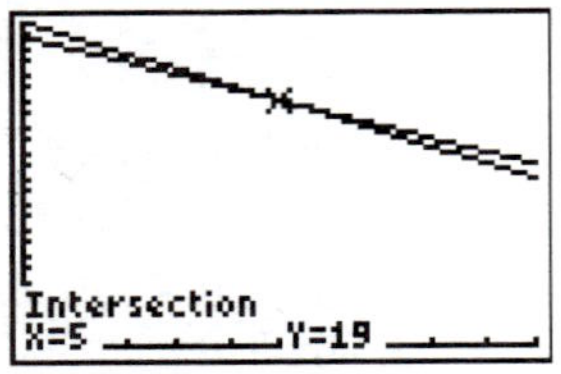

$m + c = 24$

$c = 24 - m$

$$194.50m + 159.50c = 4003.00$$
$$194.50m + 159.50(24 - m) = 4003.00$$
$$194.50m + 3828 - 159.50m = 4003.00$$
$$35m + 3828 = 4003.00$$
$$35m = 175$$
$$m = 5$$

$m + c = 24$

$c = 24 - m$

$c = 24 - 5$

$c = 19$

The solution is (5, 19). There are 5 MP3 players and 19 CD players

ELABORATE LESSON 1

Section 3.4: Buyer Beware!

Purpose: In Section 3.3 students solved systems of equations using tables, graphs, and algebraically using linear combination and substitution. In Section 3.4 students use these tools to make purchasing decisions for the company we created.

Description: This section is designed as an individual or group activity. The section and the assignment for Section 3.4 challenge students to solve systems of equations and make sure they make sense in real-world contexts.

Materials: graphing calculators

Teaching Ideas: Students may need to review the different methods for solving systems of equations. If they need this help, you may start the class with a short review of Section 3.3. This section can be done by students individually or in groups. If the section is done in groups you may choose to have each group present their work and their rationale for their business decisions.

Answers

1. Let x represent: subtotal cost of materials in dollars.

   Let y represent: total cost of materials, including tax or shipping, in dollars.

2. Cloth City: $y = x + 0.0825x$ or $y = 1.0825x$.

   Discount Fabrics: $y = x + 13 + 0.05x$ or $y = 13 + 1.05x$.

3. Solve using a table:

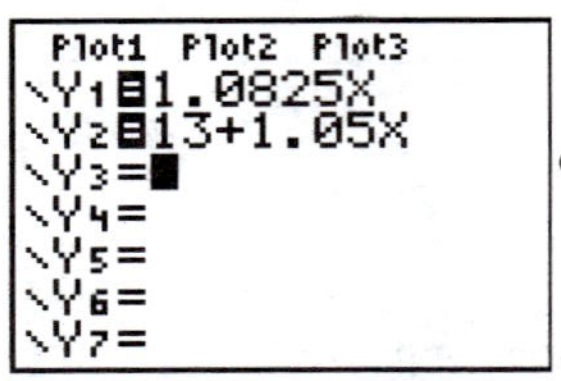

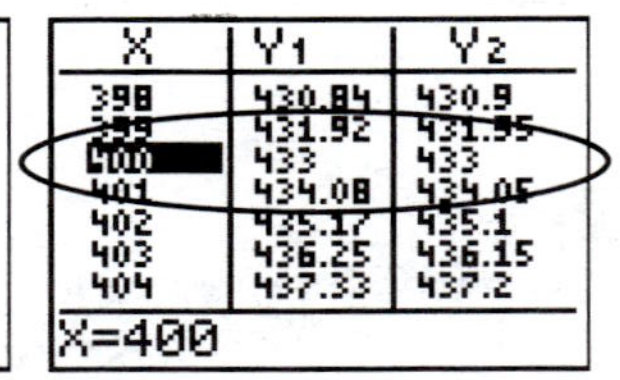

The solution is the point where the x-value yields the same y-values for both functions, which is the ordered pair (400, 433).

Solve using a graph:

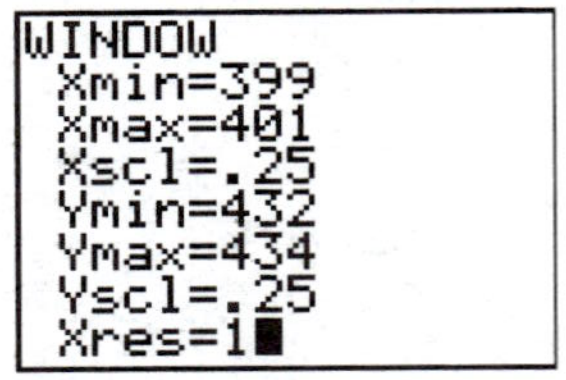

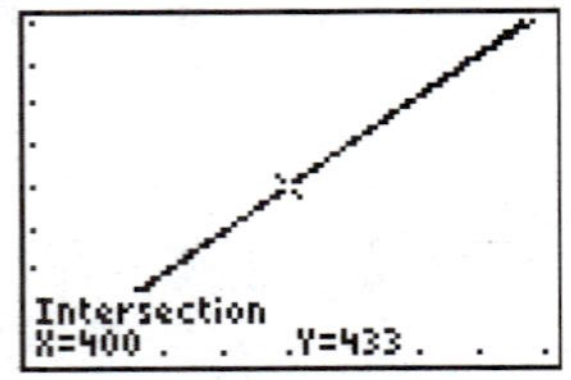

The solution is the intersection of the graphs of the two functions or the ordered pair (400, 433).

Solve using substitution:

$$\begin{cases} y = 1.0825x \\ y = 13 + 1.05x \end{cases}$$

$$1.0825x = y = 13 + 1.05x$$
$$1.0825x = 13 + 1.05x$$
$$1.0825x - 1.05x = 13 + 1.05x - 1.05x$$
$$0.0325x = 13$$
$$\frac{0.0325x}{0.0325} = \frac{13}{0.0325}$$
$$x = \$400$$

$$y = 1.0825x$$
$$y = 1.0825(\$400)$$
$$y = \$433$$

The solution is the ordered pair (x, y) or (400, 433).

4. The intersection of the two function rules is the ordered pair ($400, $433). This means that if you order $400 worth of material, both Cloth City and Discount Fabrics will charge you a total of $433 including either sales tax or shipping and handling, respectively.

5. If you plan to order less than $400 worth of materials, then you will save money using Cloth City and buying them locally. However, if you are ordering more than $400 worth of materials, then you will save money using Discount Fabrics and having the materials shipped to you.

6. $y = 37 + 1.01x$

Answers (continued)

7. 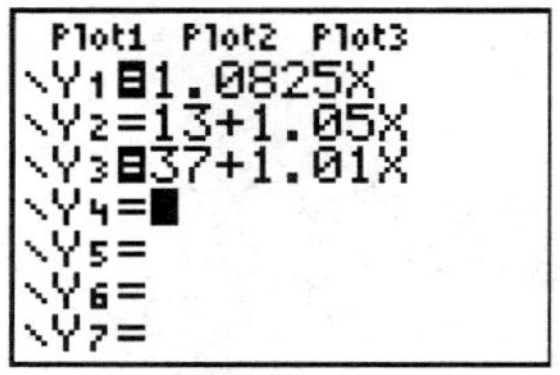

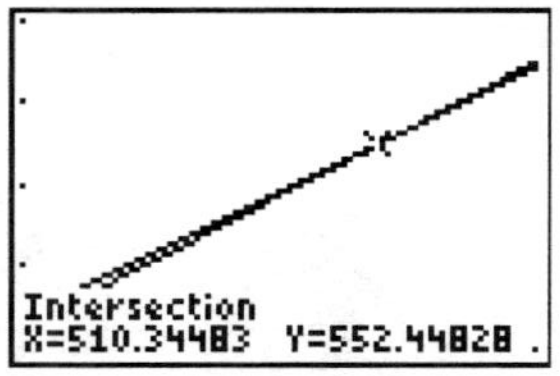

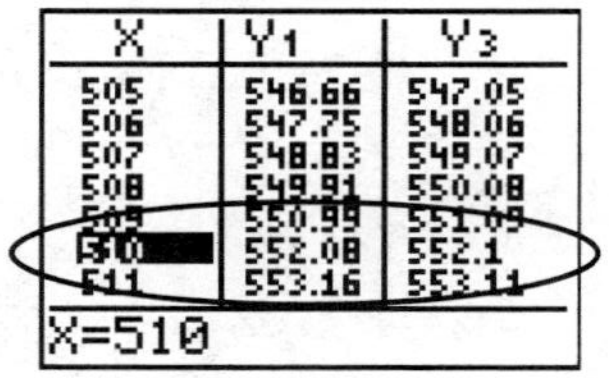

$$1.0825x = 37 + 1.01x$$
$$1.0825x - 1.01x = 37 + 1.01x - 1.01x$$
$$0.0725x = 37 \qquad y = 1.0825(510.34)$$
$$\frac{0.0725x}{0.0725} = \frac{37}{0.0725} \qquad y \approx \$552.45$$
$$x \approx \$510.34$$

If you plan to order less than \$510.34 worth of materials, then you will save money using Cloth City and buying them locally. However, if you are ordering more than \$510.34 worth of materials, then you will save money using Textile City and having the materials shipped to you.

8. Use a graph, table, and the algebraic method of your choice to compare Discount Fabrics with Textile City. Who has the better offer and why?
9. If you plan to order $300 worth of materials, who has the best offer?
10. If you plan to order $450 worth of materials, who has the best offer?
11. If you plan to order $550 worth of materials, who has the best offer?
12. If you plan to order $700 worth of materials, who has the best offer?
13. What are the intersection points where the companies with the best offer change?
14. Which of the three options is the most cost-effective way to purchase your materials?

SECTION 3.4 Assignment

For each of these situations, define variables and write a system of two equations. Solve the system using a table, a graph, and an algebraic method (linear combination or substitution). Then use words to explain the meaning of your solution in the context of the problem. Also state how the solution could be used to help decide which option to select in each situation.

1. The Zippy Photo Store charges $1.60 to develop a roll of 35mm film plus $0.10 for each print, while SuperPIX charges $1.20 for development plus $0.15 for each print. For how many prints is the cost the same?
2. Carlo is trying to decide between two used cars. The sports car costs $5000, but the insurance is $2300 per year. The SUV costs $8000, and the insurance is $800 per year. After how many years would the cost of owning either car be the same?
3. Rita is moving to a new apartment. She can rent a truck to move for $59.88 a day plus $0.49 a mile. Strong Men Movers would charge $81 a day plus $0.38 a mile. It will take one day to move. How many miles will make the costs of the two options (renting a truck or hiring Strong Men Movers) the same?
4. Flash Rent-a-Car rents a compact car for $24.95 per day plus $0.23 per mile. King Car rents the same car for $22.95 per day plus $0.31 per mile. For a one-day rental, how many miles would make the cost of renting the car from either company the same?

Answers (continued)

8. 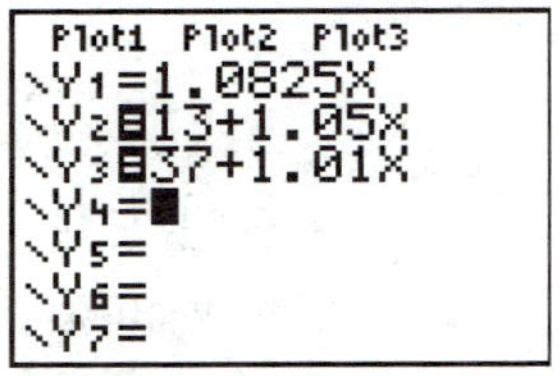

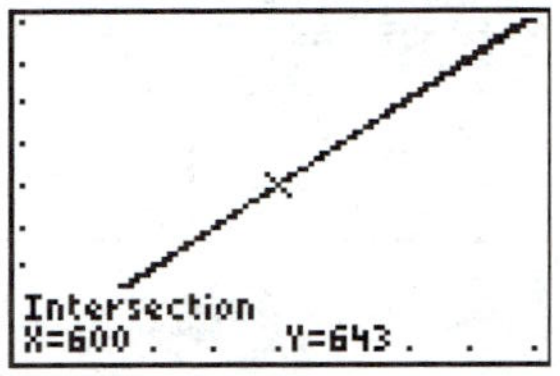

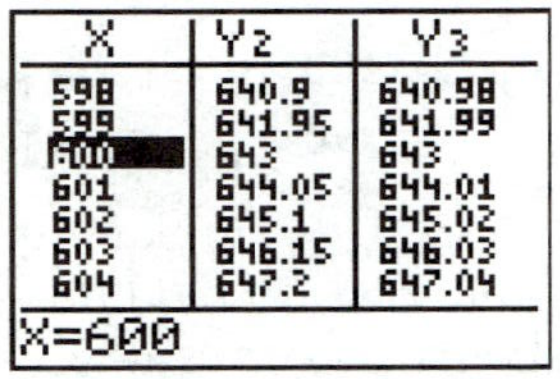

$$13 + 1.05x = 37 + 1.01x$$
$$13 + 1.05x - 1.01x = 37 + 1.01x - 1.01x$$
$$13 + 0.04x = 37$$
$$13 - 13 + 0.04x = 37 - 13 \qquad y = 13 + 1.05(600)$$
$$0.04x = 24 \qquad y = \$643$$
$$\frac{0.04x}{0.04} = \frac{24}{0.04}$$
$$x = \$600$$

If you plan to order less than $600 worth of materials, then you will save money using Discount Fabrics. However, if you are ordering more than $600 worth of materials, then you will save money using Textile City.

9. When ordering $300 worth of materials, Cloth City will charge $324.75, Discount Fabrics will charge $328, and Textile City will charge $340. Thus, Cloth City has the best offer.

10. When ordering $450 worth of materials, Cloth City will charge $487.13, Discount Fabrics will charge $485.50, and Textile City will charge $491.50. Thus, Discount Fabrics has the best offer.

11. When ordering $550 worth of materials, Cloth City will charge $595.38, Discount Fabrics will charge $590.50, and Textile City will charge $592.50. Thus, Discount Fabrics has the best offer.

12. When ordering $700 worth of materials, Cloth City will charge $757.75, Discount Fabrics will charge $748, and Textile City will charge $744. Thus, Textile City has the best offer.

13. At ($400, $433), the graphs of the functions representing Cloth City and Discount Fabrics intersect. At ($600, $643), the graphs of the functions representing Discount Fabrics and Textile City intersect. At ($510.34, $552.45), the graphs of the functions representing Cloth City and Textile City intersect. However, this intersection point is extraneous to the problem solution because between $400 worth of materials ordered and $600 worth of materials ordered, Discount Fabrics has the best offer.

14. A not-to-scale graph of the functions is shown for interpretation purposes.

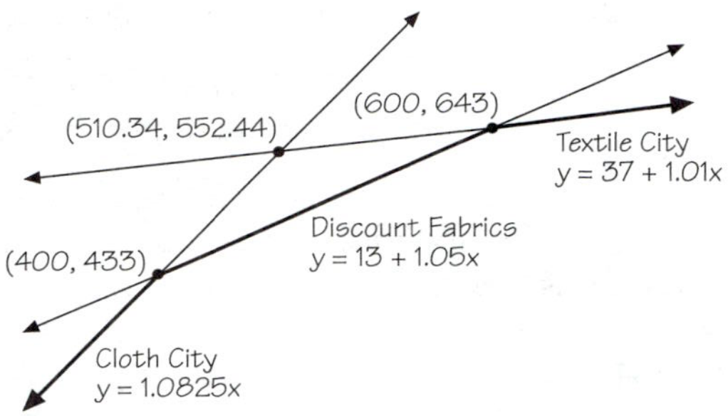

According to the graph, when $x < \$400$, Cloth City has the best offer. Thus, if you are ordering less than $400 worth of materials, you should order from Cloth City.

However, when $400 < x < 600$, Discount Fabrics has the best offer. Thus, if you are ordering between $400 and $600 worth of materials, you should order from Discount Fabrics. If you are ordering exactly $400 worth of materials, you can order from either Cloth City or Discount Fabrics since the cost is the same at each place.

Finally, when $x > 600$, Textile City has the best offer. Thus, if you are ordering more than $600 worth of materials, you should order from Textile City. If you are ordering exactly $600 worth of materials, you can order from either Textile City or Discount Fabrics since the cost is the same at each place.

Section 3.4 Assignment

Answers

1. Let p = the number of prints
   Let c = the cost of p prints

   Zippy Photo Store: $c = 1.60 + 0.10p$
   SuperPIX: $c = 1.20 + 0.15p$

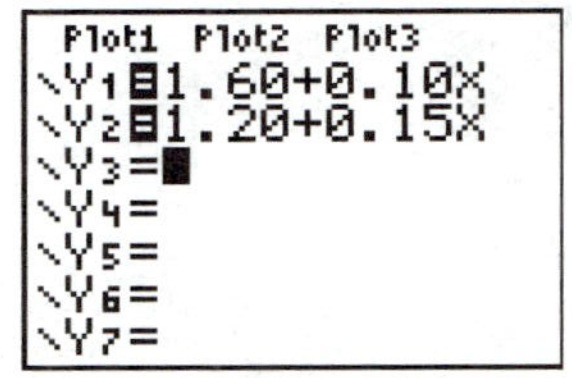

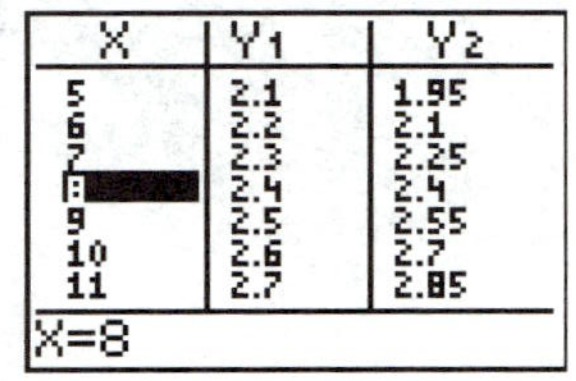

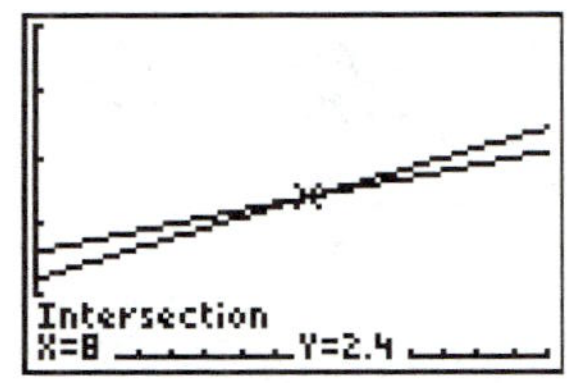

Sample solution using the linear combination method:

$$\begin{aligned} c &= 1.60 + 0.10p \\ c &= 1.20 + 0.15p \end{aligned} \rightarrow \begin{aligned} -1.5[c &= 1.60 + 0.10p] \\ c &= 1.20 + 0.15p \end{aligned} \rightarrow$$

$$\begin{aligned} -1.5c &= -2.4 + (-0.15)p \\ c &= 1.20 + 0.15p \\ \hline -0.5c &= -1.2 \\ c &= 2.4 \end{aligned} \rightarrow \begin{aligned} c &= 1.20 + 0.15p \\ 2.4 &= 1.20 + 0.15p \\ 1.2 &= 0.15p \\ 8 &= p \end{aligned}$$

The solution is (8, 2.40). This means that the cost to print 8 prints at either store is $2.40. If you have more than 8 pictures to print, using Zippy Photo Store will save you money.

Answers (continued)

2. Let y = the number of years
   Let c = the cost of owning the car for y years

   Sports car: $c = 5000 + 2300y$
   SUV: $c = 8000 + 800y$

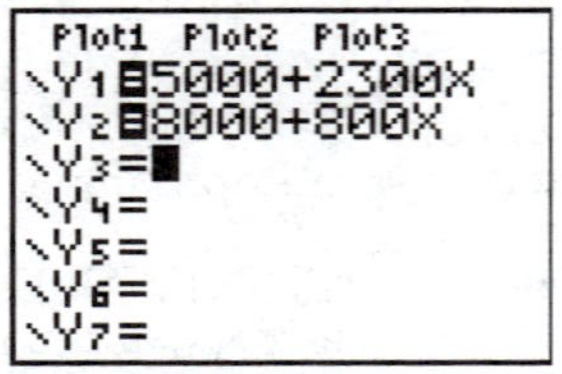

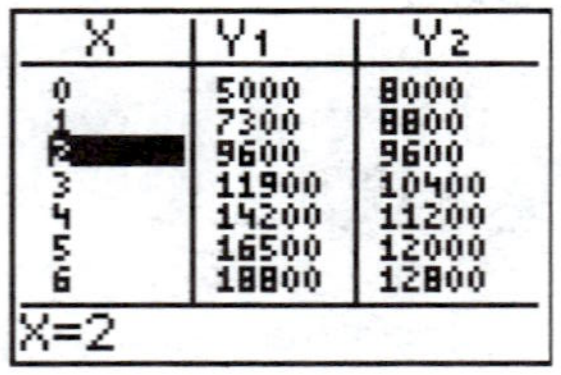

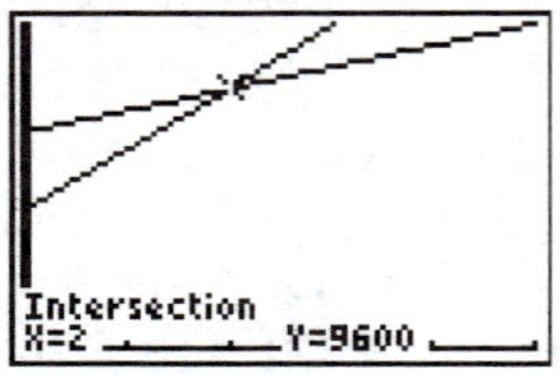

Sample solution using the substitution method:

$c = 5000 + 2300y$, $c = 8000 + 800y$ → $5000 + 2300y = 8000 + 800y$, $1500y = 3000$ → $y = 2$

$c = 5000 + 2300(2)$
$c = 5000 + 4600$
$c = 9600$

The solution is (2, 9600). This means that the cost of each car will be \$9600 after 2 years. If you plan to keep the vehicle more than 2 years, buying the SUV will save you money.

3. Let m = the number of miles
   Let c = moving cost for m miles

   Renting a truck: $c = 59.88 + 0.49m$
   Strong Men Movers: $c = 81 + 0.38m$

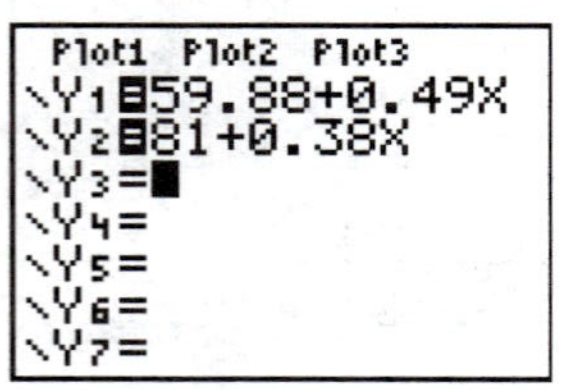

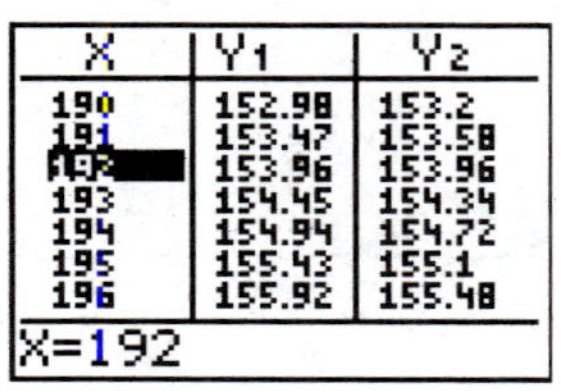

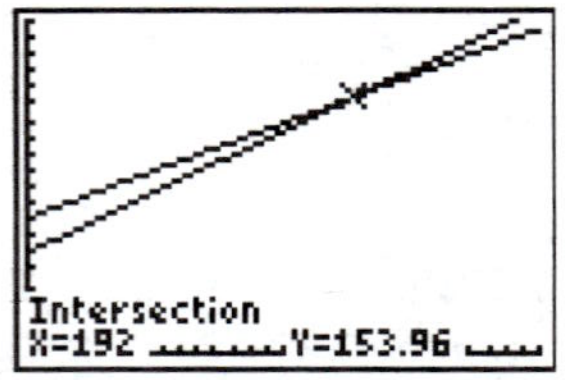

Sample solution using the substitution method:

$c = 59.88 + 0.49m$, $c = 81 + 0.38m$ → $59.88 + 0.49m = 81 + 0.38m$, $0.11m = 21.12$ → $m = 192$

$c = 59.88 + 0.49m$
$c = 59.88 + 0.49(192)$
$c = 59.88 + 94.08$
$c = 153.96$

The solution is (192, 153.96). This means that if the distance traveled is 192 miles, each option will cost \$153.96. If your move is more than 192 miles, using Strong Men Movers will save you money.

4. Let m = the number of miles
   Let c = rental cost for m miles

   Renting from Flash: $c = 24.95 + 0.23m$
   Renting from King Car: $c = 22.95 + 0.31m$

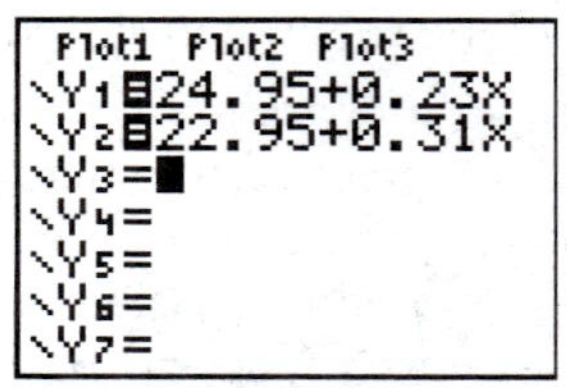

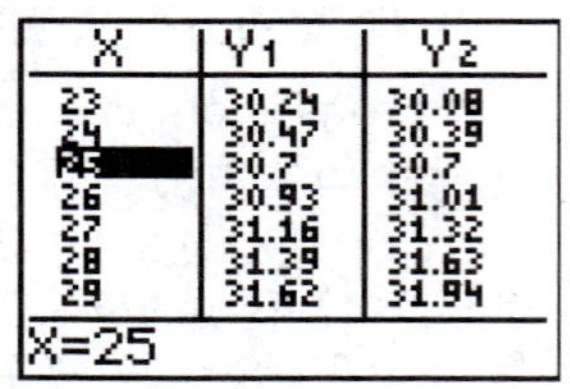

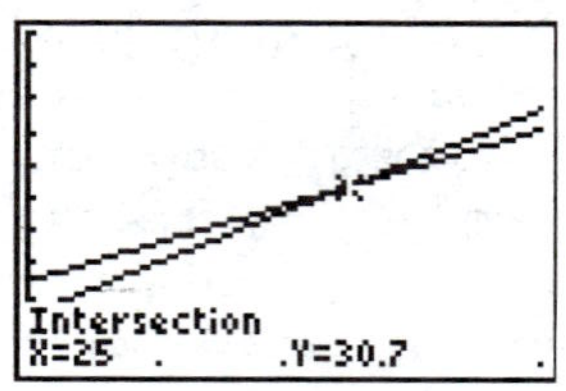

Sample solution using the substitution method:

$c = 24.95 + 0.23m$, $c = 22.95 + 0.31m$ → $24.95 + 0.23m = 22.95 + 0.31m$, $2 = 0.08m$ → $25 = m$

$c = 24.95 + 0.23m$
$c = 24.95 + 0.23(25)$
$c = 24.95 + 5.75$
$c = 30.7$

The solution is (25, 30.7). This means that if the distance traveled is 25 miles, each option will cost \$30.70. If you plan to drive more than 25 miles, renting from Flash will save you money.

SECTION 3.5 What About Those Bills?

Let's apply what you have learned thus far about solving systems of linear equations to a new situation.

1. Sometimes a businessperson needs to find creative ways to sell merchandise quickly.

   You just realized that in two days, you must pay a bill for supplies in the amount of $12,750. Your bank account is empty but you have 1000 hats in stock. You decide to have a two-day tent sale. You sell plain hats for $10 and logo hats for $12. At the end of the first day you sell a total of 500 hats and collect $5750.

   You realize that if you sell the remaining 500 hats at the same prices you will not make enough money. Thus, for the second day, you decide to sell plain hats for $11 and logo hats for $15. Assuming that you sell exactly the same number of each hat as the day before, will you make enough money to pay your bill? Justify your answer.

SUMMARY

In these sections we used the solutions to systems of linear equations to make a variety of business decisions. These decisions involve understanding costs and pay scales, choosing suppliers, and developing pricing.

SECTION 3.6 Cracking the Code

Guarding plans and ideas is important in business. If another company discovers what you plan to do, they can steal your ideas. People try to create codes to hide information. Baseball coaches have elaborate methods for giving signs to keep the opposing team from knowing their plans.

Military leaders also need to keep their plans secret. During World War II, the Germans developed a machine called Enigma that coded messages. The ability of the Allies to break this code contributed to their winning the war. Navajo code talkers contributed to the success of the armed forces in the Pacific by translating messages into the Navajo language, which was impossible for the Japanese to crack. One of the first military leaders to use coded messages was Julius Caesar. His code involved shifting the letters of the Roman alphabet. He shifted the letters three spaces forward as shown in **Figure 3.5** ($x + 3$). Each letter is assigned a number based on its position in the alphabet. So, if you add 3 to the position number of a letter you will get the position number of the coded letter. X, Y and Z (the 24th, 25th and 26th) loop around to the beginning of the alphabet, so these letters are coded A, B, and C (1, 2, and 3).

CHECK THIS!
A shift transformation moves every point the same distance in the same direction.

x	1	2	3	4	5	6	7	8	9	10	11	12	13	14	15	16	17	18	19	20	21	22	23	24	25	26
Original Letter	A	B	C	D	E	F	G	H	I	J	K	L	M	N	O	P	Q	R	S	T	U	V	W	X	Y	Z
Coded Letter	D	E	F	G	H	I	J	K	L	M	N	O	P	Q	R	S	T	U	V	W	X	Y	Z	A	B	C
$x + 3$	4	5	6	7	8	9	10	11	12	13	14	15	16	17	18	19	20	21	22	23	24	25	26	1	2	3

FIGURE 3.5. An $x + 3$ shift code.

The following quotation from Julius Caesar,

"As a rule men worry more about what they can't see than about what they can,"

would be coded as,

"Dv d uxoh phq zruub pruh derxw zkdw wkhb fdq'w vhh wkdq derxw zkdw wkhb fdq."

EVALUATE LESSON 2

Section 3.5: What About Those Bills?

Purpose: In Sections 3.1 through 3.4, students discovered multiple ways to set up and solve a system of linear equations. In this section students will have the opportunity to demonstrate their knowledge by applying those concepts and skills to solve a problem in a new context. This prepares students to examine yet another way to solve systems of equations, matrices in Sections 3.6 through 3.10, and systems of linear inequalities in Sections 3.11 through 3.15.

Description: Facing a problem in a new context, students will determine and implement a method to solve the problem. This is an open-middle performance assessment task; i.e., there is only one correct answer, but there are a variety of valid methods to obtain it. In their justification, which is the most important part of a performance assessment task, students should be able to communicate their knowledge of mathematics and demonstrate that they have mastered the concepts presented in Sections 3.1 through 3.4.

Materials: none

Teaching Ideas: This task may be assigned as group work or individual work.

Answers

1. Sample solution:

   Let p = the number of plain hats
   Let l = the number of logo hats

   The number of plain hats plus the number of logo hats is 500. So one equation is $p + l = 500$.

   Each plain hat sells for $10 and each logo hat sells for $12. Together the total amount of money collected is $5750. So the second equation is $10p + 12l = 5750$.

$$\begin{aligned} p + l &= 500 \\ 10p + 12l &= 5750 \end{aligned} \qquad \begin{aligned} p + l &= 500 \\ l &= 500 - p \end{aligned}$$

$$\begin{aligned} 10p + 12l &= 5750 \\ 10p + 12(500 - p) &= 5750 \\ 10p + 6000 - 12p &= 5750 \\ -2p &= -250 \\ p &= 125 \end{aligned} \qquad \begin{aligned} p + l &= 500 \\ l &= 500 - p \\ l &= 500 - 125 \\ l &= 375 \end{aligned}$$

The solution to the system of equations is (125, 375). This means there were 125 plain hats sold and 375 logo hats sold on the first day of the tent sale.

We need to raise $12,750 and have already collected $5750. Subtracting these numbers means that an additional $7000 is necessary to pay the bill. Raising the price of plain hats to $11.00 and the price of logo hats to $15.00 results in the following equation.

$$\begin{aligned} 125(11) + 375(15) &= \text{income} \\ 1375 + 5625 &= \text{income} \\ 7000 &= \text{income} \end{aligned}$$

Adding this $7000 to the $5750 from the first day of the sale, we have $12,750. We can pay the bill.

ENGAGE LESSON 2

Section 3.6: Cracking the Code

Purpose: Section 3.6 engages students by asking them to think about cryptography, the coding and decoding of messages. Students will use the Caesar code, which is a shift transformation, to encrypt and decrypt messages. They will be ready to explore using matrices and inverse matrices to encode and decode messages in Section 3.7.

Description: In this section students will code and decode messages using the Caesar code. The shift transformation method involves translating the letters of the alphabet either right or left to generate a coded message.

Materials: Handout 3.1

Teaching Ideas: Chapter 3 Supplemental Activities provides a number of possible activities to demonstrate the importance of coding in a wide range of settings from military to business.

As an extension activity, you may want to have students write short research papers on the Zimmermann Telegram, public key encryption, Mary Queen of Scots and the Babington Plot, or any of the many other fascinating areas of cryptography.

Have students make their own St. Cyr Cipher Strip using Handout 3.1. Students will now be able to complete the problems in Section 3.6 Assignment.

NOTE: St. Cyr Military Academy was founded by Napoleon in France. Training at St. Cyr is similar to training at West Point. During the 1880s students were taught to encode messages with the St. Cyr Cipher Strip, a variation on the Caesar code.

CHECK THIS!

Cryptography is the study of writing codes in order to provide information security.

A cryptographer is someone who creates secret code using cryptography.

Encryption is the act of altering a message or data to make it unreadable unless you know how to decrypt it.

A secret key is an encryption plan known only to the parties that exchange messages.

Cryptanalysis is the science of cracking, or deciphering, codes.

Since it is difficult to keep a key secret, you can make a shift transformation more secure by changing the number of spaces or the direction in which the letters are shifted, giving the secret key only to the person to whom you are sending the message.

In Dan Brown's book *Digital Fortress,* the two main characters send notes to each other using a variation of the Caesar code. They use a shift transformation, moving the letters one space forward, $(x + 1)$. Since both of them know the secret key, they decode each other's messages without anyone else knowing what they are saying.

I LOVE YOU is written as J MPWF ZPV.

You can make your own shift cipher device (like the one shown in **Figure 3.6**) by inserting a sliding slip of paper into a "mask" upon which the letters of the alphabet have been written. You can slide the slip of paper back and forth for different shift transformations.

Original Letters in Message

1	2	3	4	5	6	7	8	9	10	11	12	13	14	15	16	17	18	19	20	21	22	23	24	25	26
A	B	C	D	E	F	G	H	I	J	K	L	M	N	O	P	Q	R	S	T	U	V	W	X	Y	Z
D	E	F	G	H	I	J	K	L	M	N	O	P	Q	R	S	T	U	V	W	X	Y	Z	A	B	C
4	5	6	7	8	9	10	11	12	13	14	15	16	17	18	19	20	21	22	23	24	25	26	1	2	3

Y	Z	A
25	26	1

F	G	H
6	7	8

Letters in Coded Message

FIGURE 3.6.
A shift cipher device like this one is sometimes called a St. Cyr Strip.

SECTION 3.6 Assignment

1. Caesar might have used the shift three spaces forward $(x + 3)$ code to send this secret message to one of his generals. Translate the message back into its original form.

   Dwwdfn Yhuflqjhwrula dw gdzq.

2. Use your St. Cyr Cipher Strip to decode this coded message. The message was encrypted with a shift transformation of $x - 10$, ten places backward.

   Dusuiiyjo yi jxu cejxuh ev ydludjyed.

3. Use the St. Cyr Cipher Strip to encode this message with an $x + 5$, five places forward, shift transformation.

   HAVE INVENTED SELF CLEANING FOOTBALL UNIFORM.

   MEET ME AT NOON TO DISCUSS SALES TO TEXANS.

4. Write a school-appropriate note to someone in your group using the shift one forward $(x + 1)$ code. Trade notes with someone in your group and decode their message.

5. Interview at least one adult and some of your friends; ask them why encryption techniques are important in today's world. Write down their answers.

Section 3.6 Assigment

Answers

1. Decoded: Attack Vercingetorix at dawn.
2. Decoded: Necessity is the mother of invention.
3. Encoded: MFAJ NSAJSYJI XJQK HQJFSNSL KTTYGFQQ ZSNKTWR.

   RJJY RJ FY STTS YT INXHZXX XFQJX YT YJCFSX.

SECTION 3.7 The Matrix Code

Shift codes have a big problem: They are easy to crack. Players on the Wheel of Fortune game show guess letters based on how common they are. Code breakers work in much the same way. For example, if they notice that *h* has a high **frequency** (the number of times the letter is used) in a coded message, they might guess that it represents a letter such as *e* or *a* that is common in English. Once code breakers know the shift of one letter, they can decode a message quickly.

So it makes sense to find a way to code that is harder to crack. A matrix can be used to make code cracking more difficult.

> **CHECK THIS!**
> A matrix is a rectangular arrangement of numbers. The numbers are arranged in rows and columns and appear to make a rectangular shape. A row is a horizontal group of numbers. A column is a vertical group of numbers.

1. Write the message LOOK BEFORE YOU LEAP as a matrix using the pattern A = 1, B = 2, C = 3, and so on. Use 0 for the blanks between words.

   Number form of message:

   Matrix A =

You are scrambling your message by multiplying by the coding matrix $C = \begin{bmatrix} 2 & 5 \\ 1 & 3 \end{bmatrix}$. The coding matrix has 2 rows and 2 columns. You are going to use the matrix feature of a graphing calculator to multiply these two matrices. Your teacher will explain to you how to find the product of matrices A and C by entering the matrices into your graphing calculator.

2. Find the product of the two matrices and write it in the space below.

   Scrambled message:

   Matrix B =

Note that the first time O is coded it is a 30 and the next time it is 155. This means that code breaker needs more than the frequency of a letter to break the code.

3. Does the order in which you multiply the message and coding matrix to get matrix B matter and why?

The message in matrix B was coded with matrix C. That is, if A is a matrix containing the original message, the coding was done with this equation:

[C] · [A] = [B]

Scrambled message:

$$\text{Matrix B} = \begin{bmatrix} 24 & 49 & 15 & 40 & 69 & 4 & 43 & 9 & 70 & 8 \\ 43 & 76 & 25 & 62 & 113 & 6 & 71 & 14 & 115 & 16 \end{bmatrix}$$

4. How do you think you would find matrix A, the original message, when you know matrices B and C?

5. Matrix equations are solved in much the same way as algebraic equations. How do you use multiplication to solve the equation $2x = 7$?

> **CHECK THIS!**
> The additive inverse property: $a + -a = 0$.
> The multiplicative inverse property: $a \cdot \frac{1}{a} = 1$.

You can use multiplicative inverses of matrices to solve matrix equations on a graphing calculator. Your teacher will show you how to do this.

6. Will it matter in which order we multiply the inverse matrix by the coded matrix?

EXPLORE LESSON 2

Section 3.7: The Matrix Code

Purpose: Section 3.7 leads to a higher level of encryption than Section 3.6 by multiplying a matrix by a coding matrix, and then using the inverse matrix to decode the encrypted matrix. Students will be exploring the idea of matrix multiplication in an informal way in this section with a graphing calculator. They will see matrix multiplication used in a relevant situation prior to formally defining matrices and matrix operations in Section 3.8.

Description: Students will encode a message numerically and then arrange a matrix. They will use a coding matrix by which to multiply the message matrix. They will then explore undoing the coded message by multiplying it by the inverse of the coding matrix.

Materials: graphing calculator, Handout 3.2 and rubric

Teaching Ideas: Allow students to explore matrix multiplication and the undoing of the matrix multiplication informally. They will be formalizing matrix ideas in the next section. The idea of this section is to set up a relevant situation for matrices before students are introduced to solving systems of equations with matrices.

Answers

1. Matrix A =

12	15	15	11	0	2	5	6	15	18
5	0	25	15	21	0	12	5	1	16

Have students work as partners as you demonstrate how to use the calculator. They can check each other as you progress through the steps.

2nd MATRIX, EDIT, ENTER

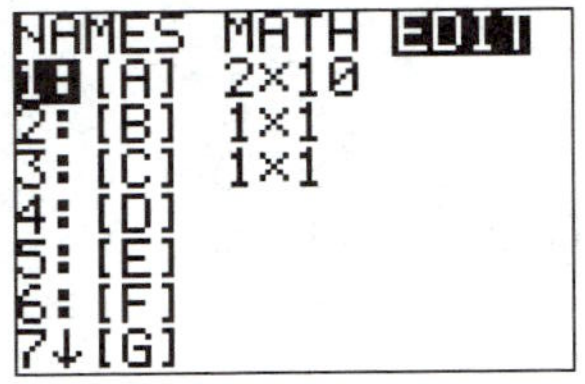

Enter the number of rows and columns.

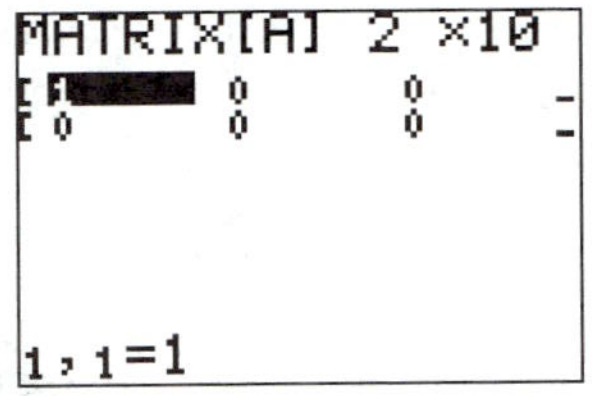

Enter the number form of the message into matrix A by arrowing to the right to fill in all of the entries. Notice how the calculator tells you in which row and which column you are entering a value.

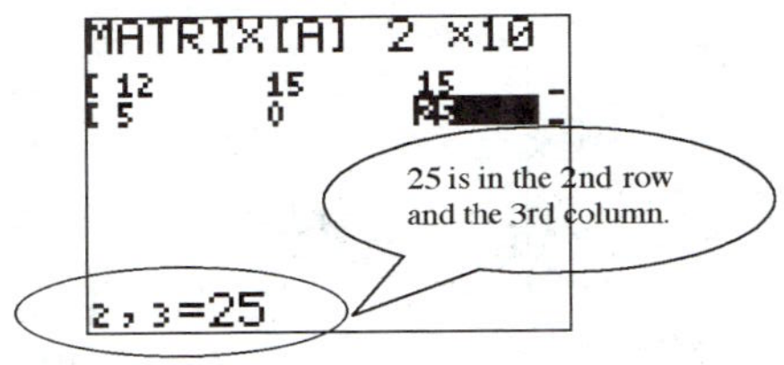

Enter the numbers from the coding matrix.

2nd MATRIX, EDIT, [C]

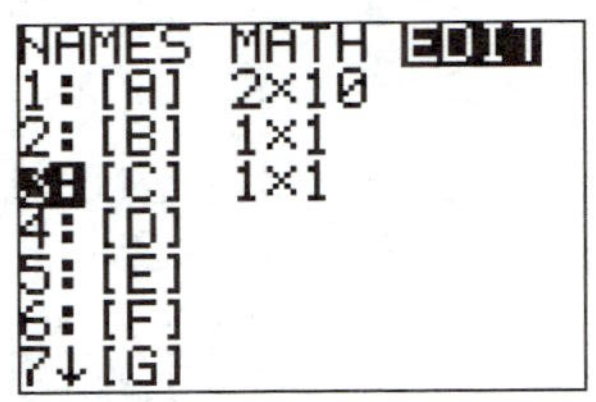

Enter the number of rows and columns.

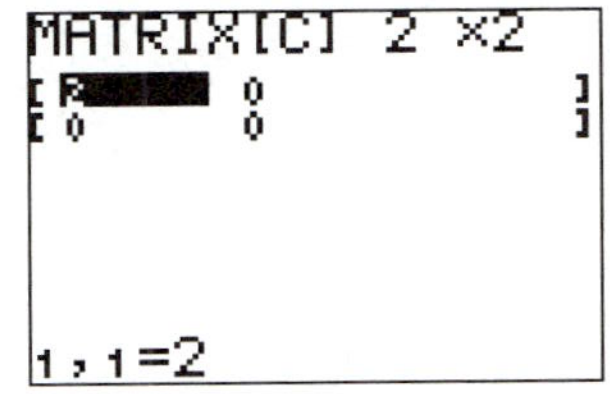

Enter the values in the coding matrix, C.

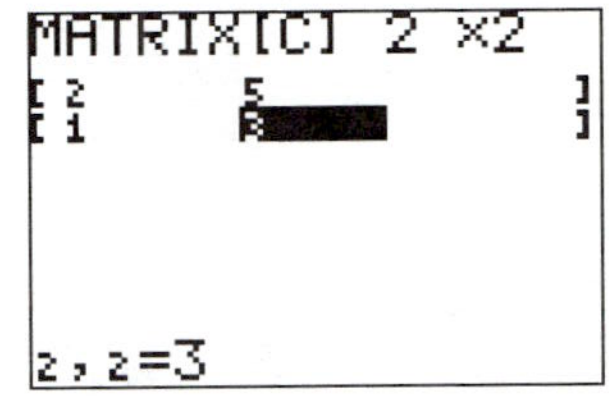

2. Demonstrate for students how to multiply the matrices.

2nd MATRIX, NAMES, [C] ENTER, *

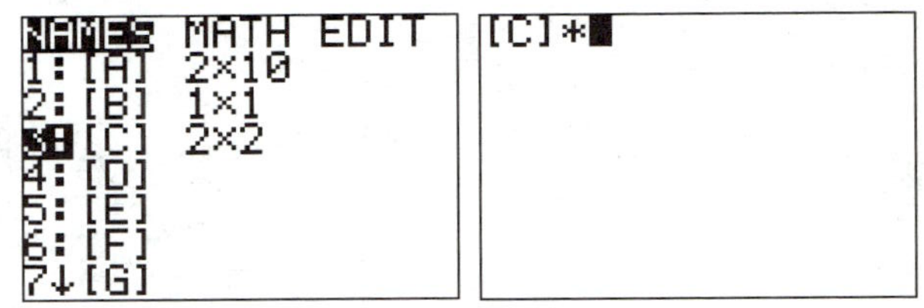

2nd MATRIX, NAMES [A] ENTER, ENTER

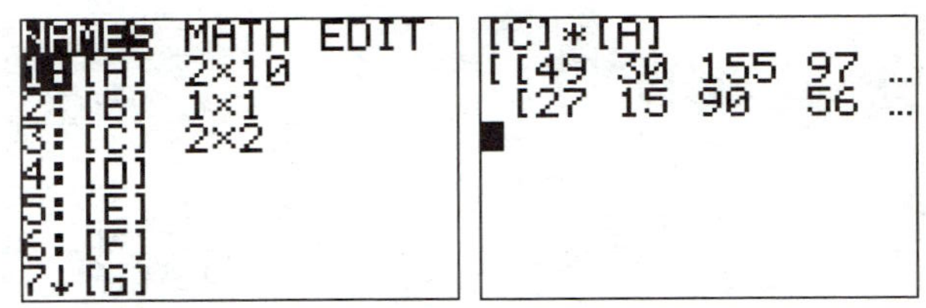

Scrambled message:

Matrix B =

49	30	155	97	105	4	70	37	35	116
27	15	90	56	63	2	41	21	18	66

Answers (continued)

3. Multiply C × A because it is impossible the other way. The number of elements in the rows and columns don't match up. We will formally explain this concept in the following section. For now, students will notice that if they multiply A × C the screen displays an error message:

4. Remind the students how matrix B was derived:

$$\begin{bmatrix}\text{Coding}\\\text{Matrix}\end{bmatrix}\cdot\begin{bmatrix}\text{Message}\\\text{Matrix}\end{bmatrix}=\begin{bmatrix}\text{Scrambled}\\\text{Matrix}\end{bmatrix}$$

$$[C] \quad \cdot \quad [A] \quad = \quad [B]$$

We need to undo the multiplying. Allow students time to try division of matrices on the calculator. They will see that they get an error message. Questions to ask students include: How else do we unmultiply? How do you divide fractions? Lead students into remembering that dividing by a fraction is equivalent to multiplying by the reciprocal or inverse of the fraction.

5. Students should know that you multiply 7 by the multiplicative inverse of 2, or 1/2.

Demonstrate for students how to multiply matrix B by the inverse of matrix C.

$$[C]^{-1}\cdot[C]\cdot[A]=[C]^{-1}\cdot[B]$$

$$[A]=[C]^{-1}\cdot[B]$$

2nd MATRIX, EDIT, [B]

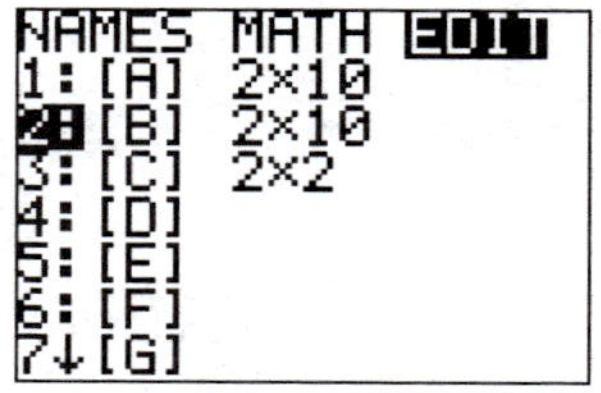

Enter number of rows, columns, and values.

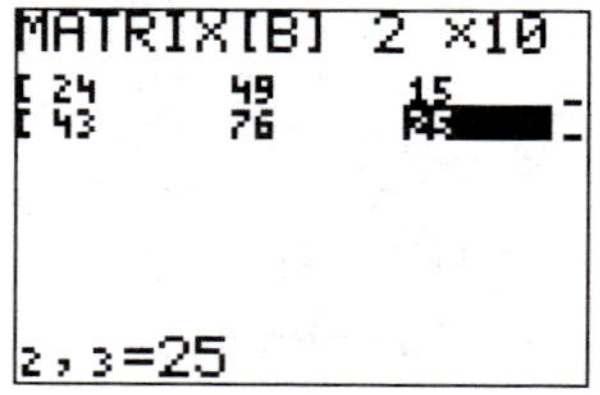

Enter new values for matrix C.

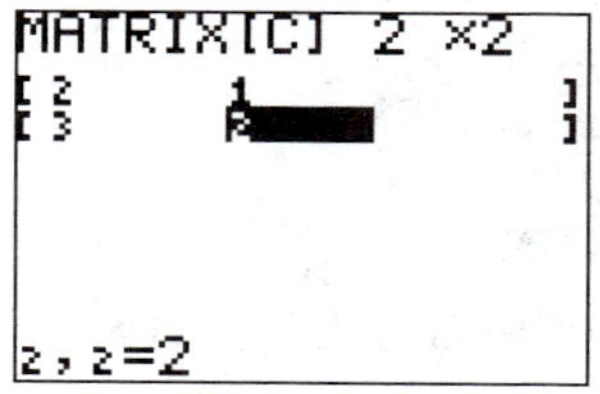

2nd QUIT, 2nd MATRIX, NAMES [C], ENTER, x^{-1}, *, 2nd MATRIX, NAMES, [B], ENTER

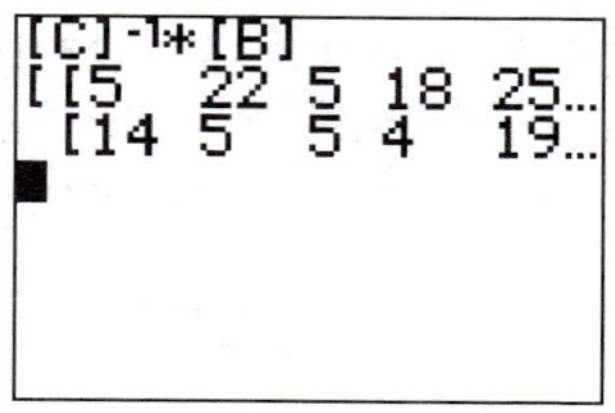

6. It does matter. Matrix C^{-1} is still a 2 × 2 matrix, and there is only one way we can multiply this matrix by the coded matrix. Allow students time to try multiplying in a different order with the calculator. This is an informal investigation; matrix multiplication will be explained formally in the following section.

7. Write your product matrix.

   Number form of the message:

   Matrix A =

8. Write your decoded message below.

SECTION 3.7 Assignment

Blank	0
A	1
B	2
C	3
D	4
E	5
F	6
G	7
H	8
I	9
J	10
K	11
L	12
M	13
N	14
O	15
P	16
Q	17
R	18
S	19
T	20
U	21
V	22
W	23
X	24
Y	25
Z	26

Create and code a secret message. Record your work on Handout 3.2. Follow the instructions below.

1. Your message must be 20 characters (letters or blanks) or less. Write your message in a table like the one below. Place one letter or a blank in each square.

2. Assign each letter in your message a number from the chart at the right. Write the number message in a table like the one below.

3. Create a 2×2 coding matrix. Write your coding matrix on the paper titled Secret Message.

   Coding matrix: $\begin{bmatrix} a & b \\ c & d \end{bmatrix}$

   To avoid problems with fractions, be sure that $ad - bc$ is equal to 1.

4. Multiply your coding matrix times your message matrix. Record the coded message on the Secret Message paper.

5. Exchange your secret message with another student. They will decode your message while you decode theirs. Write the matrix equation you used to decode the message.

6. Translate the message matrix into letters and write it on the Secret Message paper.

Answers (continued)

7. Answer:

5	22	5	18	25	2	15	4	25	0
14	5	5	4	19	0	13	1	20	8

8. DECODED MESSAGE: Everybody needs math.

Section 3.7 Assignment

Have students complete Questions 1–4 for homework. Students should be prepared to complete Questions 5 and 6 when they return to class.

SECTION 3.8

What's in a Matrix?

Tables are a common way to show data. **Figure 3.7** shows a table of the average times in minutes for the national physical fitness test endurance run in 1989 (U.S. Department of Education, 2001). The runs were 3/4 mile for ages 10 and 11 and 1 mile for ages 12 through 17:

	Boys	Girls
10–11-year-olds	7.3	8.0
12–13-year-olds	9.1	10.5
14–17-year-olds	8.6	10.7

FIGURE 3.7.
1989 times (in minutes).

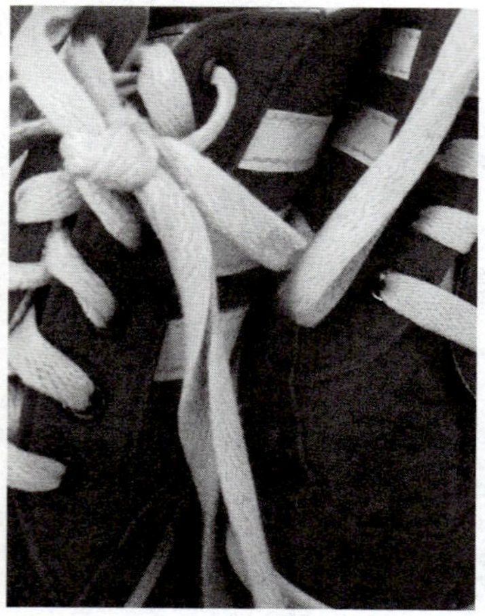

When you remove the labels, in this case the boys and the girls and their ages, you have a group of numbers called a **matrix**. The plural of this word is **matrices**. Matrices are often represented by a capital letter. Here is the matrix for the table above:

$$T = \begin{bmatrix} 7.3 & 8.0 \\ 9.1 & 10.5 \\ 8.6 & 10.7 \end{bmatrix}$$

Here T is the label for the matrix, *times*. Matrices are often described by their size or their **matrix dimensions**, the number of rows and columns. This is a 3 × 2 (read "3 by 2") matrix because it has 3 rows and 2 columns. Each entry in a matrix is called an **element or a component**. The symbol $t_{2,1}$ points to the element in the second row, first column. In this table that is 9.1.

Corresponding elements are in the same position (row and column) in each matrix.

	Boys	Girls
10–11-year-olds	7.5	8.2
12–13-year-olds	9.4	10.4
14–17-year-olds	8.8	10.9

FIGURE 3.8.
2000 times (in minutes).

If you take the matrices from the 1989 table and the 2000 table you get:

$$T = \begin{bmatrix} 7.3 & 8.0 \\ 9.1 & 10.5 \\ 8.6 & 10.7 \end{bmatrix} \qquad U = \begin{bmatrix} 7.5 & 8.2 \\ 9.4 & 10.4 \\ 8.8 & 10.9 \end{bmatrix}$$

Matrix U is the student times in 2000.

Matrices can be labor saving tools when you have a graphing calculator or a spreadsheet program. Suppose you want to compare how fast students ran in 1989 to how fast they ran in 2000 (Figure 3.7 to Figure 3.8). This involves performing a **matrix operation.** A matrix operation allows you to perform a process such as multiplication, addition, or subtraction with one or more matrices.

If you want to see how students' times have changed you could do **matrix subtraction.** Matrix subtraction allows us to subtract elements of one matrix from those of another. To subtract two matrices, subtract the corresponding elements.

$$T = \begin{bmatrix} 7.3 & 8.0 \\ 9.1 & 10.5 \\ 8.6 & 10.7 \end{bmatrix} \qquad U = \begin{bmatrix} 7.5 & 8.2 \\ 9.4 & 10.4 \\ 8.8 & 10.9 \end{bmatrix} \qquad D = T - U = \begin{bmatrix} -0.2 & -0.2 \\ -0.3 & 0.1 \\ -0.2 & -0.2 \end{bmatrix}$$

Values from matrix D, or *difference*, can then be put in a table:

	Boys	Girls
10–11-year-olds	−0.2	−0.2
12–13-year-olds	−0.3	0.1
14–17-year-olds	−0.2	−0.2

FIGURE 3.9.
Times 1989 vs. 2000 (in minutes).

This table allows you to see that student times, except for 12- and 13-year-old girls, are slower than in 1989.

In this case it made sense to use matrix subtraction. Sometimes it makes sense to add elements of matrices. This is called **matrix addition.**

For example:

$$\begin{bmatrix} 2 & 3 \\ 1 & 0 \end{bmatrix} + \begin{bmatrix} -3 & 5 \\ 7 & -3 \end{bmatrix} = \begin{bmatrix} -1 & 8 \\ 8 & -3 \end{bmatrix}$$

EXPLAIN LESSON 2

Section 3.8: What's in a Matrix?

Purpose: Students have been introduced to solving systems of equations in Sections 3.1 through 3.5 along with informally introducing matrix multiplication and inverse matrices in Section 3.7. Section 3.8 will formally define and explain matrices and matrix operations to students. They will be ready to use inverse matrices to solve systems of equations in Section 3.9.

Description: The information on matrices is presented in the textbook, but you can make transparencies from the Masters on the Teacher Resource CD-ROM.

Materials: transparencies, Matrix Mouse Hunt cards on the Teacher Resource CD-ROM

Teaching Ideas: This section is a formal explanation of matrix definitions and operations. After you explain these concepts, students will do problems involving fill-in-the-blank questions and simple matrix operations.

For matrix addition and subtraction, both matrices must be the same size. This means that they have same number of rows and columns.

There are other operations you can perform on matrices. You can multiply a matrix by a number called a **scalar**. To perform **scalar multiplication**, multiply every element in the matrix by the scalar.

EXAMPLE:

$$-5\begin{bmatrix} 3 & 0 \\ -10 & 8 \end{bmatrix} = \begin{bmatrix} -15 & 0 \\ 50 & -40 \end{bmatrix}$$

You can use the definition of equal matrices to solve **matrix equations. Equal matrices** have the same dimensions and corresponding elements are equal. To solve for x and y, write the following equations and solve each equation.

EXAMPLE:

Solve for x and y.

$$\begin{bmatrix} 2x-5 & 4 \\ 3 & 3y+12 \end{bmatrix} = \begin{bmatrix} 25 & 4 \\ 3 & y+18 \end{bmatrix}$$

$2x - 5 = 25$	$3y + 12 = y + 18$
$2x - 5 + 5 = 25 + 5$	$3y - y + 12 = y - y + 18$
$2x = 30$	$2y + 12 = 18$
$\frac{2x}{2} = \frac{30}{2}$	$2y + 12 - 12 = 18 - 12$
$x = 15$	$2y = 6$
	$y = 3$

To multiply two matrices, the number of columns in the first matrix must be the same as the number of rows in the second matrix.

Each element in row 1 of the first matrix is multiplied by an element of column 1 in the second matrix, and the products are added together. The result becomes the row 1, column 1 element in the answer matrix. The process is continued until every row of the first matrix is paired with a column of the second matrix.

EXAMPLE:

Multiply.

$$\begin{bmatrix} 2 & -1 & 5 \\ 7 & 3 & -2 \end{bmatrix} \cdot \begin{bmatrix} 0 & -3 \\ -4 & 2 \\ 5 & 9 \end{bmatrix}$$

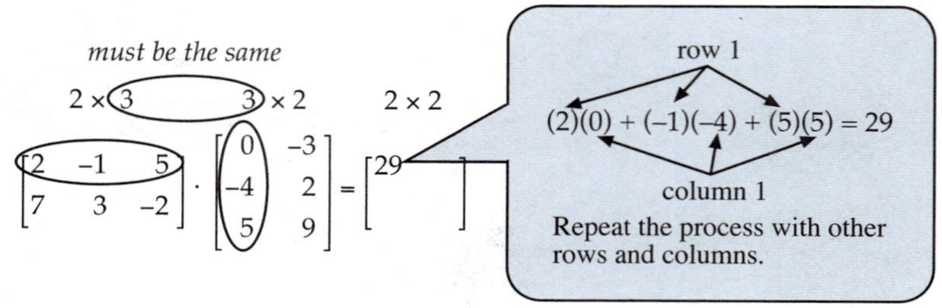

In general, when you pair row m of the first matrix with column n of the second matrix in this way, you get the row m, column n element of the product matrix.

Let's see if you get the same answer when you multiply matrix P by matrix Q as you do when you reverse the order of multiplication.

$$P = \begin{bmatrix} 2 & 3 \\ 0 & -1 \end{bmatrix} \qquad Q = \begin{bmatrix} -3 & 1 \\ 2 & 0 \end{bmatrix}$$

$$P \cdot Q = \begin{bmatrix} 2 & 3 \\ 0 & -1 \end{bmatrix} \cdot \begin{bmatrix} -3 & 1 \\ 2 & 0 \end{bmatrix} = \begin{bmatrix} 2(-3)+3(2) & 2(1)+3(0) \\ 0(-3)+(-1)(2) & 0(1)+(-1)(0) \end{bmatrix} = \begin{bmatrix} 0 & 2 \\ -2 & 0 \end{bmatrix}$$

$$Q \cdot P = \begin{bmatrix} -3 & 1 \\ 2 & 0 \end{bmatrix} \cdot \begin{bmatrix} 2 & 3 \\ 0 & -1 \end{bmatrix} = \begin{bmatrix} -3(2)+1(0) & -3(3)+1(-1) \\ 2(2)+0(0) & 2(3)+0(-1) \end{bmatrix} = \begin{bmatrix} -6 & -10 \\ 4 & 6 \end{bmatrix}$$

Multiply $P \cdot Q$ and $Q \cdot P$ with your calculator to check your answers.

Does order matter when multiplying matrices?

Yes. $P \cdot Q = \begin{bmatrix} 0 & 2 \\ -2 & 0 \end{bmatrix}$, but $Q \cdot P = \begin{bmatrix} -6 & -10 \\ 4 & 6 \end{bmatrix}$, so $P \cdot Q \neq Q \cdot P$.

Therefore, multiplication of matrices is not commutative.

Answers

Example

After demonstrating conceptually how matrices are multiplied, remind students how they can multiply matrices on a graphing calculator.

2nd MATRIX, EDIT, [A], ENTER

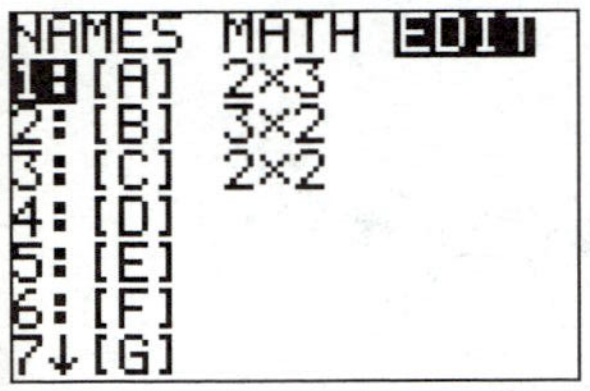

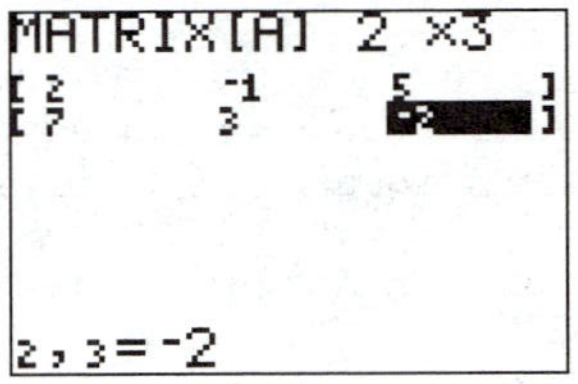

2nd MATRIX, EDIT, [B], ENTER

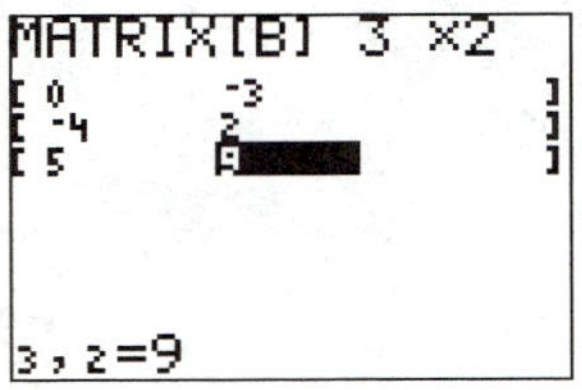

2nd MATRIX, NAMES, [A], 2nd MATRIX, NAMES, [B], ENTER, ENTER

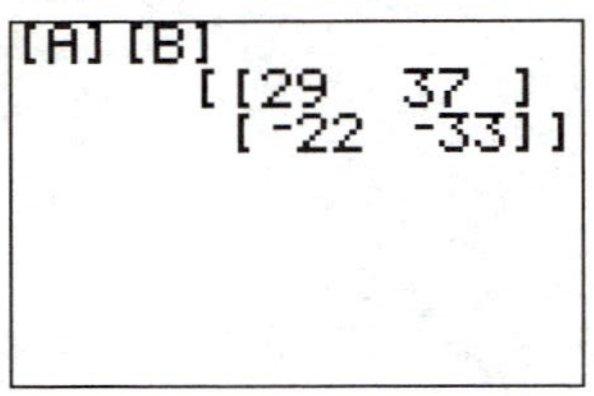

SECTION 3.8 Assignment

Fill in each blank with an appropriate word or number.

1. A __________ is a rectangular arrangement of numbers. It is arranged in _______ and __________. The plural form of this is __________. They are labeled with a ______________ and are written within brackets.

2. The ____________ of a matrix define the size of the matrix. The number of rows is always stated ________.

3. The matrix $A = \begin{bmatrix} 27 & -3 & 14 \\ 3.8 & -9 & 45 \end{bmatrix}$ is referred to as a ___ × ___ matrix.

4. Each number in a matrix is called an __________.

5. Two matrices are _______ matrices if and only if they have the same dimensions and their corresponding elements are equal. ____________ are the elements in the same position (row and column) of each matrix.

6. You can multiply a matrix by a number called a ________ by multiplying every element in the matrix by it.

7. You can use the definition of equal matrices to solve a ________.

8. If two matrices have the ________ dimensions, you can ______ or __________ the corresponding elements to find the sum or difference of the matrices.

9. To multiply two matrices, the number of __________ in the first matrix must be the same as the number of ________ in the second matrix.

Use matrices A, B, C, D, or E for Questions 10–16.

$A = \begin{bmatrix} 5 & -2 \\ 1 & 7 \end{bmatrix}$ $B = \begin{bmatrix} 12 & 4 \\ -3 & -1 \end{bmatrix}$ $C = \begin{bmatrix} 1 & 0 & 5 \\ 7 & 2 & 4 \end{bmatrix}$

$D = \begin{bmatrix} -3 & 9 & -1 \\ 5 & 0 & 12 \end{bmatrix}$ $E = \begin{bmatrix} 0 & 1 & 2 \\ 3 & 4 & 5 \\ 6 & 7 & 8 \end{bmatrix}$

10. What are the dimensions of matrix C?

11. Which matrix is a 3 × 3 matrix?

12. If matrix A is equal to $\begin{bmatrix} 5 & 3y-11 \\ 2x+9 & 7 \end{bmatrix}$, solve for x and y.

13. Multiply –3B.

14. Add C + D.

15. Simplify 5A – B.

16. Multiply A · D.

Section 3.8 Assignment

Answers

1. A *matrix* is a rectangular arrangement of numbers. It is arranged in *rows* and *columns*. The plural form of this is *matrices*. They are represented by a *capital letter* and are written within brackets.
2. The *dimensions* of a matrix define the size of the matrix. The number of rows is always stated *first*.
3. 2 × 3
4. element
5. Two matrices are *equal* matrices if and only if they have the same dimensions and their corresponding elements are equal. *Corresponding elements* are the elements in the same position (row and column) of each matrix.
6. scalar
7. You can use the definition of equal matrices to solve a *matrix equation*. In equal matrices, *corresponding* elements are equal.
8. If two matrices have the *same* dimensions, you can *add* or *subtract* the corresponding elements to find the sum or difference.
9. To multiply two matrices, the number of *columns* in the first matrix must be the same as the number of *rows* in the second matrix because the number of elements will be paired up and multiplied together.
10. 2 × 3
11. Matrix E
12. $x = -4$, $y = 3$
13. $\begin{bmatrix} -36 & -12 \\ 9 & 3 \end{bmatrix}$
14. $\begin{bmatrix} -2 & 9 & 4 \\ 12 & 2 & 16 \end{bmatrix}$
15. $\begin{bmatrix} 13 & -14 \\ 8 & 36 \end{bmatrix}$
16. $\begin{bmatrix} -25 & 45 & -29 \\ 32 & 9 & 83 \end{bmatrix}$

In order to give students ample opportunity to practice matrix operations after completing Section 3.8, do the Matrix Mouse Hunt activity.

In order to use the circuit, print each full page on colored card stock, each of which has an answer on the top half and a problem on the bottom half. You may also want to put each page into a sheet protector so that it will last longer. Tape the pages onto the walls in your classroom or out in the hall in mixed up order.

Students should be paired for this activity. Each pair will start the circuit at any of the cards and work the problem that is on the bottom half of that card. An answer sheet is provided for students to carry with them and record their answers. You may want to assign a starting problem for pairs of students so that they will be spread out. On the student answer sheet, write an answer from the card you want them to start on in the Starting Point box. Once the student pair has worked their first problem, they search the other cards to find the one that has their answer on the top half. On that card will be a new problem (again on the bottom half). The students continue to work the new problem, record the answer on their answer sheet, and find the next card that has that answer on the top. This procedure continues until they get back to the original card, which has the answer to their last problem, thus completing the circuit. Each student should turn in a completed answer sheet.

SECTION 3.9

Using Matrices to Solve Systems of Equations

A square matrix with ones along the diagonal that goes from the upper left corner to the lower right corner and with zeros in every other position is called an **identity matrix**.

2 × 2 identity matrix:

$$I = \begin{bmatrix} 1 & 0 \\ 0 & 1 \end{bmatrix}$$

3 × 3 identity matrix:

$$I = \begin{bmatrix} 1 & 0 & 0 \\ 0 & 1 & 0 \\ 0 & 0 & 1 \end{bmatrix}$$

EXAMPLE:

What is the product of the identity matrix and matrix L?

$$L = \begin{bmatrix} -5 & 4 & 3.2 \\ -1 & 0 & 6 \end{bmatrix}$$

$$\begin{bmatrix} 1 & 0 \\ 0 & 1 \end{bmatrix} \cdot \begin{bmatrix} -5 & 4 & 3.2 \\ -1 & 0 & 6 \end{bmatrix} = \begin{bmatrix} \quad & \quad & \quad \\ \quad & \quad & \quad \end{bmatrix}$$

If you multiply a matrix by the identity matrix, the product is the original matrix.

If the product of two matrices is the identity matrix then the matrices are <u>**inverses**</u> of each other.

EXAMPLE:

$$A = \begin{bmatrix} -3 & 5 \\ 5 & -8 \end{bmatrix} \qquad A^{-1} = \begin{bmatrix} 8 & 5 \\ 5 & 3 \end{bmatrix}$$

$$A^{-1} \cdot A = I$$

$$\begin{bmatrix} 8 & 5 \\ 5 & 3 \end{bmatrix} \cdot \begin{bmatrix} -3 & 5 \\ 5 & -8 \end{bmatrix} = \begin{bmatrix} 8(-3)+5(5) & 8(5)+5(-8) \\ 5(-3)+3(5) & 5(5)+3(-8) \end{bmatrix} = \begin{bmatrix} 1 & 0 \\ 0 & 1 \end{bmatrix}$$

$$A \cdot A^{-1} = I$$

$$\begin{bmatrix} -3 & 5 \\ 5 & -8 \end{bmatrix} \cdot \begin{bmatrix} 8 & 5 \\ 5 & 3 \end{bmatrix} = \begin{bmatrix} -3(8)+5(5) & -3(5)+5(3) \\ 5(8)+(-8)(5) & 5(5)+(-8)(3) \end{bmatrix} = \begin{bmatrix} 1 & 0 \\ 0 & 1 \end{bmatrix}$$

If you do not know the inverse of a matrix, you can find it with a graphing calculator. This is especially helpful for matrices that are larger than 2 × 2. Your teacher will give you instructions on how to find the inverse matrix with the graphing calculator.

EXAMPLE:

Use a graphing calculator to find the inverse of matrix W.

$$W = \begin{bmatrix} 3 & 1 \\ 4 & 2 \end{bmatrix}$$

Check to be sure that the two matrices are inverses by multiplying them on the calculator.

EXAMPLE:

Use a graphing calculator to find the inverse of matrix F.

$$F = \begin{bmatrix} 4 & 3 & 0 \\ 5 & 2 & -3 \\ -3 & -1 & 2 \end{bmatrix}$$

Multiply $F \cdot F^{-1}$ (and $F^{-1} \cdot F$) to see if the product is the identity matrix.

You have already learned several methods for solving a system of equations: graphing, tables, substitution, and linear combinations. Inverse matrices are an additional method to help you find the solution.

EXAMPLE:

$$2x + y = 9$$

$$-2x - 3y = -7$$

If we look at the coefficients of x and y in the two equations above, we can represent them with a matrix called the **coefficient matrix**.

$$A = \begin{bmatrix} 2 & 1 \\ -2 & -3 \end{bmatrix}$$

The variables are represented with a different matrix called the **variable matrix**.

$$X = \begin{bmatrix} x \\ y \end{bmatrix}$$

ELABORATE LESSON 2

Section 3.9: Using Matrices to Solve Systems of Equations

Purpose: In Sections 3.1 through 3.5, students studied several methods for solving systems of equations. In Section 3.8 they learned the basics of matrix operations. In this section they will learn an additional method for solving systems: inverse matrices.

Description: Students will be introduced to the identity matrix and inverse matrices. Then they will solve systems of equations using inverse matrices with the graphing calculator.

Materials: graphing calculator

Teaching Ideas: Help students see the value of using matrices to solve systems of equations in ways that are similar to solving individual ones. In order to do this, students should learn the definition of the identity matrix and its relationship to inverse matrices. Students will be able to find the inverse of a 2 × 2 matrix by hand. Students do not need to be able to find the inverse matrix by hand, especially for matrices larger than 2 × 2. You will demonstrate for them how to find the inverse matrix on the graphing calculator.

Answers

Example:

Using the graphing calculator:

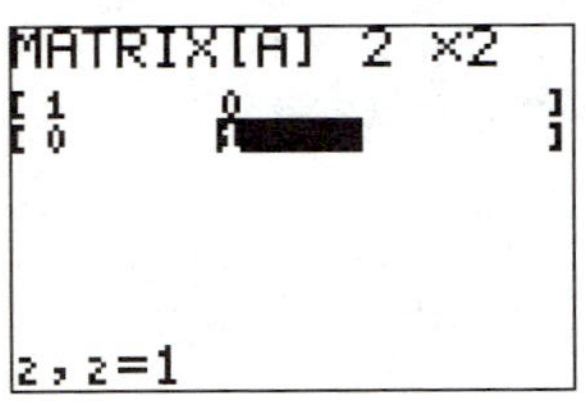

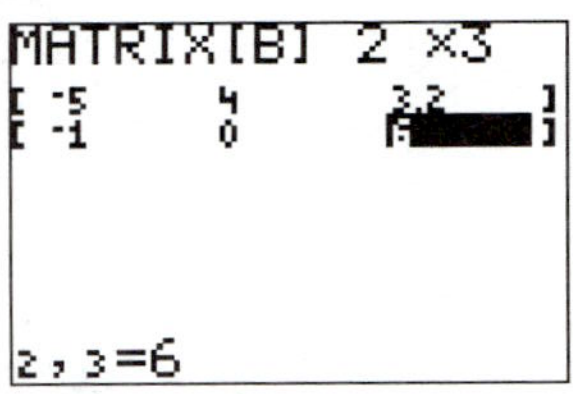

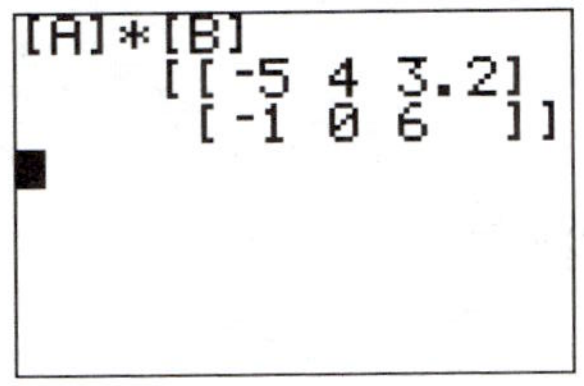

Example:

Demonstrate how to find the inverse matrix. Have students follow along as you demonstrate and check with a partner to be sure they have the correct steps.

Example:

Enter matrix W.

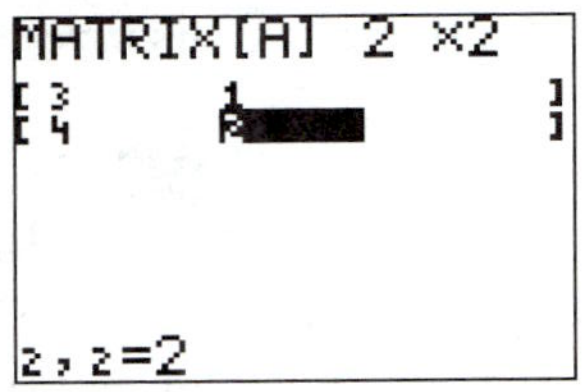

2nd MATRIX, NAMES, x^{-1}, ENTER.

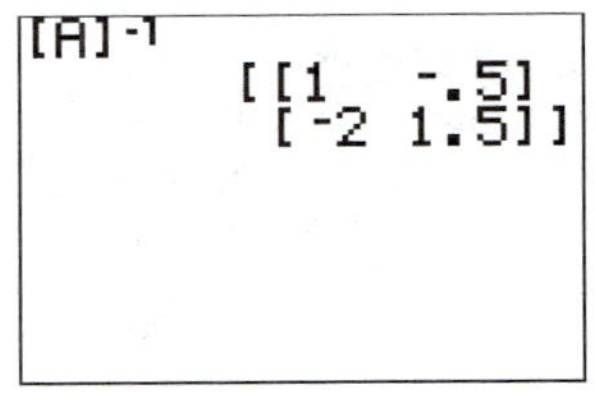

$$W^{-1} = \begin{bmatrix} 1 & -\frac{1}{2} \\ -2 & \frac{3}{2} \end{bmatrix}$$

Enter matrix A

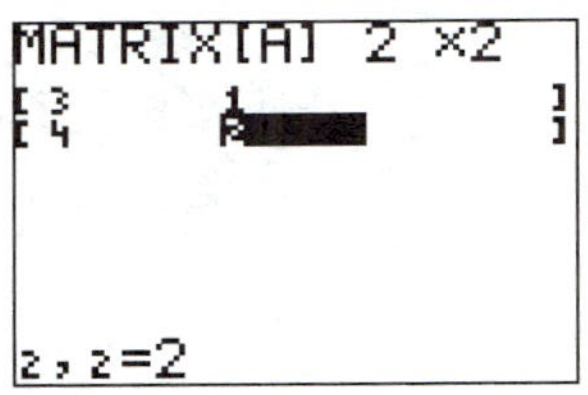

Enter A^{-1}

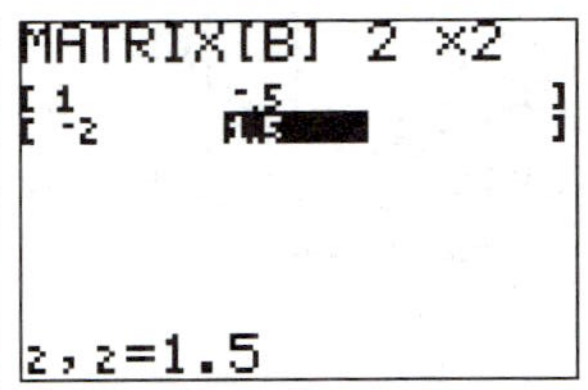

Multiply A × A^{-1}

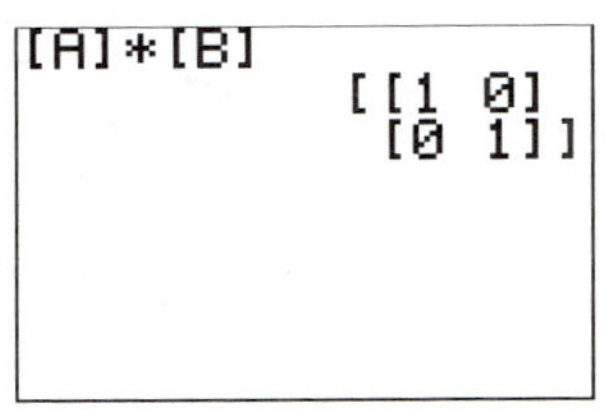

Example:

2nd MATRIX, EDIT, ENTER

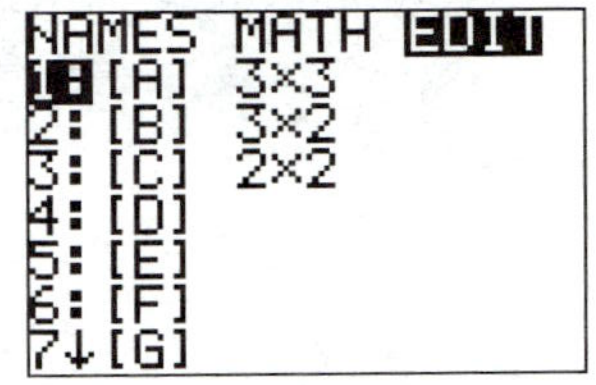

Enter the matrix

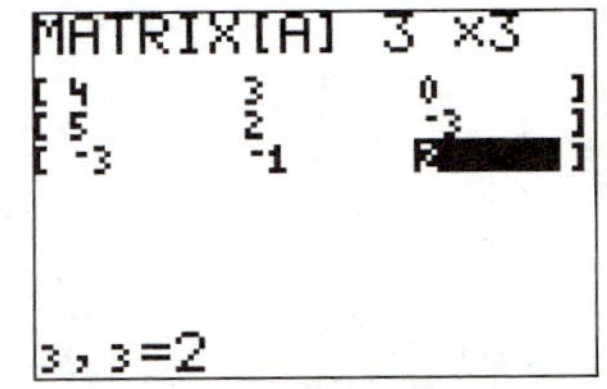

2nd QUIT, 2nd MATRIX

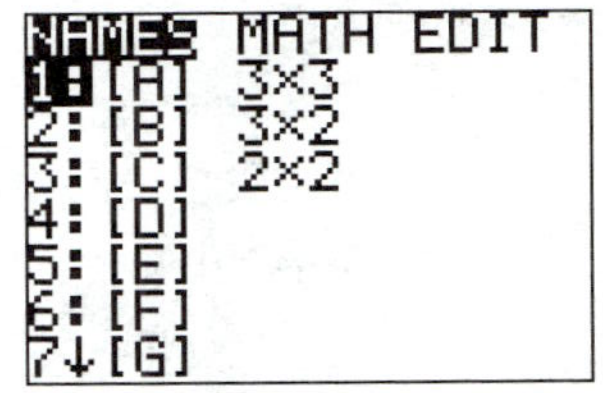

NAMES, ENTER, x^{-1}, ENTER

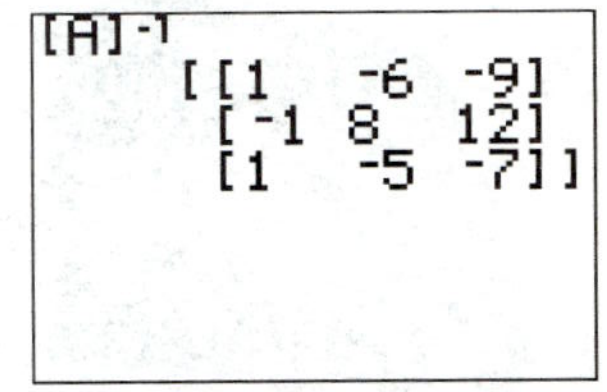

So, the inverse of matrix F is F^{-1}.

$$F^{-1} = \begin{bmatrix} 1 & -6 & -9 \\ -1 & 8 & 12 \\ 1 & -5 & -7 \end{bmatrix}$$

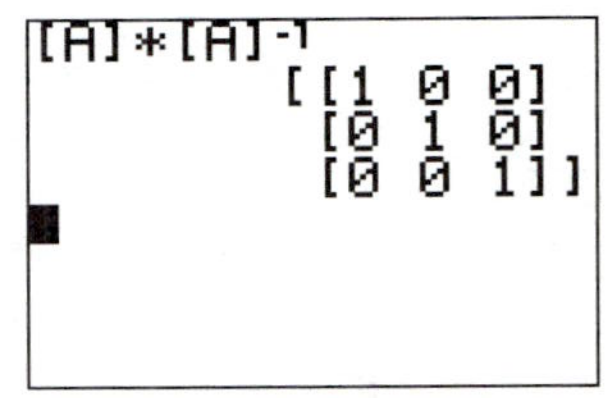

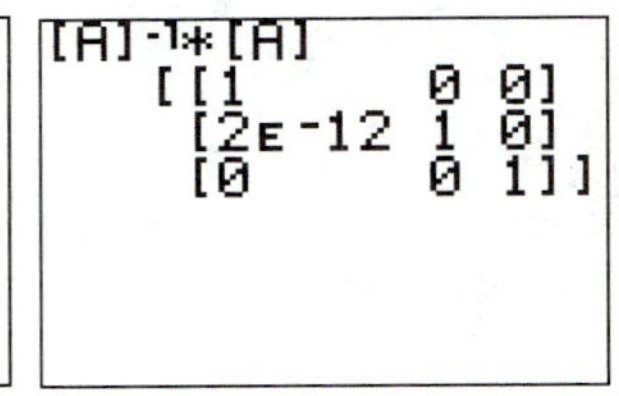

This might be a good opportunity to ask students what 2E – 12 means. The calculator representation of scientific notation 2E –12 means 2×10^{-12} or 0.000000000002, which is so small that for our purposes it is 0.

Finally, the constants on the right side of each equation also have their own matrix called the **constant matrix**.

$$B = \begin{bmatrix} 9 \\ -7 \end{bmatrix}$$

You can rewrite the equations as the product of two matrices:

$$\begin{bmatrix} 2 & 1 \\ -2 & -3 \end{bmatrix}\begin{bmatrix} x \\ y \end{bmatrix} = \begin{bmatrix} 2x+y \\ -2x-3y \end{bmatrix} \text{ and is equal to } \begin{bmatrix} 9 \\ -7 \end{bmatrix}.$$

So, the equations could be written as

$$\begin{bmatrix} 2 & 1 \\ -2 & -3 \end{bmatrix}\begin{bmatrix} x \\ y \end{bmatrix} = \begin{bmatrix} 9 \\ -7 \end{bmatrix}$$

[A][X] = [B].

Multiply each side of the matrix equation by the inverse of matrix A.

$[A]^{-1}[A][X] = [A]^{-1}[B]$

Remember that matrix multiplication is not always commutative, so you must left-multiply on both sides of the matrix equation.

Don't forget that the product of a matrix and its inverse is the identity matrix, commonly labeled matrix I.

$[I][X] = [A]^{-1}[B]$

If you multiply a matrix by the identity matrix the product is the original matrix.

$[X] = [A]^{-1}[B]$

Let's do the same steps with the actual numbers in the matrix.

$$\begin{bmatrix} 2 & 1 \\ -2 & -3 \end{bmatrix}\begin{bmatrix} x \\ y \end{bmatrix} = \begin{bmatrix} 9 \\ -7 \end{bmatrix}$$

$$\begin{bmatrix} 2 & 1 \\ -2 & -3 \end{bmatrix}^{-1}\begin{bmatrix} 2 & 1 \\ -2 & -3 \end{bmatrix}\begin{bmatrix} x \\ y \end{bmatrix} = \begin{bmatrix} 2 & 1 \\ -2 & -3 \end{bmatrix}^{-1}\begin{bmatrix} 9 \\ 7 \end{bmatrix}$$

$$\begin{bmatrix} 1 & 0 \\ 0 & 1 \end{bmatrix}\begin{bmatrix} x \\ y \end{bmatrix} = \begin{bmatrix} 2 & 1 \\ -2 & -3 \end{bmatrix}^{-1}\begin{bmatrix} 9 \\ -7 \end{bmatrix}$$

$$\begin{bmatrix} x \\ y \end{bmatrix} = \begin{bmatrix} 5 \\ -1 \end{bmatrix}$$

So, the solution to the system of equations is $x = 5$ and $y = -1$.

Inverse matrices are especially useful when solving systems of equations with three variables.

EXAMPLE:

Solve the system of equations using inverse matrices on the graphing calculator.

$$2x - 3y + z = -7$$
$$x + 2y + 3z = 0$$
$$-3x - 5y + 9z = 1$$

$$\begin{bmatrix} x \\ y \\ z \end{bmatrix} = \begin{bmatrix} 2 & -3 & 1 \\ 1 & 2 & 3 \\ -3 & -5 & 9 \end{bmatrix}^{-1}\begin{bmatrix} -7 \\ 0 \\ 1 \end{bmatrix}$$

$$\begin{bmatrix} x \\ y \\ z \end{bmatrix} = \begin{bmatrix} -2 \\ 1 \\ 0 \end{bmatrix}$$

So the solution set is $x = -2$, $y = 1$, and $z = 0$.

EXAMPLE:

This month, Franco's checking account had a total of 24 checks written and ATM withdrawals. The bank charges \$0.30 for each check and \$1.50 for each ATM withdrawal. If his service charge was \$12.00, how many of each type of activity did Franco carry out this month? Write a system of equations to represent this situation. Clearly define your variables.

This problem can be represented with the system:

$$x + y = 24$$
$$0.30x + 1.50y = 12$$

where x is the number of checks written and y is the number of ATM withdrawals.

Solve the system using inverse matrices.

$$\begin{bmatrix} x \\ y \end{bmatrix} = \begin{bmatrix} 1 & 1 \\ 0.30 & 1.50 \end{bmatrix}^{-1}\begin{bmatrix} 24 \\ 12 \end{bmatrix}$$

$$\begin{bmatrix} x \\ y \end{bmatrix} = \begin{bmatrix} 20 \\ 4 \end{bmatrix}$$

Franco wrote 20 checks and had 4 ATM withdrawals.

SECTION 3.9 Assignment

1. Are these matrices inverses of each other? Why or why not?

$$A = \begin{bmatrix} 3 & -5 \\ -3 & 6 \end{bmatrix}$$

$$B = \begin{bmatrix} 2 & \frac{5}{3} \\ 1 & 1 \end{bmatrix}$$

2. Use a graphing calculator to find the inverse of this 2 × 2 matrix.

$$C = \begin{bmatrix} -4 & -2 \\ 5 & 2 \end{bmatrix}$$

3. Use a graphing calculator to find the inverse of this matrix.

$$D = \begin{bmatrix} 1 & 1 & 0 \\ -3 & -2 & 2 \\ 3 & 1 & -3 \end{bmatrix}$$

4. A softball team bought 24 jerseys and 15 hats for a total cost of $1713. Later, they bought 5 more jerseys and 12 more hats for a total cost of $490. If the cost per item was the same for each order, what was the cost of each jersey and each hat?

5. A collection of nickels, dimes, and quarters has a value of $3.40. There are twice as many dimes as quarters, and there are 32 coins in all. How many of each kind are there?

6. A lawn and garden factory makes rakes, shovels, and clippers at a monthly cost of $6850 for a total of 2150 items. It costs $2 to make a rake, $3 to make a shovel, and $4 to make a clipper. Wholesale, a rake sells for $3, a shovel sells for $4.50, and a clipper sells for $5.50. If the total income is $9825, how many of each item did the factory make?

7. A college student earned $177.50 last week working three part-time jobs. During the 30-hour workweek, she spent twice as many hours babysitting at $5 an hour as she did tutoring at $12 an hour. If she earned $4.50 an hour working in her grandmother's garden, how many hours did she spend at each job last week?

SECTION 3.10 Opening Day

The promotions manager for the Texas Rangers is planning a special opening day giveaway. Each of the first 5000 fans will receive either a souvenir cap or a Rangers blanket. The manager knows that the caps cost $5 each and the blankets cost $12 each. What combination of caps and blankets can he purchase for a cost of $32,000?

After deciding how many caps and blankets to buy, the manager must also find the shipping costs. If the cost is $0.25 to ship each cap and $0.60 to ship each blanket, what will be the total cost for the opening day giveaway (including purchase price and shipping costs)?

Section 3.9 Assignment

Answers

1. Yes, they are inverses. When you multiply the two matrices the product is the identity matrix.

2. The inverse matrix is

$$C^{-1} = \begin{bmatrix} 1 & 1 \\ -\frac{5}{2} & -2 \end{bmatrix}$$

3. The inverse matrix is

$$D^{-1} = \begin{bmatrix} 4 & 3 & 2 \\ -3 & -3 & -2 \\ 3 & 2 & 1 \end{bmatrix}$$

4. The system of equations for this problem with j = the cost of a jersey and h = the cost of a hat is:

$$\begin{aligned} 24j + 15h &= 1713 \\ 5j + 12h &= 490 \end{aligned}$$

The solution is a jersey costs \$62 and a hat costs \$15.

5. Let n = the number of nickels, d = the number of dimes, and q = the number of quarters. The system for this problem is:

$$\begin{aligned} n + d + q &= 32 \\ 0.05n + 0.10d + 0.25q &= 3.40 \\ d &= 2q \end{aligned}$$

To translate this into matrix form, all equations must be in standard form. That means that all variables must be on the left side in a particular order and the constant must be on the right. Also, fill in any missing terms with a zero coefficient.

$$\begin{aligned} n + d + q &= 32 \\ 0.05n + 0.10d + 0.25q &= 3.40 \\ 0n + d - 2q &= 0 \end{aligned}$$

Then write the system as a product of matrices. After entering the matrices into the graphing calculator, find $A^{-1} \cdot B$.

$$\begin{bmatrix} n \\ d \\ q \end{bmatrix} = \begin{bmatrix} 1 & 1 & 1 \\ 0.05 & 0.10 & 0.25 \\ 0 & 1 & -2 \end{bmatrix}^{-1} \begin{bmatrix} 32 \\ 3.40 \\ 0 \end{bmatrix}$$

The solution is 14 nickels, 12 dimes, and 6 quarters.

6. The system of equations for this problem with r = the number of rakes, s = the number of shovels, and c = the number of clippers can be written as:

$$\begin{aligned} r + s + c &= 2150 \\ 2r + 3s + 4c &= 6850 \\ 3r + 4.5s + 5.5c &= 9825 \end{aligned}$$

The solution is 500 rakes, 750 shovels, and 900 clippers.

7. The system of equations for this problem with x = the number of hours babysitting, y = the numbers of hours tutoring, and z = the number of hours gardening is:

$$\begin{aligned} x + y + z &= 30 \\ x &= 2y \\ 5x + 12y + 4.5z &= 177.5 \end{aligned}$$

The solution is 10 hours babysitting, 5 hours tutoring, and 15 hours gardening.

EVALUATE LESSON 2

Section 3.10: Opening Day

Purpose: In previous sections students explored and formalized different ways to solve systems of equations. The purpose of this section is to allow students to demonstrate an understanding of finding the solution set for a system of linear equations.

Description: This is a performance assessment. Students should not only be able to find the answer, but also justify why they think their answer is correct. Students should be allowed to use any of the methods for solving systems of equations they have learned so far in this chapter. There are many ways to arrive at the one correct answer; therefore, how students defend and justify their choices is more important than arriving at the answer.

Materials: graphing calculator

Teaching Ideas: Students should solve this problem individually but should be allowed to use a graphing calculator. There are several ways to get to the correct solution. Allow students to choose the method they are the most comfortable with using then justify their solution. Use the rubric to evaluate the students' work.

Students may solve this problem in several ways. They may solve with a graph, substitution, linear combinations, table, or with inverse matrices.

$$\begin{aligned} x + y &= 5000 \\ 5x + 12y &= 3200 \end{aligned}$$

The promotions manager can purchase 4000 caps and 1000 blankets for \$32,000.

$$\begin{aligned} \text{The shipping cost} &= 0.25(4000) + 0.60(1000) \\ &= \$1600.00 \end{aligned}$$

$$\begin{aligned} \text{The total cost} &= \$32{,}000 + \$1600 \\ &= \$33{,}600 \end{aligned}$$

SECTION 3.11

Play Ball!

A LOOK AHEAD

In 1965 the world's first sports stadium with a domed roof was built. The Astrodome in Houston, Texas, was the home of the Houston Astros. It was originally built with glass windows as ceiling tiles. However, during the first game, outfielders discovered that the sunlight shining through the windows made it difficult to see fly balls. Thus, the glass was painted in order to let some light in yet reduce glare. Without direct sunshine, the grass died. To solve this problem, engineers created an artificial grass called Astroturf, and artificial turf was born.

After the construction of the Astrodome, other cities built domed stadiums such as the Metrodome in Minneapolis, Minnesota, or the Superdome in New Orleans, Louisiana. During the 1990s, the people of Houston decided to build a new baseball stadium in downtown Houston. The new stadium, now known as Minute Maid Park, has a retractable roof. When the weather is nice, the Astros can play baseball with an open roof. However, if it is hot or rainy, the roof can be closed, and baseball can be played indoors.

Your sporting goods company plans to sell merchandise at a local baseball stadium. In this section, you will examine information regarding prices and expenses and make decisions about selling your merchandise.

At the stadium souvenir stand, you will sell hats with the home team's logo for $12, and T-shirts with the home team's logo for $20. The stadium charges $500 to rent the space to sell items during a home game, and the team charges $1.50 per item with their logo in royalties. Each hat costs $7 in materials and labor to produce, and each T-shirt costs $9 in materials and labor to produce. There is only enough space in the stand to stock 200 total items. You want to know which combinations of hats and shirts will generate more income than expenses; that is, will create a profit.

1. Use a table like **Figure 3.10** to identify possible combinations of hats and shirts that satisfy the requirement to have no more than 200 items in stock.

Number of Hats	Number of Shirts	Total Number of Items

FIGURE 3.10. Stadium sales table.

2. Neglecting the stadium stand rent, what is the net profit (difference between income and expenses) generated by the sale of one hat?

3. Neglecting the stadium stand rent, what is the net profit generated by the sale of one shirt?

4. Use your combinations of hats and T-shirts and a table like **Figure 3.11** to identify the total income and the total expenses for each combination.

Income Generated by Hats	Income Generated by Shirts	Total Income

Expenses Generated By Hats	Expenses Generated by Shirts	Stand Rent	Total Expenses

FIGURE 3.11. Income and expenses table.

5. How many possible combinations of hats and shirts can you bring to sell at the stadium that will satisfy the restrictions of the situation?

6. Which combination of hats and shirts is best? Justify your choice.

A LOOK AHEAD

ENGAGE LESSON 3

Section 3.11: Play Ball!

Purpose: Students have been introduced to using systems of equations to model situations and solve problems. In this section students will examine a situation in which there is not one unique solution but a range of possible solutions to a particular problem. In Section 3.12 students will explore ways (tabular, graphical, and symbolic) to determine the specific range of possible solutions.

Description: Our sporting goods company has been invited to sell merchandise at a local baseball stadium. Students will read a short history of baseball stadiums in one city then be given information about sale prices and expenses for being a vendor at the stadium. They will use a table to determine possible numbers of shirts and numbers of hats that they should bring for their booth at the stadium.

Materials: pencil, paper, graphing calculator for each student

Teaching Ideas: Students should work in small groups (3 to 4 students) and have access to a calculator. Give students a limited amount of time to work on the questions in their groups (around 15–20 minutes, perhaps longer if students continue to be engaged with the problem). At the end of this time, debrief the students by discussing their answers to the questions.

Answers

1. Sample answers are shown:

Number of Hats	Number of Shirts	Total Number of Items
100	100	200
75	50	125
150	45	195
50	100	150
60	75	135
70	70	140

2. $3.50

3. $9.50

4. Sample answers based on answers from Question 1:

Income Generated by Hats	Income Generated by Shirts	Total Income
1200	2000	3200
900	1000	1900
1800	900	2700
600	2000	2600
720	1500	2220
840	1400	2240

Expenses Generated By Hats	Expenses Generated by Shirts	Stand Rent	Total Expenses
850	1050	500	2400
637.5	525	500	1662.5
1275	472.5	500	2247.5
425	1050	500	1975
510	787.5	500	1797.5
595	735	500	1830

5. There are an infinite number of combinations.

6. Answers may vary. Since this is a system of inequalities, there is not one mathematically best choice. The important part of the answer is the students' justifications of their choices.

SECTION 3.12 Not Everything is Created Equal

In the previous section we looked at the following problem.

Your sporting goods company plans to sell merchandise at a local baseball stadium. At the stadium souvenir stand, you will sell hats with the home team's logo for $12 and T-shirts with the home team's logo for $20. The stadium charges $500 to rent the space to sell items during a home game, and the team charges $1.50 per item with their logo in royalties. Each hat costs $7 in materials and labor to produce, and each T-shirt costs $9 in materials and labor to produce. There is only enough space in the stand to stock 200 total items. What possible combinations of hats and shirts will generate more income than expenses; that is, will create a profit?

You created a table of possible combinations of numbers of hats and T-shirts that satisfy this situation. Use your table and the context of the problem to answer these questions.

1. Let h represent the number of hats and s represent the number of T-shirts. Write an expression to represent the amount of income that the sale of hats (h) and T-shirts (s) will generate.

2. Write an expression to represent the amount of expenses that the sale of a number of hats (h) and a number of T-shirts (s) will generate.

INEQUALITY

An inequality is a statement that compares two variables. For example these are common inequality symbols:

Symbol	Example	Meaning (from left to right)
$<$	$x < y$	x is less than y
$>$	$x > y$	x is greater than y
$\leq$	$x \leq y$	x is less than or equal to y
$\geq$	$x \geq y$	x is greater than or equal to y

3. Use the words **income** and **expenses** to write an inequality that describes the conditions under which a profit will be generated (income will exceed expenses).

4. Substitute your expressions from Questions 1 and 2 into your inequality from Question 3.

5. Solve this inequality for s in terms of h.

6. For this situation and ordered pairs (h, s), which quadrant of the coordinate plane should we consider? Why?

7. Use your graphing calculator to graph this inequality. Describe your window, and sketch your results.

CHECK THIS!

Recall that when graphing an inequality, if the relationship includes "or equal to," use a solid line. If the relationship is "less than" or "greater than" but not "equal to," then use a dashed line.

Recall that when graphing a "less than" inequality, you need to shade the values below your graphed line.

Recall that when graphing a "greater than" inequality, you need to shade the values above your graphed line.

8. Write an inequality representing the possible combinations of hats (h) and T-shirts (s) that you can store in your souvenir stand.

Just like systems of equations, you can work with two inequalities at the same time.

SYSTEMS OF INEQUALITIES

Systems of inequalities are two (or more) inequalities that have the same variables. For example, $2x + 3y < 4$ and $3x + 4y < 5$ are a system of inequalities

9. Use your graphing calculator to graph the number of T-shirts vs. the number of hats for the inequality from Question 8 in the same window as the inequality from Questions 4 and 5. Record your window and sketch your results.

EXPLORE LESSON 3

Section 3.12: Not Everything is Created Equal

Purpose: In Section 3.11 students used a table to examine a range of possible solutions to a particular problem. In Section 3.12 students will use graphical and symbolic representations to determine the specific range of possible solutions. In Section 3.13 students will take their experiences from this section to formalize different methods of solving systems of inequalities and compare them to methods of solving systems of equations.

Description: Students will continue exploring the situation posed in the Engage phase of this lesson. From the tables they generated in Section 3.11, they will develop symbolic rules to describe the constraints of the situation. Then they will use those rules to create a graphical representation of the problem. Students will investigate different regions of the graph and make a conjecture about which region best satisfies the problem situation.

Materials: graphing calculator with the capability to graph inequalities, graph paper, blue- and yellow-colored pencils (optional)

Teaching Ideas: Students should work in small groups (3 to 4 students). Encourage them to discuss the problems with each other and to ask questions of each other before asking the teacher. If students struggle with the graphical representation of a system of inequalities, refer them back to their tables from Section 3.11, the Engage phase of this lesson.

If students struggle with identifying the feasible region, or the solution to the system of inequalities, have them use a blue-colored pencil to shade one inequality then a yellow-colored pencil to shade the other inequality. Since yellow and blue make green, the feasible region will color-code itself.

Answers

1. $12h + 20s = \text{Income}$

2. $500 + 8.50h + 10.50s = \text{Expenses}$

3. Income > Expenses

4. $12h + 20s > 500 + 8.50h + 10.50s$

5.
$$12h + 20s > 500 + 8.50h + 10.50s$$
$$12h + 20s - 10.50s > 500 + 8.50h + 10.50s - 10.50s$$
$$12h + 9.50s > 500 + 8.50h$$
$$12h - 12h + 9.50s > 500 + 8.50h - 12h$$
$$9.50s > 500 - 3.50h$$
$$\frac{9.50s}{9.50} > \frac{500 - 3.50h}{9.50}$$
$$s > \frac{500 - 3.50h}{9.50}$$

6. In this situation, it only makes sense to consider whole numbers of hats and T-shirts. Thus, we only need to consider the first quadrant.

7. 

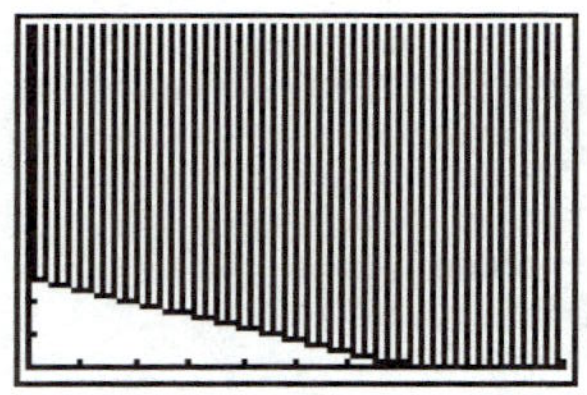

To graph a linear inequality on a TI-83 or TI-84, follow these steps:

1) If necessary, solve the linear inequality for the dependent variable. Then enter it into the function grapher as though it were a linear equation.

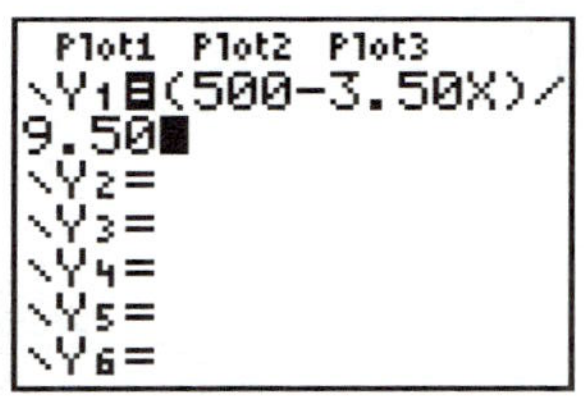

2) Use the Left-Arrow key to highlight the line type symbol just to the left of Y1. Press the Enter key until the desired line type appears. To graph "greater than", use the ◥ line type. To graph "less than", use the ◣ line type.

Sample graph sketched on paper:

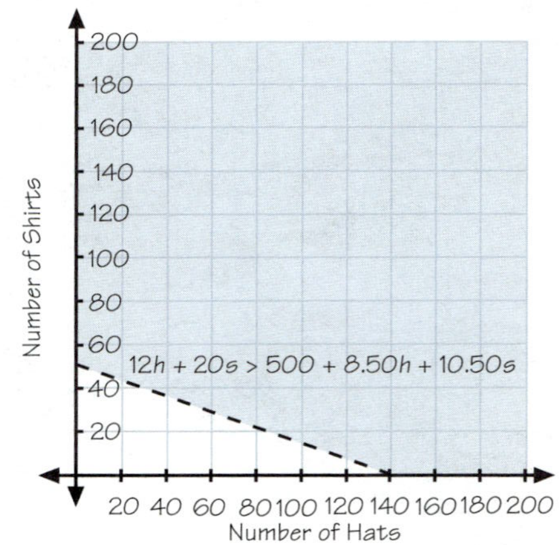

8. $h + s \leq 200$

Answers (continued)

9.
```
WINDOW
 Xmin=0
 Xmax=200
 Xscl=20
 Ymin=0
 Ymax=200
 Yscl=20
 Xres=■
```

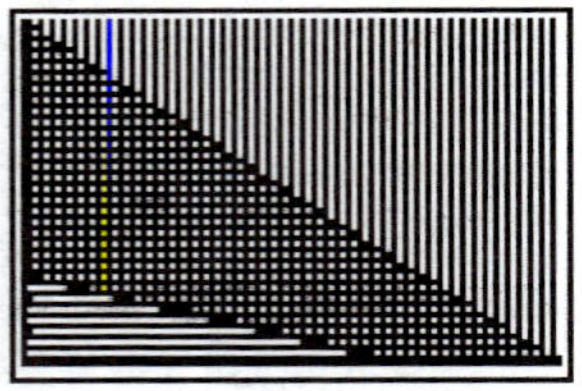

Sample graph sketched on paper:

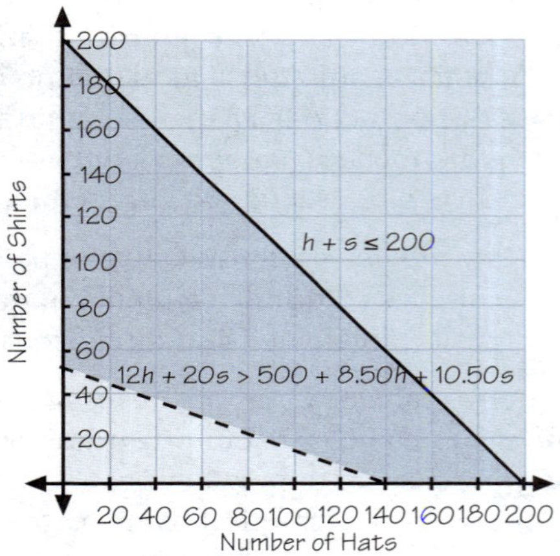

10. Describe the graphs of both inequalities.

11. In this situation, what do the x-coordinate (h-coordinate) and y-coordinate (s-coordinate) represent?

12. Choose a point that lies above and to the right of both lines. Substitute the h-value and s-value for this point into both inequalities. Which inequalities does the point satisfy?

13. Choose a point that lies in the region between both lines. Substitute the h-value and s-value for this point into both inequalities. Which inequalities does the point satisfy?

14. Choose a point that lies below and to the left of both lines. Substitute the h-value and s-value for this point into both inequalities. Which inequalities does the point satisfy?

15. Choose two more points from the region where one point satisfies both inequalities. Do these points also satisfy both inequalities? Justify your answer.

16. Make a conjecture about the region of the coordinate plane where both inequalities are true.

SECTION 3.12 Assignment

1. Graph the following system of inequalities.

$$\begin{cases} y \le 2x - 4 \\ 2x + 3y < 12 \end{cases}$$

a) Is the point (4, –3) a part of the solution to the system of inequalities? Why or why not?

b) Is the point (7, 5) a part of the solution to the system of inequalities? Why or why not?

c) Is the point (3, 2) a part of the solution to the system of inequalities? Why or why not?

d) Is the origin a part of the solution to the system of inequalities? Why or why not?

2. Graph the following system of inequalities.

$$\begin{cases} y \le 2x - 1 \\ x - 3y \ge 18 \end{cases}$$

a) At what point do the lines intersect? Is this point a part of the solution to the system of inequalities? Why or why not?

b) Describe the solution set to the system of inequalities in terms of x and y.

3. To meet shipping requirements at the local U Pack It We Ship It store, the perimeter of the base of a rectangular box must be less than 30 inches. The length must be at least the width of the base of the box.

a) Write two inequalities that describe this situation.

b) What are a reasonable domain and range? Assume width is the independent variable and length is the dependent variable.

Answers (continued)

10. Answers may vary. Students should observe that the first inequality, which was a "greater than" relationship, consists of a dashed line, and the region above the line is shaded. The second inequality, which was a "less than or equal to" relationship, consists of a solid line, and the region below the line is shaded. There is a region between the two lines that is shaded for both inequalities.

11. For any given point, the x- or h-coordinate represents the number of hats sold and the y- or s-coordinate represents the number of T-shirts sold.

12. A point above and to the right of both lines satisfies the inequality $8h + 15s > 500 + 1.50(h + s)$. For example, consider the point (140, 100) that represents selling 140 hats and 100 T-shirts.

$$12h + 20s \stackrel{?}{>} 500 + 8.50h + 10.50s$$
$$12(140) + 20(100) > 500 + 8.50(140) + 10.50(100)$$
$$3680 > 2740$$

$$h + s \stackrel{?}{\le} 200$$
$$(140) + (100) \not\le 200$$
$$240 \le 200$$

The income for 140 hats and 100 T-shirts exceeds the expenses; thus, the first inequality is true. However, there are more than a total of 200 hats and T-shirts, which violates the second inequality.

13. A point in the region between both lines satisfies both inequalities. Consider the point (100, 100), which represents 100 hats and 100 T-shirts.

$$12h + 20s > 500 + 8.50h + 10.50s$$
$$12(100) + 20(100) \stackrel{?}{>} 500 + 8.50(100) + 10.50(100)$$
$$3200 > 2400$$

$$h + s \le 200$$
$$(100) + (100) \stackrel{?}{\le} 200$$
$$200 \le 200$$

The income for 100 hats and 100 T-shirts exceeds the expenses; thus, the first inequality is true. There are a total of 200 hats and T-shirts, which satisfies the second inequality.

14. A point below and to the left of both lines satisfies the inequality $h + s \le 200$. For example, consider the point (20, 20), which represents selling 20 hats and 20 T-shirts.

$$12h + 20s > 500 + 8.50h + 10.50s$$
$$12(20) + 20(20) \stackrel{?}{>} 500 + 8.50(20) + 10.50(20)$$
$$640 \not> 880$$

$$h + s \le 200$$
$$(20) + (20) \stackrel{?}{\le} 200$$
$$20 \le 200$$

The income for 20 hats and 20 T-shirts is less than the expenses; thus, the first inequality is not true. However, there are fewer than 200 hats and T-shirts and this satisfies the second inequality.

15. If students select two points from the region between the two lines, then both inequalities will be satisfied. For example, consider the point (20, 100).

$$12h + 20s > 500 + 8.50h + 10.50s$$
$$12(20) + 20(100) \stackrel{?}{>} 500 + 8.50(20) + 10.50(100)$$
$$2240 > 1720$$

$$h + s \le 200$$
$$(20) + (100) \stackrel{?}{\le} 200$$
$$120 \le 200$$

Now consider the point (120, 60).

$$12h + 20s > 500 + 8.50h + 10.50s$$
$$12(120) + 20(60) \stackrel{?}{>} 500 + 8.50(120) + 10.50(60)$$
$$2640 > 2150$$

$$h + s \le 200$$
$$(120) + (60) \stackrel{?}{\le} 200$$
$$180 \le 200$$

16. If you graph both inequalities on the same coordinate grid, then the region where the shading overlaps is the region that visually shows the possible solutions to the system of inequalities.

Section 3.12 Assignment

Answers

1.

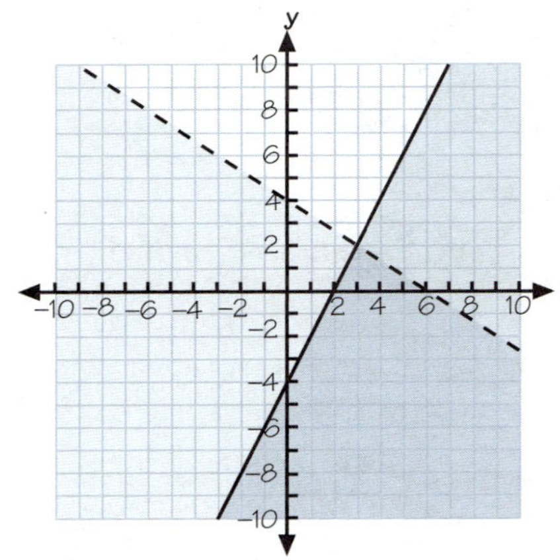

a) Yes, because it lies in the shaded region for both inequalities.

b) No, because it only lies in the shaded region for one inequality, $y \le 2x - 4$.

c) No, because it lies at the intersection of the two lines, and the second inequality, $2x + 3y < 12$, is a "less than" relationship but not an equal relationship (the symbol is "less than" instead of "less than or equal to").

d) No, because it only lies in the shaded region for one inequality, $2x + 3y < 12$.

Answers (continued)

2.

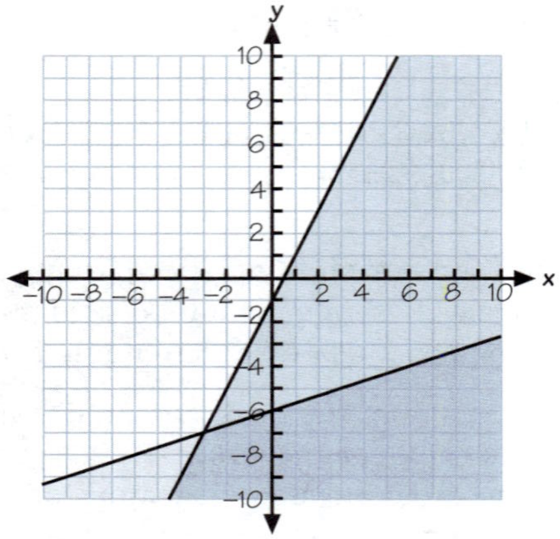

a) The lines intersect at the point (–3, –7). This point is a part of the solution to the system of inequalities because both inequalities include "or equal to" as a part of the relationship between x and y.

b) The solution set is the region to the right of the line $y = 2x - 1$ and beneath the line $x - 3y = 18$.

3. a) $2l + 2w < 30$, or $l + w < 15$

$l \geq w$

b) A reasonable domain for the width is real numbers such that $0 < w < 7.5$.

A reasonable range for the length is real numbers such that $0 < l < 15$.

c) What are some possible values for length and width that satisfy the requirements of the shipping store? Use a table like **Figure 3.12** to organize your data.

Length	Width	Perimeter

FIGURE 3.12.
Box measurement table.

d) Graph the inequalities relating length and width. Use width on the x-axis and length on the y-axis.

e) Are the lengths and widths from your table in the solution set according to your graph? How do you know?

4. The local petting zoo charges $1 for children's admission and $3 for adults. An elementary school class is going on a field trip to the zoo and has $50 budgeted for admission. The small bus they will take holds up to 30 people. What possible combinations of children and adults could attend the field trip? To find out, answer these questions.

a) Define your variables.

Let x = _____

Let y = _____

b) Write two inequalities representing this situation.

c) What are a reasonable domain and range?

d) Graph the two inequalities.

5. Use your graph to fill in each blank with a value so that the combination can attend.

a) 15 children, ______ adults

b) 3 children, _____ adults

c) ___ children, 2 adults

d) ____ children, 5 adults

SECTION 3.13 Solutions to Systems of Inequalities

In the previous section you solved a system of inequalities graphically by identifying a region of the coordinate plane where the points describe pairs that make both of the inequalities true.

Now let's look at some important ideas about systems of inequalities and compare them to what we know about solving systems of equations.

FEASIBLE REGION

The region of a graph containing points that make all of the inequalities in the system true is called the feasible region.

1. When you made a table of possible data points that satisfied all of the conditions of your problem, how did you decide which points to use?

2. How did you develop your expressions for income and expenses?

3. When you wrote your inequalities, how did you decide which relation to use: less than, greater than, or equal to?

4. When you graphed your inequalities, how did you decide which region to shade?

5. How did you decide which region satisfied both inequalities?

6. Are the lines in your graph part of the feasible region? Explain.

7. What does the point of intersection of the lines in the graph represent? Is it a part of the feasible region?

8. How does the solution of a system of inequalities compare to the solution of a system of equations?

Answers (continued)

c)

Length	Width	Perimeter
10	3	26
8	6	28
8	5	26
6	4	20
5	5	20
8.5	4	25

d)

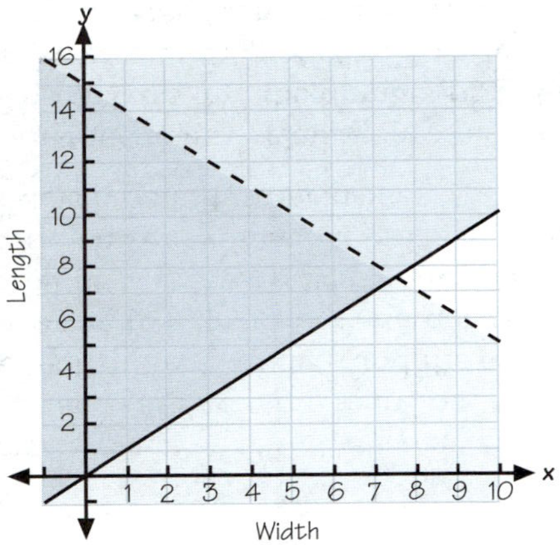

e) Yes. Each of the points lies either inside the shaded region or on a solid line.

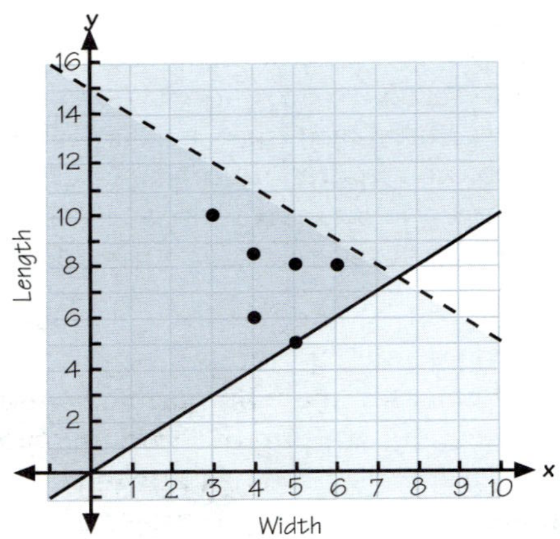

4. a) Let x = number of children

   Let y = number of adults

   b) $x + 3y \leq 50$ and $x + y \leq 30$

   c) A reasonable domain is whole numbers up to and including 30.

   A reasonable range is whole numbers up to and including 16.

   d)

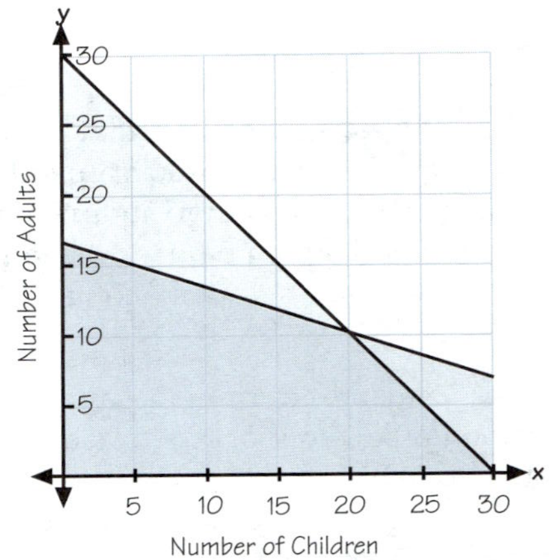

5. a) Answer: any whole number less than 11

   b) Answer: any whole number less than 15

   c) Answer: any whole number less than 28

   d) Answer: any whole number less than 25

EXPLAIN LESSON 3

Section 3.13: Solutions to Systems of Inequalities

Purpose: In Section 3.12 students graphically solved a system of inequalities. In this section students will formalize methods of solving systems of inequalities using tables and graphs and compare these to methods of solving systems of equations. This will prepare students to use linear programming in Section 3.14 to maximize income in a problem situation.

Description: This section is designed to be a teacher-led discussion of the concepts in the lesson. Use the questions to facilitate the class discussion.

Materials: graphing calculator with the capability to graph inequalities (INEQUALZ app for the TI calculator), graph paper

Teaching Ideas: Debrief students on Section 3.12 by having them share their answers to the questions. As they share, be sure to fill in any missing concept, process, or skill that could have been used to answer the questions. Then work through the questions in this section with the entire class. Once again be sure all necessary concepts, processes, or skills are addressed.

Answers

1. Answers may vary. Students will likely have discussed the points within their groups. Most of them will have used some type of educated guess and check approach, beginning with points that satisfied the simplest condition (no more than 200 items in the souvenir stand) and then seeing if that number of hats and T-shirts generated more income than expenses.

Answers (continued)

2. The expression for income is developed because \$12 per hat and \$20 per T-shirt is a rate of change. In order to find the total amount of income from selling a hat, for example, the number of hats should be multiplied by the cost per hat.

   The expression for expenses is developed because there is a fixed rate of \$500 charged without regard to the number of items sold. This fixed rate is a starting point, or y-intercept, of a linear function. The expenses of logo licensing fee, labor, and materials depend on the number of items produced and sold. Thus, the \$8.50 charge per hat is a rate of change or slope and should be multiplied by the number of hats sold, and the \$10.50 charge per T-shirt sold is a rate of change and should be multiplied by the number of T-shirts sold.

3. Answers may vary. Students may cite key words such as "no more than" or "at least." Be sure that students understand the concept of comparing two quantities and determining which one is greater or less. Often students depend on key words in a problem situation and develop faulty conclusions.

4. The first inequality relating income and expenses states that the number of T-shirts sold must be greater than or equal to a certain expression. Hence, all s-values greater than the resulting line, or above, should be shaded.

   The second inequality limits the number of hats and T-shirts stocked in the souvenir stand to 200 total. The line represents all values of h and s that sum to 200. Since the total must be less than 200, all points that are beneath the line should be shaded.

5. During the Explore phase of the lesson, Section 3.12, students select an ordered pair from a point in different regions of the graph and then substitute those values into both inequalities. The points from the region that made both inequalities true indicate which region satisfies both of the inequalities at the same time.

   This region is called the **feasible region** for the solution of the system of inequalities. All of the ordered pairs of points contained in this region will satisfy all of the inequalities in the system.

6. If the line is a part of an inequality that is "less than or equal to" or "greater than or equal to" or it is a solid line, then the line is a part of the feasible region. If the line is a part of an inequality that is "less than" or "greater than" but not "equal to" or is a broken line, then the line is not a part of the feasible region.

7. The point of intersection represents the values that make the expressions on either side of the relational symbol $>$ or $<$ equal in value. It is only a part of the feasible region when both inequalities include "or equal to" as a part of the relationship.

8. A system of equations has one unique solution that is represented by the point of intersection of the graphs of the equations. A system of inequalities, however, has multiple solutions that are represented by a feasible region of the coordinate plane.

9. Suppose a customer orders hats and T-shirts. They want no more than 60 items and no more than twice as many T-shirts as hats.

   a) What are a reasonable domain and range for this situation? Justify your answer.

   b) Write an inequality in terms of h, the number of hats, and s, the number of T-shirts, to describe the condition "they want no more than 60 items."

   c) Write an inequality in terms of h and s to describe the condition "no more than twice as many T-shirts as hats."

   d) Graph both of these inequalities together on your graphing calculator. Choose an appropriate window and sketch your results. Identify the solution to the system.

SUMMARY

Solving Systems of Inequalities

Systems of inequalities have more than one solution. So a good way to represent the solution is with a graph. To create a graph for the solution to a system of inequalities,

- identify an appropriate domain and range for the situation. Use these to set up your independent and dependent variable axes;
- graph each inequality. Use a solid line if the inequality uses the symbol $\geq$, "greater than or equal to," or the symbol $\leq$, "less than or equal to." Use a broken line if the inequality uses the symbol $>$, "greater than," or $<$, "less than";
- shade the appropriate region for each inequality. Shade the y-values that are greater than (above) the line if the inequality uses "greater than." Shade the y-values that are less than (below) the line if the inequality uses "less than"; and
- identify the feasible region. The feasible region is the part of the graph where the shaded regions of the individual inequalities overlap.

In this section you examined systems of linear inequalities.

1. You looked at tabular and graphical representations of the solution to systems.
2. You compared the solution for a system of equations to the solution for a system of inequalities.

Answers (continued)

9. a) You must have whole numbers for both the domain and range; you cannot have either negative or fractional amounts of hats or T-shirts.

   b) $h + s \leq 60$

   c) $s \leq 2h$

   d) The solution is the region bounded by the inequalities $h + s \leq 60$ and $s \leq 2h$ and the x-axis, as shown below.

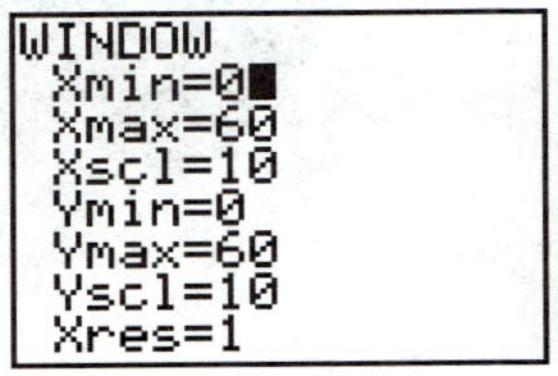

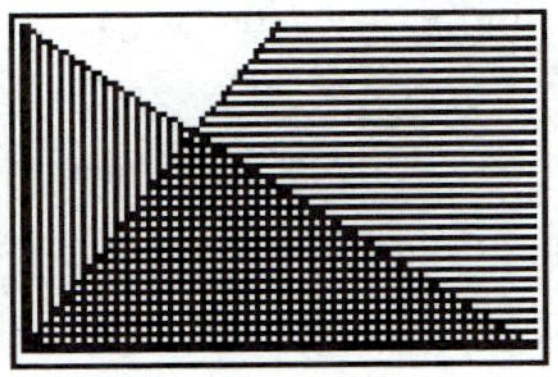

A paper-and-pencil graph might look like this:

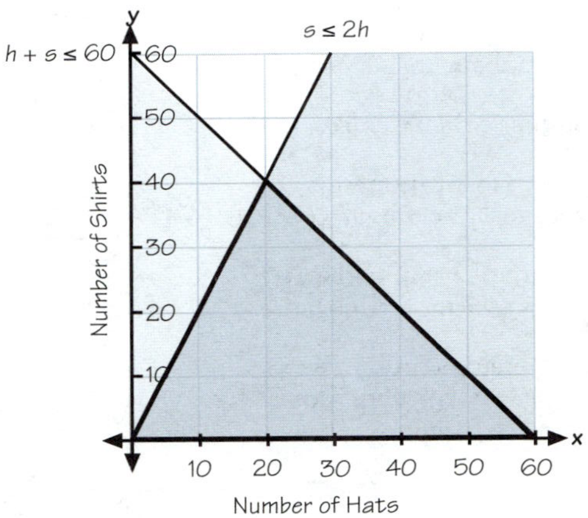

SECTION 3.13 Assignment

1. A men's medium baseball jersey requires $2\frac{1}{4}$ yards of fabric. A men's large baseball jersey requires $3\frac{3}{4}$ yards of fabric. You have 75 yards of fabric and want to make no more than 30 jerseys. How many of each type of baseball jersey can you make?

   a) Define your variables. Identify the independent and dependent variables.

   b) What are a reasonable domain and range for this situation? Justify your answer.

   c) Write two inequalities to describe the situation.

   d) Solve each inequality for your dependent variable in terms of your independent variable.

   e) Graph both of these inequalities together. Choose an appropriate window and sketch your results. Identify the solution to the system.

2. A loaf of rye bread requires 3 cups of flour and $\frac{1}{2}$ cup of sugar. A loaf of pumpernickel bread requires $2\frac{1}{2}$ cups of flour and 1 cup of sugar. You have 40 cups of flour and 10 cups of sugar. How many of each kind of loaf can you bake?

   a) Define your variables. Identify the independent and dependent variables.

   b) What are a reasonable domain and range for this situation? Justify your answer.

   c) Write two inequalities to describe the situation.

   d) Solve each inequality for your dependent variable in terms of your independent variable.

   e) Graph both of these inequalities together. Choose an appropriate window and sketch your results. Identify the solution to the system.

3. A ballpark offers two packages for birthday party favors. Package A includes 5 tickets for a ball toss game and 6 pennants for the home team. Package B includes 15 tickets for a ball toss game and 3 pennants for the home team. The ballpark management want to have at least 30 packages available. They also want to include at least 300 tickets but no more than 180 pennants. How many of each package can they assemble?

   a) Define your variables. Identify the independent and dependent variables.

   b) What are a reasonable domain and range for this situation? Justify your answer.

   c) Write three inequalities to describe the situation.

   d) Solve each inequality for your dependent variable in terms of your independent variable.

   e) Graph both of these inequalities together. Choose an appropriate window and sketch your results. Identify the solution to the system.

Section 3.13 Assignment

Answers

1. a) Let m = number of medium jerseys be the independent variable

   Let l = number of large jerseys be the dependent variable

   b) Since m is the independent variable, a reasonable domain for this situation is all whole numbers such that $m \le 30$. If all you made were medium jerseys, you could make 30 and have fabric left over. Since l is the dependent variable, a reasonable range for this situation is all whole numbers such that $l \le 20$. 20 is the maximum number of large jerseys that could be made with 75 yards of fabric.

   c) $2.25m + 3.75l \le 75$ and $m + l \le 30$

   d)
$$\begin{aligned} 2.25m + 3.75l &\le 75 \\ 2.25m - 2.25m + 3.75l &\le 75 - 2.25m \\ 3.75l &\le 75 - 2.25m \\ \frac{3.75l}{3.75} &\le \frac{75 - 2.25m}{3.75} \\ l &\le 20 - 0.6m \end{aligned}$$
$$\begin{aligned} m + l &\le 30 \\ m - m + l &\le 30 - m \\ l &\le 30 - m \end{aligned}$$

   e) The solution is the region bounded by the inequalities $2.25m + 3.75l \le 75$ and $m + l \le 30$ and the x-axis and the y-axis, as shown below.

```
WINDOW
 Xmin=0
 Xmax=30
 Xscl=5
 Ymin=0
 Ymax=30
 Yscl=5
 Xres=1
```

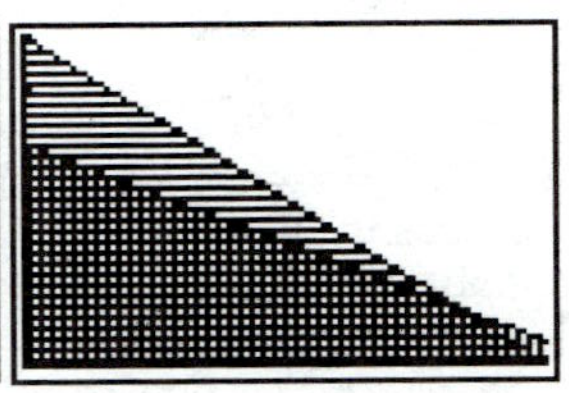

2. a) Let r = number of loaves of rye bread be the independent variable

   Let p = number of pumpernickel bread be the dependent variable

   b) Since r is the independent variable, a reasonable domain for this situation is all whole numbers such that $r \le 13$. If all you baked were loaves of rye bread, you could make 13 loaves. Since p is the dependent variable, a reasonable range for this situation is all whole numbers such that $p \le 10$. If all you baked were loaves of pumpernickel bread, you could make 10 loaves.

   c) $3r + 2.5p \le 40$ and $0.5r + p \le 10$

   d)
$$\begin{aligned} 3r + 2.5p &\le 40 \\ 3r - 3r + 2.5p &\le 40 - 3r \\ 2.5p &\le 40 - 3r \\ \frac{2.5p}{2.5} &\le \frac{40 - 3r}{2.5} \\ p &\le 16 - 1.2r \end{aligned}$$
$$\begin{aligned} 0.5r + p &\le 10 \\ 0.5r - 0.5r + p &\le 10 - 0.5r \\ p &\le 10 - 0.5r \end{aligned}$$

   e) The solution is the region bounded by the inequalities $3r + 2.5p \le 40$ and $0.5r + p \le 10$ and the x-axis and the y-axis, as shown below.

```
WINDOW
 Xmin=0
 Xmax=20
 Xscl=5
 Ymin=0
 Ymax=20
 Yscl=5
 Xres=1
```

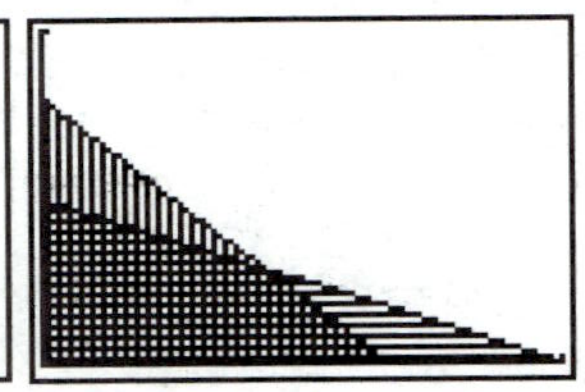

3. a) Let x = number of package A be the independent variable

   Let y = number of package B be the dependent variable

   b) Since x is the independent variable, the domain for this situation is restricted to whole numbers (you cannot have partial packages). The ballpark only has 180 pennants; therefore, if all packages were package A, then the highest number of package A's that can be assembled is 30.

   Since y is the dependent variable, the range for this situation is also restricted to whole numbers. The ballpark only has 180 pennants; therefore, if all packages were package B, then the highest number of package B's that can be assembled is 60.

   c) $x + y \ge 30$, $5x + 15y \ge 300$, and $6x + 3y \le 180$

   d)
$$\begin{aligned} x + y &\ge 30 \\ x - x + y &\ge 30 - x \\ y &\ge 30 - x \end{aligned}$$
$$\begin{aligned} 5x + 15y &\ge 300 \\ 5x - 5x + 15y &\ge 300 - 5x \\ 15y &\ge 300 - 5x \\ \frac{15y}{15} &\ge \frac{300 - 5x}{15} \\ y &\ge 20 - \frac{1}{3}x \end{aligned}$$
$$\begin{aligned} 6x + 3y &\le 180 \\ 6x - 6x + 3y &\le 180 - 6x \\ 3y &\le 180 - 6x \\ \frac{3y}{3} &\le \frac{180 - 6x}{3} \\ y &\le 60 - 2x \end{aligned}$$

   e) The solution is the region bounded by the inequalities $x + y \ge 30$, $5x + 15y \ge 300$, and $6x + 3y \le 180$ and the y-axis, as shown.

```
WINDOW
 Xmin=0
 Xmax=30
 Xscl=5
 Ymin=0
 Ymax=60
 Yscl=5
 Xres=1
```

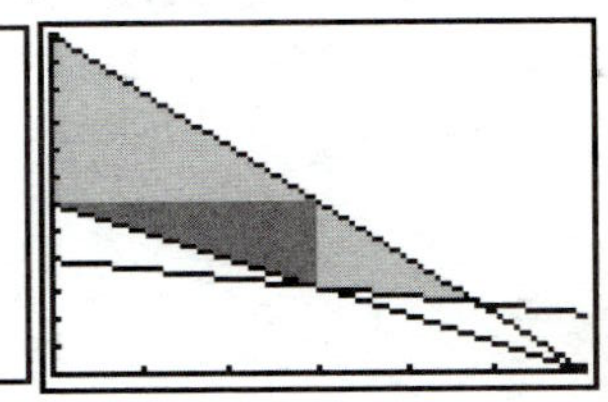

SECTION 3.14 Make the Most of It

In Sections 3.12 and 3.13, you used a system of inequalities to represent a problem about your sports apparel business. A local baseball stadium has invited you to sell hats and T-shirts at home games. Thus far, you have used linear inequalities to represent constraints on the situation. The system of inequalities yielded a feasible region with possible combinations of hats and T-shirts that satisfy all of the constraints.

Which one of those combinations can you use to generate the most income for your business? In this section you will explore a way to use the graphs of linear inequalities to answer this question.

Earlier in this chapter we have explored the following problem.

Your sporting goods company plans to sell merchandise at a local baseball stadium. At the stadium souvenir stand, you will sell hats with the home team's logo for \$12, and T-shirts with the home team's logo for \$20. The stadium charges \$500 to rent the space to sell items during a home game, and the team charges \$1.50 per item with their logo in royalties. Each hat costs \$7 in materials and labor to produce and each T-shirt costs \$9 in materials and labor to produce. There is only enough space in the stand to stock 200 total items. What possible combinations of hats and shirts will generate more income than expenses; i.e., what combinations will create a profit?

For now, let's ignore the limit of 200 items, since we are thinking about production at the factory. Consider that the cost for hats is \$2 in material and \$5 in labor. Also, the cost for T-shirts is \$4 in material and \$5 in labor. There is still a licensing fee of \$1.50 per item for the use of the team's logo.

Upon further analysis of your business accounts, you decide that you can spend no more than \$1000 in material and \$1500 in labor for one week. In order to maximize your income, how many hats and how many T-shirts should you make?

In order to solve this problem, answer these questions.

1. First, create a table like **Figure 3.13** to organize the information:

	Cost of Materials per Item	Cost of Labor per Item	Expected Income per Item
Hats			
T-shirts			
Maximum Allowed			

FIGURE 3.13.
Table of cost data.

2. Let h represent the number of hats sold and s represent the number of T-shirts sold. Write a function rule that describes the amount of income (I) in terms of the number of hats (h) and the number of T-shirts (s) sold.

3. Write a system of inequalities to represent each of these costs in terms of h and s:

   a) Cost of Materials

   b) Cost of Labor

4. What domain and range restrictions do we need to consider in the context of this problem? How do you know?

5. Solve the inequalities from Question 3 for s in terms of h.

6. Graph the two inequalities from Question 5. Consider your answer to Question 4 to set an appropriate window for your graph. Describe your window and sketch your results.

7. What is the feasible region for this system?

8. What are the vertices of the feasible region? Are these points included in the solution to the system of inequalities?

9. Use your income function to find the income that is generated by these four combinations of hats and T-shirts. Use a table like **Figure 3.14** to organize your data.

Points (h, s)	Income Function $I = 12h + 20s$

FIGURE 3.14.
Income possibilities.

ELABORATE LESSON 3

Section 3.14: Make the Most of It

Purpose: In Sections 3.12 and 3.13, students examined tabular and graphical solutions to systems of inequalities. In this section students will use systems of inequalities to solve contextual problems that ask them to maximize or minimize a given quantity, such as cost or income. This prepares students to apply their knowledge of systems of inequalities to solving a problem in a new context in Section 3.15.

Description: In this section students apply and extend their knowledge. Students will set up and solve a system of inequalities. Students will then use a simplified linear programming model in order to determine the maximum income that can be generated from selling a combination of hats and T-shirts within certain constraints.

Materials: graphing calculator with the capability to graph inequalities, graph paper (optional)

Teaching Idea: Have students solve problems using multiple representations (graph, table, symbolic manipulation). The investigation calls for students to use a graphing calculator. You may need to spend some time reviewing solving an equation or inequality for one variable in terms of other variables. You may also need to spend some time reviewing how to graph inequalities on a graphing calculator. Other graphing calculator skills that are used in this activity could include calculating the value of intersection points, calculating the value of an x-intercept (zero), and using the Trace function to identify the coordinates of a certain point.

Students should work on this section while in their groups and should be encouraged to communicate with each other by asking questions and providing assistance as necessary. This will help the teacher with technology issues that arise as well as content issues that need to be addressed.

NOTE: Linear programming is a method that is commonly used in industry. In this section students are told to find the vertices of the feasible region and calculate the quantity being minimized or maximized. The reason that this method works is that all of the objective functions (functions describing the quantity being minimized or maximized) are parallel lines. You can easily demonstrate the parallel lines using a straightedge such as a ruler or piece of spaghetti. Find the equation of one objective function and then graph it. Lay the straightedge on top of the line and move it parallel to the original line until it reaches the last point of your feasible region. This last point will always be a vertex of the feasible region.

Answers

1.

	Cost of Materials per Item	Cost of Labor per Item	Expected Income per Item
Hats	\$2	\$5	\$12
T-shirts	\$4	\$5	\$20
Maximum Allowed	\$1000	\$1500	

2. $I = 12h + 20s$

3. a) $2h + 4s \le 1000$

   b) $5h + 5s \le 1500$

4. Only whole numbers can be used to describe the numbers of hats and T-shirts produced, so we can restrict $h > 0$ and $s > 0$.

5. $2h + 4s \le 1000$; $4s \le -2h + 1000$; $s \le -\frac{1}{2}h + 250$ $5h + 5s \le 1500$; $5s \le 1500 - 5h$; $s \le 300 - h$

6.

```
Plot1 Plot2 Plot3
\Y1=(1000-2X)/4
\Y2=(1500-5X)/5
\Y3=
\Y4=
\Y5=
\Y6=
\Y7=
```

```
WINDOW
 Xmin=0
 Xmax=500
 Xscl=50
 Ymin=0
 Ymax=350
 Yscl=50
 Xres=1
```

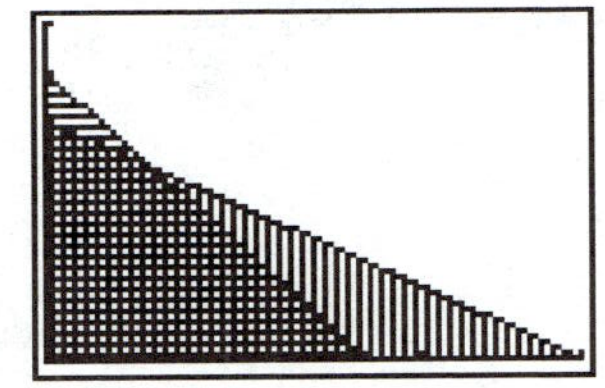

7. The region bounded by $s = 0$, $h = 0$, $2h + 4s \le 1000$, and $5h + 5s \le 1500$ as indicated by the checkered shading on the graph in Question 6.

8. There are four vertices: (0, 250), (0, 0), (300, 0), and (100, 200). Since both inequalities include "or equal to," the boundary points, including the vertex, are a part of the solution.

9.

Points (h, s)	Income Function $I = 12h + 20s$
(0, 250)	$I = 12(0) + 20(250)$ $I = \$5000$
(0, 0)	$I = 12(0) + 20(0)$ $I = \$0$
(300, 0)	$I = 12(300) + 20(0)$ $I = \$3600$
(100, 200)	$I = 12(100) + 20(200)$ $I = \$5200$

10. According to our table, how many hats and T-shirts should we produce in order to maximize our income?

11. Is it possible, within the constraints of this problem, to earn an income of $7000? Justify your answer.

12. Is it possible, within the constraints of this problem, to earn an income of $3000? Justify your answer.

13. Are there other points from the feasible region that generate more income than one of the vertices? Expand your table to record your findings.

If you want to minimize or maximize a quantity when you have two constraints that can be represented by linear inequalities, you can use a method called **linear programming**. Here are the steps:

- *Define a system of inequalities.* Use the information in the problem to write a system of inequalities that describe the constraints. A table, such as the one in Figure 3.14, will help you organize the data and write your inequalities.
- *Write an equation describing the quantity you wish to minimize or maximize.* In the example used in this section, you tried to maximize income, so you wrote an income function.
- *Graph both inequalities on the same coordinate plane.*
- *Identify a feasible region.* The feasible region is the area of overlapped shading between the inequalities. The domain and range of the situation will also help define the feasible region. In the example used in this section, you can only have positive numbers of hats and T-shirts, so the feasible region is restricted to Quadrant I.
- *Find the coordinates of the vertices of the feasible region.* Using technology such as a graphing calculator can help you to find the intersection point of two lines.
- *Substitute the coordinates of each of the vertices into the function you are trying to minimize or maximize.* You will need to substitute each ordered pair into your function and calculate the desired quantity.

- *Identify the coordinates that yield the minimum or maximum value of the function.* One of the vertices of the feasible region will yield a minimum or maximum for the quantity. This ordered pair is the solution.

Answers (continued)

10. According to the table, producing 100 hats and 200 shirts will generate the most income.

11. No. Consider the income function $7000 = 12h + 20s$. Graphing this function (bold line) over our system of inequalities reveals that this function does not pass through the feasible region.

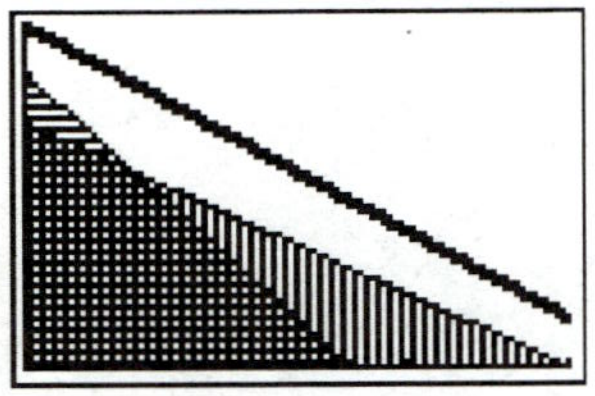

12. Yes. The income function $3000 = 12h + 20s$ passes through the feasible region.

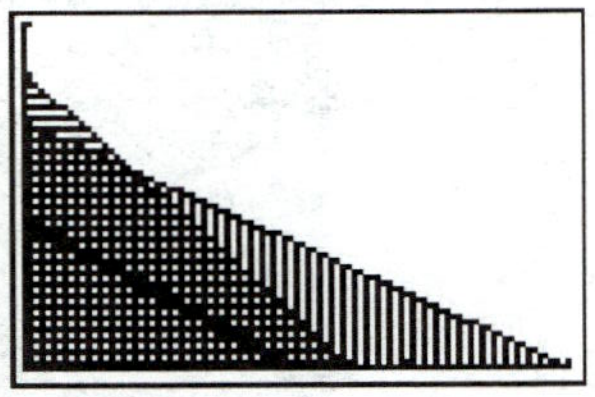

13. There are no other points yielding more than $5200 in income. Data for four more sample points are shown.

Points (h, s)	Income Function $I = 12h + 20s$
(0, 250)	$I = 12(0) + 20(250)$ $I = \$5000$
(0, 0)	$I = 12(0) + 20(0)$ $I = \$0$
(300, 0)	$I = 12(300) + 20(0)$ $I = \$3600$
(100, 200)	$I = 12(100) + 20(200)$ $I = \$5200$
(50, 40)	$I = 12(50) + 20(40)$ $I = \$1400$
(100, 100)	$I = 12(100) + 20(100)$ $I = \$3200$
(200, 100)	$I = 12(200) + 20(100)$ $I = \$4400$
(10, 150)	$I = 12(10) + 20(150)$ $I = \$3120$

SECTION 3.14 Assignment

1. Two machines, Machine A and Machine B, produce two types of chili: beef and vegetarian. One machine does the processing of the raw ingredients and mixes them together. The other machine cooks the chili and packages it. There is a profit of \$150 per case of vegetarian chili and \$100 per case of beef chili. To make a case of vegetarian chili, Machine A must run 2 hours and Machine B must run 9 hours. To make a case of beef chili, Machine A must run 4 hours and Machine B must run 2 hours. Each machine runs 24 hours a day. What combination of cases of vegetarian chili and beef chili will result in a maximum profit?

   Let x = the number of cases of vegetarian chili and y = the number of cases of beef chili.

   a) What domain and range make sense for this situation?

   b) Write an income function in terms of x and y.

   c) Write four inequalities, in terms of x and y, that describe the constraints.

   d) Graph the system of inequalities. Identify the feasible region. If you use a graphing calculator, also record your viewing window.

   e) Identify the vertices of your feasible region. Use your income function to calculate the income for each combination of cases of vegetarian chili and cases of beef chili.

   f) What combination of cases of vegetarian chili and beef chili will result in a maximum profit?

2. A trucking company can ship the hats and T-shirts produced by your sporting goods company. Their trucks can hold a maximum of 250 cubic feet in volume and 2030 pounds in weight. One box of T-shirts has a volume of 5 cubic feet, weighs 35 pounds, and will generate \$114 in profit. One box of hats has a volume of 3 cubic feet, weighs 35 pounds, and will generate \$70 in profit. What combination of boxes of hats and T-shirts can you ship to maximize your profit?

   Let x = the number of boxes of hats and y = the number of boxes of T-shirts.

   a) What domain and range make sense for this situation?

   b) Write four inequalities, in terms of x and y, that describe the constraints.

   c) Write a profit function in terms of x and y.

   d) Graph the system of inequalities. Identify the feasible region. If you use a graphing calculator, also describe your viewing window.

   e) Identify the vertices of your feasible region. Use your income function to calculate the profit for each combination of boxes of hats and boxes of T-shirts.

   f) What combination of boxes of hats and boxes of T-shirts will result in a maximum profit?

Section 3.14 Assignment

Answers

1. a) The domain describes the possible values for the number of cases of vegetarian chili, which must be a whole number. The range describes the possible values for the number of cases of beef chili, which must also be a whole number.

   b) $I = 150x + 100y$

   c)

$$\begin{aligned} x &\geq 0 & \quad & x \geq 0 \\ y &\geq 0 & & y \geq 0 \\ 2x + 4y &\leq 24 & \text{or} \quad & y \leq -\tfrac{1}{2}x + 6 \\ 9x + 2y &\leq 24 & & y \leq -\tfrac{9}{2}x + 12 \end{aligned}$$

   d)

```
WINDOW
 Xmin=0
 Xmax=12
 Xscl=1
 Ymin=0
 Ymax=13
 Yscl=1
 Xres=1
```

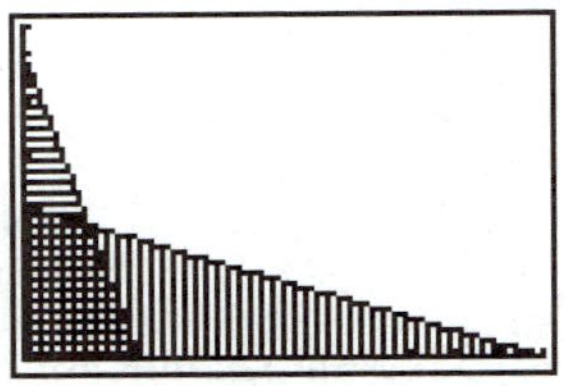

The feasible region is the region bounded by the x-axis and y-axis, and the inequalities $2x + 4y \leq 24$ and $9x + 2y \leq 24$.

   e)

Vertex	Process	Income
(1.5, 5.25)	150(1.5) + 100(5.25)	$750
(0, 6)	150(0) +100(6)	$600
$\left(2\frac{2}{3}, 0\right)$	$150\left(2\frac{2}{3}\right) + 100(0)$	$400
(0, 0)	150(0) + 100(0)	$0

   f) The vertex (1.5, 5.25) has the most income, $750. However, since our domain and range are restricted to whole numbers, consider the point (1, 5). The income generated by this combination will be 150(1) + 100(5) = $650. Therefore, 1 case of vegetarian chili and 5 cases of beef chili will yield the maximum profit.

2. a) The domain describes the possible values for the number of boxes of hats, which must be a positive whole number. The range describes the possible values for the number of boxes of T-shirts, which must also be a positive whole number.

   b)

$$\begin{aligned} x &\geq 0 & \quad & x \geq 0 \\ y &\geq 0 & & y \geq 0 \\ 3x + 5y &\leq 250 & \text{or} \quad & y \leq -\tfrac{3}{5}x + 50 \\ 35x + 35y &\leq 2030 & & y \leq -x + 58 \end{aligned}$$

   c) $P = 70x + 114y$

   d)

```
WINDOW
 Xmin=0
 Xmax=60
 Xscl=5
 Ymin=0
 Ymax=60
 Yscl=5
 Xres=1
```

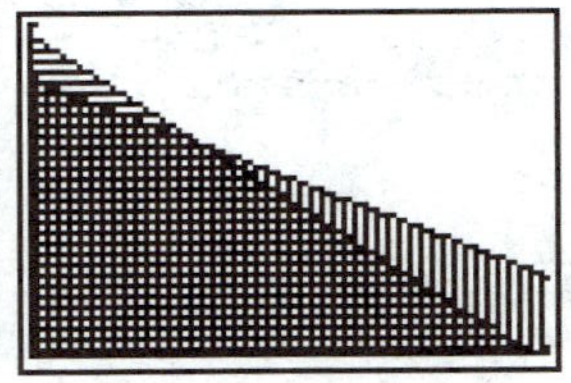

The feasible region is the region bounded by the x-axis and y-axis and the inequalities $3x + 5y \leq 250$ and $35x + 35y \leq 2030$.

   e)

Vertex	Process	Income
(20, 38)	70(20) + 114(38)	$5732
(0, 50)	70(0) + 114(50)	$5700
(58, 0)	70(58) + 114(0)	$4060
(0, 0)	70(0) + 114(0)	$0

   f) The vertex (20, 38) has the most profit, $5732. Therefore, 20 boxes of hats and 38 boxes of T-shirts will yield the maximum profit.

SECTION 3.15

Count the Trees

Let's apply what you have learned thus far about building and using systems of linear inequalities to a new situation.

1. Your friend, Judy, has started a tree farm. She grows oak and pine seedlings and then ships them to customers. Because your business has become so successful, she comes to you for advice.

   A customer wants to order a combination of oak and pine seedlings. The customer wants at least 8 seedlings to include at least 5 pine seedlings. The box your friend usually uses for shipping can hold no more than 18 pounds. One oak seedling weighs 1 pound and costs \$3 to ship. One pine seedling weighs 2 pounds but because of favorable shipping regulations only costs \$2 to ship. What combination of oak and pine seedlings should Judy advise that the customer order to minimize shipping costs? Justify your answer.

Modeling Project Growing a Business

Titangrow is an amazing new fertilizer developed by Newton Farm Products. Demand is expected to be high, but there are many issues to think about besides sales.

First of all, the company must think about making the product. A key ingredient they use in Titangrow is salmon bone meal. This ingredient is expensive: Each increase of 5% in the quantity of salmon bone meal adds 20% to the manufacturing costs.

In testing the product, the company found that when the product has 25% bone meal it is most effective. At 15% bone meal it is very effective. And at 10% bone meal it is somewhat effective. Below 10% bone meal, the product is not effective. The company has decided to create 3 products:

Titangrow Super (25%)

Titangrow Advanced (15%)

Titangrow Standard (10%)

Each 10-lb bag of Titangrow Standard costs the company \$3 to make.

Market research makes the company think that demand for Titangrow Super will be equal to Titangrow Advanced and Standard combined.

The company must also keep track of inventory at three warehouses.

Create a model for manufacturing, pricing (and profit), and keeping track of the inventory for these three products. Then create a summary report that addresses these issues for the entire product line. When you develop your model, remember that systems of equations, matrices, and linear programming are powerful business tools. Use them wisely.

EVALUATE LESSON 3

Section 3.15: Count the Trees

Purpose: In Sections 3.11 through 3.13, students explored and formalized methods of solving a system of linear inequalities. In Section 3.14 they used linear inequalities in order to solve a problem involving maximizing or minimizing a quantity such as cost. In this section students will have the opportunity to demonstrate what they have learned about using a variety of methods of solving problems using systems of linear inequalities.

Description: Students determine and implement a method to solve the problem. This is an open-middle performance assessment. There are several methods that may be used to arrive at the answer. In their justification, students should demonstrate that they have mastered the concepts presented in this lesson.

Materials: none

Teaching Ideas: Assign the section as a performance assessment task. Use the mathematics performance assessment rubric to evaluate the students' work.

Answers

1. Sample solution:

   First, define the variables and write the inequalities describing the constraints of the situation. Let x represent the number of oak seedlings and y represent the number of pine seedlings. Then we can write the following inequalities:

   Total number of seedlings: $x + y \geq 8$

   At least 5 pine seedlings: $y \geq 5$

   Weight of package: $x + 2y \leq 18$

   Next, write the cost function in terms of x and y: $C = 3x + 2y$.

   Graph each inequality and identify the feasible region.

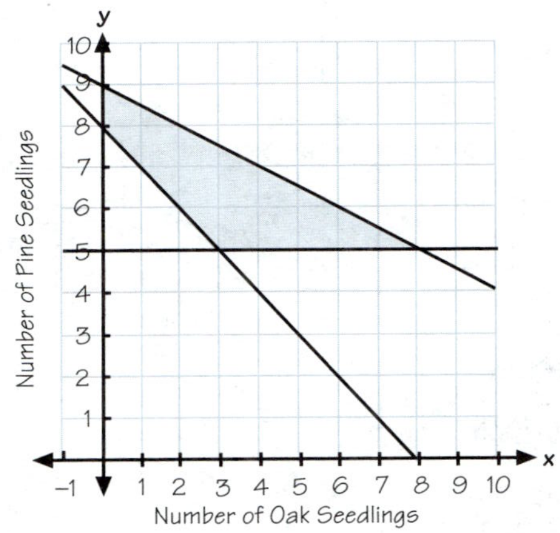

   Now identify the vertices of the feasible region. Test each point in the cost function to find the ordered pair that yields the minimum cost.

Vertex	Process	Cost
(0, 8)	3(0) + 2(8)	\$16
(0, 9)	3(0) + 2(9)	\$18
(3, 5)	3(3) + 2(5)	\$19
(8, 5)	3(8) + 2(5)	\$34

   In order to minimize shipping costs, Judy should advise the customer to purchase no oak seedlings and 8 pine seedlings.

Practice Problems

1. Suppose that one music subscription service, Dozster, charges \$11 per month plus \$0.85 per song; and a second service, Melody, charges \$8 per month plus \$1.00 per song. Let c be the monthly cost in dollars and n be the number of songs downloaded in a month.

   a) Write a linear equation that represents the monthly subscription cost for Dozster.

   b) Write a linear equation that represents the monthly subscription cost for Melody.

   c) Compare the monthly subscription cost for each company if you download 10 songs. Which is cheaper?

   d) Compare the monthly subscription cost for each company if you download 30 songs.

   e) How many songs could you download from each subscription service for \$50?

   f) Use a system of equations to find the number of downloads and the cost at which the two services are the same.

   g) Graph this system of equations and record your window.

   h) For what number of songs is Melody a better deal? For what number of songs is Dozster a better deal?

2. A coffee shop sells several different kinds of coffee. The shop also uses some of its coffees to make its own custom blends. Coffee A sells for \$6.00 a pound. Coffee B sells for \$10.00 a pound. The shop's manager wants to create a blend of the two types A and B that sells for \$7.00 a pound. The manager wants to make 10 pounds of this blend.

   Use a for the number of pounds of coffee A in the blend. Use b for the number of pounds of coffee B.

   a) Write an equation that describes the total number of pounds of coffees A and B in the blend.

   b) Write an equation that describes the total dollar value of the blend.

   c) What is the number of pounds of each coffee in 10 pounds of the blend?

3. Joel is saving his money to buy a used car. The price of the car is \$5525, but it will be reduced by \$150 for each month that the car remains unsold. Joel currently has \$3250 in his savings account and will be able to save an additional \$175 each month.

   a) Write an algebraic expression to represent each of the following:

   the price of the car in n months

   the amount of money Joel will have saved after n months

   b) Determine the number of months until Joel will be able to buy the car. Show your work.

4. Using only 34-cent stamps and 20-cent stamps, Peggy put \$3.52 postage on a package. She used twice as many 34-cent stamps as 20-cent stamps. Determine how many of each type of stamp she used.

5. Mr. Parson sells nuts at a corner stand. He sells peanuts for \$1.00 per pound and walnuts for \$2.00 per pound. He wants to make 50 pounds of peanut-walnut mix. He will maintain the cost per pound of each type of nut and sell the mix for \$1.60 per pound. How many pounds of each nut should he put in the mix? Show how you arrive at your answers.

6. A florist is offering two different package deals of roses and carnations. One package contains 20 roses and 34 carnations for \$50.40. The other package contains 15 roses and 17 carnations for \$32.70. This information can be represented by the system of equations below, where r is the cost of one rose and c is the cost of one carnation.

   $20r + 34c = 50.40$

   $15r + 17c = 32.70$

   Solve the system of equations to find the cost of one rose.

Practice Problems

Answers

1 a) $c = 11 + 0.85n$

b) $c = 8 + n$

c) Dozster: $c = 11 + 0.85(10) = 11 + 8.50 = \$19.50$.
Melody: $c = 8 + 10 = \$18$. Melody is cheaper.

d) Dozster: $c = 11 + 0.85(30) = 11 + 25.50 = \$36.50$.
Melody: $c = 8 + 30 = \$38$. Dozster is cheaper.

e) Dozster: $50 = 11 + 0.85n$; $n = 45.9$. So, you could download 45 songs.
Melody: $50 = 8 + n$; $n = 42$ songs.

f) $c - 0.85n = 11$

$c - \quad n = 8$

$0.15n = 3$

$n = 20$

$c - 20 = 8$

$c = 28$

So, if 20 songs are purchased from either service, the cost is \$28.

g)

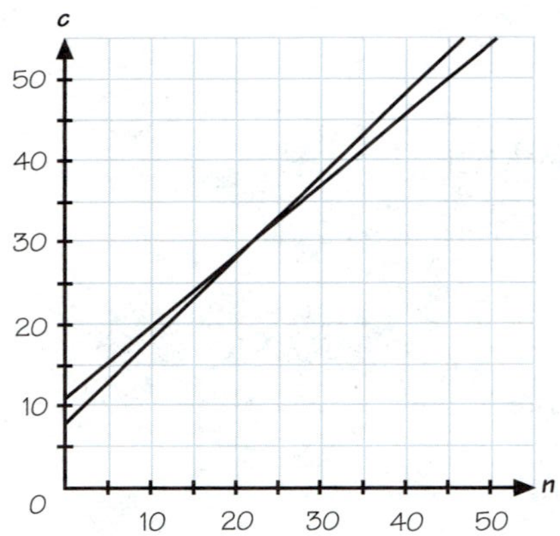

h) If fewer than 20 songs are purchased, Melody is the better deal. If you purchase more than 20 songs, Dozster is the better deal.

2. a) $a + b = 10$

b) $6a + 10b = (7)(10)$

c) $a = 7.5$ pounds; $b = 2.5$ pounds

3. a) An expression to represent the price of the car: $5525 - 150n$

An expression to represent the amount saved: $3250 + 175n$

b) Sample answer: Joel can buy the car when the price of the car is equal to the amount of money saved.

$5525 - 150n = 3250 + 175n$

$2275 = 325n$

$7 = n$

So in 7 months Joel can buy the car.

4. She used eight 34-cent stamps and four 20-cent stamps.

5. The mixture should contain 20 pounds of peanuts and 30 pounds of walnuts. Let p represent the number of pounds of peanuts and w represent the number of pounds of walnuts. Then $p + w = 50$ and $p + 2w = 50(1.60)$. The solution of this system of equations is $p = 20$ and $w = 30$.

6. One rose costs \$1.50.

7. Sketch a graph of this system of equations.

$y = -x - 3$

$y = x - 3$

8. In the following system of equations, what would be the first step in eliminating the variable y?

$x + y = 7$

$2x - y = 5$

9. This graph was made to compare the costs of renting copy machines from Company A and from Company B. What information is given by the point of intersection of the two lines?

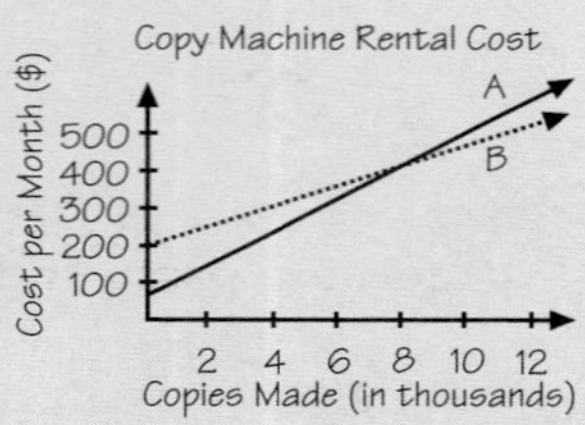

10. A clothing company receives orders from three shops. The first shop orders 25 jackets, 75 shirts, and 75 pairs of pants. The second shop orders 30 jackets, 40 shirts, and 35 pairs of pants. The third shop orders 20 jackets, 40 shirts, and 35 pairs of pants. Display this information in a matrix with rows representing the shops and columns representing the type of clothing ordered. Label the rows and columns of your matrix.

11. Through July 20, 1997, the three baseball players with the highest batting average in the National League had these batting statistics.

	AB	R	H	HR	RBI	Avg.
L. Walker (Colorado)	343	88	138	27	79	.402
Gwynn (San Diego)	372	64	147	15	84	.395
Piazza (Los Angeles)	332	56	118	19	62	.355

These are the statistics for the same players through September 30, 1997.

	AB	R	H	HR	RBI	Avg.
L. Walker (Colorado)	568	143	208	49	130	.366
Gwynn (San Diego)	592	97	220	17	119	.372
Piazza (Los Angeles)	556	104	201	40	124	.355

Find and label a matrix that displays the changes in these statistics. Notice that the batting averages for two of the three players decreased. How is a decrease shown in your matrix?

12. A state university announced a 7% raise in tuition. The current rates per semester hour are shown in the following table.

	Undergraduate	Graduate
Resident	$75.00	$99.25
Nonresident	$204.00	245.25

a) Write and label a matrix that represents this information.

b) Find a new matrix that represents the tuition rates per semester hour after the 7% raise goes into effect. Label your matrix.

c) Find a matrix that represents the dollar increase for each of the categories. Label your matrix.

d) Which matrix operation did you use in b? Which matrix operation did you use in c?

13. Nancy has jewelry shops in Westmarket, Eastmarket, and Oldmarket plazas. Her sales of cultured pearls for July are shown in the following table.

	Old	West	East
Earrings	10	8	12
Pins	6	5	4
Necklaces	3	2	2
Bracelets	4	3	2

7.

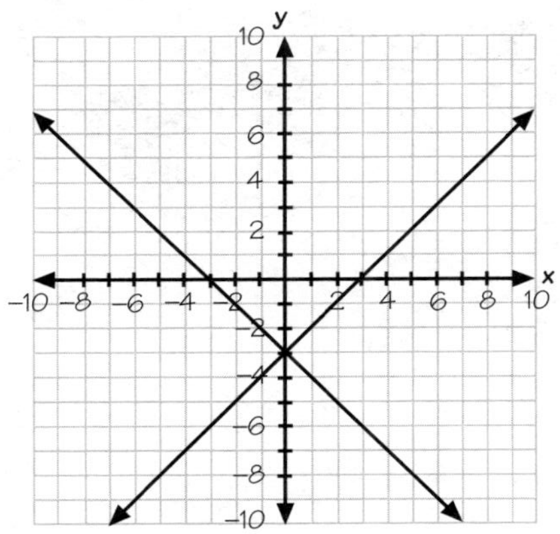

8. Add the two equations.

9. The number of copies for which the total cost per month is the same for both companies

10.

	Jackets	Shirts	Pants
Store 1	25	75	75
Store 2	30	50	50
Store 3	20	40	35

11.

	At Bats	Runs	Hits	HRs	RBIs	
L. Walker	225	55	70	22	51	−0.036
Gwynn	220	33	73	2	35	−0.023
Piazza	224	48	83	21	62	0.007

Decrease in batting average is shown with a negative sign.

12. a)

	Undergrad	Grad
Resident	\$75.00	\$99.25
Nonresident	\$204.00	\$245.25

b)

	Undergrad	Grad
Resident	\$80.25	\$106.25
Nonresident	\$218.28	\$262.42

c)

	Undergrad	Grad
Resident	\$5.25	\$6.95
Nonresident	\$14.28	\$17.17

d) Scalar multiplication in b.

Subtraction of matrices in c.

13.
$$\$\begin{bmatrix} 40 & 35 & 80 & 45 \end{bmatrix} \begin{matrix} e \\ p \\ n \\ b \end{matrix} \overset{\begin{matrix} \text{Old} & \text{West} & \text{East} \end{matrix}}{\begin{bmatrix} 10 & 8 & 12 \\ 6 & 5 & 4 \\ 3 & 2 & 2 \\ 4 & 3 & 2 \end{bmatrix}} = \$\begin{bmatrix} 1030 & 790 & 870 \end{bmatrix}$$

Earrings sell for \$40 a pair, pins for \$35 each, necklaces for \$80 each, and bracelets for \$45. Use matrix multiplication to find Nancy's total sales at each location.

14. Use matrices to find (x, y) if

$5x + 3y = 2$

$7x - 2y = 3$

15. A small company makes unfinished tables and chairs. Each table uses 40 board feet of wood, and each chair uses 20 board feet. It takes 6 hours of labor to make a table and 8 hours to make a chair. There are 2000 board feet of wood and 600 labor hours available for the next week.

a) Let T represents the number of tables made in the next week and C the number of chairs. Write inequalities that relate T and C to the total available amount of wood, and to the total available labor.

b) The inequalities $T \geq 0$ and $C \geq 0$ also describe this situation. Why?

c) Draw a graph that represents this situation.

d) Can the company produce 30 tables and 35 chairs in the next week? Defend you answer.

e) Can the company produce 50 tables and 10 chairs in the next week? Defend you answer.

16. Eric wants to buy chips and soda for the guests at his party. Each large bag of chips costs \$2.40 and each large bottle of soda costs \$2.00. He has \$48.00 to spend.

a) Write a system of inequalities that describes the limits on Eric's choices for number of bags of chips C and number of bottles of soda S that he can buy.

b) Draw a graph that represents the possible numbers of bags of chips and bottles of soda that Eric can buy.

c) If Eric buys 10 bags of chips, what is the largest number of bottles of soda that he can buy?

17. Write a system of inequalities that describes the graph in **Figure 3.15.**

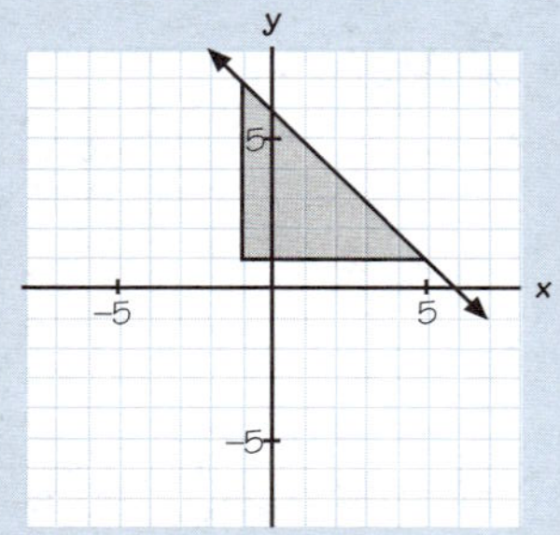

FIGURE 3.15.
Graph for Question 17.

18. A to Z Auto Service claims that their car repair rates are the lowest in town. They charge an initial fee of \$50 plus \$30 per hour. Assuming their claim is true, draw a graph that describes the rates charged by A to Z's competitors.

19. Julie does not want to spend more than \$300 on ice skating. Her skates will cost \$42, her lessons will cost a total of \$56, and the practice time will cost \$7.50 per hour. Write an inequality Julie could use to determine the maximum number of hours, h, she can practice without spending more than \$300.

20. Principal Greene has received a \$1500 grant to buy new printers at his school. He has a choice between \$50 color printers and \$130 high-resolution laser printers. He wants at least 20 new printers. If C represents the number of color printers and L represents the number of laser printers, write a set of inequalities that models his choice.

14. $\left(\frac{13}{31}, \frac{-1}{31}\right)$

15. a) $40T + 20C \le 2000$
    $6T + 8C \le 600$

b) Negative numbers for tables or chairs make no sense.

c) Sample answer:

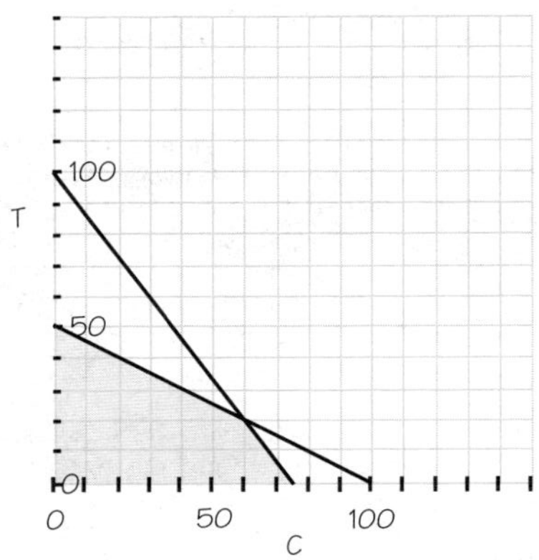

d) Yes, the shaded area in the graph shows the number of each item they can make, and 30 tables and 35 chairs are within this area.

e) No. The shaded area in the graph shows that when 50 tables are made, 0 chairs can be made.

16. a) $C \ge 0$
    $S \ge 0$
    $2.40C + 2.00S \le 48$

b) Sample answer:

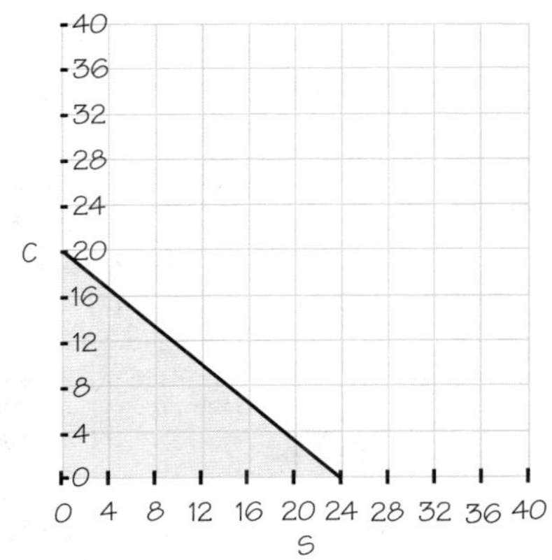

c) 12

17. $y \le 6 - x$
    $y \ge 1$
    $x \ge -1$

18.

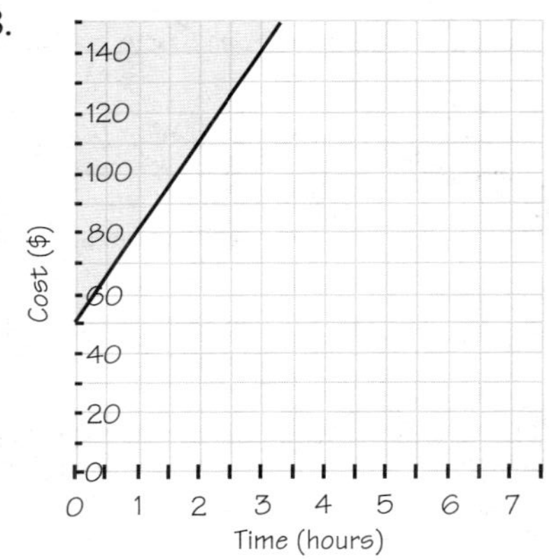

19. $42 + 56 + 7.50h \le 300$

20. $50C + 130L \le 1500$
    $C + L \ge 20$

Glossary

KEY CONCEPTS

Additive Inverse: The additive inverse of a number a is the number b, such that $a + b = 0$. So the additive inverse of 2 is –2.

Coefficient Matrix: A matrix that contains the coefficients from a system of equations.

Constant Matrix: A matrix that contains the constant from a system of equations.

Corresponding Elements: Elements in matrices that are in the same position (row and column) of each matrix.

Equal Matrices: Two matrices are equal matrices if and only if they have the same dimensions and their corresponding elements are equal.

Expense: An amount paid for a good or service. In business, expenses must be deducted from income to find profit.

Feasible Region: The area of a graph containing all the points that make all of the inequalities in the system true.

Frequency: In this chapter, frequency is the number of times a value occurs. If it occurs often the frequency is high. If it does not occur often the frequency is low.

Identity Matrix: A square matrix (the same number or rows and colums) that contains ones running diagonally from the upper left corner to the lower right corner and has zeroes in all of the other entries. Multiplying a square matrix by an identity matrix does not change the matrix.

Income: In most cases the amount earned from work. In business, income minus expenses equals profit.

Inequality: A type of statement that compares two values. For example these are common inequality symbols:

Symbol	Example	Meaning (from left to right)
$<$	$x < y$	x is less than y
$>$	$x > y$	x is greater than y
$\leq$	$x \leq y$	x is less than or equal to y
$\geq$	$x \geq y$	x is greater than or equal to y

Intersect or Intersection: In this chapter, the point (x, y) in a graph where two lines cross.

Inverse Matrix: The product of a square matrix and its inverse matrix is an identity matrix. Not every square matrix has an inverse matrix.

Linear Combination Method: A way to solve systems of equations in which one of the variables is eliminated by adding equations.

Linear Programming: The study and processes for maximizing or minimizing a function based on linear constraints. Linear programming is a tool used in business to make financial decisions.

Matrix (Matrices): An arrangement of numbers written between brackets. The numbers are arranged in rows and columns and make a rectangular shape. A row is a horizontal group of numbers. A column is a vertical group of numbers. The plural of matrix is matrices.

Matrix Addition: A process of adding two matrices. Matrix addition can only be performed on matrices with the same dimensions.

Matrix Dimensions: The number of rows and columns in a matrix. A matrix with 3 rows and 2 columns is a 3 × 2 matrix (read "3 by 2").

Matrix Element or Component: A single entry in a matrix.

Matrix Multiplication: A process of multiplying two matrices.

Matrix Operation: A process, such as addition or scalar multiplication, carried out on one or more matrices.

Matrix Subtraction: A process of subtracting two matrices. Matrix subtraction can only be performed on matrices with the same dimensions.

Multiplicative Inverse: The multiplicative inverse of a number a is the number b, such that $a \times b = 1$. So the multiplicative inverse of 2 is $\frac{1}{2}$.

Profit: The amount left over after expenses have been deducted from income. This can be a negative number (often called "loss").

Scalar Multiplication: A process where all of the elements in matrix are multiplied by a number. The number is called a scalar.

Solution to a System of Equations: A set of values that are solutions to every individual equation in a system.

Substitution Method: A way to solve systems of equations by isolating a variable in one equation then substituting into the other equation.

System of Equations: Two or more equations. A system of linear equations is a group of two or more linear equations.

Systems of Inequalities: Two (or more) inequalities that have the same variables. For example, $2x + 3y < 4$ and $3x + 4y < 5$ are a system of inequalities.

Variable Matrix: A matrix that contains the variables from a system of equations.

CHAPTER 4

Art: Transformations, Symmetry, and Proportions

PREPARATION READING

Picture Perfect

Visual Art takes many forms. A pencil drawing and a graphic design are examples of art. A computer-generated graphic, a quilt, and a painting are also forms of art. Many occupations make use of art as well. For example, architects use drawings to represent building designs.

Artists and architects use flat surfaces to create pictures that appear to be three-dimensional. In using a flat surface, such as a canvas or paper, to represent three-dimensional space, artists and architects use mathematical ideas and geometric principles to make a flat image look like the real object.

What does mathematics have to do with creating art? Have you noticed patterns that repeat in paintings, logos, and even wallpaper? Sometimes the images that make up the pattern are always the same size, sometimes they are made to look 3-D, or three-dimensional, by drawing some objects over others and changing their size.

An architect uses the geometry of one-point or multi-point perspective to help a client visualize what a proposed building will look like. A theater designer uses geometry to create the illusion of depth and distance on a small stage. An animator guides you into and around three-dimensional objects using a two-dimensional screen. A painter uses perspective and shading to add depth to drawings.

Before you begin, take a look around you and notice the lines and shapes that make up your world. Think about how you might sketch the room you are in or the front of your school.

Small children view the world differently than you do. Their sense of depth is not well developed. Just look at a picture drawn by a young child, such as **Figure 4.1.**

FIGURE 4.1.
House by Jackson Barber, *age 8.*

CHAPTER 4

Art: Transformations, Symmetry, and Proportions

Preparation Reading: Picture Perfect

Purpose: The purpose of the preparation reading is to pique students' interest in the connection between art and mathematics.

Description: This lesson begins with a comparison of a primitive sketch by a young child to a sophisticated sketch by an architect. Students are prompted to make comparisons between the two figures to start a conversation about the use of drawing techniques that may prompt a sense of relative size and relative distance. The conversations may also include discussion of how one represents a three-dimensional setting such as a building and its grounds using a two-dimensional representation such as a sketch or architectural rendering.

Materials: preparation reading

Teaching Ideas: Before reading the preparation reading, prompt students to create a K-W-L chart as shown below.

What I Know	What I Want to Know	What I Learned

Prompt students to brainstorm what they know about the relationship between mathematics and art. Prompt students to record what they know in the What I Know column of the K-W-L chart.

Prompt students to read the preparation reading. Prompt students to record familiar vocabulary in the What I Know column and unfamiliar vocabulary in the What I Want to Know column.

After students have completed the reading, prompt them to record questions they would like to have answered during this unit of study in the What I Want to Know column of the K-W-L chart. Direct students to move into groups of 2 to share their recordings from the first 2 columns. Prompt students to make any additions to their charts based on the input from their peers. Students will keep these charts throughout this unit of study, adding to the What I Learned column at the end of each lesson.

Notice that the sun is too large and the building appears flat. Now compare Figure 4.1 with an architect's drawing for a new school (**Figure 4.2**).

FIGURE 4.2 Tantasqua Regional High School.

The drawing helped community members imagine what the completed project might look like before they voted on whether or not to build the new school. A realistic drawing is an important tool in their decision.

In this chapter we will explore three questions:

- How do artists create patterns by moving or transforming an object?
- What geometric principles guide an artist's creation of an accurate drawing?
- What principles give a picture depth?

This chapter explores how the answers to these questions help artists, architects, and others develop realistic drawings.

SECTION 4.1 Transformations in Art

Repetition is a technique used often by artists. By repeating a basic element an artist can create a pleasing painting, tapestry, or other work. There are several ways to repeat a basic design element.

Consider the tapestry in **Figure 4.3.** The artist has repeated a basic condor element in two ways.

FIGURE 4.3.
Condor by Cerapio Vallejo.

Figure 4.4 shows how the condor in the upper left corner (condor 1) can slide to its right to produce another condor. A **slide** is one type of repetition used by artists.

FIGURE 4.4.
A slide produces another condor.

FIGURE 4.5. *Slide condor 1 to the right, then slide it down, and then flip it to the left.*

CHECK THIS!
In mathematics a slide is called a translation.

CHECK THIS!
In mathematics a flip is called a reflection.

But not all of the condors can be produced by sliding one of them. For example, the condor directly below condor 1 cannot be obtained by sliding condor 1. It can be obtained from condor 1 by sliding condor 1 to the right, then down, and then flipping it to the left (see **Figure 4.5**). A **flip** is another kind of repetition used by artists.

All of the condors in this tapestry can be obtained from condor 1 by translating or reflecting or a combination of the two.

ENGAGE Lesson 1

Section 4.1: Transformations in Art

Purpose: To engage students' interest in the mathematics behind repetitive patterns in art

Description: Students examine and discuss various ways in which artists use repetitive patterns to create two-dimensional designs.

Materials: none

Teaching Ideas: Place students into groups of 2 or 3. After groups have discussed the questions, ask groups to share their results.

The mirror in **Figure 4.6** demonstrates a third way artists use repetition.

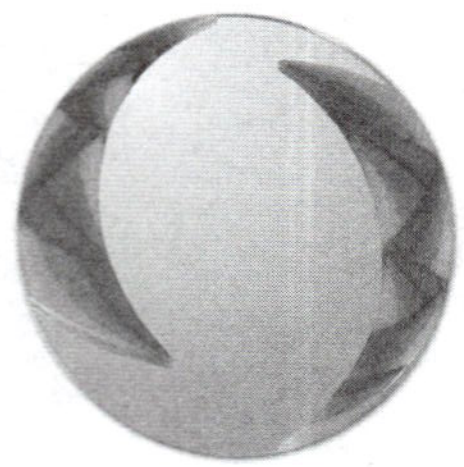

FIGURE 4.6.
Art Nouveau mirror by Cidinha.

The crescent shape on the right edge of the mirror can be obtained from the one on the left by spinning the left one 180° about a central point. **Figure 4.7** shows how one crescent tip can be rotated to get the corresponding tip of the other crescent. A **spin** is another kind of repetition used by artists.

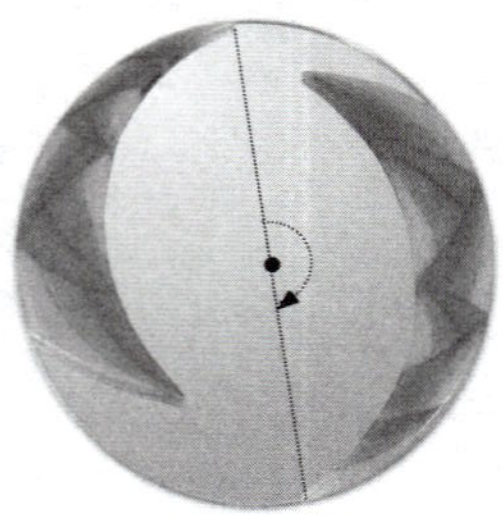

FIGURE 4.7.
Spinning one crescent tip 180° produces the other.

CHECK THIS!
In mathematics a spin is called a rotation.

Translations, reflections, and rotations are types of **transformations**. That is, each is a process that uses a basic image to create new ones. In this lesson you will learn more about transformations and how they are used in art and in mathematics.

In each of the following figures, identify a basic element and describe ways to get the other sections of the figure by translating, reflecting, and/or rotating the basic element. Your basic element should be as small a portion of the entire figure as possible.

1.

FIGURE 4.8. Image for Question 1.

2.

FIGURE 4.9. Image for Question 2.

3.

FIGURE 4.10. Image for Question 3.

4.

FIGURE 4.11. Image for Question 4.

5.

FIGURE 4.12. Image for Question 5.

6.

FIGURE 4.13. Image for Question 6.

7.

FIGURE 4.14. Image for Question 7.

8.

FIGURE 4.15. Image for Question 8.

9.

FIGURE 4.16. Image for Question 9.

10.

FIGURE 4.17. Image for Question 10.

Answers

1. Basic element: either the right or left half of the image. The other half can be obtained by reflecting the chosen element.
2. Basic element: either the right or left half of the image. The other half can be obtained by reflecting the chosen element.
3. Basic element: any of the three arrows. The other two arrows can be obtained by rotating the chosen element 120° and 240°.
4. Basic element: any of the six snowflake petals. The other five can be obtained by rotating the chosen element 60°, 120°, 180°, 240°, and 300°.
5. Sample answer: The square at the center of the cross (think of the cross as composed of five squares). Each of the four branches of the cross can be obtained by translating the chosen element to the left, to the right, upward, and downward.
6. Sample answer: The spiral in the upper left, along with two straight segments extending from it. The spiral to the right of the chosen element and the one below it can be obtained by reflecting the chosen element. The remaining spiral can be obtained by rotating the chosen element.
7. Basic element: either the left or right half of the figure. The other half can be obtained by rotating the chosen element 180°.
8. Basic element: any of the 12 cogs together with the arc between it and an adjacent cog. The remainder of the figure can be obtained by rotating the chosen element 30°, 60°, 90°, 120°, 150°, 180°, 210°, 240°, 270°, 300°, and 330°.
9. Basic element: any of the star's five sections. The remaining sections can be obtained by rotating the chosen element 72°, 144°, 216°, and 288°.
10. Basic element: the circled portion. The remainder of the top half of the figure can be obtained by translating the basic element. The left third of the bottom half can be obtained by reflecting the basic element. The rest of the bottom half can be obtained by translating and then reflecting the basic element.

SECTION 4.2 Types of Symmetry

Many works of art have one or more kinds of symmetry. Symmetry is often easy to recognize, but can be hard to define. In Section 4.1 you learned about translations, reflections, and rotations. One way to describe a symmetric image is that a copy of the image can be translated, reflected, or rotated to match the original.

In this section you will create symmetric images. The basic instructions for Questions 1–4 are:

- Use two circular coffee filters. Place one on top of the other.
- Fold the double-filter as described.
- Make a variety of cuts along each straight edge of the folded filters.
- Unfold them.
- Describe how one filter (the copy) can be translated, rotated, or reflected to obtain the other. If there is more than one way to obtain one from the other, describe all of the ways that you can find.

1. Fold the filters in half once as shown in **Figure 4.18.** Make various cuts along the edge as shown in **Figure 4.19** (but not exactly the same as in the figure).

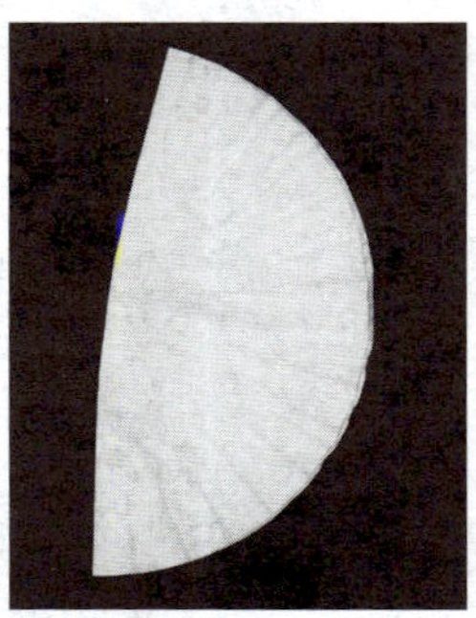

FIGURE 4.18. Folded filters.

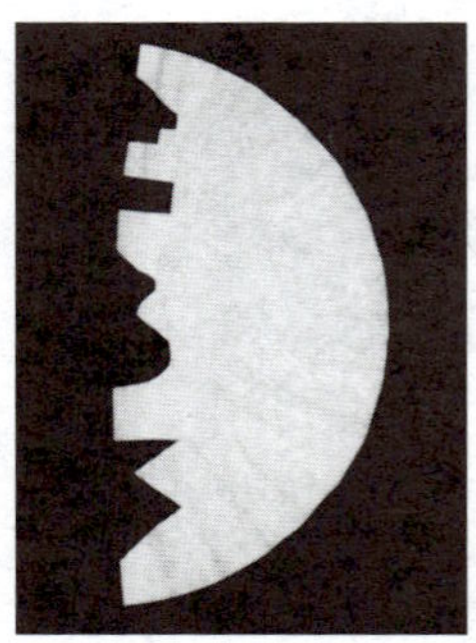

FIGURE 4.19. Folded and cut filters.

2. Fold the filters in half and then in half again. (After the two folds, you should have quarter-filters). Make various cuts along the edge.
3. Fold the filters in half, then in half again, and then in half again. (After the three folds, you should have an eighth-filters). Make various cuts along the edge.
4. Fold the filters in half, and then fold into thirds. Make various cuts along the edge.

An image has **reflection symmetry** (also called line symmetry) when a copy of it can be reflected through a line to match the original image. The line is called a **line of symmetry**. For example, **Figure 4.20** shows the line of symmetry for a butterfly image.

FIGURE 4.20. Reflection symmetry.

An image has **rotation symmetry** when a copy of it can be rotated less than 360° about a central point to match the original image. When you describe rotation symmetry, try to identify the central point and to state the size of the smallest possible rotation for which the copy and original match. For example, the flower in **Figure 4.21** has 72° rotation symmetry.

FIGURE 4.21. 72° rotation symmetry.

5. For each of the objects you created in Questions 1–4, identify all lines of symmetry and the size of any rotation symmetry.
6. Does the flower in Figure 4.21 have any symmetry other than rotation symmetry? Explain.

EXPLORE Lesson 1

Section 4.2: Types of Symmetry

Purpose: Students create designs that demonstrate reflection, rotation, and related symmetries.

Description: Students use coffee filters to create objects that demonstrate reflections and rotations.

Materials: circular coffee filters, scissors

Teaching Ideas: Place students in groups of 2 or 3. Give each group approximately 8 coffee filters and a pair of scissors. Prompt students to follow the directions in Section 4.2 to create objects with various kinds of symmetry. Have groups share their results with other groups.

When students make cuts in two edges of folded filters, be sure that they use different patterns of cuts on the two edges. Also be sure that when students answer the questions, they allow slight differences in the patterns of the cuts that are due to imperfections in the folds and cuts.

At the end of the lesson, prompt students to add thoughts to the What I Learned column of the K-W-L chart

Answers

1. You can reflect one of the cut filters (turn it over) to obtain the other.
2. You can reflect one of the cut filters (turn it over) to obtain the other. You can also rotate one of the filters 180° to obtain the other. Sample result:

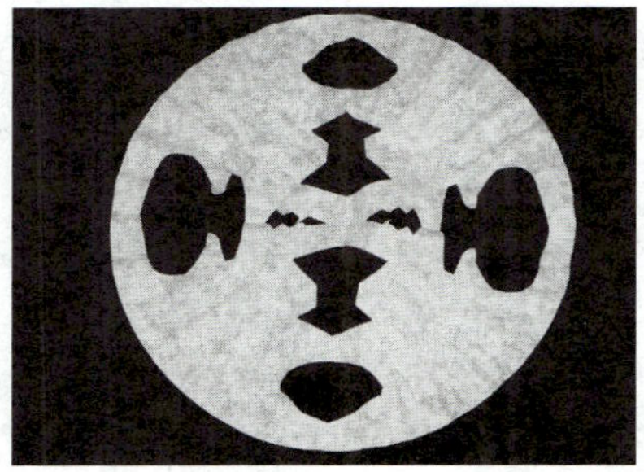

3. You can reflect one of the cut filters (turn it over) to obtain the other. You can also rotate one of the filters 90° to obtain the other. Sample result:

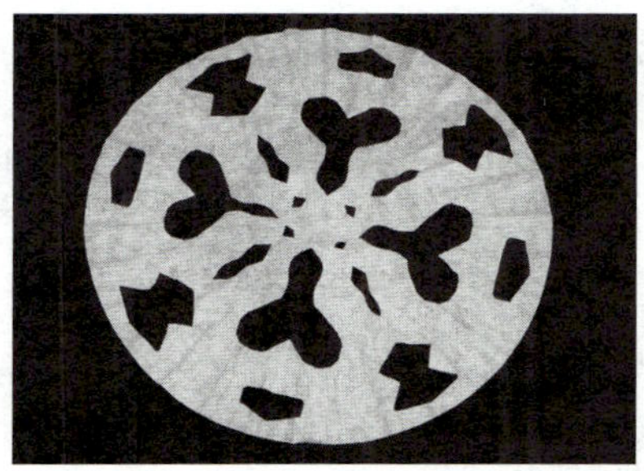

4. You can reflect one of the cut filters (turn it over) to obtain the other. You can also rotate one of the filters 120° to obtain the other. Sample result:

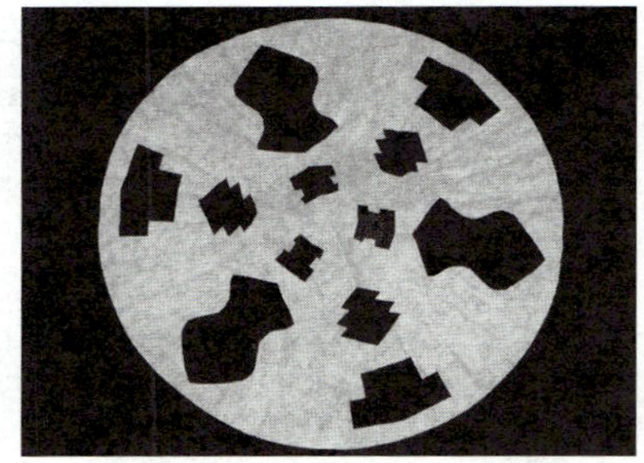

5. In Question 1, there is a single line of symmetry along the fold.

   In Question 2, there are two lines of symmetry, one along each fold. There is also 180° rotation symmetry.

   In Question 3, there are four lines of symmetry, one along each fold. There is also 90° rotation symmetry.

   In Question 4, there are three lines of symmetry, one along each fold. There is also 120° rotation symmetry.

6. Yes. There are five lines of symmetry: one bisecting each of the five petals.

SECTION 4.3

Transformations, Isometries, and Symmetry

Many computer software programs have drawing features that allow you to create and edit images. For example, you can copy an image and paste it elsewhere. Many programs let you reflect or rotate an image. All of these operations are mathematical transformations. In this section you will take a closer look at how some of these operations work.

When a geometric object is transformed into another one, the original is called the **pre-image**. The new object is called the **image**. The vertices of the pre-image are labeled with capital letters, such as ABC for a triangle. The vertices of the image are often labeled with the same letters, but with an apostrophe after each. The apostrophe is read as **prime**. The transformation is often indicated with an arrow.

CHECK THIS!

The notation $\Delta ABC \rightarrow \Delta A'B'C'$ says that triangle ABC is transformed into triangle A-prime B-prime C-prime. ΔABC is the pre-image. $\Delta A'B'C'$ is the image.

Draw a triangle on a sheet of paper and label its vertices A, B, and C. Be sure the triangle is not special; that is, don't draw an equilateral triangle, an isosceles triangle, or a right triangle. Draw the triangle so that one vertex is pointing upward; use A for the label of this vertex.

1. Place a sheet of patty paper over the triangle and trace it. Slide the patty paper a few inches to the right. On the original paper, mark the locations of the new vertices and connect them. Label the new triangle's vertices A′, B′, and C′. Discuss the type of transformation that A′B′C′ is of ABC. Also discuss the properties of the original that changed and those that are the same.

2. On the paper with the original triangle, draw a vertical line to the right or the left of the triangle. Place the patty paper triangle back on the original triangle. Fold the patty paper along the line. On the original paper, mark the locations of the new vertices and connect them. Label the new triangle's vertices A′, B′, and C′. Discuss the type of transformation that A′B′C′ is of ABC. Also discuss the properties of the original that have changed and those that are the same.

3. On the paper with the original triangle, mark a point somewhere outside of the triangle. Place the patty paper triangle back on the original triangle. Place the point of your pencil on the point that you marked and spin the patty paper about 90°. On the original paper, mark the locations of the new vertices and connect them. Label the new triangle's vertices A′, B′, and C′. Discuss the type of transformation that A′B′C′ is of ABC. Also discuss the properties of the original that have changed and those that are the same.

4. Does your original triangle have any lines of symmetry? That is, does it have reflection symmetry? If not, what type of triangle does?

5. Does your original triangle have rotation symmetry? If not, what type of triangle does?

Translations, reflections, and rotations are all special types of transformations called isometries. An **isometry** is a transformation that does not change size or shape. (Later in this chapter you will examine a transformation that is not an isometry.)

Isometries change the location of an object. Some of them change an object's orientation. You can orient a simple geometric figure by labeling the vertices A, B, C, and so on as you move from vertex to vertex in a clockwise direction. When an isometry changes the orientation, as you move from A′ to B′ to C′, and so on, you move in a counterclockwise direction.

6. Which transformations change orientation: translation, reflection, or rotation?

Each of these three isometries has specific characteristics that are used to identify it.

- A translation has a distance and a direction. For example, you might slide ΔABC two inches to the right. Translations change location, but do not change size, shape, or orientation.
- A reflection has a line (sometimes called a mirror). For example, you might reflect ΔABC about line $\overleftrightarrow{CD}$. Reflections change location and orientation, but do not change size or shape.
- A rotation requires a point and an angle. For example, you might rotate ΔABC 30° counterclockwise about point D. Rotations change location, but do not change size, shape, or orientation.

EXPLAIN Lesson 1

Section 4.3: Transformations, Isometries, and Symmetry

Purpose: Explain the attributes of symmetries, transformations, and isometries.

Description: Students use patty paper to describe the effect of symmetries, transformations, and isometries on geometric figures.

Materials: patty paper

Teaching Ideas: If your students are familiar with basic graphic operations on a computer, you might begin with a whole-class discussion of the following questions:

1. When you "cut and paste" a figure on a computer, what happens?
2. How might you use patty paper to illustrate what happens?
3. When you "copy and paste" a figure on a computer, what happens?
4. How might you use patty paper to illustrate what happens?
5. When you "rotate" a figure on a computer, what happens?
6. How might you use patty paper to illustrate what happens?
7. When you "reflect" a figure on a computer, what happens?
8. How might you use patty paper to illustrate what happens?

Answers to these questions vary. The goal is to have the students identify that the new images are congruent to the original image.

Place students into small groups. Have students complete the questions in these groups. As students complete each question, prompt them to compare and contrast the newly completed question with previously completed questions.

When students have completed the questions, debrief their conclusions in a whole-class setting. You might tell students that when animated art is created for Websites, these principles are applied to the pixels that make up the image to be animated.

At the end of the lesson, prompt students to add thoughts to the What I Learned column of the K-W-L chart. Reproducible versions of this chart are on the Teacher Resources CD-ROM.

Answers

1. This is a translation. The location of the original triangle is changed, but its size and shape are the same.
2. This is a reflection. The location of the original triangle is changed, but its size and shape are the same. Vertices B′ and C′ have switched sides, so the orientation of the triangle is changed.
3. This is a rotation. The location of the original triangle is changed, but its size and shape are the same. Students might also observe that vertex A is "pointing in a different direction;" however, this does not change the triangle's orientation.
4. No. A triangle with at least two sides equal (an isosceles triangle) has a line of symmetry that runs halfway between the two equal sides and perpendicular to the remaining side.
5. No. A triangle with three sides equal (an equilateral triangle) has rotation symmetry of 60° about a central point.
6. Only reflection

CHECK THIS!

Mathematicians consider a counterclockwise rotation positive and a clockwise rotation negative. For example, a rotation 30° clockwise is considered a rotation of –30°.

Sometimes two or more isometries are combined. For example, in **Figure 4.22**, ΔABC is translated to the right to get $\Delta A'B'C'$.

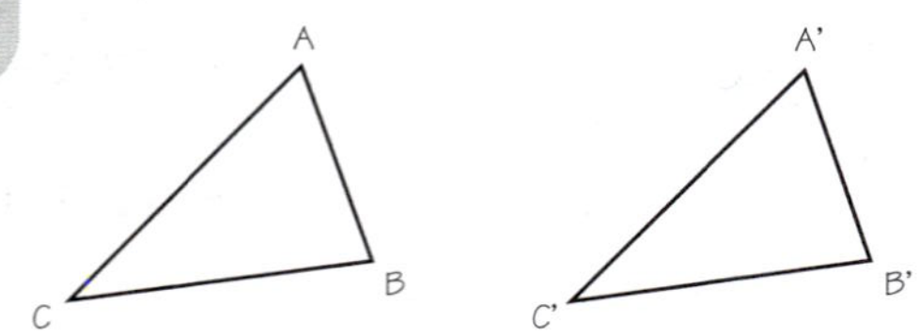

FIGURE 4.22.
A translation.

In **Figure 4.23**, $\Delta A'B'C'$ is reflected across the line shown to obtain $\Delta A''B''C''$.

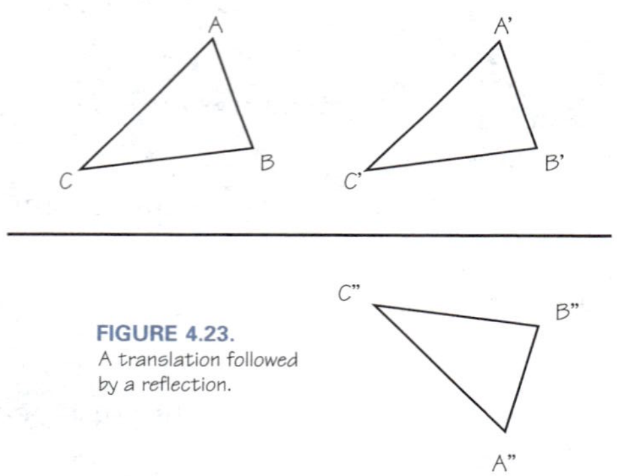

FIGURE 4.23.
A translation followed by a reflection.

GLIDE REFLECTION

A glide reflection is a composite of a translation and a reflection.

$\Delta A''B''C''$ is a called a **glide reflection** of ΔABC.

7. What properties of a figure are changed by a glide reflection?

8. Use patty paper to create a composite of two reflections through a pair of parallel lines. That is, draw a line to the right of ΔABC and reflect the triangle through it. Then draw a second line parallel to the first and to the right of the image. Reflect the image through this second line. How is the final image related to ΔABC?

9. Use patty paper to create a composite of two reflections through a pair of intersecting lines. That is, draw a line to the right of ΔABC and reflect the triangle through it. Then draw a second line that intersects the first and that is to the right of the image and below it. Reflect the image through this second line. How is the final image related to ΔABC?

SECTION 4.3 Assignment

1. **Figure 4.24** is a design element. Identify the smallest basic portion of it that can be used to obtain the rest and describe isometries that can be used to build up the design.

FIGURE 4.24.
A design element.

FIGURE 4.25.
Fabric design by Debra L. Hayden.

2. **Figure 4.25** shows how artist Debra Hayden used a basic element similar to the one in Figure 4.24 to create a larger work. Describe at least two ways that she could have done this.

3. Describe the symmetries of the basic element in Figure 4.24.

4. Discuss how an artist could use transformations to create **Figure 4.26** from a basic element of the figure.

FIGURE 4.26.
Image for Question 4.

5. **Figure 4.27** is a capital letter A in the Arial font. Some capital letters are symmetric. Identify those that have symmetry and describe the symmetry of each.

FIGURE 4.27.
Image for Question 5.

Answers (continued)

7. Size and shape are unchanged. Location and orientation are changed.
8. It is a translation of ΔABC. The distance of the translation appears to be double the distance between the parallel lines.
9. It is a rotation of ΔABC. The size of the rotation appears to be double the size of the angle between the intersecting lines.

Section 4.3 Assignment

1. Sample answer: The basic element is an eighth of the figure, such as the portion shown here:

   The portion below it can be obtained by reflecting across the line running from upper left to lower right. Then this entire piece can generate the rest of the figure by rotating 90°, 180°, and 270°.

2. Translations alone can be used. For example, the basic portion in the upper left (one-sixth of the entire figure) can be translated to the right to get a third of the figure. Then this third can be translated downward to get another third and downward twice as far to get the final third.

   Reflections alone can be used. For example, the basic portion in the upper left (one sixth of the figure) can be reflected across a vertical line running through the middle of the figure to get a third of the figure. Then this third can be reflected across a horizontal line a third of the way from the top of the figure to get the middle third. This portion can then be reflected across a horizontal line a third of the way from the bottom to get the bottom third.

3. It has rotation symmetry of 90°. It has four lines of symmetry: one is vertical, one is horizontal, and the other two are diagonal.
4. Sample answer: The leaf on the lower left can be used to create the entire figure. A pair of translations of this leaf produces the other two lower leaves. Three glide reflections of this leaf produce the three top leaves.
5. The answer depends on the font because some letters are slightly asymmetric in some fonts. The following are typical answers:

   A: reflection symmetry about a vertical line

   B: reflection symmetry about a horizontal line

   C: reflection symmetry about a horizontal line

   D: reflection symmetry about a horizontal line

   E: reflection symmetry about a horizontal line

   H: reflection symmetry about horizontal and vertical lines; rotational symmetry of 180°

   I: reflection symmetry about horizontal and vertical lines; rotational symmetry of 180°

   K: reflection symmetry about a horizontal line

   M: reflection symmetry about a vertical line

   N: rotation symmetry of 180°

   O: if perfectly circular, it has an infinite number of lines of symmetry and rotational symmetry of any number of degrees; if an oval, it has vertical and horizontal lines of symmetry and rotational symmetry of 180°

   S: rotational symmetry of 180°

   T: reflection symmetry about a vertical line

   U: reflection symmetry about a vertical line

   V: reflection symmetry about a vertical line

   W: reflection symmetry about a vertical line

   X: reflection symmetry about vertical and horizontal lines; rotational symmetry of 180° or perhaps 90°, depending on how it is written

   Y: reflection symmetry about a vertical line

   Z: rotational symmetry of 180°

6. You may hear people talk about point symmetry. The plant in **Figure 4.28** demonstrates this type of symmetry. **Point symmetry** occurs when every point of a figure has an image directly across the figure's center, as shown in **Figure 4.29.**

FIGURE 4.28. Image for Question 6.

FIGURE 4.29. Image for Question 6.

a) Do any letters of the alphabet demonstrate point symmetry? If so, which ones?

b) Point symmetry is the same as another type of symmetry that you have studied in this lesson. Explain.

7. **Figures 4.30, 4.31, and 4.32** are oil company logos. Describe the symmetry of each.

FIGURE 4.30. Shell Oil logo.

FIGURE 4.31. Gulf Oil logo with letters.

FIGURE 4.32. Gulf Oil logo without letters.

SECTION 4.4

Algebraic Representations of Symmetries and Isometries

All of the isometries that you have examined change an object's location. People who use isometries in their work often need to describe locations precisely. When precise location is important, people use coordinates. For example, each pixel on a computer screen has coordinates that identify its location and that are used to transform the pixel or an object composed of many pixels in various ways.

In this section you will use coordinates to describe isometries exactly.

PART A: TRANSLATIONS ON A COORDINATE PLANE

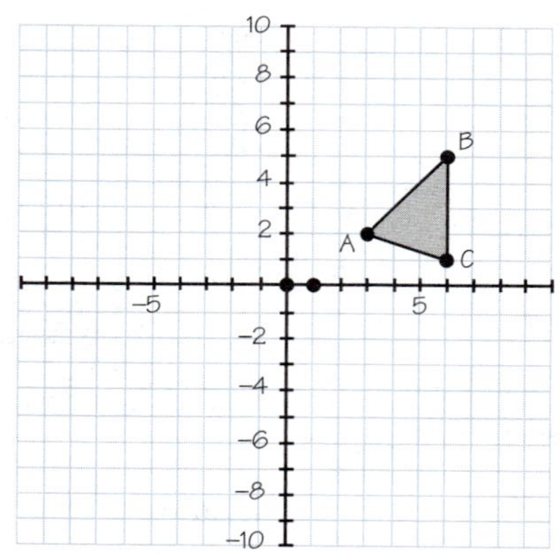

FIGURE 4.33. Triangle ABC.

A **translation** is a transformation that slides an object a given distance and in a given direction. Thus, every translation has a distance and a direction.

1. On a sheet of grid paper, plot ΔABC as shown in **Figure 4.33.** Record the coordinates of the vertices of ΔABC, the pre-image, in a table like **Figure 4.34.**

2. Trace ΔABC and the coordinate axes onto a sheet of patty paper. Label the copy, the image, A′B′C′.

Pre-Image			Image		
A	B	C	A′	B′	C′

FIGURE 4.34. Table for Question 1.

Answers (continued)

6. a) H, I, N, O, S, X, Z

   b) It is the same as 180° rotational symmetry. Some students may say that it is the same as having a pair of perpendicular lines of symmetry, which is also correct.

7. The Shell logo has reflection symmetry about a vertical line. The Gulf logo with letters has no symmetry, but with the letters removed it has reflection symmetry about vertical and horizontal lines, rotation symmetry of 180°, and therefore also has point symmetry.

ELABORATE Lesson 1

Section 4.4: Algebraic Representations of Symmetries and Isometries

Purpose: Students use algebraic notation to describe isometries.

Description: Students use graph paper to develop algebraic explanations to describe isometries of figures drawn on a coordinate plane. Algebraic generalizations are made for translations, reflections across the y-axis, reflections across the x-axis, reflections over lines, and 90° rotations.

Materials: patty paper, graph paper

Teaching Ideas: Place students in pairs. Have students complete Parts A–E in these groups. As students complete each part, prompt the students to compare and contrast the newly completed part with previously completed parts.

When students have completed Parts A–E, debrief their conclusions in a whole-class setting.

At the end of the lesson, prompt students to add thoughts to the What I Learned column of the K-W-L chart.

Answers

Part A: Translations on a Coordinate Plane

1.

Pre-Image			Image		
A	B	C	A′	B′	C′
(3, 2)	(6, 5)	(6, 1)	(0, 4)	(3, 7)	(3, 3)

3. Place the copy on top of the original so that the axes are aligned. Slide the copy 3 units left and 2 units up. Record the coordinates of the image in your table.
4. Did the triangle change size or shape?
5. Return the image to its original position and repeat 3 and 4 two more times using different values for the horizontal and the vertical translation.
6. Write equations that describe a translation of (x, y) horizontally h units and vertically k units. Use x' and y' for the coordinates of the image.
7. How does one represent a vertical or horizontal direction and a distance of a translation on a coordinate plane?

A **reflection** is a transformation that flips an object across a line.

PART B: REFLECTION ACROSS THE Y-AXIS

8. Refer back to your original ΔABC. Record the coordinates of the vertices of ΔABC, the pre-image, in a table like **Figure 4.35**.

Pre-Image			Image		
A	B	C	A′	B′	C′

FIGURE 4.35.
Table for Question 8.

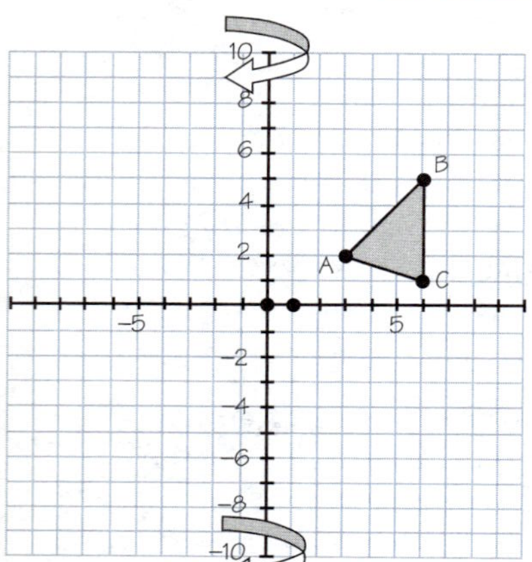

FIGURE 4.36.
Graph for Question 9.

9. Place the copy of ΔABC from Part A on top of the original so that the axes are aligned. Reflect the triangle across the y-axis, the line of reflection, by flipping the paper and realigning the axes (**Figure 4.36**). Record the coordinates of the image in your table.
10. Did the triangle change size or shape?
11. Describe the relationship between the line of reflection and vertices A and A′. Are these relationships true for the other vertices of the triangle?
12. Repeat 8–11 two times drawing the pre-image ΔABC in a different position, the fourth quadrant for example, each time.

13. Write equations that describe a reflection of (x, y) across the y-axis.
14. Do the y-coordinates change? If so, how?
15. Why are x-coordinates of the image opposites of those of the pre-image?

PART C: REFLECTION ACROSS THE X-AXIS

16. Refer back to your original ΔABC. Record the coordinates of the vertices of ΔABC, the pre-image, in a table like **Figure 4.37.**

Pre-Image			Image		
A	B	C	A′	B′	C′

FIGURE 4.37.
Table for Question 16.

17. Place the copy of ΔABC from Part A on top of the original so that the axes are aligned. Reflect the triangle across the x-axis, the line of reflection, by flipping the copy of ΔABC and realigning the axes (**Figure 4.38**). Record the coordinates of the image in your table.

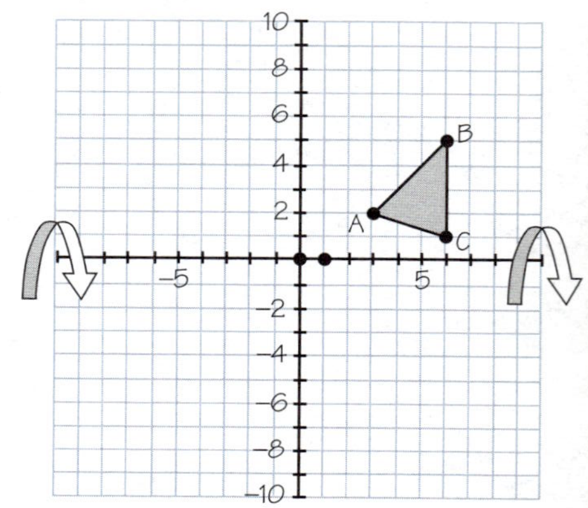

FIGURE 4.38.
Graph for Question 17.

18. Did the triangle change size or shape?
19. Describe the relationship between the line of reflection and vertices A and A′. Are these relationships true for the other vertices of the triangle?
20. Repeat 16–19 two times drawing the pre-image ΔABC in a different position (the fourth quadrant for example) each time.
21. Write equations that describe a reflection of (x, y) across the x-axis.
22. Do the x-coordinates change? If so, how?
23. Why are the y-coordinates of the image the opposite of those of the pre-image?

Answers (continued)

4. No, the triangle changed location.

6. $x' = x + h$; $y' = y + h$.

7. Vertical or horizontal direction is implied by the sign on the amount of change. The distance is indicated by the amount of the change. For example, +3 implies a translation to the right of 3 units. –3 implies a translation to the left of 3 units. +2 implies a translation of 2 units up while –2 implies a translation of 2 units down.

Part B: Reflection across the *y*-axis

8.

Pre-Image			Image		
A	B	C	A′	B′	C′
(3, 2)	(6, 5)	(6, 1)	(–3, 2)	(–6, 5)	(–6, 1)

10. No, the triangle changed location and orientation.

11. The vertices are equidistant from the line of reflection, and the segment connecting the vertices is perpendicular to the line of reflection. Yes, these relationships are true for the other vertices.

NOTE: The following questions might prove useful if students are having trouble determining the relationships between A and A′ and the line of reflection.

- What measurement tools could you use to investigate the relationships between A and A′ and the line of reflection?
- What do you notice about a segment connecting A and A′ and the line of reflection when you fold the paper along the line of reflection?

13. $x' = -x$; $y' = y$.

14. The image is reflected across the y-axis, so the distance of each point from the x-axis remains unchanged.

15. The x-coordinates of the pre-image and the image have the same absolute value because the line of reflection is the y-axis. The corresponding vertices between the pre-image and the image are equidistant from the line of reflection.

Part C: Reflection across the *x*-axis

16.

Pre-Image			Image		
A	B	C	A′	B′	C′
(3, 2)	(6, 5)	(6, 1)	(3, –2)	(6, –5)	(6, –1)

18. No, the triangle changed location and orientation.

19. The vertices are equidistant from the line of reflection, and the segment connecting the vertices is perpendicular to the line of reflection. Yes, these relationships are true for the other vertices.

NOTE: The following questions might prove useful if students are having trouble determining the relationships between A and A′ and the line of reflection.

- What measurement tools could you use to investigate the relationships between A and A′ and the line of reflection?
- What do you notice about a segment connecting A and A′ and the line of reflection when you fold the paper along the line of reflection?

21. $x' = x$; $y' = -y$.

22. The image is reflected across the x-axis, so the distance of each point from the y-axis remains unchanged.

23. The y-coordinates of the pre-image and the image have the same absolute value because the line of reflection is the x-axis. The corresponding vertices between the pre-image and the image are equidistant from the line of reflection.

PART D: 90° ROTATIONS ON A COORDINATE PLANE

A **rotation** is a transformation that spins an object about a given center a given number of degrees.

24. Draw triangle ΔABC in the first quadrant of a new sheet of grid paper. Record the coordinates of the vertices of ΔABC, the pre-image, in a table like **Figure 4.39.**

Angle of Rotation	Pre-Image			Image		
	A	B	C	A′	B′	C′

FIGURE 4.39.
Table for Question 24.

25. Trace ΔABC and the coordinate axes onto another sheet of patty paper. Label the image A′B′C′.

26. Place the image on top of the original so that the coordinate axes are aligned. Rotate the copy 90° about the origin by turning the copy counterclockwise (**Figure 4.40**). Record the coordinates of the image in your table.

CHECK THIS!

In geometry, a rotation is counterclockwise if the angle is positive and clockwise if the angle is negative.

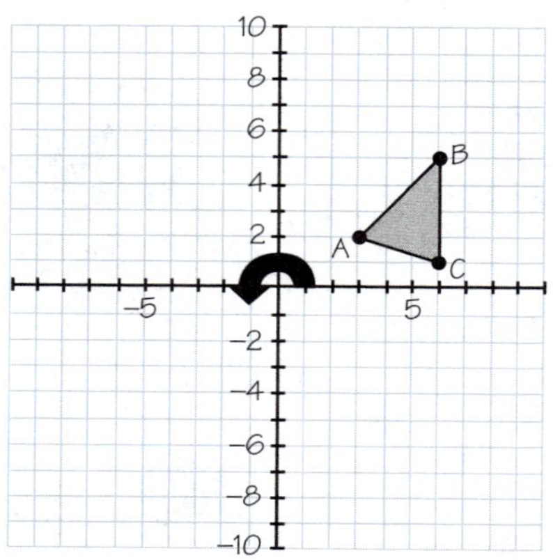

FIGURE 4.40.
Graph for Question 26.

27. Did the triangle change size or shape?

28. Repeat 26 and 27 for rotations of 180°, 270°, and 360°. Record the results in your table.

29. Describe a rotation by multiples of 90° in words.

Coordinate descriptions of rotations that are not multiples of 90° use trigonometry, so you will not do them now. Trigonometry is discussed in Chapter 8.

PART E: GLIDE REFLECTIONS USING THE X-AXIS

Since glide reflections are a composite of a translation and a reflection they require a distance, a direction, and a line of reflection.

Conduct an investigation into glide reflections using the x-axis as the line of reflection. Proceed as you did in Parts A–D. Begin with a triangle drawn in one of the quadrants. Select a distance to translate it horizontally and a distance to translate it vertically. After you have translated it, reflect the image across the x-axis and record the results. Do the same for more triangles until you think you can answer Question 30.

30. Write equations that describe a glide reflection of the point (x, y) h units horizontally, k units vertically, and across the x-axis.

Part D: 90° Rotations on a Coordinate Plane

24.

Angle of Rotation	Pre-Image			Image		
	A	B	C	A′	B′	C′
90°	(3, 2)	(6, 5)	(6, 1)			
180°	(3, 2)	(6, 5)	(6, 1)			
270°	(3, 2)	(6, 5)	(6, 1)			
360	(3, 2)	(6, 5)	(6, 1)			

27. No, the triangle only changed location.

28. Answer:

Angle of Rotation	Pre-Image			Image		
	A	B	C	A′	B′	C′
90°	(3, 2)	(6, 5)	(6, 1)	(−2, 3)	(−5, 6)	(−1, 6)
180°	(3, 2)	(6, 5)	(6, 1)	(−3, −2)	(−6, −5)	(−6, −1)
270°	(3, 2)	(6, 5)	(6, 1)	(2, −3)	(5, −6)	(1, −6)
360	(3, 2)	(6, 5)	(6, 1)	(3, 2)	(6, 5)	(6, 1)

29. Every 90°, the absolute values of the x- and y-coordinates reverse and their signs correspond to the quadrant where the image is.

Part E: Glide Reflections using the *x*-axis

30. $x' = x + h;\ y' = -(y + k)$.

SECTION 4.4 Assignment

1. What are four types of isometries?

2. What are the basic attributes of translations?

3. What are the basic attributes of reflections?

4. What are the basic attributes of rotations?

5. Draw ΔABC on your paper. If you reflect the triangle across any one of its sides, what shape is formed by combining the original triangle and the new triangle?

6. In **Figure 4.41**, which of these do not produce an image of the rectangle with two of its vertices lying on the y-axis?

   A. A reflection across the x-axis
   B. A translation right 3 units
   C. A translation up 2 units
   D. A reflection across the y-axis

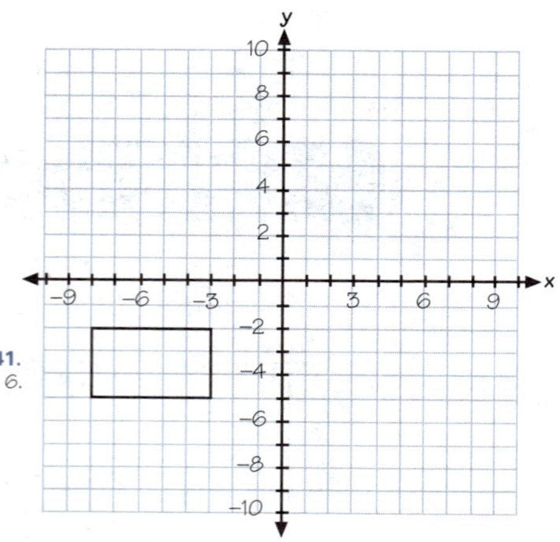

FIGURE 4.41.
Graph for Question 6.

7. In **Figure 4.42**, if triangle ΔABC is reflected across the x-axis, what are the coordinates of A'?

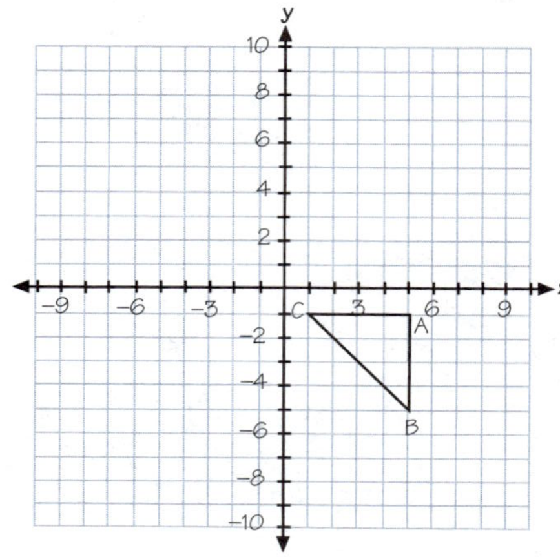

FIGURE 4.42.
Graph for Question 7.

8. In **Figure 4.43**, the vertices of a pre-image are at (–2, –2), (–1, –2), (–1, 0), and (–2, 0).

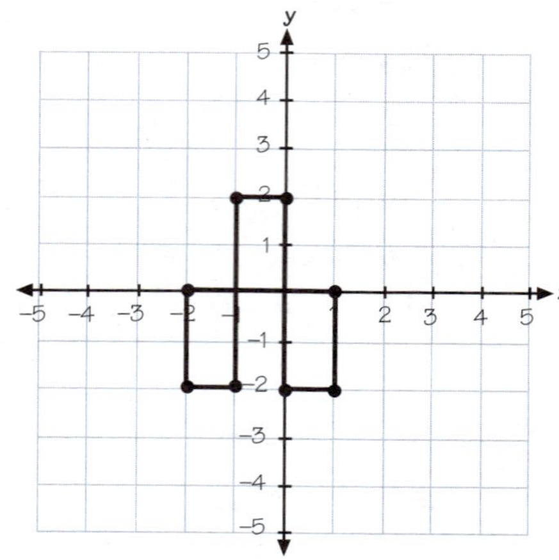

FIGURE 4.43.
Glide reflections on a coordinate plane.

a) Write equations that describe the use of a glide reflection to obtain another part of the figure.

b) Write equations that describe the use of a translation to obtain the remaining part of the figure.

Section 4.4 Assignment

Answers

1. Translation, reflection, rotation, glide reflection
2. Translations preserve size, shape, and orientation, but change location according to $(x, y) \rightarrow (x + h, y + k)$.
3. Reflections preserve size and shape, but they reverse orientation. Each point in the pre-image and its corresponding point in the image are equidistant from the line of reflection. The segment connecting the corresponding points in the image and pre-image is perpendicular to the line of reflection.
4. Rotations preserve size, shape, and orientation. Rotations about the origin in 90° increments cause absolute values of the x- and y-coordinates to reverse. The sign of a coordinate corresponds to the quadrant where the image is.
5. A quadrilateral with the side used for the reflection as one of its diagonals
6. A, C, and D
7. (5, 1)
8. a) $x' = x + 1$; $y' = -y$.

   b) $x' = x + 2$; $y' = y$.

SECTION 4.5 Design a Logo

Create a graphic design for a corporate logo that incorporates translations, reflections, rotations, or glide reflections. The design should include elements of translation, rotation, and reflection symmetry.

Transfer your logo to a coordinate plane. Label points on the pre-image. Use algebraic notation to verify the accuracy of your use of symmetry to create your logo.

Write a short paragraph describing how and why each element of symmetry is incorporated into the design.

SECTION 4.6 It's a Matter of Perspective

Artists, architects, illustrators, and animators are trained to draw objects and scenes. Their drawings need to be accurate and realistic. Small children have not received training to draw objects and scenes. A child's drawing may appear to be less accurate and realistic than the drawing of a trained artist.

Examine the picture assigned to your group by your teacher. Use **Figures 4.44 to 4.47** to answer the following questions:

1. What do you notice about this picture?
2. What do you notice about the sizes of the different objects in the picture?
3. What do you notice about the sizes of the different objects and their placement in the picture?

FIGURE 4.44. *Cave drawings similar to those found at Hamangala.*

FIGURE 4.45. *Architectural rendering of a building.*

EVALUATE Lesson 1

Section 4.5: Design a Logo

Purpose: In Sections 4.1 through 4.4, students investigated and formalized different types of symmetries and transformations. In Section 4.5 students will have the opportunity to demonstrate their knowledge of symmetries and transformations and the applications of mathematics that are used in graphic design, preparing them to investigate symmetry and transformations as they apply to functions in Chapter 5.

Description: Students will be asked to create a graphic design that includes different transformations and symmetries. They will also need to be able to describe, using mathematically appropriate vocabulary, how their design incorporates each of the types of transformations and symmetries.

This is an open-middle performance assessment task; in other words, there is only one correct answer but there is a variety of valid methods to obtain it. In their justification, which is the most important part of a performance assessment task, students should be able to communicate their knowledge of mathematics and demonstrate that they have mastered the concepts presented in Sections 4.1 through 4.4.

Materials: protractor, ruler, tracing paper (such as patty paper), tools students may need to make logo

Teaching Ideas: This task may be assigned as group work or individual work. The mathematics performance assessment rubric should be used to evaluate students' work.

Answers will vary. Use the performance assessment rubric to determine the extent to which students have accomplished the task posed by this section.

ENGAGE Lesson 2

Section 4.6: It's a Matter of Perspective

Purpose: In Section 4.6 students explore works of art to identify attributes common to paintings or sketches that make use of perspective. In the following phases of this lesson, students will explore the idea of drawing in perspective.

Description: In the last lesson students explored symmetry and transformations that are isometries, transformations that do not change the size and shape of a geometric figure. In this lesson students will explore dilation: a transformation that is not an isometry because it changes the size of a figure (but not its shape). Students may have varied levels of exposure to artwork and the idea of perspective in paintings and sketches. As students compare works of art, they make note of which best represent a real-life situation. These comparisons allow for a common experience with works of art for all the students. Students are asked to make observations about the relative size of objects in the paintings to learn that objects close to the viewer should be larger than objects farther away. This serves as the first part of understanding perspective. At the close of this section, students should be able to describe whether or not a painting or sketch makes use of perspective. This will prepare students for further exploration of techniques used to create perspective.

Materials: chart paper, chart markers, sheet of chart paper titled perspective

Teaching Ideas: Divide the class into groups. Assign each group one figure to analyze. Prompt each student group to answer the three questions about their given figure with as much detail as possible.

As students discuss their answers to each question, prompt each group to compare the objects in the foreground of the figure with the objects in the background.

Students may or may not be familiar with the terms foreground and background, so point to objects in the foreground of the figure as you ask about the foreground.

After students complete their discussions, direct the students into new groups. Each group should have 4 members, representing each of the figures. Each member should be prepared to discuss a different figure. Prompt the students to discuss their observations about the figure from their original group. After each student has shared the discussion from his or her original group, prompt the students to compare and contrast the figures. Key questions to ask during student discussions are "Which paintings would be most like a photograph of the same scene?" and "Why?" Those paintings that make use of perspective will be most like a photograph. Prompt each group to summarize their observations on chart paper. Use the observations to create a class summary of the group discussions.

Discuss the different meanings of perspective with students. Use the class summary of the group discussions to create a list of observable attributes of perspective on the sheet of chart paper titled perspective.

Prompt students to work in pairs or individually to complete Questions 4–9.

Answers

For Figure 4.44

1. What do you notice about this picture?

   Answers may vary. Students should see that while the picture is primitive they can see that it represents people and animals

2. What do you notice about the relative size of the different objects in the picture?

   Answers may vary. The animals are larger than the humans. The humans are all approximately the same size.

Answers (continued)

3. What do you notice about the relative size of the different objects and their placement in the picture?

   Answers may vary. The placement of the animals makes them appear to be floating above the humans. The hunters appear to be riding on the backs of the animals.

For Figure 4.45

1. What do you notice about this drawing compared to the cave drawing?

   Answers may vary. Students should describe that the architectural drawing looks more realistic than the cave drawing.

2. What do you notice about the sizes of the different objects in the picture?

   Answers may vary. The people in the foreground seem larger than the people near the car.

3. What do you notice about the sizes of the different objects and their placement in the picture?

   Answers may vary. The trees get smaller as they get farther away; same with the people.

FIGURE 4.46. *Perspective Absurdities* by William Hogarth.

FIGURE 4.47. Ancient Egyptian art.

4. Which pictures most accurately depict a real-life scene? Why?
5. Which pictures least accurately depict a real-life scene? Why?

The term perspective has several different meanings:

a) A view: "From what perspective are you looking down into the parking lot?"

b) A mental outlook: "What is your perspective on standardized dress codes for high school students?"

c) The relationship of parts to the whole: "What is the perspective on the soccer team's success this year? What role did the players play? What role did their training play? What role did the coach play?"

d) The technique of representing objects in three-dimensional space with a two-dimensional plane: "How does the artist use perspective to make a painting that looks like real life?"

6. Which paintings seem to make use of perspective?

Sometimes an artist may use perspective to get your attention. The painting shown in Figure 4.46 contains several intentional mistakes.

PERSPECTIVE

Perspective is the technique of representing objects from three-dimensional space in a two-dimensional plane. Through the use of perspective, an artist's two-dimensional drawing or painting imitates the appearance of a three-dimensional object or scene.

7. Work together with your group to identify the real-world inaccuracies in the painting by William Hogarth (Figure 4.46). Describe each error in detail.

8. Describe what the artist needs to do to correct each item you identified in Question 7.

Question 9 will help you identify elements or principles involved in perspective drawing.

FIGURE 4.48. *The Bathers at Asnieres* by George Seurat.

9. Look at *The Bathers* by George Seurat (**Figure 4.48**). The artist has tried to represent a three-dimensional subject on a two-dimensional plane. The two-dimensional plane is the surface of the canvas or paper.

   a) Imagine that your group is in a hot-air balloon floating over the subjects in this picture. Draw a bird's-eye view of what you would see. You don't have to sketch every detail. Just draw simple outlines of some of the objects in each picture and label them. (For example, you might use ovals labeled as dog, man lying down, man wearing hat, or sailboat to indicate objects.)

   b) Explain what clues in the original picture helped you decide where to place particular objects in your drawing.

   c) Discuss how the artist tried to convey depth.

Answers (continued)

For Figure 4.46

1. What do you notice about this painting?

   Answers may vary. They should notice that some of the things in the picture do not look possible.

2. What do you notice about the size of the different objects in the picture?

   Answers may vary. The person walking on the hill seems too large when compared to the wagon crossing the bridge.

3. What do you notice about the size of the different objects and their placement in the picture?

   Answers may vary. Some objects that are closer to the viewer appear to be larger than those objects that are further from the viewer, but some objects in the distance appear larger than objects in the foreground of the painting.

For Figure 4.47

1. What do you notice about this painting?

   Answers may vary. They should notice that the picture does not seem to try to render people accurately.

2. What do you notice about the size of the different objects in the picture?

   Answers may vary. Half of the people are one size. Others are a different size. Some appear to have on masks.

3. What do you notice about the size of the different objects and their placement in the picture?

   Answers may vary. All subjects appear to be in the same plane. There is no background.

4. Answers may vary. Students will probably answer Figures 4.45 and 4.46. Students looking closely at Figure 4.46, the Hogarth painting, may decide it is not accurate. If a student raises this, remind the class to keep this thought in mind.

5. Answers may vary. Students will probably answer Figures 4.44 and 4.47.

6. Answers may vary. Figures 4.45 and 4.47. Some students may recognize that the Hogarth painting (Figure 4.46) uses perspective in a funny and impossible way. They may suggest that the artist is doing this on purpose.

7. (1) The woman leaning out of the second-story window is lighting the pipe of the person walking on a distant hill. The trees on the distant hill overlap the sign on the garage. The fishing pole closest to the viewer hangs over the fishing pole that is farthest from the viewer.

   (2) The swan on the lake is larger than the cattle/sheep that are closer. The man walking on the distant hill is larger than the coach crossing the bridge. The trees on the distant hill get larger as they get farther away. The crow in the tree is about the same size as the smallest sheep, which is closer to the viewer, and the crow is too large relative to the tree.

   (3) The shadow under the bridge falls in the opposite direction from the shadows of the trees on the distant hill.

   (4) The lines on the building don't appear to be uniform.

   Students may find additional inaccuracies.

8. (1) Do not allow a distant object to overlap or interact with an object in the foreground.

   (2) Change the size of several objects so that distant objects are smaller than objects in the foreground.

   (3) Create shadows that appear on the same side of all objects.

   (4) Make sure the parallel lines on the building appear to imitate a real-life perspective.

9. a) Note that this drawing does not show all the elements in the picture.

   b) Possible clues include relative sizes of objects, positions of objects in the picture, overlapping of objects, and shading.

   c) Seurat put some objects in front of other objects. The objects that are in front are closer to the viewer. In addition, he made objects that were farther away smaller. Objects in the distance show less detail.

SECTION 4.7

Location and Size

PART 1: WHAT'S IN THE FRONT?

In Section 4.6 you identified several factors that are important to the proper use of perspective in art. One of these is the location of objects with respect to each other: objects close to the viewer should overlap those that are behind them. Another factor is the relative size of objects. In this part of this section you will take a closer look at overlapping. In the second part you will take a closer look at size.

Within your group, use centimeter grid paper to create 4 cubes as described in the table in **Figure 4.49.**

Cube	Side Length (cm)
A	2
B	4
C	6
D	8

FIGURE 4.49.
Measurements of cubes.

Use masking tape or string to mark a 35 centimeter by 35 centimeter square region on a flat surface. Label the sides of the square north, south, east, and west.

Position the cubes according to the diagram shown in **Figure 4.50.**

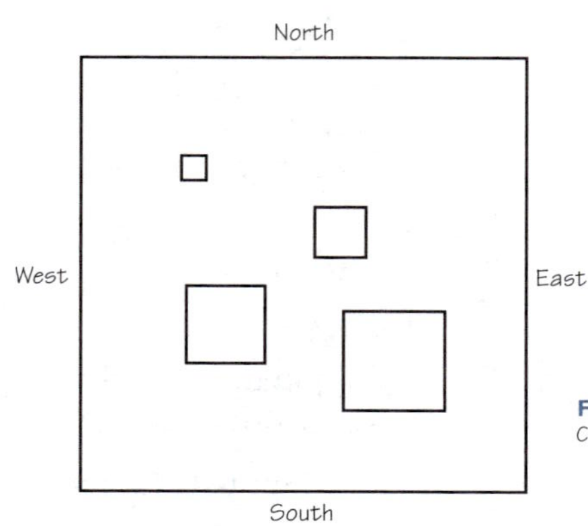

FIGURE 4.50.
Cube placement.

1. Lower yourself so that you are eye level with the cubes. Draw a two-dimensional sketch of what you see from the north, the south, the east, and the west.

2. How are the sketches alike?

3. How are the sketches different?

4. How does overlapping reveal which cube is in front of another cube?

5. **Figure 4.51** shows the top view of a table on which three cubes are sitting. Draw four eye level views of what you would see if you were sitting at positions A, B, C, and D.

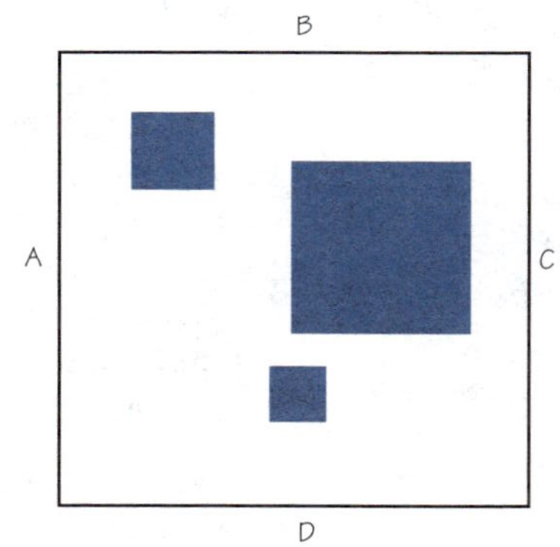

FIGURE 4.51.
Top view of three cubes on a table.

6. What statements can you make about how to draw images of objects so that the person looking at the image knows which objects are closer than other objects?

PART 2: THE RULER METHOD

In Part 1 you investigated the principle of overlapping: objects close to the viewer should overlap those that are behind them. You have also seen that objects far away from the viewer should be drawn smaller than objects close to the viewer. In this part you will take measurements of the size an object appears to be to a viewer.

- Hold a cube directly in front of you so all you see is a square.
- Move the cube toward you. What happens?

EXPLORE Lesson 2

Section 4.7: Location and Size

Purpose: In Section 4.6 students discussed the importance of perspective in making artwork appear realistic. In this section students explore using overlapping and diminution to represent three-dimensional objects in two-dimensional drawings.

Description: In Part 1 of this section students create models of cubes and arrange the cubes based on top views. Students then shift their perspective on the cubes by viewing the arrangements from the north, the south, the east, and the west. Following this, students create sketches of various views without the aid of a physical model.

In Part 2 students take measurements of the apparent height of a student who is several meters away. Then they use these data to calculate ratios that lead to an analysis using similar triangles in Section 4.8.

Materials: centimeter grid paper printed on cardstock, scissors, masking tape or string, meter stick

Teaching Ideas: Given the length of the section it may be done in two class periods.

For Part 1, group the students into pairs. Provide each pair of students with centimeter grid paper printed on cardstock, 2 pairs of scissors, a meter stick, and masking tape or string. Prompt the students to use the centimeter grid paper to build the cubes described in Figure 4.49. Prompt students to place the cubes on a flat surface as described in Figure 4.50. Ask students to justify their sketches of the views from the north, the south, the east, and the west.

For Part 2, place students in groups of four. The remaining instructions for Part 2 are in the student text.

Answers

1. Sample answers:

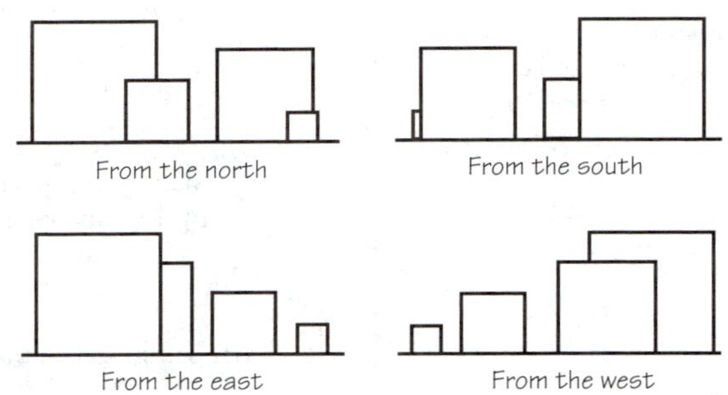

2. Answers may vary. All four sketches contain the same four cubes.
3. Answers may vary. Which cube appears to be closer to the eye of the viewer varies as the location of the viewer varies. Other cubes' locations also vary.
4. Answers may vary. If a cube is in the foreground, that cube overlaps any cubes that appear to be behind it or in the background.
5. Answer:

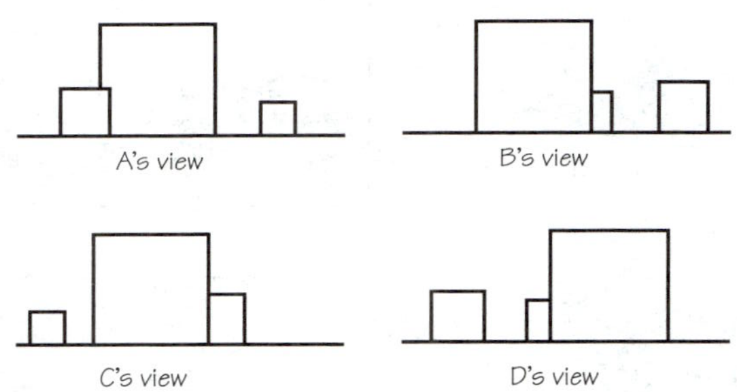

6. Sample answer: Objects that are closer to the viewer may overlap objects that are farther from the viewer.

- Move the cube away from you. What happens?
- What general statement might we make about objects as they move farther away?

You will work with your group to take some measurements that illustrate numerically why objects appear to grow smaller as they move farther away.

Decide which member of your group will do each of the following:

- hold the ruler (holder)
- stand at a distance (stander)
- take measurements (measurer)
- record measurements (recorder)

The instructions that follow are written from the holder's point of view.

7. Hold a ruler in your hand, and extend your arm in front of you at eye level so that the ruler is vertical and your arm is extended fully (see **Figure 4.52**).

FIGURE 4.52. Photograph showing how to hold the ruler.

8. The measurer should carefully measure the distance from your eye to the ruler.
9. Instruct a member of your group (the stander) to stand several meters away from you. The measurer should use a meter stick or tape measure to measure the distance between your eye and the other person.
10. Use the ruler in your extended hand to measure the apparent height of the other person (the height you would draw the person if the paper or canvas coincided with the ruler). See **Figure 4.53.**

FIGURE 4.53. Measuring the apparent height of a person.

11. Record the measurements in a table like **Figure 4.54.**

Measure	Distance (eye to ruler)	Distance (eye to person)	Apparent Height of Person	Actual Height of Person
1				
2				
3				
4				

FIGURE 4.54. Table for Question 11.

12. Complete the ratios in the table in **Figure 4.55.**

Measure	$\frac{\text{Distance Eye to Ruler}}{\text{Apparent Height}}$	$\frac{\text{Distance Eye to Person}}{\text{Actual Height}}$
1		
2		
3		
4		

FIGURE 4.55. Table for Question 12.

13. What do you notice about the ratios of the measures?
14. **Figure 4.56** shows a viewer looking at a person. The picture plane represents the canvas or paper on which an artist draws or paints. Use the figure to explain your conclusion about ratios in Question 13.

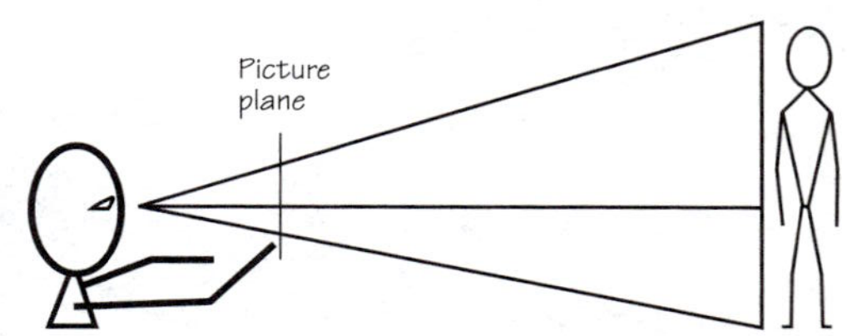

FIGURE 4.56. Image for Question 14.

Answers (continued)

8. Typical answers will be between 1.5 and 2.5 feet.
9. Answers will depend on space available.
13. The ratios are approximately equal to each other.
14. Sample answer: The smaller and larger triangles are similar. Therefore, the ratio of the distance to the picture plane (ruler) to the apparent height (in the picture plane) must equal the ratio of the distance to the person being viewed to the actual height of the person.

SECTION 4.7 Assignment

1. Look at Figure 4.45 on p. 215. Describe at least one object that overlaps another. Which of the two objects is closer to the viewer?

2. Refer to Hogarth's painting (see Figure 4.46 on p. 216). Describe where Hogarth violates the principle of overlapping.

3. **Figures 4.57 and 4.58** show the top view of two different tables on which three cubes are sitting.

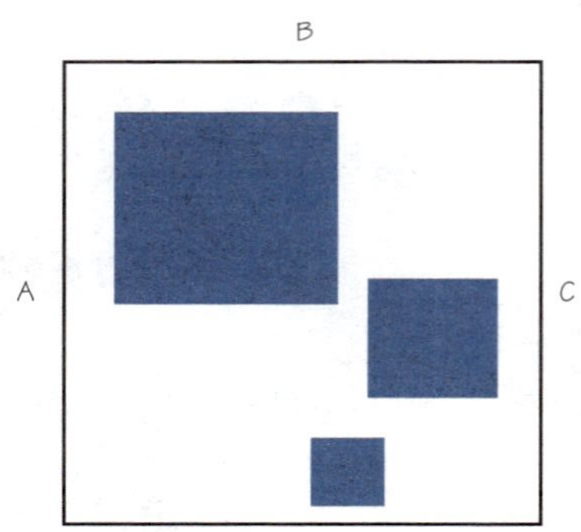

FIGURE 4.57.
Top view of three cubes on Table 1.

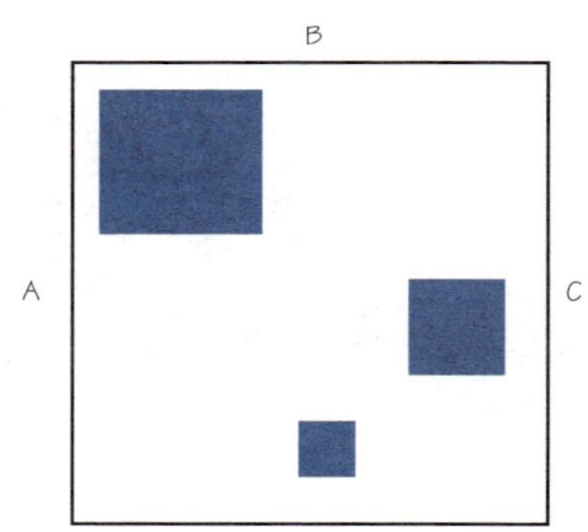

FIGURE 4.58.
Top view of three cubes on Table 2.

   a) Look at the top view presented in Figure 4.57. Draw what you would see if you were sitting at position D.

   b) Look at the top view presented in Figure 4.58. Draw what you would see if you were sitting at position D.

4. Martin built a cityscape using rectangular prisms. He made the following statements about his cityscape as viewed from eye level.

   a) From the southern view, I see a model building that is 2 inches tall. Behind it is a model building that is 6 inches tall. To the right of the 2-inch model, I see a 4-inch tall model building.

   b) From the eastern view, I see a model building that is 4 inches tall. To its right, I see a model building that is 3 inches tall. A 6-inch tall building stands behind this building.

   c) From the northern view, I see a model building that is 4 inches tall that stands behind a 3-inch tall building. I also see a 6-inch tall building.

   d) From the western view, I see a 6-inch tall building. I also see a 2-inch tall building with a 4-inch tall building behind it.

   Sketch the four side views to scale and the top view that match Martin's description. Explain your thinking.

5. Tommy used cubes to build five models of buildings that are each 1 cube wide. The models are described in **Figures 4.59 and 4.60.**

Building	Height (cubes)	Length (cubes)
A	1	4
B	2	2
C	4	1
D	3	1
E	1	2

FIGURE 4.59. Table for Question 5.

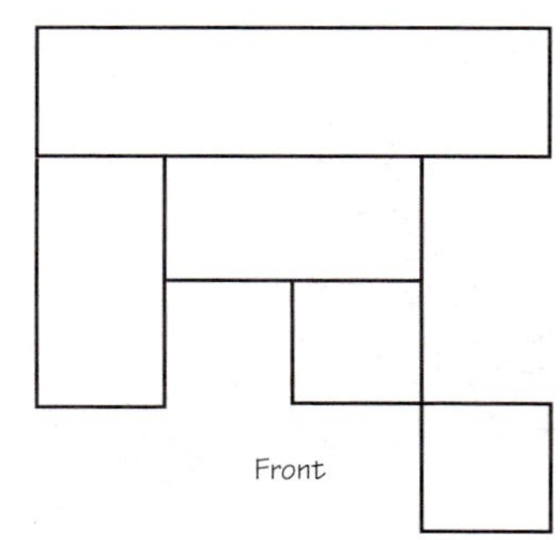

FIGURE 4.60. Top view.

   How many possible arrangements of these five models match the top view? Sketch the front, right, and back views of each arrangement.

Section 4.7 Assignment

Answers

1. Answers may vary. Students may answer that trees overlap the building.

2. Sample answers: The sign hangs behind the distant trees. The line from the fishing pole closest to the viewer hangs behind the fishing pole farther from the viewer.

3. a) Sample views from A, B, C, D:

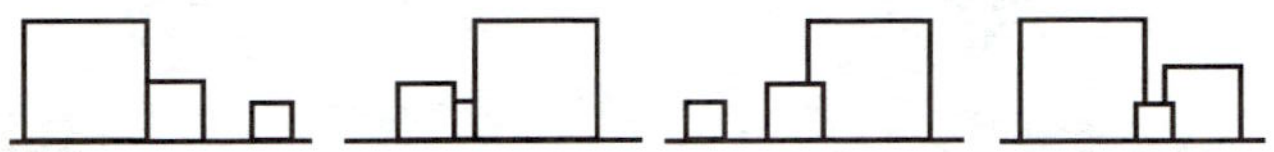

b) Sample views from A, B, C, D:

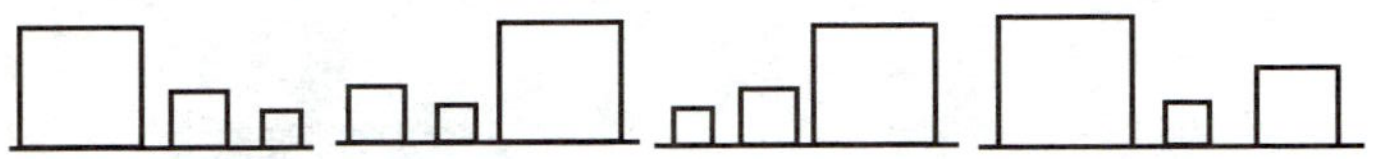

4. Sample answer:

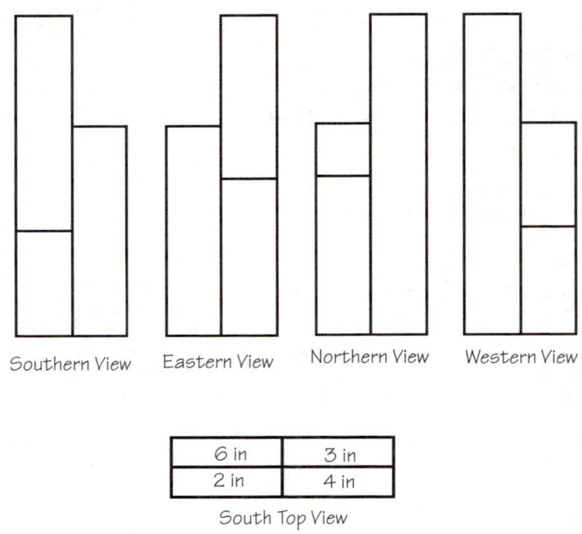

5. Sample answers:

Arrangement 1

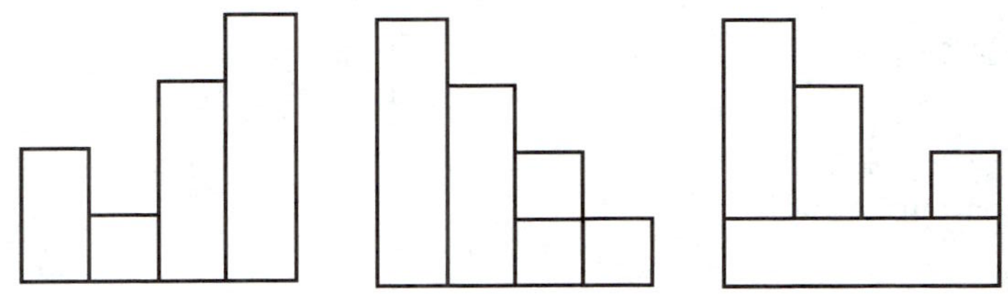

Arrangement 2

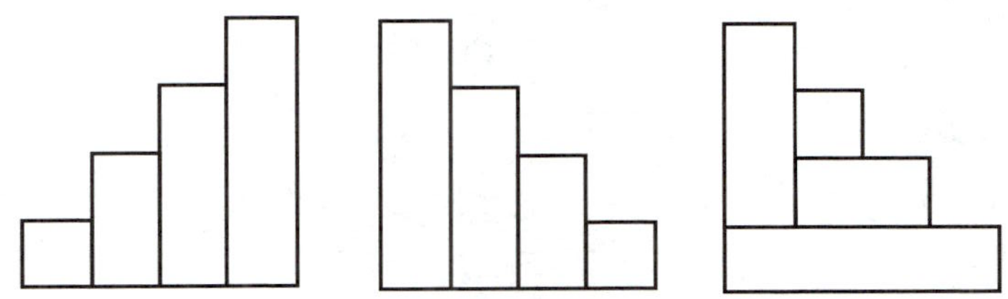

Arrangement 3

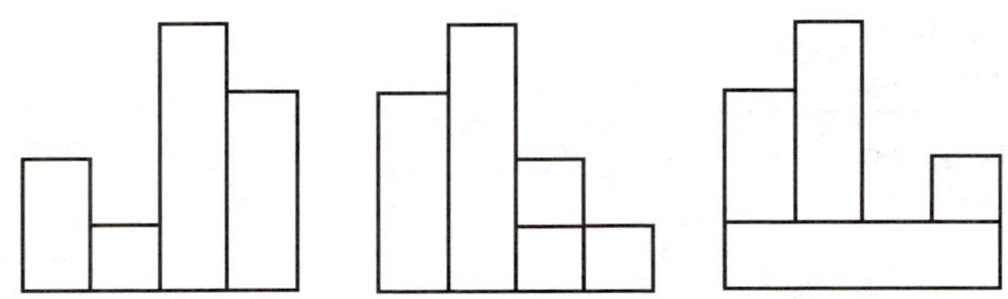

Arrangement 4

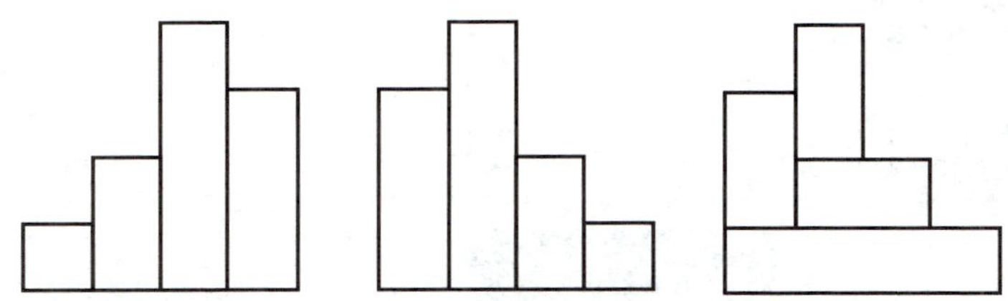

6. **Figure 4.61** is a sketch of your hand held up 2 feet in front of your eye.

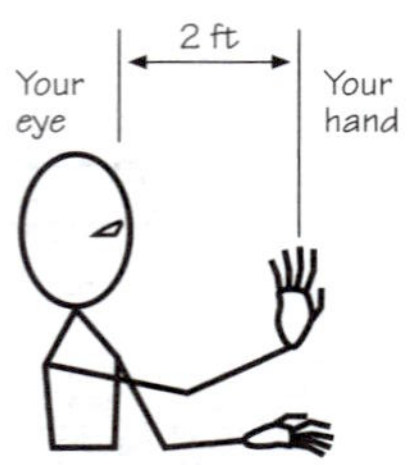

FIGURE 4.61. Figure for Question 6.

Suppose your hand is 7 inches long and just blocks out a friend who is 5′ 6″ tall.

a) Make a sketch showing your eye, your hand, and your friend. Show all of the given measurements in your figure.

b) How far from your eye is your friend standing? Explain how you got your answer.

SECTION 4.8 Principles of Perspective: Diminution and Scale

In the last section you probably noticed that the farther an object is away from you the smaller it appears. This basic principle is called **diminution**.

Diminution

CHECK THIS!

The word diminution means the act of reducing or decreasing. This word comes from the word diminish.

Look at the Seurat painting (Figure 4.48) and think about how he used diminution. For example, Seurat used the principle of diminution when he made people in the boat smaller than the people on the land. But he had to decide how much smaller to make them. The key to this decision lies in a technique called **scaling**. To help you see how scaling works you first need to know what is meant by **lines of sight**. You also need to recall a few facts about **similar triangles**.

CHECK THIS!

Lines of sight are imaginary lines running from your eye to an object at which you are looking.

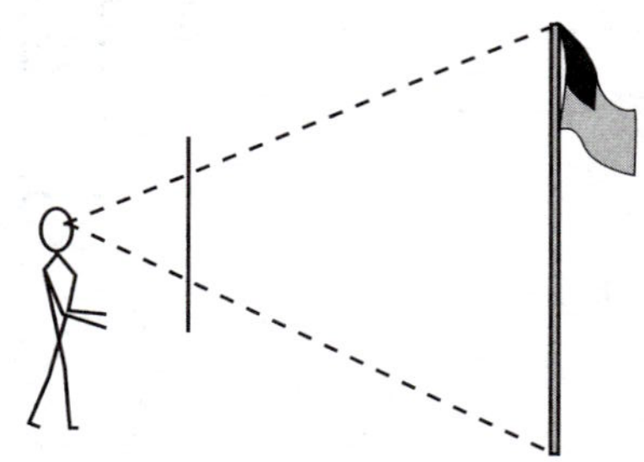

FIGURE 4.62. Lines of sight to top and bottom of a flagpole.

Suppose a person is viewing a flagpole as shown in **Figure 4.62.** The dashed lines represent the lines of sight from the viewer's eye to the top and bottom of the flagpole. The vertical line in front of the viewer represents a canvas (the picture plane). The lines of sight intersect the picture plane to indicate where the top and bottom of the flagpole should be in a sketch made on the canvas.

Figure 4.62 shows two triangles. One of them is the small triangle with portions of the two lines of sight and the picture plane as sides. Another is the large triangle with the lines of sight and the flagpole as sides. These two triangles appear to be similar. That is, they appear to be the same shape, but of different sizes.

Answers (continued)

6. a) Sample sketch:

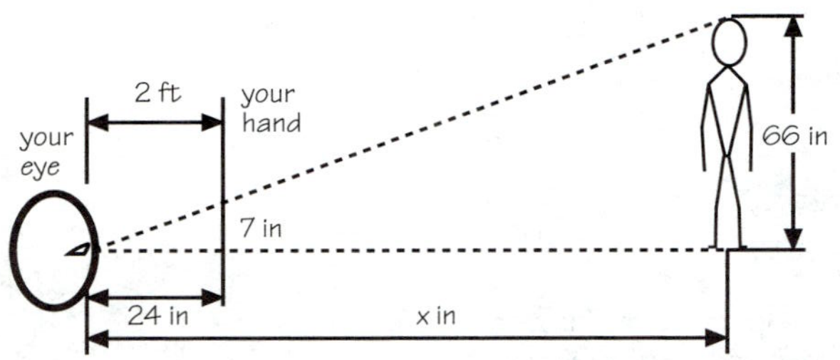

b) About 226 inches or 18′ 10″ away. The ratio $\frac{\text{distance to hand}}{\text{hand length}}$ should equal the ratio $\frac{\text{distance to friend}}{\text{friend's height}}$.

EXPLAIN Lesson 2

Section 4.8: Principles of Perspective: Diminution and Scale

Purpose: In Section 4.7 students explored the issues of overlapping and sizing in perspective drawing. In this section students explain the principles of diminution and scale that are used to draw three-dimensional objects in two dimensions. In Section 4.9 they will expand the concept of diminution to one- and two-point perspective drawings.

Description: Students apply their results from Section 4.7 about the relationships between the apparent heights of objects and their distances from the viewer. These relationships form the foundation for understanding scale. As discussions about scale occur, students refine their understanding of the role of ratio and proportion in art.

Materials: rulers, protractors

Teaching Ideas: Explain to the class the conventions for naming and identifying similar figures. Prompt students to label and name the similar triangles that are represented in Figure 4.62. Prompt students to describe the ratios and proportions that may be written based on the viewer's line of sight. Use the ratios generated by the students to formally define scale.

Prompt students to work in groups of 2–3 to complete the questions. When students have completed the work, have a volunteer from each group rotate to another group. Prompt the members of each group to share strategies.

At the end of the lesson, prompt students to add thoughts to the What I Learned column of the K-W-L chart.

Mathematicians use the following conventions when discussing triangles:

- Vertices of triangles are labeled with capital letters.
- A side of a triangle is labeled with the same letter (but in lower case) as its opposite angle. For example, in **Figure 4.63**, side *a* is opposite angle A. A side can be named with the letters of its endpoints. Thus, side *a* is also referred to as $\overline{BC}$.
- An angle is named with the letter of its vertex. If two or more angles shared a vertex, three letters are used to name an angle. For example, ∠A can also be called ∠BAC.

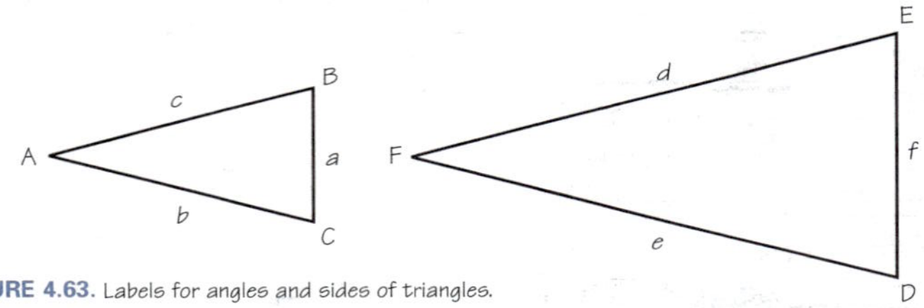

FIGURE 4.63. Labels for angles and sides of triangles.

In Figure 4.63, triangle ABC appears to be similar to triangle FED. Mathematicians write this statement with symbols: ΔABC~ΔFED. When two triangles are similar, corresponding angles are congruent and ratios of corresponding sides are equal. For example, in Figure 4.63, if ΔABC~ΔFED, then

- $\angle A \cong \angle F$; $\angle B \cong \angle E$; $\angle C \cong \angle D$
- $\frac{AB}{FE} = \frac{BC}{ED} = \frac{AC}{FD}$.

Mathematicians have proved a theorem that is helpful in deciding whether two triangles are similar. It says that if two triangles have two pairs of congruent angles, then the triangles must be similar. The proof is based on the fact that the sum of the measure of any triangle's angles is always 180°.

CHECK THIS!

Although having all pairs of angles congruent makes triangles similar, the same is not true of other polygons. For example, a square and a rectangle have four pairs of congruent angles, but they are not similar.

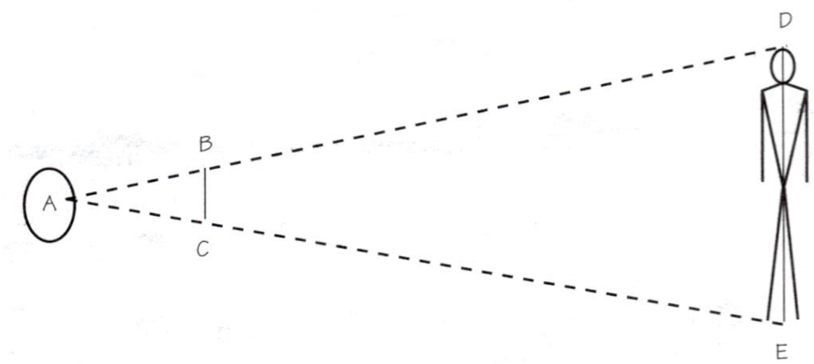

FIGURE 4.64. Image for Question 1.

1. **Figure 4.64** shows a viewer's eye at A, lines of sight to the top (D) and bottom (E) of a person, and the points where the lines of sight cross the picture plane (B and C). What triangles are similar? Why?

2. In Question 1, what ratios must be equal?

The **scale** of an object in a picture is the ratio of the image height in the picture (the apparent height) to the actual height of the object. That is, the scale is the fraction $\frac{\text{apparent height}}{\text{actual height}}$.

3. In Figure 4.64, what ratio represents the scale?

In **Figure 4.65**, a line representing the distance from the viewer to the person being viewed has been added to Figure 4.64. This line creates an altitude in each triangle.

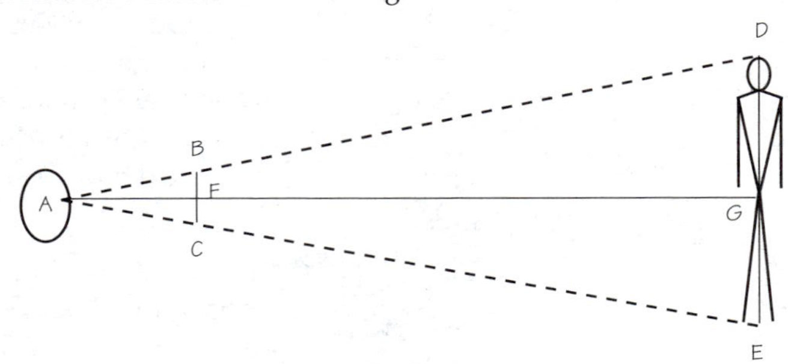

FIGURE 4.65. Figure for Question 4.

CHECK THIS!

An altitude of a triangle connects a vertex to the opposite side and is perpendicular to the opposite side.

Answers

1. Answer: $\Delta ABC \sim \Delta ADE$. They are similar because they share the angle at A and because the picture plane and the plane of the person being viewed are parallel.

2. Answer: $\frac{AB}{AD} = \frac{BC}{DE} = \frac{AC}{AE}$

3. Answer: $\frac{BC}{DE}$

Mathematicians have proved that corresponding altitudes of similar triangles have the same ratio as corresponding sides.

4. In Figure 4.65, what ratio of altitudes must equal the ratio $\frac{BC}{DE}$?

5. Suppose you are drawing on a canvas that is 2 feet in front of you. You use proper perspective to sketch an image 6 inches tall of an object that is really 3 feet tall.

   a) What is your drawing's scale?

   b) How far away are you from the object?

6. Suppose an object is 6 feet tall. You are 18 feet away from it, and you are sketching it on a canvas that is 2 feet in front of you. How tall should the object be in your picture?

In scaling an object to its proper size, an artist uses a mathematical transformation called **dilation**. A dilation uses a fixed point to project every point of a geometric figure. To an artist, the fixed point is the artist's eye. To a mathematician, the fixed point is the dilation's center.

Figure 4.66 shows a dilation of ΔABC from point E that doubles the distance between E and each point of ΔABC. ΔA'B'C' is the image that results. For example, the distance from E to A' is twice the distance from E to A. Since the distance from E to ΔABC is doubled, the sides of ΔABC are scaled by a factor of 2.

In a dilation, a segment's **scale factor** is the ratio $\frac{\text{length in image}}{\text{length in pre-image}}$.

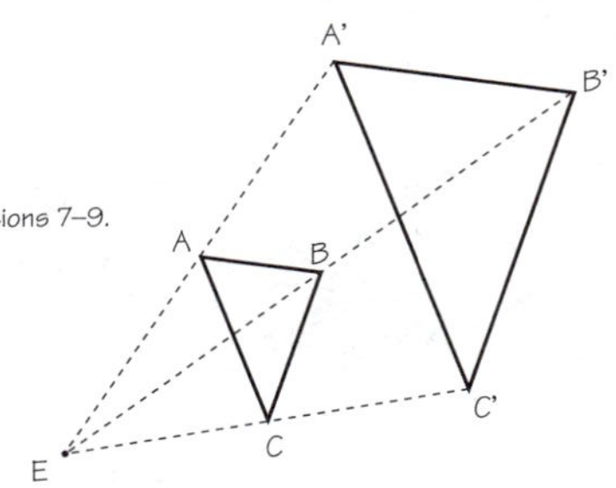

FIGURE 4.66.
Image for Questions 7–9.

CHECK THIS!

Doubling the lengths of the sides of a triangle does not double the triangle's area. Doubling the sides of a triangle quadruples its area.

7. Use rulers and protractors to confirm that ΔABC and ΔA'B'C' are similar and that dilation doubles the lengths of ΔABC's sides.

8. Describe a dilation using point E that triples the lengths of ΔABC's sides. Trace ΔABC and point E onto a sheet of paper and perform the dilation.

9. Describe a dilation using point E that halves the lengths of ΔABC's sides. Trace ΔABC and point E onto a sheet of paper and perform the dilation.

Answers (continued)

4. Answer: $\frac{AF}{AG}$

5. a) Answer: $\frac{6}{36} = \frac{1}{6}$

   b) Answer: If x is the distance to the object, $\frac{6}{36} = \frac{24}{x}$. So, $x = 144$ inches or 12 feet.

6. Answer: $\frac{x}{6} = \frac{2}{18}$, so $\frac{12}{18} = \frac{2}{3}$ feet or 8 inches.

7. Students should confirm that corresponding angles have equal measures and that the ratios of lengths of sides of the larger triangle to lengths of sides of the smaller triangle equal 2.

8. Sample description: Draw line segments from E to A, B, and C. Extend the segments to three times their original length and connect the endpoints to form a triangle.

9. Sample description: Draw line segments from E to A, B, and C. Use a ruler to find the midpoint of each segment and connect the midpoints to form a triangle.

SECTION 4.8

Assignment

1. The Washington Monument is 169 meters high. How far should you stand from it to block it out with your hand? Assume your hand is 18 cm in length and the picture plane is 50 cm from your eye. Sketch the situation first. (Be careful with the units.)

2. Suppose you are sketching a tree on a piece of paper taped to an easel (see **Figure 4.67**). The paper is 2 feet from your eye. The tree is 10 feet high and 40 feet from your eye. How tall should you make the image of the tree on the paper?

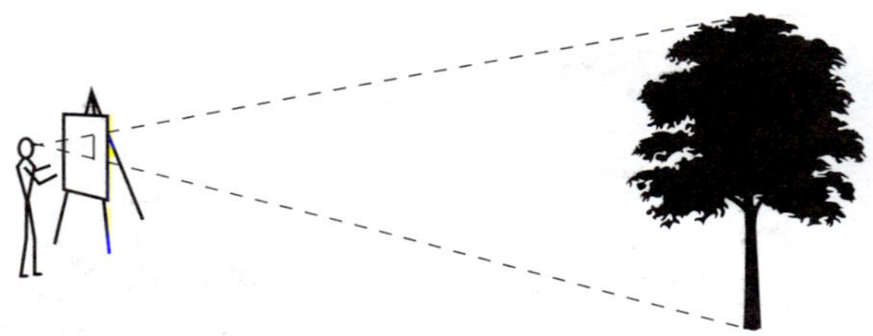

FIGURE 4.67. Lines of site for drawing a tree.

3. At many amusement parks and beaches, there are stores that take a photo, enlarge it, and give you a larger-than-life portrait suitable for framing. They also give you a small wallet-sized copy. **Figure 4.68** shows examples using stick pictures (not to the same scales).

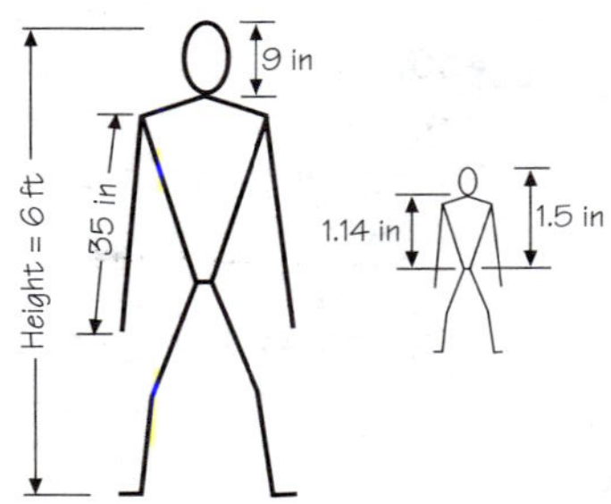

FIGURE 4.68. Poster-sized and wallet-sized pictures.

a) Is enough information given to find the height of the smaller figure? If not, make up a reasonable value for the information you need.

b) Assume the leg length of the large portrait is 35 inches (so the torso length is 28 inches). Find the leg length and head height for the smaller figure.

c) What fraction of the large figure's height is in its legs? What fraction of the small figure's height is in its legs?

4. Trace **Figure 4.69** onto your paper. Sketch a side-doubling dilation of ΔABC using E as the center.

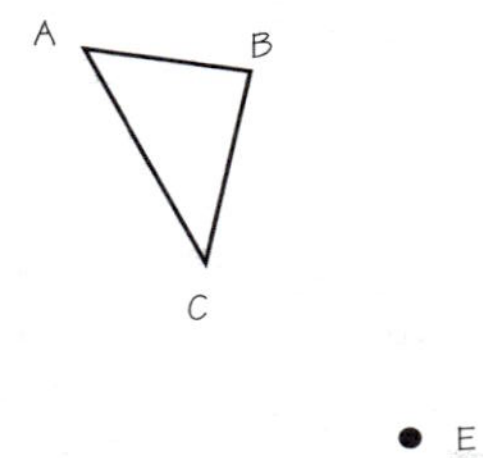

FIGURE 4.69. Image for Question 4.

5. In Question 4, does the dilation produce a larger similar triangle if E is inside ΔABC? Does it matter where point E is located? Explain.

6. a) Since two triangles are similar if they have two pairs of congruent angles, two right triangles are similar if they have just one pair of congruent acute angles. Explain.

b) **Figure 4.70** shows a right triangle ABC with right angle at A and an altitude drawn from vertex A to the hypotenuse $\overline{BC}$. List all of the similar triangles. Be sure you have the vertices in proper correspondence.

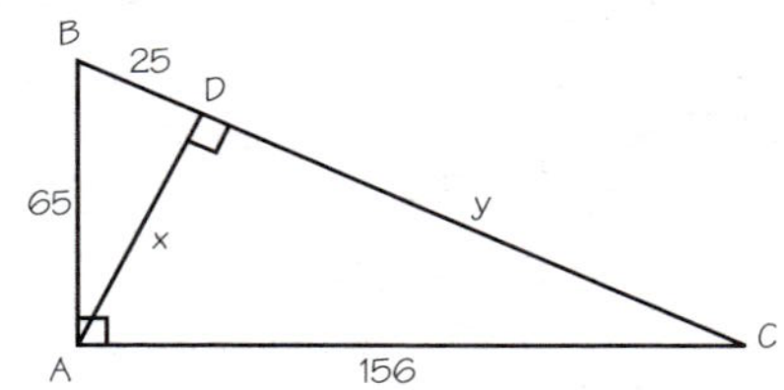

FIGURE 4.70. Image for Question 6.

c) Use the fact that corresponding sides of similar triangles are proportional to find lengths x and y in Figure 4.70.

d) In Figure 4.70, $\angle B$ measures approximately 67.4°. Find the measures of all the angles of the triangles.

Section 4.8 Assignment

Answers

1. Sample answer:

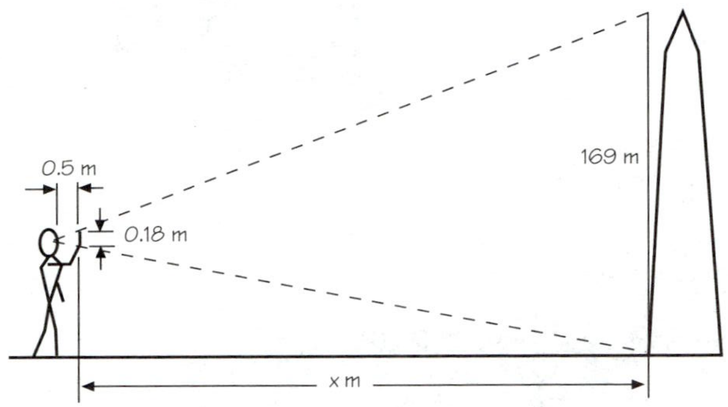

Solve a proportion $0.5/0.18 = x/169$; $x \approx 470$ m.

2. Answer: 6 inches

3. a) Yes. The head size for the smaller figure is 0.36 inches ($1.5 - 1.14 = 0.36$). Then the height of the smaller figure can be found by solving the proportion $0.36/9 = h/72$. So the height of the smaller figure is 2.88 inches.

   b) The leg length of the smaller figure can be found using the proportion $0.36/9 = x/35$, so the length is 1.4 inches. The head height of the smaller figure is about 0.36 inches.

   c) 35/72 of the height is legs. The ratio is the same for the smaller figure.

4. Sample sketch (not drawn to size):

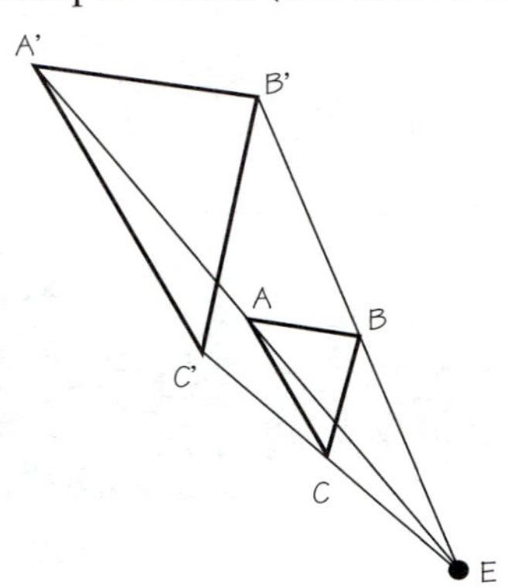

5. No matter where E is located, (experiment with putting E inside the triangle), ΔA'B'C' is similar to ΔABC, and its sides are twice as long as those of ΔABC.

6. a) Sample answer: Since right triangles automatically have a pair of equal angles (the right angles), only one additional pair is needed.

   b) Answer: ΔABC~ΔDBA~ΔDAC

   c) $\frac{x}{25} = \frac{25}{65}$, so $x = 60$; $y/156 = 60/65$, so $y = 144$.

   d) $m\angle ABC = m\angle BDA = m\angle CDA = 90°$

   $m\angle C = m\angle BDA = 22.6°$

   $m\angle B = m\angle DAC = 67.4°$

SECTION 4.9

A Parallel Universe (One-Point and Two-Point Perspective)

When you hold a cube at eye level so one face is parallel to your plane of vision, the cube looks like a square. If the cube is transparent, then you see the otherwise invisible sides and edges of the cube (see **Figure 4.71**).

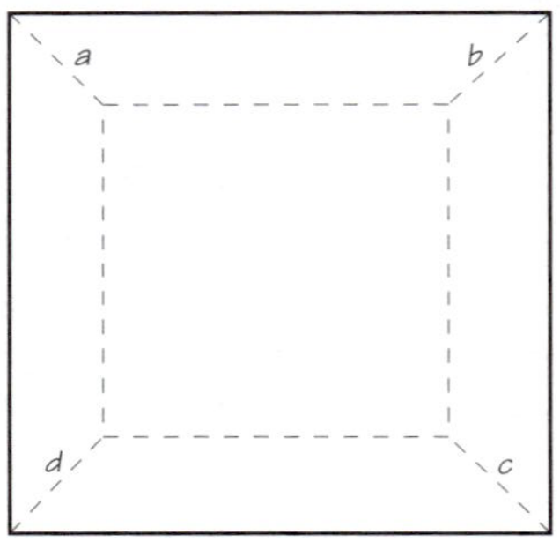

FIGURE 4.71. Transparent cube.

1. The dotted lines represent the edges of the hidden faces. The edges labeled *a*, *b*, *c*, and *d* are perpendicular to the plane of vision.

   a) What do you notice about the vertical edges?

   b) How do the front and back faces compare?

   c) Why does the back face appear smaller than the front face of the cube?

   d) What do you notice about the top, bottom, and side faces?

   e) What do you notice about the edges labeled *a*, *b*, *c*, and *d*, which are perpendicular to the plane of vision?

FIGURE 4.72. The inside of a hallway.

2. Refer to the picture of a hallway in **Figure 4.72.**

   a) There are several lines that run the length of the hallway. Select two of these lines. Place a ruler or edge of a paper along each of them. Describe where the two lines intersect.

   b) Repeat a using two lines that run the length of the ceiling. How does their intersection point compare to the intersection point of the floor lines?

In reality, the walls of the hallway in Figure 4.72 are parallel. Yet, in the picture the distance between the walls appears to narrow. If you imagine the walls were extended, they would converge to a single point. This is an example of the artistic principle of **convergence**: The apparent distance between parallel horizontal lines narrows. The single point to which the lines converge is called a **vanishing point**.

The vanishing point for horizontal lines is at the eye level of the viewer (see **Figure 4.73**). This eye level is called the **horizon line**—the line on which the earth and the sky appear to meet. The vanishing point for the railroad ties in Figure 4.73 is a point on the horizon.

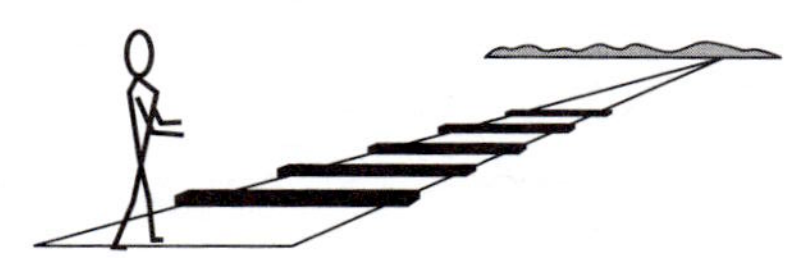

FIGURE 4.73. The vanishing point is on the horizon at the eye level of the viewer.

ELABORATE Lesson 2

Section 4.9: A Parallel Universe (One-Point and Two-Point Perspective)

Purpose: In previous sections students studied the principles of overlapping and diminution. In this section students elaborate on their understanding of the principles of overlapping and diminution and the concept of scale using a new context: one-point and two-point perspective.

Description: Students apply their discussions from Sections 4.6–4.8 to the ideas of one-point and two-point perspective. Discussion returns to the visualization of a square as students examine a true rendering of a transparent cube and as they examine the view of a cube when the line of sight is centered on a vertical edge of the cube. This discussion allows students to differentiate between one-point perspective and two-point perspective. Students use straightedges to reveal the use of one-point perspective and two-point perspective in works of art as they identify the vanishing point(s) found in photographs, paintings, and sketches.

Materials: ruler or straightedge (a small 3″ × 5″ index card works well), tracing paper or patty paper, copies of Handouts 4.1, 4.2, and 4.3

Teaching Ideas: As a whole class complete Question 1. Upon completion of Question 1, place students in groups of 2. Assign Questions 2–6 to half of the groups. Assign the remaining groups Questions 7–10. When all groups have completed their work, regroup students into new groups of 2 with one member who examined one-point perspective and one member who examined two-point perspective in each group.

Prompt students to work in these pairings to create a summary of how overlapping, diminution, and scaling are related to one-point and two-point perspective.

At the end of the lesson, prompt students to add thoughts to the What I Learned column of the K-W-L chart.

Answers

1. a) They are parallel and they look like they are parallel and vertical in the drawing.

   b) Both appear as squares. The back face appears smaller than the front face.

   c) Diminution. The faces are the same size in real life, but the images are different sizes because the back face is farther from the viewer.

   d) Although they are in reality squares, they look like trapezoids.

   e) Although they really are parallel to one another, they do not appear to be parallel. If you extend each line toward the center of the cube, they cross at a point in the middle of the squares.

2. a) The lines intersect in the middle of the back of the hallway.

   b) These lines also intersect in the middle of the back hallway.

A similar effect is visible when you look down the parallel rails of a railroad track. They also appear to converge in the distance (see **Figure 4.74**).

FIGURE 4.74.
Railroad tracks illustrate convergence.

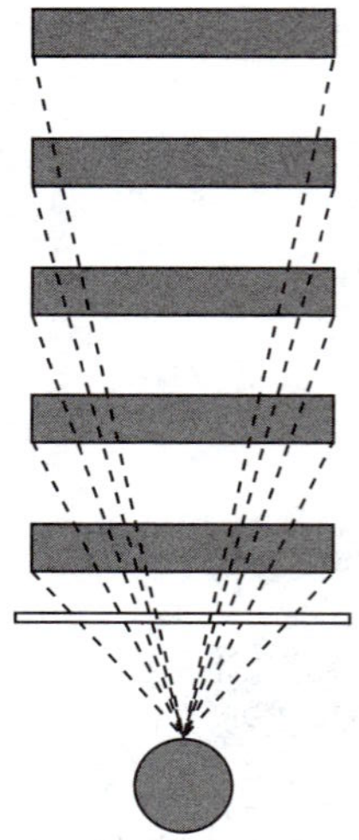

FIGURE 4.75.
Top view of person viewing a series of railroad ties.

3. **Figure 4.75** is a top view of a person observing a railroad track and a series of equally spaced railroad ties. The lines of sight are drawn from the viewer to the ends of the first five ties.

   a) To simplify the picture, the rails connecting the ties are not shown. Describe the relationship between the two rails in real life.

   b) The **projection** of the first tie in the picture plane is almost as wide as the picture (see Figure 4.74). Compare the length of the projection of the fourth tie to the length of the projection of the first tie. How do they differ?

   c) Describe the projection of the railroad tie that is as far as the eye can see from the viewer. (What does it look like? Where is it located in the picture plane?)

   d) Each tie connects the parallel rails. What does your answer to c tell you about the images of the rails in the picture?

CHECK THIS!

In this chapter the projection of a real object is its image on the picture plane. The image is formed by lines of sight from the viewer to points on the object.

4. **Figure 4.76** is a side view of the same person viewing the first four railroad ties.

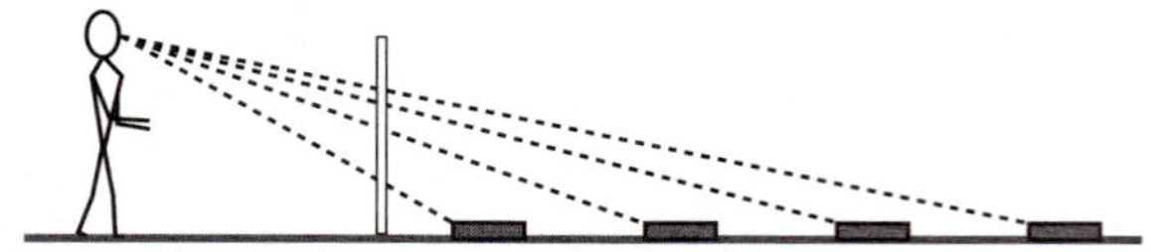

FIGURE 4.76.
Side view observing a series of equally spaced railroad ties.

   a) The projection of the first tie appears near the bottom of the picture plane (see Figure 4.74). How is the location of the projection of the fourth tie different from the location of the projection of the first tie?

   b) Describe the location in the picture plane of the tie (not shown) that is farthest from the viewer. Assume the ground is level.

Lines of sight may be drawn from a viewer to one or more objects in a scene. A line of sight that is parallel to the ground (assuming the ground is level) is said to be at the eye level of the viewer (see **Figure 4.77**).

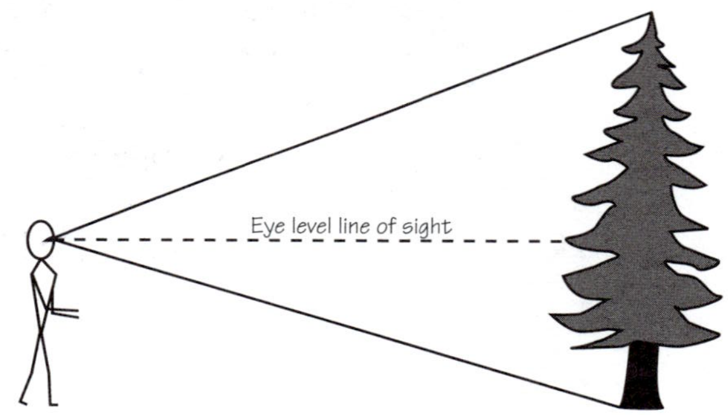

FIGURE 4.77.
Sketch illustrating line of sight at the eye level of the viewer.

5. **Figure 4.78** shows an incomplete picture. The artist has drawn in the **horizon**, the horizontal line at the viewer's eye level where the earth and sky appear to meet. She has added two dots at the bottom of her picture to mark the edges of a roadway that she plans to sketch.

   a) Where in the picture do you expect to find the portion of roadway that is as far as the eye can see: above the horizon, below the horizon, or on the horizon? How large will it appear?

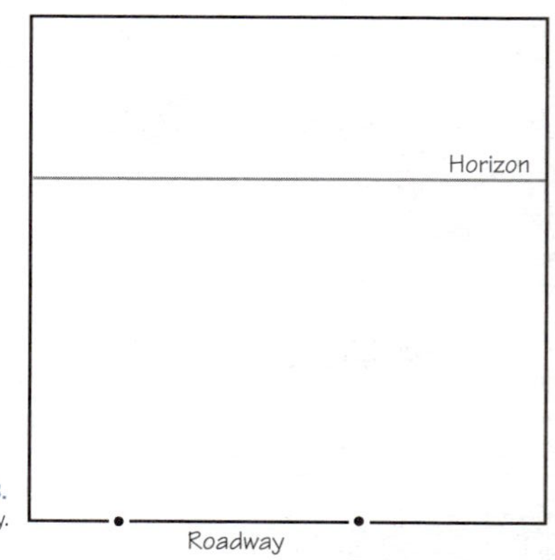

FIGURE 4.78.
Incomplete picture of roadway.

Answers (continued)

3. a) They are parallel.

   b) The projection of the fourth tie is significantly shorter than the projection of the first tie.

   c) The farthest tie would appear as a point. Assuming the ground is level, the point will be directly in front of the viewer at eye level. The ties appear shorter and smaller as their distance from the viewer increases; that's what diminution is all about. The width approaches zero for extremely large distances.

   d) The rails must appear to meet at a point.

4. a) It is located higher in the picture plane.

   b) It appears as a point at eye level to the artist/viewer.

5. a) It will appear as a point on the horizon.

b) Trace a copy of Figure 4.78. Sketch the roadway for the artist. What assumptions did you make about the viewer's location in relation to your roadway?

FIGURE 4.79.
Print Shop Interior by Jan van der Straet.

6. a) On Handout 4.1, which is the same as **Figure 4.79**, draw lines along all edges that represent lines parallel to the center aisle. Extend the lines to identify the locations of any corresponding vanishing points.

   b) What is true of the vanishing points for sets of parallel lines from various objects in Figure 4.79? Explain.

Next, you will apply what you have learned about the principles of diminution and convergence to drawings with two vanishing points.

THE CORNER STONE

7. Hold a cube so that the top and bottom are parallel to the ground and one face is parallel to your plane of vision. Keeping the bottom of the cube parallel to the ground, rotate the cube so that no face is parallel to your plane of vision.

   a) Trace a copy of **Figure 4.80** onto paper.

   b) Which pairs of edges in Figure 4.80 are, in reality, parallel? How is the principle of convergence apparent in the sketch of the rotated cube?

   c) For the cube in Figure 4.80, there are two vanishing points. Draw them on your copy of Figure 4.80.

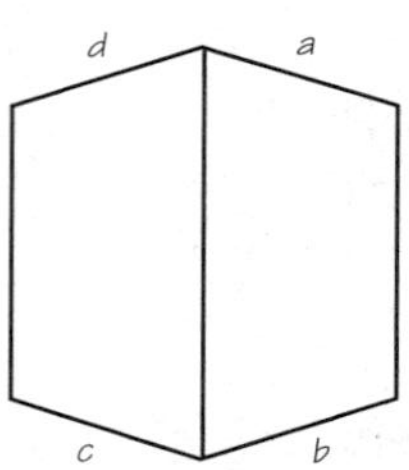

FIGURE 4.80.
View of cube looking directly at a vertical edge.

   d) The vanishing point is on the horizon for horizontal lines that increase in distance from the viewer. You have drawn two vanishing points in c. Are there two horizons? Explain.

8. a) Locate the vanishing points and horizon in **Figures 4.81 and 4.82.** Draw the vanishing points and horizon on Handout 4.2 provided by your teacher.

FIGURE 4.81.
Photograph for Question 8.

FIGURE 4.82.
Painting for Question 8.

   b) How many vanishing points are there? How do you know?

Answers (continued)

b) Sample answers:

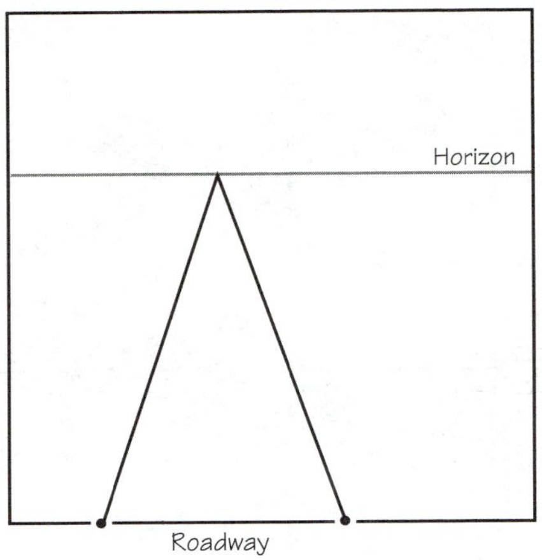

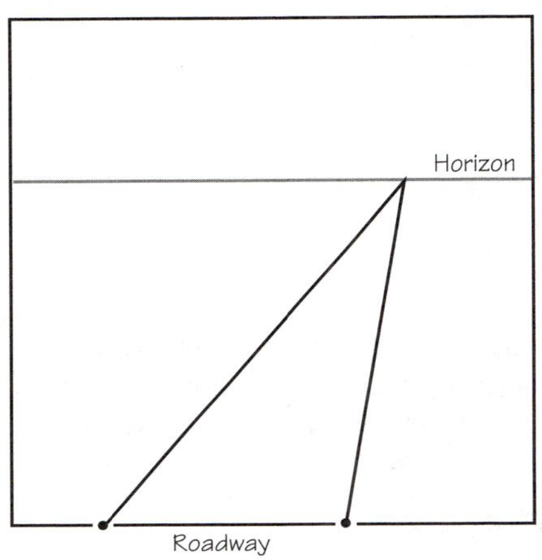

The lines representing the sides of the roadway should meet at a point on the horizon. The viewer's location determines where the vanishing point should be placed on the horizon. In the first sample picture, the viewer is standing in the middle of the road and looking straight down the road. In the second picture, the viewer is standing closer to one side of the road.

6. a) Answer:

b) They are the same point because they represent parallel lines in the scene.

7. b) Edges a and b and edges c and d are, in reality, parallel. The distance between edges a and b appears to narrow. The same is true for the distance between edges c and d.

c) The points of intersection of the extended edges are the vanishing points.

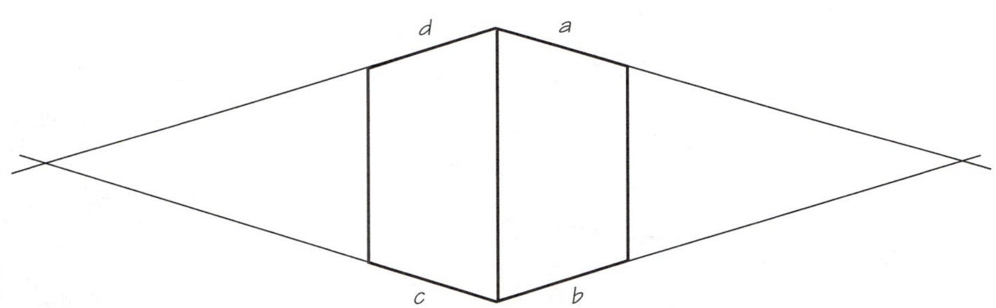

d) No, there is only one horizon. Parallel lines that are parallel to the ground (and increase in distance from the viewer) converge to a point on the horizon that is at eye level of the viewer. The two sets of parallel edges in the cube are perpendicular to each other but parallel to the ground. Hence the two sets of parallel lines converge to vanishing points that are at the eye level of the viewer. Here there is only one horizon, and both vanishing points are on the horizon.

Answers (continued)

8. a) Answer:

b) There are at least two apparent vanishing points because the rectangular objects do not have faces parallel to the plane of vision. (The front face in Figure 4.83 is only slightly turned from the plane of vision.)

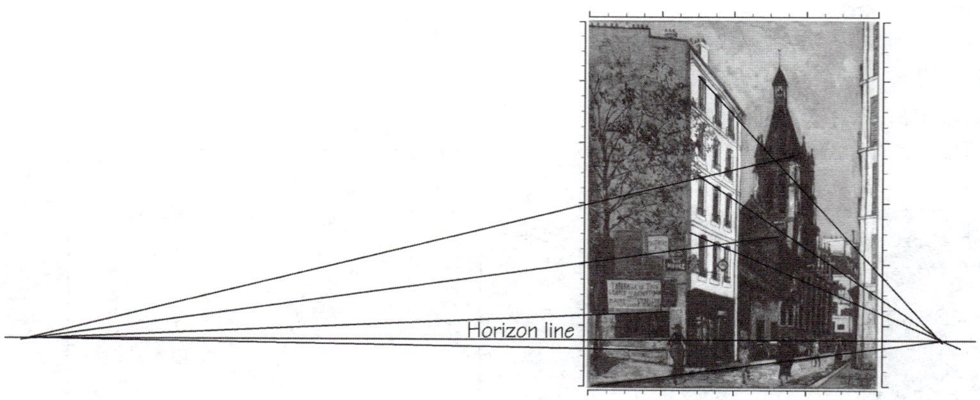

9. Practice drawing a rectangular box using two vanishing points. Use pencil and paper or a geometric drawing utility. Begin with a drawing that looks similar to **Figure 4.83.** Notice the box is drawn entirely below the horizon.

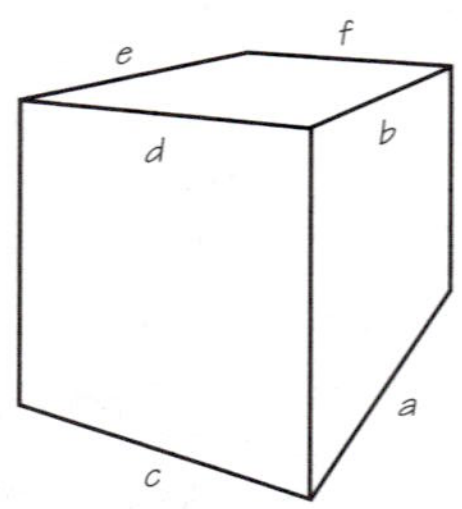

FIGURE 4.83.
Perspective drawing of a box using two vanishing points.

a) Why does edge e appear to converge to the same vanishing point as edges a and b?

b) Why does edge f appear to converge to the same vanishing point as edges c and d?

c) In Figure 4.83, you see two sides and the top of the box. Why is the artist able to see the top of the box in Figure 4.83 but not in Figure 4.80?

d) Describe what happens to the drawing of the box when you bring the two vanishing points closer together.

e) Describe what happens to the drawing of the box when you move the two vanishing points farther apart.

10. Revisit the painting by William Hogarth (Figure 4.46 on p. 216). How many vanishing points do you observe? Describe where they are located and explain how you found them.

SECTION 4.9 Assignment

1. Draw lines of convergence and identify the vanishing points for **Figure 4.84.**

FIGURE 4.84.
Le Pont de l'Europe 1876 by Gustave Caillebotte.

FIGURE 4.85.
The Flatiron Building.

2. a) Find the vanishing point for **Figure 4.85.** Use Handout 4.3 provided by your teacher.

   b) Is the vanishing point in the center of the picture, to the right of center, or to the left of center? Is the vanishing point inside or outside of the bounds of the picture?

3. Revisit Figure 4.79 on p. 236. Draw the horizon line using a different color pen from the one used to draw the lines of convergence.

4. Revisit Figure 4.85. Draw the horizon line using a different color pen from the one you used in Question 2.

Answers (continued)

9. a) Edge *e* is parallel to edges *a* and *b* in the rectangular figure.

   b) Edge *f* is parallel to edges *c* and *d* in the rectangular figure.

   c) The eye level of the artist is higher than the sides of the box in Figure 4.84.

   d) It seems to bring the object or scene closer to the viewer.

   e) It seems to move the objects farther from the viewer.

10. There are several vanishing points. For the tall building to the right in the picture, the roof line and window frames on the left face seem to converge to a point to the right of the painting; the window and roof lines for the right face seem to converge to a point to the left of the painting. For the shorter building to the right in the painting, the slats in the left face seem to converge to a point to the left that is different from any other vanishing point; the slats in the right face seem to converge to a point on the left. The man in the lower right corner is standing on bricks or tiles. The line in the bricks seems to converge to a point in front of the picture. There could be more vanishing points!

Section 4.9 Assignment

Answers

1. Answer for Figure 4.85:

2. a) Answer:

   b) The vanishing point is to the left of center. It is outside the bounds of the picture.

3. Answer:

4. Answer:

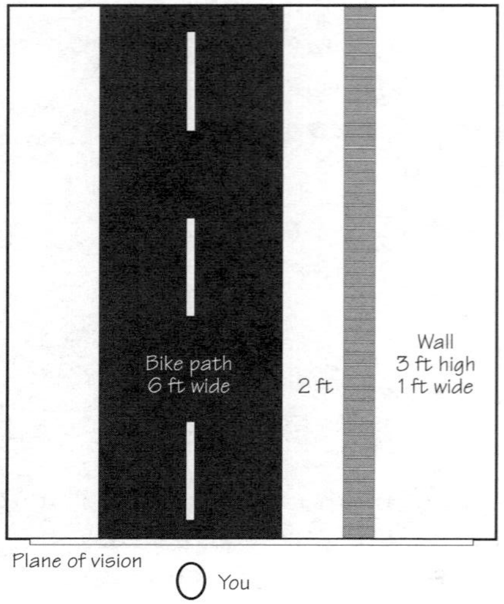

FIGURE 4.86. Aerial view.

5. **Figure 4.86** shows an aerial view of a scene.

   a) Draw a perspective view of the same scene with one vanishing point.

   b) Suppose power poles are in line with the viewer's line of sight to the vanishing point on the horizon. Describe what the viewer sees. What principle(s) of perspective is (are) responsible for this phenomenon?

6. Optical illusions can be created when the principle of diminution or the principle of convergence is violated.

   a) Describe the effect of not reducing the height of a series of objects and not reducing the spacing between the objects as other objects converge to the vanishing point or points (see **Figure 4.87**).

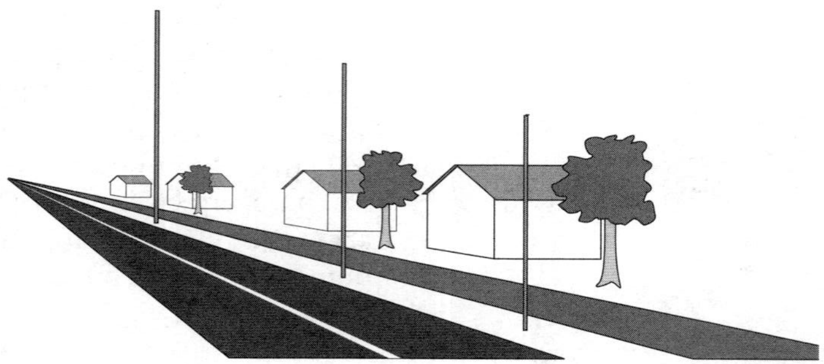

FIGURE 4.87. Drawing for 6a.

   b) Describe the effect of not reducing the height but allowing the spacing between the objects to agree with principles of perspective (see **Figure 4.88**).

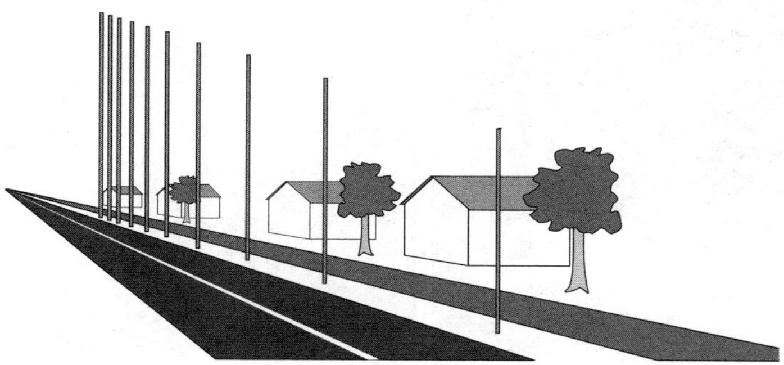

FIGURE 4.88. Drawing for 6b.

   c) Explain how the B.C. cartoon in **Figure 4.89** violates a principle of perspective drawing.

FIGURE 4.89. B.C. cartoon.

7. **Figure 4.90** shows a sketch of a room containing three rectangles representing a chest, a large wardrobe, and an air purification unit. The drawings of the objects are incomplete. Only the sides facing the viewer are sketched.

   a) Trace Figure 4.90 onto paper. Complete the perspective drawings of the chest, wardrobe, and air purification unit so that they appear to have depth. What principles of perspective did you use to guide your drawing?

   b) In your completed drawing, you should be able to see the top of one of the rectangles and the bottom of another. For the third object, you should be able to see neither its top nor its bottom. What determines whether the viewer can see the top or the bottom of an object in a perspective drawing? Under what conditions are both the top and the bottom of an object not visible to the viewer?

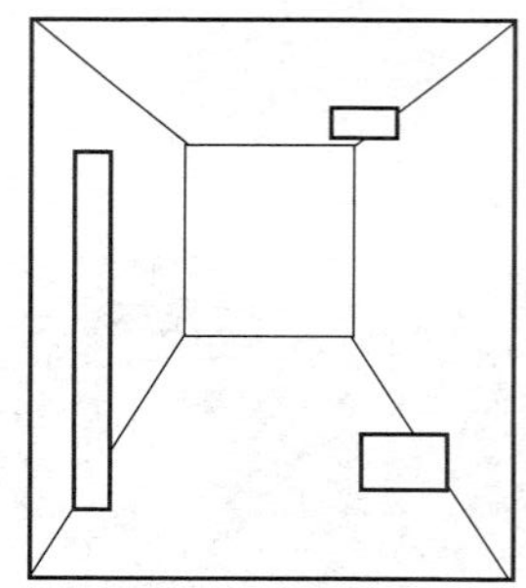

FIGURE 4.90. Incomplete drawing of room containing three objects.

Answers (continued)

5. a) Sample answer:

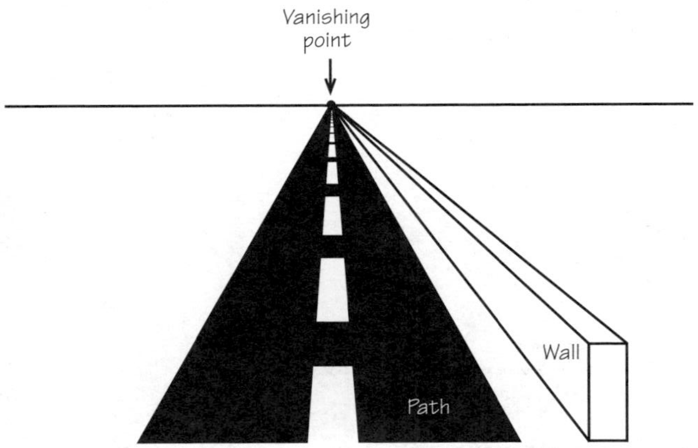

b) Because of diminution, the pole closest to the viewer appears to be the largest. Because of overlapping, the first pole completely covers the view of all the other poles.

6. a) The objects do not appear to be equal in height even though when you measure them they are the same height.

b) The objects appear to grow larger in relation to their distance from the viewer.

c) It illustrates the principle of diminution in reverse. The objects appear to diminish in direct relation to the distance from the viewer. (In other words, closer objects appear smaller and more distant objects appear larger.)

7. a) Sample sketch:

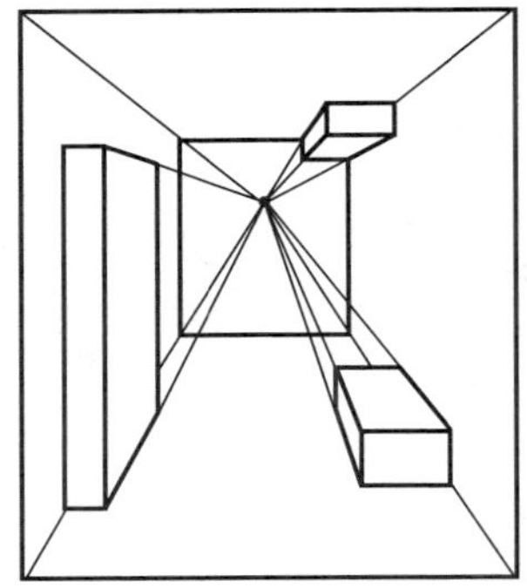

Sample answer: Determine the vanishing point by extending the floor and ceiling lines and marking their point of intersection. All other lines that, in reality, are parallel converge to this point. (The only exception to this principle is when the lines are parallel to the picture plane.) Extend lines from the corners of the rectangles to the vanishing point. Locate where you want to place the end of the rectangle and draw a vertical line that extends to the convergence lines. If the top or bottom of the rectangle is visible, add a horizontal line to complete the drawing of the end of the rectangular solid farthest from the viewer. Connect the front and back of the boxes along the convergence lines. Erase any lines that would lie behind the objects (principle of overlapping).

b) If the top of the object lies below the horizon line, then the viewer can see the top. If the bottom of an object lies above the horizon line, then the viewer can see the bottom. If the object straddles the horizon line, then the viewer can see neither the top nor the bottom.

8. **Figure 4.91** shows the front view of a cottage. Draw the cottage in perspective with two vanishing points.

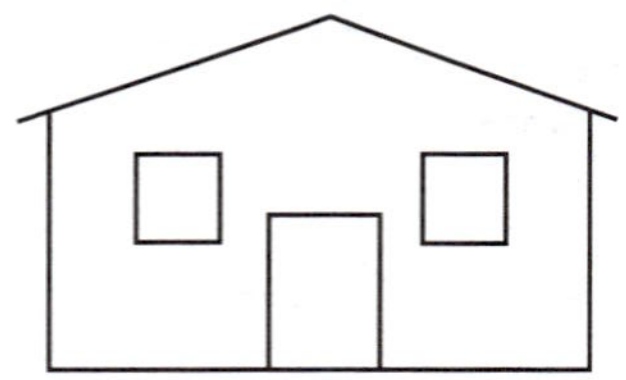

FIGURE 4.91.
Front view of cottage.

9. Eight-year old Jackson Barber drew the picture in **Figure 4.92.** It does not adhere to the principles of perspective. Draw the same scene on your own paper using principles of perspective and two vanishing points.

FIGURE 4.92.
Child's drawing that does not conform to principles of perspective.

SECTION 4.10 Setting Your Sights

FIGURE 4.93.
Trees lining a roadway.

1. In **Figure 4.93** you see a photograph of a roadway.

   a) What do you notice about the lines that form the sides of the road in the picture?

   b) Next, focus on the trees on the right side of the road. Suppose you draw a line connecting the treetops (assume the trees are approximately the same height) and another line connecting the tree bottoms. What would you notice about these lines?

   c) Use the rulers that surround the picture to identify the location of the vanishing point for the lines joining the tops of the trees and the bottoms of the trees. Then imagine drawing a horizontal line through that vanishing point. What does this horizontal line represent?

   d) The shoulders of the road extend beyond the white lines on both edges of the road. Pretend a ladder is placed from the edge of the shoulder to the top of each tree, forming a triangle. Describe the triangles that result from doing this with each tree.

Answers (continued)

8. Sample sketch:

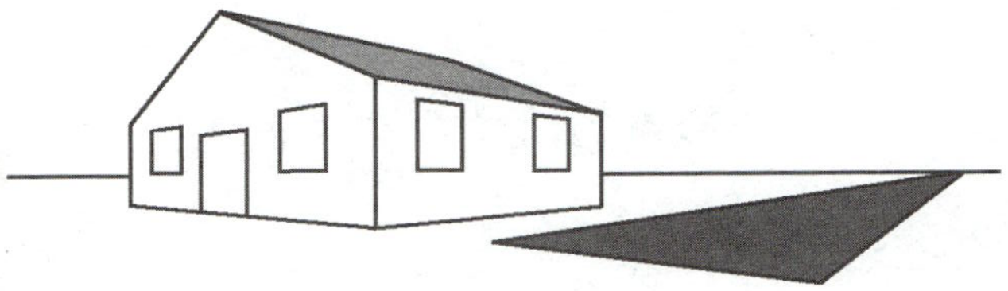

9. Sample sketch:

EVALUATE Lesson 2

Section 4.10: Setting Your Sights

Purpose: In Sections 4.6 through 4.9, students investigated and formalized different principles of perspective. In Section 4.10 students demonstrate their knowledge of perspective and the applications of mathematics that are used to create works of art.

Description: Students examine a photograph, a painting, and a drawing then identify elements of perspective that are found in each object. This is an open-middle performance assessment task; in other words, there is only one correct answer, but there is a variety of valid methods to obtain it. In their justification, which is the most important part of a performance assessment task, students should be able to communicate their knowledge of mathematics and demonstrate that they have mastered the concepts presented in Sections 4.6 through 4.9.

Materials: ruler, copies of Handouts 4.4 and 4.5 for each student

Teaching Ideas: This task may be assigned as group work or individual work. The mathematics performance assessment rubric should be used to evaluate students' work.

Answers

1. a) As you look down the road, the apparent distance between the two sides of the road narrows. The lines converge.

   b) The apparent height of the trees gets smaller as you look down the road. So the line connecting treetops converges with the line connecting tree bottoms.

   c) If you use the rulers as a coordinate system, then the equation for the horizontal line is $y = 2$. This is the horizon: the line where the sky and land seem to meet.

   d) The triangles appear to be similar, but decreasing in size. Eventually, because of diminution, the triangles vanish to a point. The vanishing point for the row of triangles is the same as the vanishing point for the white lines on the edges of the road.

e) Sketch a top view of the road. (Use circles to indicate the trees.) What should be true of the lines that form the sides of the road? What about the lines connecting the treetops and tree bottoms?

FIGURE 4.94. *Chemin de la Machine, Louveciennes* by Alfred Sisley.

2. a) On Handout 4.4, which is the same as **Figure 4.94**, draw lines marking the sides of the road and along the treetops on either side of the road. Draw lines along the roofs of buildings on the right side of the road.

   b) What is true of these lines? Explain.

3. **Figure 4.95** shows two roads meeting at an intersection. You are an architect, and your client wants to build an office at the corner opposite where the viewer is standing. Draw Figure 4.95 on your own paper, or use Handout 4.5 provided by your teacher.

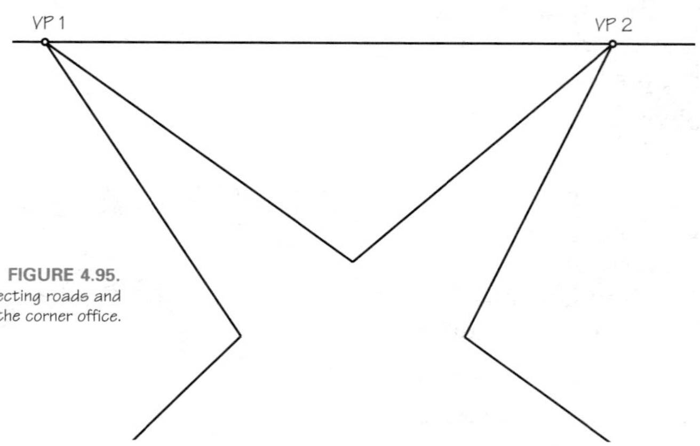

FIGURE 4.95. Two intersecting roads and the site of the corner office.

a) Make a detailed drawing showing what the office will look like when it is constructed. Include windows, doors, and a sign on the building. (For an extra challenge, try drawing a sloped roof rather than a flat roof.)

b) What is the effect if the vanishing points are brought closer together or moved farther apart? Draw sketches to illustrate your answer.

c) What would have to change in the context to increase the apparent distance between the vanishing points?

Answers (continued)

e) Sample sketch:

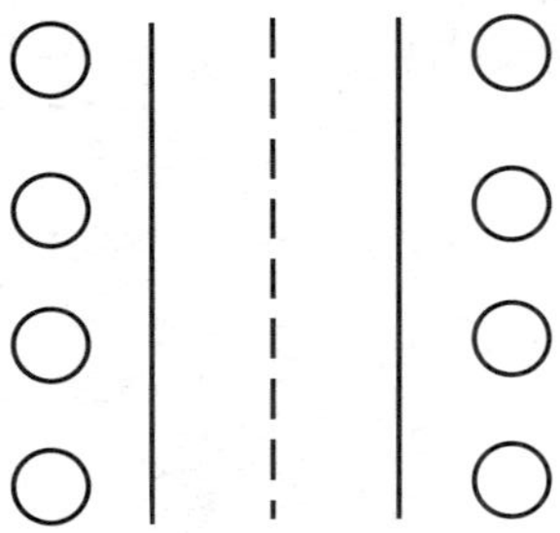

The lines that form the sides of the road should be parallel. The lines indicating the treetops and tree bottoms should lie directly on top of each other. In real life, these lines are parallel but at different heights. In a two-dimensional perspective viewed from above, the lines coincide.

2. a) Answer:

b) They have the same vanishing point because the edges of the roofs, the sides of the road, and the tops of the trees are all essentially parallel to each other.

3. a) Sample sketch:

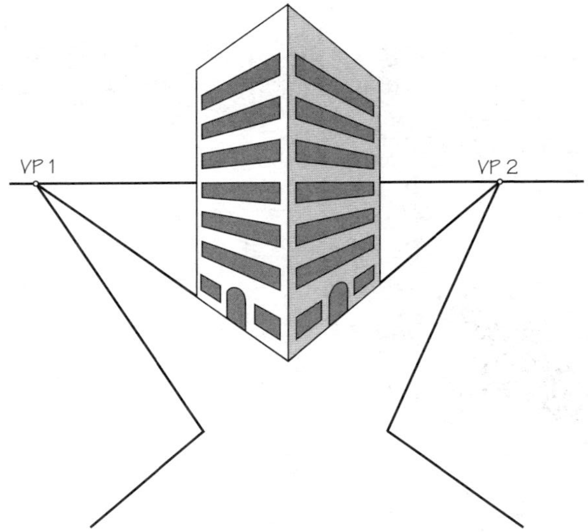

b) When you increase the distance between the vanishing points, the angles at the front corner get larger. In reality, right angles look larger as they are viewed from a lower point of view. So separating the vanishing points can make it seem that you are looking from a greater distance. See the examples below.

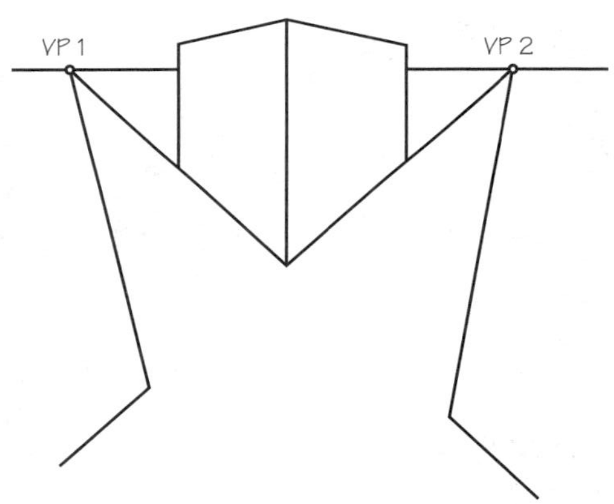

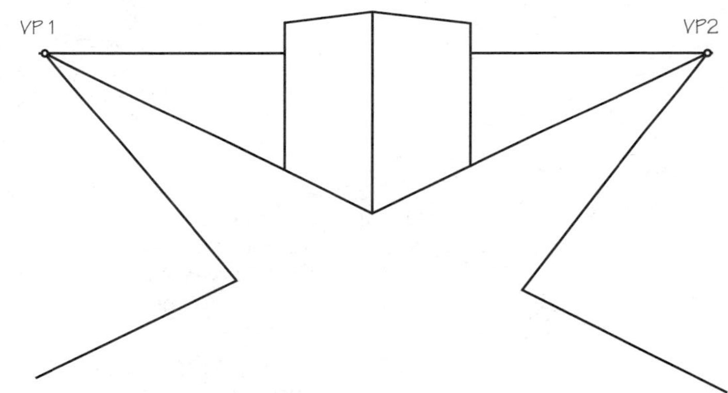

c) Increase the distance between the viewer and the object.

Modeling Project It's a Masterpiece!

In this chapter you have seen a number of ways artists use mathematics to create art. Now it is *your* turn. Use tools you have developed in this chapter to create an original piece of art. This drawing or painting should use the principles of symmetry, overlapping, and diminution. While the subjects of the art can be anything you would like to draw, including an architectural drawing of a building or objects around your home and school, your drawing should include notes that describe how you used mathematics to transform your objects in order to create patterns and give the drawing depth. These notes should include the types of transformations you used to create your work and details like the scale factors you chose and why.

For a twist on this project, revisit the Hogarth painting (Figure 4.46 on p. 216) and create your own impossible drawing. Instead of your notes explaining how you used mathematics to create the image, your notes should explain how you could use mathematics to "fix" your drawing.

Finally, be creative and have fun. Be ready to share your drawing with the class and answer questions about your work.

Practice Problems

1. In each of these figures, identify a basic element and describe ways to get the other sections of the figure by applying one or more of the transformations you studied in this chapter. Your basic element should be as small a portion of the entire figure as possible.

a) 

b) 

c) 

d) 

e) 

f) 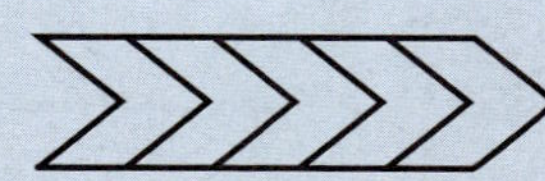

Practice Problems

Answers

1. a) Sample answer: The basic element marked in the figure can be translated repeatedly to the left to get the rest of the figure.

b) Sample answer: The basic element marked in the figure can be translated to the left to get the third element. Glide reflections of the basic element can be used to get the second and fourth.

c) Sample answer: Glide reflections of the bird marked in the figure will produce birds on either side of it and every other bird thereafter. The basic element can be translated to produce the other birds.

d) Sample answer: The basic element marked in the figure can be translated to the right to get the other elements in the first row. Translating this row downward produces the third, fifth, and so on rows. The element below and to the right of the basic element can be obtained from the basic element by rotating it 180°, and translations can then be used to get the rest of the second row, the fourth row, and so on.

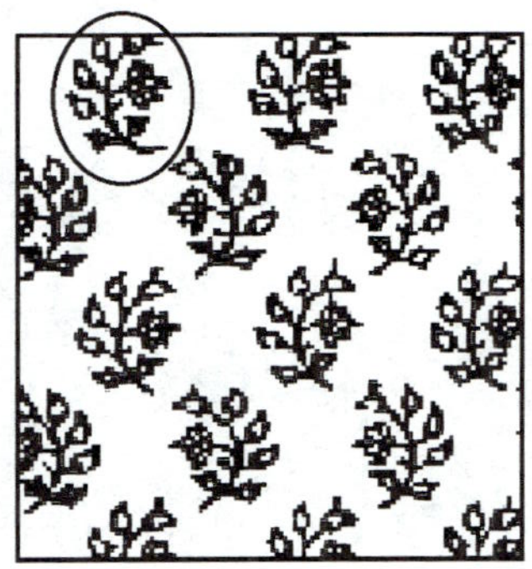

e) Sample answer: The basic element marked in the figure can be reflected to the right, reflected downward, and rotated 180° to obtain the pattern that is a sixth of the entire figure. This sixth can then be translated to obtain the rest of the figure.

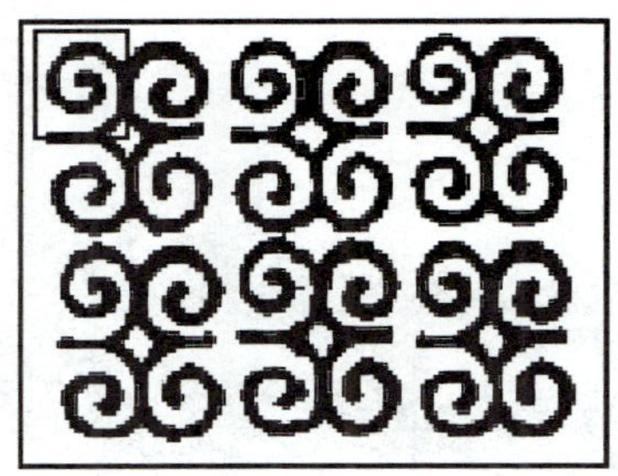

f) Sample answer: The basic element marked in the figure can be reflected downward to obtain the pattern that is a fifth of the entire figure. This fifth can then be translated to the right four times to obtain the rest of the figure.

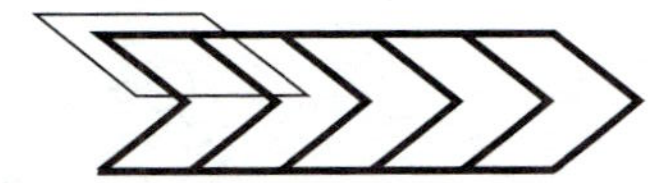

2. The natural world has many examples of symmetry (although it is usually not perfect). Describe the lines of symmetry and the types of symmetry in each of these photos of living things.

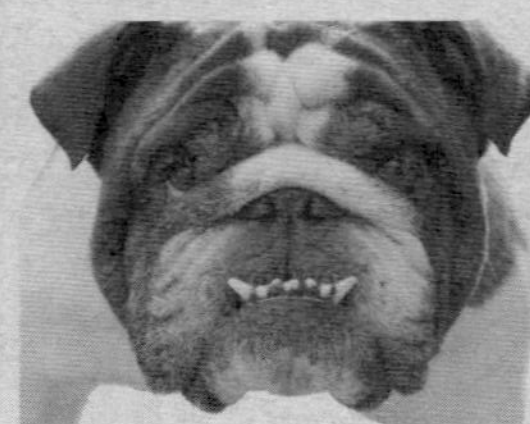

3. Three spheres are arranged on a table.

   a) **Figure 4.97** shows the top view. Draw the views from A and B. What principle of perspective did you use in your drawing?

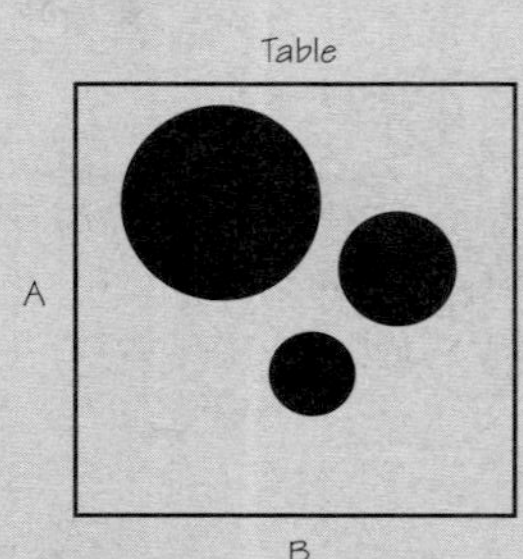

FIGURE 4.97. Three spheres arranged on a table.

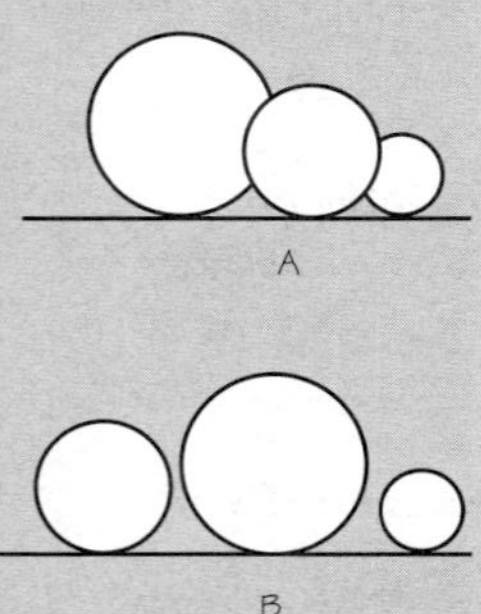

FIGURE 4.98. Side views of the rearranged spheres.

   b) Suppose the spheres are rearranged. Views from sides A and B are shown in **Figure 4.98**. Draw an overhead view of the table. Explain how you arrived at your answer.

4. In each of these questions, suppose a ruler is held at arm's length so the plane of vision is 27 inches from the eye of the viewer.

   a) You are 100 yards away from the goalpost on a football field. The image height of the goalpost is 1.6 inches. How high are the goalposts?

   b) Close to shore on the opposite side of a lake is a tower that is 50 feet tall. The image height of the tower is 1 inch. How wide is the lake?

   c) A certain basketball player stands 86 inches tall. You are standing under the basket and he is attempting a free throw, which means he is about 18 feet from you. How tall is his image?

5. **Figure 4.99** shows the top view of a person looking at five railroad ties.

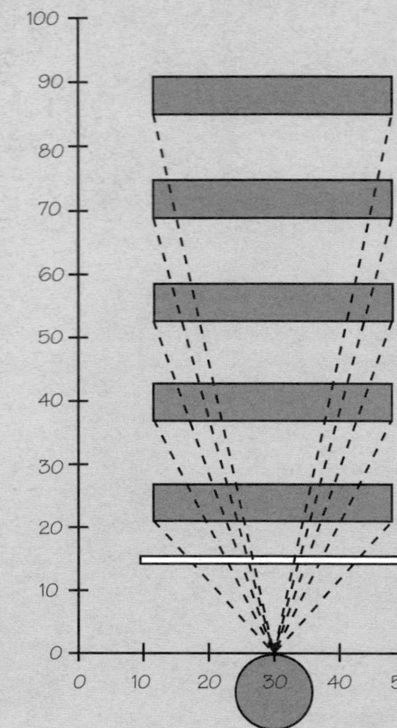

FIGURE 4.99. Top view of person looking at railroad ties.

   a) Measure the length of each projection in the picture plane. Then measure the distance between each tie and the viewer's eye. Record your measurements in a table like **Figure 4.100**.

Tie Number	1	2	3	4	5
Distance to Viewer's Eye					
Projection Length					

FIGURE 4.100. Table for recording railroad tie lengths and distances.

   b) Based on your measurements, describe the relationship between the length of the projection of each tie and its distance from the viewer's eye.

   c) Graph your data. Label the horizontal axis distance to eye and label the vertical axis projection length. Describe the shape of your plot.

Answers (continued)

2. Insect has one vertical line of symmetry running vertically through its body.

   Starfish has 5 lines of symmetry each running though the middle of an arm. The starfish also has rotational symmetry. It can be rotated $360°/5 = 72°$ to coincide with itself.

   The dog's face has one vertical line of symmetry.

   Four leaf clover has four lines of symmetry two of which bisect pairs of opposite petals and two that pass between petals. The clover also has rotational symmetry: It can be rotated $360°/4 = 90°$ to coincide with itself.

3. a) The principle of overlapping was used.

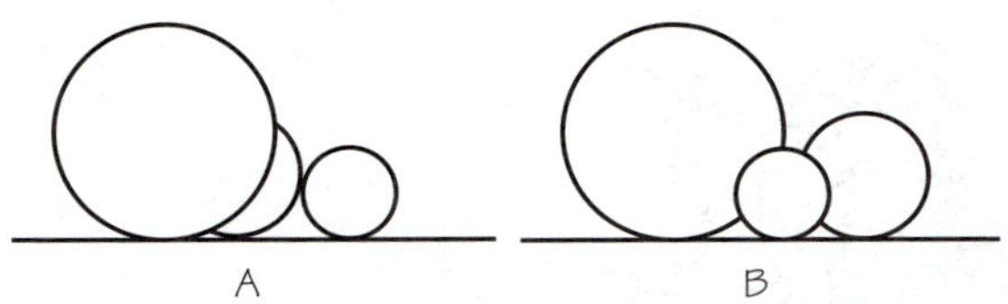

   b) The following picture shows how to determine the placement of the spheres on the table.

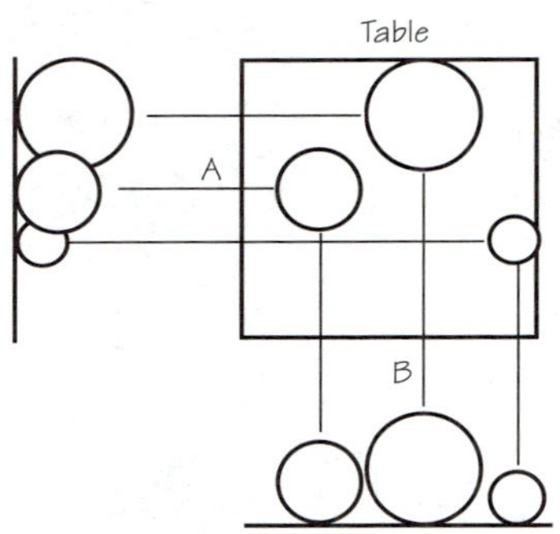

4. a) Approximately 18 feet high. Solve one of the following proportions:

   Proportion #1:

   (1.6 in)/(27 in) = (x ft)/(300 ft). Here the solution will be $x \approx 17.78$ ft or approximately 17′ 9″.

   Proportion #2:

   (1.6 in)/(x in) = (27 in)/(3600 in). Here the solution will be $x = 213.4$ in or about 17′ 9″.

   b) The lake is approximately 1350 feet across. Solve the proportion:

   1 in/27 in = 50 ft/x.

   c) His image is 10.75 inches tall. Solve the proportion:

   x/27 in = (86 in)/(216 in).

5. a) Sample answer based on tie 1 being the railroad tie closest to the viewer's eye and tie 5 being the farthest. The scale is millimeters.

Tie Number	1	2	3	4	5
Distance to Viewer's Eye	21	37	53	69	85
Projection Length	25	14	10	7.3	6.5

   b) The relationship appears to be an inverse variation. As the distance increases, the projection length decreases.

   c) Sample graph:

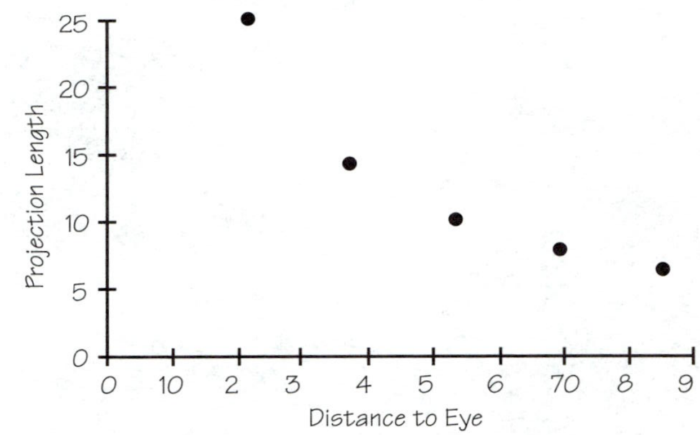

   If you connect the plotted points with a smooth curve, the curve gets closer to the horizontal axis as the distance to the eye increases. In addition, the curve bends upward.

6. a) Copy **Figure 4.101** on your own paper and label the unlabeled sides and vertices according to mathematical convention.

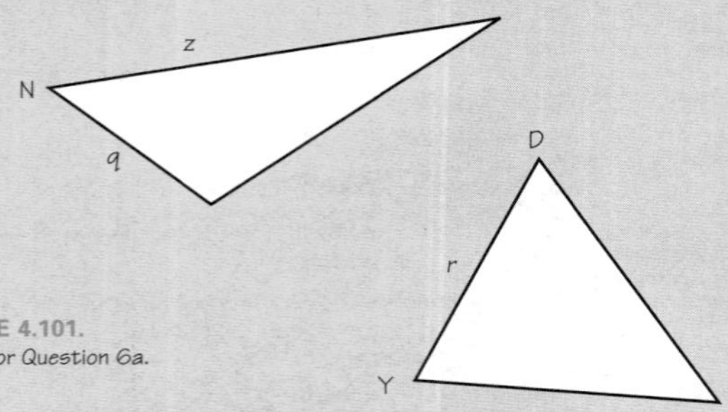

FIGURE 4.101.
Figure for Question 6a.

b) For each of the triangles in **Figure 4.102**, how long is the side opposite angle V?

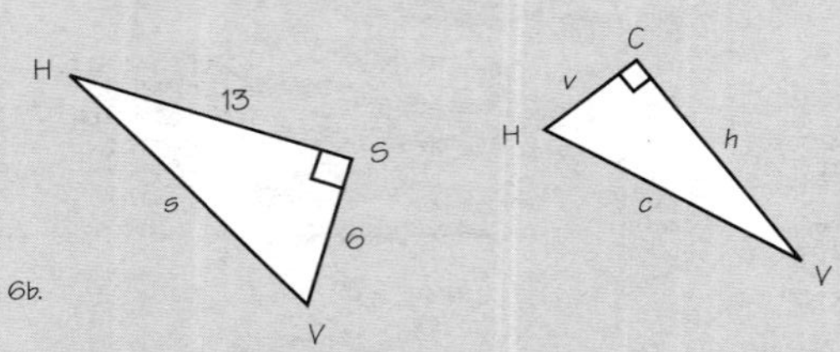

FIGURE 4.102.
Figure for Question 6b.

7. Three equal-height poles are to be placed 100 feet apart, increasing in distance from the viewer. Which drawing (see **Figure 4.103**) seems to be the best representation? Explain your answer.

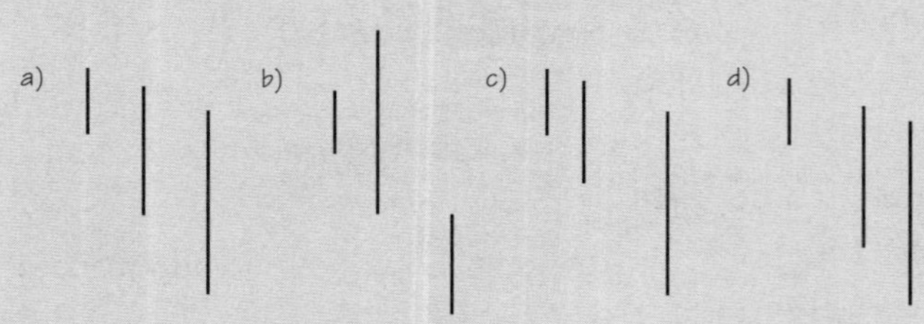

FIGURE 4.103.
Four different drawings of three poles.

8. **Figure 4.104** is a perspective view of a hallway.

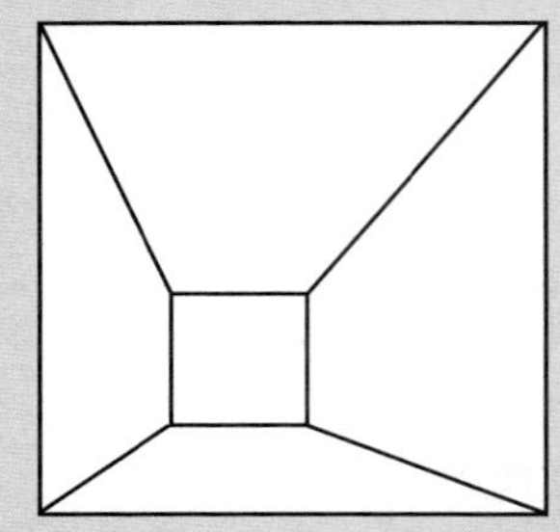

FIGURE 4.104.
Perspective view of hallway.

a) Trace a copy of Figure 4.104 on your own paper. Determine the vanishing point and the horizon line.

b) Raise the vanishing point and move it to the right. Redraw the hallway using the new vanishing point.

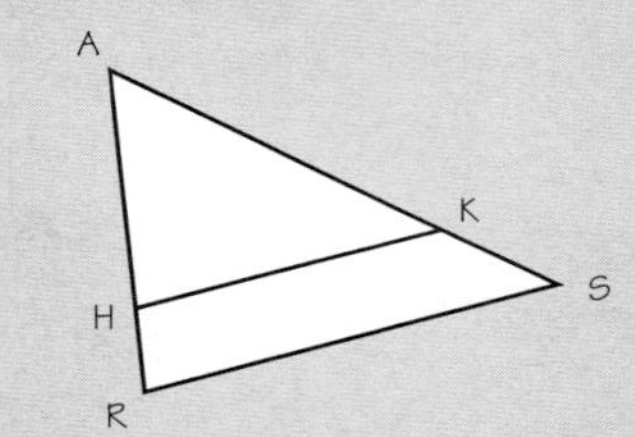

FIGURE 4.105.
Figure for Question 9.

9. In **Figure 4.105**, RS is parallel to HK.

a) Which angles have equal measure?

b) What is the relationship between triangles ARS and AHK? Explain.

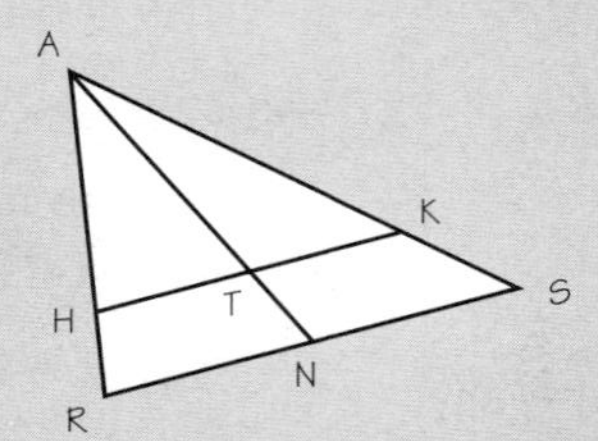

FIGURE 4.106.
Figure for Question 9c.

c) **Figure 4.106** was created by adding the line segment AN to Figure 4.105. Segment AN bisects segment HK. Explain why it must also bisect RS.

Answers (continued)

6. a) Answer:

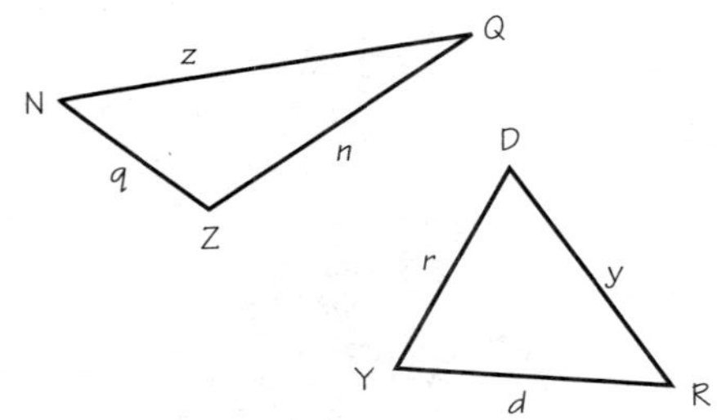

b) 13 and v

7. The correct perspective is view c. According to the principle of diminution, the poles should appear smaller and closer together the farther they are from the viewer. This is only true for c.

8. a) Drawing of vanishing point and horizon line:

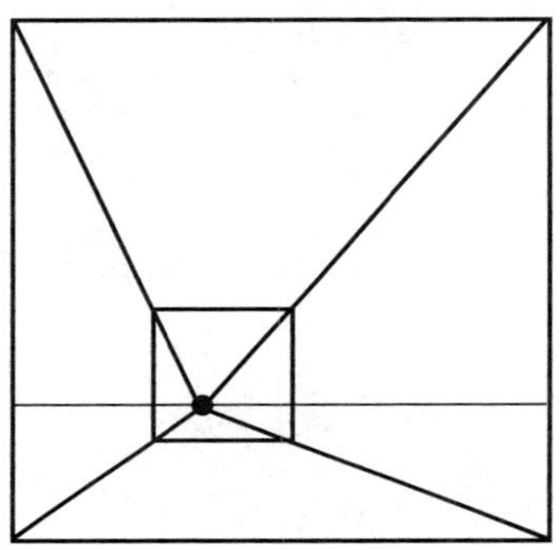

b) Sample picture:

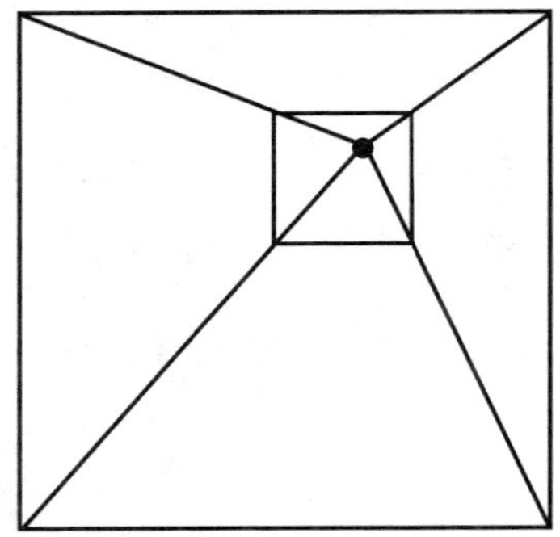

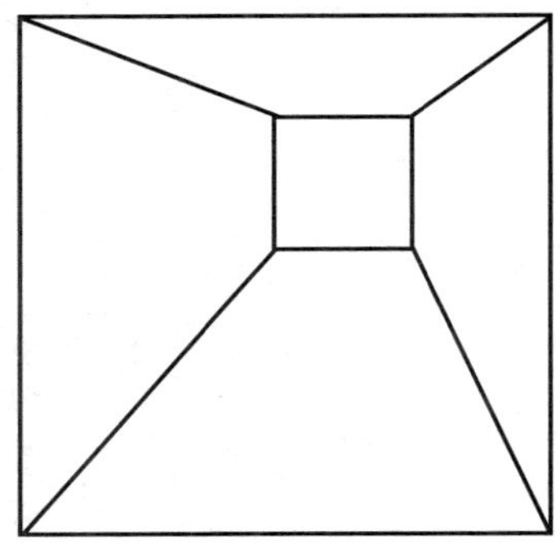

9. a) m∠AHK = m∠ARS, m∠AKH = m∠ASR, m∠A = m∠A

b) Triangles ARS and AHK are similar triangles. Corresponding angles have equal measure.

c) Triangle ATH is similar to triangle ANR, so HT/RN = AT/AN. Similarly, triangle ATK is similar to triangle ANS, so TK/NS = AT/AN. Therefore, HT/RN = TK/NS, so HT/TK = RN/NS. Hence, HT/TK = RN/NS. But HT/TK = 1:1. So, RN = NS and segment AN bisects RS.

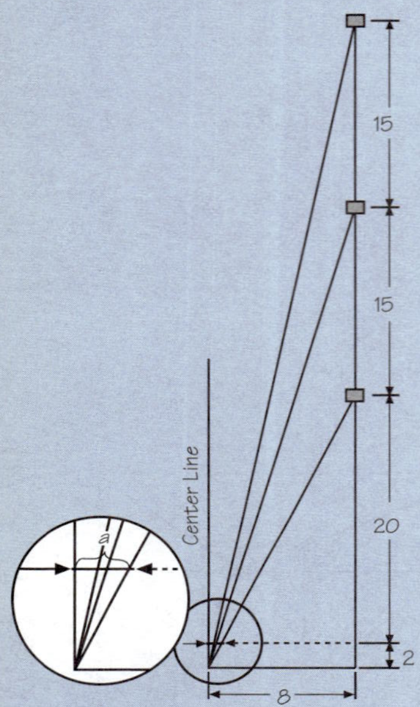

FIGURE 4.107.
Top view of hallway support poles, including artist's viewing position (not to scale).

10. Suppose you want to draw accurately a series of poles that support a cover above a walkway. **Figure 4.107** shows a top view of the poles in relation to the artist and the plane of vision. The artist chooses a viewing position 8 feet to the left of the line of poles and 22 feet from the first pole that is to be drawn in the picture. The 9-foot poles are 15 feet apart. The distance between the artist's eye and the picture plane is 2 feet.

    a) The center line is an imaginary line from the artist's eye to the vanishing point. In a top view, this line is parallel to the line of poles. Describe the relationship in a perspective drawing between (1) the center line and (2) a line drawn through the poles at the artist's eye height. If drawn, how would the center line appear in the picture?

    b) In the drawing, what is the scale at the first pole? At the second pole? At the third pole?

    c) What is the image size of Pole 1? What are the image sizes of Pole 2 and Pole 3?

    d) Determine the horizontal distance a, the distance between the image of Pole 1 and the center line. How can you use the scale at Pole 1 to determine this distance?

    e) Determine the horizontal distance b, the distance between the image of Pole 2 and the imaginary center line.

    f) Determine the horizontal distance c, the distance between the image of Pole 3 and the imaginary center line.

Glossary

KEY CONCEPTS

Convergence: Artistic principle asserting that lines or edges of objects that in reality are parallel appear to come together as they recede from the observer.

Corresponding angles: Angles in corresponding positions within similar figures. Corresponding angles are equal in measure.

Dilation: A transformation in which the distance to every point's image is exactly k times its original distance from some specified, fixed point. The fixed point is called the center of the dilation; k is the scale factor for the dilation.

Diminution: The phenomenon by which an object appears smaller as its distance from the observer increases.

Glide reflection: A composite of translation and reflection. A glide reflection reverses a figure's orientation.

Horizon Line: The imaginary line on the plane of vision (or picture plane) containing the vanishing points of all horizontal (in reality, parallel to the ground) converging lines. It is always on the same level as the viewer's eyes (the eye level).

Hypotenuse: The side opposite the right angle of a right triangle. It is the longest side in the right triangle.

Image: In transformations, a new object created by transforming a pre-image.

Isometry: A transformation that preserves size and shape.

Line of Sight: An imaginary line between the viewer's eye and a point on a three-dimensional object being observed.

Line of Symmetry: A line that divides a figure into two identical mirror images.

One-Point Perspective: Occurs with only one family of horizontal parallel lines not parallel to the plane of vision; all the lines converge to a single vanishing point located on the horizon line.

Overlapping: A technique to achieve a sense of depth and space in drawings by showing which objects are in front and which are in back.

Perspective: A technique of representing objects from three-dimensional space in a two-dimensional plane, like a picture. Through the use of perspective, an artist's two-dimensional drawing or painting imitates the appearance of a three-dimensional object or scene.

Plane of Vision (or Picture Plane): An imaginary plane between the observer and the three-dimensional object(s) being viewed. The image of the object is projected onto the plane. In this chapter the picture plane is always vertical.

Point Symmetry: A figure has point symmetry when it fits exactly on itself after a rotation of 180°.

Pre-image: When a geometric object is transformed into another one, the original is called a pre-image.

Projection: The projection of a real object is its image on the picture plane. The image is formed by lines of sight from the viewer to points on the object.

Pythagorean Formula: A formula relating the lengths of three sides of a right triangle: $a^2 + b^2 = c^2$, where c is the length of the hypotenuse and a and b are the lengths of the other two sides (legs).

Answers (continued)

10. a) Because these two lines are horizontal with the ground and parallel to each other, in a perspective drawing they intersect at the vanishing point. The center line is a line of sight from the artist's eye to the vanishing point. In a perspective drawing, it would appear as a point—the vanishing point.

    b) Scale = (distance from artist's eye to picture plane)/(distance from artist's eye to object)

    At first pole: $2/22 = 1/11$ or approximately 0.09.

    At second pole: $2/37$ or approximately 0.054.

    At third pole: $2/52 = 1/26$ or approximately 0.038.

    c) Image size of Pole 1: $(1/11)(9) \approx 0.82$ ft or approximately 9.8 inches.

    Image size of Pole 2: $(2/37)(9) = 18/37 \approx 0.49$ ft or approximately 5.8 inches.

    Image size of Pole 3: $(1/26)(9) = 9/26 \approx 0.35$ ft or approximately 4.2 inches.

    d) Multiply the horizontal distance (8 ft) by the scale at Pole 1 to give $(0.09)(8) = 0.72$ ft or about 8.6 inches.

    e) $(2/37)(8) = 16/37 \approx 0.43$ ft or approximately 5.2 inches.

    f) $(1/26)(8) = 8/26 \approx 0.31$ ft or approximately 3.7 inches.

Reflection: A transformation that flips a figure across a line. A reflection reverses a figure's orientation.

Reflection (line) Symmetry: When a figure fits exactly on itself after reflection through a line. The line divides the figure into two identical halves.

Rotation: A transformation that spins a figure about a given center a given number of degrees.

Rotation Symmetry: When a figure fits exactly on itself after it spins less than 360° about a center.

Scale: The ratio of an image size (height or width) in a picture to the actual size of the object.

Scale Factor: In a dilation, the ratio of the new lengths to the old lengths.

Similar Triangles: Two triangles are similar when they have equal corresponding angles and equal ratios of corresponding sides.

Transformation: A process that creates a new figure from an original. The original is called the pre-image; the new figure is called the image.

Translation: A transformation that slides a figure a given distance in a given direction.

Two-Point Perspective: Occurs with exactly two families of horizontal parallel lines not parallel to the plane of vision; all the lines converge to one of two vanishing points located on the horizon line.

Vanishing Point: The imaginary point to which parallel lines not in the plane of vision appear to converge.

5

CHAPTER

Motion: Quadratic Functions

PREPARATION READING

It's Show Time

You may have watched scenes like these in the movies:

The star jumps off a roof and lands in the back of a passing truck.

During a chase the star speeds down the street on a motorcycle, through a cross street and just misses a truck.

Auto thrill shows also do stunts using cars and trucks. Jeff Lattimore's stunt in Chittwood's Thrill Show is "The Leap for Life." Jeff has performed it for years. In his stunt Jeff climbs a ladder and stands on an eight-foot stool. Then the ladder is removed leaving Jeff in the air just as a car comes speeding at him. Jeff jumps right before the car hits the stool and knocks it out from under him. Jeff lands safely on the ground.

Since this stunt is complicated, trial-and-error would be a poor method. A mistake could result in loss of life. Using mathematics and understanding the laws of physics are key to planning stunts.

In this chapter you'll see functions that model motion. You'll use mathematical models to predict where an object will be at a specific time. Then you'll use the models to plan and test a small-scale stunt in your class.

CHAPTER 5

Motion and Quadratics

Preparation Reading: It's Show Time

Description: The reading introduces the major context of the chapter: different ways of representing motion. Students will read about a stuntman whose life depends on the accuracy of mathematical models used to calculate the exact positions of moving objects. In the rest of the chapter, students will investigate types of motion that lead to linear, quadratic, and square root functions.

Materials: none

Teaching Ideas: Students should work in small groups (3 to 4 students). To help insure that all students read the preparation reading, have students read silently then record three important facts or things that they learned. Ask students to share their individual responses within the group, then ask each group to share two of the group's responses with the entire class. After the reading students should want to know more about using mathematics to model and predict motion.

SECTION 5.1

Walking the Walk

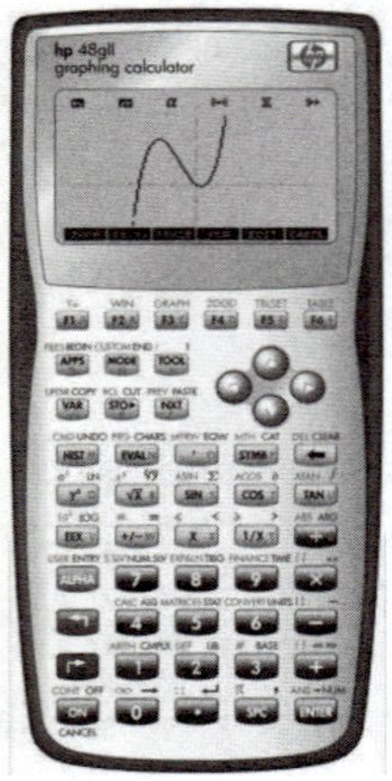

In this section you will look at distance-versus-time graphs. For these graphs, you will consider time as the independent variable and distance as the dependent variable.

A motion detector can track a student walking toward or away from it. The motion detector along with a graphing calculator can be used to create a distance-versus-time graph of a student walk.

Imagine a student walking away from a motion detector.

1. As time increases, what happens to the student's distance from the motion detector?

2. Imagine a distance-versus-time graph of the walk. Would you expect the graph to be a straight line or a curved line? Why?

3. Would you expect the graph to increase or decrease from left to right? Why?

Use a motion detector and a graphing calculator to collect data and create a distance-versus-time graph for a classmate who walks away from the motion detector.

Sketch your graph.

4. How can you tell from the graph that the student was walking away from the motion detector?

5. Was the student walking at a constant rate? How can you tell from the graph?

Again use a motion detector and a graphing calculator to collect data and create a distance-versus-time graph for a second classmate who walks away from the motion detector.

Sketch your graph.

A sample graph is shown in **Figure 5.1.**

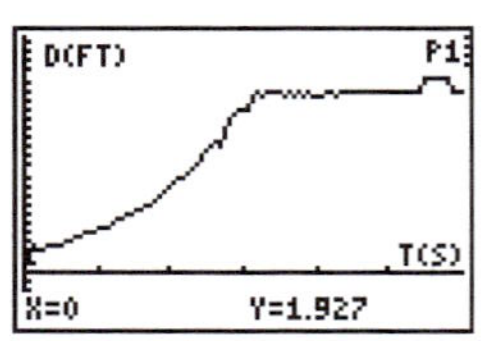

FIGURE 5.1.
A sample distance-versus-time graph.

6. How can you tell from the graph that the second student was walking away from the motion detector?

7. Was the second student walking at a constant rate? How can you tell from the graph?

Now you will analyze data taken from student walks in front of a motion detector. When you track the walk of another student, you can see how you would have to walk to produce the same distance-versus-time graph.

Cara walked in a straight line in front of a motion detector. The motion detector recorded her distance from it every 0.1 second for 6 seconds. Cara's distance-versus-time graph appears in **Figure 5.2.**

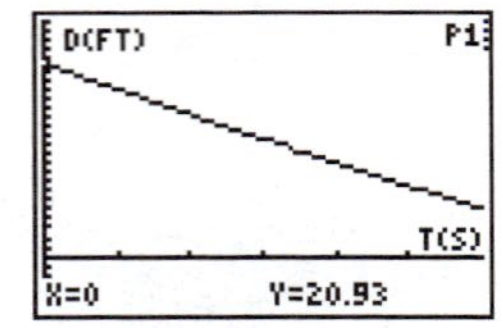

FIGURE 5.2.
Cara's distance-versus-time graph.

8. Based on Figure 5.2, did Cara walk toward or away from the motion detector? How can you tell from her graph?

9. Did she begin moving as soon as the motion detector began collecting data or did she pause first? How can you tell from her graph?

10. After she began moving, did she walk at a constant rate, or did she speed up or slow down? Explain how you decided.

11. What plan could you give to a walker so that he or she could walk a graph that is the same as Cara's?

12. Using your group's plans, have one group member walk in front of a motion detector. Is the graph of the walk close to the graph of Cara's walk?

13. How can you modify your plans so that the walk will more closely model Cara's walk?

Paul walked in a straight line in front of a motion detector. The motion detector recorded his distance from it every 0.1 second for 6 seconds. Paul's distance-versus-time graph appears in **Figure 5.3.**

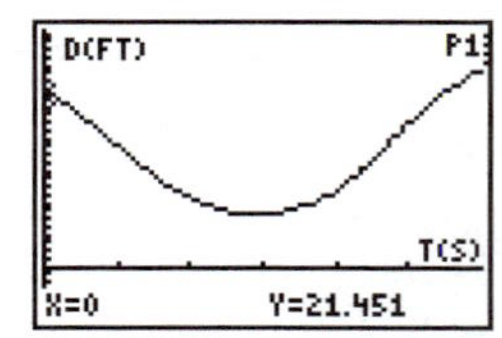

FIGURE 5.3.
Paul's distance-versus-time graph.

14. Based on Figure 5.3, did Paul walk toward or away from the motion detector? How can you tell from his graph?

15. After he began moving, did he walk at a constant rate, or did he speed up or slow down? Explain how you decided.

ENGAGE LESSON 1

Section 5.1: Walking the Walk

Purpose: In previous chapters students explored functional relationships and used linear functions and systems of linear functions to model data. In Section 5.1, students will use a motion detector to collect data that model linear motion. They will also collect data that model non-linear motion. In the next phase of this lesson, students will explore one type of non-linear relationship, quadratic functions.

Description: In this section students will investigate distance-versus-time graphs using data collected from a motion detector such as a CBR. They will interpret distance-versus-time graphs, answering questions about increasing or decreasing distance with respect to time and increasing or decreasing speeds with respect to time.

Materials: motion detector, graphing calculator, overhead graphing calculator display

Teaching Ideas: Students should work in small groups (3 to 4 students). The data collection portion of this phase of the lesson should be done as a whole-class demonstration. Student volunteers should be solicited as walkers for the data collection. If students have never used a motion detector to collect data in this manner, you may need to run several practice trials before settling on one set of data to use with the class to debrief the activity.

Answers

NOTE: The responses to these questions may vary. The correctness of the responses is not important. These questions are posed at this point in the lesson in order to get students thinking about the walking activity they are about to do and make conjectures about what will happen.

1. Sample answer: As time increases, the student's distance from the motion detector increases.
2. Sample answer: The graph will be a straight line if the student walks at a constant rate. It will be curved if the student speeds up or slows down.
3. Sample answer: The graph should increase from left to right because as time increases, the student's distance from the motion detector increases.

NOTE: Set up a class demonstration for this walk. Privately ask the student volunteer to walk at a steady pace away from the motion detector. The answers to the next two questions are based on this walk.

Graphs should be similar to one of the graphs below.

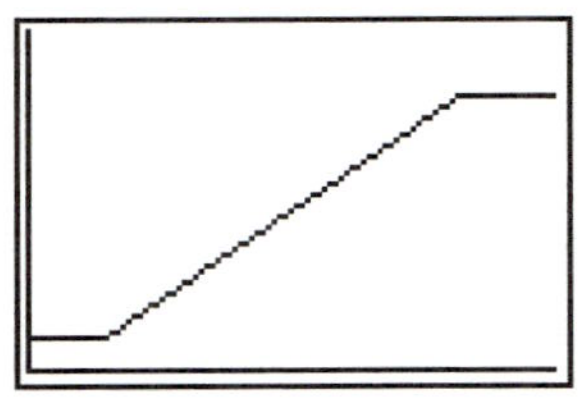

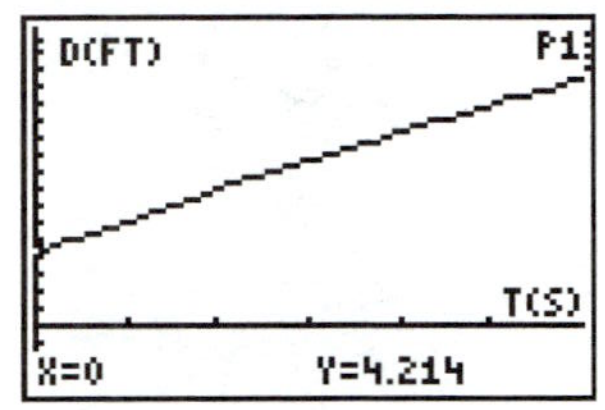

4. Sample answer: As time increased, the student's distance from the motion detector increased.
5. Sample answer: We can tell that the student was walking at a constant rate because for every second, the same amount of distance was covered. Use the TRACE feature on the calculator to verify this distance-versus-time relationship in the graph.

NOTE: Again, set up a class demonstration for this walk. Privately ask the student volunteer to begin walking slowly away from the motion detector and increase speed as he gets farther away. The answers to the next two questions are based on this walk.

6. Sample answer: As time increased, distance away from the motion detector increased.
7. Sample answer: We can tell that the second student was not walking at a constant rate because for every second, the amount of distance covered was not the same. Use the TRACE feature on the calculator to verify this distance-versus-time relationship in the graph.
8. Sample answer: Cara walked toward the motion detector. As time increased, distance from the motion detector decreased.
9. She began as soon as the motion detector started collecting data. If she paused first, the graph would have a short horizontal section. That would indicate that her distance from the motion detector was not changing during the corresponding period of time.
10. Sample answer: She walked at a constant velocity. The portion of the graph that corresponds to Cara's motion appears approximately linear.
11. Students should use the TRACE feature to find the starting point $(0, y)$ then trace to find the y-value at $x = 1$ $(1, y)$. They can then use the two points to calculate rate (slope).

    Sample answer: Start about 21 feet from the motion detector. Walk toward the motion detector at a rate of about 2.5 feet per second.
12. NOTE: Either have a representative from each group walk as a class demonstration or have students do the walk in their group.
13. Sample answer: Change the starting point or rate.
14. At first he walked toward the motion detector then walked away. Distance was decreasing for 3 seconds then began to increase.
15. Sample answer: At first he walked at a constant rate, then slowed down to change direction. After he changed direction he increased his rate.

16. What rules could you give to a walker so he or she could walk a graph like Paul's?

17. Using your group's rules, have one group member walk in front of a motion detector. Is the graph of the walk like the graph of Paul's walk?

18. How can you modify your rules so that the walk will more closely model Paul's walk?

19. What differences in Cara's and Paul's walks caused the differences in the two graphs?

CHECK THIS!

The rate of change of distance over a certain period of time is called speed.

SECTION 5.2

Linear versus Non-Linear

In the last section you saw graphs created by students walking in a straight line in front of a motion detector. Cara's graph was close to a straight line while Paul's graph was curved.

In this section you will compare data that create a straight line (a linear relationship) with data that create a curved graph (a quadratic relationship).

CHECK THIS!

A linear relationship has a constant rate of change and is represented by a function of the form $f(x) = mx + b$.

CHECK THIS!

A quadratic relationship has a variable rate of change and is represented by a function of the form $f(x) = ax^2 + bx + c$.

Victoria walked in front of a motion detector. Some of the data are shown in **Figure 5.4.**

Time, t (seconds)	Distance, d (feet)
0	2
1	5
2	8
3	11
4	14
5	17

FIGURE 5.4.
Data from Victoria's walk.

Answers (continued)

16. Sample answer: Start about 21.5 feet from the motion detector. Walk toward the motion detector at a rate of about 5 feet per second. Gradually slow down then begin walking away from the motion detector. When walking away gradually increase your rate to 5 feet per second. Be sure that students understand that this graph does not require an up or down motion.
17. NOTE: Either have a representative from each group walk as a class demonstration or have students do the walk in their group.
18. Sample answer: Change the starting point or rate.
19. Although both students walked in a straight line in front of the motion detector, the graphs differed due to the rates of the walks. A constant rate created a linear graph while a variable rate created a non-linear graph. Also Paul changed directions while Cara did not.

EXPLORE LESSON 1

Section 5.2: Linear versus Non-Linear

Purpose: In Section 5.1 students used distance-versus-time graphs to distinguish between linear and non-linear data. In this section students will investigate characteristics of one type of non-linear data, quadratic relationships, using a finite differences approach. In Section 5.3 students will formalize what a quadratic relationship is and use different representations of the parent function.

Description: In this section students will use finite differences in order to explore the constant rate of change that exists in a linear function. Then students will look for patterns in the finite differences of non-linear data. They will discover that for linear functions, the first differences are constant, and that for quadratic functions, the second differences are constant.

Materials: graphing calculator

Teaching Ideas: Students should work in small groups (3 to 4 students). This section is designed for students to deepen their understanding of slope as a constant rate of change in linear functions. Monitor calculator use and assist as necessary.

Ricardo also walked in front of a motion detector. Some of the data recorded by the motion detector for his walk are shown in **Figure 5.5.**

Time, t (seconds)	Distance, d (feet)
0	25
1	15
2	9
3	7
4	9
5	15

FIGURE 5.5.
Data from Ricardo's walk.

1. What do you see in the data for Victoria's and Ricardo's walks?

2. Use your graphing calculator to create a distance-versus-time scatter plot of Victoria's data. Record your window and sketch your graph.

3. Use your graphing calculator to create a distance-versus-time scatter plot of Ricardo's data. Record your window and sketch your graph.

4. What do you see in the graphs for Victoria's and Ricardo's walks?

5. What do you think might be a cause of the differences in the graphs?

One tool to help develop representation of a function is differences in consecutive table values.

Let's look at the first and second differences in the two tables.

CHECK THIS!

First difference is the change from one table value to the next.

CHECK THIS!

Second difference is the change from one first difference to the next.

6. Find the first and second differences in the table values for Victoria's walk. Record them in a table like the one in **Figure 5.6.**

	Time (seconds)	Distance (feet)	First Difference	Second Difference
	0	2		
+1			—	
	1	5		—
+1			—	
	2	8		—
+1			—	
	3	11		—
+1			—	
	4	14		—
+1			—	
	5	17		

FIGURE 5.6.
Differences for Victoria's walk.

7. What patterns do you see?

The function that models Victoria's walk where d is distance and t is time is $d = 3t + 2$. Recall that an equation in this form ($y = mx + b$) has a linear graph where m is the slope of the line and b is the y-intercept.

CHECK THIS!

The parent function for the linear family is y = x. The graph of y = x is the line shown below.

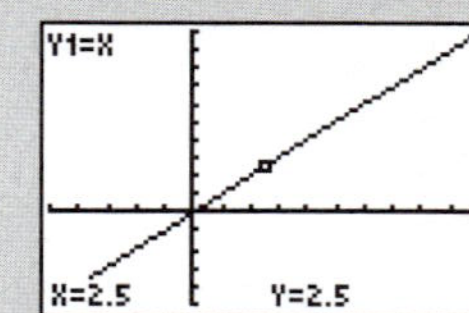

8. Look at the table with differences for Victoria's walk. Where do you see the number 3? What does 3 represent here?

9. Look at the table with differences for Victoria's walk. Where do you see the number 2? What does 2 represent here?

10. How do you think the first and second differences in a table of data can be represented by a linear function?

11. Explain how you think first differences relate to the function rule.

12. Find the first and second differences in the table values for Ricardo's walk. Record them in a table like the one in **Figure 5.7.**

13. What patterns do you see?

	Time (seconds)	Distance (feet)	First Difference	Second Difference
	0	25		
+1			—	
	1	15		—
+1			—	
	2	9		—
+1			—	
	3	7		—
+1			—	
	4	9		—
+1			—	
	5	15		

FIGURE 5.7.
Differences for Ricardo's walk.

Answers

1. Sample answer: Victoria's distance from the motion detector increased at a constant rate. Ricardo's distance decreased then increased. The increase and decrease of distance for Ricardo were not at a constant rate.

2.

```
WINDOW
 Xmin=0
 Xmax=6
 Xscl=1
 Ymin=0
 Ymax=20
 Yscl=1
 Xres=█
```

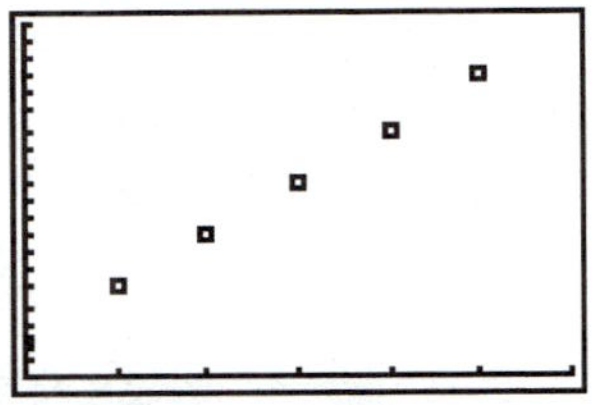

3.

```
WINDOW
 Xmin=0
 Xmax=6
 Xscl=1
 Ymin=0
 Ymax=25
 Yscl=1
 Xres=█
```

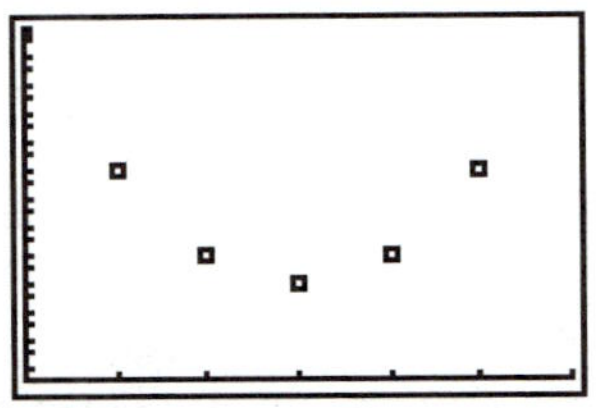

4. Sample answer: Victoria's graph appears to be linear. Ricardo's graph appears to be shaped like a parabola.

5. Sample answer: Victoria walked at a constant rate in the same direction. Ricardo did not walk at a constant rate and changed directions.

6.

	Time (seconds)	Distance (feet)	First Difference	Second Difference
	0	2		
+1	1	5	+3	
+1	2	8	+3	0
+1	3	11	+3	0
+1	4	14	+3	0
+1	5	17	+3	0

7. Sample answer: The first differences are all +3. The second differences are all 0.

8. The number 3 is the first difference. In this situation the number 3 represents a rate of 3 feet per second.

9. The number 2 is paired with 0 in the table. It represents the number of feet Victoria was away from the motion detector when the motion detector started to collect data.

10. The first differences are constant.

11. First differences represent the rate of change for the function.

12. Answer:

	Time (seconds)	Distance (feet)	First Difference	Second Difference
	0	25		
+1	1	15	−10	
+1	2	9	−6	+4
+1	3	7	−2	+4
+1	4	9	+2	+4
+1	5	15	+6	+4

13. Sample answer: The first differences are different. The second differences are all +4.

The equation for the function that models Ricardo's walk where d is distance and t is time is $d = 2t^2 - 12t + 25$. Recall that $y = ax^2 + bx + c$ where $a \neq 0$ is a **quadratic function**. Quadratic functions are often used to model relationships in the real world such as distance versus time and area versus length. The graph of a quadratic function is U-shaped and is called a **parabola**. The point at which the graph changes between increasing and decreasing is called the **vertex**.

CHECK THIS!

The parent function for the quadratic family is $y = x^2$. The graph of $y = x^2$ is the parabola shown below.

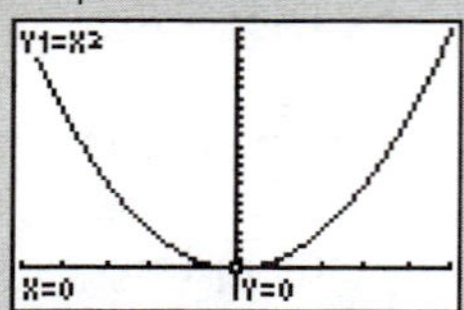

14. Look at the table for Ricardo's walk. What is the relationship between the second differences and the number 2, the coefficient of t^2, in $d = 2t^2 - 12t + 25$?

15. Look at the table for Ricardo's walk. Where do you see the number 25? What does 25 represent here?

16. Look at the table for Ricardo's walk. How does the first, first difference compare to the coefficients of t^2 and t in $d = 2t^2 - 12t + 25$?

17. How can the first and second differences in a table of data be represented by a quadratic function?

18. How do the first and second differences relate to the function rule that models the data in a quadratic relationship?

19. In **Figure 5.8,** one of the tables represents a linear function while the other one shows a quadratic function. Without graphing, find which one is linear and which one is quadratic. How did you decide?

Table 1

x	y
0	0
1	2
2	4
3	6
4	8

Table 2

x	y
0	0
1	2
2	8
3	18
4	32

FIGURE 5.8. Two tables of x- and y-values.

20. Use the first and second differences in Table 1 to write a function rule.

21. Use the first and second differences in Table 2 to write a function rule.

SECTION 5.2 Assignment

INTERPRETING MOTION GRAPHS

The first graph in **Figure 5.9** was recorded when Jasmyne walked in front of a motion detector. The program recorded time in seconds and distance in feet. The TRACE feature of her calculator was used to find the distance for 0.9, 2.9, and 4.9 seconds (the other three graphs).

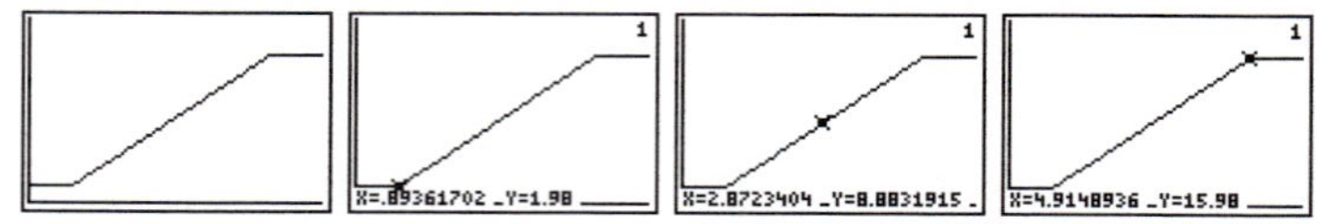

FIGURE 5.9. Graph of Jasmyne's walk.

1. How far was Jasmyne from the motion detector when the program began running?

2. How many seconds did Jasmyne stand still before moving?

3. How much time did Jasmyne spend walking?

4. How far did she walk?

5. During the time that Jasmyne was walking, did she walk at a constant rate, increase her speed, or decrease her speed?

6. During the time that Jasmyne was walking, at what rate did she walk?

7. The two graphs in **Figure 5.10** were recorded by a motion detector when a student walked in front of it. Assuming the scales are the same for both graphs, in which case was the student walking faster? Explain your answer.

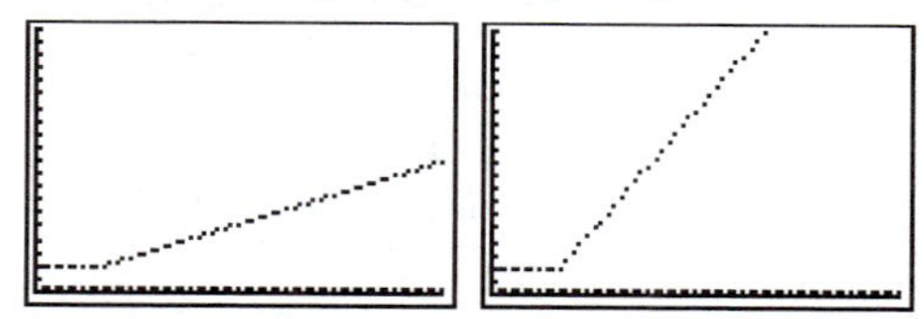

Walk 1. Walk 2.

FIGURE 5.10. Two walks recorded by a motion detector.

Answers (continued)

14. The number 2 is one-half the second differences.
15. The number 25 is paired with 0 in the table. It represents the number of feet Ricardo was away from the motion detector when the motion detector started to collect data.
16. The first, first difference is –10; this is the sum of 2 and –12.
17. In a quadratic function the first differences are not constant. The second differences are constant.
18. The first of the first differences is equal to $a + b$ and the second difference is equal to $2a$.
19. In Figure 5.8, Table 1 represents a linear function since first differences are constant. Table 2 represents a quadratic function since second differences are constant.
20. The function rule is $y = 2x$.
21. The function rule is $y = 2x^2$.

Section 5.2 Assignment

Answers

1. 1.98 ft
2. 0.9 sec
3. 4.9 – 0.9 = 4 sec
4. 15.98 – 1.98 = 14 ft
5. Jasmyne walked at a constant rate.
6. 14 ft/4 sec = 3.5 ft/sec
7. The student walked faster in Walk 2 than in Walk 1. The slope of the line segment representing the time the student was in motion is greater in Walk 2 than in Walk 1.

A motion detector can be used to create a distance-versus-time graph for a toy car.

8. Imagine a toy car moving along a straight line in front of a motion detector. The motion produces a distance-versus-time graph that is a straight line. What does this tell you about the toy car's rate?

9. How could you determine the toy car's rate?

10. Suppose that the distance-versus-time graph has a negative slope. What does that tell you about the car's motion?

11. Suppose that a toy car is traveling along a straight line in front of a motion detector. The motion produces a distance-versus-time graph that is curved. What does this tell you about the car's rate? Explain.

Suppose a movie stunt calls for a person to step off a roof as soon as a truck reaches a white line painted across a road. To design such a stunt, you need to calculate details about the person's fall. For example, you need to know how long it would take for the person to land on the ground.

12. Imagine that a stunt person steps off a roof 25 feet above the ground and falls vertically to the ground. Sketch a possible height-versus-time graph for the falling body. Add scales and labels to your axes.

13. Explain why the stunt person will fall as you have shown in your graph.

14. How might you collect data that would help you decide whether your graph does a good job of describing the stunt person's fall?

Jackie and Jermaine drew the graphs in **Figure 5.11** as part of their answers to Question 12.

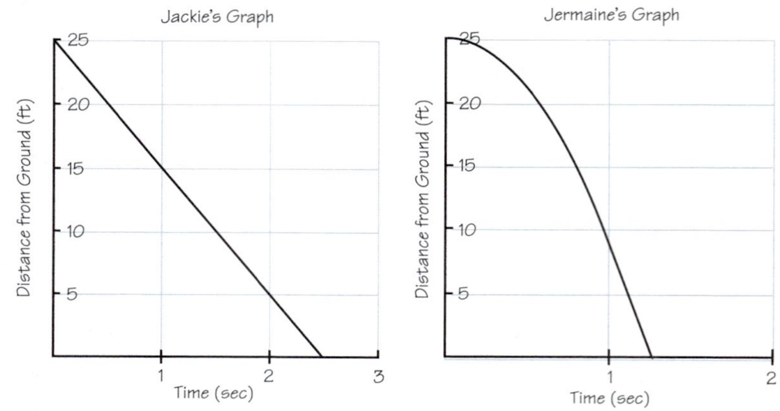

FIGURE 5.11. Jackie's and Jermaine's graphs.

15. Suppose these graphs show the motion of a walker in front of a motion detector. Write a plan for a walker so that the graph would look like Jackie's.

16. Write rules for a walker so that the graph would look like Jermaine's.

A motion detector tracked Rashida as she jogged. Some of the data are shown in **Figure 5.12.** Create a scatter plot of the distance-versus-time data. Be sure to add labels and scales to your axes.

Time (seconds)	Distance from Sensor (feet)
0	2
1	3.15
2	6.8
3	12.95
4	21.6
5	32.75
6	46.4
7	62.55
8	81.2

FIGURE 5.12. Distance-versus-time data for Rashida's walk.

17. Sketch and describe the shape of Rashida's graph.

18. Did Rashida jog at a steady pace, speed up, or slow down? How can you tell from the graph?

19. Use first and second differences to find a rule. Test your rule over your scatter plot. Write your rule and sketch the graph.

Answers (continued)

8. The car is traveling at a constant rate.
9. Determine the slope of the line.
10. The car is traveling toward the motion detector.
11. The car's rate changes during the trip. The car either speeds up or slows down.
12. Sample answer:

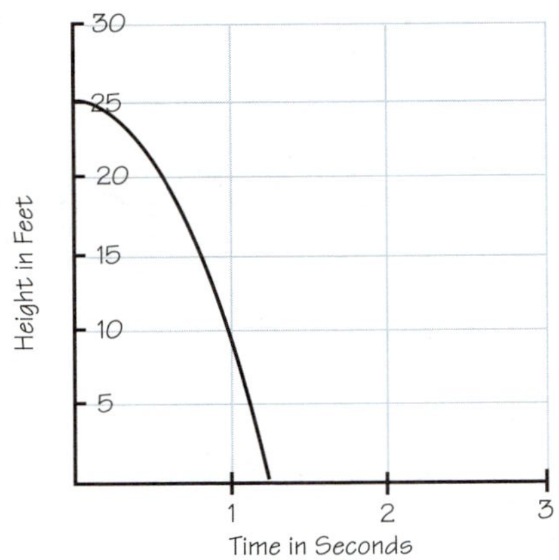

13. Sample answer: The stunt person would fall slowly at first and then pick up speed during his fall.
14. Sample answer: You could make a videotape of stunt people falling from buildings. Then you could slow the video down and pause at equal intervals of time to measure the stunt person's height above the ground each time the video was paused. However, students will not be able to collect data of this sort. Instead, you might drop various objects and use a motion detector to gather distance-versus-time data on their falls.
15. Sample answer: The walker should stand about 26 feet from the motion detector and then walk at a constant rate toward the sensor. The person taking the readings should begin the motion detector program when the walker is 25 feet from the sensor.
16. Sample answer: The walker should stand about 25 feet from the motion detector. When the program begins, the walker should walk toward the sensor. At first the walker should walk slowly and then gradually increase her rate until she reaches the motion detector.
17.

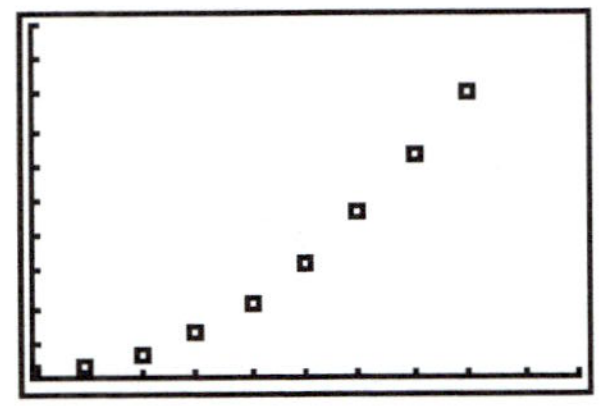

It is a curve, opening upwards and increasing. It looks like it might be part of a parabola.

18. The first part of the graph is not very steep. As time goes on, the graph becomes steeper, meaning she started slowly then gradually increased her speed.
19. The second differences are $2.5 = 2a$ so $a = 1.25$.

    The first of the first differences is 1.15, so $a + b = 1.15$, which means $b = -0.1$.

    The starting point or y-intercept is 2 so $c = 2$.

    Substitute into $y = ax^2 + bx + c$ to get $y = 1.25x^2 - 0.1x + 2$.

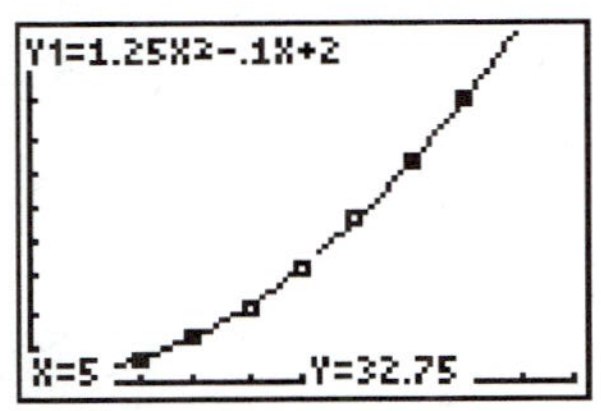

SECTION 5.3

Characteristics of Quadratic Functions

In the last section you explored linear and quadratic relationships. In this section you will formalize how and why those relationships work the way they do.

Recall the student walks in front of a motion detector.

1. Describe a walk that creates a linear relationship.
2. Describe a walk that creates a non-linear relationship.
3. Describe a walk that creates a linear relationship with a positive slope.
4. Describe a walk that creates a linear relationship with a negative slope.
5. Describe a walk that has a graph similar in shape to a parabola.

As you discovered in the last section, the use of first and second differences is a good tool to develop a function rule.

CHECK THIS! Recall that in a linear relationship the first differences are constant.

CHECK THIS! Recall that in a quadratic relationship the second differences are constant.

Let's take a closer look at linear and quadratic relationships.

6. In a linear relationship, how are the first differences related to the function rule?
7. In a linear relationship, how are the first differences related to the graph of the function rule?
8. In a quadratic relationship, how are the first differences related to the graph of the function rule?
9. In a quadratic relationship, how are the second differences related to the function rule?

x	$y = ax^2 + bx + c$	First Differences		Second Differences
0	$a \cdot 0^2 + b \cdot 0 + c = c$			
		$a + b + c - c =$	$a + b$	
1	$a \cdot 1^2 + b \cdot 1 + c = a + b + c$			$2a$
		$4a + 2b + c - (a + b + c) =$	$3a + b$	
2	$a \cdot 2^2 + b \cdot 2 + c = 4a + 2b + c$			$2a$
		$9a + 3b + c - (4a + 2b + c) =$	$5a + b$	
3	$a \cdot 3^2 + b \cdot 3 + c = 9a + 3b + c$			$2a$
		$16a + 4b + c - (9a + 3b + c) =$	$7a + b$	
4	$a \cdot 4^2 + b \cdot 4 + c = 16a + 4b + c$			

FIGURE 5.13. Quadratic differences table.

10. Notice that the first of the first differences in **Figure 5.13** is equal to $a + b$ and the second difference is equal to $2a$. How can you use these facts to develop a quadratic model?
11. Use these facts to justify the quadratic model, $d = 2t^2 - 12t + 25$, for Ricardo's walk in Section 5.2.
12. Use first and second differences to find a function rule for each data set.

a)

x	y
0	2
1	10
2	24
3	44
4	70

b)

x	y
0	4
1	6
2	8
3	10
4	12

SUMMARY

Using differences can help write a function rule for linear or quadratic data. Assume that the difference between consecutive x-values in the table is constant. If the relationship is **linear**, then the first differences in the y-values will be constant, indicating a constant rate of change. Linear relationships can be expressed in the form $y = mx + b$ where m and b are real numbers.

If the first differences in the y-values are not constant, then there is not a constant rate of change and the relationship is non-linear. If the second differences, or the differences between the first differences, are constant then the relationship is **quadratic**. Quadratic relationships can be expressed in the form $y = ax^2 + bx + c$ where a, b, and c are real numbers.

EXPLAIN LESSON 1

Section 5.3: Characteristics of Quadratic Functions

Purpose: In Section 5.2 students used finite differences to investigate linear and quadratic functions. In Section 5.3 students will use finite differences to write algebraic rules that describe quadratic relationships. This prepares students to investigate another tool for writing equations to describe quadratic relationships, transformations, and curve fitting in Section 5.4.

Description: In this phase of the lesson, students will use finite differences to identify and write quadratic equations to describe sets of data.

Materials: graphing calculators

Teaching Ideas: Debrief students after Section 5.2 by having them share their answers to the questions. As they share, be sure to fill in any missing concept, process, or skill that could have been used to answer the questions. Then work through the questions in this section with the entire class. Once again be sure all necessary concepts, processes, or skills are addressed.

Answers

1. The student walks a constant rate (speed) away from or toward the motion detector.
2. The student walks a variable rate (speed) in front of the motion detector.
3. The student walks a constant rate (speed) away from the motion detector.
4. The student walks a constant rate (speed) toward the motion detector.
5. The student either increases or decreases speed as he or she walks in front of the motion detector.
6. The first differences indicate the rate of change or slope of the linear function. The slope is m in the function $y = mx + b$.
7. First differences can be shown as a rate of change on the graph of a linear function.

   See example below for $y = 3x + 2$.

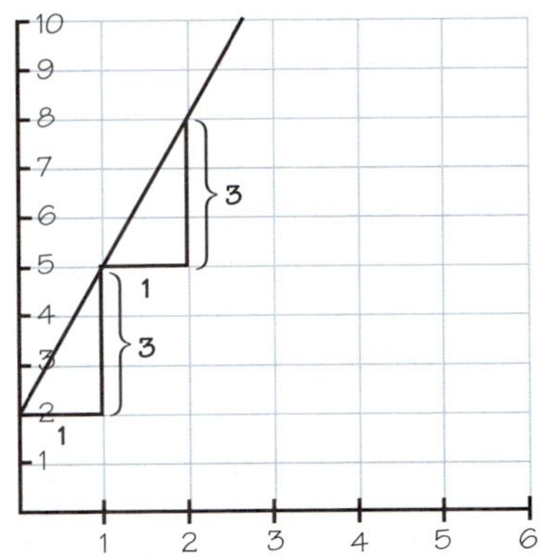

8. In a quadratic relationship, the first differences are not constant but do show a constant rate of change.

   See example below for $y = x^2 + 2$. Here the constant rate of change is 2.

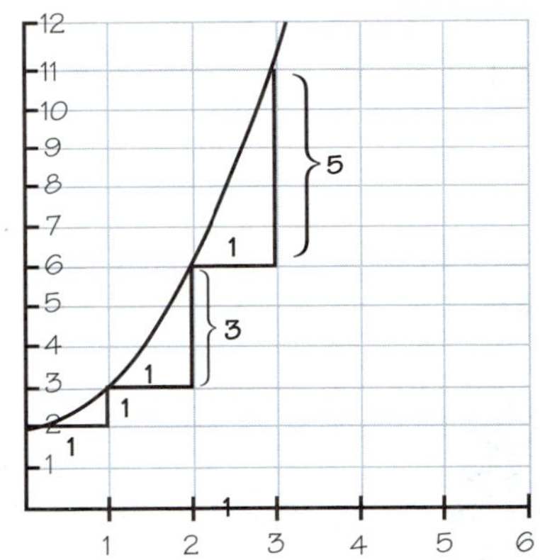

9. Students should realize that the second differences are double the coefficient of the x^2 term.

   Pursue this relationship abstractly by examining the table for the equation $y = ax^2 + bx + c$.

   When x-values are substituted into polynomial form, $y = ax^2 + bx + c$, y-values can be calculated in terms of a, b, and c. This allows us to make generalizations relating first and second differences to the coefficients a, b, and c.

10. To determine the coefficient of x^2 in $y = ax^2 + bx + c$, divide the second difference by 2. Then determine b so that $a + b$ is equal to the first of the first differences.

11. Ricardo's Walk

Time (seconds)	Distance (feet)
0	25
1	15
2	9
3	7
4	9
5	15

Since the first of the first differences is equal to -10, $a + b$ must equal -10. Since the second differences are all 4, a must equal 2. Since $a + b = -10$ and $a = 2$, the value of b must be -12. At 0 seconds Ricardo was 25 feet from the motion detector so $c = 25$. Substituting into the general form of a quadratic function, $ax^2 + bx + c$ gives the rule for Ricardo's walk, $d = 2t^2 - 12t + 25$.

12. a) $y = 3x^2 + 5x + 2$

    b) $y = 2x + 4$

Using Finite Differences to Write Linear Functions:

Δx	Time (seconds)	Distance (feet)	Δy
	0	2	
+1	1	5	+3
+1	2	8	+3
+1	3	11	+3
+1	4	14	+3
+1	5	17	+3

FIGURE 5.14.
Finite differences for linear functions.

$y = mx + b$ where

- b is the y-value when $x = 0$
- m is the ratio of $\frac{\Delta y}{\Delta x}$.

Using Finite Differences to Write Quadratic Functions:

- Use the polynomial form $y = ax^2 + bx + c$.
- If the x-interval is equal to 1, the value for a is half of the constant second difference ($2a$ = second difference).
- If the first x-value is 0, the first, first difference is equal to $a + b$.
- The y-value when $x = 0$ is the value of c.

CHECK THIS!

A quadratic is a type of polynomial, a mathematical expression of the form $a_nx^n + a_{n-1} + a_{n-2}x^{n-2} + \ldots + a_1x^1 + a_0$. Polynomials of one, two, or three terms are called monomial, binomial, or trinomial, respectively. The polynomial form of a quadratic function is the same as the general form, $y = ax^2 + bx + c$.

SECTION 5.3 Assignment

1. Use first and second differences to find a quadratic function for **Figure 5.15.**

x	y
0	6
1	5
2	10
3	21
4	38

FIGURE 5.15.
Table of x- and y-values.

2. Make a quadratic function and create a table.

   a) First choose values for a, b, and c.

   b) Next substitute them into $y = ax^2 + bx + c$. Write the results in a table like the one shown in **Figure 5.16.**

   c) Use first and second differences to check your work.

x	y
0	
1	
2	
3	
4	

FIGURE 5.16.
Table for Question 2.

The table in **Figure 5.17** shows the distance a car travels from the time the driver decides to stop to the time the car comes to a complete stop. These data are for an alert driver, a well-maintained car, and dry road conditions.

3. Find a function rule to model the relationship between each distance and speed.

   a) Reaction distance: ________________

   b) Braking distance: ________________

   c) Stopping distance: ________________

4. Complete Figure 5.17.

Speed	Reaction Distance	Braking Distance	Stopping Distance
0	0	0	0
10	10	5	15
20	20	20	40
30	30	45	75
40	40	80	120
50			
60			
70			

FIGURE 5.17.
Car stopping distance data.

Section 5.3 Assignment

Answers

1. $y = 3x^2 - 4x + 6$

2. Answers will vary.

If students are to compute the equations manually, they need to develop or be told the first and second difference rules for tables where $x = 0$, 10, and 20.

The first, first difference is $100a + 10b$.

The second difference is $200a$.

3. Sample answer:

   a) The linear equation for reaction distance is $y = x$.

   b) The quadratic equation for braking distance is $y = 0.05x^2$.

   c) Stopping Distance = Braking Distance + Reaction Distance = $0.05x^2 + x$.

4. Answer:

Speed	Reaction Distance	Braking Distance	Stopping Distance
50	50	125	175
60	60	180	240
70	70	245	315

5. According to Figure 5.17, a driver should allow 120 feet to stop safely when driving 40 mph. About how many car lengths is this? In your experience, how much space do most drivers allow?

6. How might an advocate for reducing highway speeds from 70 mph to 55 mph use stopping distances to support his or her case?

7. The Figure 5.17 data are for an alert driver. How might tuning the radio or talking on a cell phone or to other passengers affect these data?

8. Use your knowledge of quadratic models to complete the table in **Figure 5.18.**

Speed	Stopping Distance for Large Trucks	Stopping Distance for Trains
0	0	0
10	21	900
20	54	2020
30	99	3360
40	156	4920
50		
60		
70		

FIGURE 5.18. Stopping distance data for large trucks and trains.

9. Find a function rule to model the relationship between each distance and speed.

   a) Stopping distance for large trucks: ____________________

   b) Stopping distance for trains: ____________________

10. The average speed of trains traveling through town is about 40 mph. What is the stopping distance for that speed in feet? What is the stopping distance in miles?

11. If the train comes around a bend and the conductor sees a car stalled on the tracks about half a mile ahead, what is the fastest speed the train can travel and be able to stop to avoid a collision?

SECTION 5.4 Quadratics on the Move

In Chapter 4 you explored translations, reflections, and dilations. These transformations can be applied to the graphs of functions. Since graphs of functions are done on a coordinate plane, you can use algebraic rules for these transformations.

Knowing about function transformations will help you later in this chapter. If you know how to translate, reflect, and dilate a parabola then you can quickly fit a quadratic function to a set of data.

TRANSLATIONS OF QUADRATIC GRAPHS

1. Use grid paper to sketch the graph of $y = x^2$. Then complete a table like **Figure 5.19.**

x	$y = x^2$
−3	
−2	
−1	
0	
1	
2	
3	

FIGURE 5.19. Table for Question 1.

The effect of changing $y = x^2$ to $y = x^2 + k$.

2. Complete the table in **Figure 5.20** for $y = x^2$ and $y = x^2 + 3$, then graph both functions on your graphing calculator. Make a sketch of both graphs.

x	Pre-Image	Image
	$y = x^2$	$y = x^2 + 3$
−3		
−2		
−1		
0		
1		
2		
3		

FIGURE 5.20. Table for Question 2.

3. Write a sentence that describes the transformation.

Answers (continued)

5. Answers may vary: about 10 to 12 car lengths.
6. Answers may vary. One example is to show the increase in stopping distances from 55 mph to 70 mph.
7. Sample answer: Reaction time will be slower, requiring more reaction distance.
8. Answer:

Speed	Stopping Distance for Large Trucks	Stopping Distance for Trains
50	225	6700
60	360	8700
70	399	10,920

9. a) Large truck: $y = 0.06x^2 + 1.5x$

   b) Train: $y = 1.1x^2 + 79x$
10. 4920 feet = 0.93 miles (there are 5280 feet in 1 mile)
11. Just under 25 mph

ELABORATE LESSON 1

Section 5.4: Quadratics on the Move

Purpose: In Sections 5.2 and 5.3, students used finite differences to write quadratic equations that describe a set of data. In this section students will use what they learned in Chapter 4 about geometric transformations to perform function transformations. This concept will enable students to use function transformations to fit a model to any set of data, whether it is linear or non-linear, without using regression.

Description: In this section students will apply what they learned about geometric transformations in Chapter 4 to functions. They will then use function transformations to fit a quadratic function to a given set of data.

Materials: graphing calculator

Teaching Ideas: Emphasize the use of multiple representations in this section. Be sure that students understand the graphical, tabular, and algebraic representation of the parent function $y = x^2$. Then guide them through the application of these representations to transformations. The Card Match activity provided as supplementary material can be used to reinforce transformations.

Answers

1.

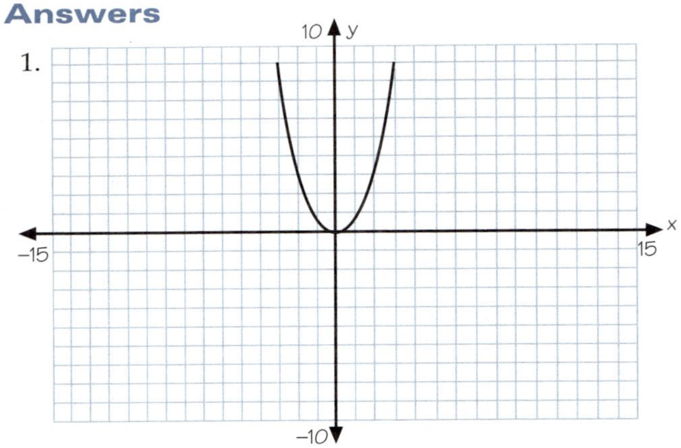

x	$y = x^2$
–3	9
–2	4
–1	1
0	0
1	1
2	4
3	9

2. Answer:

x	Preimage	Image
	$y = x^2$	$y = x^2 + 3$
–3	9	12
–2	4	7
–1	1	4
0	0	3
1	1	4
2	4	7
3	9	12

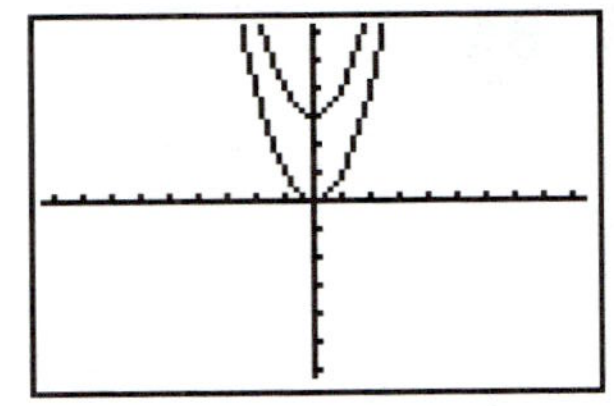

3. The graph of $y = x^2$ is translated vertically up 3 units.

4. Complete the table in **Figure 5.21** for $y = x^2$ and $y = x^2 - 2$, then graph both functions on your graphing calculator. Make a sketch of both graphs.

x	Pre-Image	Image
	$y = x^2$	$y = x^2 - 2$
−3		
−2		
−1		
0		
1		
2		
3		

FIGURE 5.21. Table for Question 4.

5. Write a sentence that describes the transformation.

6. Describe what you think the effect will be on the graph of a quadratic function when a positive or negative number is added to the function after squaring the x-value.

The effect of changing $y = x^2$ to $y = (x + h)^2$.

7. Complete the tables in **Figure 5.22** for $y = x^2$ and $y = (x - 3)^2$, then graph both functions on your graphing calculator. Make a sketch of both graphs.

x	Pre-Image
	$y = x^2$
−3	
−2	
−1	
0	
1	
2	
3	

x	Image
	$y = (x - 3)^2$
0	
1	
2	
3	
4	
5	
6	

FIGURE 5.22. Tables for Question 7.

8. Write a sentence that describes the transformation.

9. Describe the effect on the graph of a quadratic function when a positive number is subtracted from the x-value before squaring.

10. Describe the effect on the graph of a quadratic function when a negative number is subtracted from the x-value before squaring.

The effect of changing $y = x^2$ to $y = -x^2$.

11. Complete the table in **Figure 5.23** for $y = x^2$ and $y = -x^2$, then graph both functions on your graphing calculator. Make a sketch of both graphs.

x	Pre-Image	Image
	$y = x^2$	$y = -x^2$
−3		
−2		
−1		
0		
1		
2		
3		

FIGURE 5.23. Table for Question 11.

12. Write a sentence that describes the transformation.

The effect of changing $y = (x + 2)^2$ to $y = -(x + 2)^2$.

13. Complete the table in **Figure 5.24** for $y = (x + 2)^2$ and $y = -(x + 2)^2$, then graph both functions on your graphing calculator. Make a sketch of both graphs.

x	Pre-Image	Image
	$y = (x + 2)^2$	$y = -(x + 2)^2$
−5		
−4		
−3		
−2		
−1		
0		
1		

FIGURE 5.24. Table for Question 13.

14. Write a sentence that describes the transformation.

15. Describe the effect on the graph of a quadratic function when the opposite sign is applied to the function values.

The effect of changing $y = x^2$ to $y = ax^2$.

16. Complete the table in **Figure 5.25** for $y = x^2$ and $y = 2x^2$, then graph both functions on your graphing calculator. Make a sketch of both graphs.

x	Pre-Image	Image
	$y = x^2$	$y = 2x^2$
−3		
−2		
−1		
0		
1		
2		
3		

FIGURE 5.25. Table for Question 16.

17. Write a sentence that describes the transformation.

Answers (continued)

4. Answer:

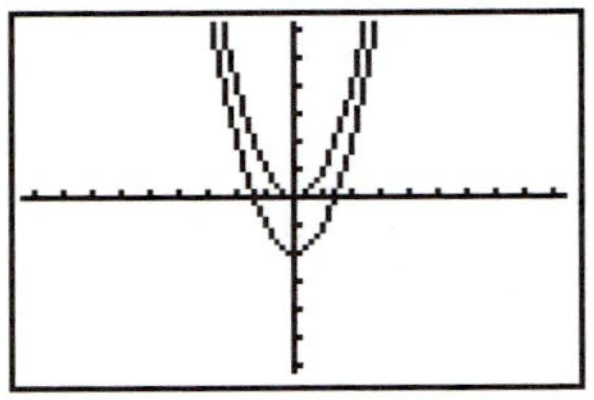

x	Pre-Image	Image
	$y = x^2$	$y = x^2 - 2$
−3	9	7
−2	4	2
−1	1	−1
0	0	−2
1	1	−1
2	4	2
3	9	7

5. The graph of $y = x^2$ is translated vertically down 2 units.

6. Students should discover that in the transformation $y = f(x) + k$, the graph is translated (shifted) vertically up k units if k is positive and down k units if k is negative. Discuss with students that the geometric term "translate" is synonymous with the term "shift" and that they should become used to interchanging the words.

7. Answer:

x	Pre-Image
	$y = x^2$
−3	9
−2	4
−1	1
0	0
1	1
2	4
3	9

x	Image
	$y = (x - 3)^2$
0	9
1	4
2	1
3	0
4	1
5	4
6	9

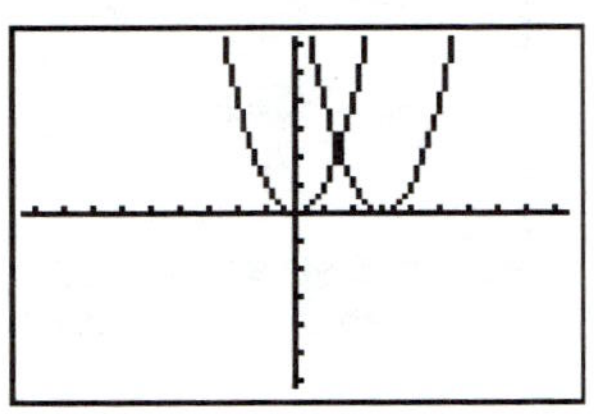

8. The graph of $y = x^2$ is translated horizontally 3 units to the right.

9. When generalizing this rule with students, pay particular attention to the way this transformation is normally written: The transformation $y = (x - a)^2$ shifts the graph a units to the right if a is positive. Stress that since the subtraction sign is part of the form $x - a$ in the function $y = (x - 3)^2$, a is positive.

10. When generalizing this rule with students, pay particular attention to the way this transformation is normally written: The transformation $y = (x - a)^2$ shifts the graph a units to the left if a is negative. If necessary, rewrite $x + 2$ as $x - (-2)$ to show that in this problem a is negative.

11.

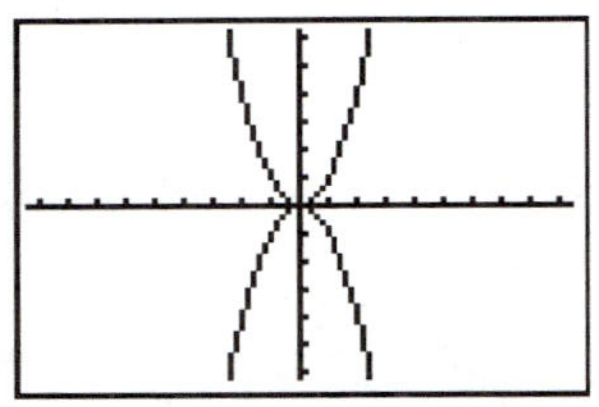

x	Pre-Image	Image
	$y = x^2$	$y = -x^2$
−3	9	−9
−2	4	−4
−1	1	−1
0	0	0
1	1	−1
2	4	−4
3	9	−9

12. The graph is the reflection of $y = x^2$ reflected across the x-axis.

13.

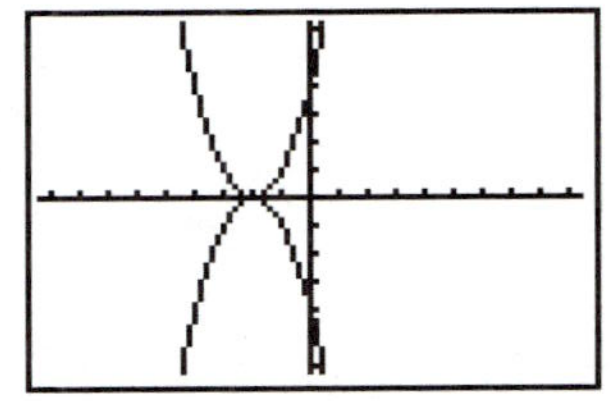

x	Pre-Image	Image
	$y = (x + 2)^2$	$y = -(x + 2)^2$
−5	9	−9
−4	4	−4
−3	1	−1
−2	0	0
−1	1	−1
0	4	−4
1	9	−9

14. The graph of $y = (x + 2)^2$ is reflected across the x-axis.

15. The graph is reflected across the x-axis.

16.

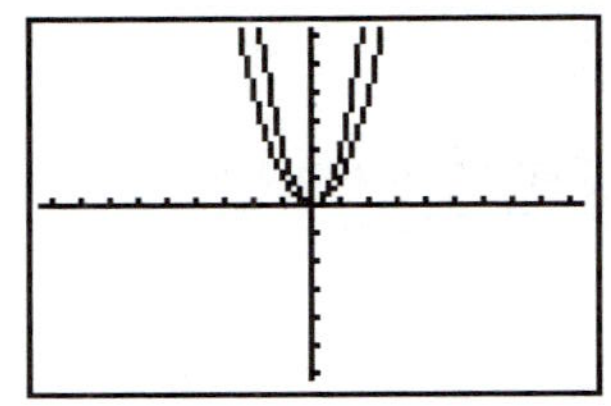

x	Pre-Image	Image
	$y = x^2$	$y = 2x^2$
−3	9	18
−2	4	8
−1	1	2
0	0	0
1	1	2
2	4	8
3	9	18

17. The graph of $y = x^2$ is vertically stretched by a factor of 2.

x	Pre-Image	Image
	$y = x^2$	$y = \frac{1}{2}x^2$
–3		
–2		
–1		
0		
1		
2		
3		

FIGURE 5.26.
Table for Question 18.

18. Complete the table in **Figure 5.26** for $y = x^2$ and $y = \frac{1}{2}x^2$, then graph both functions on your graphing calculator. Make a sketch of both graphs.

19. Write a sentence that describes the transformation.

20. Describe the effect on the graph of a quadratic function when a scale factor greater than 1 is applied to the function values.

21. Describe the effect on the graph of a quadratic function when a scale factor less than 1, but greater than 0, is applied to the function values.

$y = a(x - h)^2 + k$		
	Verbal Description	Visual
$\lvert a \rvert > 1$	Vertical stretch of the graph of the function	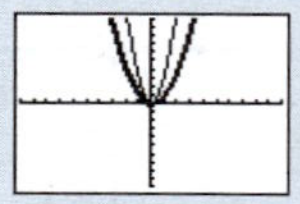
$\lvert a \rvert < 1$	Vertical compression of the graph of the function	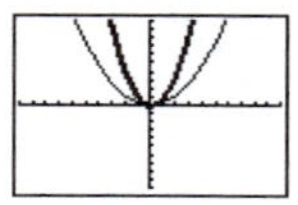
$-a$	Reflection of the graph of the function across the x-axis	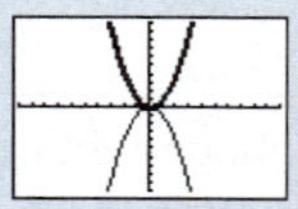
h	Horizontal translation of the graph of the function	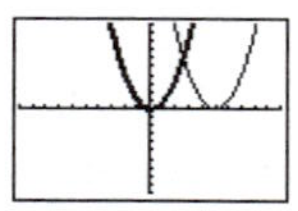
k	Vertical translation of the graph of the function	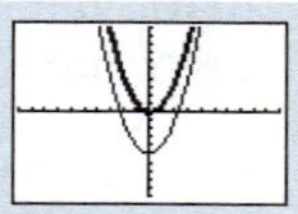

FIGURE 5.27.
Transformation summary.

Transforming functions is a useful tool when you want to fit a function to data. For example, recall Paul's parabola-shaped walk from Section 5.1 shown in **Figure 5.28.**

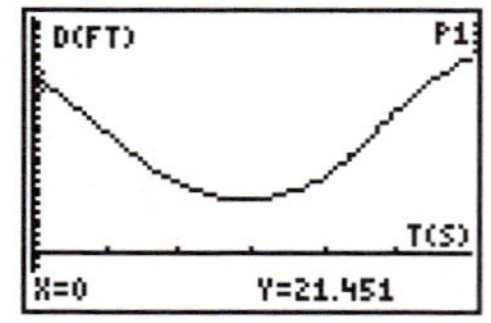

FIGURE 5.28.
Paul's walk.

To fit a function to these data, first compare the graph of the parent function for quadratics, $y = x^2$, to the scatter plot of Paul's walk shown in **Figure 5.29.**

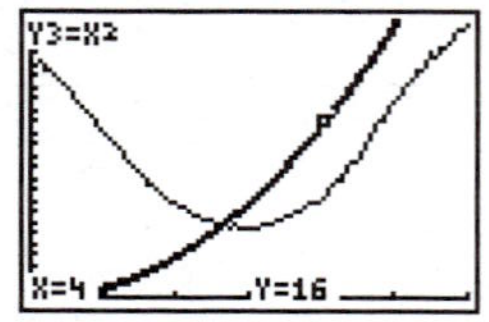

FIGURE 5.29.
Parent function and Paul's walk.

22. The graph of $y = x^2$ must be translated to the right to resemble the walk more closely. Write a function rule that translates the parabola to the right.

23. Now the graph of $y = (x - 3)^2$ must be shifted vertically to more closely resemble the walk. Write a function rule that has the necessary vertical shift.

24. Next the graph of $y = (x - 3)^2 + 7$ must be stretched vertically to resemble the walk more closely. Write a function rule that has the necessary vertical stretch.

Answers (continued)

18.

x	Pre-Image	Image
	$y = x^2$	$y = \frac{1}{2}x^2$
–3	9	$4\frac{1}{2}$
–2	4	2
–1	1	$\frac{1}{2}$
0	0	0
1	1	$\frac{1}{2}$
2	4	2
3	9	$4\frac{1}{2}$

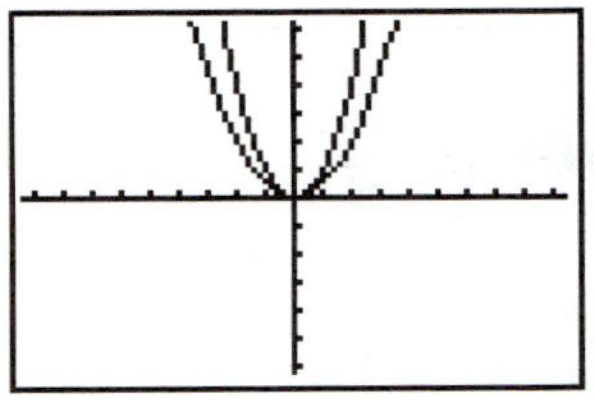

19. The graph of $y = x^2$ is vertically compressed by a scale factor of $\frac{1}{2}$.

20. The graph is vertically stretched by the scale factor.

21. The graph is vertically compressed by the scale factor.

22. Use the trace feature on the calculator to approximate the vertex. Since the x-value is approximately 3, graph $y = (x - 3)^2$.

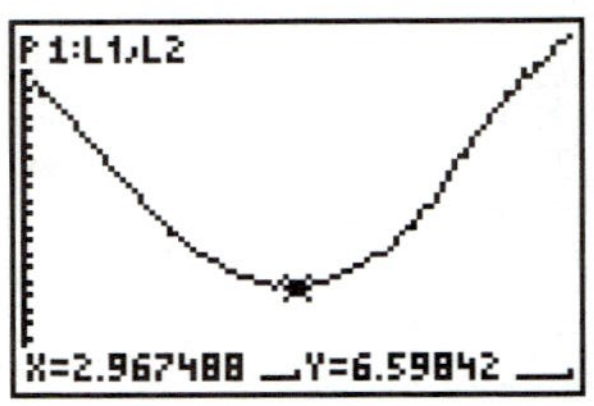

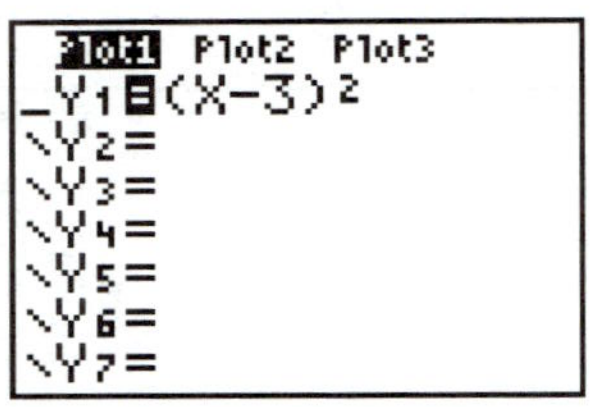

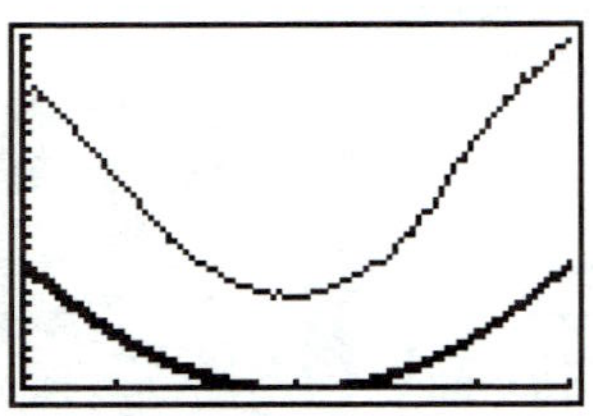

23. Since the y-value in the trace is approximately 7, graph $y = (x - 3)^2 + 7$.

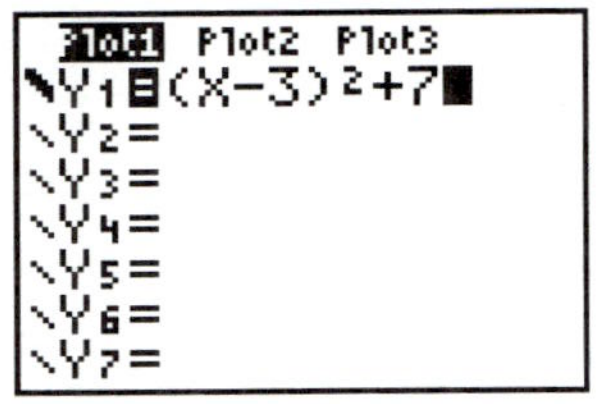

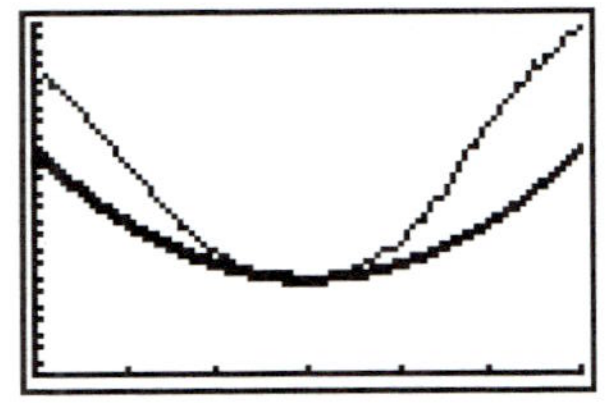

24. Once again, use the trace feature to get an approximate y-value when x is equal to 5 on the scatter plot as well as on the graph of $y = (x - 3)^2 + 7$.

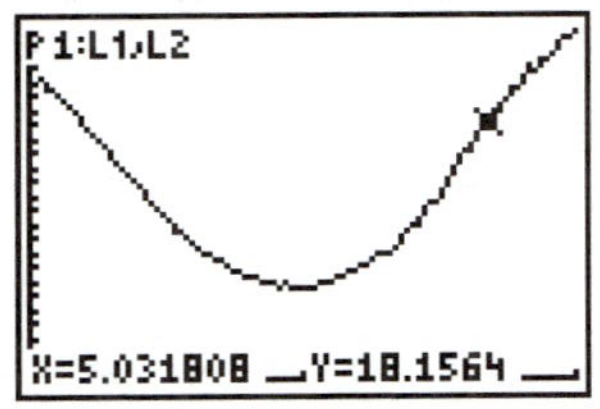

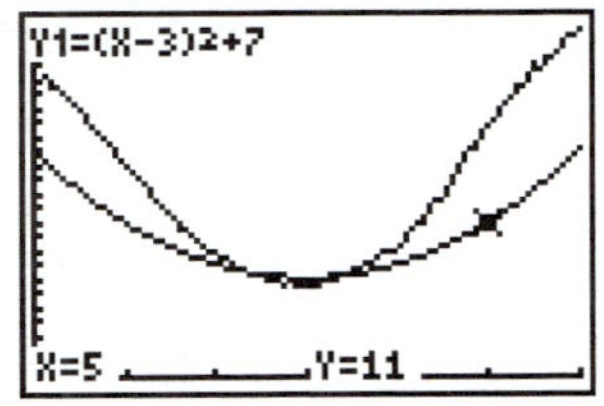

Since the y-value for the scatter plot is greater than the y-value for the function $y = (x - 3)^2 + 7$, a scale factor greater than one must be applied to the rule. The graph of $y = 2(x - 3)^2 + 7$ is shown below as a sample fit.

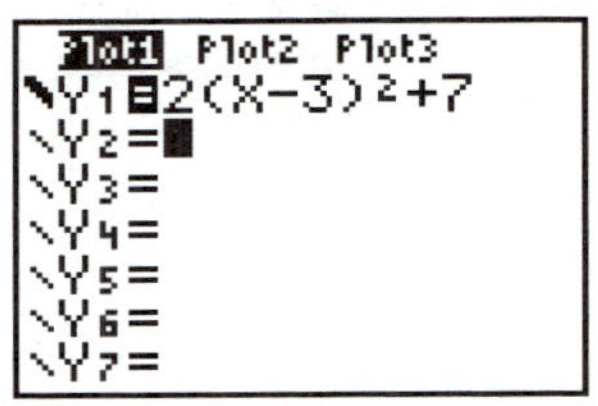

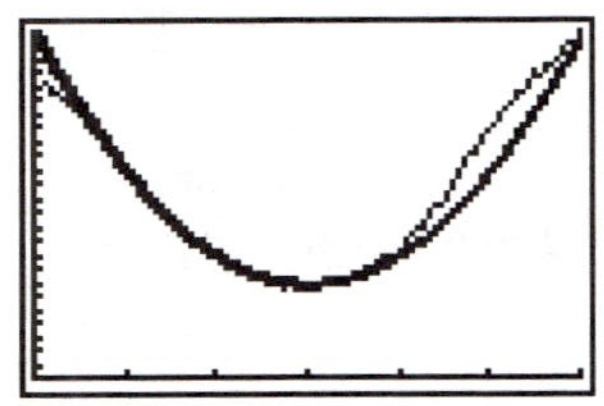

SECTION 5.4 Assignment

The graph of the parent function $y = x^2$ is graphed in the window shown in **Figure 5.30.**

```
WINDOW
 Xmin=-10
 Xmax=10
 Xscl=1
 Ymin=-10
 Ymax=10
 Yscl=1
 Xres=■
```

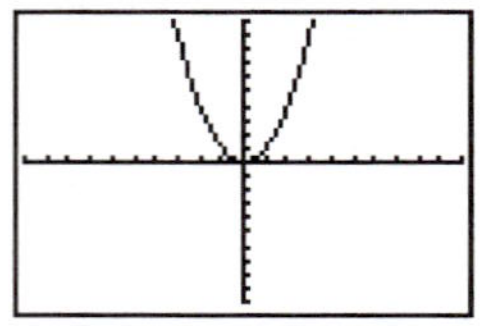

FIGURE 5.30. The parent function $y = x^2$.

Each graph below was graphed using the window above, and is a transformation on $y = x^2$. The vertex of the parabola is labeled in each case. For each graph, write a function rule **and** a verbal description of the transformation.

1. 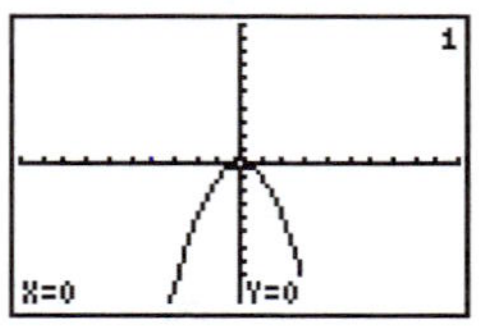

2. 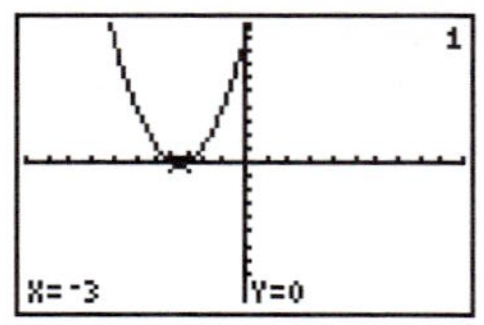

3. 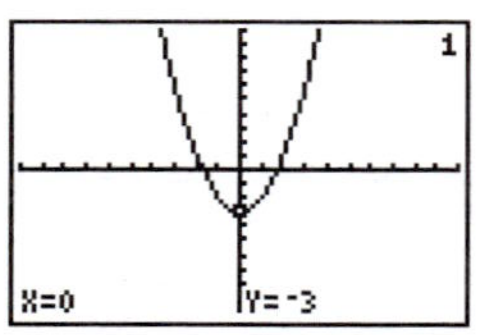

4. 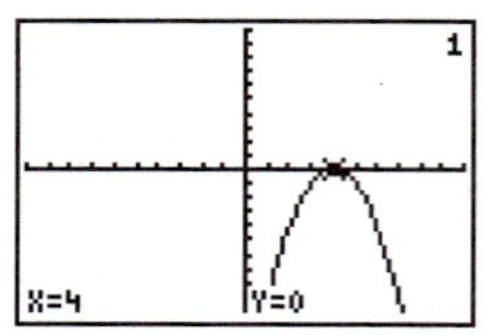

5. 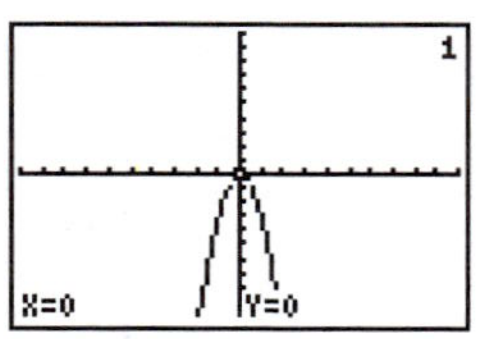

6. 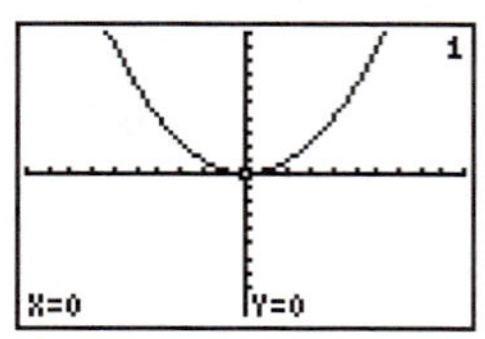

Questions 7–11 each represent a transformation of $y = x^2$. For each, sketch a graph **and** a write a verbal description of the transformation.

7. $y = x^2 - 4$

8. $y = (x - 4)^2$

9. $y = 2(x - 3)^2$

10. $y = \frac{1}{3}x^2$

11. $y = (x - 5)^2 + 3$

12. The scatter plot in **Figure 5.31** was made with the window shown.

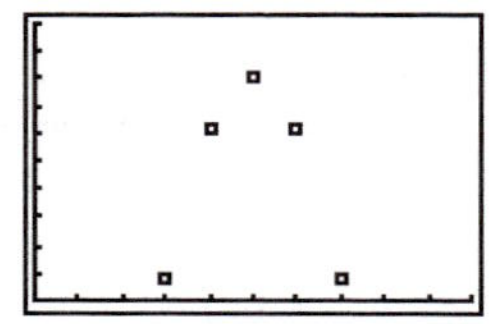

```
WINDOW
 Xmin=0
 Xmax=10
 Xscl=1
 Ymin=0
 Ymax=10
 Yscl=1
 Xres=■
```

FIGURE 5.31. Figure for Question 12.

13. Use transformations to develop a function that fits the data in the scatter plot. Explain your process.

Section 5.4 Assignment

Answers

1. $y = -x^2$, a reflection of $y = x^2$ across the x-axis
2. $y = (x + 3)^2$, a translation of $y = x^2$ horizontally three units to the left
3. $y = x^2 - 3$, a translation of $y = x^2$ vertically three units down
4. $y = -(x - 4)^2$, a reflection of $y = x^2$ across the x-axis followed by a translation horizontally four units to the right
5. $y = -3x^2$, a reflection of $y = x^2$ across the x-axis followed by a vertical stretch by a scale factor of 3
6. $y = \frac{1}{3}x^2$, a vertical compression of $y = x^2$ by a scale factor of $\frac{1}{3}$
7. Answer: A translation of $y = x^2$ vertically four units down

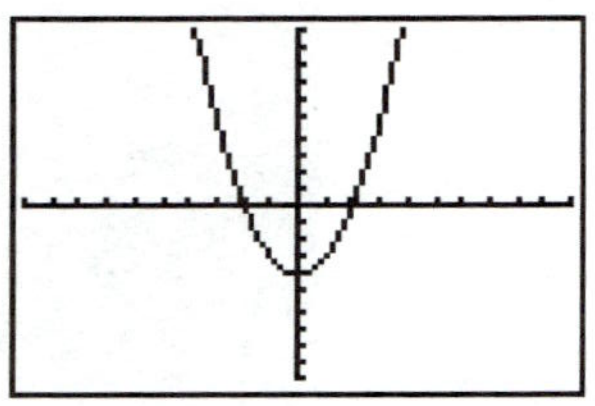

8. Answer: A translation of $y = x^2$ horizontally four units to the right

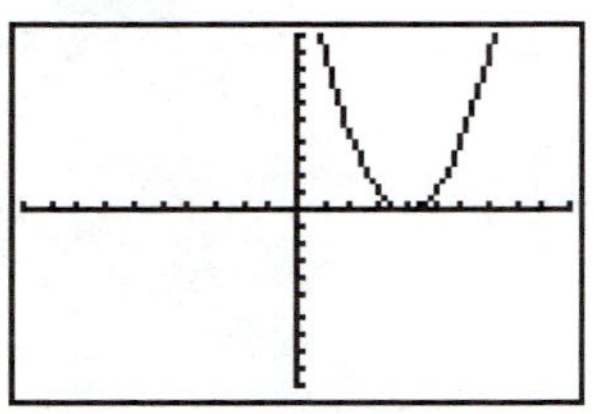

9. Answer: A translation of $y = x^2$ horizontally three units to the right followed by a vertical stretch by a scale factor of 2

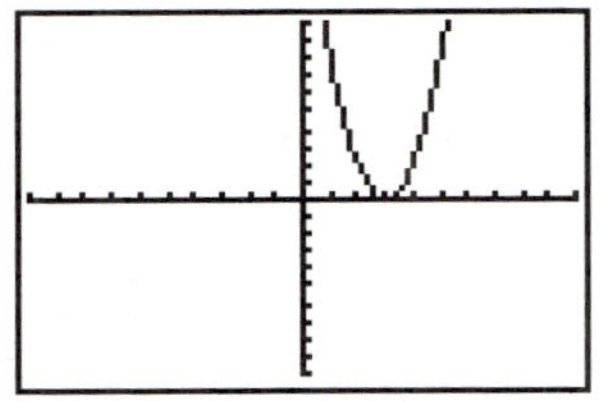

10. Answer: A vertical compression of $y = x^2$ by a scale factor of $\frac{1}{3}$

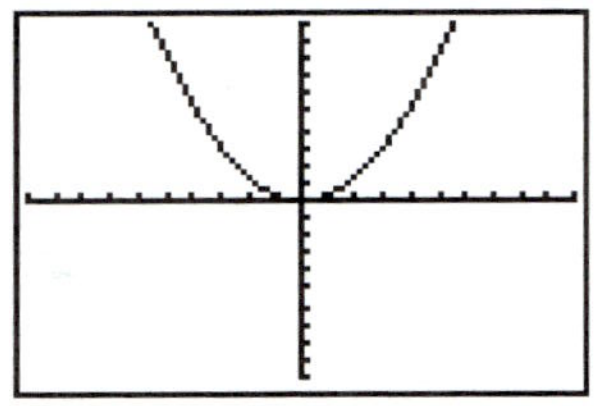

11. Answer: A translation of $y = x^2$ horizontally five units to the right followed by a vertical shift three units up

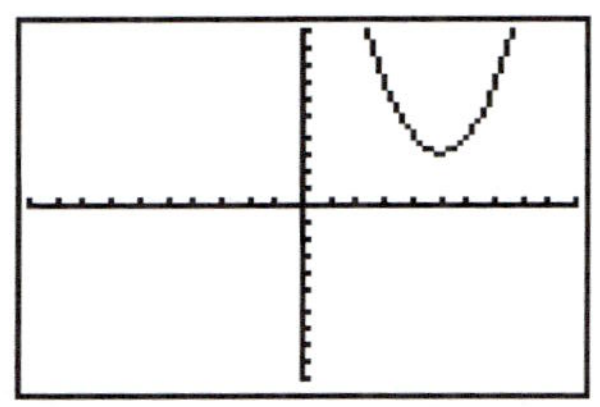

13. Sample answer: Start with the parent function $y = x^2$. Reflect using $y = -x^2$. Move five units to the right $y = -(x - 5)^2$. Translate vertically up eight units $y = -(x - 5)^2 + 8$. Vertically stretch by a scale factor of 1.8, $y = -1.8(x - 5)^2 + 8$.

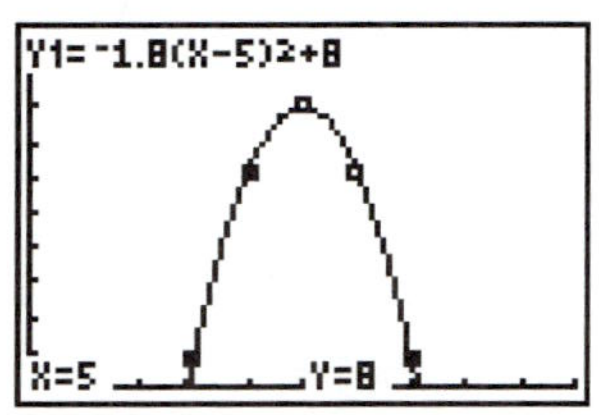

SECTION 5.5

Oh, That Traffic!

Let's apply to a new situation what you have learned about quadratic functions.

Sometimes motion isn't as speedy as at other times.

A mass transit organization in a large city is measuring the speed at which vehicles travel on the freeways. By measuring the speed at specific times of day over a long period of time, they can calculate average speeds during the day. This information can be put into a graph (see **Figure 5.32**) that helps commuters decide how much time they should allow to get to work.

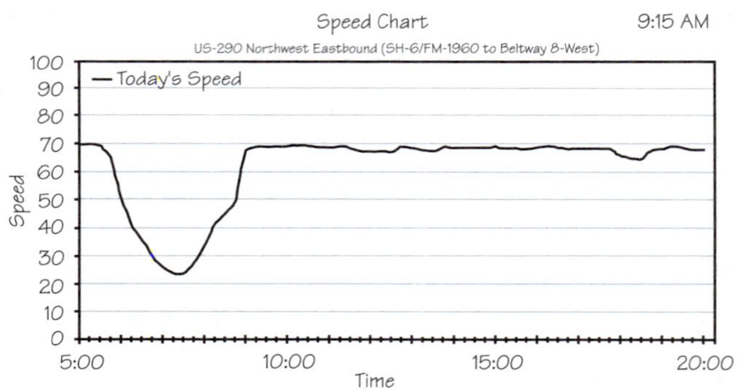

FIGURE 5.32. Travel times for a local freeway.

For one part of a freeway in this city, the speeds at half-hour intervals during the morning rush hour have been recorded (see **Figure 5.33**). The start of rush hour, 5:00 a.m., is time interval 0. The time interval increases by 1 for each successive half hour; i.e., 5:30 a.m. is time interval 1, 6:00 a.m. is time interval 2, etc.

Time	Time Interval	Average Speed (mph)
5:00 a.m.	0	70
5:30 a.m.	1	50
6:00 a.m.	2	35
6:30 a.m.	3	25
7:00 a.m.	4	20
7:30 a.m.	5	20
8:00 a.m.	6	25
8:30 a.m.	7	35
9:00 a.m.	8	50
9:30 a.m.	9	70

FIGURE 5.33. Morning rush hour data.

1. Use Figure 5.33 to make a scatter plot of speed versus time interval. Then build a model to predict the speed in terms of the time interval and graph your model over the scatter plot.

2. Ms. Sanchez and Mr. Williams travel this part of the freeway on their way to work in the morning. Ms. Sanchez reaches this part of the freeway at 6:10 a.m., and Mr. Williams reaches this part of the freeway at 8:10 a.m. If the traffic moves at average speed, who will travel faster? Justify your answer.

EVALUATE LESSON 1

Section 5.5: Oh, That Traffic!

Purpose: In Sections 5.1 through 5.4, students developed the concept of a quadratic function and learned to transform the parent function to fit certain circumstances. In this section students will have the opportunity to demonstrate their knowledge by applying those concepts and skills to solve a problem in a new context. This prepares students to examine multiple approaches to solving quadratic equations and modeling data using quadratic functions in the rest of this chapter.

Description: Facing a problem in a new context, students will determine and implement a method to solve the problem. This is an open-middle performance assessment task; consequently there is only one correct answer, but there are a variety of valid methods to obtain it. In their justification, which is the most important part of a performance assessment task, students should be able to communicate their knowledge of mathematics and demonstrate that they have mastered the concepts presented in Sections 5.1 through 5.4.

Materials: none

Teaching Ideas: This task may be assigned as group work or individual work. The mathematics performance assessment rubric should be used to evaluate students' work.

Answers

1. Sample solution:

   Let x = time interval

   Let y = average speed

   A scatter plot of speed versus time interval will look like this:

To build a model, use a finite differences approach, calculating the first and second differences for the table of data.

Time Interval	Average Speed	First Differences	Second Differences
0	70	–20	5
1	50	–15	5
2	35	–10	5
3	25	–5	5
4	20	0	5
5	20	5	5
6	25	10	5
7	35	15	5
8	50	20	
9	70		

Since the second differences are constant, this relationship can be represented with a quadratic function of the form $y = ax^2 + bx + c$. So we know the following to be true:

- $2a$ = second difference
- c = y-value when $x = 0$
- $a + b$ = first, first difference

Substituting what you know,

- $2a = 5$, so $a = 2.5$
- $c = 70$
- $a + b = -20$, so $b = -20 - a = -20 - 2.5 = -22.5$

To find the quadratic model that best fits these data, substitute $a = 2.5$, $b = -22.5$, and $c = 70$ into $y = ax^2 + bx + c$ to get $y = 2.5x^2 - 22.5x + 70$.

Graphing this function over the scatter plot will test its accuracy:

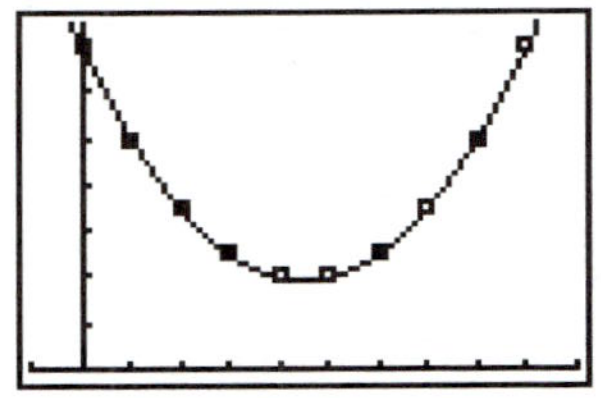

2. Ms. Sanchez arrives at this part of the freeway at 6:10 a.m., which is between time interval 2 and time interval 3. A time interval of 1 is 30 minutes. So 5 minutes is $\frac{1}{6}$ of a time interval, and 10 minutes is $\frac{1}{3}$ of a time interval, so 6:10 a.m. is time interval $2\frac{1}{3}$.

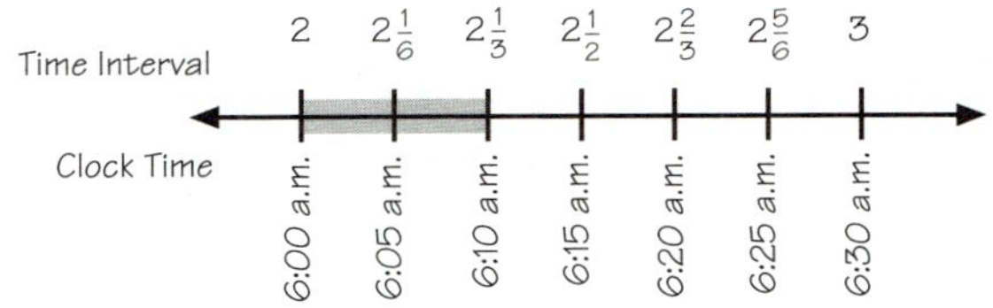

The speed when $x = 2\frac{1}{3}$ will be what Ms. Sanchez should expect to be traveling. Using the TRACE feature of a graphing calculator reveals this speed to be about 31.1 miles per hour.

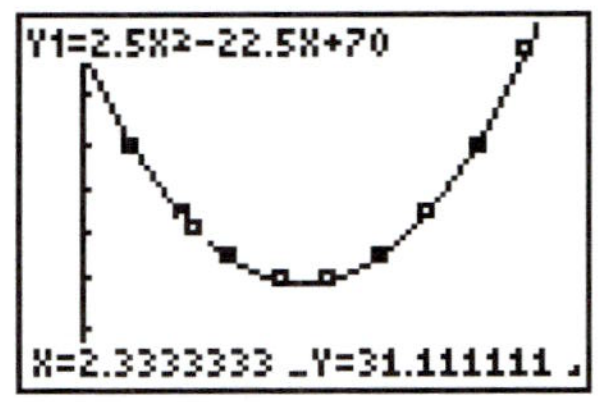

Mr. Williams arrives at this part of the freeway at 8:10 a.m., which is between time interval 6 and 7.

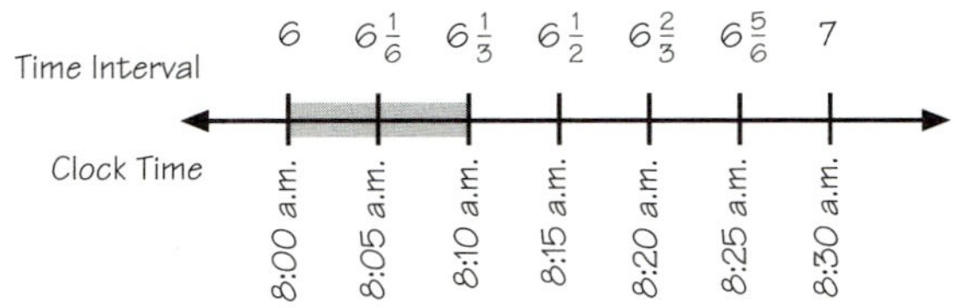

Answers (continued)

So Mr. Williams will arrive at 8:10 a.m., or time interval $6\frac{1}{3}$. Use the TRACE feature of the graphing calculator to find the speed Mr. Williams can expect to be traveling, which is about 27.8 miles per hour.

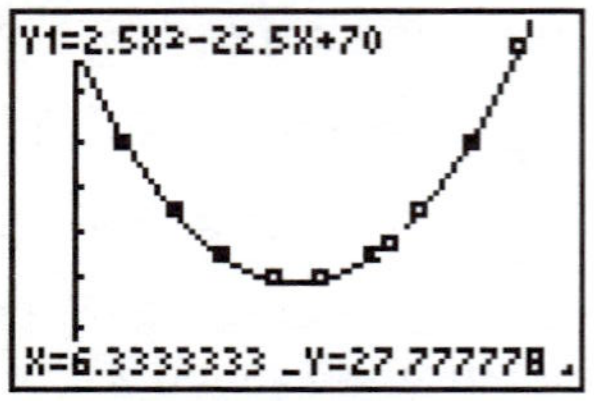

Since Ms. Sanchez will be traveling about 31.1 mph and Mr. Williams about 27.8 mph, Ms. Sanchez will be traveling faster.

SECTION **5.6**

3...2...1...Blast Off!

On Earth, gravity pulls objects toward the ground. This force, as you may have learned in your science class, changes an object's speed. If the object is moving away from Earth, gravity slows it down. If the object is moving toward Earth, gravity speeds it up. This change in speed is called acceleration.

CHECK THIS!

Acceleration is the change in an object's speed per unit of time.

When astronauts try to escape Earth's gravity and travel into space, they must overcome this force. One way they do that is to use a rocket. In this section you will observe a model rocket launch. While your rocket won't fly into space, you will observe changes in its speed as a result of the force of gravity.

MODEL ROCKET LAUNCH SIMULATION

Begin by setting the MODE and the WINDOW on your graphing calculator as shown in **Figure 5.34.**

```
Normal Sci Eng
Float 0123456789
Radian Degree
Func Par Pol Seq
Connected Dot
Sequential Simul
Real a+bi re^θi
Full Horiz G-T
```

```
WINDOW
 Tmin=0
 Tmax=5
 Tstep=.1
 Xmin=0
 Xmax=5
 Xscl=1
 Ymin=0
 Ymax=110
 Yscl=10
```

FIGURE 5.34. Settings for graphing calculator.

You've now changed your calculator setting from FUNCTION mode (what we normally use to graph, in terms of x and y) to PARAMETRIC mode. In parametric mode, we graph two functions, x_T and y_T, both in terms of *time*.

1. Press the Y= key on your graphing calculator. Enter each of these into your calculator.

   $X_{1T} = 2.5$

   $Y_{1T} = 1.6T^2 + 80T$

2. Press the GRAPH key on your calculator. Sketch the graph produced on the axes.

3. Is this graph what you expected? Why or why not?

4. In Section 5.1 you examined distance-versus-time walking graphs. One of the graphs had a parabolic shape. How does the flight of the rocket compare to the walk that produced a parabolic graph?

ENGAGE LESSON 2

Section 5.6: 3...2...1...Blast Off!

Purpose: In previous sections students investigated and defined quadratic functions and how to perform transformations on the parent function, $y = x^2$. In Section 5.6 students will investigate a problem concerning rocketry involving motion that can be described by a quadratic function. This sets the stage for students to learn multiple methods to solve quadratic equations in Sections 5.7 through 5.12.

Description: In this section of the lesson the teacher will demonstrate a rocket launch. Students will estimate and describe features of the rocket launch, such as the maximum height of the rocket, how long the rocket is in the air, and what a graph of the rocket's motion might look like. Then the teacher will guide students through a graphing calculator simulation of the rocket launch.

This lesson encompasses Sections 5.6 through 5.14 and introduces three methods of solving quadratic equations. Each method will have its own Explore-Explain cycle. Hence, this lesson is a modified 5E lesson where there is one Engage, three cycles of Explore-Explain, one Elaborate, and one Evaluate.

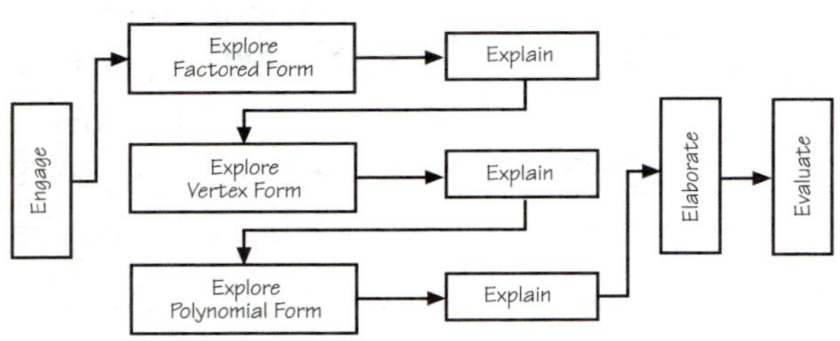

After Exploring and Explaining (formalizing) using the method of solving by factoring when quadratic equations are given in factored form, students will practice using this method in Assignment 5.8. After Exploring and Explaining using the method of inverse operations, including completing the square, students will practice using this method in Assignment 5.10. After Exploring and Explaining using the method of the Quadratic Formula, students will practice using this method in Assignment 5.12. Section 5.13, the Elaborate phase of this lesson, will enable students to make connections among the three forms of a quadratic equation and empower students to select an appropriate method to solve a quadratic equation of any given form. In Section 5.14 students will have the opportunity to solve a contextual problem involving a quadratic function using the method of their choice, including a defense of the chosen method.

Materials: rocket (preassembled using the directions below from the following materials: film canister with a lid that snaps inside the canister, cardstock, tape), effervescent tablet(s) such as Alka-Seltzer™, water, tennis ball

Teaching Ideas: This section is designed to be a teacher-led whole-class demonstration. The teacher or a student volunteer should perform the actions. The teacher will pose critical questions, allowing students time to record their thoughts in their notebooks and discuss either within their groups or with the whole class.

The first part of this section consists of a model rocket launch. For the rocket launch, it is recommended that about half of an effervescent tablet be used as rocket fuel. The teacher should construct the rocket ahead of time using the pattern found on the Teacher Resource CD-ROM. When selecting a film canister, it is important that the canister be one where the lid snaps inside the canister rather than closing over the lip of the canister. Also, it is **highly** recommended that you have a flat container, such as an empty plastic tub, under the rocket to serve as a launch pad to catch the expelled liquid fuel as the rocket launches. Although the rockets are extremely safe, it is recommended that good laboratory practice be followed and all participants wear safety glasses.

The second part of this section consists of a mathematical simulation of a rocket launch. The graphing calculator simulation of the rocket launch involves using the parametric mode. Students DO NOT need to have an understanding of parametric functions in order to observe or to emulate the rocket launch simulation. Parametric mode is the most efficient way to simulate the height of the rocket as a function of time in a way that is similar to the rocket's behavior in the real world. The teacher can set up the graphing calculator simulation prior to the class demonstration. Alternatively, depending on the class, the teacher can ask the students to perform the simulation by walking them through the calculator steps.

Model Rocket Launch Demonstration:

a) Show students the preassembled rocket. Explain to them that you are going to launch the rocket and that they should make observations about the height of the rocket, time that the rocket is in the air, etc., after the rocket is placed and launched. Students should have their notebooks out, ready to record what they observe in their notes.

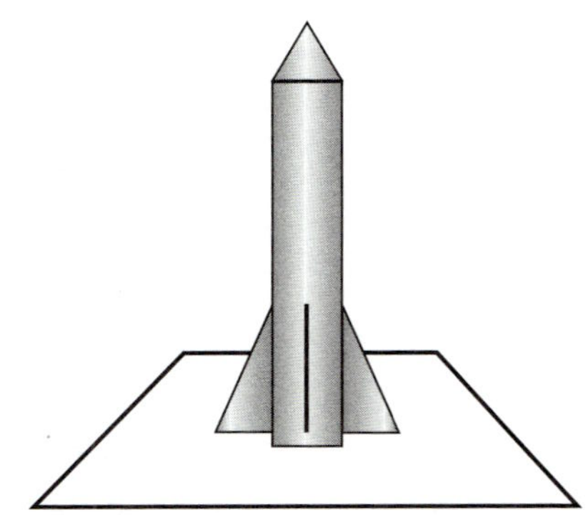

b) Fill the film canister about $\frac{1}{3}$ full of water.

c) Add about half of an effervescent tablet and cap the film canister. Quickly set the rocket in place on the launch pad.

Answers (continued)

d) Stand back and watch it blast off!

Use the following questions to debrief the launch. Do not answer them, but ask students to record their answers.

1. About how high did the rocket travel?
2. About how long was the rocket in the air?
3. Where did the rocket land in relation to the launch pad?
4. Sketch a graph of the rocket's height as a function of time.

Answers to the questions will vary. The goal is not to have students arrive at exact, correct answers; rather, the goal is to get students to record the picture in their mind's eye of what a rocket's path looks like.

Next, use a graphing calculator to show the students what a simulation of the model rocket launch might look like using technology.

Answers

Model Rocket Launch Simulation

1. If the calculator your students are using can do a "bubble trace," set that option for this graph as shown.

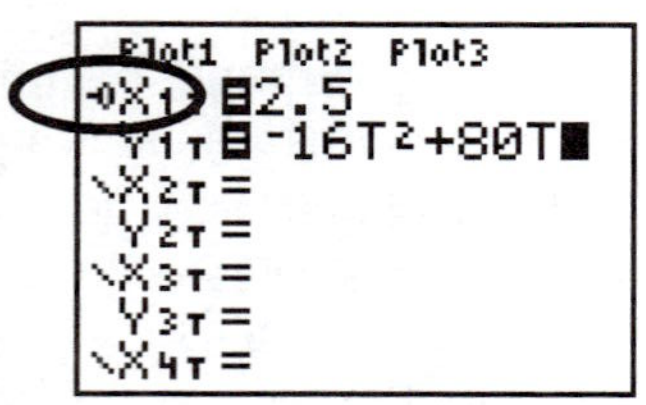

2. 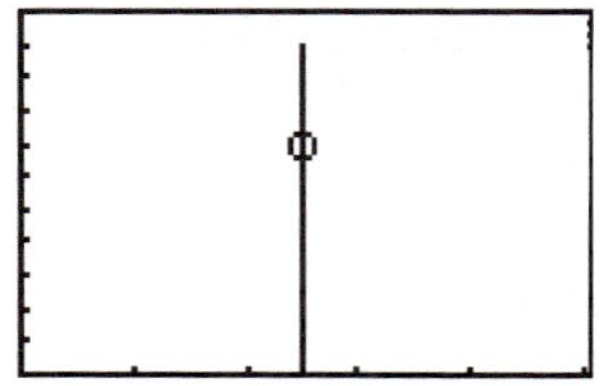

3. The students may have been expecting to see a parabola since the function was quadratic. Ask the students why the graph goes up and then comes back down.

   At this time, explain to the students that this graph represents the rocket's height at any time, *t*. The graph is actually simulating the PATH of the rocket going up and down. The $X_{1t} = 2.5$ merely indicates where to draw the graph on the screen.

   Students frequently get confused with parabolic free-fall motion problems. Often in a mathematics class, students are given a quadratic function that indicates the height of a falling body as a function of time. Students confuse the graph of height vs. time with the actual path of a falling object.

   To reinforce the representation on this parametric graph, throw a tennis ball STRAIGHT up in the air and watch it fall back to the ground. Then run the rocket launch simulation again. Equate the tennis ball's path with the graph simulating the rocket launch.

4. Answers may vary. Students should notice that the rocket rose quickly at first but slowed down, then fell back to the ground at an increasing rate of speed. Likewise, when the student was walking a parabolic-shaped graph, he or she started out walking very fast, slowed down, then turned around and walked at an increasing rate of speed.

SECTION 5.6 Assignment

1. Based on the information provided, graph a parabola.
   - Line of symmetry is $x = 2$.
   - The vertex has a y-coordinate of -1.
   - $(3, 0)$ is an x-intercept.
   - $(0, 3)$ is the y-intercept.

CHECK THIS!

A line of symmetry in a parabola is a line that passes through the vertex as shown in Figure 5.35.

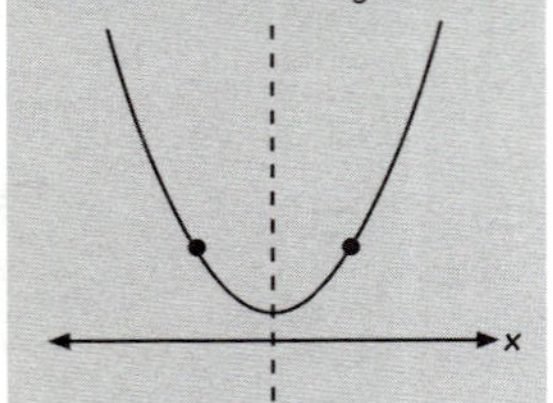

FIGURE 5.35.

Each point on the parabola can be reflected across the line of symmetry to match a point on the other side of the parabola.

2. What are the coordinates of the vertex?
3. What are the coordinates of the other x-intercept?
4. Name one other point that is on the parabola and explain how you found this point.
5. Write an equation of this parabola.

In Chapter 1 equations of the form $y - y_1 = m(x - x_1)$ are introduced as a way of describing linear models. This form is called point-slope form.

The **vertex form** of a quadratic function is $y = a(x - h)^2 + k$. This form can be modified by subtracting k from both sides to get $y - k = a(x - h)^2$.

6. Compare the point-slope form for linear functions with the modified vertex form for quadratics. How are they similar? How are they different?

SECTION 5.7 There Could Be Several Factors

Quadratic functions can be used to model many things that occur in the real world. In Section 5.6 you used a graphing calculator to simulate a rocket launch.

Aerospace engineers use quadratic functions to model the motion of a rocket, much like you did in Section 5.6. Then they use these quadratic functions to get quadratic equations to help solve problems.

Recall that we used $Y_{1T} = -16T^2 + 80T$, where T represents time and Y represents the height of a rocket, to describe the motion of the rocket (see **Figure 5.36**).

FIGURE 5.36.
Modeling a rocket's height.

$Y_{1T} = -16T^2 + 80T$ can be rewritten as $y = -16x^2 + 80x$ if we let x be time and y be the height of the rocket.

1. Write an equation that can be solved to find the times when the rocket was 64 feet above the ground.
2. Rewrite this equation in $ax^2 + bx + c = 0$ form, and completely factor the polynomial.

CHECK THIS!

Factoring Polynomials

Step 1: Find the largest factor common to all the terms of the polynomial.

Step 2: Use the distributive law to write the polynomial as a product where one factor is the largest common factor.

Section 5.6 Assignment

Answers

1. 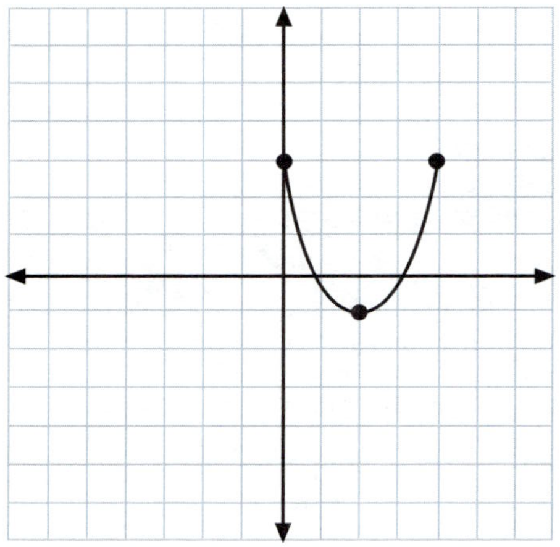

2. (2, –1)
3. (1, 0)
4. Answers will vary. Sample answer: (4, 3); this is the symmetrical point to the y-intercept that was given.
5. Any one of the following is a correct representation for this equation:

   $y = (x - 2)^2 - 1$

   $y = (x - 1)(x - 3)$

   $y = x^2 - 4x + 3$
6. Answers may vary. Similarities the students may notice include that both equations have a constant being subtracted from y, both equations have a constant being subtracted from x, and both equations have a constant multiplier to the quantity containing x. Differences the students may notice include that the linear equation has no exponent, but the quadratic equation has an exponent of 2; the linear equation has a multiplier that is called "slope," but the quadratic equation has a general multiplier, "a."

EXPLORE LESSON 2

Section 5.7: There Could Be Several Factors

Purpose: In Section 5.6 students used a rocket launch and graphing calculator simulation in order to demonstrate a real-world application of quadratic functions. In Section 5.7 students will be faced with a situation requiring them to set up and solve a quadratic equation in factored form. In Section 5.8 students will formalize the mechanics of solving a quadratic equation by factoring.

Description: Based on their experiences with the rocket launch simulation, students will be posed several questions that lead to the development of quadratic equations in factored form. These questions will develop the need to solve quadratic equations by factoring.

Materials: graphing calculator

Teaching Ideas: Students should work in small groups (3 to 4 students). This section is designed for students to make connections between the rocket launch simulation and solving quadratic equations.

In this lesson students will be required to factor trinomials. If your students need more help with factoring, you may need to spend some time reviewing different methods of factoring.

Answers

1. $-16x^2 + 80x = 64$
2. $-16x^2 + 80x - 64 = 64 - 64$

   $-16x^2 + 80x - 64 = 0$

   $-16x^2 + 80x - 64 = 0$

   $-16(x^2 - 5x + 4) = 0$

   $-16(x - 4)(x - 1) = 0$

EXAMPLE:

Factor $18x^5 - 9x^3 + 27x^2$.

Although there are other factors common to the three terms (18, 9, and 27), such as $3x$, the largest common factor is $9x^2$.

So you factor the polynomial using $9x^2$.

$18x^5 - 9x^3 + 27x^2$

$= (9x^2)(2x^3) + (9x^2)(3)$

$= 9x^2(2x^3 - x + 3)$

3. Use your graphing calculator to graph $y = ax^2(2x^3 - x + 3)$. Sketch the graph and record your viewing window.

4. At which x-values does the graph cross the x-axis?

5. Use your graphing calculator to make a table of values for $y = -16x^2 + 80x - 64$. At what x-values is $y = 0$? How do these values relate to the x-values where the graph crosses the x-axis?

6. How do the x-values from Questions 4 and 5 relate to the factors in the factored form from Question 2?

7. What happens when you substitute each x-value into the factored form?

8. What is the meaning of $y = 0$ in this situation?

9. Write an equation that can be solved to find the times when the rocket was on the ground.

10. Completely factor the polynomial you wrote in Question 9.

11. Use your graphing calculator to graph the function in factored form. Sketch the graph and record your viewing window.

12. At which x-values does the graph cross the x-axis?

13. Use your graphing calculator to make a table of values for $y = -16x^2 + 80x$. At what x-values is $y = 0$? How do these values relate to the x-values where the graph crosses the x-axis?

14. How do the x-values from Questions 11 and 12 relate to the factors in the factored form from Question 9?

15. Substitute the x-values from Question 13 into the factored form and simplify. Describe your results.

16. Write a sentence that describes the relationship between the solutions to a quadratic equation of the form $ax^2 = bx + c = 0$ and the factors of $ax^2 + bx + c$.

Answers (continued)

3.

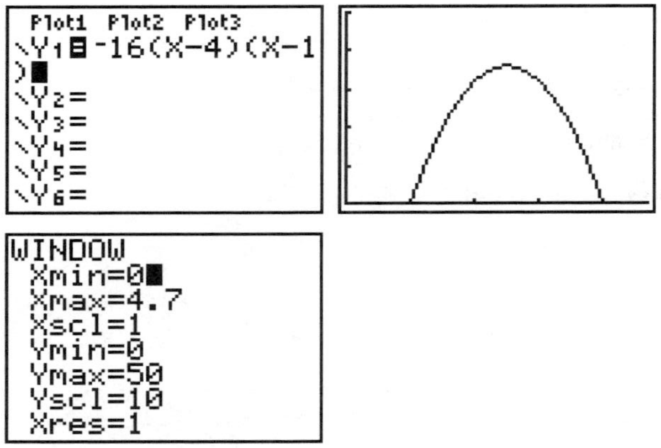

4. The graph crosses the x-axis when $x = 1$ or $x = 4$.

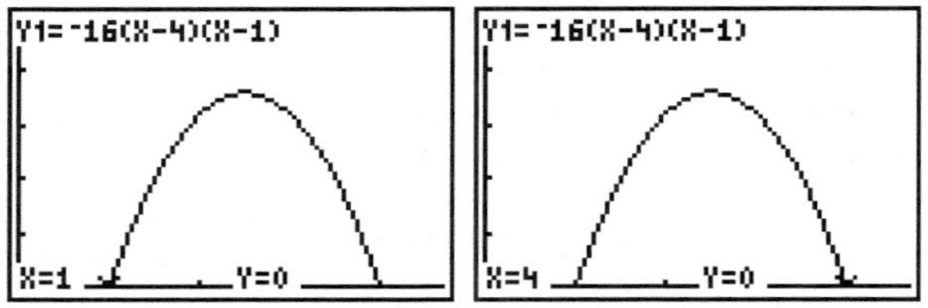

5. When $x = 1$ and $x = 4$, $y = 0$. These are the same x-values where the graph crosses the x-axis.

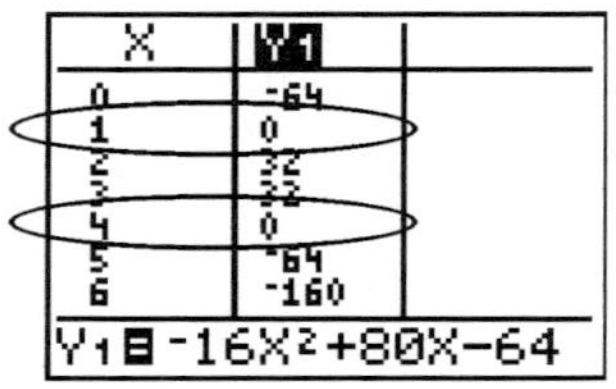

6. The x-values are the opposite to the constants in the factors of the polynomial in the factored equation $-16(x - 4)(x - 1) = 0$.

7. When $x = 4$ is substituted into the factored expression $-16((4) - 4)((4) - 1)$, the second factor becomes $4 - 4$, which is 0, and the product of the three factors becomes zero.

When $x = 1$ is substituted into the factored expression $-16((1) - 4)((1) - 1) = 0$, the third factor becomes $1 - 1$, which is 0, and the product of the three factors becomes zero.

8. When $y = 0$, the height of the rocket is 0, and the rocket is on the ground.

9. $-16x^2 + 80x = 0$

10. $-16x^2 + 80x = 0$

$-16x(x - 5) = 0$

11.

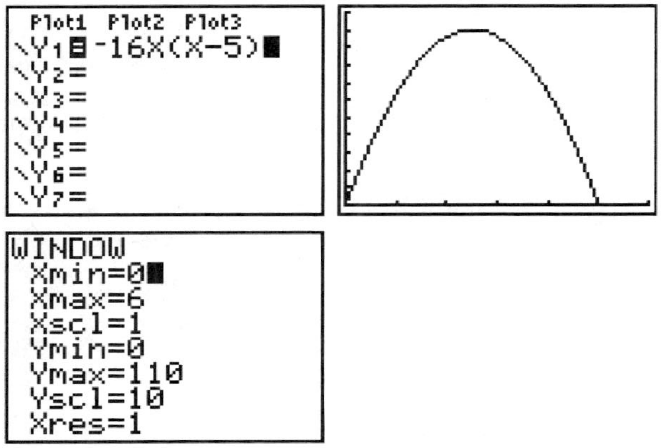

12.

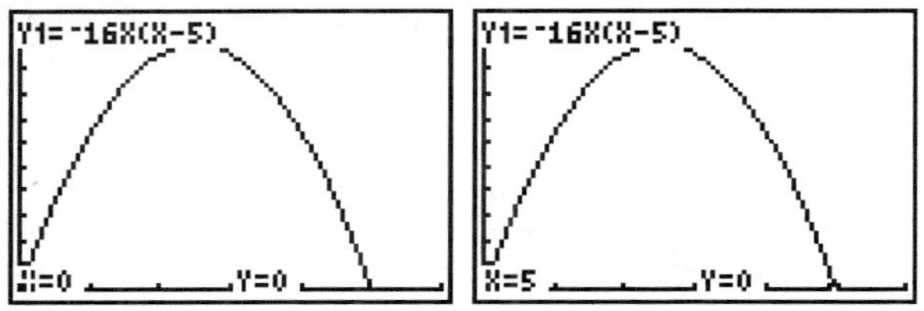

The graph crosses the x-axis when $x = 0$ or $x = 5$.

13. When $x = 0$ and $x = 5$, $y = 0$. These are the same x-values where the graph crosses the x-axis.

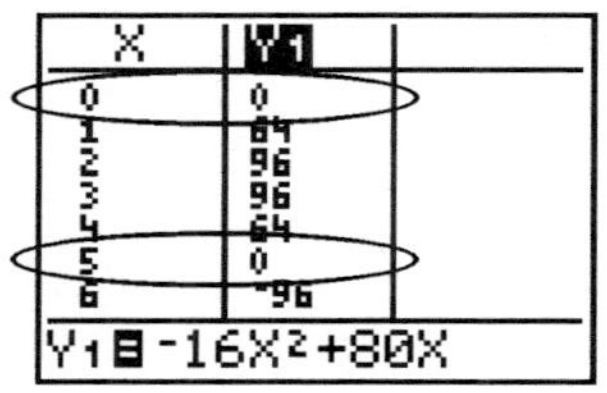

14. The x-values are the opposite to the constants in the factors of the polynomial in the factored equation $-16x(x - 5) = 0$.

15. When $x = 0$ is substituted into the factored expression $-16(0)((0) - 5)$, the first factor becomes 16(0), which is 0, and the product of the two factors becomes zero.

When $x = 5$ is substituted into the factored expression $-16(5)((5) - 5)$, the second factor becomes $5 - 5$, which is 0, and the product of the two factors becomes zero.

16. Answers may vary. Sample answer: The solutions to a quadratic equation are values that make one of the factors equal to zero.

SECTION 5.8

Solving Quadratic Equations by Factoring

$y = ax^2 + bx + c$ is called the **general form** of a quadratic function. This form does not provide a lot of information about the graph of the function. Other forms are more useful because they display some important features of the graph. One form of a quadratic function is called the **factored form** and is written $y = a(x - r_1)(x - r_2)$, where r_1 and r_2 are integers.

1. Consider $y = -0.5(x - 1)(x - 5)$, which is written in factored form.

   a) Use a graphing calculator to graph the function. Sketch the graph and record your viewing window.

   b) Where does the graph cross the x-axis?

   c) Substitute the x-values where the graph crosses the x-axis into the function. What y-values do they yield?

Factored form $y = a(x - r_1)(x - r_2)$ is a useful way to write a quadratic function. The numbers r_1 and r_2 are the x-coordinates of the points where the graph crosses the x-axis. These x-coordinates are called the **roots** of the quadratic equation $-0.5x(x - 1)(x - 5) = 0$. They are also known as ***x*-intercepts** of the graph or **zeros** of the function $y = -0.5x(x - 1)(x - 5)$.

   d) Why is an x-intercept also called a zero?

   e) How can you tell if a point is an x-intercept from just the coordinates of the point?

In the factored form of a quadratic, the only things that change are the numbers. In general, it can be written like this:

$y =$ ____ $(x -$ ____ $)(x -$ ____ $)$

The challenge is to fill in the blanks with the correct numbers.

2. Write a quadratic function in factored form that has x-intercepts at -4 and $+2$. Use your graphing calculator to graph the function, sketch the graph, and record your window.

3. Adjust your answer so that the vertex is located at the point $(-1, -3.6)$. Explain how you determined this answer.

4. a) Change your answer into polynomial form ($y = ax^2 + bx + c$).

   b) Use your graphing calculator to display the graph of both the factored form and the polynomial form. Compare the two graphs and explain any similarities or differences.

   c) If you are graphing two functions, but only one graph appears, what does this mean?

You can use factored form to solve quadratic equations. You may have noticed that substituting a zero of a function causes one of the factors to be 0. Multiplying any number or factor by 0 yields a value of 0. So, if you have a quadratic in polynomial form that equals 0, such as $x^2 - 6x + 5 = 0$, you can use factoring to solve the equation.

Here are the steps to use factored form to solve quadratic equations.

- Factor the polynomial completely.

  $x^2 - 6x + 5 = 0$

  $(x - 5)(x - 1) = 0$

- Set each factor equal to 0.

  $x - 5 = 0 \qquad x - 1 = 0$

- Solve each equation for x.

  $x - 5 = 0 \qquad x - 1 = 0$

  $x = 5 \qquad x = 1$

This method of solving a quadratic equation uses the **Zero-Product Property**. This property states that if the product of two real numbers is zero, then one or both of the numbers must be equal to zero. In other words, if $ab = 0$, then either $a = 0$ or $b = 0$.

EXPLAIN LESSON 2

Section 5.8: Solving Quadratic Equations by Factoring

Purpose: In Section 5.7 students explored the relationship between the factors of a quadratic equation and the x-intercepts of the graph. In this section students will formalize this relationship and use it to solve quadratic equations by factoring. In the next few sections students will explore and formalize other methods of solving quadratic equations so that by Section 5.13 students will be able to select and defend a method of solving a quadratic equation in a problem-solving situation.

Description: In this section students will use their graphing calculators to formalize the connection between the zeros of a function and the factors contained in the factored form of a quadratic equation. They will learn to use factoring to solve quadratic equations, and formalize the Zero-Product Property.

Materials: graphing calculators

Teaching Ideas: Debrief students after Section 5.7 by having them share their answers to the questions. As they share, be sure to fill in any missing concept, process, or skill that could have been used to answer the questions. Then work through the questions in this section with the entire class. Once again be sure all necessary concepts, processes, or skills are addressed.

Answers

1. a)

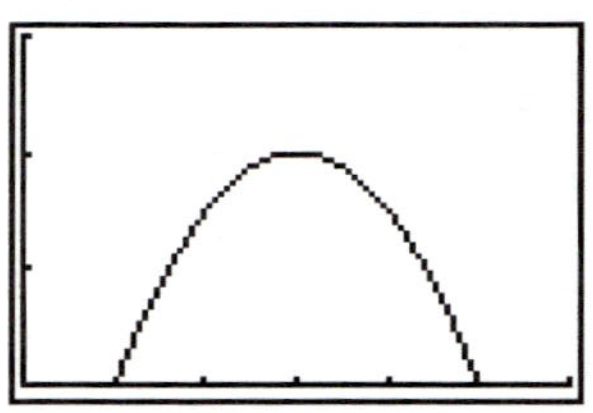

b) The graph crosses the x-axis at $x = 1$ and $x = 5$.

c) $y = -0.5(x - 1)(x - 5)$ $\quad$ $y = -0.5(x - 1)(x - 5)$
$y = -0.5(1 - 1)(1 - 5)$ $\quad$ $y = -0.5(5 - 1)(5 - 5)$
$y = -0.5(0)(-4)$ $\quad$ $y = -0.5(4)(0)$
$y = 0$ $\quad$ $y = 0$

d) Using the x-value from an x-intercept produces a y-value of 0 for the equation.

e) If the y-coordinate for a point on the graph is 0, then the point is an x-intercept.

2. Answers will vary, but will be of the form: $y = a(x + 4)(x - 2)$, with some value for a. The sample graph uses $a = 2$.

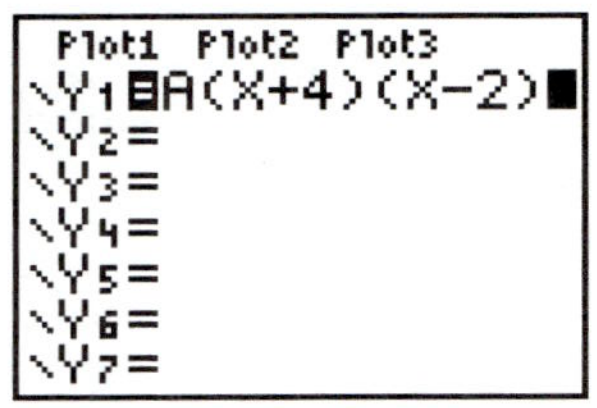

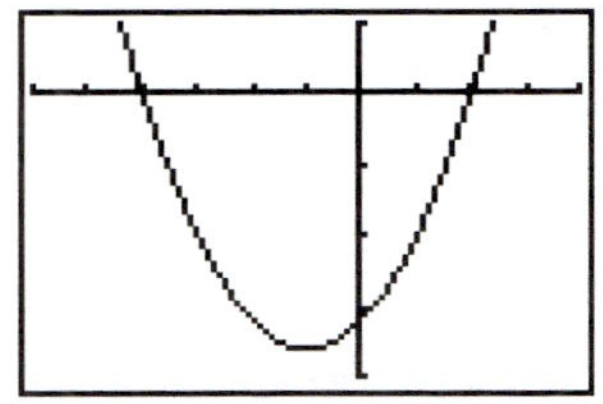

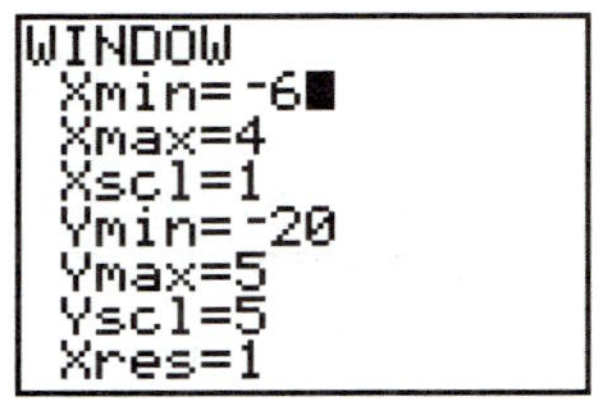

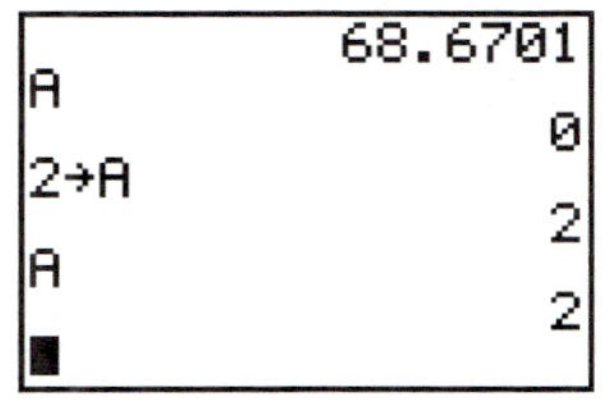

NOTE: You can use constants such as A in the function grapher or table generator. To set or change the value of A, go to your home screen, type the value you wish to use, press the STO key, ALPHA-A, then press ENTER.

3. $y = 0.4(x + 4)(x - 2)$

Either students will use trial-and-error by adjusting the value for a to match the vertex point coordinates or will solve the following equation:

$$-3.6 = a(-1 + 4)(-1 - 2)$$
$$-3.6 = a(3)(-3)$$
$$-3.6 = -9a$$
$$\frac{-3.6}{-9} = \frac{-9a}{-9}$$
$$0.4 = a$$

4. a) $y = 0.4x^2 + 0.8x - 3.2$

b)

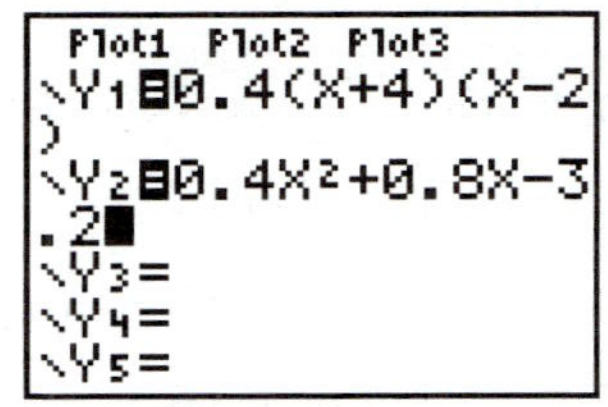

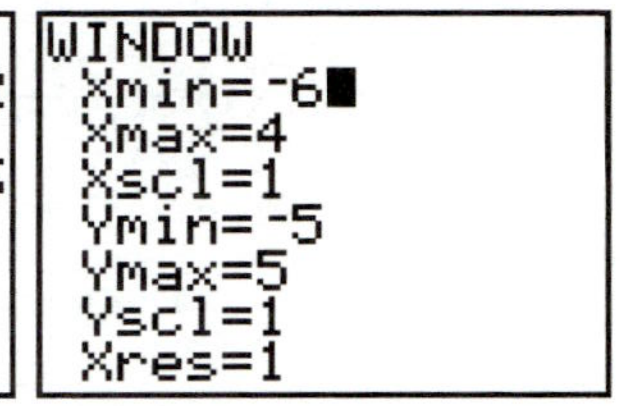

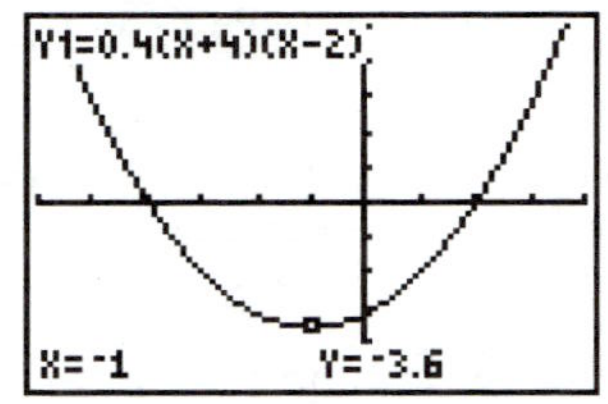

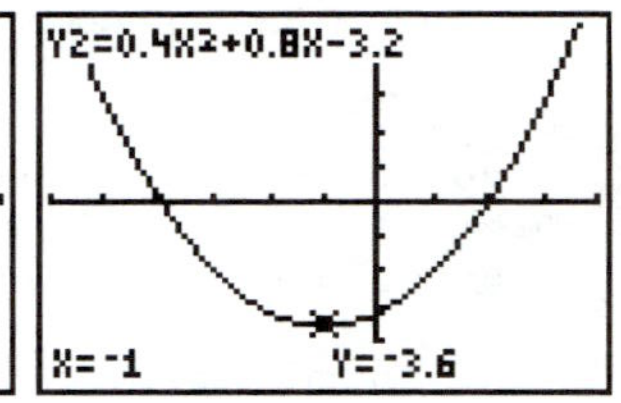

Both graphs cross the x-axis at -4 and 2 and have a vertex at $(-1, -3.6)$.

c) When only one graph appears, it suggests that the two expressions are algebraically equivalent because the second graph is being drawn right on top of the first. (In this case, that's a good thing!)

If a polynomial is factored and the polynomial is equal to zero, then at least one of its factors must also be equal to zero. In other words, if $(x-5)(x-1)=0$, then $x-5=0$ or $x-1=0$. As long as one of these two factors is equal to 0, then that value of x will solve the equation.

Let's take another look at our rocketry example from Section 5.7.

Recall that the function describing our rocket launch simulation from Section 5.6 was $y=-16x^2+80x$. We wrote an equation that could be solved to find the times when the rocket was 64 feet above the ground, $-16x^2+80x=64$.

5. Rewrite this equation in $ax^2+bx+c=0$ form, and completely factor the polynomial.

6. Use the Zero-Product Property to solve the equation.

7. The height (in feet) of a rocket, y, after x seconds of flight can be found using $y=-16x^2+80x$. At what time(s) will the rocket be 84 feet above the ground?

8. What types of problems are best solved with the Zero-Product Property?

SUMMARY

If a quadratic that can be factored equals 0, then the equation can be solved using the Zero-Product Property.

- The Zero-Product Property states that for all real numbers a and b, if $ab=0$, then $a=0$ or $b=0$.

To solve an equation by factoring:

- First, write the quadratic equation in polynomial form, $ax^2+bx+c=0$.
- Second, completely factor the polynomial.
- Third, set each factor equal to zero.
- Fourth, solve each of these equations for x.
- Since each of these x-values makes at least one factor equal to zero, by the Zero-Product Property, each of these x-values solves the equation.

SECTION 5.8 Assignment

1. One way to see factored form is with an area model that represents the various parts of a quadratic expression. The pieces needed to represent $2x^2+7x+3$ are shown in **Figure 5.37**.

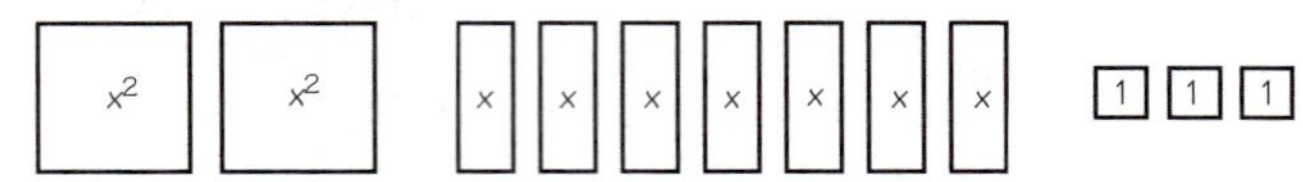

FIGURE 5.37. Area model for a quadratic.

a) What are the dimensions for each type of piece?

b) $2x^2+7x+3$ can be in factored form by arranging the pieces to form a rectangle. On a piece of paper, trace the pieces in Figure 5.37, and then arrange them to form a rectangle. Sketch the results.

c) The length and width of the rectangle represent the factors. What are the factors of $2x^2+7x+3$?

d) Write $y=2x^2+7x+3$ in factored form.

e) Solve $2x^2+7x+3=0$ using the Zero-Product Property.

f) Write $y=4x^2+12x+9$ in factored form.

g) Solve $4x^2+12x+9=0$ using the Zero-Product Property.

2. A flare is launched from a life raft. The relationship between height and time is modeled by $h=192t-16t^2$ where h is height in feet and t is time in seconds. When is the flare 512 feet above the raft?

3. A toy rocket is fired in the middle of a field. The relationship between height and time is modeled by $h=-16t^2+256t$ where h is height in feet and t is time in seconds. When does the rocket's height equal 540 feet?

Answers (continued)

5.

$$-16x^2 + 80x - 64 = 64 - 64$$
$$-16x^2 + 80x - 64 = 0$$
$$-16x^2 + 80x - 64 = 0$$
$$-16(x^2 - 5x + 4) = 0$$
$$-16(x - 4)(x - 1) = 0$$

6. If $-16(x - 4)(x - 1) = 0$, then one of the three factors must be equal to zero. Set each factor equal to zero, then solve for x.

   1. $-16 = 0$
      no solution
   2. $x - 4 = 0$
      $x = 4$
   3. $x - 1 = 0$
      $x = 1$

   So, if $x = 1$ or $x = 4$, then $y = 64$. When the rocket has been in the air for 1 second or 4 seconds, the height of the rocket will be 64 feet above the ground.

7. To solve this problem, first write and completely factor an equation:

$$-16x^2 + 80x = 84$$
$$-16x^2 + 80x - 84 = 0$$
$$-4(4x^2 - 20x + 21) = 0$$
$$-4(2x - 7)(2x - 3) = 0$$

   Next, set each factor equal to zero, then solve that equation. It is not necessary to set constant factors equal to zero because there is no x-value that will make $4 = 0$.

$$2x - 7 = 0 \qquad 2x - 3 = 0$$
$$2x = 7 \qquad 2x = 3$$
$$x = \frac{7}{2} \qquad x = \frac{3}{2}$$

8. Answers may vary. Students should make connections between the roots of a quadratic equation and the solutions. Therefore, students should use factoring and the Zero-Product Property when they encounter problems that give information about the roots of the quadratic equation. Also, if a polynomial will factor, the Zero-Product Property will yield a solution more quickly than other methods that will be explored in subsequent sections.

Section 5.8 Assignment

Answers

1. a) The large square is x by x; the small square is 1 by 1; the rectangles are x by 1.

   b) Sample answer:

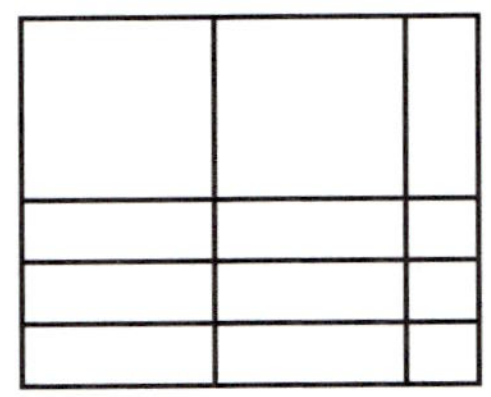

   c) $(x + 3)$ and $(2x + 1)$

   d) $y = (x + 3)(2x + 1)$

   e)

$$2x^2 + 7x + 3 = 0$$
$$(2x + 1)(x + 3) = 0$$

$$2x + 1 = 0$$
$$2x + 1 - 1 = 0 - 1$$
$$2x = -1$$
$$\frac{2x}{2} = \frac{-1}{2}$$
$$x = \frac{1}{2}$$

$$x + 3 = 0$$
$$x + 3 - 3 = 0 - 3$$
$$x = -3$$

   f) $y = (2x + 3)^2$

   g) $4x^2 + 12x + 9 = 0$
   $(2x + 3)^2 = 0$

   Since this polynomial factors to a binomial squared, there is only one root of the equation, and it is a double root.

$$2x + 3 = 0$$
$$2x + 3 - 3 = 0 - 3$$
$$2x = -3$$
$$\frac{2x}{2} = \frac{-3}{2}$$
$$x = -\frac{3}{2}$$

2.

$$192t - 16t^2 = 512$$
$$-16t^2 + 192t - 512 = 0$$
$$-16(t^2 - 12t + 32) = 0$$
$$-16(t - 8)(t - 4) = 0$$

$$t - 8 = 0 \qquad t - 4 = 0$$
$$t = 8 \qquad t = 4$$

   The flare's height is exactly 512 feet at 4 and 8 seconds.

3.

$$-16t^2 + 256t = 540$$
$$-16t^2 + 256t - 540 = 0$$
$$-4(4t^2 - 64t + 135) = 0$$
$$-4(2t - 27)(2t - 5) = 0$$

$$2t - 27 = 0 \qquad 2t - 5 = 0$$
$$2t = 27 \qquad 2t = 5$$
$$t = \frac{27}{2} \qquad t = \frac{5}{2}$$

   The rocket will be 540 feet above the ground at 2.5 seconds and again 13.5 seconds after launch.

4. A rectangle has a length, l, and a width, w. The area of a rectangle can be found using $A = lw$. The perimeter of a rectangle can be found using $P = 2l + 2w$. The perimeter of a certain rectangle is 48 centimeters.

   a) Use the perimeter formula to write a function rule for the length of the rectangle in terms of its width.

   b) Write a function for the area of the rectangle in terms of its width.

   c) If the rectangle has an area of 143 square centimeters, what are its length and width?

SECTION 5.9 So What's Your Point?

Quadratic functions can be used to model many things that occur in the real world. In Section 5.6 you used a graphing calculator to simulate a rocket launch.

Suppose that a rocket launch can be simulated with $y = -16x^2 + 128x$, where x represents the time since launch in seconds and y represents the height of the rocket in feet.

1. Use your graphing calculator to graph the function. Sketch your graph and record your window.

2. Use the TRACE feature to estimate the coordinates of the vertex.

3. Use your graphing calculator to find the location of the vertex. How did you find your answer?

4. Recall that the vertex form of a quadratic function is $y = a(x - h)^2 + k$. Use the vertex you found in Question 3 to write a vertex form for this function.

5. Use transformations to fit an appropriate function, in vertex form, to the parabola. Use your graphing calculator to confirm your answer.

6. Use algebra to show that your function from Question 5 is equivalent to $y = -16x^2 + 128x$.

7. Here is a quadratic expression: $2(x + 3)^2 + 2$.

   a) List the operations you would use, in order, to evaluate this expression. Then evaluate the expression for $x = 4$.

   b) Knowing the order of operations can help in solving equations. Use your answer from a to list the opposite operations (i.e., opposite of multiplication is division).

Answers (continued)

4. a)

$$P = 2l + 2w$$
$$48 = 2l + 2w$$
$$48 - 2w = 2l + 2w - 2w$$
$$48 - 2w = 2l$$
$$\frac{48-2w}{2} = \frac{2l}{2}$$
$$24 - w = l$$

b) $A = lw$

$A = (24 - w)w$

c)

$$A = (24 - w)w$$
$$143 = 24w - w^2$$
$$143 - 24w + w^2 = 24w - 24w - w^2 + w^2$$
$$w^2 - 24w + 143 = 0$$
$$(w - 11)(w - 13) = 0$$

$$(w - 11)(w - 13) = 0$$
$$w - 11 = 0 \quad w - 13 = 0$$
$$w = 11 \quad w = 13$$

11 cm and 13 cm are the dimensions of a rectangle with an area of 143 square centimeters and a perimeter of 48 cm. One is the length; the other is the width.

EXPLORE LESSON 2

Section 5.9: So What's Your Point?

Purpose: In Sections 5.7 and 5.8, students explored situations and formalized methods to solve quadratic equations by factoring. In Section 5.9 students will explore situations that will require a quadratic equation to be written in vertex form. In Section 5.10 students will formalize the mechanics of solving a quadratic equation by first writing it in vertex form, then applying inverse operations, to include taking the square root.

Description: Students will be posed with several questions that lead to the writing of quadratic equations in vertex form. These questions will develop the need to solve quadratic equations by using inverse operations.

Materials: graphing calculator

Teaching Ideas: Students should work in small groups (3 to 4 students). This section is designed for students to make connections between the rocket launch simulation and the vertex of the graph of the parabola that describes the motion of the rocket. These connections will lead to a practical use for the vertex form of a quadratic function, which was developed in Section 5.4.

In this lesson students will be required to identify the vertex of a parabola and to use the vertex form of a quadratic equation. Students may need to review using the graphing calculator to identify the vertex of a parabola or using the vertex form, depending on their skill and fluency levels.

Answers

1.

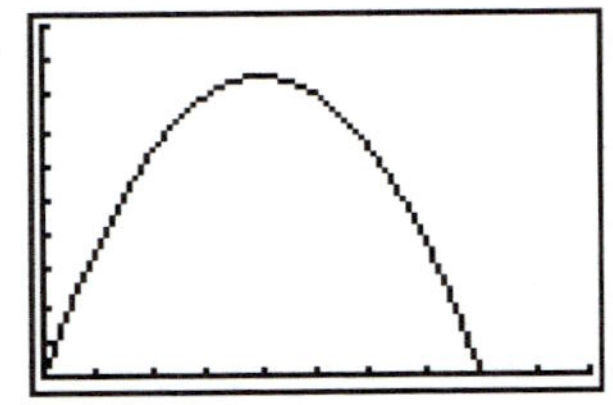

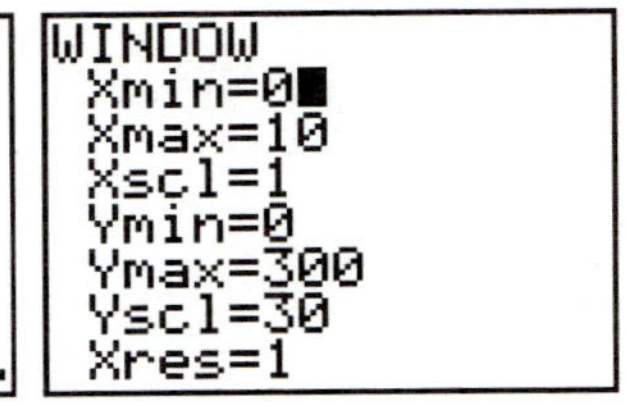

2. The vertex is near the point (4, 256).

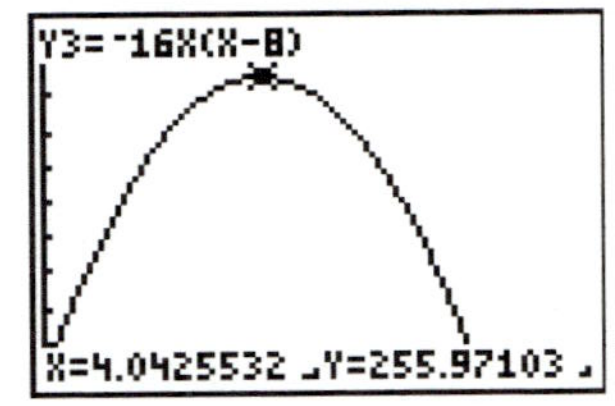

3. The vertex of the parabola is the point (4, 256). Explanations may vary, but a sample explanation could be:

Use the 2nd-CALC feature of the graphing calculator to calculate the maximum value of the parabola. First, press 2nd-CALC, then select "maximum."

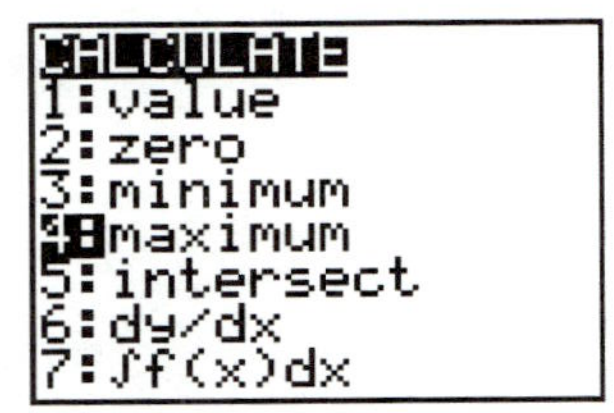

Answers (continued)

Next, use the right and left arrow keys to select the left bound to the left of where you think the vertex is, then the right bound to the right of where you think the vertex is. Make a guess near the vertex, then press ENTER.

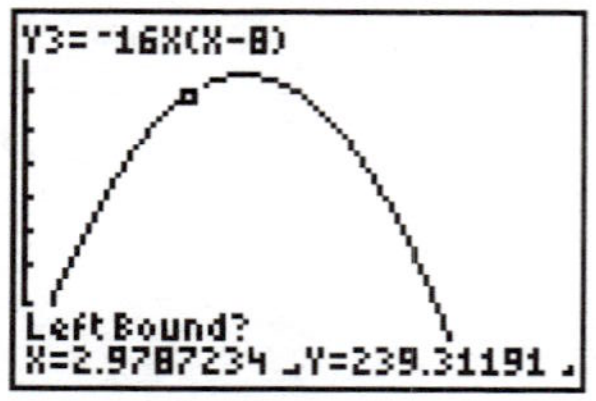

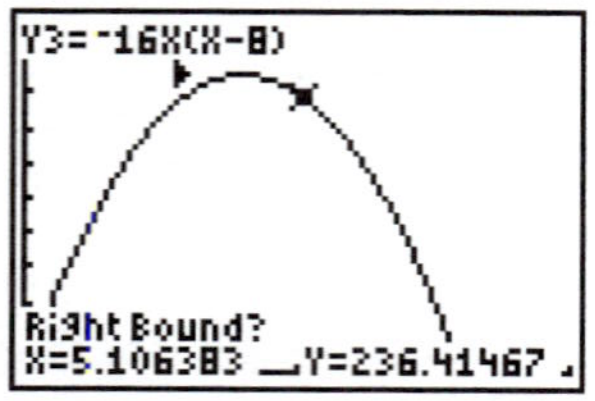

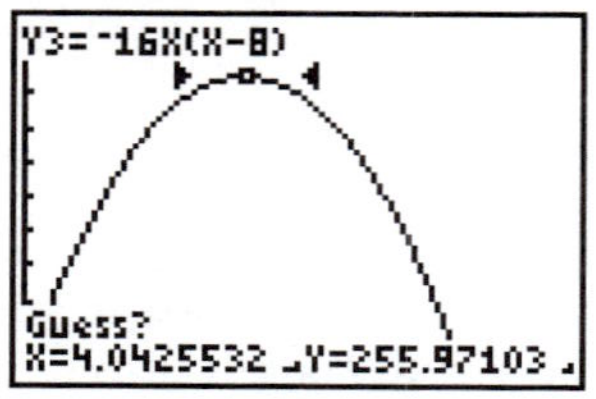

The calculator yields the value (3.99999, 256), which rounds to (4, 256).

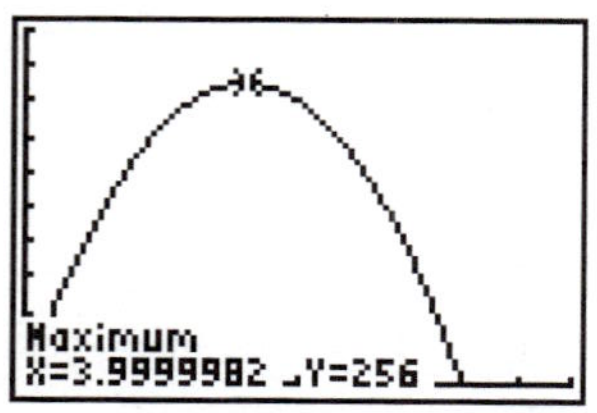

4. Since the vertex is (4, 256), $h = 4$ and $k = 256$. Substituting these values into the vertex form equation yields $y = a(x-4)^2 + 256$.

5. Since the parabola opens downward (is concave down), $a < 0$. The original equation contains a value of –16, so it makes sense to try $a = -16$. Thus, $y = -16(x-4)^2 + 256$ would be a good parabola to try initially. The graphing calculator confirms that this parabola matches the original parabola.

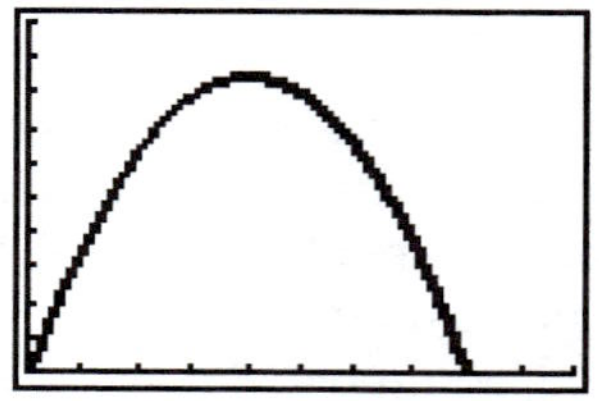

6. $y = -16(x-4)^2 + 256$
$y = -16(x^2 - 8x + 16) + 256$
$y = -16x^2 + 128x - 256 + 256$
$y = -16x^2 + 128x$

7. a) Addition $(4 + 3 = 7)$
Exponents $(7^2 = 49)$
Multiplication $(2(49))$
Addition $(98 + 2 = 100)$

b) Addition $(4 + 3 = 7)$ → Subtract 3
Exponents $(7^2 = 49)$ → Take the square root
Multiplication $(2(49))$ → Divide by 2
Addition $(98 + 2 = 100)$ → Subtract 2

c) If you want to solve $2(x + 3)^2 + 2 = 20$ you can start with 20 and apply the operations that you listed in b in reverse order. What numbers do you get? (Hint: there are two answers!)

d) To formalize your work, apply the same steps to both sides of $2(x + 3)^2 + 2 = 20$, one at a time, in the same order you used in c. Write down what the equation becomes after each of the steps.

8. a) A ball is shot straight up from a cannon. After 6 seconds, it reaches a height of 576 feet. Using $a = -16$, write a function in vertex form that can be used to model this rocket's flight.

b) Use this function to find the height of the ball 4 seconds after launch.

c) At what time will the ball be 320 feet above the ground?

SECTION 5.10

Solving Quadratic Equations Using Inverse Operations

There are many ways to solve a quadratic equation. The best method depends on the form of the equation. Section 5.9 uses vertex form.

The Tower of the Americas in San Antonio, Texas, was built for the 1968 World's Fair. At 750 feet it is taller than both the Washington Monument in Washington, DC and the Space Needle in Seattle, Washington.

1. A penny is dropped from the top of the Tower of the Americas. The height of the penny, h, after t seconds can be described by $h = -16t^2 + 750$. At about what time will the penny be 100 feet from the ground?

One way to look at the vertex form, $y = a(x - h)^2 + k$, is by transformations from Section 5.4. Vertex form is the result of starting with the parent function, $y = x^2$, stretching or compressing it by a factor of a, reflecting it if $a < 0$, and then translating it horizontally h units and vertically k units. In each translation the only point that needs to be tracked is the vertex. The dilation (compression or stretch) affects how quickly or slowly the graph grows as the values for x increase or decrease.

Another way to look at the vertex form is through order of operations. They are shown as an arrow diagram in **Figure 5.38.**

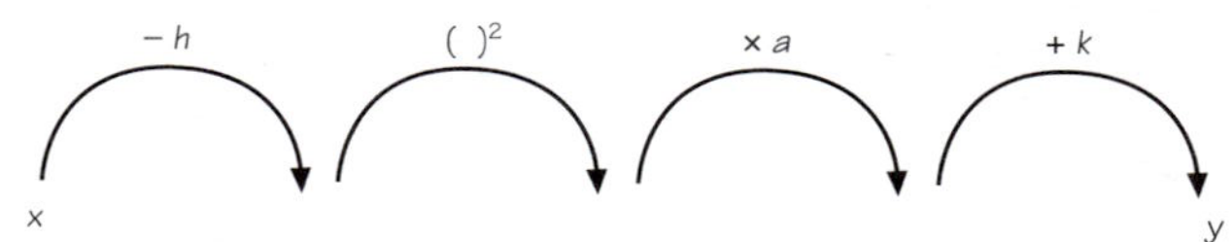

FIGURE 5.38.
Arrow diagram.

Because the operations can be done in sequence, one at a time, vertex form can be useful for solving equations using inverse operations.

Answers (continued)

c) Subtract 2: $20 - 2 = 18$
Divide by 2: $18 \div 2 = 9$
Take the square root: $\sqrt{9} = \pm 3$
Subtract 3: $+3 - 3 = 0$
$-3 - 3 = -6$

d) Original equation: $2(x + 3)^2 + 2 = 20$
Subtract 2: $2(x + 3)^2 + 2 - 2 = 20 - 2$
$2(x + 3)^2 = 18$
Divide by 2: $\frac{2(x+3)^2}{2} = \frac{18}{2}$
$(x + 3)^2 = 9$
Take the square root: $\sqrt{(x+3)^2} = \sqrt{9}$
$x + 3 = \pm 3$
Subtract 3: $x + 3 - 3 = \pm 3 - 3$
$x = \pm 3 - 3$

$x = 3 - 3 \quad x = -3 - 3$
$x = 0 \quad x = -6$

8. a) $y = -16(x - 6)^2 + 576$

b) Substitute $x = 4$ into the equation and simplify.

$y = -16(x - 6)^2 + 576$
$y = -16((4) - 6)^2 + 576$
$y = -16(-2)^2 + 576$
$y = 512$

c) Substitute $y = 320$ into the equation and solve for x:

$y = -16(x - 6)^2 + 576$
$320 = -16(x - 6)^2 + 576$
$320 - 576 = -16(x - 6)^2 + 576 - 576$
$-256 = -16(x - 6)^2$
$\frac{-256}{-16} = \frac{-16(x-6)^2}{-16}$
$16 = (x - 6)^2$
$\sqrt{16} = \sqrt{(x-6)^2}$
$\pm 4 = x - 6$

$\pm 4 = x - 6$
$+4 = x - 6 \quad -4 = x - 6$
$4 + 6 = x \quad -4 + 6 = x$
$10 = x \quad 2 = x$
$x = 2, 10$

EXPLAIN LESSON 2

Section 5.10: Solving Quadratic Equations Using Inverse Operations

Purpose: In Section 5.9 students used the vertex form of a quadratic equation to explore the operations that are being performed in evaluating a quadratic expression in vertex form and how the inverse operations can be used to solve a quadratic equation. In this section students will formalize these relationships and apply them to solve quadratic equations using inverse operations. In the next few sections students will explore and formalize other methods of solving quadratic equations so that by Section 5.13, students will be able to select and defend a method of solving a quadratic equation in a problem-solving situation.

Description: In this section students will formalize the relationship between the vertex form of a quadratic equation and the operations that are being used in that form. Students will also generalize the inverse operations in order to develop an algorithm that can always be used to solve quadratic equations once they are in vertex form. This algorithm will give students another tool with which to solve quadratic equations. Students will also look at contextual problems and solve them.

Materials: graphing calculators

Teaching Ideas: Debrief students after Section 5.9 by having them share their answers to the questions. As they share, be sure to fill in any missing concept, process, or skill that could have been used to answer the questions. Then work through the questions in this section with the entire class. Once again be sure all necessary concepts, processes, or skills are addressed.

Answers

1. Substitute $h = 100$, and solve for t:

$h = 16t^2 + 750$
$100 = -16t^2 + 750$
$100 - 750 = -16t^2 + 750 - 750$
$-650 = -16t^2$
$\frac{-650}{-16} = \frac{-16t^2}{-16}$
$40.625 = t^2$
$\sqrt{40.625} = \sqrt{t^2}$
$\pm 6.37 \approx t$

2. $y = a(x - h)^2 + k$, if you know the values of y, a, h, and k, describe how to solve the equation for x. (Hint: Look back at work in Question 1.)

Every year for New Year's Eve, a fireworks display is launched from the Tower of the Americas. Suppose that a firework is launched from the top of the tower. Its height, h, above the ground can be found from the time, t, since launch using $h(t) = -16t^2 + 64t + 750$.

Planners need to know how high the firework will go before it begins to descend. Also, they must know how long the firework will take to reach this height. One way that we can answer these questions is to rewrite the function in vertex form using a method called **completing the square**.

Before using completing the square to answer these questions, let's consider a simpler problem.

Algebra	Tiles
a) Represent the function using tiles. $f(x) = x^2 + 4x + 1$	
b) Group the x-terms using parentheses. $f(x) = (x^2 + 4x) + 1$	
c) Find $\frac{1}{2}$ of the coefficient of x. Square this value, then add it to your x-terms inside the parentheses. $f(x) = (x^2 + 4x + \_\_\_) + 1$ $\frac{4}{2} = 2$ $2^2 = 4$	
d) You want to manipulate the equation, not change its value. So, take the value from step c) that you added and subtract it outside the parentheses. Factor your trinomial, and combine like terms to write the vertex form. *To add 4, then, we also need to subtract 4.* $f(x) = (x^2 + 2x + 4) + 1$ $f(x) = (x + 2)^2 - 4 + 1$ *Combine like terms.* $f(x) = (x + 2)^2 - 3$	

FIGURE 5.39. Table for Question 3.

3. Write $f(x) = x^2 + 4x + 1$ in vertex form. To do this, use algebra tiles to represent the polynomial. In your notebook, make a table like **Figure 5.39** to record the steps and sketch the tiles.

4. Write $g(x) = 2x^2 + 4x + 3$ in vertex form. To do this, use algebra tiles to represent the polynomial. In your notebook, make a table like **Figure 5.40** to record the steps and sketch the tiles.

Algebra	Tiles
a) Group the x-terms: $g(x) = (2x^2 + 4x) + 3$	
b) We just followed a procedure for completing the square when $a = 1$, so let's factor 2 out of both of the x-terms. $g(x) = 2(x^2 + 2x) + 3$	
c) Find $\frac{1}{2}$ of the coefficient of x. Square this value, then add it to your x-terms inside the parentheses. $g(x) = 2(x^2 + 2x + \_\_\_) + 3$ $\frac{2}{2} = 1$ $1^2 = 1$	
d) You want to manipulate the equation, not change its value. So, take the value from step c) that you added and subtract it outside the parentheses. Factor your trinomial, and combine like terms to write the vertex form. *To add 1, we need to subtract 1. Don't forget, however. We are adding TWO groups of 1, or 2, and need to subtract 2 to compensate.* $g(x) = 2(x + 1)^2 - 2 + 3$ $g(x) = 2(x + 1)^2 + 1$	

FIGURE 5.40. Table for Question 4.

Answers (continued)

2.

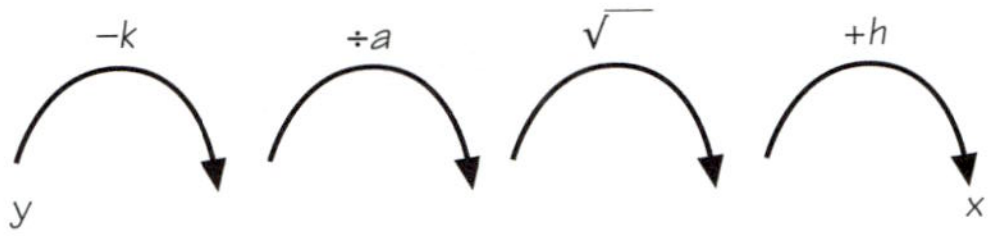

Start with y; subtract k, divide by a, take the square root of that answer, and then add h.

3.

Algebra	Tiles
a) Represent the function using tiles. $f(x) = x^2 + 4x + 1$	
b) Group the x-terms using parentheses. $f(x) = (x^2 + 4x) + 1$	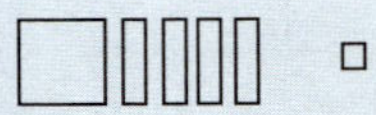
c) Find $\frac{1}{2}$ of the coefficient of x. Square this value, then add it to your x-terms inside the parentheses. $f(x) = (x^2 + 4x + \_\_\_) + 1$ $\frac{4}{2} = 2$ $2^2 = 4$	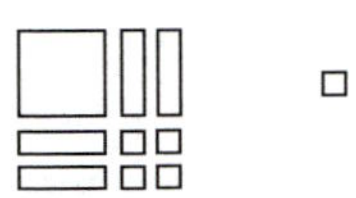
d) You want to manipulate the equation, not change its value. So, take the value from step c) that you added and subtract it outside the parentheses. Factor your trinomial, and combine like terms to write the vertex form. *To add 4, then, we also need to subtract 4.* $f(x) = (x^2 + 2x + 4) + 1$ $f(x) = (x + 2)^2 - 4 + 1$ *Combine like terms.* $f(x) = (x + 2)^2 - 3$	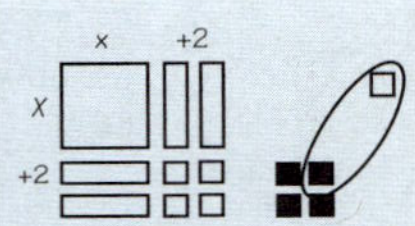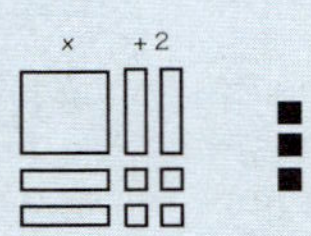

4.

Algebra	Tiles
a) Group the x-terms: $g(x) = (2x^2 + 4x) + 3$	
b) We just followed a procedure for completing the square when $a = 1$, so let's factor 2 out of both of the x-terms. $g(x) = 2(x^2 + 2x) + 3$	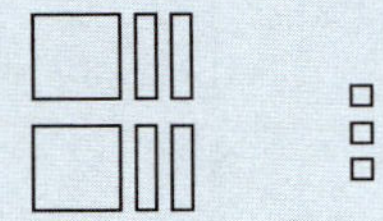
c) Find $\frac{1}{2}$ of the coefficient of x. Square this value, then add it to your x-terms inside the parentheses. $g(x) = 2(x^2 + 2x + \_\_\_) + 3$ $\frac{2}{2} = 1$ $1^2 = 1$	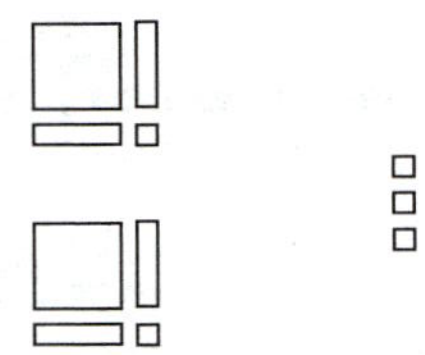
d) You want to manipulate the equation, not change its value. So, take the value from step c) that you added and subtract it outside the parentheses. Factor your trinomial, and combine like terms to write the vertex form. *To add 1, we need to subtract 1. Don't forget, however. We are adding TWO groups of 1, or 2, and need to subtract 2 to compensate.* $g(x) = 2(x + 1)^2 - 2 + 3$ $g(x) = 2(x + 1)^2 + 1$	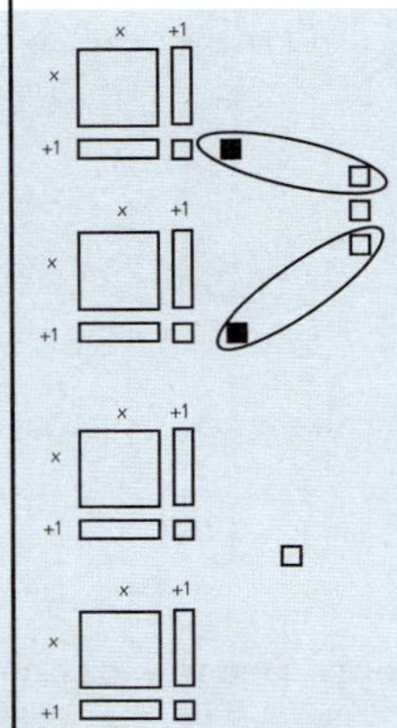

Completing the square can be thought of in two ways.

Geometrically, you are physically completing a square. The tiles reveal this process. In Question 3 we were completing the square for $x^2 + 4x$. We built what we could out of algebra tiles then asked ourselves, "What number of unit tiles do I need to complete the square?" The answer is 4 tiles (see **Figure 5.41**). The factors of the resulting polynomial are $(x + 2)$ and $(x + 2)$.

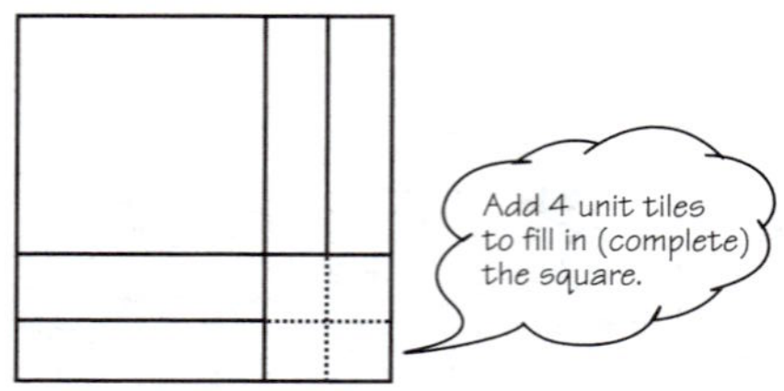

FIGURE 5.41. Completing the square geometrically.

Algebraically, you are completing a binomial perfect square. In Question 3, we were completing the square for $x^2 + 4x$. We want to write this expression as a binomial squared; that is, a polynomial of the form $(x + \#)^2$.

$x^2 + 4x +$ _____

$x^2 + 4x + \left(\frac{4}{2}\right)^2$

$x^2 + 4x + 4$

$(x + 2)(x + 2)$

$(x + 2)^2$

Solving a quadratic equation by completing the square has two parts.

Part 1. Rewrite the polynomial in vertex form using completing the square.

Part 2. Use inverse operations to isolate the variable.

EXAMPLE:

Solve $x^2 - 3x = 10$ by completing the square.

Step 1: The left-hand side needs to be written in vertex form. Since $-3 \div 2 = -1.5$, and $(-1.5)^2 = 2.25$, add 2.25 to both sides of the equation.

$x^2 - 3x + 2.25 = 10 + 2.25$

Step 2: Rewrite the left-hand side as a binomial squared.

$(x^2 - 3x + 2.25) = 12.25$

$(x - 1.5)^2 = 12.25$

Step 3: Use inverse operations to solve for x.

Take the square root of each side:

$\sqrt{(x-1.5)^2} = \sqrt{12.25}$

$x - 1.5 = \pm 3.5$

Add 1.5 to each side of the equation.

$x - 1.5 = \pm 3.5$

$x - 1.5 + 1.5 = \pm 3.5 + 1.5$

Simplify.

$x = \pm 3.5 + 1.5$

$x = 3.5 + 1.5 \qquad x = -3.5 + 1.5$

$x = 5 \qquad x = 2$

SUMMARY

If a quadratic equation is in vertex form, $y = a(x - h)^2 + k$, then it can be solved by using inverse operations.

- Start with y.
- Subtract k.
- Divide by a.
- Take the square root.
- Add h.

Sometimes a quadratic equation can be solved by a method called **completing the square**. To solve $x^2 + bx + c = 0$ using this method:

- Write the equation in the form $x^2 + bx = -c$.
- Add $\left(\frac{b}{2}\right)^2$ to both sides:

$$x^2 + bx + \left(\frac{b}{2}\right)^2 = -c + \left(\frac{b}{2}\right)^2$$

- Rewrite the left side as a perfect square:

$$\left(x + \frac{b}{2}\right)^2 = -c + \left(\frac{b}{2}\right)^2$$

- Take the square root of each side and solve for x.

SECTION 5.10 Assignment

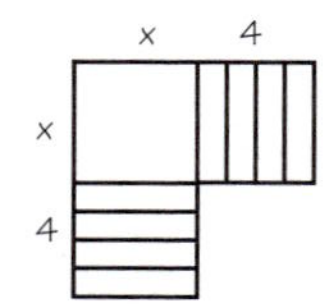

FIGURE 5.42.
An incomplete area model.

1. **Figure 5.42** is an incomplete area model. Complete the square.

   a) What algebraic expression does Figure 5.42 represent?

   b) How many small squares (area of each = 1) do you need to add to complete the square?

   c) What algebraic expression does the model then represent?

   d) What algebraic expression represents the sum of the pieces being used to complete the square?

2. Complete each square, then write the polynomial as a binomial squared.

   a) $x^2 - 16x +$ _____

   b) $x^2 + 7x +$ _____

3. Write $h(x) = 3x^2 - 12x - 1$ in vertex form. Show your work in a table like **Figure 5.43**.

Algebra	Sketch of Tiles
a) Group the x-terms using parentheses.	
b) Factor out a 3 from both the x^2 and x-terms.	
c) Find $\frac{1}{2}$ of the coefficient of x. Square this value, then add it to your x-terms inside the parentheses.	
d) You want to convert the equation, not change its value. So, take the value from step c) that you added, multiply it by the 3 that you factored out (distribute, distribute!), and subtract it outside the parentheses. Factor your trinomial, and combine like terms to write the vertex form.	

FIGURE 5.43.
Table for Question 3.

Section 5.10 Assignment

Answers

1. a) $x^2 + 8x$

   b) 16

   c) $(x + 4)^2$

   d) $x^2 + 8x + 16$

2. a) 64; $(x - 8)^2$

   b) $\frac{49}{4}$; $\left(x + \frac{7}{2}\right)^2$

3.

Algebra	Sketch of Tiles
a) Group the x-terms using parentheses. $h(x) = (3x^2 - 12x) - 1$	
b) Factor out a 3 from both the x^2 and x-terms. $h(x) = 3(x^2 - 4x) - 1$	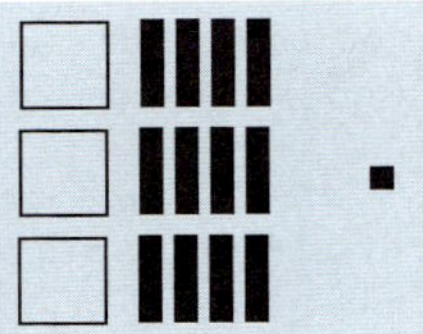
c) Find $\frac{1}{2}$ of the coefficient of x. Square this value, then add it to your x-terms inside the parentheses. $h(x) = 3(x^2 - 4x + \_\_\_) - 1$; $\frac{-4}{2} = -2$; $(-2)^2 = 4$	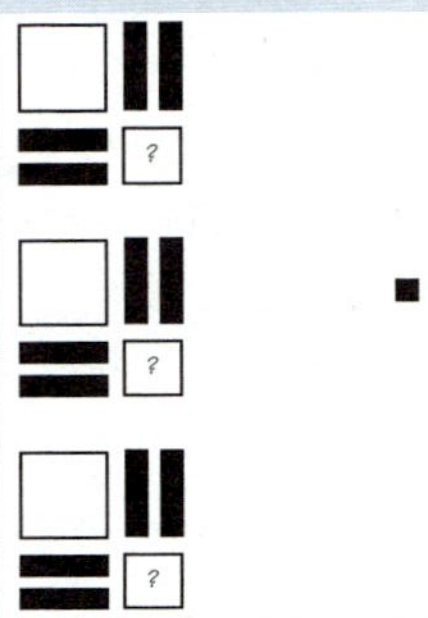
d) You want to convert the equation, not change its value. So, take the value from step c) that you added, multiply it by the 3 that you factored out (distribute, distribute!), and subtract it outside the parentheses. Factor your trinomial, and combine like terms to write the vertex form. *To add 4, we need to subtract 4. Don't forget, however. We are adding THREE groups of 4, or 12, and need to subtract 12 to compensate.* $h(x) = 3(x^2 - 4x + 4) - 12 - 1$; $h(x) = 3(x + 2)^2 - 13$	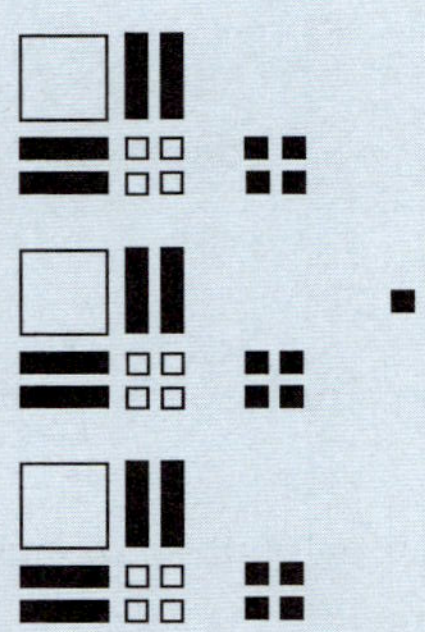

4. Use completing the square to write $f(x) = \frac{1}{2}x^2 + 2x + 3$ in vertex form.

For Questions 5–7, solve by completing the square and inverse operations.

5. $x^2 + 4x = 12$

6 $x^2 + 8x - 11 = 0$

7. $5x^2 - 10x - 30 = 0$

8. A football is kicked into the air. It hits the ground 60 yards downfield. At its highest point, the ball is 7 yards above the ground. Assume the relationship between height and distance is quadratic.

a) Sketch a graph of the relationship between height and distance.

b) Label the vertex with its coordinates. What does this point mean here?

c) Label the x-intercepts with their coordinates. What do these points mean here?

d) Substitute the coordinates of the vertex into the vertex form $y = a(x - h)^2 + k$.

e) To find the value of a, choose a known point on the parabola and substitute the values for x and y. Solve for a.

f) Write a function rule to model the relationship between the height of the ball and the distance downfield.

g) Assume the ball is being kicked for a 51-yard field goal. The height of the field goal crossbar is 10 feet. Will the ball make it through the goal posts? Show how you found your answer.

SECTION 5.11 The Quadratic Safety Net

You have seen three ways to write a quadratic function:

Polynomial Form: $y = ax^2 + bx + c$

Factored Form: $y = a(x - r_1)(x - r_2)$

Vertex Form: $y = a(x - h)^2 + k$

You have also seen two ways to solve quadratic equations. The first method, factoring, involves applying the Zero-Product Property to solve each factor for x. If the quadratic is in polynomial form, it must first be factored in order to use this method. The second method, using inverse operations, involves applying the inverse operations to the equation to isolate x, which includes taking the square root to "undo" the quadratic nature of the equation. If the quadratic is in polynomial form, it must first be expressed as a binomial squared, which usually involves completing the square.

1. Consider $y = x^2 + 4x + 3.75$.

a) Does the polynomial $x^2 + 4x + 3.75$ factor? How can you tell? If so, what are the factors?

b) Use your graphing calculator to graph the function $y = x^2 + 4x + 3.75$. Sketch your graph and record your window.

c) What are the zeros of the function? How do you know?

d) Based on your answers so far, what is the solution to $x^2 + 4x + 3.75 = 0$?

e) Now, consider $y = x^2 + 5x - 3$. Can the function be written in factored form? If so, write it in factored form. If not, why not?

f) Graph $y = x^2 + 5x - 3$ on your graphing calculator. Sketch your graph and record your window.

Answers (continued)

4. $f(x) = \frac{1}{2}x^2 + 2x + 3$

$f(x) = \frac{1}{2}(x^2 + 4x) + 3$

$f(x) = \frac{1}{2}\left(x^2 + 4x + \left(\frac{4}{2}\right)^2\right) + 3 - \frac{1}{2}\left(\frac{4}{2}\right)^2$

$f(x) = \frac{1}{2}(x + 2)^2 + 3 - 2$

$f(x) = \frac{1}{2}(x + 2)^2 + 1$

5. $x^2 + 4x + \_\_\_ = 12 + \_\_\_$

$x^2 + 4x + \left(\frac{4}{2}\right)^2 = 12 + \left(\frac{4}{2}\right)^2$

$x^2 + 4x + 4 = 12 + 4$

$(x + 2)^2 = 16$

$\sqrt{(x+2)^2} = \sqrt{16}$

$x + 2 = \pm 4$

$x + 2 = \pm 4$

$x + 2 = 4 \qquad x + 2 = -4$

$x + 2 - 2 = 4 - 2 \qquad x + 2 - 2 = -4 - 2$

$x = 2 \qquad x = -6$

6. $x^2 + 8x - 11 + 11 = 0 + 11$

$x^2 + 8x = 11$

$x^2 + 8x + \left(\frac{8}{2}\right)^2 = 11 + \left(\frac{8}{2}\right)^2$

$x^2 + 8x + 16 = 11 + 16$

$(x + 4)^2 = 27$

$\sqrt{(x+4)^2} = \sqrt{27}$

$x + 4 = \pm\sqrt{27}$

$x = -4 \pm \sqrt{27}$

7. $5x^2 - 10x = 30$

$5(x^2 - 2x) = 30$

$5\left(x^2 - 2x + \left(\frac{2}{2}\right)^2\right) = 30 + 5\left(\frac{2}{2}\right)^2$

$5(x^2 - 2x + 1) = 30 + 5$

$5(x - 1)^2 = 35$

$\frac{5(x-1)^2}{5} = \frac{35}{5}$

$(x - 1)^2 = 7$

$\sqrt{(x-1)^2} = \sqrt{7}$

$x - 1 = \pm\sqrt{7}$

$x = 1 \pm \sqrt{7}$

8. a)

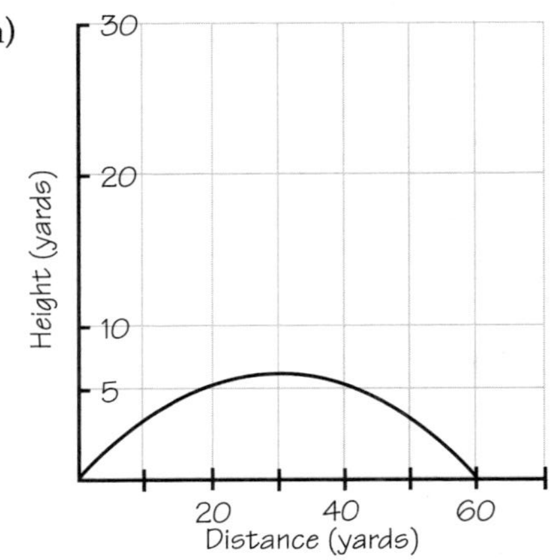

b) Vertex: (30, 7). At 30 yards downfield from where the ball is kicked, it is 7 yards high.

c) x-intercepts (0, 0) is where the ball is kicked. (60, 0) is where the ball lands downfield relative to where it is kicked.

d) $h(d) = a(d - 30)^2 + 7$

e) $-\frac{7}{900} = a$; solutions may vary, depending on which point the students choose. For the point (60, 0), the following is the solution to find a.

$h(d) = a(d - 30)^2 + 7$

$0 = a(60 - 30)^2 + 7$

$0 = a(30)^2 + 7$

$0 = a(900) + 7$

$0 - 7 = 900a + 7 - 7$

$-7 = 900a$

$\frac{-7}{900} = \frac{900a}{900}$

$-\frac{7}{900} = a$

f) Any of the following equations is a correct representation of the path of the football.

$h(d) = \frac{-7}{900}(d - 30)^2 + 7$

$h(d) = \frac{-7}{900}d(d - 60)$

$h(d) = \frac{-7}{900}d^2 + \frac{7}{15}d$

g) Yes, the ball will clear the goal posts. At a distance of 51 yards downfield from where the ball is kicked, it will be 3.57 yards or 10.71 feet off of the ground. The height of the ball can be found by substituting $d = 51$ into the equation that models the height of the ball and simplifying the expression. For example,

$h(d) = \frac{-7}{900}(d - 30)^2 + 7$

$h(51) = \frac{-7}{900}(51 - 30)^2 + 7$

$h(51) = \frac{-7}{900}(21)^2 + 7$

$h(51) = \frac{-7}{900}(441) + 7$

$h(51) = -3.43 + 7$

$h(51) = 3.57$

EXPLORE LESSON 2

Section 5.11: The Quadratic Safety Net

Purpose: In Sections 5.7 through 5.10, students explored and formalized two methods to solve quadratic equations: factoring and inverse operations, including completing the square. In Section 5.11 students will explore situations that use polynomial form to describe quadratic functions and motivate a need for the Quadratic Formula. In Section 5.12 students will formalize the Quadratic Formula and its implications for determining solutions of quadratic functions.

Description: Previously, students graphed quadratic functions in polynomial form, then used the graphing calculator to identify the zeros of the function. These zeros were related to the factored form of a quadratic function. In this section students will explore a situation where the zeros of a function are not easily found numerically. This will build the need for an alternate method to find zeros of a quadratic function when the numbers are not clean (i.e., when the numbers are irrational). This alternate method is the Quadratic Formula.

Materials: graphing calculator

Teaching Ideas: Students should work in small groups (3 to 4 students). This section is designed for students to recognize that factoring will work in some cases, but not all. Students should recognize that sometimes quadratic equations have messy solutions, and we need some kind of rule that will help us to solve any quadratic equation. This rule is the Quadratic Formula.

The Quadratic Formula will be stated here, but the derivation will be left to the Explain phase in Section 5.12.

In this lesson students will be required to use the graphing calculator to find roots of quadratic equations. Depending on their skill level, some students may require extra assistance with the technology.

Answers

1. a) Answers may vary. In order to determine the factors of the polynomial, we need to find two numbers whose sum is 4 and whose product is 3.75.

   Students may think that the polynomial does not factor since the constant term is 3.75, and factoring is traditionally taught using integers instead of decimals. For the purposes of this phase of the lesson, this is an acceptable response. Students will have the opportunity to revisit their answer later in this section.

   Mathematically speaking, the polynomial $x^2 + 4x + 3.75$ factors to $(x + 2.5)(x + 1.5)$.

   b)

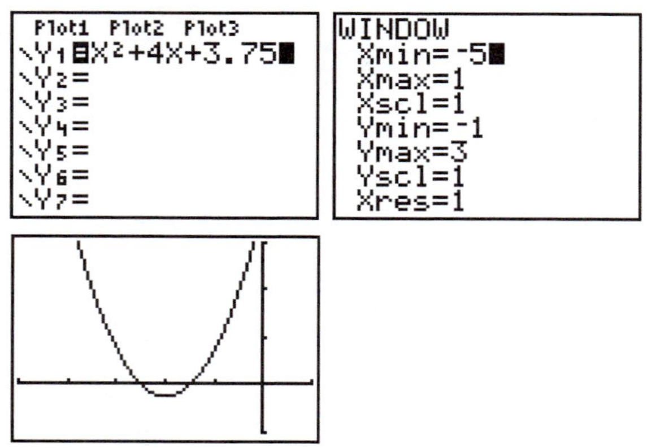

   c) The roots of the function are $x = -1.5$ and $x = -2.5$. They can be found using the TRACE feature, or the CALC feature of a graphing calculator.

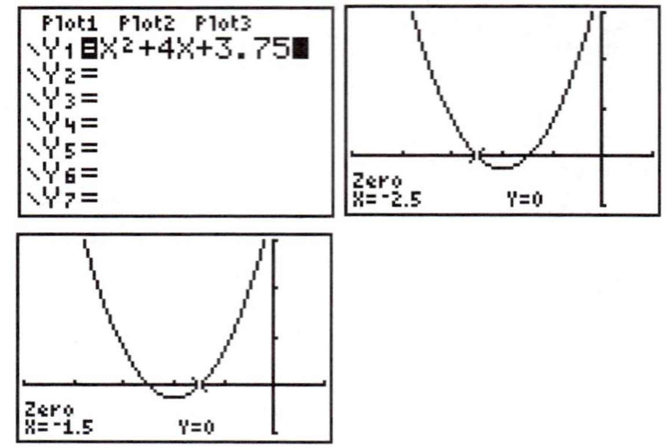

   d) Since the roots of the function are $x = -1.5$ and $x = -2.5$, those x-values are also the solution to the equation $x^2 + 4x + 3.75 = 0$.

   e) No. There are no factors of –3 that will sum to 5.

   f)

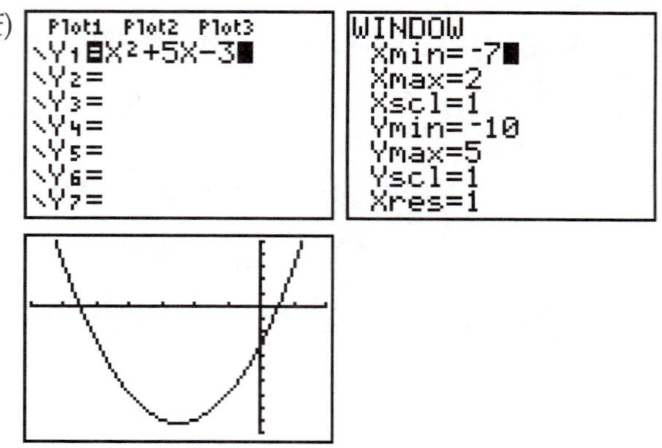

g) Find the zeros of $y = x^2 + 5x - 3$. Explain your process.

h) Based on your answers, what is the solution to $x^2 + 5x - 3 = 0$? Are these exact or approximate? How do you know?

i) Solve $x^2 + 5x - 3 = 0$ without using a graphing calculator. Leave your answer in exact form.

SECTION 5.12 Solving Quadratic Equations Using the Quadratic Formula

The method of completing the square, followed by applying inverse operations, can be used to solve any quadratic equation. However, the arithmetic and algebra are time comsuming. So, mathematicians have developed a formula to solve any quadratic equation.

Instead of solving quadratic equations one at a time by completing the square, you can solve $ax^2 + bx + c = 0$ for x. This will give you a formula that can be used every time. You can complete the square once and for all!

Begin with $ax^2 + bx + c = 0$. Your goal is to solve this equation for x using completing the square, followed by inverse operations to isolate x.

Step 1: Factor out the a from the x-terms:

$$a\left(x^2 + \frac{b}{a}x\right) + c = 0$$

Step 2: Complete the square:

Take half of the coefficient of x. Square this value.

$$\left(\frac{1}{2}\cdot\frac{b}{a}\right)^2 = \frac{b^2}{4a^2}$$

Add this to the left side of the equation.

$$a\left(x^2 + \frac{b}{a}x + \frac{b^2}{4a^2}\right) + c = 0 + \boxed{???}$$

But wait! You cannot add something to one side of an equation without adding it to the other! To balance this equation, we have to add the same thing to the right side.

Look closely at the left side of the equation. We added $\frac{b^2}{4a^2}$ to the left side but we placed it inside the parentheses. When the a is distributed to each term in the parentheses, we actually added $\frac{b^2}{4a^2}$ to the left side. So we need to add this term to the right side: $a\left(x^2 + \frac{b}{a}x + \frac{b^2}{4a^2}\right) + c = 0 + \frac{a}{1}\cdot\frac{b^2}{4a^2}$.

Answers (continued)

g) The zeros of the function are $x \approx -5.54$ and $x \approx 0.54$. They can be found using the TRACE feature, or the CALC feature of a graphing calculator.

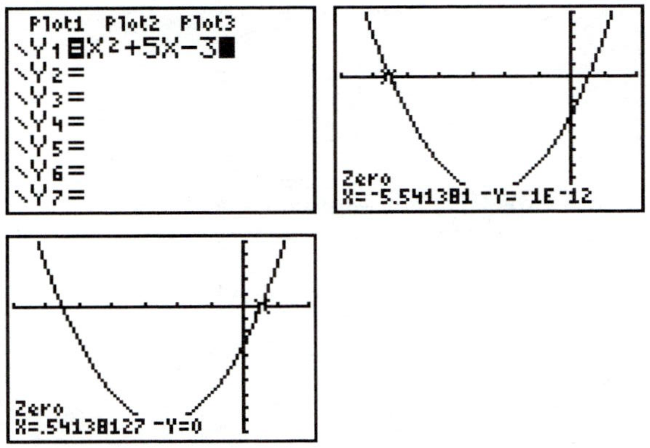

h) The approximate solutions to the equation are $x \approx -5.54$ and $x \approx 0.54$. These are approximate values because they have been rounded.

i) The equation could be solved using the method of completing the square then using inverse operations. An exact answer will be in radical form.

$$x^2 + 5x - 3 = 0$$
$$(x^2 + 5x) - 3 = 0$$
$$(x^2 + 5x + (2.5)^2) - 3 - (2.5)^2 = 0$$
$$(x^2 + 5x + 6.25) - 3 - 6.25 = 0$$
$$(x + 2.5)^2 - 9.25 = 0$$
$$(x + 2.5)^2 = 9.25$$
$$x + 2.5 = \sqrt{9.25}$$
$$x = -2.5 \pm \sqrt{9.25}$$

$$x = -2.5 \pm \sqrt{9.25}$$

$x = -2.5 + \sqrt{9.25}$ $\quad$ $x = -2.5 - \sqrt{9.25}$

$x \approx 0.54$ $\quad$ $x \approx -5.54$

For comparison with Question 8, show students that the answers from Questions 8 and 9 are approximately equal.

EXPLAIN LESSON 2

Section 5.12: Solving Quadratic Equations Using the Quadratic Formula

Purpose: In Section 5.11 students revisited the methods they previously used to solve quadratic equations. In both cases they encountered messy mathematics that elicited the need for a better way to solve quadratic equations. In this section students will formalize the Quadratic Formula and use it to solve quadratic equations that are in polynomial form. In Section 5.13 students will examine the methods that they have used to solve quadratic equations so far, and be able to select and defend a method of solving a quadratic equation in a problem-solving situation.

Description: In this section students will formalize the Quadratic Formula, then use it to solve quadratic equations in polynomial form.

Materials: graphing calculators

Teaching Ideas: Debrief students after Section 5.11 by having them share their answers to the questions. As they share, be sure to fill in any missing concept, process, or skill that could have been used to answer the questions. Then work through the questions in this section with the entire class. Once again be sure all necessary concepts, processes, or skills are addressed.

In this section, show students the derivation of the Quadratic Formula. It is not necessary that they be able to generate the Quadratic Formula on their own, but they should have a general idea of where it comes from; that mathematicians have developed a specific formula to solve any quadratic equation.

Simplify the right side.

$$a\left(x^2+\frac{b}{a}x+\frac{b^2}{4a^2}\right)+c=0+\frac{\not{a}}{1}\cdot\frac{b^2}{4a\cdot\not{a}}$$

$$a\left(x^2+\frac{b}{a}x+\frac{b^2}{4a^2}\right)+c=\frac{b^2}{4a}$$

Factor the left side.

$$a\left(x+\frac{b}{2a}\right)^2+c=\frac{b^2}{4a}$$

Step 3: Subtract c from both sides.

$$a\left(x+\frac{b}{2a}\right)^2=\frac{b^2}{4a}-c$$

Step 4: Divide by a (or, multiply by the reciprocal of a).

$$\frac{1}{\not{a}}\cdot\not{a}\left(x+\frac{b}{2a}\right)^2=\frac{1}{a}\cdot\left(\frac{b^2}{4a}-c\right)$$

$$\left(x+\frac{b}{2a}\right)^2=\frac{b^2}{4a^2}-\frac{c}{a}$$

Step 5: Find a common denominator on the right side:

The common denominator can be found by identifying the least common multiple of $\frac{b^2}{4a^2}$ and $\frac{c}{a}$. In other words, we need to multiply the second fraction by $\frac{4a}{4a}$ in order to have a common denominator of $4a^2$.

$$\left(x+\frac{b}{2a}\right)^2=\frac{b^2}{4a^2}-\frac{4a}{4a}\cdot\frac{c}{a}$$

$$\left(x+\frac{b}{2a}\right)^2=\frac{b^2}{4a}-\frac{4ac}{4a^2}$$

$$\left(x+\frac{b}{2a}\right)^2=\frac{b^2-4ac}{4a^2}$$

Step 6: Take the square root of both sides. (Don't forget that there are a positive and a negative square root!)

$$\sqrt{\left(x+\frac{b}{2a}\right)^2}=\sqrt{\frac{b^2-4ac}{4a^2}}$$

$$x+\frac{b}{2a}=\pm\sqrt{\frac{b^2-4ac}{4a^2}}$$

$$x+\frac{b}{2a}=\pm\frac{\sqrt{b^2-4ac}}{\sqrt{4a^2}}$$

$$x+\frac{b}{2a}=\pm\frac{\sqrt{b^2-4ac}}{2a}$$

Step 7: Solve for x.

Subtract $\frac{b}{2a}$ from both sides.

$$x=-\frac{b}{2a}\pm\frac{\sqrt{b^2-4ac}}{2a}$$

$$x=\frac{-b\pm\sqrt{b^2-4ac}}{2a}$$

QUADRATIC FORMULA

If $ax^2+bx+c=0$, then $x=\frac{-b\pm\sqrt{b^2-4ac}}{2a}$.

Now we can use the Quadratic Formula to solve any quadratic equation.

EXAMPLE 1:

Use the Quadratic Formula to solve $3x^2+4x+1=0$ for x.

Step 1: Define a, b, and c.

Let $a=3$, $b=4$, and $c=1$.

Step 2: Substitute the values for a, b, and c into the Quadratic Formula.

$$x=\frac{-b\pm\sqrt{b^2-4ac}}{2a}\qquad x=\frac{-4\pm\sqrt{4^2-4(3)(1)}}{2(3)}$$

Step 3: Simplify using the order of operations (the square root sign acts as a grouping symbol).

$$x=\frac{-4\pm\sqrt{4^2-4(3)(1)}}{2(3)}$$

$$x=\frac{-4\pm\sqrt{16-12}}{6}$$

$$x=\frac{-4\pm\sqrt{4}}{6}$$

$$x=\frac{-4\pm2}{6}$$

$$x=\frac{-4+2}{6}\qquad x=\frac{-4-2}{6}$$

$$x=\frac{-2}{6}\qquad x=\frac{-6}{6}$$

$$x=-\frac{1}{3}\qquad x=-1$$

EXAMPLE 2:

Use the Quadratic Formula to solve $4x^2-6x-9=0$ for x.

Step 1: Define a, b, and c.

Let $a=4$, $b=-6$, and $c=-9$.

Step 2: Substitute the values for a, b, and c into the Quadratic Formula.

$$x=\frac{-b\pm\sqrt{b^2-4ac}}{2a}$$

$$x=\frac{-(-6)\pm\sqrt{(-6)^2-4(4)(-9)}}{2(4)}$$

Step 3: Simplify using the order of operations (the square root sign acts as a grouping symbol).

$$x = \frac{-(-6) \pm \sqrt{(-6)^2 - 4(4)(-9)}}{2(4)}$$

$$x = \frac{6 \pm \sqrt{36 + 144}}{8}$$

$$x = \frac{6 \pm \sqrt{180}}{8}$$

$$x = \frac{6 \pm 6\sqrt{5}}{8}$$

$$x = \frac{6 \pm 6\sqrt{5}}{8}$$

$$x = \frac{6(1 \pm \sqrt{5})}{8}$$

$$x = \frac{3(1 \pm \sqrt{5})}{4}$$

$$x = \frac{3}{4} \pm \frac{3\sqrt{5}}{4}$$

SUMMARY

A quadratic equation in the form $ax^2 + bx + c = 0$ can be solved with the Quadratic Formula.

- If $ax^2 + bx + c = 0$, then $x = \frac{-b \pm \sqrt{b^2 - 4ac}}{2a}$.

To solve an equation with this formula:

- First, identify the values of a, b, and c.
- Second, substitute the values for a, b, and c into the Quadratic Formula.
- Third, simplify using the order of operations.
  - ✓ The square root sign acts as a grouping symbol.
  - ✓ Simplify the square root completely.
  - ✓ Reduce all fractions.

SECTION 5.12 Assignment

1. Use the Quadratic Formula to solve each of these:

   a) $3x^2 + 10x + 7 = 0$

   b) $2x^2 + 5x - 9 = 0$

2. Solve each of these using the most appropriate method.

   a) $y = x^2 + 6x + 9$

   b) $y = x^2 - 2.5x - 6$

   c) $y = -2x^2 - x - 0.12$

   d) $y = x^2 - 21x + 14$

   e) $y = \frac{1}{2}x^2 - 2x + 2$

   f) $y = -3x^2 + 9x + 12$

3. A stone is thrown from a catapult. The function $h(x) = -16x^2 + 80x$ describes the height (h) of the stone as a function of time.

   a) How high is the stone at 3.5 seconds? Explain how you found your answer.

   b) When will the stone be 36 feet high? Explain how you found your answer.

Section 5.12 Assignment

Answers

1. a) $x = \frac{-10 \pm \sqrt{100 - 4(3)(7)}}{2(3)}$ $\quad x = \frac{-10 \pm 4}{6}$

$x = \frac{-10 \pm \sqrt{100 - 84}}{6}$ $\quad x = \frac{-10 + 4}{6}$ $\quad x = \frac{-10 - 4}{6}$

$x = \frac{-10 \pm \sqrt{16}}{6}$ $\quad x = \frac{-6}{6}$ $\quad x = \frac{-14}{6}$

$x = \frac{-10 \pm 4}{6}$ $\quad x = -1$ $\quad x = -\frac{7}{3}$

b) $x = \frac{-5 \pm \sqrt{25 - 4(2)(-9)}}{2(2)}$

$x = \frac{-5 \pm \sqrt{25 + 72}}{4}$

$x = \frac{-5 \pm \sqrt{97}}{4}$

2. a) $x = -3$

b) $x = -1.5, 4$

c) $x = -0.2, -0.3$

d) $x = \frac{21 \pm \sqrt{385}}{2}$

e) $x = 2$

f) $x = -1, 4$

3. a) Students should realize that they can find this answer by substituting 3.5 for x since time is the independent variable.

$h(3.5) = -16(3.5)^2 + 80(3.5)$
$h(3.5) = -16(12.25) + 80(3.5)$
$h(3.5) = -196 + 280$
$h(3.5) = 84$

The rocket was 84 feet high at 3.5 seconds.

b) Since height is the dependent variable, substitute 36 for $h(x)$, and solve for x.

$36 = -16x^2 + 80x$
$36 - 36 = -16x^2 + 80x - 36$
$0 = -16x^2 + 80x - 36$

While the Quadratic Formula can be used, since this equation factors, it is much quicker to factor the polynomial and apply the Zero-Product Property.

$0 = -16x^2 + 80x - 36$
$0 = -4(4x^2 - 20x + 9)$
$0 = -4(2x - 9)(2x - 1)$

$2x - 9 = 0 \qquad 2x - 1 = 0$
$2x = 9 \qquad 2x = 1$
$x = \frac{9}{2} \qquad x = \frac{1}{2}$

The stone will be 36 feet high 0.5 seconds and again 4.5 seconds after launch.

SECTION 5.13

It's a Wrap!

In this chapter you have explored three ways to write quadratic functions:

Polynomial form: $y = ax^2 + bx + c$

Factored form: $y = a(x - r_1)(x - r_2)$

Vertex form: $y = a(x - h)^2 + k$

You have used three methods to solve quadratic equations:

✓ Factor the polynomial, then apply the Zero-Product Property.

✓ Complete the square, then apply inverse operations.

✓ Use the Quadratic Formula.

Which method is best? It depends on the situation. In this section you will look at problems involving motion. You will consider why one form of a quadratic function is better than others and why one method of solution is quicker than others.

THREE WAYS TO REPRESENT QUADRATIC FUNCTIONS

1. Enter each of these into your graphing calculator:

   $p(x) = 2x^2 - 12x + 16$

   $f(x) = 2(x - 2)(x - 4)$

   $v(x) = 2(x - 3)^2 - 2$

   a) Graph the functions on your calculator. Sketch the graphs and record your window.

   b) What can you conclude about the three functions? On what evidence do you base your conclusion?

   c) Prove that $p(x)$ and $f(x)$ are equivalent.

   d) Prove that $p(x)$ and $v(x)$ are equivalent.

   e) What is the vertex of this parabola? Which function most closely reveals the vertex?

   f) What are the x-intercepts of this parabola? Which function most closely reveals them?

   g) What is the y-intercept of this parabola? Which function most closely reveals it?

2. For each function in **Figure 5.44,** fill in the missing representations.

Graph	Factored Form	Vertex Form	Polynomial Form
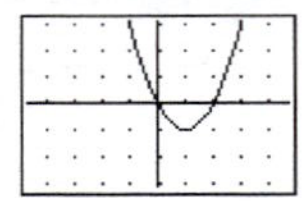	$y = x(x - 2)$	$y = (x - 1)^2 - 1$	$y = x^2 - 2x$
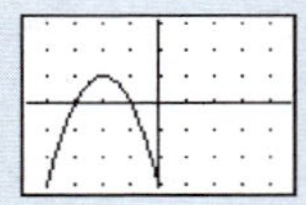			
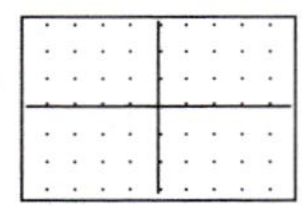	$y = (x - 2)(x + 1)$		
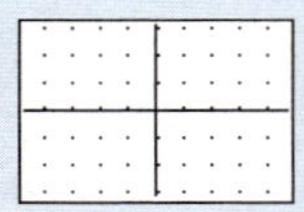		$y = 2(x - 2)^2 - 2$	
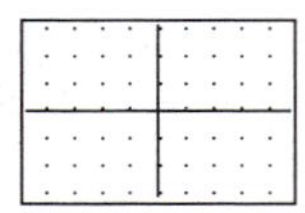			$y = -\frac{1}{2}x^2 + 2$

FIGURE 5.44. *Table for Question 2.*

3. When is it best to solve equations with factoring and the Zero-Product Property?

4. When is it best to solve equations with inverse operations?

5. When is it best to solve equations with the Quadratic Formula?

ELABORATE LESSON 2

Section 5.13: It's a Wrap!

Purpose: In Sections 5.7 through 5.12, students developed three different ways to solve quadratic equations. In Section 5.13 students will examine quadratic functions in polynomial, vertex, and factored forms, convert between the three representations, and place each representation in a context. This prepares students to select a method for solving quadratic equations in a problem-solving context in Section 5.14.

Description: In this lesson students will be given a quadratic function or equation in one representation and will be asked to generate the other representations of a quadratic function or equation. They will also be given problem situations and asked to identify which representation is most appropriate, select a method for solving the problem, and solve.

Materials: graphing calculator

Teaching Ideas: Students should work in small groups (3 to 4 students). This section is designed for students to make connections between the multiple representations of a quadratic function. Students will also select an appropriate method to solve quadratic equations.

Answers

1. Teachers may need to remind students that $p(x) =$, $f(x) =$, and $v(x) =$ are all equivalent to $y =$.

a)

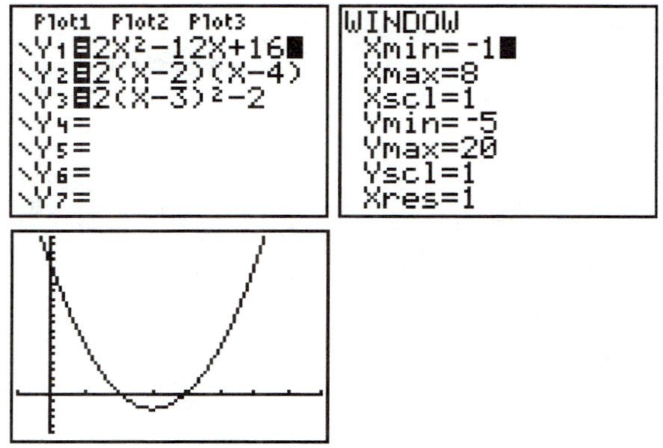

b) All three functions have the same graph. This indicates that the three functions are all equivalent representations of the same relation.

c) $p(x) = 2x^2 - 12x + 16$ $\quad f(x) = 2(x-2)(x-4)$

$$\begin{aligned} &= 2(x^2 - 4x - 2x + 8) \\ &= 2(x^2 - 6x + 8) \\ &= 2x^2 - 12x + 16 \end{aligned}$$

d) $p(x) = 2x^2 - 12x + 16$ $\quad v(x) = 2(x-3)^2 - 2$

$$\begin{aligned} &= 2(x-3)(x-3) - 2 \\ &= 2(x^2 - 6x + 9) - 2 \\ &= 2x^2 - 12x + 18 - 2 \\ &= 2x^2 - 12x + 16 \end{aligned}$$

e) The vertex of the parabola is (3, –2). $v(x)$, which is a function in vertex form, reveals this information.

f) The x-intercepts are (2, 0), and (4, 0). $f(x)$, which is a function in factored form, reveals this information.

g) The y-intercept is (0, 16). $p(x)$, which is a function in polynomial form, reveals this information.

Answers (continued)

2.

Graph	Factored Form	Vertex Form	Polynomial Form
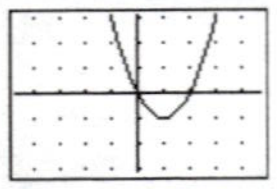	$y = x(x-2)$	$y = (x-1)^2 - 1$	$y = x^2 - 2x$
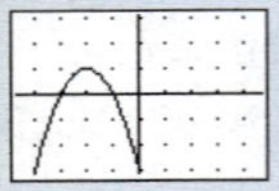	$y = -(x-1)(x-3)$	$y = -(x+2)^2 + 1$	$y = -x^2 - 4x - 3$
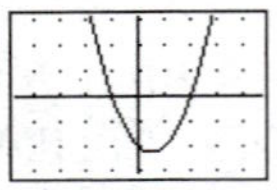	$y = (x-2)(x+1)$	$y = (x-0.5)^2 - 2.25$	$y = x^2 - x - 2$
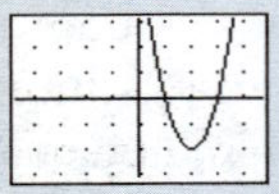	$y = 2(x-1)(x-3)$	$y = 2(x-2)^2 - 2$	$y = 2x^2 - 8x + 6$
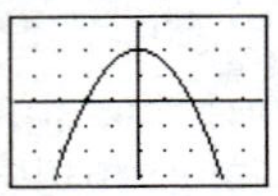	$y = -\frac{1}{2}(x-2)(x+2)$	$y = -\frac{1}{2}(x-0)^2 + 2$	$y = -\frac{1}{2}x^2 + 2$

3. The Zero-Product Property yields quick solutions when equations are given in factored form, or if an equation given in polynomial form is easily factored. In a word problem, if information is given about roots of the equation or zeros of the function, then an equation in factored form can be written and used to solve the problem.

4. When an equation is given in vertex form, inverse operations can yield a quick solution. An equation given in polynomial form could be rewritten in vertex form using completing the square, although sometimes this process can become tedious. In a word problem, if information is given about conditions leading to a maximum or minimum value, an equation in vertex form can be written and used to solve the problem.

5. When an equation is given in polynomial form and does not quickly factor, the Quadratic Formula can be used to generate the solutions. The Quadratic Formula can yield integer, rational, or irrational solutions and will work for all quadratic equations. In a word problem, if information is given about acceleration, starting speed, or a starting point, then an equation in polynomial form can be written and used to solve the problem.

6. Use the table in **Figure 5.45** to record your findings about the forms of quadratic functions and the methods to solve quadratic equations.

	General Form of Equation	Critical Attributes	Preferred Method of Solution
Factored Form			
Vertex Form			
Polynomial Form			

FIGURE 5.45.
Table for Question 6.

SECTION 5.13 Assignment

1. In **Figure 5.46**, fill in the blank columns.

Graph	Factored Form	Algebra Tiles	Vertex Form	Polynomial Form
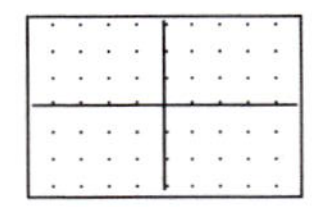				$y = -3x^2 + 6x$
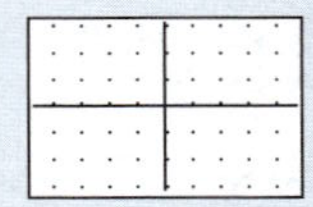				

FIGURE 5.46.
Table for Question 1.

2. Change each of these into the specified form.

   a) $y - 5 = 0.5(x + 2)^2$ into polynomial form

   b) $y = 2x^2 - 7x + 6$ into factored form

   c) $y = -2(x + 2)(x - 1)$ into polynomial form

   d) $y = 1.4(x - 1)(x - 5)$ into vertex form

   e) $y - 8 = -2(x + 4)^2$ into factored form

   f) $y = x^2 - 2x - 15$ into vertex form

3. Sketch the graph of each function without using a graphing calculator. In each case explain how the form of the function helped you determine the shape of the graph.

   a) $y = -0.5(x + 1)(x - 3)$

   b) $y = 2(x + 2)^2 - 4$

   c) $y = x^2 + 3x - 2$

Answers (continued)

6.

	General Form of Equation	Critical Attributes	Preferred Method of Solution
Factored Form	$y = a(x - r_1)(x - r_2)$	Roots are $(r_1, 0)$ and $(r_2, 0)$ If $\lvert a \rvert > 1$, then the graph $y = x^2$ is vertically stretched. If $\lvert a \rvert < 1$, then the graph is vertically compressed. If $a < 0$, then the parabola is down. If $a > 0$, then the parabola is up. Roots are quickly identified.	Zero-Product Property
Vertex Form	$y = a(x - h)^2 + k$	Vertex is (h, k) If $\lvert a \rvert > 1$, then the graph is vertically stretched. If $\lvert a \rvert < 1$, then the graph is vertically compressed. If $a < 0$, then the parabola is concave down. If $a > 0$, then the parabola is up. Transformations are quickly identified.	Inverse Operations
Polynomial Form	$y = ax^2 + bx + c$	y-intercept is $(0, c)$ If $\lvert a \rvert > 1$, then the graph is vertically stretched. If $\lvert a \rvert < 1$, then the graph is vertically compressed. If $a < 0$, then the parabola is down. If $a > 0$, then the parabola is up.	Quadratic Formula

Section 5.13 Assignment

Answers

1.

Graph	Factored Form	Algebra Tiles	Vertex Form	Polynomial Form
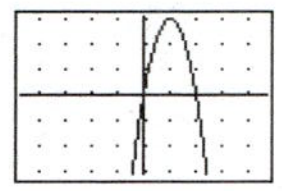	$y = -3x(x - 2)$	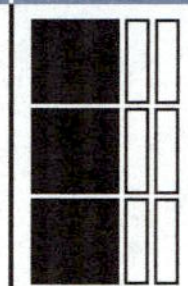	$y = -3(x - 1)^2 + 3$	$y = -3x^2 + 6x$
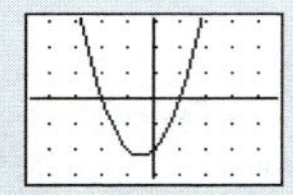	$y = (x - 1)(x + 2)$		$y = (x + 0.5)^2 - 2.25$	$y = x^2 + x - 2$

2\. a)
$$y - 5 = 0.5(x + 2)^2$$
$$y - 5 = 0.5(x^2 + 4x + 4)$$
$$y - 5 = 0.5x^2 + 2x + 2$$
$$y - 5 + 5 = 0.5x^2 + 2x + 2 + 5$$
$$y = 0.5x^2 + 2x + 7$$

b)
$$y = 2x^2 - 7x + 6$$
$$y = (2x - 3)(x - 2)$$

c)
$$y = -2(x + 2)(x - 1)$$
$$y = -2(x^2 + 2x - 1x - 2)$$
$$y = -2(x^2 + x - 2)$$
$$y = -2x^2 - 2x + 4$$

d)
$$y = 1.4(x - 1)(x - 5)$$
$$y = 1.4(x^2 - 6x + 5)$$
$$y = 1.4[(x^2 - 6x + 9) + (-9 + 5)]$$
$$y = 1.4[(x - 3)^2 - 4]$$
$$y = 1.4(x - 3)^2 - 1.4(4)$$
$$y = 1.4(x - 3)^2 - 5.6$$

e)
$$y - 8 = -2(x + 4)^2$$
$$y - 8 = -2(x^2 + 8x + 16)$$
$$y - 8 = -2x^2 - 16x - 32$$
$$y - 8 + 8 = -2x^2 - 16x - 32 + 8$$
$$y = -2x^2 - 16x - 24$$
$$y = -2(x^2 + 8x + 12)$$
$$y = -2(x + 2)(x + 6)$$

f)
$$y = x^2 - 2x - 15$$
$$y = (x^2 - 2x) - 15$$
$$y = (x^2 - 2x + 1) - 15 - 1$$
$$y = (x - 1)^2 - 16$$

3\. a) Since this equation is in factored form, first plot the zeros of the function. In this case a is negative, so the parabola is concave down (opens downward). Then find the line of symmetry (halfway between the zeros). Substitute this x-value into the equation to find the y-value of the vertex. Connect the points with a smooth curve.

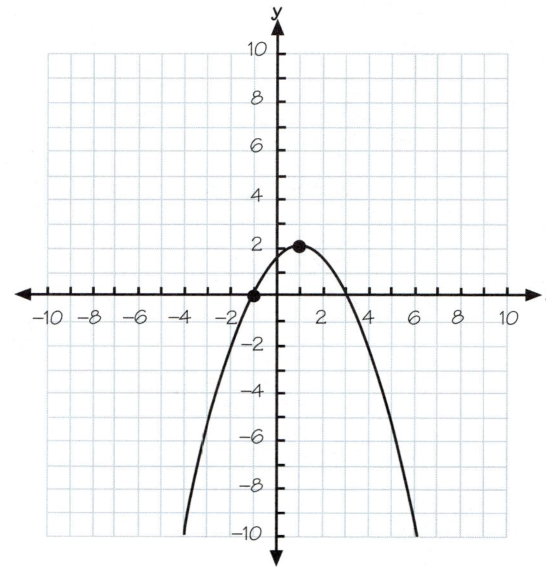

Answers (continued)

b) Since this equation is in vertex form, first plot the vertex. Since a is positive, the parabola will be concave up (opens upward). From the vertex, moving one unit in the x-direction will result in moving 2 units upward (instead of 1 unit in the parent function) since there is a vertical stretch of 2. Connect the points with a smooth curve.

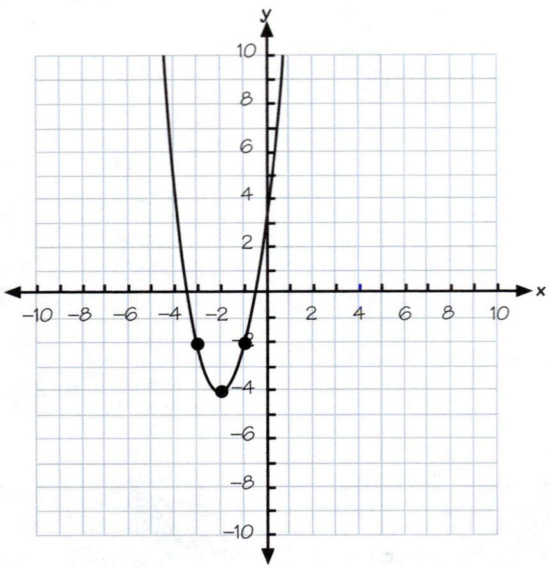

c) Since this equation is in polynomial form, first plot the y-intercept. Since a is positive, the parabola will be concave up (opens upward). Select 2 x-values that are small, then substitute those into the polynomial to find the y-values. Plot these points. Connect the points with a smooth curve.

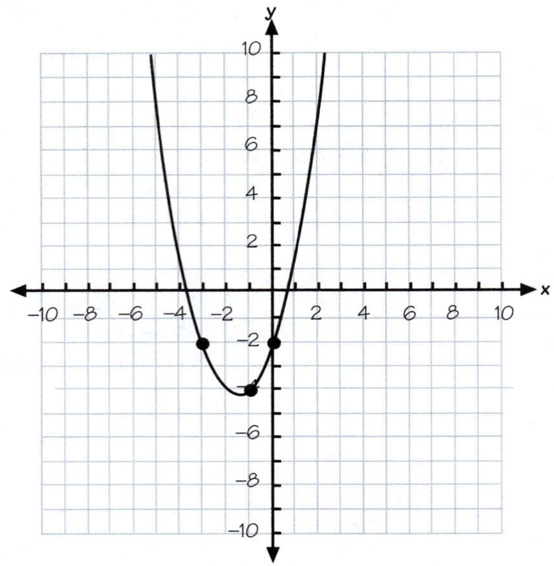

4. A toy rocket blasts off from the ground and peaks in 3 seconds at 144 feet. Assume the relationship between height and time is quadratic, and let the value of a be -16.

   a) Write a function in vertex form for the height of the rocket at time t.

   b) Write a function in factored form for the height of the rocket at time t.

   c) Write a function in polynomial form for the height of the rocket at time t.

   d) Sketch a graph of the relationship between height and time. If you use a graphing calculator, record your window.

   e) When does the rocket hit the ground? Why?

   f) Find the coordinates of the vertex and the x-intercepts. Label these on your graph.

   g) How high was the rocket 2 seconds after launch? Show how you found this.

   h) When will the rocket again be the same height as your answer in g? Explain how you found this.

5. The freshman class is deciding whether or not to raise the price of tickets to the spring dance. $p = 75 + 100d - 20d^2$ can be used to estimate profit, p, by raising the ticket price d dollars.

   a) What profit is expected if they do not raise the price (i.e., if $d = 0$)?

   b) What is the least amount they can raise the price if they want to make $155?

   c) What method did you use to answer b? Why did you choose that method?

SECTION 5.14 Build a Better Sandbox . . .

A company builds sandboxes as shown in **Figure 5.47.** The series number determines the surface area covered with sand.

FIGURE 5.47. A series of sandboxes.

Series Number	Sandbox & Seating Tiles	Written Description
1	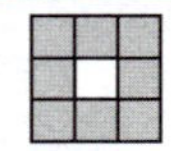	A 3 x 3 sandbox has 1 square unit of its surface area covered with sand.
2	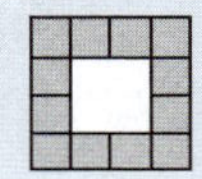	A 4 x 4 sandbox has 4 square units of its surface area covered with sand.
3	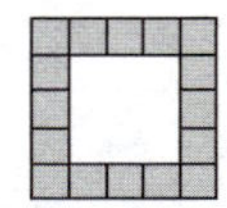	A 5 x 5 sandbox has 9 square units of its surface area covered with sand.
4	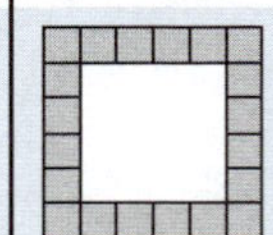	A 6 x 6 sandbox has 16 square units of its surface area covered with sand.

1. Find a function rule that describes the total surface area of the sandbox, including the part that is covered with sand and the surrounding seating tiles, in terms of the series number. Justify your answer.

2. Find the series number of the sandbox with 196 square units of surface area. Justify your answer.

Answers (continued)

4. a) $y = -16(x - 3)^2 + 144$

   b) $y = -16x(x - 6)$

   c) $y = -16x^2 + 96x$

   d) 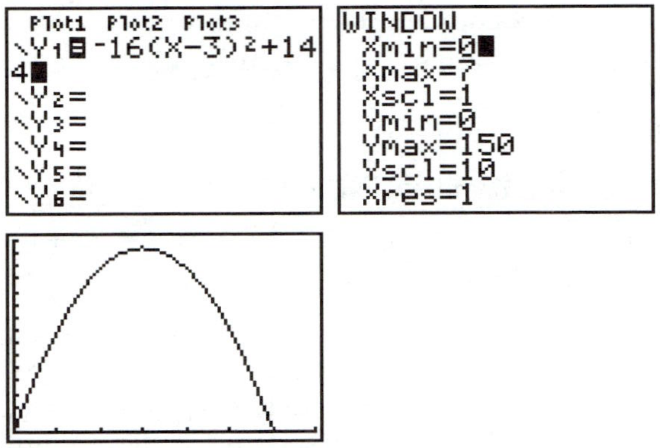

   e) The rocket hits the ground after 6 seconds, since that is the second x-value when $y = 0$.

   f) Vertex: (3, 144)
   Roots: (0, 0) and (6, 0)

   g) The rocket was 128 feet above the ground.

   Explanations may vary. One way to find the answer is to substitute $x = 2$, then simplify the expression.

   $y = -16((2) - 3)^2 + 144$
   $y = -16(-1)^2 + 144$
   $y = -16 + 144$
   $y = 128$

   h) After 4 seconds, the rocket will also be 128 feet above the ground.

   Explanations may vary. One way to find the answer is to substitute $y = 128$, then solve the equation for x.

   $128 = -16x^2 + 96x$
   $0 = -16x^2 + 96x - 128$
   $0 = -16(x^2 - 6x + 8)$
   $0 = -16(x - 2)(x - 4)$

   $0 = -16(x - 2)(x - 4)$
   $x - 2 = 0 \quad x - 4 = 0$
   $x = 2 \quad x = 4$

5. a) $75

   $p = 75 + 100(0) - 20(0)^2$
   $p = 75 + 0 + 0$
   $p = 75$

   b) $1 (They could also raise the price $4 to generate a profit of $155, but the least amount is $1.)

   $155 = 75 + 100d - 20d^2$
   $0 = 75 - 155 + 100d - 20d^2$
   $0 = -20d^2 + 100d - 80$
   $0 = -20(d^2 - 5d + 4)$
   $0 = -20(d - 1)(d - 4)$

   $d - 1 = 0 \quad d - 4 = 0$
   $d = 1 \quad d = 4$

   c) Answers may vary. The importance of this question is not to emphasize that one method of solving a quadratic equation is better than another but to get students to think about why they would want to choose one method over another, beyond the perception that "it's easier." When students know and understand multiple representations of quadratic functions and multiple methods of solving quadratic equations, then they are empowered to choose the most appropriate method to solve any problem involving quadratic functions.

EVALUATE LESSON 2

Section 5.14: Build a Better Sandbox...

Purpose: In Sections 5.6 through 5.13, students discovered multiple ways to solve quadratic equations. In this section students will have the opportunity to demonstrate their knowledge by applying those concepts and skills to solve a problem in a new context. This prepares students to delve deeper into quadratic functions in Sections 5.15 through 5.19.

Description: Facing a problem in a new context, students will determine and implement a method to solve the problem. This is an open-middle performance assessment task; in other words, there is only one correct answer, but there are a variety of valid methods to obtain it. In their justification, which is the most important part of a performance assessment task, students should be able to communicate their knowledge of mathematics and demonstrate that they have mastered the concepts presented in Sections 5.6 through 5.13.

Materials: none

Teaching Ideas: This task may be assigned as group work or individual work. The mathematics performance assessment rubric should be used to evaluate students' work.

Answers

1. Sample solution: First organize the information in a table.

Series Number	Process Column	Surface Area with Sand	Process Column	Total Surface Area
1	1 × 1	1	1 + 4(1 + 1)	9
2	2 × 2	4	4 + 4(2 + 1)	16
3	3 × 3	9	9 + 4(3 + 1)	25
4	4 × 4	16	16 + 4(4 + 1)	36
n	$n \times n$	n^2	$n^2 + 4(n + 1)$	$n^2 + 4n + 4$

If n represents the series number, then $n^2 + 4n + 4$ can be used to find the total surface area of the sandbox, including the part covered with sand and the seating tiles around the perimeter of the sandbox.

2. The method of solution that the student chooses for this portion will depend on the form of the expression that they developed in Question 1. In this example, we found a rule in polynomial form that factors, so we will solve by factoring.

To find the series number that has a total surface area of 196 square units, set the expression for total surface area equal to 196, then solve for n.

$$n^2 + 4n + 4 = 196$$
$$(n + 2)^2 = 196$$
$$\sqrt{(n+2)^2} = \sqrt{196}$$
$$n + 2 = \pm 14$$

$$n + 2 = \pm 14$$

$$n + 2 = 14 \qquad n + 2 = -14$$
$$n + 2 - 2 = 14 - 2 \qquad n + 2 - 2 = -14 - 2$$
$$n = 12 \qquad n = -16$$

$n = 12$ and $n = -16$ are the roots of the equation $n^2 + 4n + 4 = 196$. However, in the context of this problem, the series number must be positive so $n = -16$ has no meaning. Therefore, the series number of 12 will have a total surface area of 196 square units.

SECTION 5.15 Motion with a Pendulum

A **pendulum** is a string with a mass (called a "bob") attached at the end.

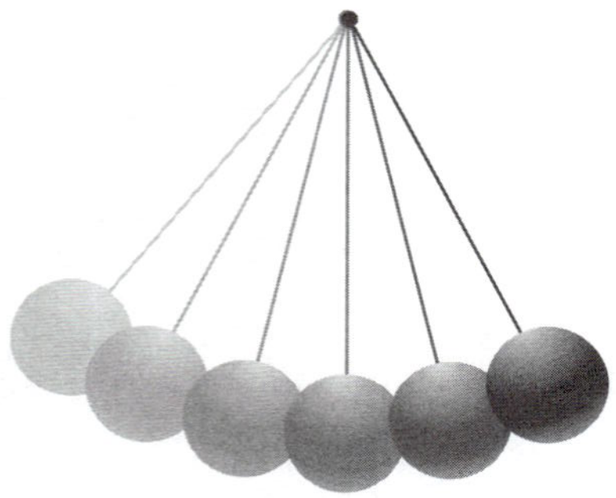

Galileo

One of the first people to investigate the motion of a pendulum was Galileo, a famous astronomer. He noticed a chandelier moving in a cathedral in Pisa, Italy. He explored the relationships among the length of a pendulum, the weight on the end of the pendulum, the angle of release, and the time it takes for one swing. Galileo's work led to the invention of the pendulum clock.

CHECK THIS!

The period of a pendulum is the time it takes for one complete back-and-forth motion of the pendulum.

1. What types of motion have you investigated so far in this chapter?
2. Give examples of events where the motion is like that of a pendulum.

SECTION 5.15 Assignment

1. In **Figure 5.48**, a circle has been drawn with points evenly spaced around it. Straight lines connect all pairs of points. In order to predict the number of lines that have been drawn, you might want to start with a simpler problem and look for a pattern. Fill in the table in **Figure 5.49.** Hint: You may want to draw a circle, put the number of points on it, connect the points with lines, and then count them.

Number of Points on the Circle	Number of Lines Drawn
2	
3	
4	
5	
6	
7	

FIGURE 5.49. Table for Question 1.

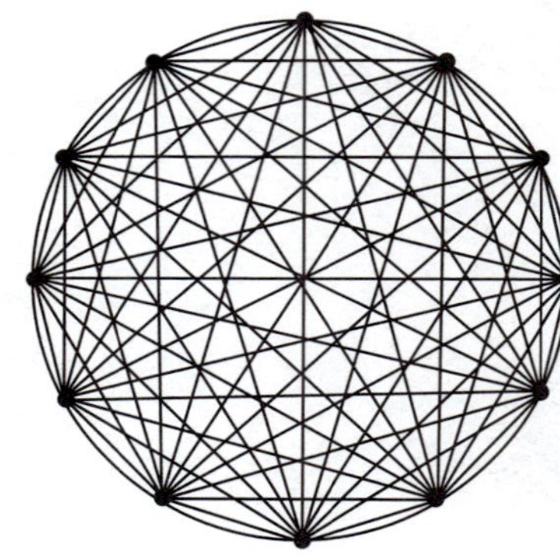

FIGURE 5.48. A circle with drawn points.

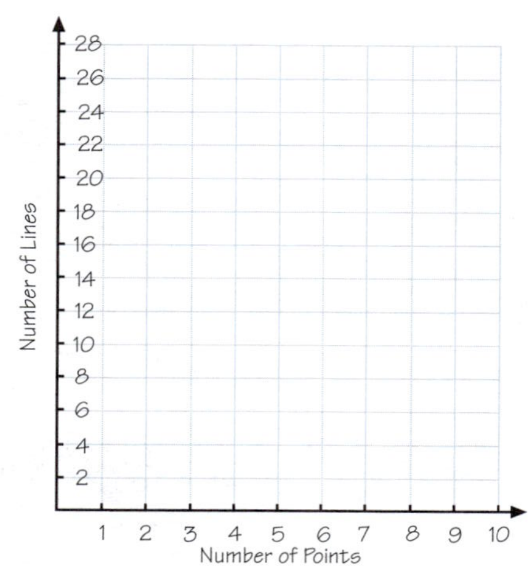

FIGURE 5.50. Graph for Question 2.

2. Graph the ordered pairs from the table on a graph like **Figure 5.50.**
3. Decide if the relationship between the number of lines and the number of points is linear or quadratic. Write a function rule that models this relationship.
4. Use your model to predict how many lines you would draw if there were 12 points on the circle.
5. Is your prediction close to the actual number of lines?

ENGAGE LESSON 3

Section 5.15: Motion with a Pendulum

Purpose: The purpose of this section is to engage students in thinking about the motion of a pendulum so they will be able to connect this type of motion with the square root function in Section 5.19. By observing demonstrations of changing the angle of release or the amount of weight on the pendulum, they should notice that neither weight nor angle affects the period of a pendulum.

Description: In this section the teacher will allow students to discuss the different types of motion they have studied up to this point, define a pendulum, and then demonstrate the motion of a pendulum. As a class, students will collect data by varying only the weight or angle of the pendulum. The length of the pendulum will remain the same.

Materials: string, washers, stopwatch, protractor, markers, large sheet of grid paper, large paper clip

Teaching Ideas: This section is an investigation of the motion of pendulums. The main idea that students should understand is that neither the angle of release nor the weight on the end of a pendulum affects the period of the pendulum. They will investigate reflections across the line $y = x$ and inverse functions in the following two sections, so that they will be prepared for the data collection experiment in Section 5.19. This section has been written as a teacher demonstration, but you may choose to allow students to do the experiment in small groups.

Prior to class you will need to set up a pendulum. If you set it up near the chalkboard, make sure that the pendulum swings freely without hitting the board. If you have a ring stand you may use it to hold the pendulum also. Draw a horizontal line at the lowest point of the swing and another line at approximately 10° off of center to use as guides for the demonstration. You may want to have a protractor attached at the top of the string to convince students that the angle is the same for each trial. Tie a paper clip to the end of the string and put 2 washers on the paper clip. Additional washers will be added to the paper clip in the second demonstration.

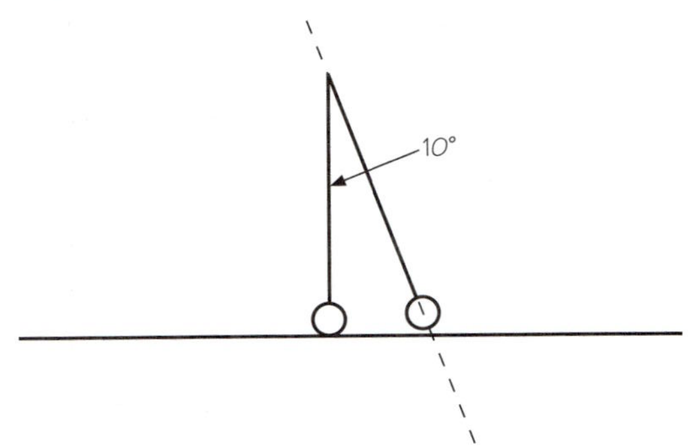

For this pendulum investigation, we are going to keep it simple. We will ignore air resistance, friction, and gravity. We will keep the arc less than 15° since the model we will be investigating in Section 5.19 only works for arcs less than 15°.

Before beginning the demonstration allow students time to discuss and answer the questions.

Answers

1. Answers will vary, but hopefully students will mention walking at a constant rate with the motion detector and launching a rocket. It is acceptable for students to extend beyond the chapter and add other types of motion they are familiar with such as driving a car or bouncing on a trampoline.

2. Answers may vary. Some examples of pendulums are a grandfather clock, a trapeze artist, a swing at the playground, or a ride at an amusement park.

Demonstrate swinging a pendulum. Ask students the following facilitation questions:

- What do you observe as I swing the pendulum?

  Answers may vary. Students may notice that the pendulum comes back almost to the same height after each swing. They may say that the time to complete one period stays approximately the same. Or, they may simply notice that the pendulum swings back and forth. Because of gravity, the pendulum will eventually come to a stop.

- What variables are involved in this motion?

  Answers may vary. Students will probably mention the length of the pendulum, the angle at which the pendulum starts off center, or the weight on the end of the pendulum.

- Do you think the length of the period will change if you start from a different angle?

  Answers may vary. Accept all answers students give. An investigation will follow.

- Do you think the length of the period will change if you add more weight?

  Answers may vary. Accept all answers students give. An investigation will follow.

- Do you think the length of the period will change if the string changes in length?

  Answers may vary. Accept all answers students give. This investigation is in Section 5.19.

Answers (continued)

For the demonstration you will need the help of four students. One student, the counter, will count the number of complete swings of the pendulum. A second student, the timer, will time 10 swings of the pendulum. A third student, the calculator, will calculate the average time for each swing. A fourth student will hold the bob at the designated spot and release it. Tell students that you are investigating whether or not the time for each swing decreases, increases, or stays the same when you change the angle from where the pendulum is released from small to medium to large. As students complete the experiment, record the results in a table on the overhead projector. An example of collected data is shown.

Size of Angle of Swing	Time for 10 Swings	Average Time per Swing
Small angle (~5°)	13.76 sec	1.4 sec
Medium angle (~10°)	13.94 sec	1.4 sec
Large angle (~15°)	14.15 sec	1.4 sec

- What do you observe?

  The average time per swing is approximately the same no matter what the angle of release is.

For the next demonstration you may choose four different students. In addition to the four helping assignments you used in the previous demonstration, you will need another student to add additional weights. If you use washers, the student will simply need to add the desired number of washers. As students complete the experiment, record the results in a table on the overhead projector. An example of collected data is shown.

Number of Washers	Time for 10 Swings	Average Time per Swing
2	14.12 sec	1.4 sec
4	14.46 sec	1.4 sec
6	14.59 sec	1.5 sec

- What do you observe?

  The average time per swing is approximately the same no matter what the mass of the bob is.

If students want to investigate changing the length of the pendulum at this time, let them know they will be doing that investigation in Section 5.19. Try not to give any hints about what changing the length might do in order to build suspense.

Section 5.15 Assignment

Answers

1.

Number of Points on the Circle	Number of Lines Drawn
2	1
3	3
4	6
5	10
6	15
7	21

2.

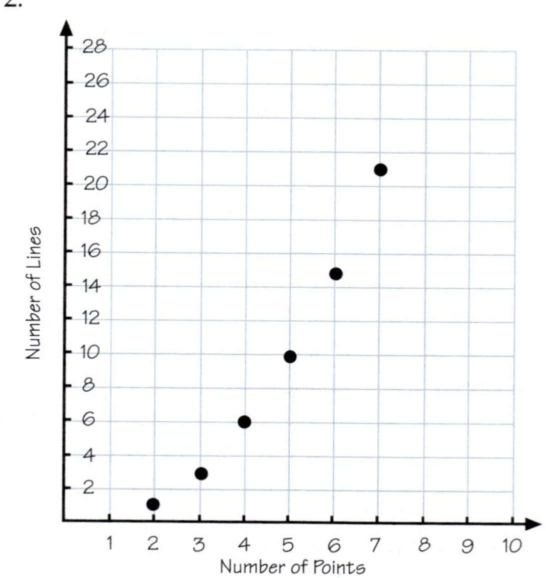

3. After investigating the graph, students should decide that the relationship is quadratic by investigating the first and second differences. They may use various methods to find the best model. Quadratic regression gives the equation $f(x) = 0.5x^2 - 0.5x$. Students may choose to use a transformation equation: $f(x) = \frac{1}{2}\left(x - \frac{1}{2}\right)^2 - \frac{1}{8}$.
   They may think logically. Each of the x points has $x - 1$ lines originating from that point. In order to count each line only once, divide by 2.
   So, $f(x) = \frac{x(x-1)}{2}$. Some students may want to use the table to predict a pattern. This model is appropriate for integers greater than or equal to 2 since fractions of a point are not realistic in this situation.
4. All models from Question 4 should give the same answer, 66 lines.
5. Answer will vary. If student answered yes, they should be able to provide how close they were to 66.

SECTION 5.16

Reflections with Patty Paper

A REFLECTION WITH PATTY PAPER

1. Connect the points listed below in order on coordinate graph paper. When you see the word STOP, do not connect the point before the word STOP to the point after the word STOP.

   {(–8, 0), (–9, –2), (–3, –2), (–4, 0), (–8, 0), (–8, 8), (–4, 8), (–4, 0), STOP, (–8, 8), (–6, 10), (–4, 8), STOP}

2. Graph the line $y = x$ on the coordinate axes.
3. What do you notice about the coordinates of the points that lie on the graph of $y = x$?
4. Trace your drawing from Question 1, the line $y = x$, and the coordinate axes onto a sheet of patty paper.
5. Fold the patty paper on the line $y = x$.
6. Trace the reflection of the drawing onto the patty paper.
7. What do you notice about the reflection?
8. Trace the reflection of the drawing onto the coordinate grid. Draw the reflection with a different color pencil.
9. List the coordinates of the corresponding points of the reflected drawing in a table like **Figure 5.51.**

FIGURE 5.51.
Table for Question 9.

Coordinates of Points in Original Drawing	Corresponding Coordinates of Points in Reflected Drawing
(–8, 0)	
(–9, –2)	
(–3, –2)	
(–4, 0)	
(–4, 8)	
(–6, 10)	
(–8, 8)	

10. What do you notice about the coordinates?
11. What are the x-intercepts on the original drawing?
12. Which points in the reflection correspond to the x-intercepts in the original drawing?
13. What happens if you reflect the reflection back over the line $y = x$?

ANOTHER REFLECTION

14. Graph the line $y = 2x + 4$ on the coordinate axes.
15. Draw the graph of $y = x$.
16. Trace $y = 2x + 4$, $y = x$, and the coordinate axes onto a sheet of patty paper.
17. Fold the patty paper along the line $y = x$.
18. Trace the reflection of $y = 2x + 4$ onto the patty paper.
19. Place the patty paper back on the coordinate grid. Trace the reflection of $y = 2x + 4$ onto the coordinate grid. Draw the reflection with a different color pencil.
20. What do you notice about the reflection?
21. List three points that lie on $y = 2x + 4$ and list the coordinates of corresponding points on the reflected line.
22. What do you notice about the coordinates of corresponding points?
23. What is the slope of the line $y = 2x + 4$?
24. What is the slope of the reflected line?
25. What is the y-intercept of the line $y = 2x + 4$?
26. What is the x-intercept of the reflected line?
27. What is the x-intercept of $y = 2x + 4$?
28. What is the y-intercept of the reflected line?
29. What is the equation of the reflected line?

EXPLORE LESSON 3

Section 5.16: Reflections with Patty Paper

Purpose: The purpose of this activity is to give students the opportunity to investigate informally the inverse of a relation defined as the reflection over the line $y = x$ using three different sets of relations: a simple drawing of a rocket, a linear function, and the quadratic parent function. Students are relating functions they have worked with previously to their reflections over the line $y = x$. This will allow students to relate the reflection of the quadratic function to its inverse, the square root function, in Section 5.18.

Description: In this section students will graph a simple drawing of a rocket, a linear function, and the quadratic parent function and graph the reflection of each across the line $y = x$. The main concept students should notice is that the coordinates of corresponding points are reversed.

Materials: graph paper, patty paper

Teaching Ideas: This section is an informal investigation of inverses. Students should notice that when a relation or function is reflected across the line $y = x$, the domain and range are reversed. Avoid using the word inverse until Section 5.18.

Answers

3. The x- and y-coordinates for each point on the graph of the line $y = x$ are equal. Examples include (3, 3) and (–4, –4).

7. Student answers may vary. Accept all of the possible answers. For example, students might respond, "The reflection of the rocket drawing is the same size as the original drawing," or, "It has been reflected across the line $y = x$," or, "Instead of pointing straight up, it is pointing to the right."

8. Answers may vary.

9.

Coordinates of Points in Original Drawing	Corresponding Coordinates of Points in Reflected Drawing
(–8, 0)	(0, –8)
(–9, –2)	(–2, –9)
(–3, –2)	(–2, –3)
(–4, 0)	(0, –4)
(–4, 8)	(8, –4)
(–6, 10)	(10, –6)
(–8, 8)	(8, –8)

10. The x- and y-coordinates are reversed.

11. The x-intercepts are (–8, 0) and (–4, 0).

12. The x-intercepts in the original become the y-intercepts, (0, –8) and (0, –4), in the reflection.

13. The rocket drawing would go back to its original position.

Before proceeding to the next portion of this activity, walk around and visually inspect students' work. It should look like the graph below.

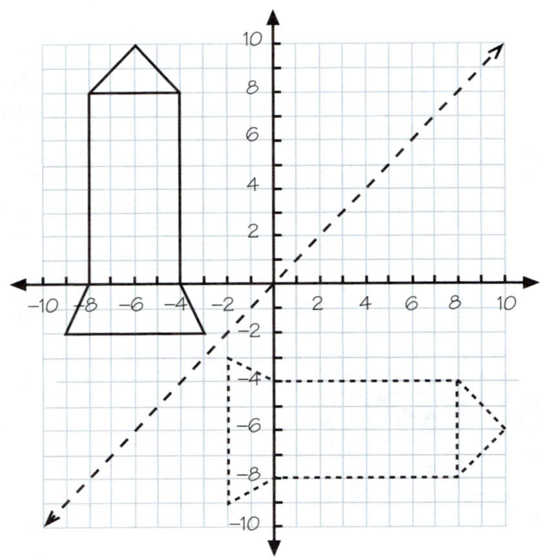

20. Answers may vary. Students may simply say that the reflection is a mirror image, or they might notice that the slopes of the two lines are reciprocals. The y-intercept is the opposite of $\frac{4}{2}$.

21.

Coordinates of Points on the Line $y = 2x + 4$	Coordinates of Points on the Reflected Line
(0, 4)	(4, 0)
(–2, 0)	(0, –2)
(1, 6)	(6, 1)

22. The coordinates of corresponding points are reversed.

23. The slope is 2.

24. The slope of the reflected line is $\frac{1}{2}$.

25. The y-intercept is the point (0, 4).

26. The x-intercept of the reflected line is (4, 0).

27. The x-intercept of $y = 2x + 4$ is (–2, 0).

28. The y-intercept of the reflected line is (0, –2).

29. The equation of the reflected line is $y = \frac{1}{2}x - 2$.

30. What happens if you reflect the reflection back over the line $y = x$?

A FINAL REFLECTION

31. Graph $y = x^2$ on a coordinate grid.
32. Graph $y = x$.
33. Trace $y = x^2$, $y = x$, and the coordinate axes onto a sheet of patty paper.
34. Fold the patty paper along the line $y = x$.
35. Trace the reflection of $y = x^2$ onto the patty paper.
36. Trace the reflection of $y = x^2$ onto the coordinate grid. Draw the reflection with a different color pencil.
37. What do you notice about the reflection?
38. Is the reflection a function? How do you know?
39. List three points that lie on $y = x^2$ and list the coordinates of corresponding points on the reflected graph.
40. What do you notice about the coordinates of corresponding points?
41. What are the domain and range of $y = x^2$?
42. What are the domain and range of the reflection of $y = x^2$?
43. What happens if you reflect the reflection back over the line $y = x$?

SECTION 5.16 Assignment

1. Plot the following relation.

   {(2, 2), (2, 4), (–1, 4), (–3, 1), (–6, –1), (–5, –4), (–3, –6), (1, –5), (2, –2), (4.5, 0.5)}

2. What is the domain of this relation?
3. What is the range of this relation?
4. Plot the graph of $y = x$.
5. Plot the reflection of the relation from Question 1 over the line $y = x$.
6. List the coordinates of corresponding points as shown in the table in **Figure 5.52.**

Original Coordinates	Reflected Coordinates
(2, 2)	
(2, 4)	
(–1, 4)	
(–3, 1)	
(–6, –1)	
(–5, –4)	
(–3, –6)	
(1, –5)	
(2, –2)	
(4.5, 0.5)	

FIGURE 5.52.
Table for Question 6.

7. What is the domain of the reflected relation?
8. What is the range of the reflected relation?
9. What do you notice about the domain and range of the relation and its reflection?
10. If a point from the relation lies on the line $y = x$, where does the reflection of that point lie?

Answers (continued)

30. The line is reflected back to its original position.

Walk around the room visually inspecting students' work, assessing informally if they have correctly followed the instructions. Each student's graph should look like the following graph.

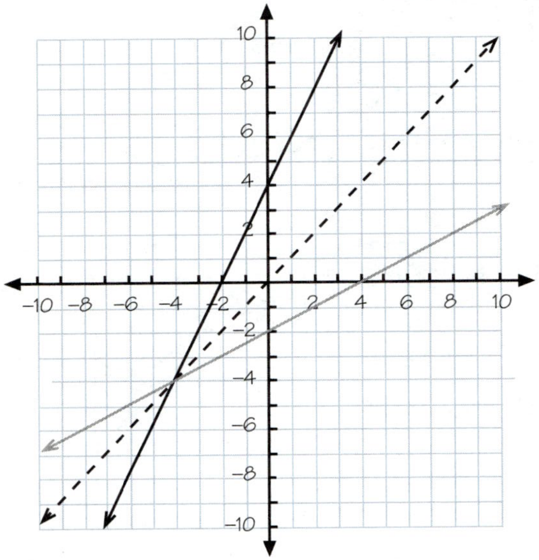

37. Answers may vary. The parabola is reflected across the line $y = x$. The parabola is pointing sideways instead of upwards. The parabola stayed the same shape, but in a different direction.

38. No, the reflection of the parabola is not a function, because for a given x-value, there are two possible y-values when $x > 0$.

39.

Coordinates of Points on the Graph of $y = x^2$	Corresponding Coordinates of Points in Graph of $y = x^2$ Reflected Over the Line $y = x$
(0, 0)	(0, 0)
(−3, 9)	(9, −3)
(2, 4)	(4, 2)

40. Again, the coordinates of corresponding points are reversed.

41. The domain is all real numbers, or, in interval notation, $(-\infty, \infty)$. The range is $\{y \mid y \geq 0\}$ or $[0, \infty)$.

42. The domain is $\{x \mid x \geq 0\}$, or in interval notation, $[0, \infty)$. The range is all real numbers or $(-\infty, \infty)$.

43. The parabolic shape is reflected back to its original position.

Walk around the room visually inspecting students' work, assessing informally if they have correctly followed the instructions. Each student's graph should look like the following graph.

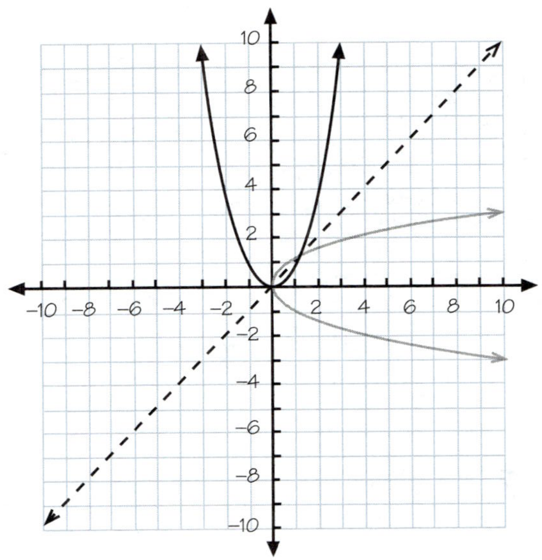

Section 5.16 Assignment

Answers

2. {2, −1, −3, −6, −5, −3, 1, 2, 4.5}

3. {2, 4, 1, −1, −4, −6, −5, −2, 0.5}

Questions 4 and 5:

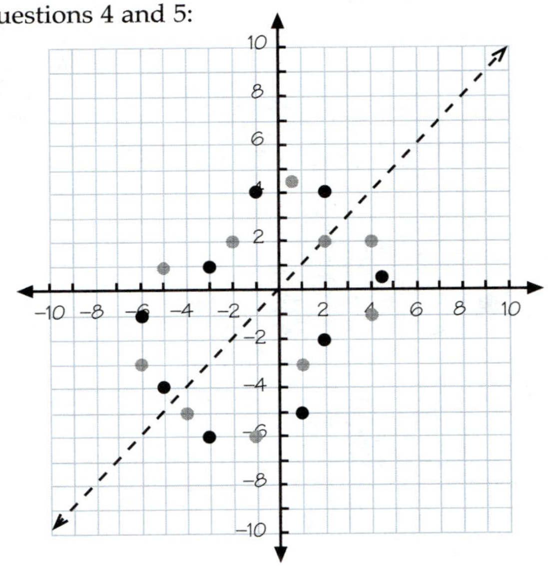

6.

Original Coordinates	Reflected Coordinates
(2, 2)	(2, 2)
(2, 4)	(4, 2)
(−1, 4)	(4, −1)
(−3, 1)	(1, −3)
(−6, −1)	(−1, −6)
(−5, −4)	(−4, −5)
(−3, −6)	(−6, −3)
(1, −5)	(−5, 1)
(2, −2)	(−2, 2)
(4.5, 0.5)	(0.5, 4.5)

7. {2, 4, 1, −1, −4, −6, −5, −2, 0.5}

8. {2, −1, −3, −6, −5, −3, 1, 2, 4.5}

9. The domain and range of the relation and its reflection are switched. The domain becomes the range, and the range becomes the domain.

10. If a point in the relation lies on the line $y = x$, the reflection of that point also lies in the same place on the line. For example, (2, 2) is reflected to (2, 2) because the x- and y-coordinates are the same number.

SECTION 5.17 Inverses

The line $y = x$ acts like a mirror for a relation and its reflection across the line. When you reflect a relation across the line $y = x$, the reflection is called the **inverse**. The x- and y-coordinates of each point in a relation are reversed when you graph the reflection.

CHECK THIS!

A relation is a set of ordered pairs.

1. You may also graph a relation and its inverse on the graphing calculator. Your teacher will give you instructions on how to graph the inverse with the graphing calculator. Enter the domain of the following data set (10 members) into List 1 of your graphing calculator. Enter the range (also 10 members) into List 2.

   {(3, 2), (3, 3), (3, 4), (4, 1), (4, 5), (6, 1), (6, 5), (7, 2), (7, 3), (7, 4)}

2. Examine the scatter plot with List 1 serving as the domain and List 2 serving as the range.
3. Then activate a second scatter plot with List 1 acting as the range and List 2 acting as the domain.
4. Graph the line $y = x$. What do you observe?

CHECK THIS!

- The inverse of a relation is the reflection of the relation over the line y = x.
- When a relation is reflected over the line y = x, the x- and y-coordinates of corresponding points are reversed.
- The domain of the relation becomes the range of its inverse, and the range of the relation becomes the domain of its inverse.
- All relations have inverses.
- The term inverse is used two ways in mathematics: one to describe an opposite operation, the other to describe a type of relation.

An inverse reverses or *undoes* a previous action: A pitcher throws a ball; the catcher catches it and throws it back to the pitcher. Taking your shoes off and putting them back on is an example of an inverse. If you graph a relation, reflect it over the line $y = x$, and then reflect it back, you will have the same relation that you started with.

5. Where in mathematics have you used inverse operations?
6. Not every action can be reversed. List some examples of actions that cannot be reversed.
7. Graph $f(x) = x^2$. Graph the inverse. List five points from each graph in a table.
8. Is $f(x) = x^2$ a function?
9. What are the domain and range for $f(x) = x^2$?
10. Is the inverse of $f(x) = x^2$ a function?
11. What are the domain and range for the inverse?

Not all functions have inverses that are also functions. If you substitute $x = -2$ into $f(x) = x^2$ you get $f(-2) = (-2)^2 = 4$. If you try to undo 4 there are two possible values, 2 or −2 as shown in **Figure 5.53.**

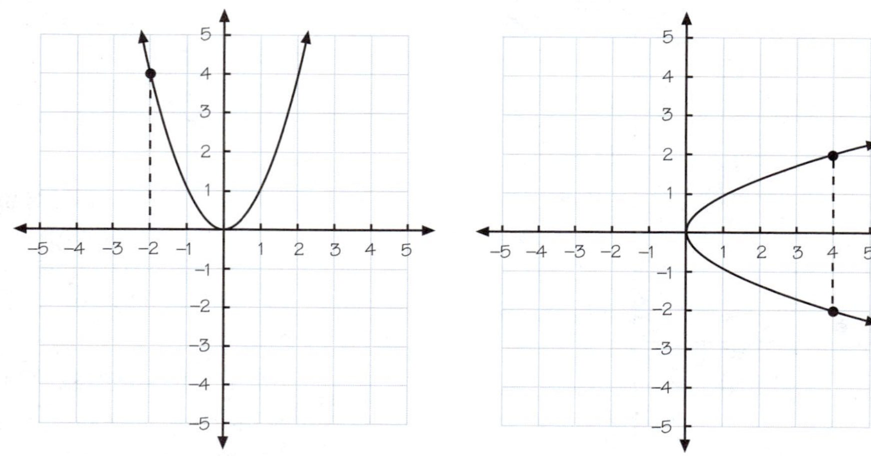

FIGURE 5.53. Inverse of a function.

EXPLAIN LESSON 3

Section 5.17: Inverses

Purpose: Students will develop the definition of inverse in this section. They will formalize the square root function as the inverse of the quadratic function. Students will then be able to do a data collection activity in Section 5.19 involving a square root function.

Description: Students will manipulate data sets in a graphing calculator in order to formalize the concept of inverse. After formalizing the definition of the inverse and the square root parent function as the inverse of a quadratic function, they will use a graphing calculator to investigate transformations to the square root parent function.

Materials: graphing calculator

Teaching Ideas: Continue to encourage students to use patty paper or a graphing calculator to check for the correct inverse if they are unsure.

Answers

1. Remind students how to enter numbers into a list: STAT, Edit, then enter the members.

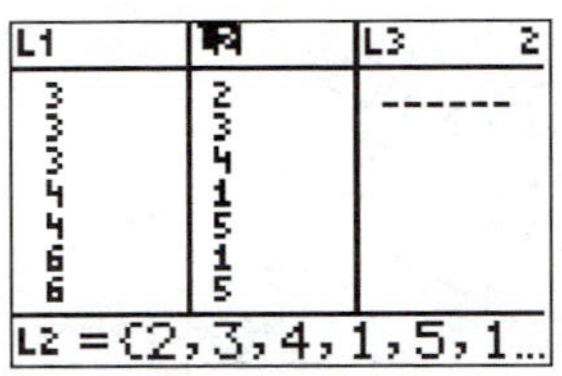

2. Have students turn on STAT PLOT and check to be sure that L1 and L2 are entered as the domain and range for the graph.

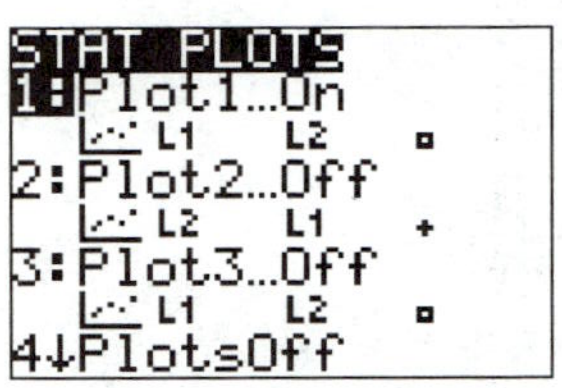

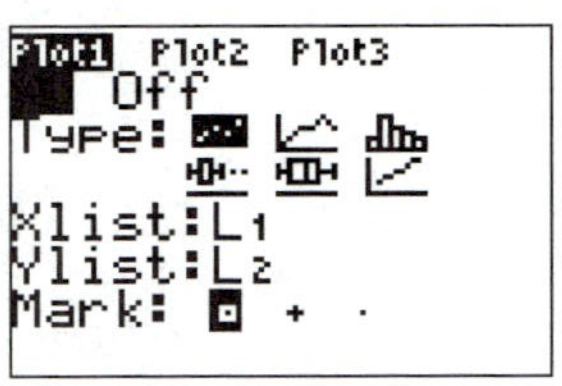

Use an appropriate WINDOW, then graph the relation. (If you use a window where the y-axis is 2/3 as long as the x-axis, the graph will appear proportional.)

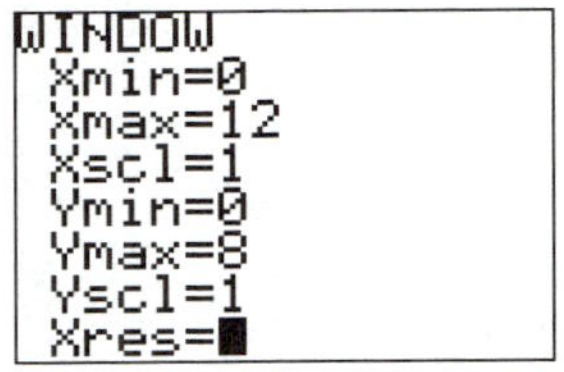

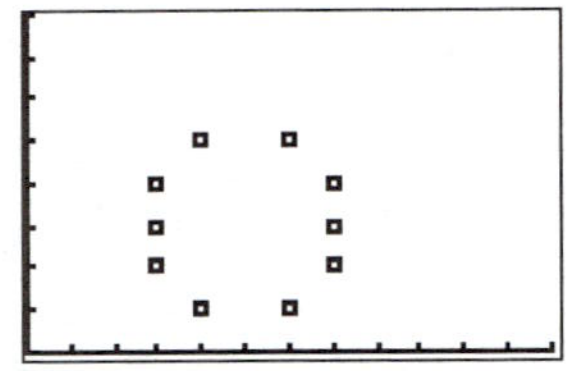

3. Have students turn on STAT PLOT 2. This time L2 is the domain and L1 is the range. Use a different mark for this graph.

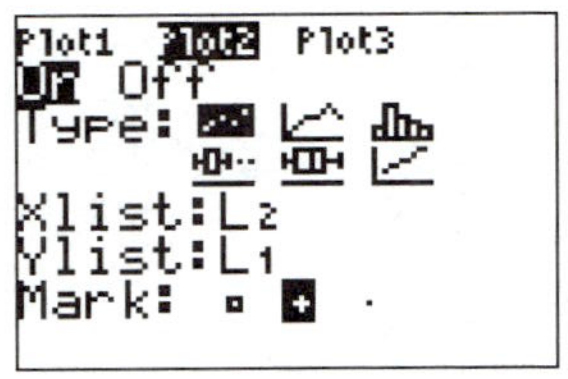

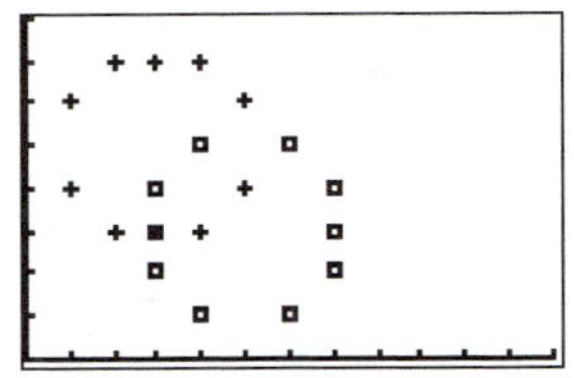

4. 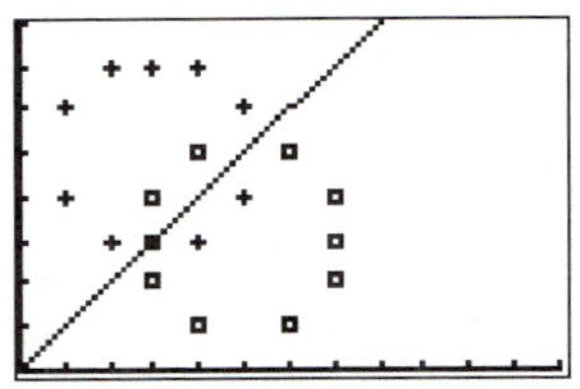

By switching the domain and range, you have graphed the inverse relation. Also, the relation and its inverse are reflected across the line $y = x$.

This data set generates a very obvious shape. The students should be able to see clearly the reflection across the line $y = x$.

5. When you solve an equation such as $x + 3 = 6$, to undo the addition of 3 you add the additive inverse, –3, to both sides of the equation.

When you solve an equation such as $2x = 8$, to undo the multiplication by 2 you multiply by the multiplicative inverse, $\frac{1}{2}$, on both sides of the equation.

In Chapter 4 you solved systems of equations using matrices by multiplying both sides of the matrix equation by the inverse matrix.

6. Answers will vary. Some examples might be:

If you run your car run into a telephone pole, it may be impossible to undo the damage.

It would be impossible to "undial" after dialing a wrong number.

A homerun pitch is impossible to "unthrow."

Rockets cannot be "unlaunched."

7.

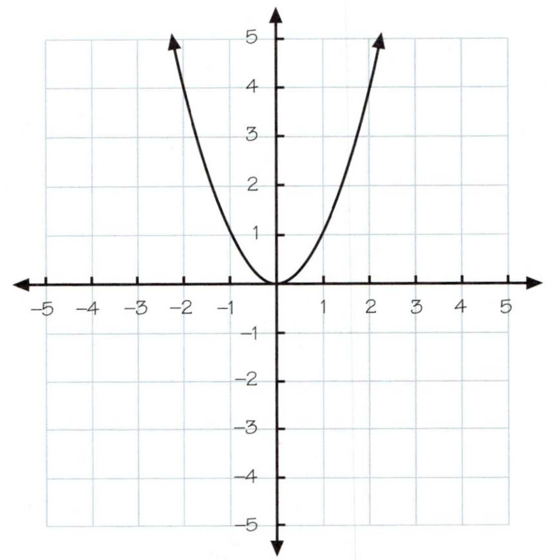

x	y
–2	4
–1	1
0	0
1	1
2	4

Answers (continued)

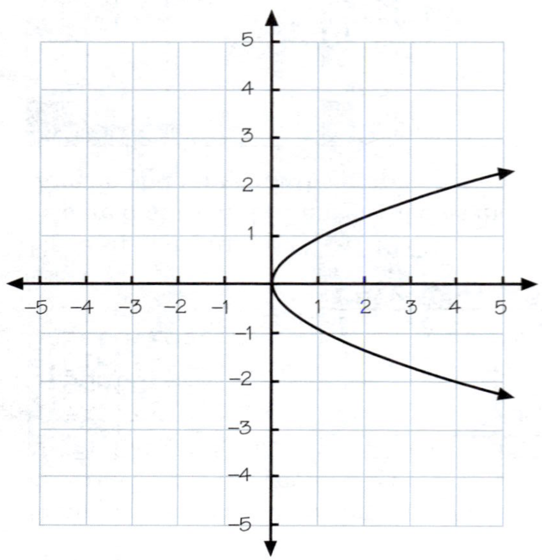

x	y
4	2
4	−2
1	1
1	−1
0	0

8. Yes, for every x-value there is only one y-value. The graph passes the vertical line test.

9. The domain is $\{x \mid x \text{ is a real number}\}$. The range is $\{y \mid y \geq 0\}$.

10. No, there are ordered pairs with the same x-value, but different y-values. The graph does not pass the vertical line test.

11. The domain is $\{x \mid x \geq 0\}$. The range is $\{y \mid y \text{ is a real number}\}$.

In order for $f(x) = x^2$ to have an inverse that is also a function, mathematicians restrict the domain of $f(x) = x^2$ to $x \geq 0$. This gives you only the right branch of the parabola. Then the inverse is a function. The inverse of $f(x) = x^2$ needs to undo the action of squaring a number. The operation that reverses squaring is the square root. The mathematical notation for an inverse function is $f^{-1}(x)$.

If $f(x) = x^2$ for $x \geq 0$, then $f^{-1}(x) = \sqrt{x}$ for $x \geq 0$ (see **Figure 5.54**).

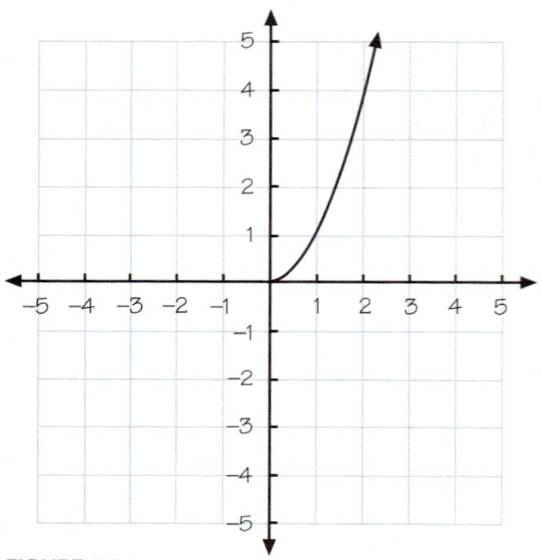

$f(x) = x^2, x \geq 0, y \geq 0$

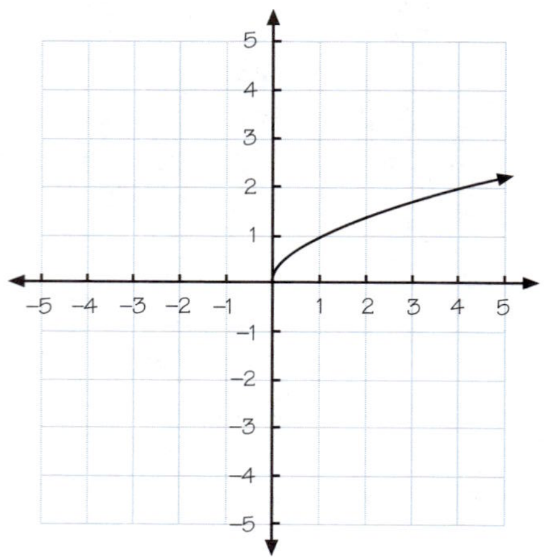

$f(x) = \sqrt{x}\ , x \geq 0, y \geq 0$

FIGURE 5.54.
Inverse of $f(x) = x^2$ for $x > 0$.

Graphing the function and its inverse on the same coordinate axes demonstrates the inverse relationship (see **Figure 5.55**).

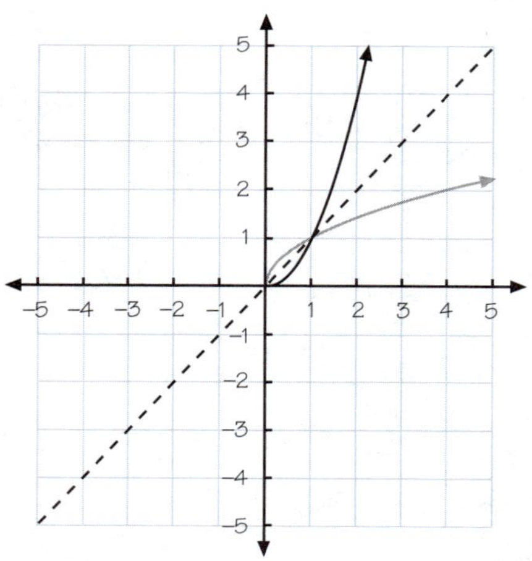

FIGURE 5.55.
Plot of the inverse relationship of Figure 5.54.

12. Use your graphing calculator to graph the parent function $y = \sqrt{x}$. Then graph the additional functions in **Figure 5.56** and fill in the blanks.

Function	Describe Changes to the Parent Function	Domain	Range
$y = \sqrt{x}$	Parent function	$x \geq 0$	$y \geq 0$
$y = \sqrt{x} + 2$		$x \geq 0$	$y \geq 2$
$y = \sqrt{x} - 3$		$x \geq 0$	$y \geq -3$

FIGURE 5.56.
Table for Question 12.

What do you observe?

13. Use your graphing calculator to graph the functions in **Figure 5.57** and fill in the blanks.

Function	Describe Changes to the Parent Function	Domain	Range
$y = \sqrt{x}$	Parent function	$x \geq 0$	$y \geq 0$
$y = \sqrt{x+3}$		$x \geq -3$	$y \geq 0$
$y = \sqrt{x-2}$		$x \geq 2$	$y \geq 0$

FIGURE 5.57.
Table for Question 13.

What do you observe?

14. Use your graphing calculator to graph the functions in **Figure 5.58** and fill in the blanks.

Function	Describe Changes to the Parent Function	Domain	Range
$y = \sqrt{x}$	Parent function	$x \geq 0$	$y \geq 0$
$y = 0.5\sqrt{x}$		$x \geq 0$	$y \geq 0$
$y = 2\sqrt{x}$		$x \geq 0$	$y \geq 0$

FIGURE 5.58.
Table for Question 14.

What do you observe?

15. Use your graphing calculator to graph the functions in **Figure 5.59** and fill in the blanks.

Function	Describe Changes to the Parent Function	Domain	Range
$y = \sqrt{x}$	Parent function	$x \geq 0$	$y \geq 0$
$y = -\sqrt{x}$		$x \geq 0$	$y \leq 0$
$y = -0.5\sqrt{x}$		$x \geq 0$	$y \leq 0$

FIGURE 5.59.
Table for Question 15.

What do you observe?

Answers (continued)

12.

Function	Describe Changes to the Parent Function	Domain	Range
$y = \sqrt{x}$	Parent function	$x \geq 0$	$y \geq 0$
$y = \sqrt{x} + 2$	Parent function is shifted vertically up 2 units	$x \geq 0$	$y \geq 2$
$y = \sqrt{x} - 3$	Parent function is shifted vertically down 3 units	$x \geq 0$	$y \geq -3$

```
Plot1 Plot2 Plot3
\Y1=√(X)
\Y2=√(X)+2
\Y3=√(X)-3
\Y4=
\Y5=
\Y6=
\Y7=
```

```
WINDOW
 Xmin=-2
 Xmax=5
 Xscl=1
 Ymin=-4
 Ymax=5
 Yscl=1
 Xres=1
```

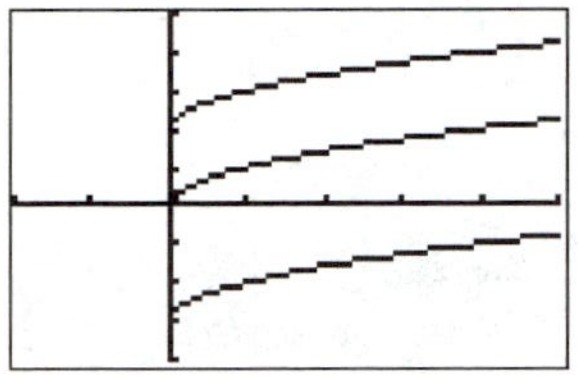

Adding or subtracting a constant shifts the graph of the parent function vertically up or down.

13.

Function	Describe Changes to the Parent Function	Domain	Range
$y = \sqrt{x}$	Parent function	$x \geq 0$	$y \geq 0$
$y = \sqrt{x+3}$	Parent function is shifted horizontally to the left 3 units	$x \geq -3$	$y \geq 0$
$y = \sqrt{x-2}$	Parent function is shifted horizontally to the right 2 units	$x \geq 2$	$y \geq 0$

```
Plot1 Plot2 Plot3
\Y1=█(X)
\Y2=√(X+3)
\Y3=√(X-2)
\Y4=
\Y5=
\Y6=
\Y7=
```

```
WINDOW
 Xmin=-5
 Xmax=5
 Xscl=1
 Ymin=-5
 Ymax=5
 Yscl=1
 Xres=1
```

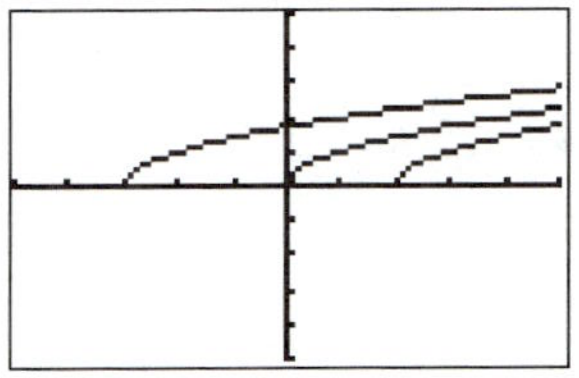

Adding or subtracting a constant beneath the radical shifts the parent function horizontally left or right.

14.

Function	Describe Changes to the Parent Function	Domain	Range
$y = \sqrt{x}$	Parent function	$x \geq 0$	$y \geq 0$
$y = 0.5\sqrt{x}$	Parent function is compressed vertically	$x \geq 0$	$y \geq 0$
$y = 2\sqrt{x}$	Parent function is stretched vertically	$x \geq 0$	$y \geq 0$

```
Plot1 Plot2 Plot3
\Y1=√(X)
\Y2=0.5√(X)
\Y3=2√(X)
\Y4=█
\Y5=
\Y6=
\Y7=
```

```
WINDOW
 Xmin=-1
 Xmax=5
 Xscl=1
 Ymin=-1
 Ymax=5
 Yscl=1
 Xres=█
```

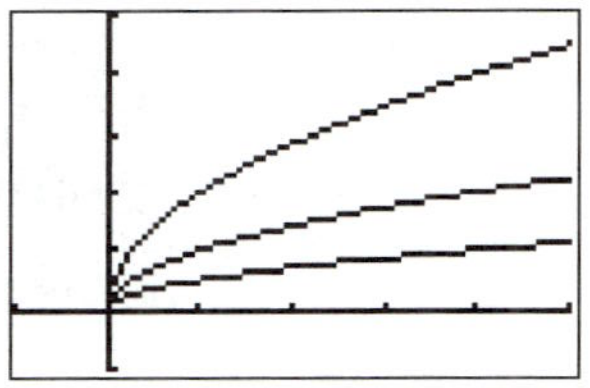

Changing the coefficient, a, for $y = a\sqrt{x}$ vertically stretches or compresses the graph of the parent function. If $0 < a < 1$, the graph is compressed vertically. If $a > 1$, the graph is stretched vertically.

15.

Function	Describe Changes to the Parent Function	Domain	Range
$y = \sqrt{x}$	Parent function	$x \geq 0$	$y \geq 0$
$y = -\sqrt{x}$	Parent function is reflected over the x-axis	$x \geq 0$	$y \leq 0$
$y = -0.5\sqrt{x}$	Parent function is compressed vertically and reflected over the x-axis	$x \geq 0$	$y \leq 0$

```
Plot1 Plot2 Plot3
\Y1=√(X)
\Y2=-√(X)
\Y3=-0.5√(X)
\Y4=
\Y5=
\Y6=
\Y7=
```

```
WINDOW
 Xmin=-1
 Xmax=3
 Xscl=1
 Ymin=-3
 Ymax=3
 Yscl=1
 Xres=1
```

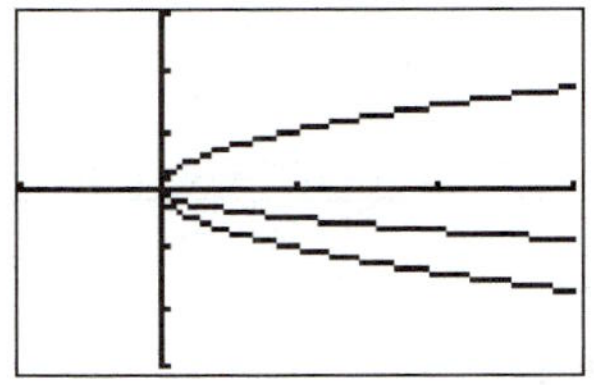

If the coefficient, a, is less than 0, the function, $y = a\sqrt{x}$, is reflected over the x-axis.

16. How are the transformations of $y = \sqrt{x}$ similar to what you learned about transformations of $y = x^2$?

17. Use a graphing calculator to make a table for $y = \sqrt{x-4}$, why does the word "ERROR" appear for values of x that are less than 4?

18. Graph $f(x) = x^2 + 2$ for $x \geq 0$. Graph its inverse. What is the equation of the inverse, $f^{-1}(x)$?

SUMMARY

The graph of the inverse of the graph of a function is the reflection of the function over the line $y = x$. The domain of the function becomes the range of its inverse. The range of the function becomes the domain of its inverse.

The inverse of the quadratic parent function where $x \geq 0$ is the square root function. Transformations of the square root function have effects like those of the quadratic parent function.

- $f(x) = \sqrt{x} + c$ results in a vertical translation.
- $f(x) = \sqrt{x+c}$ results in a horizontal translation.
- $f(x) = a\sqrt{x}$ results in a vertical compression, vertical stretch, or a reflection over the x-axis.

SECTION 5.17 Assignment

1. Plot this relation and its inverse on a grid. Use a different color to distinguish the function from its inverse.

   {(–8, 0), (–7, 1), (–4, 2), (1, 3), (5, 3.5), (8, 4)}

2. Graph the inverse of the relation in **Figure 5.60** on the same grid. (Hint: You can fold your paper along the line $y = x$ to check your answer.)

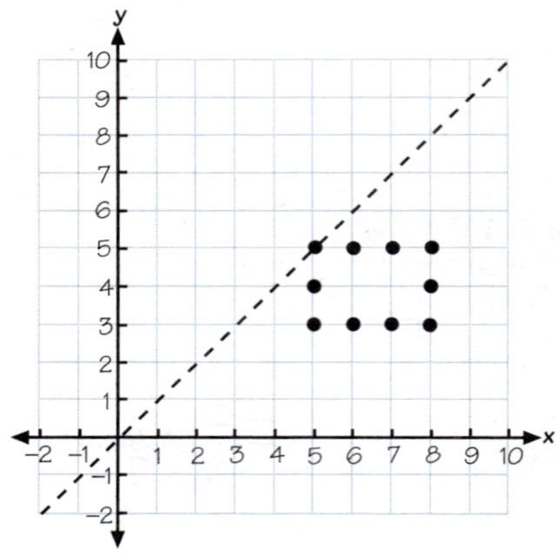

FIGURE 5.60.
Scatter plot for Question 2.

For Questions 3–8, is the statement true or false?

3. If the point (L, D) is a member of some function, then (D, L) is a member of its inverse.

4. If a function's graph intersects the graph of the line $y = x$ at a certain point, its inverse must intersect the line $y = x$ at the same point.

5. The inverse of any function is also a function.

6. A function could intersect the line $y = x$ at (5, 8).

7. The graphs of a function and its inverse are mirror images across the line $y = x$.

8. The domain of any set serves as the range of its inverse.

Answers (continued)

16. The transformations are similar because adding or subtracting constants generates horizontal or vertical shifts depending on where the constants are placed. Changing the coefficient of the function generates a vertical stretch or compression. Having a negative coefficient causes a reflection over the x-axis.

17.

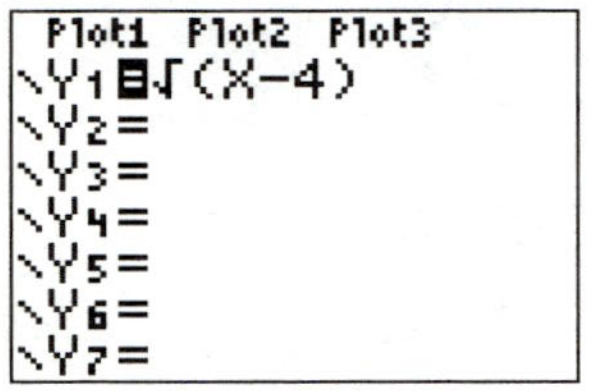

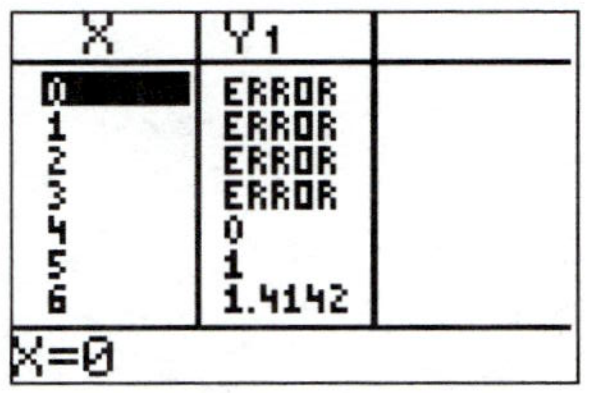

The square root of a negative number is not a real number. For example, $\sqrt{1-4} = \sqrt{-3}$, which is not a real number.

18.

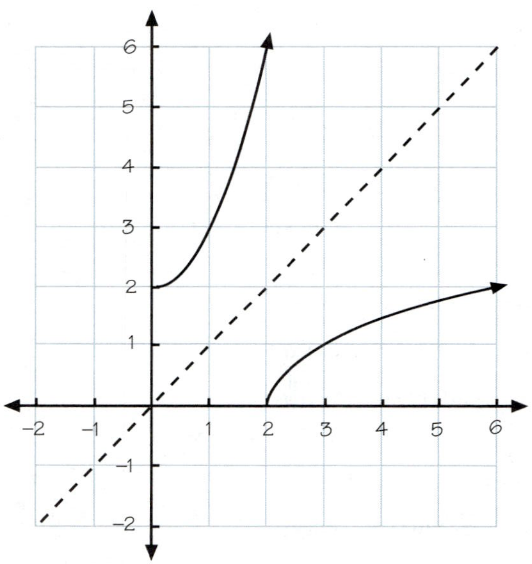

The parent function, $f(x) = \sqrt{x}$, has been shifted 2 units horizontally to the right, so the equation of the inverse is $f^{-1}(x) = \sqrt{x-2}$. Check your answer with the graphing calculator.

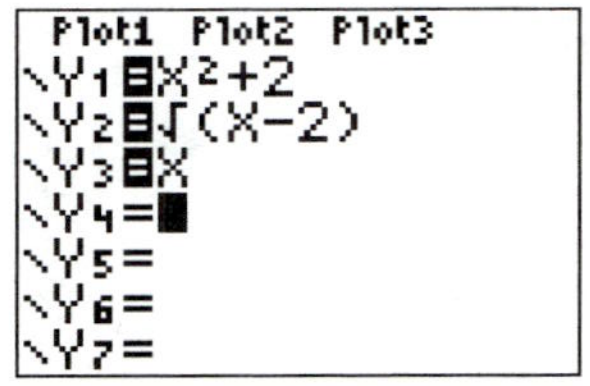

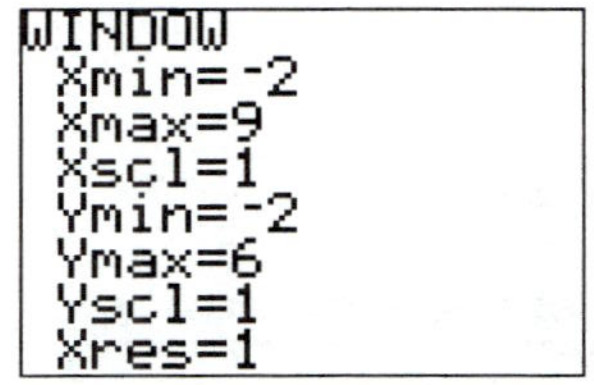

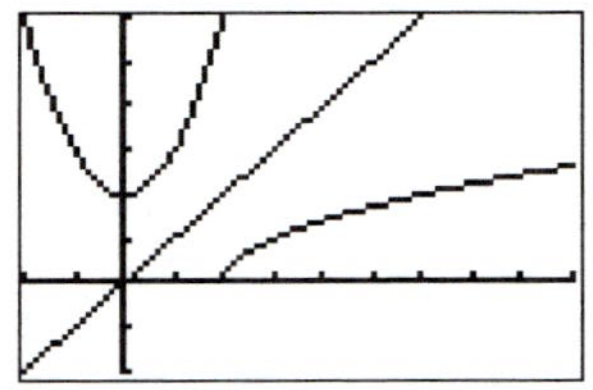

Section 5.17 Assignment

Answers

1.

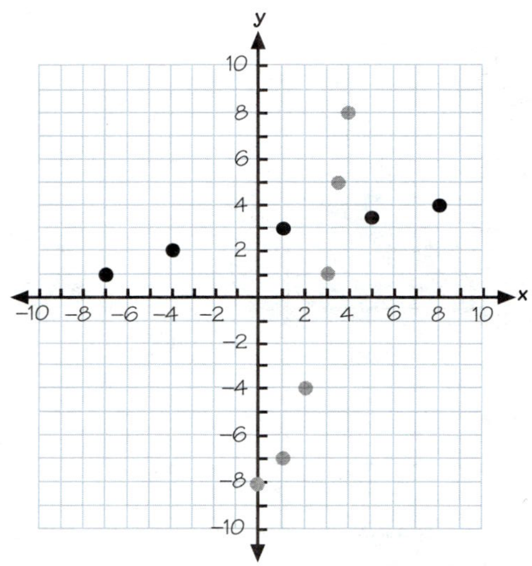

2.

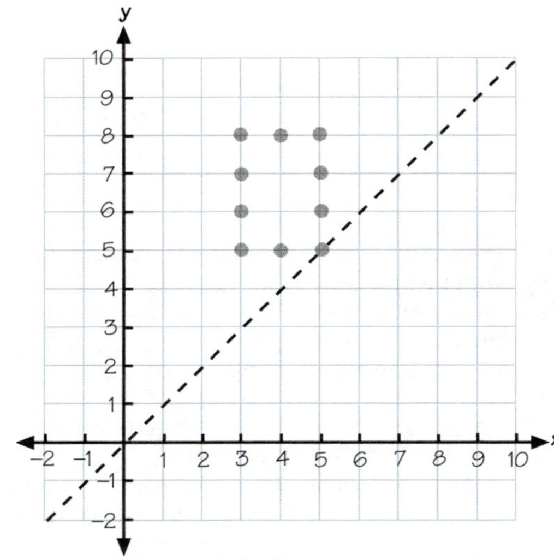

3. True

4. True

5. False

6. False

7. True

8. True

9. Graph $f(x) = \sqrt{x} + 3$ on a grid.

   a) State the domain.

   b) State the range.

   c) Using your knowledge of inverses, state the domain and range of the inverse (f^{-1}) of $f(x) = \sqrt{x} + 3$.

   d) Sketch the graph of the inverse on the same grid.

   e) Now find the inverse function.

10. Graph the following transformations of the square root function.

    a) $f(x) = -2\sqrt{x}$

    b) $f(x) = \sqrt{x} + 2$

    c) $f(x) = \sqrt{x+2}$

SECTION 5.18 Just a Swingin'

Set up an experiment as shown in **Figure 5.61** and time the period of the pendulum for different lengths of string. Here are the steps.

Step 1: On a large sheet of chart paper draw a vertical line.

Step 2: Use a protractor to measure 10° off center of the vertical line and draw another line.

Step 3: Tape the sheet of chart paper to the edge of a desk or table.

Step 4: Tape a pencil to the surface of the desk at the point where the vertical line and the angled line intersect.

Step 5: Make sure you have enough space for a pendulum to swing freely.

Step 6: Tie a washer to the end of the string.

Step 7: Attach the string to the pencil so the length from the pencil to the center of the washer is 15 centimeters.

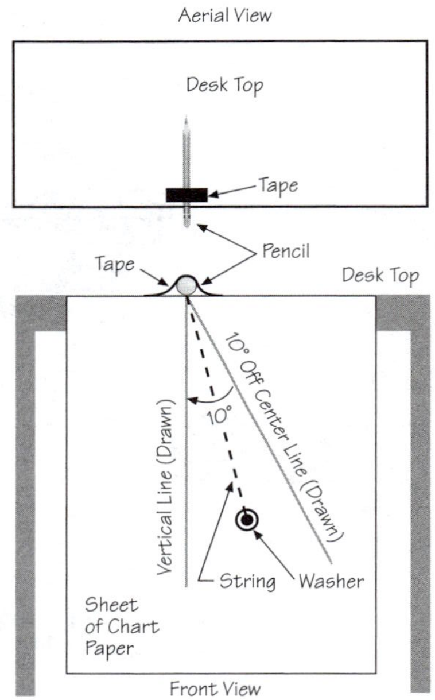

FIGURE 5.61.
Pendulum experiment.

Step 8: Pull the washer back to the 10° line. Let the washer drop freely: do not push or throw it. The washer should swing back and forth in a plane. If it appears to be swinging in a circular fashion, repeat the trial.

Length of Pendulum (cm)	Time for 10 Swings (sec)	Period (sec)
15		
25		
35		
45		
55		
65		

Step 9: Using a stopwatch or a digital watch, find the time it takes for the pendulum to make 10 full swings. Record your measurement in a table like **Figure 5.62**.

FIGURE 5.62.
Table for pendulum experiment.

Answers (continued)

9. a) The domain is $x \geq 0$ or $[0, \infty)$.

   b) The range is $y \geq 3$ or $[3, \infty)$.

   c) The domain is $x \geq 3$ or $[3, \infty)$.

   The range is $y \geq 0$ or $[0, \infty)$.

   d)

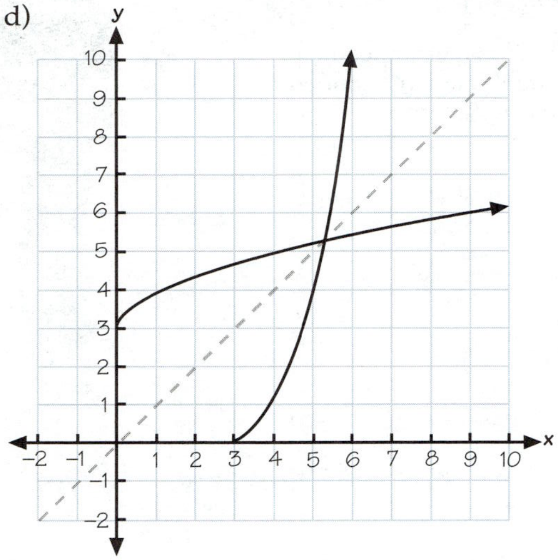

   e) Since this is the quadratic function shifted 3 units to the right, $f^{-1}(x) = (x - 3)^2$, for $x \geq 3$.

10. a)

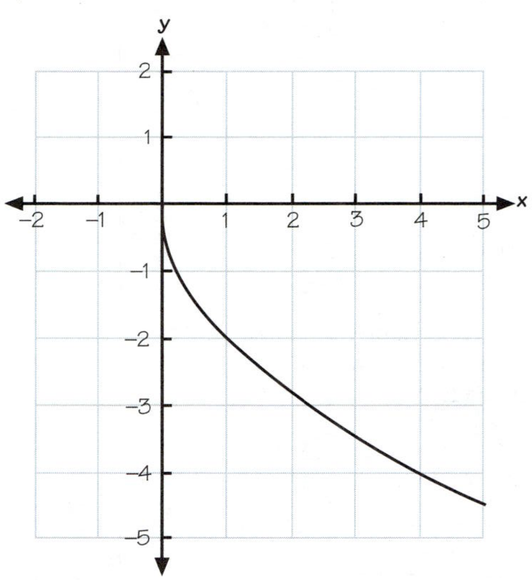

b)

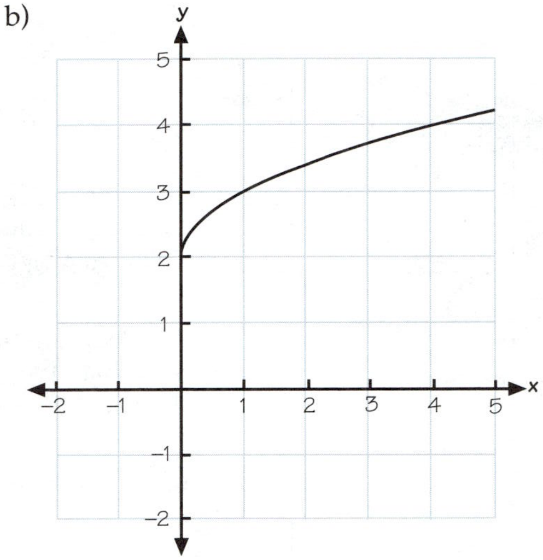

c)

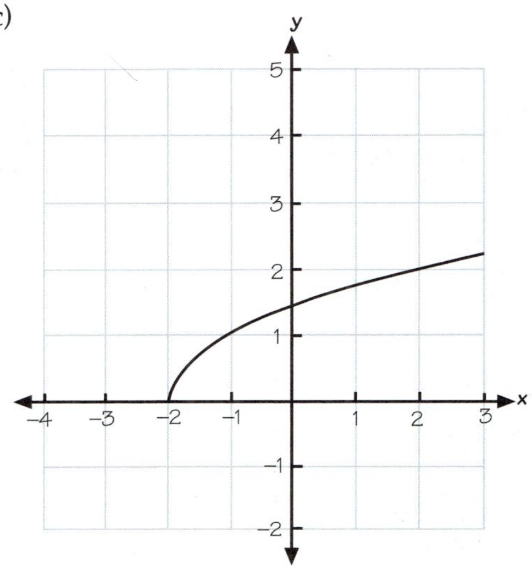

ELABORATE LESSON 3

Section 5.18: Just a Swingin'

Purpose: The purpose of this section is for students to collect data, relate the data to the square root parent function, and generate a model for the data. This connects what they learned in Sections 5.16 and 5.17 to a real-world application. They will be prepared to solve a problem involving a marble rolling down an incline in Section 5.19.

Description: Students collect data on the length of a pendulum and the period of the pendulum. They find a model using the square root function for the data collected. The only variable that is changing is the length of the pendulum; other variables, angle of release and mass of the bob, are remaining constant.

Materials: washers, digital watch or stopwatch, tape, pencil, protractor, length of string about 1 meter, meter sticks, large sheets of chart paper

Teaching Ideas: Divide students into groups of 4. Each group will need to assign jobs: timer, counter, releaser, and calculator. Each student should have their own table filled in with the collected data.

Step 10: Calculate the time for one period and record it in the table.

Step 11: Continue measuring the time required for 10 back-and-forth swings as you change the length of the string as shown in the table. You may try other lengths if you wish.

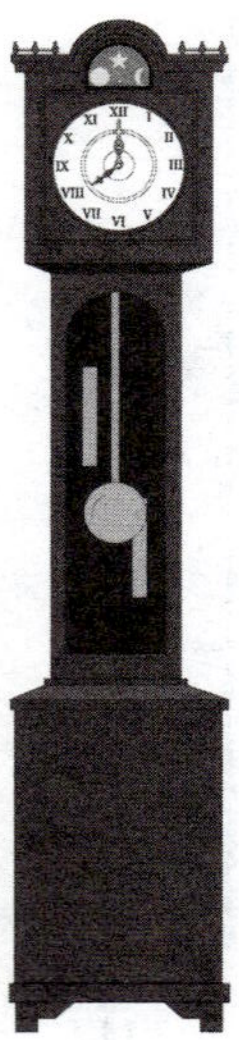

1. How does changing the length of the string affect the period of the pendulum?
2. Graph the results from your experiment on a coordinate grid, and then make a scatter plot on a graphing calculator.
3. Which parent function (linear, quadratic, or square root) best fits your data?
4. Write a function rule to fit your data. Test your function with the graphing calculator.
5. The relationship between the length of a simple pendulum and the period of the pendulum is described by

   $$\textit{Period of a pendulum} = \frac{2\pi}{\sqrt{\textit{Acceleration due to gravity}}} \cdot \sqrt{\textit{Length of pendulum}}.$$

   If the acceleration due to gravity is 980 cm/sec^2, is this equation similar to the function you found?

   $$P = \frac{2\pi}{\sqrt{\frac{980 \text{ cm}}{\text{sec}^2}}} \cdot \sqrt{L} \text{ cm}$$

   $$P = \frac{2\pi}{\sqrt{980}}\sqrt{L} \text{ sec}$$

   $$P = 0.2\sqrt{L} \text{ sec}$$

6. Use your function to predict the period for a pendulum that is 200 cm in length.
7. A clockmaker wants the swing of a pendulum to match a clock's ticking of the seconds. In other words, he wants the period of the pendulum to be 2 seconds. How long should the pendulum be?
8. Ava notices that her grandfather clock is slow. To adjust it should she make the pendulum wire longer or shorter?

SECTION 5.18 Assignment

1. Sketch the inverse of each graph in **Figure 5.63.** The graph of $y = x$ is shown as a dotted line to help you sketch the inverse.

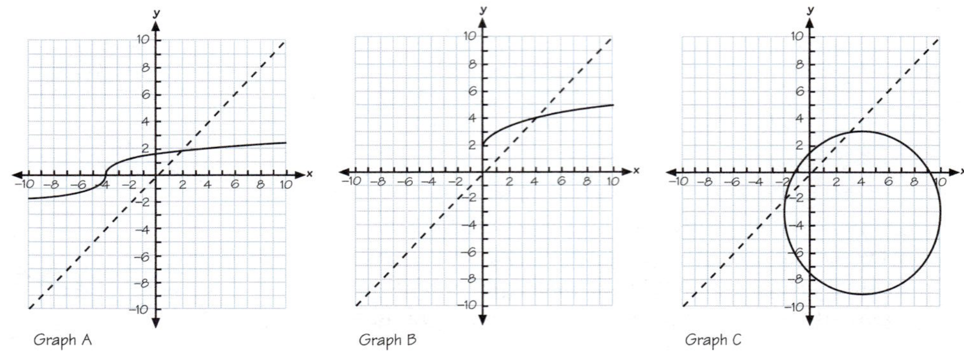

FIGURE 5.63.
Three graphs for Question 1.

2. Look at the inverses you sketched in Question 1. Decide if each inverse is a function or a relation.

   Graph A: Function or relation?

   Graph B: Function or relation?

   Graph C: Function or relation?

3. The graph of $y = (x + 1)^2$ is shown in **Figure 5.64.** Answer these questions based on this graph.

   a) State the domain and range of the function.

   b) Sketch the graph of the inverse.

   c) State the domain and range of the inverse.

   d) What two functions can be entered into your calculator in order to see the graph of the inverse?

   e) Is the inverse of $y = (x + 1)^2$ a function? Justify your answer.

   f) How could you restrict the domain of $y = (x + 1)^2$ so its inverse would be a function?

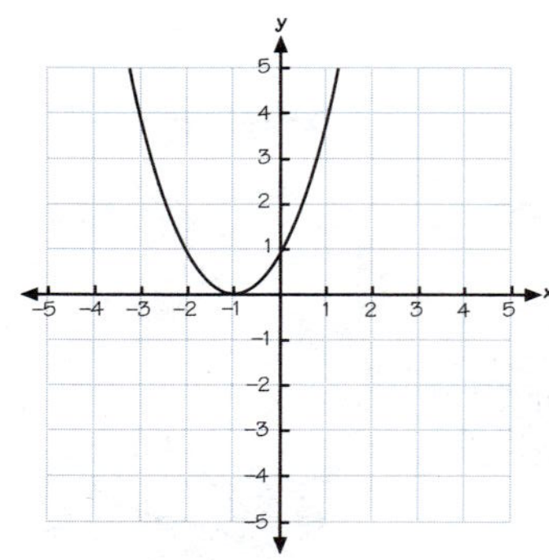

FIGURE 5.64. Graph of $y = (x + 1)^2$.

NOTE: Remember! The period is the time required for the pendulum to return to its original position.

(In Step 11, sample data is listed as a guide for teachers.)

Length of Pendulum (cm)	Time for 10 Swings (sec)	Period (sec)
15	7.72	0.772
25	9.97	0.997
35	11.68	1.168
45	13.28	1.328
55	14.68	1.468
65	15.96	1.596

Answers

1. As the pendulum gets longer, the period of each swing increases.
2. Using the sample data from the table, the scatter plot would look like the following graph.

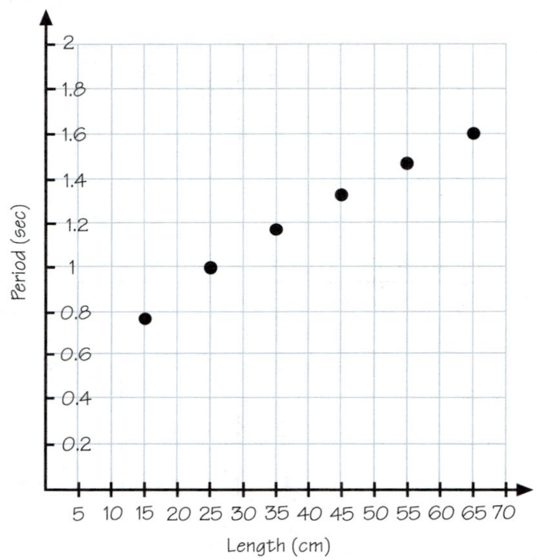

Students enter the data in the lists and graph the scatter plot with the calculator.

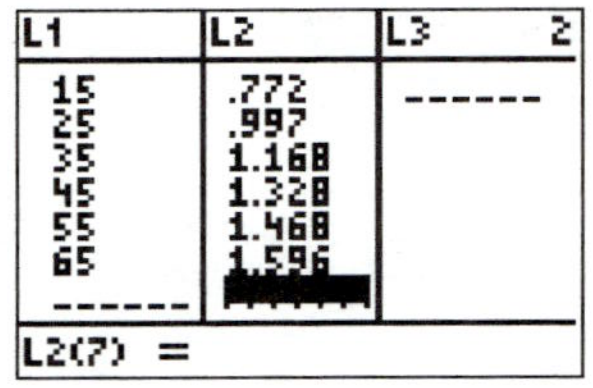

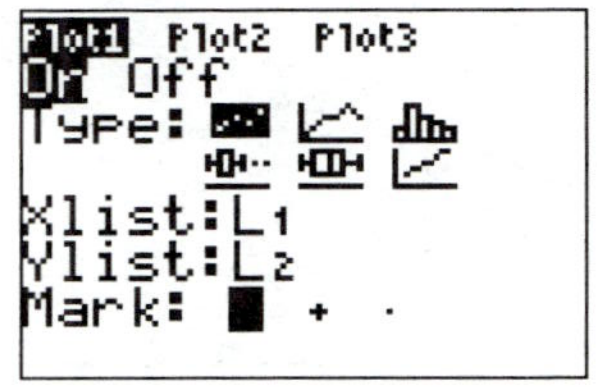

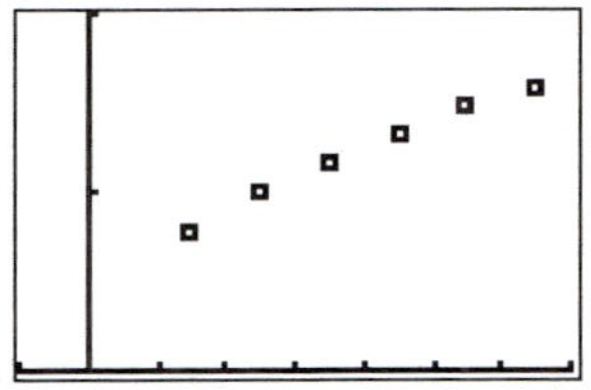

3. Answers will vary at this point. The data set appears as though it might be any of the three if you look only at the graph. If you check the first and second differences the data do not appear to be either linear or quadratic. Encourage students to investigate if the best model might be the square root function.
4. If students find the linear regression or quadratic regression, they will have fairly good fits. If the ordered pair (0, 0) is added to the list, which means when the length is 0 cm and the period is 0 sec, the graph will appear to be more of a square root function. This idea makes sense to students. As the length gets closer to 0, the period will get closer to 0 also. Linear and quadratic regression models do not contain the point (0, 0). Since the function has not been shifted horizontally or vertically, it is appropriate to change the coefficient to vertically compress the function. Allow students time to guess and check with different coefficients. They are trying to fit a curve to the data set using transformations. An example is shown below. As the coefficient gets smaller the graph is compressed until $y = 0.2\sqrt{x}$ appears to fit the data well.

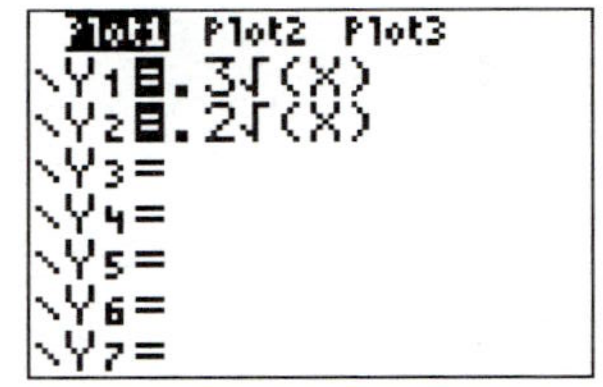

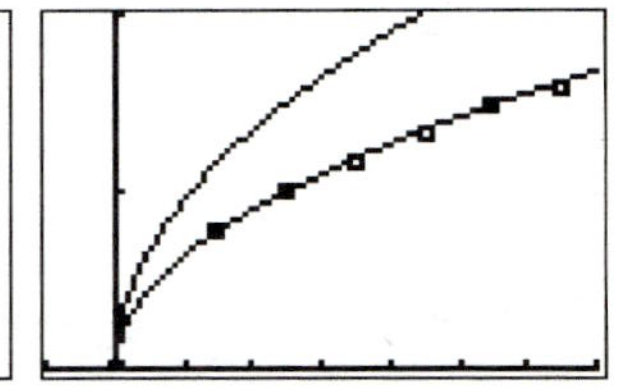

5. The model derived from the sample data is very close to this equation. Student models should be similar. The important idea here is for students to work in groups collecting data and attempting to find a model to use to predict.
6. Students could solve this problem in several ways. They could investigate the table, trace the graph, or solve algebraically. The length of the period would be approximately 2.83 seconds. You might want to have one group actually build a pendulum with a length of 200 cm in a doorframe, measure the time for 10 swings, and calculate the experimental period.
7. Students could solve this problem in several ways. They could investigate the table, trace the graph, or solve algebraically. The pendulum length should be 100 cm, or about 39.4 inches. Again, students might want to actually test their answer by building a pendulum with a length of 100 cm.
8. She needs to make the wire swing faster, so she should shorten the wire. A shorter wire decreases the period.

Section 5.18 Assignment

Answers

1. Graph A

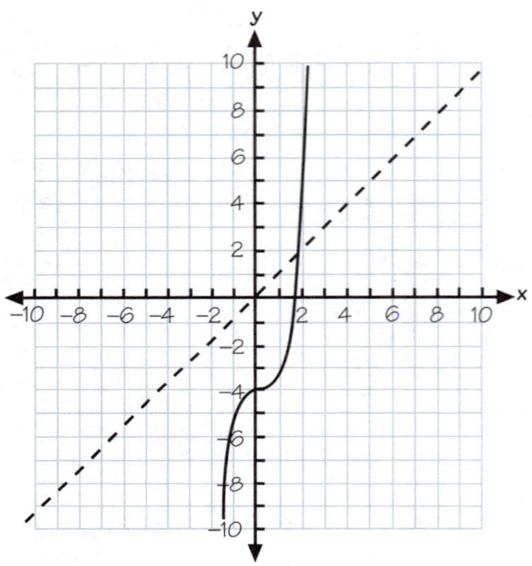

Graph B

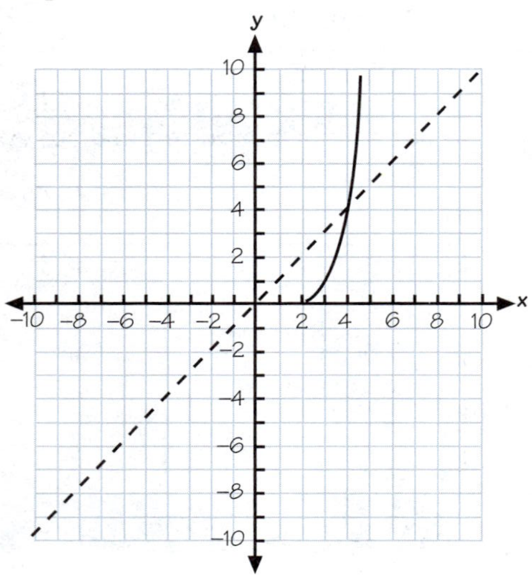

Graph C

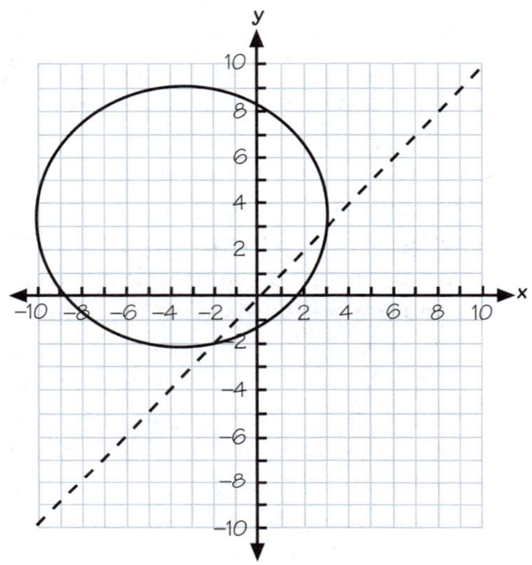

2. Graph A: Function

   Graph B: Function

   Graph C: Relation

3. a) Domain: x is any real number, or $(-\infty, \infty)$

   Range: $y \geq 0$, or $[0, \infty)$

   b)

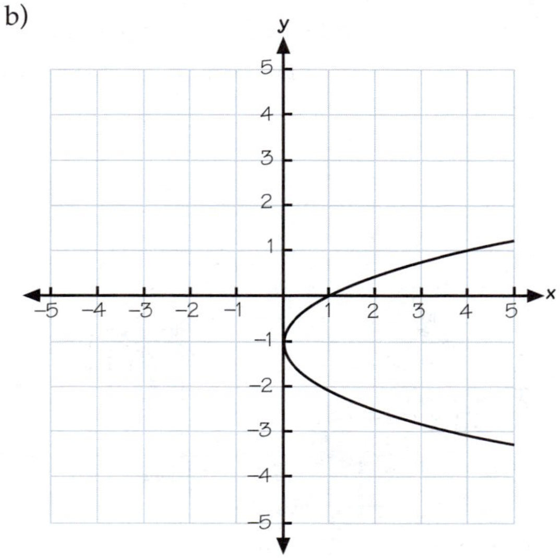

   c) Domain: $x \geq 0$, or $[0, \infty)$

   Range: y is any real number, or $(-\infty, \infty)$

   d) Students can use what they know about transformations to the parent function to find these two functions.

   $y_1 = \sqrt{x} - 1$

   $y_2 = -\sqrt{x} - 1$

   e) The inverse graph is not a function because it does not pass the vertical line test. There are multiple y-values for the same x-value.

   f) If you restrict the domain of $y = (x + 1)^2$ to $x \geq -1$, then the inverse is a function.

4. Graph each function.

   a) $f(x) = 2\sqrt{x}$

   b) $f(x) = \sqrt{x} - 3$

   c) $f(x) = \sqrt{x - 3}$

The distance you can see from the top of a building depends on the building's height. If h is the height in meters of your viewing place, the distance you can see in kilometers on a clear day is modeled by

$S(h) = 3.532\sqrt{h}$.

5. From a height of 100 meters how far can you see?

6. Draw a graph for $S(h) = 3.532\sqrt{h}$. How can the graph of $S(h)$ be obtained from the graph of the square root parent function?

7. If you can see 14 km from the top of a ride at an amusement park, how tall is the ride?

   a) List examples of situations where this viewing model might not be accurate.

SECTION 5.19 Rolling Stone

An object's acceleration on an incline is constant. The distance (d) moved from rest is related to the time (t) that a marble has been rolling by a constant k, which is unique to each incline (see **Figure 5.65**).

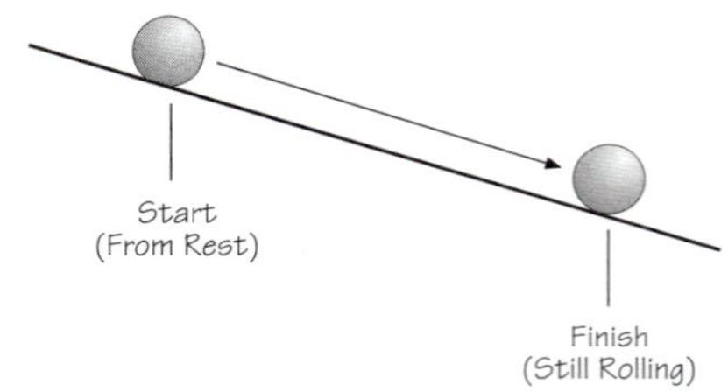

FIGURE 5.65.
A rolling marble.

Trial	Distance (d)	Time (t)
A	0.25 m	1.77 sec
B	0.40 m	2.23 sec
C	0.55 m	2.65 sec
D	0.70 m	2.96 sec
E	0.85 m	3.25 sec
F	1.00 m	3.54 sec

FIGURE 5.66.
Math students' data.

1. A group of math students measured the distance, d, a marble rolled down a plane and the time, t, it took for the marble to roll each distance. They recorded their data in the table in **Figure 5.66.**

   a) Make a scatter plot of the group's data.

   b) Find a function that fits the data.

Trial	Distance (d)	Time (t)
G	0.25 m	1.50 sec
H	0.40 m	1.95 sec
I	0.55 m	2.25 sec
J	0.70 m	2.60 sec
K	0.85 m	2.87 sec
L	1.00 m	3.05 sec

FIGURE 5.67.
Data for Question 2.

2. Another group of students measured the distance and time for a marble to roll down a different plane. Their data are shown in **Figure 5.67.**

   a) Make a scatter plot of the data.

   b) Find a function that fits the data.

Answers (continued)

4. a)

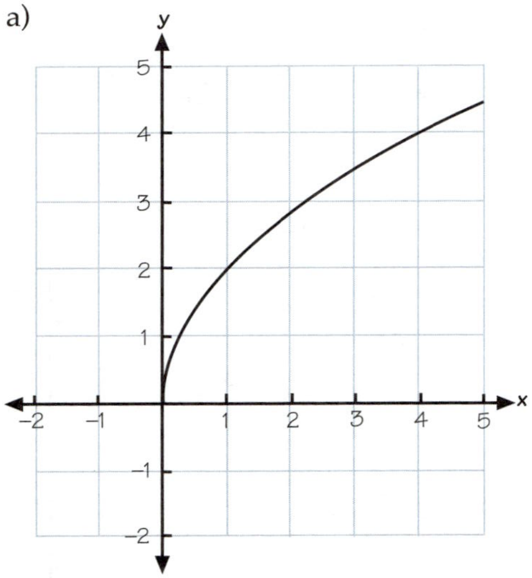

b)

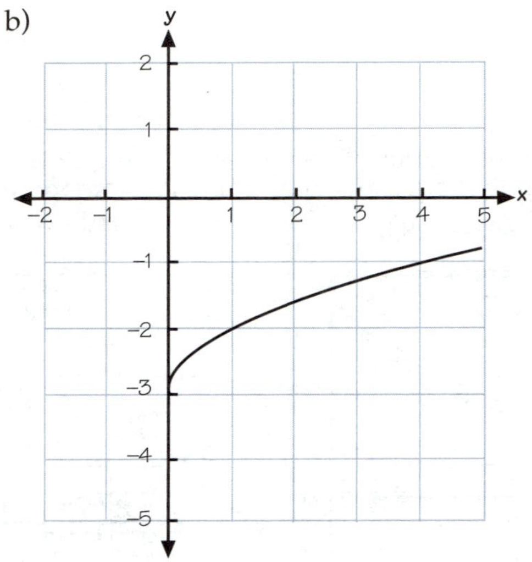

c)

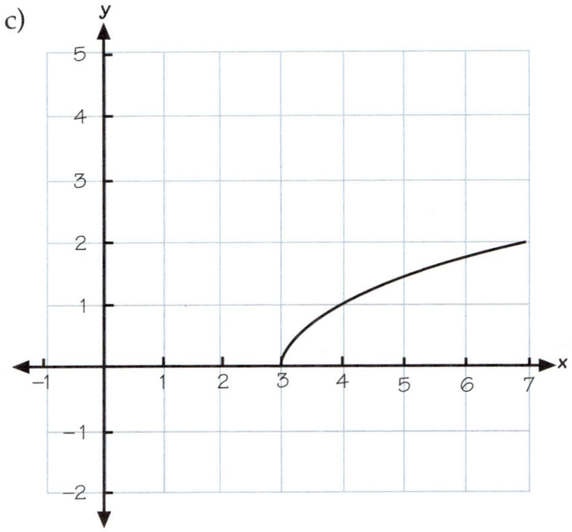

5. 35.32 km. Students may solve by using a table or by substituting into the function.

6.

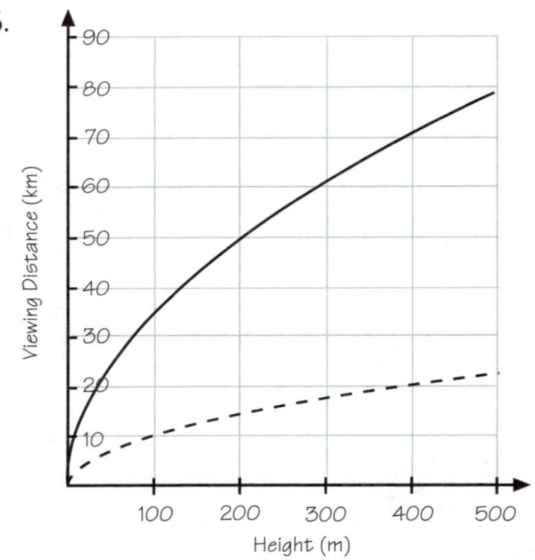

Since the coefficient is 3.532, the parent function has been stretched vertically.

7. 15.7 m. Students may solve by using a table, a graph, or solving algebraically.

a) The model might not be accurate if the weather is cloudy, rainy, or snowy. It might not be accurate if there are objects that obscure your vision such as mountains, trees, or other buildings.

EVALUATE LESSON 3

Section 5.19: Rolling Stone

Purpose: The purpose of this section is for students to use the square root function to predict the motion of a marble.

Description: This section evaluates students' understanding of the square root function and uses curve fitting to determine whether the square root function accurately models the motion of a marble rolling down an inclined plane.

Materials: none

Teaching Ideas: Students can either do this evaluation in groups or individually.

Answers

1. a)

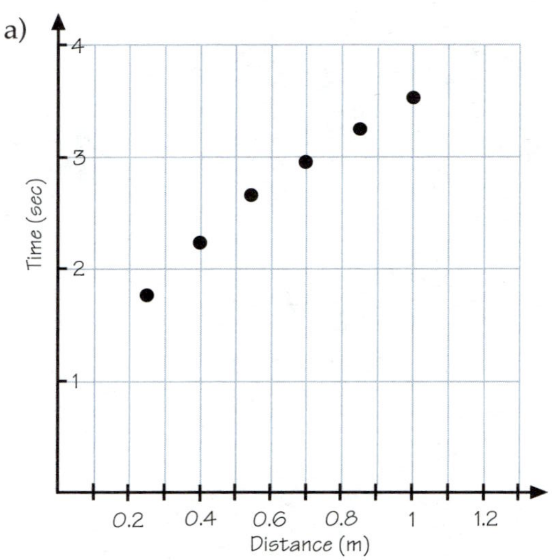

Answers (continued)

b) Using curve fitting with transformation on the graphing calculator, students should generate a model similar to the one below.

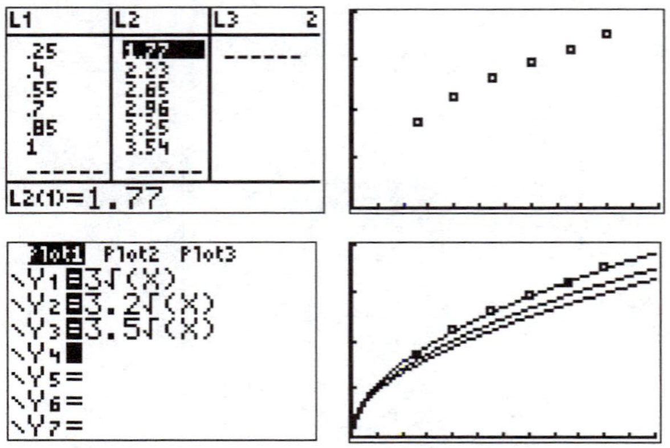

The square root function $y = 3.5\sqrt{x}$ is a good model for this set of data since the graph should include (0, 0). If the marble rolls 0 meters it would take 0 seconds.

2. a)

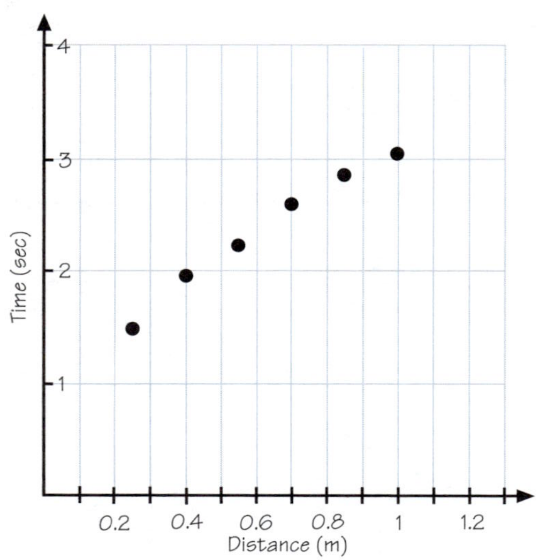

b)

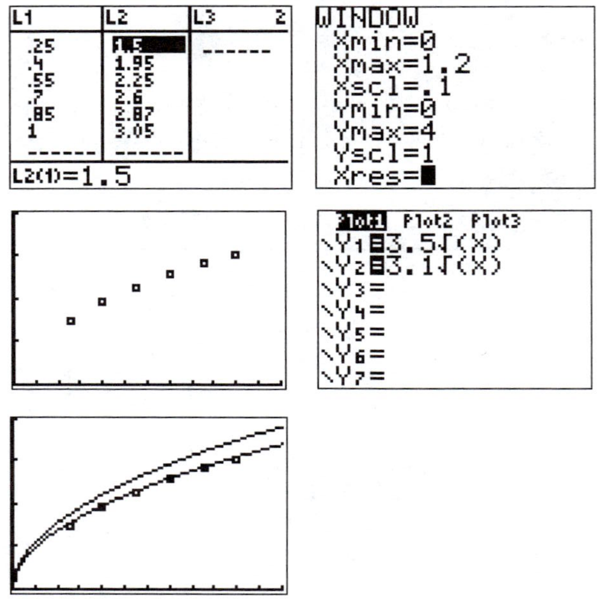

The square root function $y = 3.1\sqrt{x}$ is a good model for this set of data.

3. The two groups of students want to set up a marble stunt with their inclines as shown in **Figure 5.68.** At what distance should each marble be released for them to meet at the bottom exactly 4 seconds later?

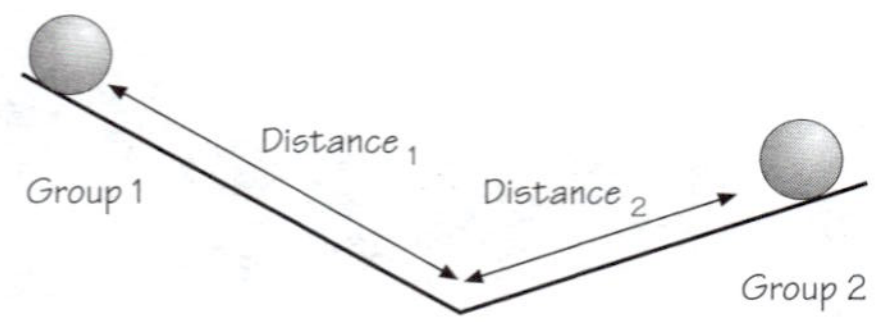

FIGURE 5.68. Inclines for Question 3.

SECTION 5.19 Assignment

The questions below are based on the relationship $k = \frac{d}{t^2}$.

1. If your value for k is 0.5 m/s^2, write a function for distance (d) in terms of time (t).

2. Use your function from Question 1 to calculate the distance your marble would move during the times in the table in **Figure 5.69.** Complete the table.

FIGURE 5.69. Data table for Question 2.

Time (t)	0 s	0.4 s	0.6 s	0.8 s	1.0 s	1.2 s
Distance (d)						

3. Let time (in seconds) serve as your domain and distance (in meters) as your range. Draw a scatter plot of your data from Question 2.

4. Graph the function you wrote in Question 1.

5. What is the parent function for this graph?

6. Now switch your domain and range so that distance (now in meters) serves as your domain and time (now in seconds) as your range. Draw a scatter plot using the data from Question 2.

7. What is the parent function for this graph?

8. Are the two graphs inverses of each other?

9. Find an equation for the inverse of your function from Question 1. Hint: Solve for t.

10. Graph the equation you found in Question 9.

Answers (continued)

3. Students may solve this problem in several ways. They may use tables, graphs, or solve algebraically.

   Solving with tables:

   Investigate the table for each model to find the distance corresponding to a time of 4 seconds in List 1 and List 2.

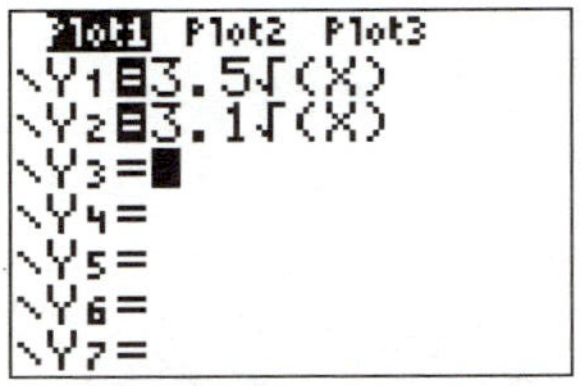

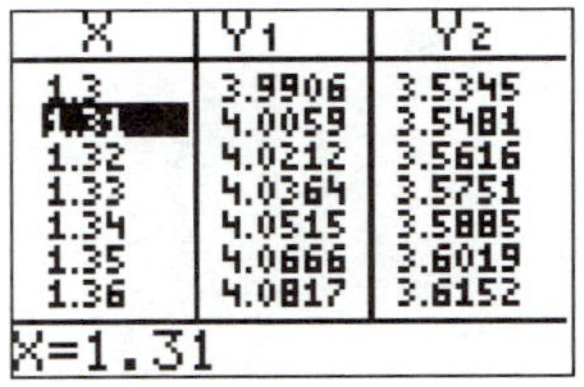

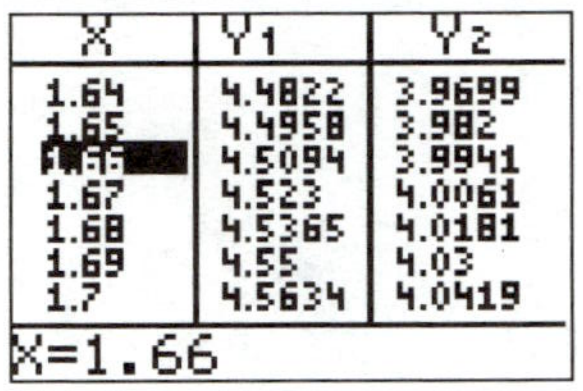

   Group 1 should place the marble at 1.31 meters, and group 2 should place their marble at 1.66 meters.

   Solving by graphing:

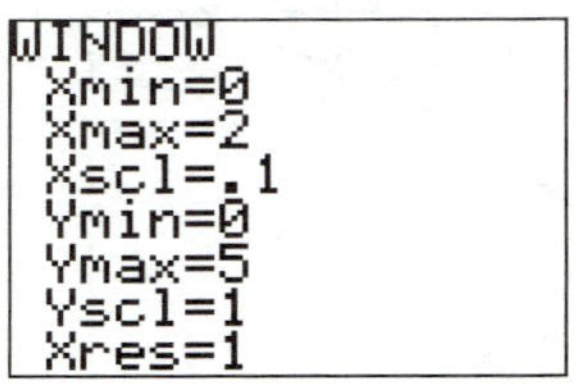

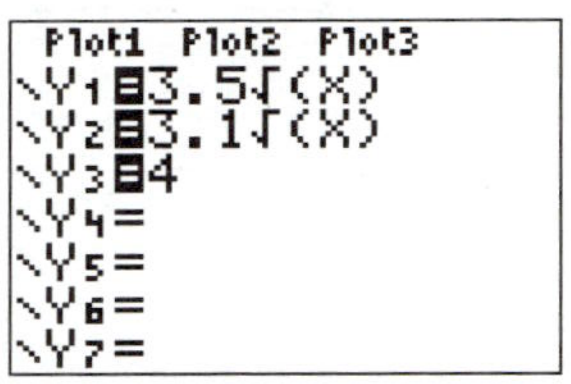

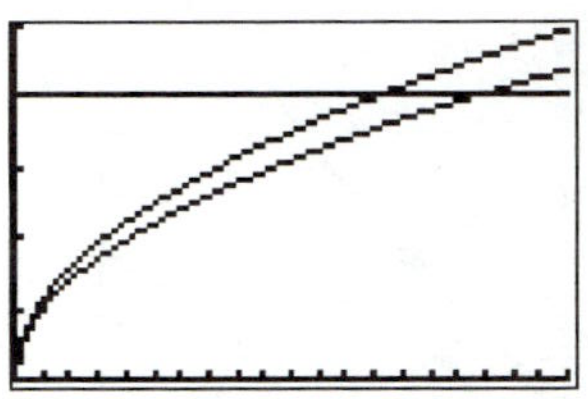

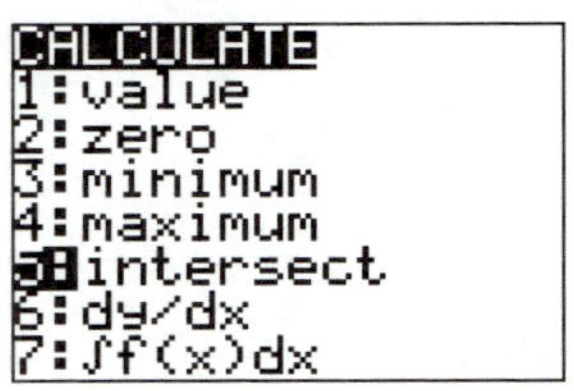

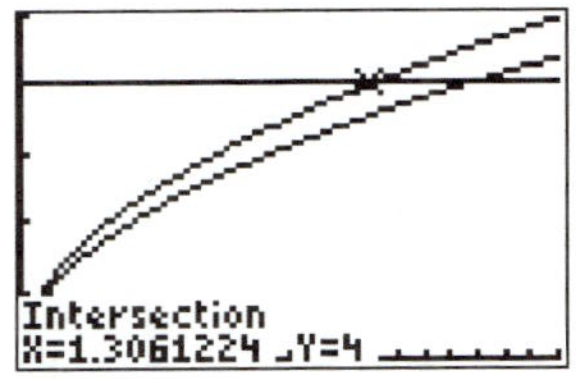

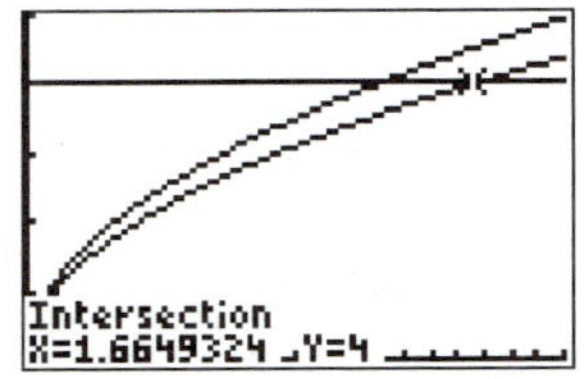

   Solving algebraically:

   $t = 3.5\sqrt{d}$ $\qquad$ $t = 3.1\sqrt{d}$

   $4 = 3.5\sqrt{d}$ $\qquad$ $4 = 3.1\sqrt{d}$

   $\frac{4}{3.5} = \sqrt{d}$ $\qquad$ $\frac{4}{3.1} = \sqrt{d}$

   $\left(\frac{4}{3.5}\right)^2 = d$ $\qquad$ $\left(\frac{4}{3.1}\right)^2 = d$

   $1.31 = d$ $\qquad$ $1.66 = d$

Section 5.19 Assignment

Answers

1. $d = 0.5t^2$

2.

Time (t)	0 s	0.4 s	0.6 s	0.8 s	1.0 s	1.2 s
Distance (d)	0 m	0.08 m	0.18 m	0.32 m	0.5 m	0.72 m

3.

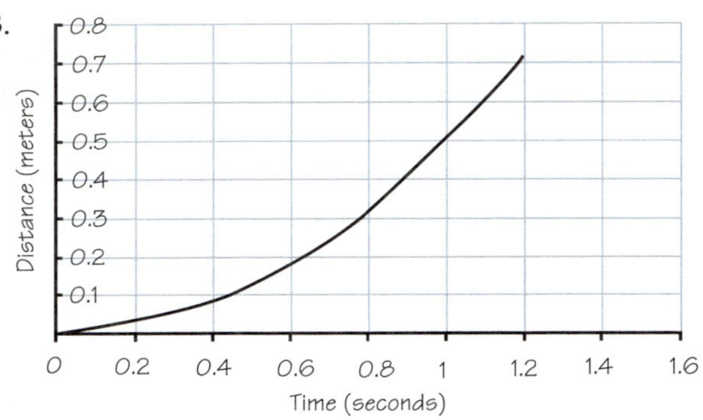

4.

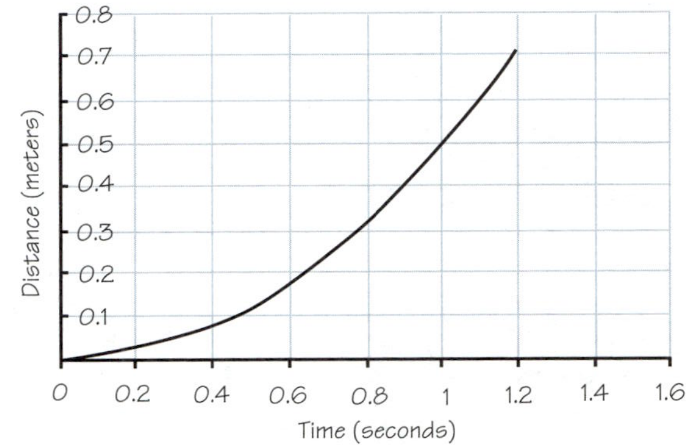

5. quadratic parent function

6.

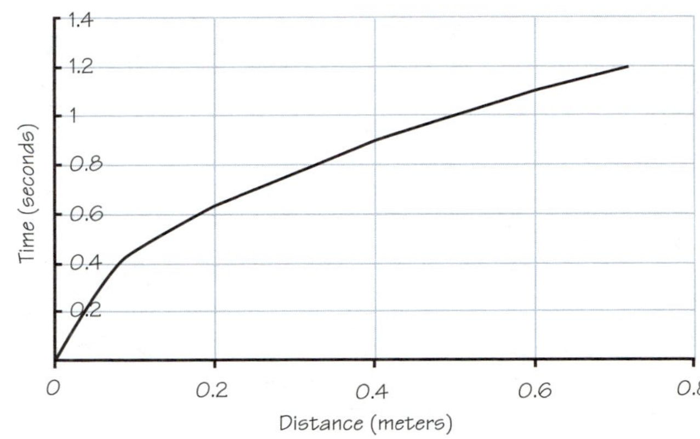

7. square root parent function

8. Yes, the x- and y-coordinates are reversed. The domain and range are switched. The two graphs are reflections over the line $y = x$.

9. $t = \sqrt{\frac{d}{0.5}} = \frac{\sqrt{d}}{\sqrt{0.5}}$ or $t = 1.414\sqrt{d}$

Answers (continued)

10.

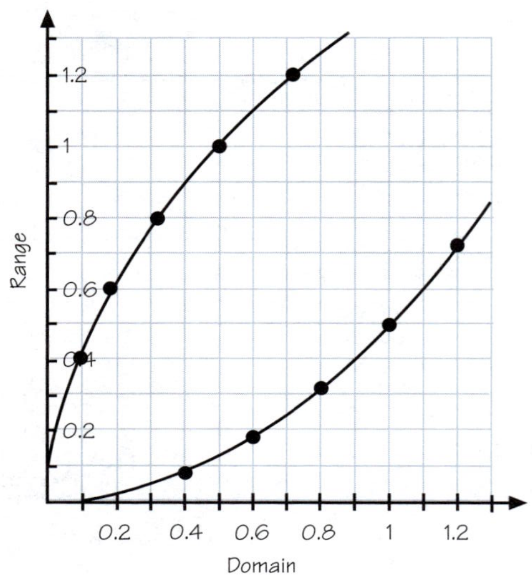

Modeling Project Designing The Hero's Fall

In this project you will design a stunt for a western movie in which the hero falls off a rooftop onto the roof of a stagecoach being drawn by a team of horses. In order to make your plans, you'll need to choose the height of the building, the speed of the stagecoach, and the dimensions of the stagecoach. Make sure your choices are as realistic as possible.

In order to make your stunt work so that the hero does not break any bones, it is important for your hero to know when to jump.

CHECK THIS!

Recall that when an object falls freely, its height can be modeled by the function

$$h = \frac{1}{2}gt^2 + h_o$$

h is the height, *g* is the acceleration due to gravity (hint: remember this is a constant), and *t* is time

1. What function will you use to model the motion of the hero during the fall? Explain your reasoning.
2. How long will it take the hero to reach the height of the roof of the stagecoach? Explain your plan.
3. In this situation, height is calculated in feet and time in seconds. However, the speed of the stagecoach is 20 miles per hour. Convert the speed of the stagecoach to feet per second.
4. The hero should begin the fall the instant the center of the stagecoach (not including the horses) reaches a mark in the road. Where should you place this mark so that the hero will land safely on top of the stagecoach? (Show the distance between the mark and the hero's drop point.) Explain your answer.

Practice Problems

1. Answer these questions about the function $y - 3\sqrt{x-2} + 1$.

   a) What is the domain?

   b) What is the range?

2. $h(t) = -5t^2 + 80$ describes the height in meters of a falling body after t seconds. Find the intercepts of the function's graph and interpret them in this context. Explain the role of the constants –5 and 80.

3. Using quadratic models sometimes requires solving quadratic equations. Among the methods you may have used to solve quadratic equations in previous courses are:

   A graphing calculator procedure such as graphing and zooming;

   A spreadsheet procedure such as zooming on a table;

   Factoring;

   The Quadratic Formula;

   Completing the square.

   The first two methods give approximate solutions and the last three give exact solutions. Solve $6x^2 + 17x = 10$ by at least two methods.

4. Solve $2x^2 - 11x + 12 = 0$ using the factoring method.

5. The table in **Figure 5.70** contains data on the distance it takes a car to stop while traveling at various speeds.

Speed (mph)	20	30	40	50	60	70
Stopping Distance (ft)	42	74	116	173	248	343

FIGURE 5.70. Stopping distances for different speeds.

   a) Create a mathematical model that predicts the stopping distance for any speed. Explain why you think your model is a good one.

Modeling Project

Designing the Hero's Fall

Sample Answer

This project can be done in groups or by individual students. The sample answer is based on the following assumptions:

a) The height of the building is 30 feet.

b) The roof of the stagecoach is 10 feet long and 9 feet high.

c) The stagecoach will maintain a constant speed of 20 mph at least 50 feet before and during the fall.

1. The hero's initial height is 30 feet and the initial velocity is 0 feet/second. Substitute these values into the function that represents freefall motion.

$h = \frac{1}{2}gt^2 + h_0$

$h = \frac{1}{2}\left(-32\frac{\text{ft}}{\text{sec}^2}\right)t^2 + 30 \text{ ft}$

$h = -16t^2 + 30$

2. The height of the stagecoach roof is 9 feet. So solve the equation $9 = -16t^2 + 30$.

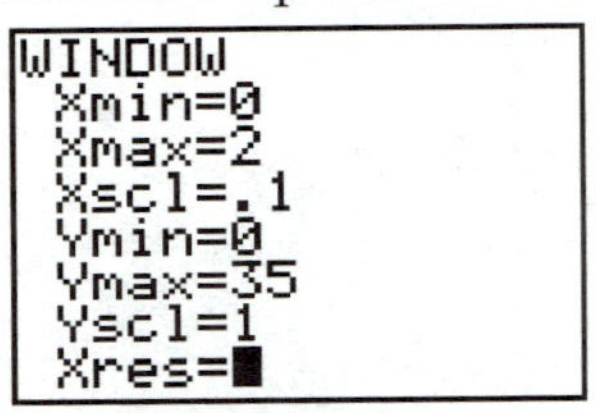

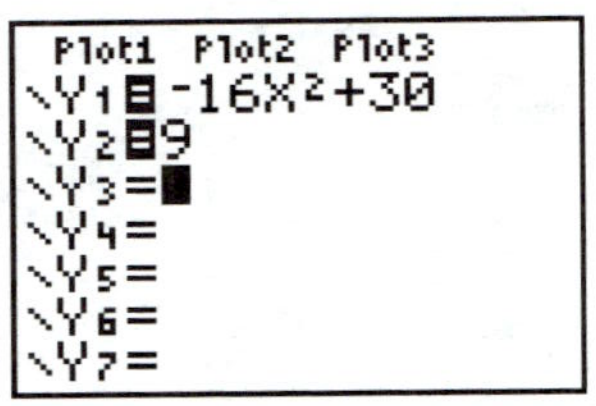

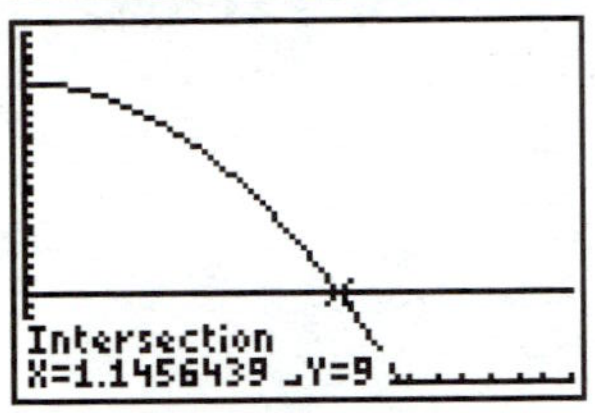

The hero will reach the height of the stagecoach roof in 1.146 seconds.

3. The stagecoach will be traveling at 29.3 feet per second.

4. Since the hero will be 9 feet above the ground in 1.146 seconds, the mark should be set so that it will take the stagecoach 1.146 seconds to get to the drop point. Multiply 1.146 seconds by the speed of the stagecoach (29.3 feet per second) to get the distance that the stagecoach will cover in 1.146 seconds. The stagecoach will travel 33.58 feet in 1.146 seconds. So set the mark at 33.58 feet.

Practice Problems

Answers

1. a) x = all numbers greater than or equal to 2 ($x \geq 2$)

   b) y = all numbers less than or equal to 1 ($y \leq 1$)

2. The vertical intercept is at 80. The x-intercepts are at (–3, 0) and (2, 0).

3. The solutions are $\frac{1}{2}$ and $\frac{-10}{3}$.

4. 1.5, 4

5. a) The model is described by the equation: $d = 0.08v^2 - 1.23v + 36.4$. When applying linear regression, the residuals look like a quadratic. The residuals from the quadratic model is more random (looks slightly like a sideways "S").

b) A police officer at an accident scene estimates that a car needed 300 feet to stop. Write an equation for this situation, and use one of the methods you learned in this chapter to solve it. Describe the method you used.

6. Write each expression in the form stated:

a) $2(x + 4)(x - 3)$ as a general quadratic expression.

b) $3x^2 + 11x + 6$ in factored form.

7. Change the following general quadratic equations into the specified form:

a) $y = 0.5x^2 + 1.5x - 5.0$ in factored form.

b) $y = 2x^2 + 8x + 3$ in vertex form.

8. The graph of $y = 0.5x^2 - 1.5x - 2$ is shown in **Figure 5.71.**

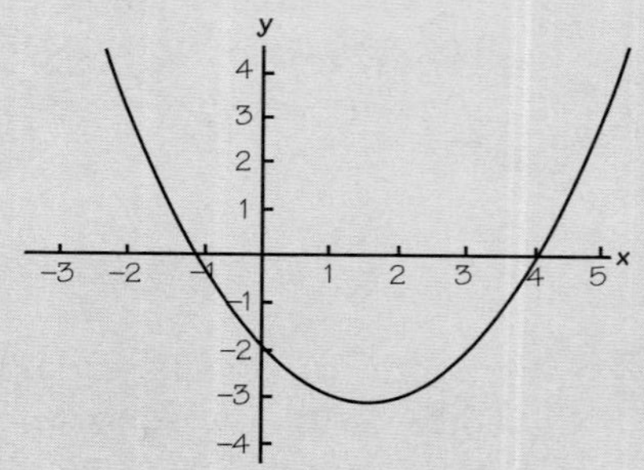

FIGURE 5.71.
Graph of $y = 0.5x^2 - 1.5x - 2$.

a) What are the x-intercepts?

b) What are the coordinates for the vertex?

c) How can you tell from looking at the original equation that the quadratic will look like a "U" (and not upside down)?

9. Given the function $y - 3 = 0.5(x + 1)^2$:

a) Explain the role that each of the three constants play in determining the graph of the function.

b) Sketch the graph of the function.

c) Stretch the graph vertically by a factor of 3 and move it 4 units to the right and 1 unit down. What equation describes the new graph?

10. Solve each of these equations. Explain the method you used.

a) $0.4(x + 3)(x - 2) = 0$

b) $1.6(x + 1)^2 - 3 = 10$

c) $2x^2 + 8x - 5 = 0$

11. A property of quadratic relationships that was explored in this chapter is that second differences are constant. You can prove that to be the case, as well as develop ways to construct the equation from patterns in the table of differences.

a) Substitute each value of x in **Figure 5.72** into $y = ax^2 + bx + c$.

b) Record the *expression* obtained for y in the column labeled y-value.

x-value	y-value	First Differences	Second Differences
0			
1			
2			
3			
4			
5			

FIGURE 5.72.
Table for Question 11.

c) Calculate first and second differences the same way you would if there were numbers in the table. Record your answers in the table.

d) Explain how the work done shows that with quadratic functions, second differences are constant.

e) Explain how to find the values for a, b, and c from the table.

Answers (continued)

b) The equation that describes this situation is: $300 = 0.08v^2 - 1.23v + 36.4$, according to the model in a. Student answers will vary; one possible method is to make the equation equal 0 by subtracting 300, and then use the Quadratic Formula with $a = 0.08$, $b = -1.23$ and $c = -263.6$. Solving the equation by any method outlined in the chapter (table, graph, completing the square, Quadratic Formula), the model predicts that the car was going around 65.6 miles/hr.

6. a) $2x^2 + 2x - 24$

   b) $(3x + 2)(x + 3)$

7. a) $y = 0.5(x + 5)(x - 2)$

   b) $y = 2(x + 2)^2 - 1$

8. a) According to the graph, the x-intercepts are at –1 and 4. Evaluating the equation with those x-values will verify that $y = 0$ at those locations.

   b) Again, according to the graph, the vertex location is at (1.5, –3). Halfway between the x-intercepts is at $x = 1.5$; substituting that value into the equation yields the y-coordinate of –3. (Another option is to complete the square.)

   c) The coefficient for the quadratic term is positive.

9. a) The 0.5 is a dilation factor, which means that the graph will rise half as high as the standard graph of $y = x^2$. (All the y-values will be half as much.) The 1 is a horizontal shift 1 unit to the left of the origin, and the 3 is a vertical shift up 3 units from the origin, locating the vertex at (–1, 3).

   b)

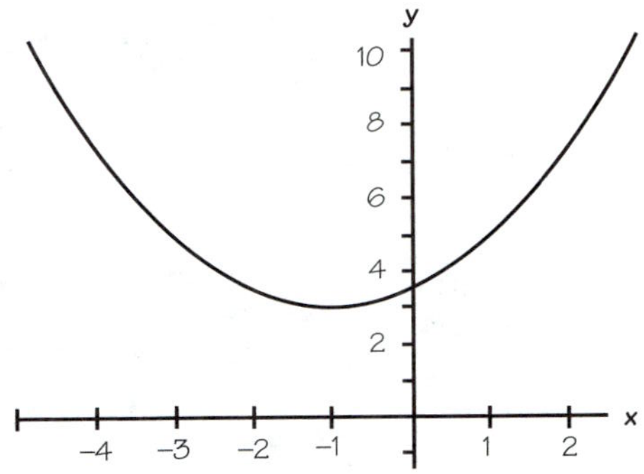

   c) The new vertex location will be (3, 2), and the new dilation factor will be 1.5. So the vertex form equation describing the new quadratic would be: $y - 2 = 1.5(x - 3)^2$.

10. a) $x = -3$ and $x = 2$. Take each factor, set it equal to 0, and solve the resulting equations. Or graph the equation $y = 0.4(x + 3)(x - 2)$ and inspect where the x-intercepts of the graph are located.

    b) The two answers are $x \approx 1.85$ and $x \approx -3.85$. Set up an arrow diagram and work the operations in reverse: 10 + 3, divide by 1.6, square root. Take –1 and add (or subtract) the result of the square root operation.

    c) The two answers are $x \approx 0.55$ and $x \approx -4.55$. Use the Quadratic Formula with $a = 2$, $b = 8$, and $c = -5$.

11. b)

x-value	y-value	First Differences	Second Differences
0	c		
		$a + b$	
1	$a + b + c$		$2a$
		$3a + b$	
2	$4a + 2b + c$		$2a$
		$5a + b$	
3	$9a + 3b + c$		$2a$
		$7a + b$	
4	$16a + 4b + c$		$2a$
		$9a + b$	
5	$25a + 5b + c$		

c) See the table in b.

d) The last column is a constant expression, $2a$. So for any value of a used in the quadratic equation, the column of second differences would have a constant value equal to $2 * a$.

e) Take the number from the column of second differences and divide by 2 to get the value for a. The value for y that is associated with $x = 0$ is the value for c. The first number in the column of first differences is equal to $a + b$; subtract a from that number to get the value for b.

f) Use your answer to e to find a function that fits the data in **Figure 5.73.**

x	y
0	2
1	9
2	20
3	35
4	54
5	77

FIGURE 5.73.
Table for Question 11f.

12. From the top of the Eiffel Tower, which is 300 meters tall, how far can you see?

13. A tsunami, a series of waves created by an undersea earthquake, hit Asia in 2004. Hundreds of thousands of people died. Now monitoring stations determine the speed and force of these waves. A key variable in the damage a wave can do is ocean depth. Ocean depth can be found using the function $s = \sqrt{gd}$ where g is 9.8 meters per second squared (the force of gravity), s is the speed of the wave in meters per second, and d is the depth of the ocean. If a wave is moving at 99 meters per second, how deep is the ocean?

Glossary

KEY CONCEPTS

Binomials: Algebraic expressions involving two terms, usually with one of them containing a variable.

Coefficient: The real number in a mathematical expression. In $3x^2$ the 3 is the coefficient.

Completing the Square: The process of placing a quadratic in vertex form. Completing the square is used to solve quadratic equations.

Dilation (graph): A transformation that stretches or compresses a graph.

Equivalent: When two expressions produce identical values and solutions for all numbers.

Factored Form of a Quadratic: A quadratic of the form $y = a(x - r_1)(x - r_2)$.

First Differences: A pattern of numbers produced by subtracting successive values from a data table in which the x-values increase by a constant amount. If first differences are constant then the equation describing the data is linear.

General Form of a Quadratic: A quadratic of the form $y = ax^2 + bx + c$.

Inverse of a Relation: A reflection of the relation over the line $y = x$.

Inverse Operations: Pairs of operations that undo each other. Addition and subtraction are inverse operations.

Polynomial Form of a Quadratic: A quadratic of the form $y = ax^2 + bx + c$.

Quadratic Formula: A formula used to solve quadratic equations of the form $ax^2 + bx + c = 0$. The solutions are:

$$x = \frac{-b \pm \sqrt{b^2 - 4ac}}{2a}.$$

Quadratic Function : A function of the form $y = ax^2 + bx + c$. The highest power of x in a quadratic is 2.

Roots of an Equation: Values that make an equation true.

Second Differences: The pattern of numbers produced by subtracting successive first differences. If the second differences are constant, then the equation describing the data is quadratic.

Square Root: A factor of a number that, when squared, gives the original number.

$\sqrt{\ }$ (Square Root Symbol): The principal square root of a number. The principal square root of a positive number is the positive square root.

Square Root Function: A function in the form $f(x) = \sqrt{x}$ is called a square root function and is defined for all $x \geq 0$.

Transformation (graph): In this chapter, a rule that changes one graph into another.

Translation (graph): A transformation that affects either the x- or y-values of an equation by adding or subtracting a number. Translations produce a "shift" or "slide" effect on a graph.

Vertex Form of a Quadratic: A quadratic of the form $y = a(x - h)^2 + k$.

Vertex of a Parabola: A point on the graph of a quadratic where the left-side and right-side (symmetric) parts meet. The vertex is either the highest or lowest point of the parabola.

x-intercept: A point where the graph of a function crosses the x-axis. The x-coordinate of the intercept is called a zero of the function.

Answers (continued)

f) The column of second differences was a constant 4; therefore, $a = 2$. The value for c is 2 (by inspecting the table where $x = 0$). The first number in the first difference column is 7; $7 - 2 = 5$. Therefore, the equation used was: $y = 2x^2 + 5x + 2$.

12. 61.18 km. Students may solve by using a table or by substituting into the function.

13. Approximately 1000 meters

Zero-Product Property: If $ab = 0$, then $a = 0$ or $b = 0$ or both $a = 0$ and $b = 0$. For example, if $(x + 4)(x - 3) = 0$ the Zero-Product Property states that either $x + 4 = 0$ or $x - 3 = 0$ or both. Thus, $x = -4$ and $x = 3$ are solutions to this equation.

Zeros of a Function: Values for which a function is 0. They are also the x-coordinates of the points where the function's graph crosses the x-axis.

CHAPTER 6

Growth & Decay: Exponential Functions

PREPARATION READING

"For the Benefit and Enjoyment of the People"

As American pioneers settled the West in the 1800s, many of them noticed the beauty of the Great Plains, Rocky Mountains, and Pacific coast. As early as the 1830s, some people began to express concern about the impact people were having on this land and the Native Americans who lived there.

In 1872, President Ulysses S. Grant signed legislation from Congress that set aside Yellowstone Country in Wyoming and Montana Territories "as a public park or pleasuring-ground for the benefit and enjoyment of the people." Yellowstone became the first piece of land in the world to become a national park and was the start of the United States National Park Service.

The National Park Service is now almost 400 parks, historic sites, and sanctuaries. States, counties, and cities also have park services that oversee state and local parks. These parks range from campgrounds and playgrounds to cultural sites.

Park managers must make many decisions about how to balance the needs of human park visitors with the need to preserve the natural environment. In a large park such as Yellowstone, park managers must be able to survey and define each part of the park. They must also keep a careful count of the population of wildlife such as bison, wolves, moose, and fish. Managing disease and giving medicine to sick animals is important in protecting endangered species.

On the human side, park managers must deal with resource management. Renewable resources such as forests must be managed in order to balance the needs of the lumber industry with the need to preserve the park's natural setting. Non-renewable resources such as coal and oil reserves must also be carefully managed.

CHAPTER 6

Growth: Exponential and Logarithmic Functions

Preparation Reading: "For the Benefit and Enjoyment of the People"

Description: The reading introduces the major context of the chapter, different aspects of park management. Students read about Yellowstone National Park and some of the issues that park managers face when making decisions about the best way to fulfill the park's mission of being "for the benefit and enjoyment of the people." In the rest of the chapter, students investigate data leading to the development of exponential growth, exponential decay, and logarithmic functions.

Materials: none

Teaching Ideas: Students should work in small groups (3 to 4 students). To help ensure that all students read the preparation reading have them read silently then record three important facts or things that they learned. Ask students to share their individual responses within the group, then ask each group to share two of the group's responses with the entire class. After the reading students should want to know more about using mathematics to model and solve problems relating to biology and ecology in a context of park management.

In a popular park such as Yellowstone, there are many visitors in a year. Park visitors can have a large impact on the park from traffic and noise pollution to the large amounts of garbage that they leave behind.

In this chapter you'll examine some of the issues that park managers face. You'll use mathematical models to solve problems such as wildlife population control and resource management. You will explore and define exponential and logarithmic functions and use them to model real-world situations. Then you will use these models to solve practical problems.

SECTION 6.1 Paper Cutting

Just about everything in our world changes over time. Sometimes the change happens quickly, such as a teenager's growth spurt over a summer or an increase of 200,000 bacteria in a container of water. Other changes occur slowly, for example, the erosion of a mountain peak or an increase in Earth's surface temperature.

Patterns of growth can be used to classify behavior. In this chapter, patterns for describing change in quantities are explored. The challenge is to understand what kind of growth patterns a quantity is undergoing and how those patterns can be modeled. Once you determine the type of growth pattern, you can apply the models to a host of other situations such as wildlife populations, radioactive decay, resource depletion, and resource harvest.

1. Cut a piece of paper in half. Lay the two sheets on top of each other. Cut each of the two sheets of paper in half again. Continue cutting, counting the number of layers of paper when the pieces are stacked up, and filling in the table in **Figure 6.1.**

Number of Cuts	Process	Number of Layers of Paper
0		
1		
2		
3		
4		
5		
6		

FIGURE 6.1. A table for paper cutting.

CHECK THIS!

Repeated multiplication can be written with a base and an exponent. For example, $5 \times 5 \times 5$ can be rewritten as 5^3. The factor 5 is the base. The number that tells how many times the base is multiplied is the exponent.

ENGAGE LESSON 1

Section 6.1: Paper Cutting

Purpose: In Chapters 2 and 5 students learned about linear and quadratic models. The purpose of this section is to give students an informal opportunity to investigate exponential functions with paper cutting. In Section 6.2 students explore the population growth of moose in a state park.

Description: Students cut a sheet of paper in half, then cut those sheets in half again, and continue the process. Students record the number of sheets they have in a table. This first model demonstrates how quickly the number of layers increases. The second model relates the number of cuts and the thickness of the stack.

Materials: scissors, 8 1/2″ × 11″ sheet of paper for each student, graphing calculator

Teaching Ideas: Have students follow the directions as printed in the text. You may wish to model cutting the paper and define what layers of paper means.

Answers

1.

Number of Cuts	Process	Number of Layers of Paper
0	$1 = 1 \times 2^0$	1
1	$1 \times 2 = 1 \times 2^1$	2
2	$(1 \times 2) \times 2 = 1 \times 2^2$	4
3	$(1 \times 2 \times 2) \times 2 = 1 \times 2^3$	8
4	$(1 \times 2 \times 2 \times 2) \times 2 = 1 \times 2^4$	16
5	$(1 \times 2 \times 2 \times 2 \times 2) \times 2 = 1 \times 2^5$	32
6	$(1 \times 2 \times 2 \times 2 \times 2 \times 2) \times 2 = 1 \times 2^6$	64

Students will probably fill in the number of cuts and the number of layers of paper columns first. When students complete those two columns and begin to discuss the process column, lead them to the idea that 1 is equal to 2^0, which follows the pattern of exponents in the table.

2. Use a copy of the grid in **Figure 6.2** to make a scatter plot of your data.

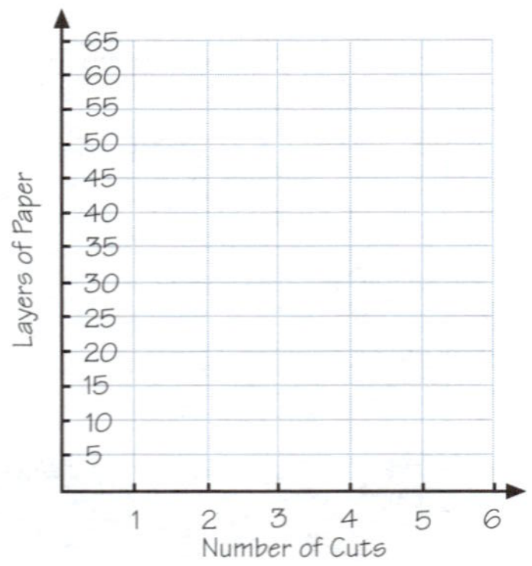

FIGURE 6.2. Layers of paper versus number of cuts.

3. Write a function for the number of layers of paper if you cut the paper n times.

4. Is this function continuous or discrete? Why?

5. What are the domain and range of this function?

6. Use a graphing calculator to make a scatter plot of the data. Then sketch the graph of the function on the scatter plot and record your window.

7. A ream of paper is almost 2 inches thick. If a ream is 500 sheets of paper, about how thick is a piece of paper?

8. Complete the table in **Figure 6.3.**

Number of Cuts	Process	Thickness (in inches)
0		
1		
2		
3		
4		
5		
6		

FIGURE 6.3. Paper cuts, process, and thicknesses.

9. Write a function rule for the thickness of the stack in inches if you cut the paper n times.

10. How are the function that models the thickness of the stack and the function you found for the number of layers after n cuts related?

11. Sketch a graph for the thickness of the stack, and label the axes.

12. How is this graph transformed from the graph that you drew in Question 6?

13. Complete the tables in **Figures 6.4 and 6.5** using the data from Questions 1 through 12.

Number of Cuts	Number of Layers
0	
1	
2	
3	
4	
5	
6	

FIGURE 6.4. Paper cuts and layers.

Number of Cuts	Thickness (inches)
0	
1	
2	
3	
4	
5	
6	

FIGURE 6.5. Paper cuts and thicknesses.

14. Recall that linear functions can be written from patterns that show repeated addition. The two functions in this section are exponential functions. Exponential functions can be written from patterns that show repeated multiplication or division. Show how the tables in Figures 6.4 and 6.5 display a pattern of repeated multiplication.

CHECK THIS!
Functions can be written using either "y =" or "f(x) =" notation.

15. A general form of an exponential equation is $f(x) = a \cdot b^x$. From the tables in Figures 6.4 and 6.5, what determines the value of a? How is a related to the graph of the function?

16. What determines the value of b?

Answers (continued)

2.

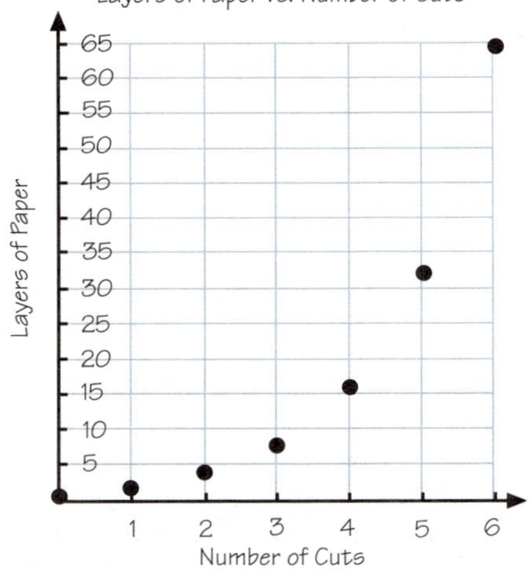

3. $f(n) = 1 \times 2^n$ or $f(n) = 2^n$

4. For this situation, the function is discrete since we cannot cut the paper a fractional number of times.

5. The domain is all whole numbers; the range is powers of 2.

6.

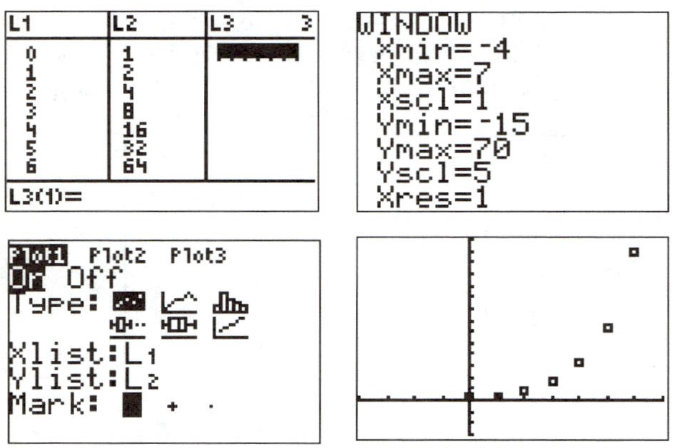

7. A sheet of paper is approximately 2 inches ÷ 500 sheets thick, or the thickness per sheet is approximately 0.004 inches.

8.

Number of Cuts	Process	Thickness (in inches)
0	0.004×2^0	0.004
1	0.004×2^1	0.008
2	0.004×2^2	0.016
3	0.004×2^3	0.032
4	0.004×2^4	0.064
5	0.004×2^5	0.128
6	0.004×2^6	0.256

9. $f(n) = 0.004 \times 2^n$

10. The function for the thickness of the stack is the function for the number of layers of paper multiplied by 0.004, the thickness of one sheet of paper.

11.

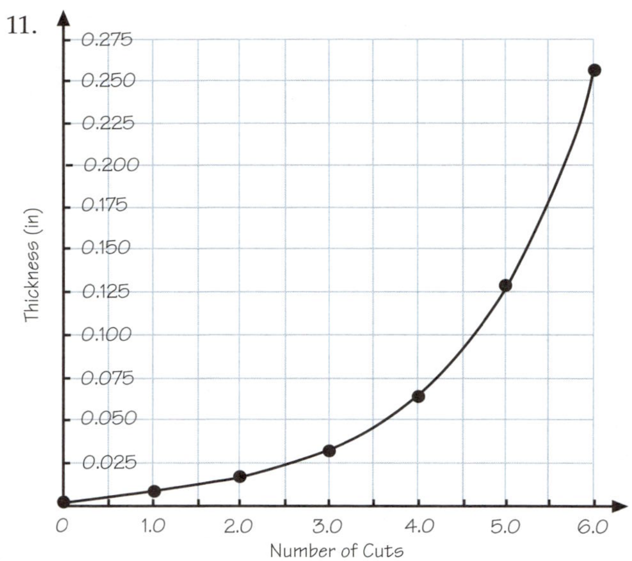

Here is a nice opportunity to talk about continuous and discrete functions. Since you cannot really cut a piece of paper half a time, the relationship shown by the line connecting the data points is not really valid.

12. The graph is vertically compressed by a factor of 0.004. At this point compare the graphs of both functions on the graphing calculator, along with their respective tables. Ask students why one graph appears to be missing on the calculator screen. Even though the domains are the same for the two functions, the range of the function for the thickness of the stack is [0.004, 0.256] when the domain is [0, 6].

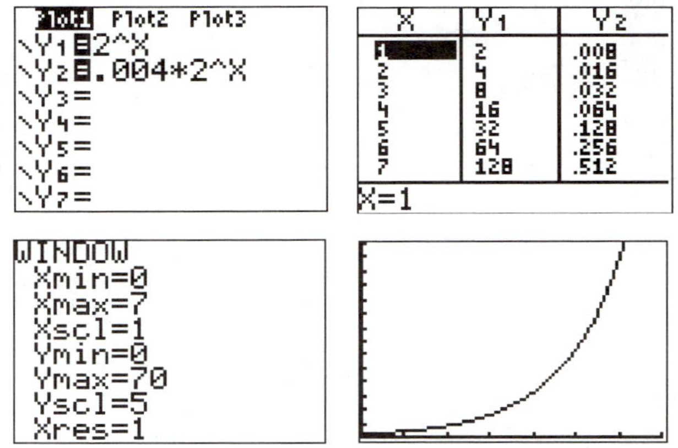

If you change the viewing window, students will see the thickness of the stack begin to grow after approximately 10 cuts.

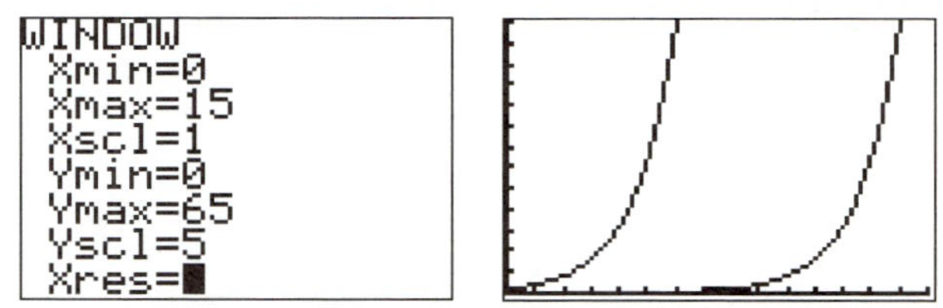

13. See answer for Question 14.

Answers (continued)

14.

Number of Cuts	Number of Layers
0	1
1	2
2	4
3	8
4	16
5	32
6	64

(× 2 between successive rows)

Number of Cuts	Thickness (inches)
0	0.004
1	0.008
2	0.016
3	0.032
4	0.064
5	0.128
6	0.256

(× 2 between successive rows)

Point out to students that this repeated multiplication (or division) is called a recursive process. The formulas for the functions developed in Questions 2 and 9 are explicit formulas.

15. The "0" term determines the value of a. On the graph, a appears as the y-intercept. The questions in this section refer only to the exponential parent function. In Section 6.9 students will investigate vertical shifts to the exponential parent function.

16. The number that you are repeatedly multiplying or dividing by determines the value of b.

SECTION 6.1 Assignment

1. If you cut the paper 22 times, how many layers do you have? Write an equation to help you solve this problem.

2. If you cut the paper 22 times and stack it up, how tall is the stack? Write an equation to help you solve this problem.

3. A box of paper is 5 reams deep. A ream has 500 sheets of paper. About how many cuts would you need to make a stack at least as thick as a box of paper?

4. The Eiffel Tower is approximately 1050 feet tall. If you had a big enough piece of paper, how many cuts would you need to have a stack that equals or exceeds this height?

SECTION 6.2 Moose Population

When animals move to a new place, it is called migration. In 1988, Adirondack State Park in upstate New York had a moose population of about 17 moose. The park is not fenced, and moose migrate into the park from other regions. Assume that the only change in population is due to migration and that the migration rate is two moose per year.

1. Make a table like **Figure 6.6** and use it to predict the number of moose in the year 2000 (12 years later).

Year	Year Number	Number of Moose
1988	0	
1989	1	
1990	2	
1991	3	
1992	4	
1993	5	
1994	6	
1995	7	
1996	8	
1997	9	
1998	10	
1999	11	
2000	12	

FIGURE 6.6. Table for predicting the number of moose in the park.

2. Make a scatter plot of your data.

3. Find a function rule that models the moose migration. Check it with the graphing calculator. Explain why your function is a good model for the moose migration.

Section 6.1 Assignment

Answers

1. Students may solve this problem in several ways. They may write the equation $x = 2^{22}$ and use a calculator to calculate $x = 4{,}194{,}304$. They may check their answer using either the table or graphing features of the graphing calculator.

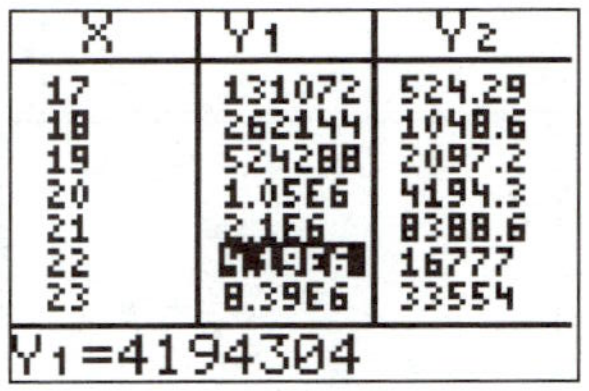

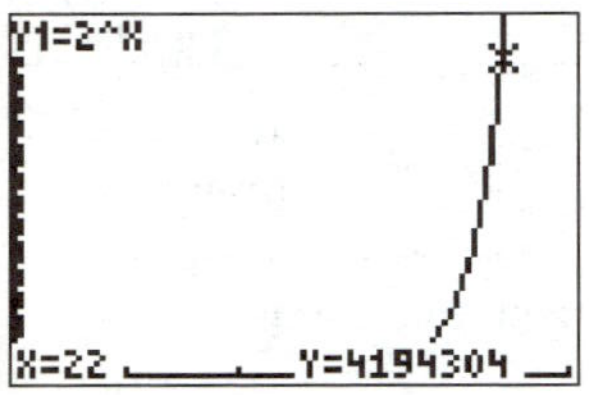

2. Students may solve this problem several ways. They may use an equation, a table, or a graph. The stack is approximately 16,777 inches (or 1398 feet) tall.

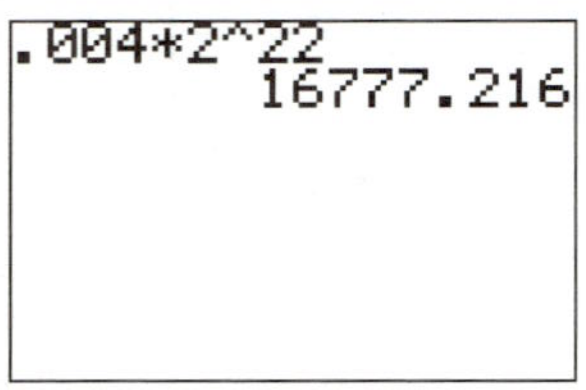

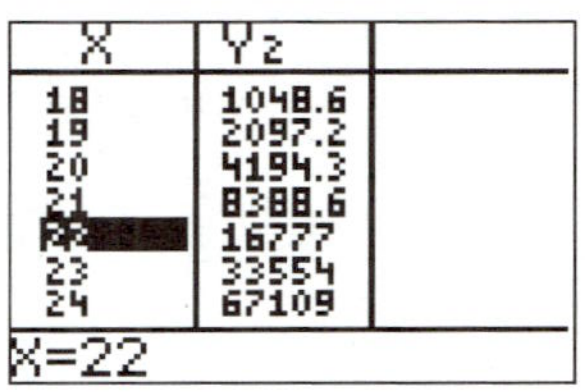

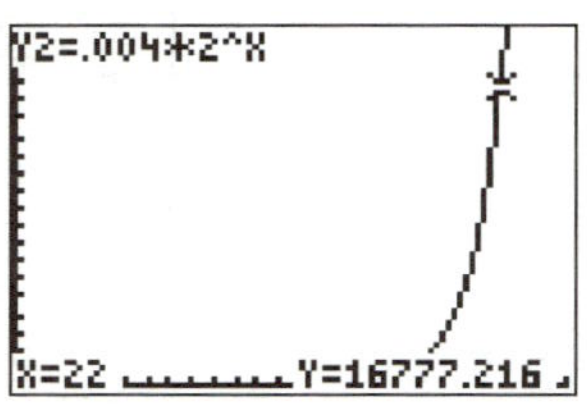

3. First, students must find out how tall the box of paper is.

   $5 \times 500 = 2500$ sheets

   2500 sheets × 0.004 inches per sheet = 10 inches tall

   At this point in their studies, most students will not be able to solve this equation, $10 = 0.004 \times 2^x$, algebraically. Students should look at the table for $y = 0.004 \times 2^x$. After 11 cuts the stack of paper is only 8.192 inches tall. After 12 cuts the paper is 16.384 inches tall. So you would need to have 12 cuts to make a stack of paper at least as thick as a box of paper. Another way to solve the problem is to let $y_2 = 10$ and find the intersection point.

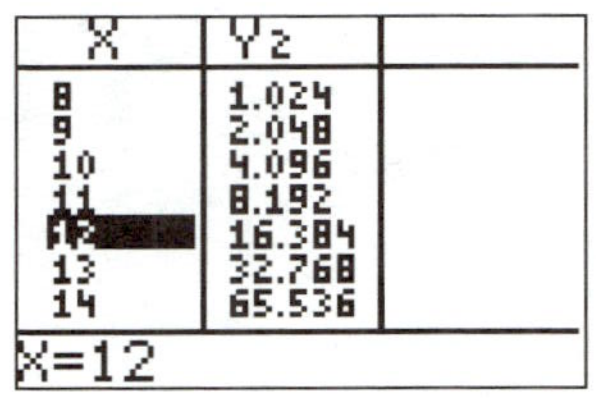

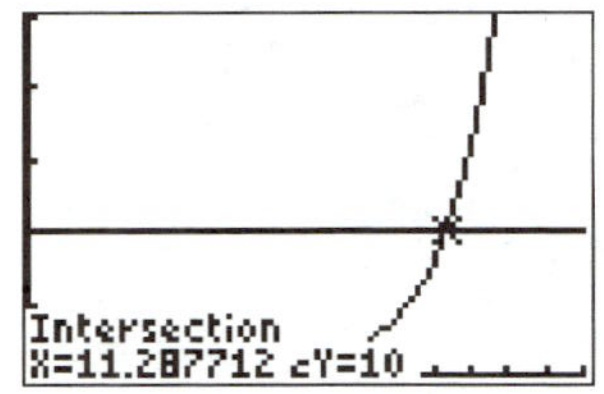

4. After changing 1050 feet to inches, the height of the Eiffel Tower is 12,600 inches. By looking at the table or the graph, students find that you would need at least 22 cuts to match or exceed the height of the Eiffel Tower.

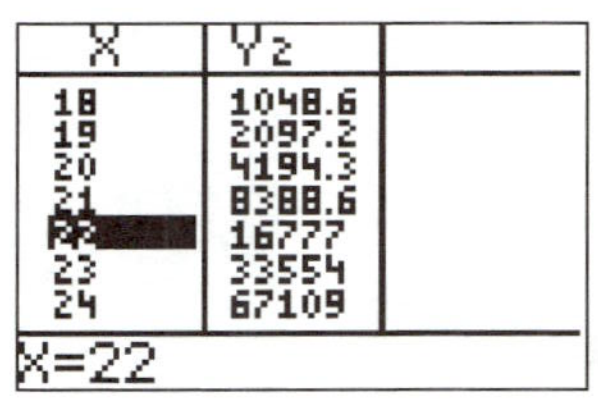

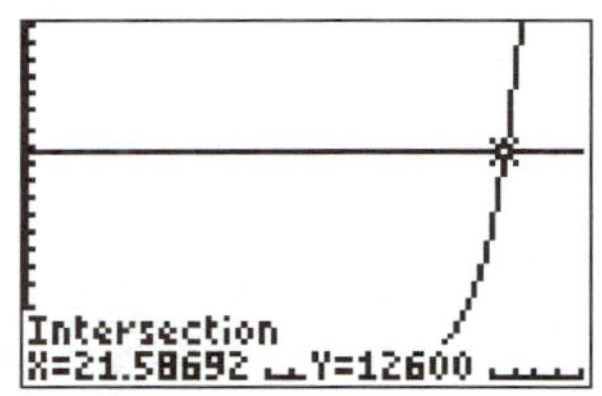

EXPLORE LESSON 1

Section 6.2: Moose Population

Purpose: In Section 6.1 students investigated paper cutting as an exponential model. In this section students will compare linear and exponential models with a moose population. In Section 6.3 we will formalize the definition of an exponential function and compare the exponential model to both linear and quadratic models.

Description: Students build both linear and exponential models for two types of population growth using number patterns and tables. The first situation involves migration of 2 moose into the park per year (linear). The second situation involves starting with 17 moose and doubling the population each year (exponential).

Materials: graphing calculators

Teaching Ideas: The big idea that students should get from this section is that linear models are addition models and exponential models are multiplication models.

Answers

1.

Year	Year Number	Number of Moose
1988	0	17
1989	1	19
1990	2	21
1991	3	23
1992	4	25
1993	5	27
1994	6	29
1995	7	31
1996	8	33
1997	9	35
1998	10	37
1999	11	39
2000	12	41

(Each successive Number of Moose: + 2)

Answers (continued)

2.

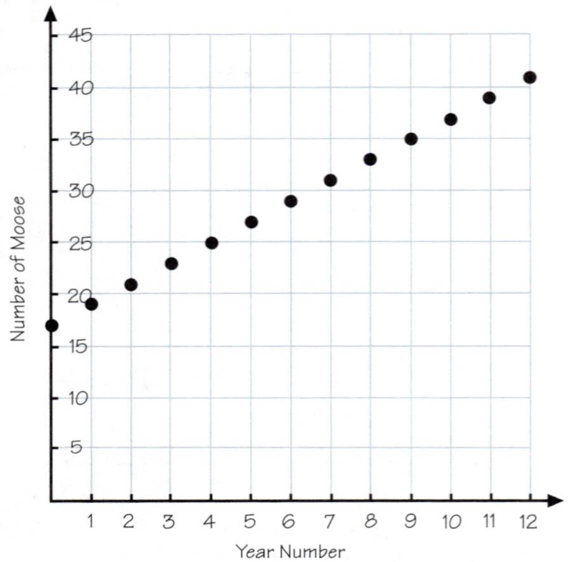

3. The data set is linear for this situation. The difference between the numbers of moose in successive years is 2. The rate of change is 2 moose per year. The initial value, or starting point, at year 0 is 17. An equation for the model is $y = 2x + 17$.

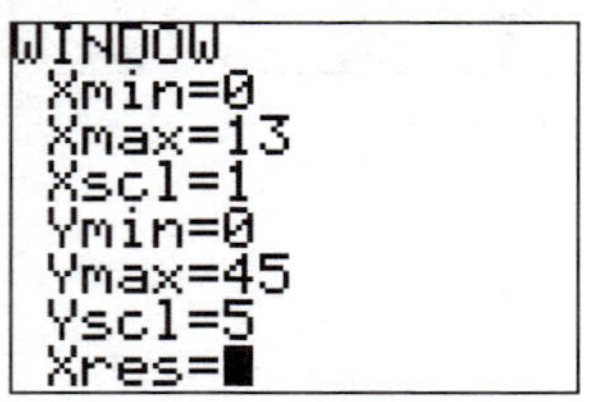

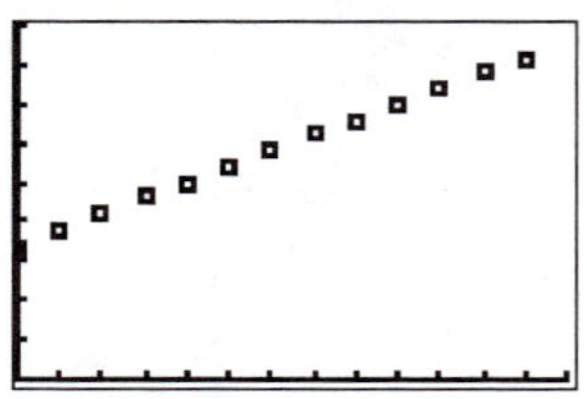

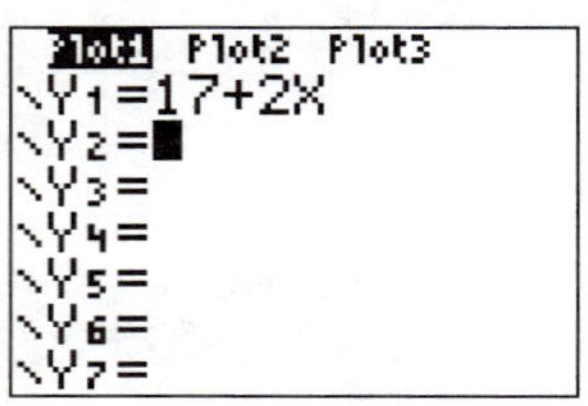

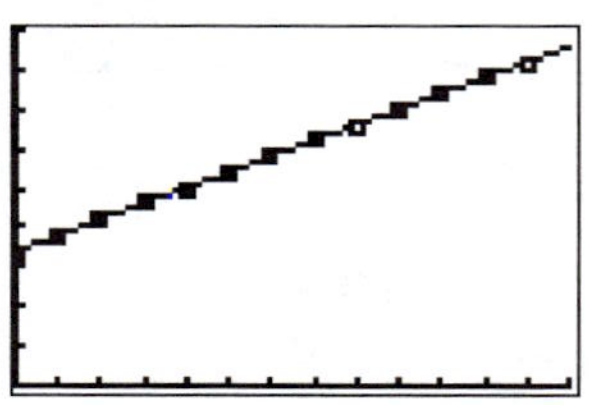

4. In Questions 1–3, the only reason for population change is the migration of moose into the park. Growth also occurs due to reproduction. Babies eventually have their own babies. If you assume that the moose population is 17 in 1988 and doubles each year due to migration and reproduction, make a table like **Figure 6.7** and use it to predict the moose population in 2000.

Year	Year Number	Process Column	Number of Moose
1988	0		
1989	1		
1990	2		
1991	3		
1992	4		
1993	5		
1994	6		
1995	7		
1996	8		
1997	9		
1998	10		
1999	11		
2000	12		

FIGURE 6.7. Table for predicting the moose population in 2000.

5. Make a scatter plot of the data in Question 4.

6. Find a function rule that models the moose migration in Question 4. Check your function with a graphing calculator. Explain why your function is a good model.

SECTION 6.2 Assignment

For each of these tables, decide if the data show exponential growth, linear growth, or neither. If the growth set is linear or exponential, write a function rule for the pattern. Make a scatter plot with a graphing calculator and check your function to see if it fits.

1.

Year	Population
0	3
1	12
2	48
3	192
4	768

2.

Year	Population
0	3
1	12
2	21
3	30
4	39

3.

Year	Population
0	3
1	12
2	27
3	48
4	75

4.

Year	Population
0	3
1	6
2	12
3	24
4	48

Answers (continued)

4.

Year	Year Number	Process Column	Number of Moose
1988	0	17×2^0	17
1989	1	17×2^1	34
1990	2	17×2^2	68
1991	3	17×2^3	136
1992	4	17×2^4	272
1993	5	17×2^5	544
1994	6	17×2^6	1088
1995	7	17×2^7	2176
1996	8	17×2^8	4352
1997	9	17×2^9	8704
1998	10	17×2^{10}	17,408
1999	11	17×2^{11}	34,816
2000	12	17×2^{12}	69,632

× 2 × 2 × 2 × 2 × 2 × 2 × 2 × 2 × 2 × 2 × 2 × 2

5.

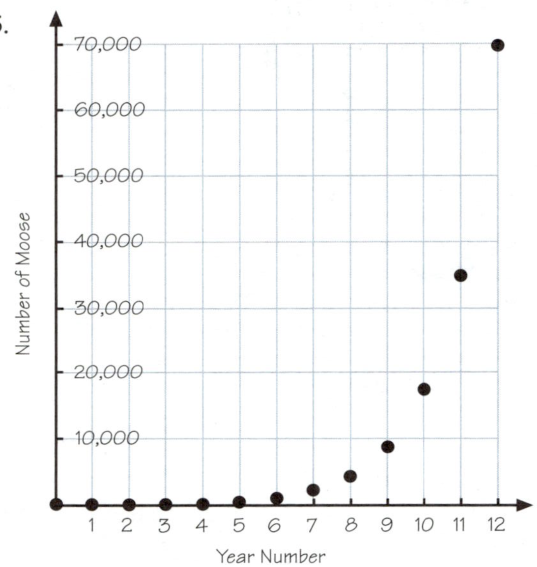

6. The data set for the moose migration situation in Question 4 is exponential. Each year the moose population is multiplied by 2 to find the population for the following year. The "0" term determines the value of a. So for this exponential model, $a = 17$. The common multiplier is 2, so $b = 2$. The exponential model is $y = 17 \times 2^x$. Check the model with the graphing calculator.

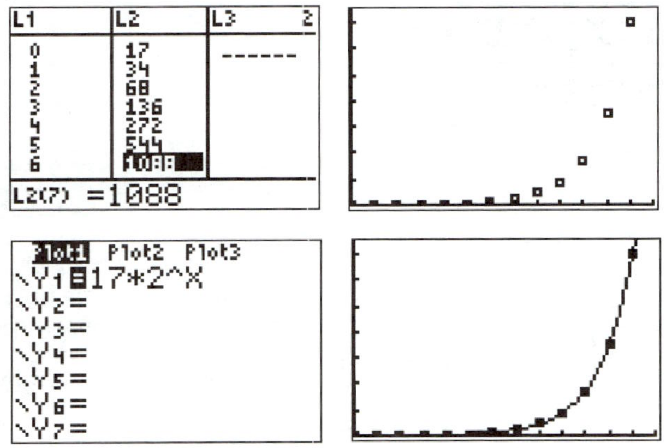

Section 6.2 Assignment

Answers

1. The growth for the data set in this table is exponential. The common multiplier is 4. The starting point is 3. The equation of the model is $f(x) = 3 \cdot 4^x$.

2. The growth for the data set in this table is linear. Each year the population increases by adding 9 to the previous year. The rate of change is 9 per year, and the starting point is 3. The equation of the linear model is $f(x) = 9x + 3$.

3. The growth for the data set in this table is neither exactly linear nor exponential. The difference between y-values is not constant, so the data set cannot be exactly modeled by a linear function. There is not a common multiplier, so the data set cannot be perfectly modeled by an exponential function.

4. The growth for the set of data shown in this table is exponential. The common multiplier for the population values is 2. The starting point, or initial value, is 3. The exponential model for this situation is $f(x) = 3 \cdot 2^x$.

SECTION 6.3 Exponential Functions

Here are the types of function models that you have seen so far in this text:

- A linear function is an addition function. As the x-values in the domain increase by a constant amount, the y-values increase by a constant amount. The first differences of function values are the same. An example of linear data and a function rule is shown in **Figure 6.8.**

x	f(x)
0	5
1	7
2	9
3	11
4	13

(x: +1, +1, +1, +1; f(x): +2, +2, +2, +2)

FIGURE 6.8. A linear set of data and its function rule.

The rate of change is 2. The linear function is $f(x) = 2x + 5$.

- In a quadratic function the second differences are constant. An example of quadratic data and a function rule is shown in **Figure 6.9.**

x	f(x)	First Differences	Second Differences
0	1		
1	3	2	
2	9	6	4
3	19	10	4
4	33	14	4

(x: +1, +1, +1, +1)

FIGURE 6.9. A quadratic set of data and its function rule.

The quadratic function is $f(x)= 2x^2 + 1$.

- An exponential function is a multiplication function. As the x-values in the domain increase by a constant amount, the y-values are multiplied by a common multiplier. An example of exponential data and a function rule is shown in **Figure 6.10.**

x	f(x)
0	4
1	12
2	36
3	108
4	324

(x: +1, +1, +1, +1; f(x): × 3, × 3, × 3, × 3)

FIGURE 6.10. An exponential set of data and its function rule.

The common multiplier is 3. The exponential function is $f(x) = 4 \cdot 3^x$.

In this same set of exponential data, notice that the first and second differences are not constants, as shown in **Figure 6.11.**

x	f(x)	First Differences	Second Differences
0	4		
1	12	8	
2	36	24	16
3	108	72	48
4	324	216	144

FIGURE 6.11. An exponential set of data with first and second differences.

CHECK THIS!

The exponential parent function is written in the form $f(x) = b^x$. The constant b is the common multiplier, the value by which you repeatedly multiply or divide. In the paper cutting in Section 6.1, you started with 1 layer when there were 0 cuts. In your investigation of the thickness of the stack after a certain number of cuts, you started with a thickness of 0.004 inches when there were 0 cuts. In the modeling situations in this chapter, the initial value occurs when the independent variable is 0. Thus, in the exponential function $f(x) = a \cdot b^x$ that is used to model these situations, the constant a is the initial value of the dependent variable.

EXPLAIN LESSON 1

Section 6.3: Exponential Functions

Purpose: In Section 6.2 students compared linear and exponential growth models. In this section the concepts of exponential functions are formally explained. Students will be ready to solve problems with real-world applications in Section 6.4.

Description: Students review linear and quadratic models. Then they discuss exponential models. Students find successive quotients to decide if an exponential model is appropriate. They investigate what happens to the graph of an exponential function as the value of a changes in $f(x) = a \cdot b^x$.

Materials: graphing calculators

Teaching Notes: The big idea continues to be that exponential functions are a multiplication model. Reinforce for students the idea that the successive quotients should all be equal, and the common multiplier is found by finding successive quotients.

In previous sections you learned how to find a common multiplier. Look at the problem below and find the common multiplier. To find the common multiplier, you will use successive quotients. A successive quotient is the ratio of a y-value to the previous y-value.

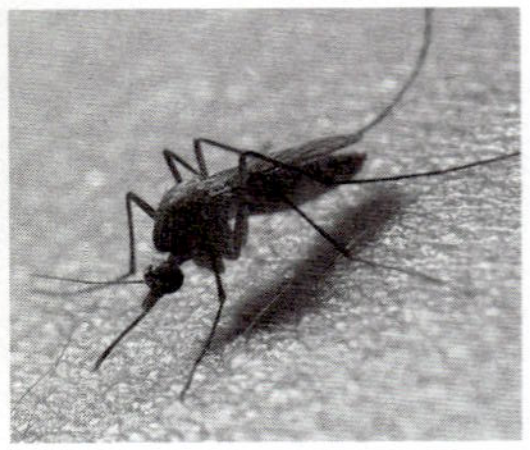

A mosquito control scientist collected the data in **Figure 6.12** over a five-month period. Rainfall was very high during this summer, and the county did not have enough money to spray in order to control the mosquito population. Each month the scientist collected 5000 mosquitoes and tested them for West Nile Virus.

Month	Month Number	Number of Mosquitoes Found Testing Positive for West Nile Virus	Successive Quotients
May	0	400	—
June	1	600	$\frac{600}{400} = 1.5$
July	2	900	$\frac{900}{600} = 1.5$
August	3	1350	$\frac{1350}{900} = 1.5$
September	4	2025	$\frac{2025}{1350} = 1.5$

FIGURE 6.12. Mosquito data.

1. Are the successive quotients the same?
2. How does finding the successive quotients help you find the common multiplier?
3. Find a function rule for the data in the table.
4. Check your function with a graphing calculator.
5. How are the three exponential functions listed below alike, and how do they differ?

$f(x) = 1 \cdot 2^x$ $\quad f(x) = 3 \cdot 2^x$ $\quad f(x) = 0.5 \cdot 2^x$

6. Use a graphing calculator to graph all three at the same time. Sketch the result on your paper.
7. How does changing a in $f(x) = a \cdot b^x$ change the graph?

CHECK THIS!

In the general exponential function $f(x) = a \cdot b^x$, if $b > 0$ and $a > 1$, the parent graph of $f(x) = b^x$ is stretched vertically. If the value of a is between 0 and 1, the graph is compressed vertically.

SECTION 6.3 Assignment

Decide if a linear, quadratic, or exponential function best describes the data in each table. Find the function.

1.

x	f(x)
0	0.75
1	1.875
2	4.6875
3	11.719
4	29.297

2.

x	f(x)
–2	8
–1	2
0	0
1	2
2	8

3.

x	f(x)
0	2
1	2.44
2	2.88
3	3.456
4	4.1472

4.

x	f(x)
–3	–11
–2	–8
–1	–5
0	–2
1	1

Answer Questions 5–10 for $f(x) = 3(2.25)^x$.

5. Graph the function.
6. Fill in the table in **Figure 6.13** with 5 values for the function.

x	f(x)
0	
1	
2	
3	
4	

FIGURE 6.13. Table of values.

Answers

1. The successive quotients are the same.
2. The successive quotients are the common multiplier. For example, $900 \times 1.5 = 1350$.
3. An exponential function is $f(x) = 400 \cdot (1.5)^x$. The initial value is 400 and the common multiplier is 1.5. At this point you might want to point out to students that this model represents a growth rate of 50%. When $b = 1$, the population remains the same. So 1.5 represents the original population plus an additional 50% $(1 + 0.5)$.
4.

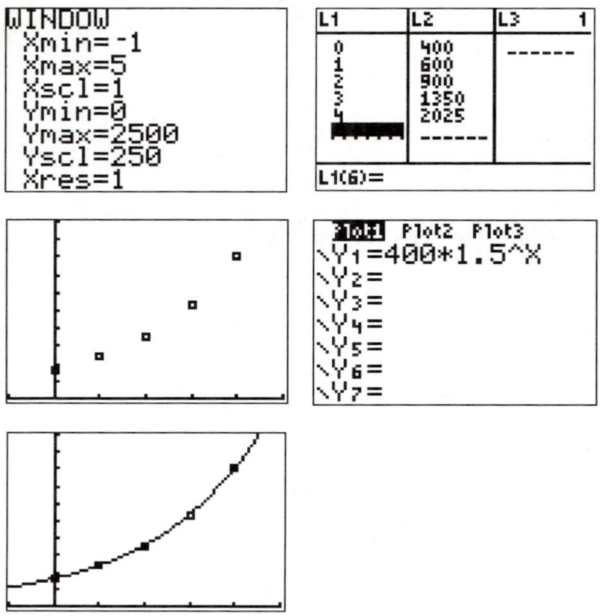

5. All three have the same common multiplier, 2. The functions have different starting points.

6.

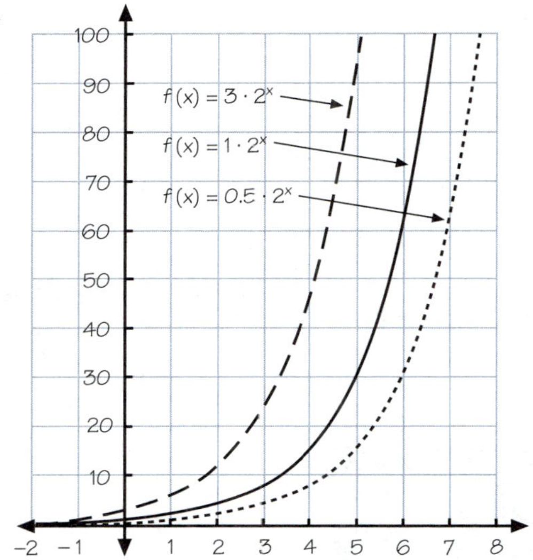

7. Values of a larger than 1 result in a vertical stretch of the graph of $f(x) = b^x$. Values of a between 0 and 1 result in a vertical compression of $f(x) = b^x$. (At this point we are using only positive values for a and b.)

Section 6.3 Assignment

Answers

1. exponential, $f(x) = 0.75\ (2.5)^x$
2. quadratic, $f(x) = 2x^2$
3. exponential, $f(x) = 2(1.2)^x$
4. linear, $f(x) = 3x - 2$
5.

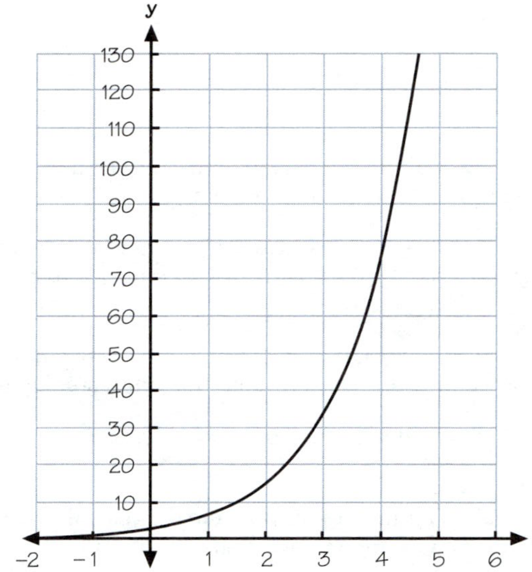

6.

x	f(x)
0	3
1	6.75
2	15.188
3	34.172
4	76.887

7. What is the y-intercept of the graph?
8. What is the common multiplier?
9. Is $f(x)$ a continuous or discrete function?
10. Is $f(x) = 3(2.25)^x$ a vertical stretch or vertical compression of $f(x) = 2.25^x$?

Use **Figure 6.14** to answer Questions 11–15.

x	$f(x)$	Successive Quotients
0	0.5	
1	1.75	
2	6.125	
3	21.438	
4	75.031	

FIGURE 6.14.
Table for Questions 11–15.

11. Find the successive quotients for the set of data.
12. Why does an exponential function fit the data?
13. What is the initial value of the function?
14. What is the common multiplier?
15. Find an exponential function that fits the data.

SECTION 6.4

M&M's® Growth

First and second differences and successive quotients are not always exactly equal in real-world data. In this activity you will gather data and fit a function.

Start with 4 M&M's in a cup. Shake the cup, roll out the M&M's, and then count the number with "m" showing. Multiply that number by 2 and add that many M&M's to the cup. Keep a record of the population and repeat the process 5 times. Before you start, answer Question 1.

1. What would you expect the population to be at each roll? Write your predictions in a table like **Figure 6.15.**

Trial	Predicted Population
0	
1	
2	
3	
4	
5	

FIGURE 6.15.
Table for predicting the M&M's population.

2. Add a column for the actual population to your table. Conduct the experiment, and fill in your actual population column.
3. How close are your results to what you predicted?
4. Combine the data with the rest of the class, and enter these data into lists in a calculator. Use transformations to create a function to fit the data.
5. Are the results more reasonable with all of the data? Take successive quotients to confirm your answer.
6. Find a function to fit the data. Check it by graphing the function on a scatter plot. Sketch the graph.

Answers (continued)

7. (0, 3)

8. 2.25

9. continuous

10. Vertical stretch

11.

x	$f(x)$	Successive Quotients
0	0.5	—
1	1.75	3.5
2	6.125	3.5
3	21.438	3.5
4	75.031	3.5

12. The successive quotients are all equal.

13. (0, 0.5)

14. 3.5

15. $f(x) = 0.5(3.5)^x$

ELABORATE LESSON 1

Section 6.4 M&M's Growth

Purpose: In the previous sections students have investigated exponential functions. In this section they will apply what they have learned to several real-world examples. In Section 6.5 you will be able to evaluate their ability to solve another real-world problem involving exponential functions.

Description: In this section students will experiment with the growth of an M&M population. When the "m" lands face up, students will count the number of m's, multiply by 2, and add that many additional M&M's. The model should be exponential, but not as "nice" as models in previous sections. When the entire class data set is investigated, the model should become a better fit.

Materials: M&M's, graphing calculator, cup

Teaching Notes: If students have trouble seeing the "m" on yellow M&M's, remove the yellow M&M's from the experiment. A shoe box lid, or other shallow lid, can be used to keep the M&M's from rolling off the desktop. This activity can be presented as a simplified population model of a species, such as the lion, which, on average, has a litter of one male and one female in each time period.

Answers

1. The population should double each time.

Trial	Predicted Population
0	4
1	8
2	16
3	32
4	64
5	128

2. Help students as they do the experiment. If they roll out 4 M&M's and 2 have "m"s showing, then they will add 2 × 2 or 4 M&M's to get a total population of 8 M&M's. This activity may work better if the yellow M&M's are removed since the "m" is more difficult to see on the yellow M&M's. Sample data are provided.

Trial	Predicted Population	Actual Population
0	4	4
1	8	8
2	16	18
3	32	34
4	64	68
5	128	138

3. Answers will vary. Students should be able to say how they made their predictions and why they believe their estimates were close or far from the actual data.

4. Students should use their knowledge of exponential functions and transformations to guess a model. They will find a more exact model in Question 6.

Sample class data for a class size of 20:

Trial	Population
0	80
1	156
2	322
3	632
4	1260
5	2548

5. The model should fit better with the increased number of data points.

6. Sample equation: $f(x) = 80(2)^x$

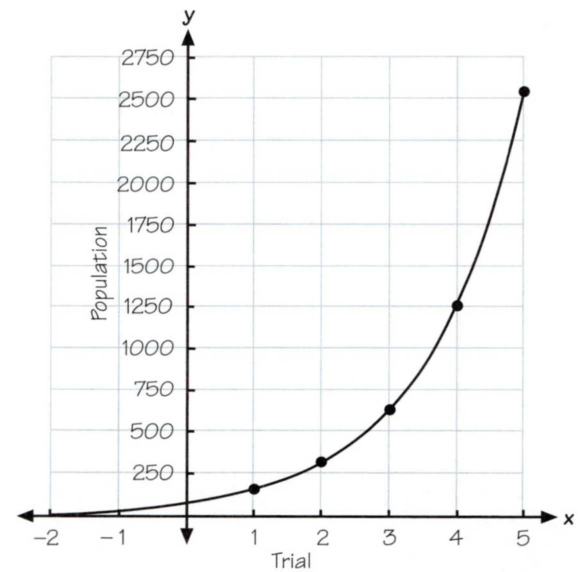

Your class's data will be different depending upon the number of M&M's with which the entire class starts.

SECTION 6.4 Assignment

In 1995 there were about 60,000 centenarians in the United States. A centenarian is a person at least 100 years old. Imagine that a researcher continued to collect data through 2002 as shown in **Figure 6.16.**

Year	Year Number	Number of Centenarians	Successive Quotients
1995	0	60,000	
1996	1	64,200	1.07
1997	2	68,694	1.07
1998	3	73,503	1.07
1999	4	78,648	1.07
2000	5	84,153	1.07
2001	6	90,044	1.07
2002	7	96,346	1.07

FIGURE 6.16. *Centenarian data from 1995–2002.*

1. Make a scatter plot of the data.
2. Explain which type of function (linear, quadratic, or exponential) should fit these data.
3. Find a function that fits the data well.
4. A researcher claimed that the number of centenarians would reach 232,000 by the year 2015. Verify that the researcher is correct based on the function you found in Question 3.
5. When will the number of centenarians reach 1,000,000 based on your model?
6. A scientist is trying to project deer population in a forest. The sample data are shown in **Figure 6.17.** Find successive differences, and then find the successive quotients. Write them in a copy of the table. Round quotients to the nearest hundredth.

Year	Year Number	Population	Successive Differences	Successive Quotients
2000	0	1500		
2001	1	1545		
2002	2	1591		
2003	3	1639		
2004	4	1688		
2005	5	1739		
2006	6	1791		
2007	7	1845		

FIGURE 6.17. *Sample data for a deer population.*

7. Sketch a scatter plot of the data.
8. Are the data linear or exponential? Why?
9. Determine the growth rate.
10. Find a function that models the growth.
11. Use your function to predict the population in the year 2010.

Section 6.4 Assignment

Answers

1.

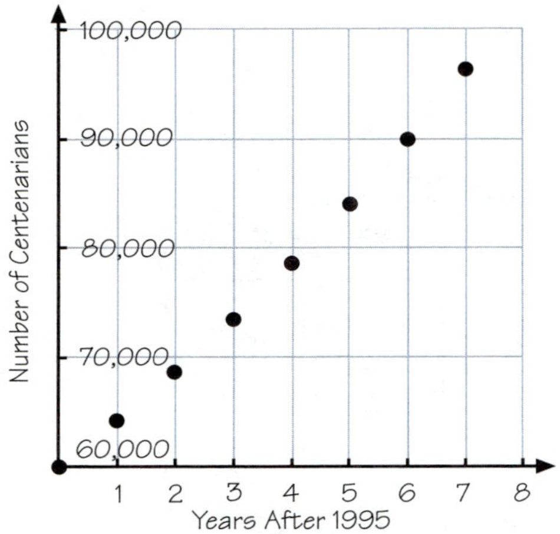

2. Even though the set of data appears linear in the scatter plot, neither the first nor the second differences are constants. The successive quotients are equal, 1.07. The best function to model the data is exponential.

3. $f(x) = 60000 \cdot 1.07^x$

4. 2015 is 20 years after 1995. So the value of the exponent in the model is 20. $60000 \cdot 1.07^{20} =$ 232,181.

5. Using the table feature of the graphing calculator students will see that the number of centenarians passes 1,000,000 forty-two years after 1995, which would be the year 2037.

6.

Year	Year Number	Population	Successive Differences	Successive Quotients
2000	0	1500	—	
2001	1	1545	45	1.03
2002	2	1591	46	1.03
2003	3	1639	48	1.03
2004	4	1688	49	1.03
2005	5	1739	51	1.03
2006	6	1791	52	1.03
2007	7	1845	54	1.03

7.

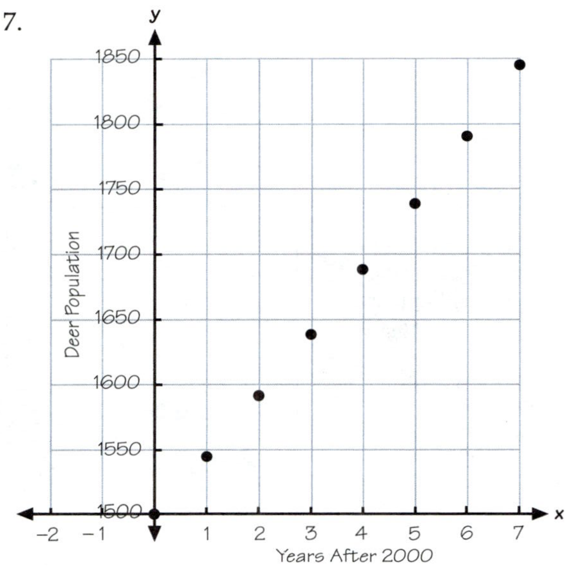

8. An exponential model fits the graph better than a linear model since all of the successive quotients are the same. (This may not be evident from the graph.)

9. The growth rate is 3% since we are multiplying by 1.03 each year.

10. $y = 1500(1.3)^x$

11. $1500(1.03)^{10} = 2016$

SECTION 6.5

Moose Population in National Parks

Let's apply to a new situation what you have learned thus far about exponential functions.

Managers from four parks in New York, New England, and Canada are meeting to discuss how they can protect the moose population. Each of them has been recording the population of moose in their parks. The tables in **Figure 6.18** contain data that they brought to the meeting.

FIGURE 6.18. Moose population data from four parks.

Adirondack Park (New York)		
Year	Year Number	Population
2000		50
2001		55
2002		61
2003		67
2004		73

Green Mountain National Forest (Vermont)		
Year	Year Number	Population
2000		2
2001		4
2002		8
2003		16
2004		32

Algonquin Provincial Park (Ontario, Canada)		
Year	Year Number	Population
2000		10
2001		15
2002		23
2003		34
2004		51

White Mountain National Forest (New Hampshire)		
Year	Year Number	Population
2000		50
2001		54
2002		58
2003		62
2004		66

If the moose population in each park continues to grow at the same rate, which park will have the most moose in 2015? Explain your answer.

SUMMARY

In this lesson you learned that exponential functions are good mathematical models for certain growth situations. You also reviewed other mathematical models:

- A linear model is an addition model.
- A quadratic model has second differences that are constant.
- An exponential model is a multiplication model.

Exponential models have these properties:

- Successive quotients are constant.
- The successive quotient is the common multiplier or the base of the exponential function.
- The parent function form for the exponential family is $f(x) = b^x$.
- A vertical compression or vertical stretch occurs when the parent function $f(x) = b^x$ is multiplied by a constant, a, $f(x) = a \cdot b^x$. This affects the y-intercept, the initial value, or a, in many modeling situations.

EVALUATE LESSON 1

Section 6.5: Moose Population in National Parks

Purpose: In Sections 6.1 through 6.4, students developed the concept of an exponential growth function and learned to transform the parent function, $y = b^x$, to fit certain circumstances. In this section students have the opportunity to demonstrate their knowledge by applying those concepts and skills to solve a problem in a new context.

Description: Facing a problem in a new context, students determine and implement a method to solve the problem. Students are asked to compare four situations and evaluate the situations in order to make a decision. This open-middle performance assessment task has only one correct answer, but there is a variety of valid methods to obtain that answer. In their justification, which is the most important part of a performance assessment task, students should be able to communicate their knowledge of mathematics and demonstrate that they have mastered the concepts presented in Sections 6.1 through 6.4.

Materials: graphing calculators

Teaching Ideas: This task may be assigned as group work or individual work. The mathematics performance assessment rubric should be used to evaluate students' work.

Sample Answer

Define "Year Number = 0" to be the year 2000. Let year number be the independent variable and population be the dependent variable. Use number patterns in the table to find an appropriate function rule.

Adirondack Park (New York)			
Year	Year Number	Population	Successive Quotient
2000	0	50	
2001	1	55	55 ÷ 50 = 1.1
2002	2	61	61 ÷ 55 = 1.1
2003	3	67	67 ÷ 61 = 1.1
2004	4	73	73 ÷ 67 = 1.1

Successive quotients are 1.1, so this is an exponential function with $b = 1.1$. The initial value is 50 moose at year 0, so $a = 50$. The function rule is $y = 50(1.1)^x$.

Green Mountain National Forest (Vermont)			
Year	Year Number	Population	Successive Quotient
2000	0	2	
2001	1	4	4 ÷ 2 = 2
2002	2	8	8 ÷ 4 = 2
2003	3	16	16 ÷ 8 = 2
2004	4	32	32 ÷ 16 = 2

Successive quotients are 2, so this is an exponential function with $b = 2$. The initial value is 2 moose at year 0, so $a = 2$. The function rule is $y = 2(2)^x$.

Algonquin Provincial Park (Ontario, Canada)			
Year	Year Number	Population	Successive Quotient
2000	0	10	
2001	1	15	15 ÷ 10 = 1.5
2002	2	23	23 ÷ 15 = 1.5
2003	3	34	34 ÷ 23 = 1.5
2004	4	51	51 ÷ 34 = 1.5

Successive quotients are 1.5, so this is an exponential function with $b = 1.5$. The initial value is 10 moose at year 0, so $a = 10$. The function rule is $y = 10(1.5)^x$.

White Mountain National Forest (New Hampshire)			
Year	Year Number	Population	Successive Quotient
2000	0	50	
2001	1	54	54 ÷ 50 = 1.08
2002	2	58	58 ÷ 54 = 1.07
2003	3	62	62 ÷ 58 = 1.07
2004	4	66	66 ÷ 62 = 1.06

Successive quotients are not the same, but are close. Looking more closely, there is a constant first difference of +4. This suggests that the moose population in White Mountain National Forest follows a linear model of $y = 4x + 50$.

Use each function rule to calculate the population in 2015, when the year number is 15.

Adirondack Park:
$y = 50(1.1)^x = 50(1.1)^{15} = 209$ moose

Green Mountain National Forest:
$y = 2(2)^x = 2(2)^{15} = 65{,}536$ moose

Algonquin Provincial Park:
$y = 10(1.5)^x = 10(1.5)15 = 4379$ moose

White Mountain National Forest:
$y = 4(15) + 50 = 110$ moose

Green Mountain National Forest will have the most moose in 2015.

SECTION 6.6

Fractional Parts and Area

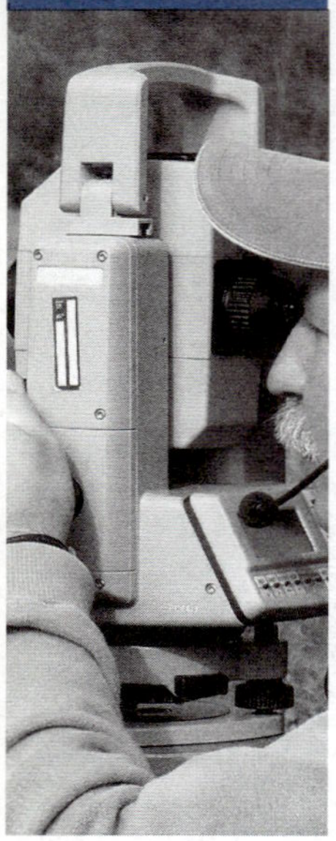

Park managers often have to divide land into sections. One way to divide land is the rectangular survey system. This system uses rectangles or squares that are divided into equal parts. Each part is then given a name based on its location in relationship to the original rectangle or square.

In this section you will model the rectangular survey system by examining how the visible area of a sheet of paper changes as the paper is folded. You will look for patterns and make predictions.

First let's look at the part of a paper that is formed when a sheet of paper is folded.

You need an 8.5-inch by 11-inch sheet of paper. Before any folds are made, the fraction of the sheet of paper is 1 (the whole sheet).

1. Fold the sheet in half once. This fold splits the paper into regions bounded by the fold lines. Unfold the paper and count the number of regions. What fraction of the original sheet of paper is each new region?

2. Continue folding the paper and complete a table like **Figure 6.19.**

Number of Folds	Fraction of the Original Sheet of Paper
0	
1	
2	
3	
4	
5	
6	

FIGURE 6.19.
Table for paper folding data.

3. What patterns do you see in the table?

4. Based on the table values, does this relationship appear to be linear? How do you know?

5. Based on the table values, does this relationship appear to be quadratic? How do you know?

6. What is a reasonable domain here? Why?

7. What is a reasonable range here? Why?

8. Sketch what you would expect a graph of the fraction of the original sheet of paper versus the number of folds to look like. Why did you sketch the graph the way you did?

Now let's look at the area of the part of a paper that is formed when a sheet of paper is folded.

You will need a new 8.5-inch by 11-inch sheet of paper to fold. Before any folds are made, calculate the area in square inches of the sheet of paper.

9. What is the area of your sheet of paper?

10. Fold the sheet of paper one time; this fold splits the paper into regions bounded by the fold lines. What is the area of each new region?

11. Continue folding the paper and complete a table like the one in **Figure 6.20.**

Number of Folds	Area of a Region
0	
1	
2	
3	
4	
5	
6	

FIGURE 6.20.
Table for region area data.

12. What patterns do you see in the table?

13. Does the relationship between area and number of folds appear to be linear? How do you know?

ENGAGE LESSON 2

Section 6.6: Fractional Parts and Area

Purpose: In Sections 6.1 through 6.5, students investigated situations that could be modeled using exponential growth functions. In Lesson 2 students will investigate situations that can be modeled using exponential decay functions. In Section 6.6 students will use paper folding to investigate the relationship between the area of a fractional region of a piece of paper and the number of folds it takes to create the region. This will allow students to explore exponential decay functions in Section 6.7.

Description: One dilemma faced by land managers and park managers, especially in western states, is how to describe the location of a piece of land. In this section students will investigate this dilemma using the problem-solving strategy of "make a model." Students will continue their paper cutting investigation from Section 6.1. This time, however, students will look for patterns in the area of the smallest region created by a certain number of folds as a fractional part of the original area. Then they will actually compute the area of the smallest region in square inches and look for number patterns in the data.

Materials: graphing calculator, two sheets of letter-size (8.5 by 11 inches) paper per student, rulers

Teaching Ideas: Students should work in small groups (3 to 4 students). Each student should individually fold the piece of paper and gather his or her own data. However, encourage students to discuss the problem together and to share their insights into solving the problem.

In particular, students will be asked to find the area of a sheet of letter-size paper. Do not tell the students how to do this. Provide them with rulers (if necessary), and encourage students to share their strategies for finding the area. Students will then be asked to find the area of a fractional region of the paper. Again, do not tell students how to find the area. If students are stuck, prompt them to use what they know about finding area and their answers to Questions 1–8.

In this phase of the lesson, it is OK for students to have approximations and conjectures. The students will formalize their findings later on and may be revising their answers to the investigation questions in this section.

Answers

1. One half

2.

Number of Folds	Fraction of the Original Sheet of Paper
0	1
1	$\frac{1}{2}$
2	$\frac{1}{4}$
3	$\frac{1}{8}$
4	$\frac{1}{16}$
5	$\frac{1}{32}$
6	$\frac{1}{64}$

3. Answers may vary. Each time the paper is folded the fractional part becomes smaller.

4. This relationship is not linear because there is not a constant rate of change.

5. This relationship is not quadratic since second differences are not constant.

6. Sample answer: A reasonable domain is whole numbers between one and six because the domain is the number of folds.

7. Sample answer: A reasonable range is fractions between zero and one because the sheet of paper starts as one whole and becomes smaller with each fold.

8. Sample answer: As the number of folds increases, the fraction of the paper gets very small but will never equal zero.

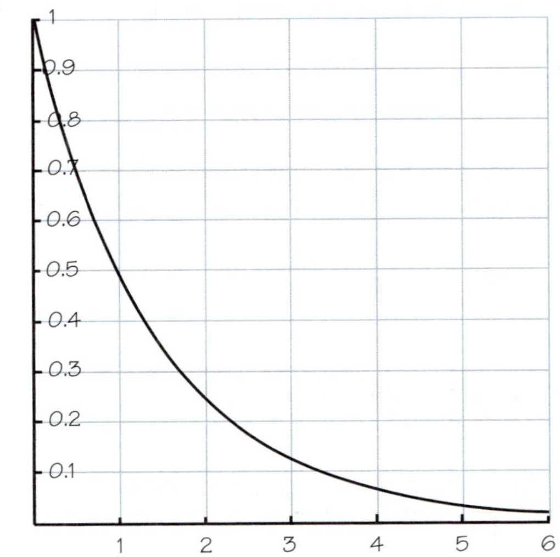

9. 93.5 square inches

10. Each new region is half the area of the original sheet of paper or 46.75 square inches.

11. Answer:

Number of Folds	Area of a Region
0	93.5 in^2
1	46.75 in^2
2	23.375 in^2
3	11.6875 in^2
4	5.84375 in^2
5	2.921875 in^2
6	1.4609375 in^2

12. Answers may vary. Each time the paper is folded, the area of the region becomes smaller.

13. This relationship is not linear because there is not a constant rate of change.

14. Does the relationship appear to be quadratic? How do you know?

15. What is a reasonable domain here? Why?

16. What is a reasonable range here? Why?

17. Sketch what you expect a graph of the area of a region versus the number of folds to look like. Why did you sketch the graph the way you did?

SECTION 6.7 Area of a Region

In the last section you used paper folding to break down a piece of paper into smaller regions, then you found the area of each region.

Examine the first table that you made in Section 6.6 (see **Figure 6.21**).

Number of Folds	Fraction of the Original Sheet of Paper
0	1
1	$\frac{1}{2}$
2	$\frac{1}{4}$
3	$\frac{1}{8}$
4	$\frac{1}{16}$
5	$\frac{1}{32}$
6	$\frac{1}{64}$

FIGURE 6.21. Table of values.

Number of Folds (n)	Process	Fraction of the Piece of Paper (f)
0	1	1
1	$1 \cdot \frac{1}{2}$	$\frac{1}{2}$
2	$1 \cdot \frac{1}{2} \cdot \frac{1}{2} = \frac{1}{2^2}$	$\frac{1}{4}$
3	$1 \cdot \_ \cdot \_ \cdot \_ = \_$	$\frac{1}{8}$
4	$1 \cdot \_ \cdot \_ \cdot \_ \cdot \_ = \_$	$\frac{1}{16}$
5	$1 \cdot \_ \cdot \_ \cdot \_ \cdot \_ \cdot \_ = \_$	$\frac{1}{32}$
6	$1 \cdot \_ \cdot \_ \cdot \_ \cdot \_ \cdot \_ \cdot \_ = \_$	$\frac{1}{64}$

FIGURE 6.22. Table of values with process column.

1. Use a graphing calculator to make a fraction of the original sheet versus the number of folds scatter plot. Record your window and sketch your graph.

2. What do you notice about the scatter plot?

3. How does each fraction of a sheet relate to the one before it?

4. Using the answer to Question 3, copy the table and fill in the blanks in the process column (see **Figure 6.22**).

5. What stays the same in the process column?

6. What changes in the process column?

7. Write a function rule that models this situation.

Answers (continued)

14. This relationship is not quadratic since second differences are not constant.

15. Sample answer: A reasonable domain is whole numbers between one and six because the domain is the number of folds.

16. Sample answer: A reasonable range is numbers between zero and 95 because the sheet of paper starts as one whole with an area of 93.5 square inches and becomes smaller with each fold.

17. Sample answer: As the number of folds increases the area of the region of paper gets very small but will never equal zero.

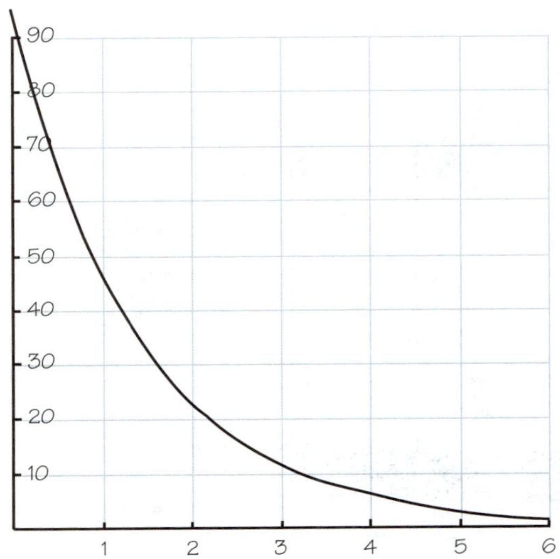

EXPLORE LESSON 2

Section 6.7: Area of a Region

Purpose: In Section 6.6 students used paper folding to investigate the area of a fractional region of a rectangle. In this section students will explore ways to develop an algebraic model based on their understanding of exponential functions from Sections 6.1 through 6.5. In Section 6.8 students will formalize the concept of exponential decay and the behavior of exponential decay functions so that they can use them to solve problems in Sections 6.9 and 6.10.

Description: In this section students will use successive quotients in order to explore the constant rate of decay that exists in an exponential decay function. Then students will build exponential functions based on their data. However, this time, the base of the exponential function will be between 0 and 1, resulting in a different kind of graph. Students will make conjectures based on their observations, which will be formalized during Section 6.8.

Materials: graphing calculator

Teaching Ideas: Students should work in small groups (3 to 4 students). This section is designed for students to build on their understanding of exponential functions from Sections 6.1 through 6.5.

Monitor calculator use and assist as necessary.

Answers

1.

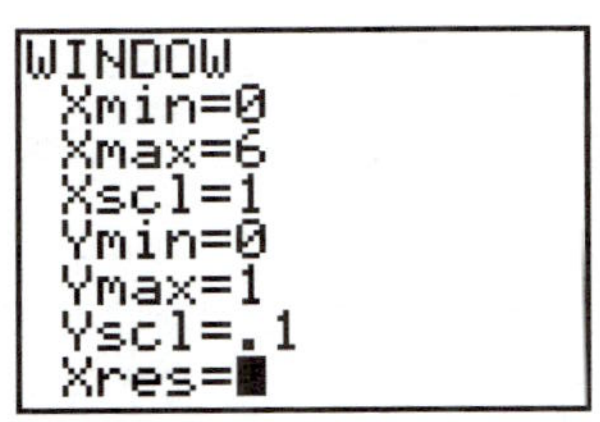

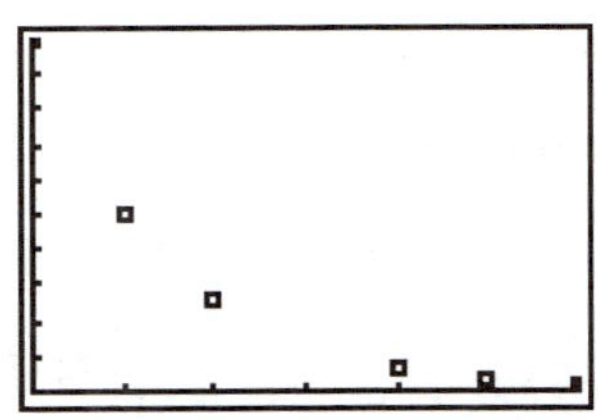

2. Answers may vary. The graph is not linear and appears to approach zero.

3. Each fraction is one-half of the one before it.

4.

Number of Folds (n)	Process	Fraction of the Piece of Paper (f)
0	1	1
1	$1 \cdot \frac{1}{2}$	$\frac{1}{2}$
2	$1 \cdot \frac{1}{2} \cdot \frac{1}{2} = \frac{1}{2^2}$	$\frac{1}{4}$
3	$1 \cdot \frac{1}{2} \cdot \frac{1}{2} \cdot \frac{1}{2} = \frac{1}{2^3}$	$\frac{1}{8}$
4	$1 \cdot \frac{1}{2} \cdot \frac{1}{2} \cdot \frac{1}{2} \cdot \frac{1}{2} = \frac{1}{2^4}$	$\frac{1}{16}$
5	$1 \cdot \frac{1}{2} \cdot \frac{1}{2} \cdot \frac{1}{2} \cdot \frac{1}{2} \cdot \frac{1}{2} = \frac{1}{2^5}$	$\frac{1}{32}$
6	$1 \cdot \frac{1}{2} \cdot \frac{1}{2} \cdot \frac{1}{2} \cdot \frac{1}{2} \cdot \frac{1}{2} \cdot \frac{1}{2} = \frac{1}{2^6}$	$\frac{1}{64}$

5. Sample answer: There is always a $1 \cdot \frac{1}{2}$.

6. The number of times we multiply by $\frac{1}{2}$.

7. $y = 1 \cdot \left(\frac{1}{2}\right)^x$

8. In the context of the paper folding activity, what does each number mean in your rule from Question 7?

9. Graph your function rule on your scatter plot. If the graph is not a good fit, change your rule. Sketch your graph.

CHECK THIS!

To find successive quotients, divide each value by the one before it.

10. What will be the fraction of the piece of paper after 9 folds?

11. If the fraction of the sheet of paper is $\frac{1}{256}$, how many times has it been folded?

12. Find the successive quotients for fraction of the sheet of paper and record them in a table like **Figure 6.23**.

Number of Folds	Fraction of the Original Sheet of Paper	Successive Quotients
0	1	
1	$\frac{1}{2}$	
2	$\frac{1}{4}$	
3	$\frac{1}{8}$	
4	$\frac{1}{16}$	
5	$\frac{1}{32}$	
6	$\frac{1}{64}$	

FIGURE 6.23.
Table of values with column for successive quotients.

13. What do you notice about the successive quotients?

14. Where do you see the successive quotients in your function from Question 7?

Now look at the second table you made in Section 6.6 (see **Figure 6.24**).

Number of Folds	Area of a Region
0	93.5 in^2
1	46.75 in^2
2	23.375 in^2
3	11.6875 in^2
4	5.84375 in^2
5	2.921875 in^2
6	1.4609375 in^2

FIGURE 6.24.
Table of values from Section 6.6.

15. Use a graphing calculator to make an area versus number of folds scatter plot. Record your window and sketch your graph.

16. What do you notice about the scatter plot?

17. How does each area relate to the area of the region before it?

Let's use a process column to help find a function.

18. Copy the table in **Figure 6.25** and fill in the blanks in the process column.

Number of Folds	Process	Area of a Region
0	93.5	93.5 in^2
1	$93.5 \cdot \frac{1}{2}$	46.75 in^2
2	$93.5 \cdot \frac{1}{2} \cdot \frac{1}{2} = 93.5 \cdot \frac{1}{2^2}$	23.375 in^2
3	$93.5 \cdot - \cdot - \cdot - = 93.5 \cdot -$	11.6875 in^2
4	$93.5 \cdot - \cdot - \cdot - \cdot - = 93.5 \cdot -$	5.84375 in^2
5	$93.5 \cdot - \cdot - \cdot - \cdot - \cdot - = 93.5 \cdot -$	2.921875 in^2
6	$93.5 \cdot - \cdot - \cdot - \cdot - \cdot - \cdot - = 93.5 \cdot -$	1.4609375 in^2

FIGURE 6.25.
Table of values with blanks in the process column.

19. What stays the same in the process column?

20. What changes in the process column?

21. Write a function rule that models this situation.

22. Graph your function rule on your scatter plot. If the graph is not a good fit, use transformations to change your rule. Sketch your graph.

23. What will be the area of the region of paper after 10 folds?

24. If the area of the region is 0.365234375 square inches, how many times has the paper been folded?

Answers (continued)

8. The 1 represents the whole sheet of paper. The $\frac{1}{2}$ represents taking one-half of the previous region with each successive fold.

9. 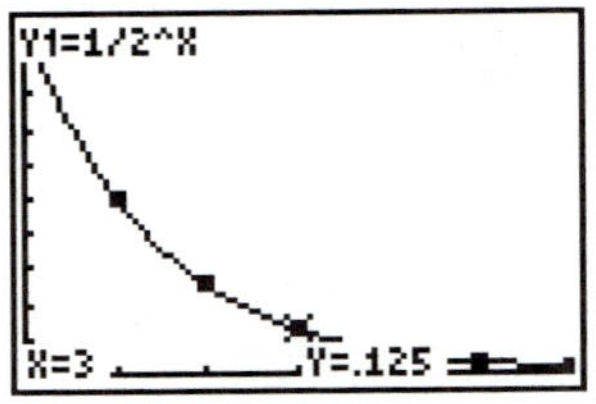

10. $\left(\frac{1}{2}\right)^9 = \frac{1}{512} \approx 0.00195$

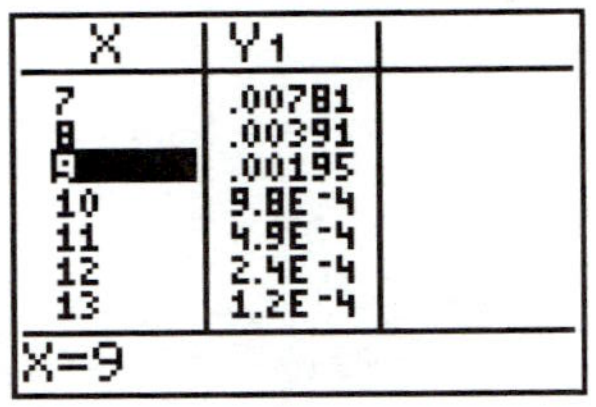

11. Since $\frac{1}{256} = 0.00390625$, the paper was folded 8 times.

Number of Folds	Fraction of the Original Sheet of Paper	Successive Quotients
0	1	
1	$\frac{1}{2}$	$\frac{1}{2} \div 1 = \frac{1}{2}$
2	$\frac{1}{4}$	$\frac{1}{4} \div \frac{1}{2} = \frac{1}{2}$
3	$\frac{1}{8}$	$\frac{1}{8} \div \frac{1}{4} = \frac{1}{2}$
4	$\frac{1}{16}$	$\frac{1}{16} \div \frac{1}{8} = \frac{1}{2}$
5	$\frac{1}{32}$	$\frac{1}{32} \div \frac{1}{16} = \frac{1}{2}$
6	$\frac{1}{64}$	$\frac{1}{64} \div \frac{1}{32} = \frac{1}{2}$

13. They all are equal to $\frac{1}{2}$.

14. The $\frac{1}{2}$ is the part of the rule that is raised to a power.

15. 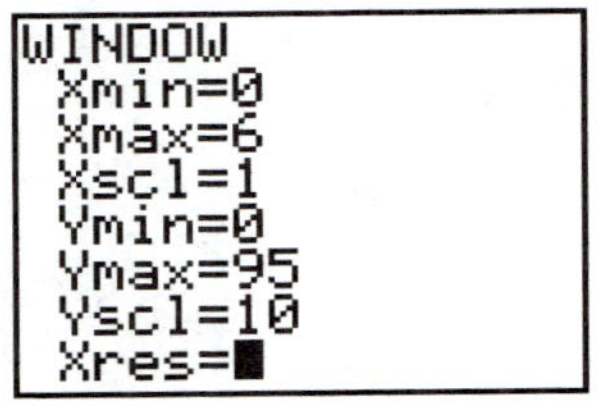

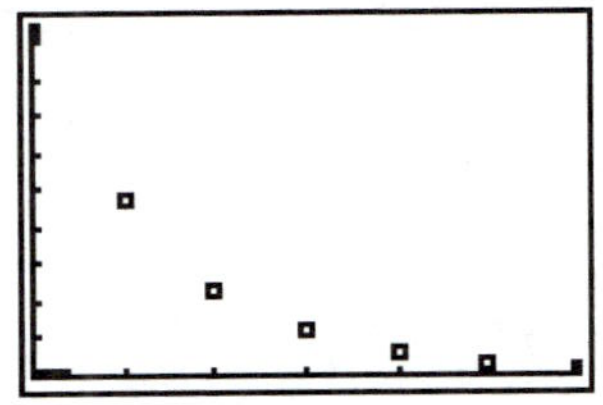

16. Answers may vary. The graph is not linear and appears to approach zero.

17. Each area is $\frac{1}{2}$ the area of the region before it.

18.

Number of Folds	Process	Area of a Region
0	93.5	93.5 in^2
1	$93.5 \cdot \frac{1}{2}$	46.75 in^2
2	$93.5 \cdot \frac{1}{2} \cdot \frac{1}{2} = 93.5 \cdot \frac{1}{2^2}$	23.375 in^2
3	$93.5 \cdot \frac{1}{2} \cdot \frac{1}{2} \cdot \frac{1}{2} = 93.5 \cdot \frac{1}{2^3}$	11.6875 in^2
4	$93.5 \cdot \frac{1}{2} \cdot \frac{1}{2} \cdot \frac{1}{2} \cdot \frac{1}{2} = 93.5 \cdot \frac{1}{2^4}$	5.84375 in^2
5	$93.5 \cdot \frac{1}{2} \cdot \frac{1}{2} \cdot \frac{1}{2} \cdot \frac{1}{2} \cdot \frac{1}{2} = 93.5 \cdot \frac{1}{2^5}$	2.921875 in^2
6	$93.5 \cdot \frac{1}{2} \cdot \frac{1}{2} \cdot \frac{1}{2} \cdot \frac{1}{2} \cdot \frac{1}{2} \cdot \frac{1}{2} = 93.5 \cdot \frac{1}{2^6}$	1.4609375 in^2

19. The $\frac{1}{2}$ and the 93.5 stay the same.

20. The number of times we multiply by $\frac{1}{2}$ changes.

21. $y = 93.5 \cdot \left(\frac{1}{2}\right)^x$

22. 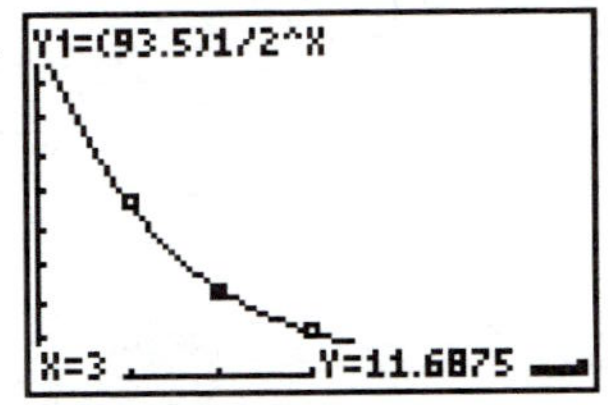

23. The area will be 0.09131 square inches.

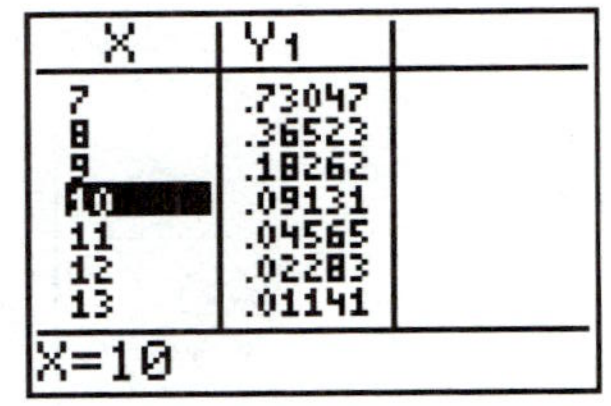

24. The paper was folded 8 times.

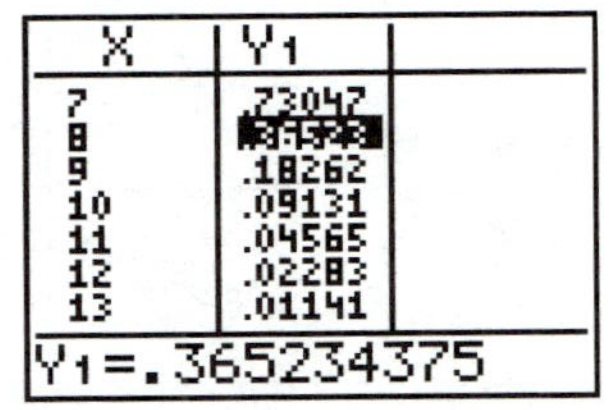

25. Calculate the successive quotients for area of a region and record them in a table like **Figure 6.26.**

Number of Folds	Area of a Region	Successive Quotients
0	93.5 in^2	
1	46.75 in^2	
2	23.375 in^2	
3	11.6875 in^2	
4	5.84375 in^2	
5	2.921875 in^2	
6	1.4609375 in^2	

FIGURE 6.26.
Table of values for calculating the successive quotients.

26. What do you notice about the successive quotients?

27. Where do you see the successive quotients in your function?

28. Write a sentence about the relationship of successive quotients to the function rule for an exponential function.

SECTION 6.7 Assignment

1. Rhonda is drawing a design in art class. She started by drawing a triangle. Next she drew a new triangle by connecting the midpoints of the sides of the first triangle. She shaded a new triangle. Next she found the midpoints of the sides of the unshaded triangles and connected them. Once again she shaded some new triangles. Her first three drawings are shown. Based on this pattern, complete the term value column in a table like **Figure 6.27.**

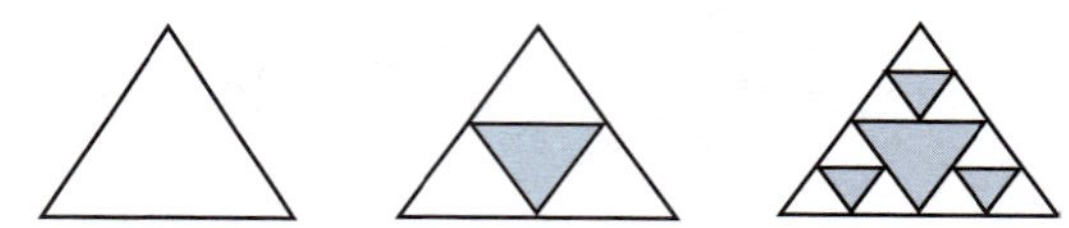

Term Number	Picture	Process Column	Term Value (Number of *Unshaded* Triangles)
0	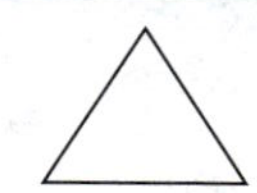	1	1
1		1 · 3	3
2	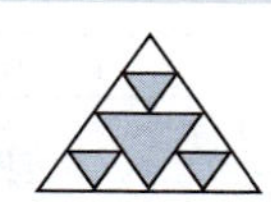		
3			
4			
n			

FIGURE 6.27.
Table for finding the number of unshaded triangles.

Answers (continued)

25.

Number of Folds	Area of a Region	Successive Quotients
0	93.5 in^2	
1	46.75 in^2	46.75 / 93.5 = 0.5
2	23.375 in^2	23.375 / 46.75 = 0.5
3	11.6875 in^2	11.6875 / 23.375 = 0.5
4	5.84375 in^2	5.84375 / 11.6875= 0.5
5	2.921875 in^2	2.921875 / 5.84375 = 0.5
6	1.4609375 in^2	1.4609375 / 2.921875 = 0.5

26. They all are equal to 0.5 or $\frac{1}{2}$.

27. The $\frac{1}{2}$ is the part of the rule that is raised to a power.

28. The successive quotients give the part of the rule that is raised to a power; it is the base (b) of an exponential function in the form $y = a \cdot b^x$.

Section 6.7 Assignment

1. See answer to Question 5.

2. Use your graphing calculator to create a term value versus term number scatter plot. Record your window and sketch your graph.
3. What do you notice about the scatter plot?
4. How does each term value relate to the one before it?
5. Fill in the process column of your table.
6. What stays the same in the process column?
7. What changes in the process column?
8. Write a function rule that models this situation.
9. Graph your function rule on your scatter plot. If the graph is not a good fit, change your rule. Sketch your graph.
10. What will be the term value for the 9th term?
11. If there are 6561 new shaded triangles, what term number is this?
12. Next, Rhonda decided to examine the area of the smallest triangles she drew. She started by defining the area of the triangle for term number 0 as one square unit. So the area of the smallest triangle in her second figure is $\frac{1}{4}$ square units. Based on this pattern, complete the term value column in a table like **Figure 6.28.**
13. Use a graphing calculator to make a term value versus term number scatter plot. Record your window and sketch your graph.
14. What do you notice about the scatter plot?
15. How does each term value relate to the one before it?
16. Fill in the process column of your table.

Term Number	Picture	Process Column	Term Value (Area of the Smallest Triangle)
0	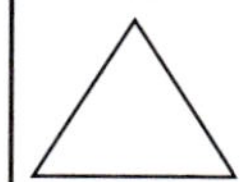	1	1
1	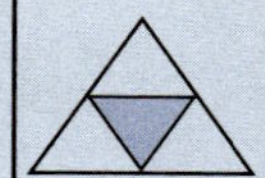	$1 \cdot \frac{1}{4}$	$\frac{1}{4}$
2			
3			
4			
n			

FIGURE 6.28. Table for finding the area of the smallest triangle.

17. What stays the same in the process column?
18. What changes in the process column?
19. Write a function rule that models this situation.
20. Graph your function rule on your scatter plot. If the graph is not a good fit, change your rule. Sketch your graph.
21. What will be the term value for the 8th term in this pattern?
22. If the area of the smallest triangle is 0.000061 square units, what term number is this?

Answers (continued)

2.

```
WINDOW
 Xmin=0
 Xmax=6
 Xscl=1
 Ymin=0
 Ymax=100
 Yscl=10
 Xres=■
```

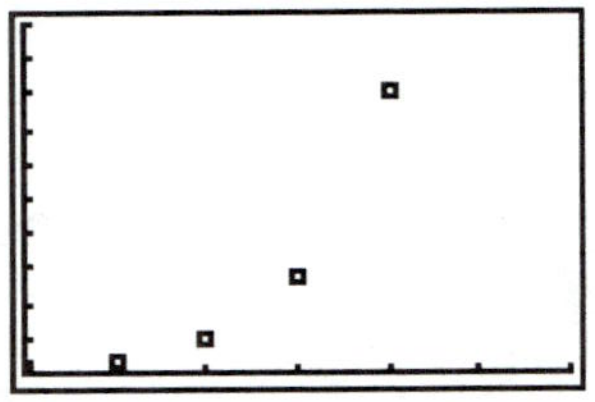

3. Answers may vary. The graph is not linear and appears to be increasing quickly.

4. Each value is 3 times the value before it.

5.

Term Number	Picture	Process Column	Term Value (Number of *Unshaded* Triangles)
0	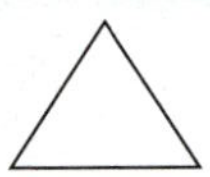	$1 = 1(3)^0$	1
1		$1 \cdot 3 = 1(3)^1$	3
2	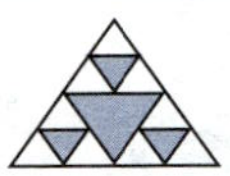	$1 \cdot 3 \cdot 3 = 1(3)^2$	9
3		$1 \cdot 3 \cdot 3 \cdot 3 = 1(3)^3$	27
4		$1 \cdot 3 \cdot 3 \cdot 3 \cdot 3 = 1(3)^4$	81
n		$1(3)^n$	

6. Sample answer: There is always a $1 \cdot 3$.

7. The number of times we multiply by 3

8. $y = 1 \cdot 3^n$

9.

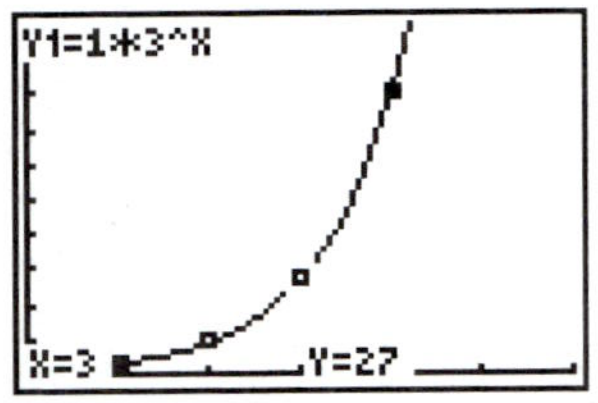

10. Answer: The value of the 9th term will be 19,683.

X	Y1
6	729
7	2187
8	6561
9	19683
10	59049
11	177147
12	531441

Y1=19683

11. Term number 8.

X	Y1
6	729
7	2187
8	6561
9	19683
10	59049
11	177147
12	531441

X=8

12. Answer: See answer to Question 16.

13.

```
WINDOW
 Xmin=0
 Xmax=6
 Xscl=1
 Ymin=0
 Ymax=1
 Yscl=.1
 Xres=■
```

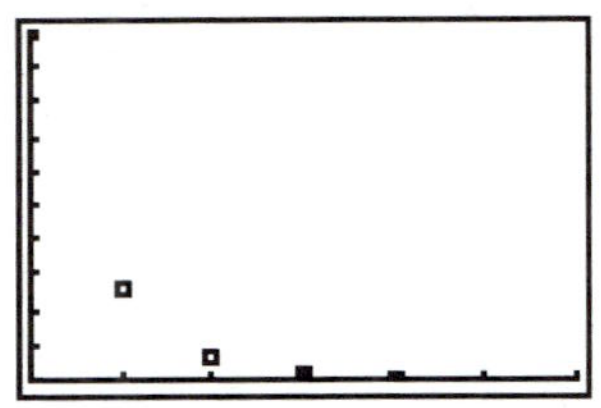

14. Answers may vary. The graph is not linear and appears to approach zero.

15. Each fraction is one fourth of the one before it.

16.

Term Number	Picture	Process Column	Term Value (Area of the Smallest Triangle)
0	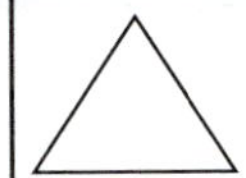	1	1
1	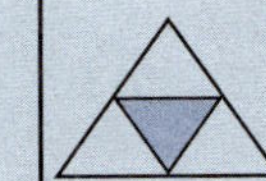	$1 \cdot \frac{1}{4}$	$\frac{1}{4}$
2		$1 \cdot \frac{1}{4} \cdot \frac{1}{4} = \frac{1}{4^2}$	$\frac{1}{16}$
3		$1 \cdot \frac{1}{4} \cdot \frac{1}{4} \cdot \frac{1}{4} = \frac{1}{4^3}$	$\frac{1}{64}$
4		$1 \cdot \frac{1}{4} \cdot \frac{1}{4} \cdot \frac{1}{4} \cdot \frac{1}{4} = \frac{1}{4^4}$	$\frac{1}{256}$
n		$\frac{1}{4^n}$	

17. Sample answer: There is always a $1 \cdot \frac{1}{4}$.

18. The number of times we multiply by $\frac{1}{4}$ changes.

19. $y = 1 \cdot \frac{1}{4^x}$ or $y = \left(\frac{1}{4}\right)^x$

20.

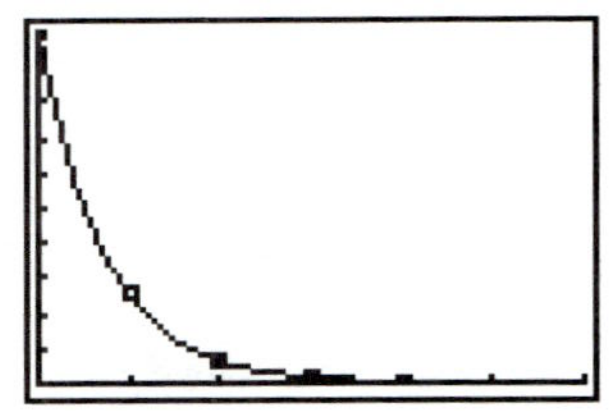

Answers (continued)

21. The term value for the 8th term is approximately 0.0000038.

22. Term number 7 has an area of 0.000061 square units.

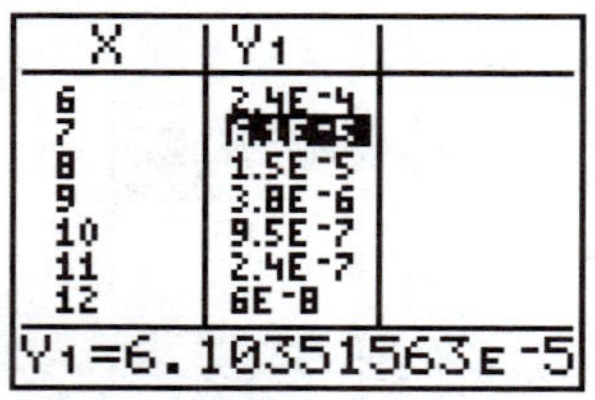

SECTION 6.8

Characteristics of Exponential Functions

In the last sections you began to explore exponential functions. In this section you will see how and why those functions work the way they do.

Recall the paper cutting in Section 6.1. Each time the paper was cut, the number of layers of paper increased. This is an example of **exponential growth**.

1. How did the table show exponential growth?

2. How did the graph show exponential growth?

Recall the paper folding in Sections 6.6–6.7. Each time the paper was folded, the size of the region decreased. The pattern you observed is an example of **exponential decay**.

3. How did the table of data show exponential decay?

4. How did the graph show exponential decay?

5. How are the function rules for the relationship between the fraction of the paper and number of folds and the relationship between area and number of folds alike?

6. How do the function rules for the relationship between the fraction of the paper and number of folds and the relationship between area and number of folds differ?

Plant Height

ΔX	Number of Days	Height (cm)	ΔY
	0	2	
+1	1	5	+3
+1	2	8	+3
+1	3	11	+3
+1	4	14	+3
+1	5	17	+3

Recall that linear relationships have a constant rate of change. In **Figure 6.29**, the height of a plant increases at a constant rate of 3 centimeters per day.

FIGURE 6.29.
Plant height.

7. In this linear relationship, what is the repeated operation?

8. What is a function rule that fits these data?

Let's compare the table for a linear relationship to the table for an exponential relationship. The table in **Figure 6.30** shows an exponential relationship between time and the amount of dissolved carbon dioxide (CO_2) in a can of soda.

Number of Hours	Dissolved CO_2
0	21.87
1	7.29
2	2.43
3	0.81
4	0.27
5	0.09

FIGURE 6.30.
Table of values for an exponential relationship.

9. How do these data show that this relationship is not linear or quadratic?

10. Is this relationship an example of exponential growth or exponential decay? How do you know?

11. Find successive quotients for the dissolved gas and record them in a table like **Figure 6.31.**

Number of Hours	Dissolved CO_2	Successive Quotients
0	21.87	
1	7.29	
2	2.43	
3	0.81	
4	0.27	
5	0.09	

FIGURE 6.31.
Table for recording successive quotients.

If a relationship is exponential and the rate of change of the independent variable is 1, then successive quotients of the dependent variable are constant.

12. What operation is repeated in an exponential relationship?

13. What is the function rule that fits these data?

EXPLAIN LESSON 2

Section 6.8 Characteristics of Exponential Functions

Purpose: In Section 6.7 students began to investigate and make conjectures about characteristics of exponential functions that led to an exponential decay situation. In Section 6.8 students will verify these conjectures, and compare linear, quadratic, exponential growth, and exponential decay functions. This prepares students to apply exponential decay to a different context in Sections 6.9 and 6.10.

Description: In this phase of the lesson students will refresh their memories about linear and quadratic functions, where they use finite differences to identify and write function rules to describe sets of data. Then they will examine the use of successive quotients to write exponential functions for both growth and decay situations. Students will use different representations of exponential functions including graphs, tables, and symbolic representations to model situations using exponential growth and decay functions.

Materials: graphing calculators

Teaching Ideas: Debrief students after Section 6.7 by having them share their answers to the questions. As they share, be sure to fill in any missing concept, process, or skill that could have been used to answer the questions. Then work through the questions in this section with the entire class. Once again, be sure all necessary concepts, processes, or skills are addressed.

Depending on students' calculator proficiency, you may need to review calculator skills of creating scatter plots, graphing function rules, and finding an appropriate window.

You may also need to review function transformations prior to or during this section, depending on students' recall of prior knowledge.

Answers

1. The values increase but not at a constant rate.
2. It is nonlinear and increases from left to right.
3. The values decrease but not at a constant rate.
4. It is nonlinear and decreases from left to right.
5. They both have $\frac{1}{2}$ raised to a power.
6. The relationship fraction of the piece of paper versus number of folds has 1 as an initial value, and the relationship area of a region versus number of folds has 93.5 as an initial value. This difference represents a transformation of the function.

   NOTE: $y = 93.5 \cdot \left(\frac{1}{2}\right)^x$ is a vertical stretch of $y = 1 \cdot \left(\frac{1}{2}\right)^x$.
7. Addition
8. $y = 3x + 2$

   NOTE: Help students recall how to find the rule using initial value and rate of change. The initial value is 2, since the height is 2 on day 0, and the rate of change is 3 centimeters per day. Also review graphing the rule over the scatter plot for verification.
9. The first differences are not constant; therefore, this relationship is not linear. The second differences are not constant; therefore, this relationship is not quadratic.
10. This relationship is an example of exponential decay because as the number of hours increases, the amount of dissolved gas decreases.
11.

Number of Hours	Dissolved CO_2	Successive Quotients
0	21.87	
1	7.29	$7.29 \div 21.87 = \frac{1}{3}$
2	2.43	$2.43 \div 7.29 = \frac{1}{3}$
3	0.81	$0.81 \div 2.43 = \frac{1}{3}$
4	0.27	$0.27 \div 0.81 = \frac{1}{3}$
5	0.09	$0.09 \div 0.27 = \frac{1}{3}$

12. Multiplication. In this example, each value for dissolved gas is multiplied by $\frac{1}{3}$ to get the next value.
13. $y = 22 \cdot \left(\frac{1}{3}\right)^x$

    NOTE: Compare finding this rule to finding the rule for a linear function.

The general function rule that models an exponential relationship is

$$y = \textbf{(initial value)} \cdot \textbf{(common multiplier)}^x$$

$$y = a \cdot b^x$$

CHECK THIS!

The common multiplier in an exponential function is also called a common ratio because it is the ratio of successive values.

14. What must be the value of b, the common multiplier (successive quotient), in an exponential decay function?

15. What must be the value of b, the common multiplier (successive quotient), in an exponential growth function?

16. How can you find the base (b) of an exponential function in the form $y = a \cdot b^x$ from a table of values?

17. How can you find the initial value (a) for $y = a \cdot b^x$ from a table of values?

Wildlife scientists use models to predict animal populations. The data in **Figure 6.32** are for a bison population in a park.

Year	Year Number	Bison Population
1998	0	600
1999	1	570
2000	2	542
2001	3	514
2002	4	489
2003	5	464
2004	6	441
2005	7	419

FIGURE 6.32.
Bison population data.

18. Do these data show a linear relationship? How do you know?

19. Do these data show an exponential relationship? How do you know?

20. Is this relationship an example of exponential growth or exponential decay? How do you know?

21. Use your graphing calculator to make a population versus year number scatter plot. Record your window and sketch your graph.

22. Write a function rule that models this situation.

23. Graph your function rule on your scatter plot. If the graph is not a good fit, change your rule. Sketch your graph.

24. Predict the bison population in 2007.

Answers (continued)

14. The common multiplier must be between 0 and 1.
15. The common multiplier must be greater than 1.
16. The base is the common multiplier (successive quotients). To find the base divide each y-value by the y-value before it.
17. The value of a is the y-value when $x = 0$, assuming there is no vertical shift.

    NOTE: Vertical shifts will be discussed in Section 6.9 as a transformation on an exponential decay function.

18. This relationship is not linear because first differences are not constant.
19. Answer: Find successive quotients. Since successive quotients are constant, the table represents an exponential relationship.

Year	Year Number	Bison Population	Successive Quotients
1998	0	600	–
1999	1	570	0.95
2000	2	542	0.95
2001	3	514	0.95
2002	4	489	0.95
2003	5	464	0.95
2004	6	441	0.95
2005	7	419	0.95

20. This example models exponential decay, since the constant multiplier is between 0 and 1.
21.

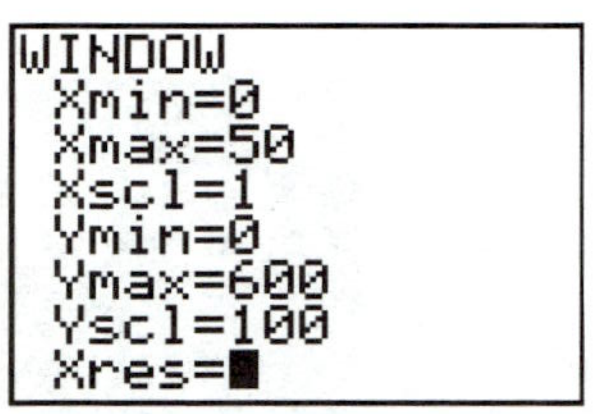

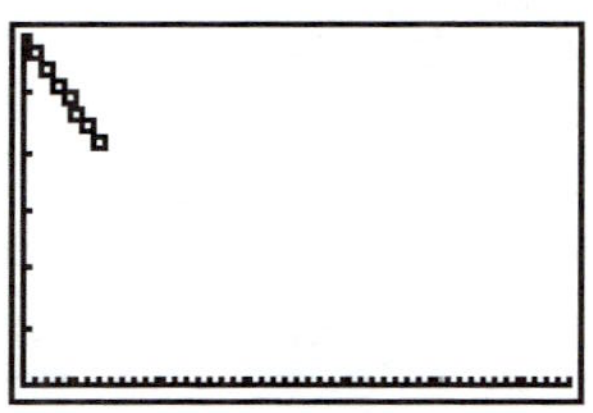

22. $y = 600 \cdot 0.95^x$
23.

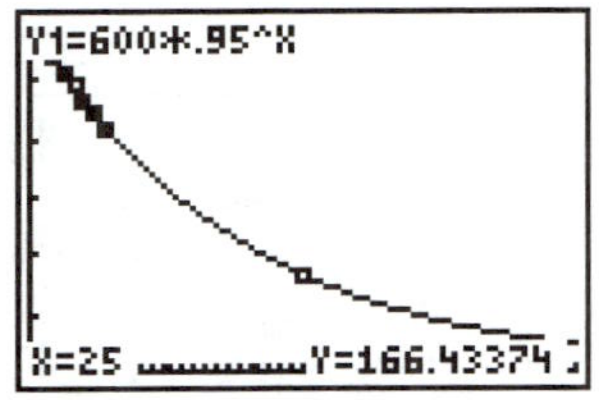

24. About 378

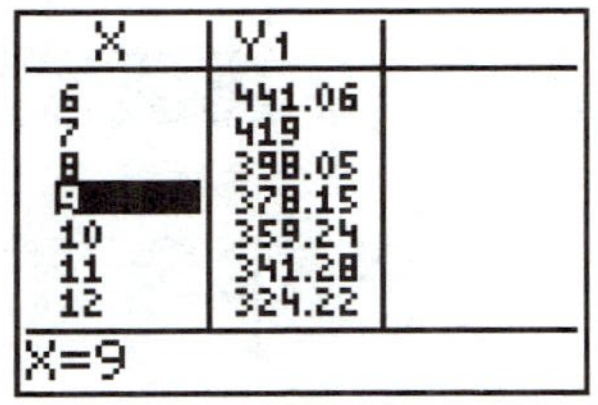

SECTION 6.8

Assignment

1. State if each of these tables shows a linear, quadratic, or exponential relationship. If the pattern is exponential, state if it is growth or decay.

a)

x	y
0	100
1	50
2	25
3	12.5
4	6.25
5	3.125

b)

x	y
0	1
1	3
2	9
3	27
4	81
5	243

c)

x	y
0	1250
1	1350
2	1458
3	1574.6
4	1700.6
5	1836.7

d)

x	y
0	0
1	2
2	8
3	18
4	32
5	50

e)

x	y
0	1
1	$\frac{1}{3}$
2	$\frac{1}{9}$
3	$\frac{1}{27}$
4	$\frac{1}{81}$
5	$\frac{1}{243}$

f)

x	y
0	2
1	5
2	8
3	11
4	14
5	17

2. Match each window and graph to one of the tables in Question 1, and write the function rule that fits the data.

a)

```
WINDOW
 Xmin=0
 Xmax=6
 Xscl=1
 Ymin=0
 Ymax=100
 Yscl=10
 Xres=■
```

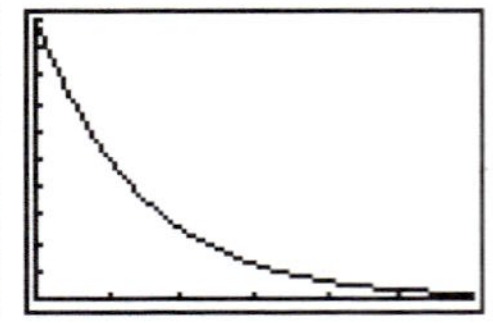

b)

```
WINDOW
 Xmin=0
 Xmax=6
 Xscl=1
 Ymin=0
 Ymax=100
 Yscl=10
 Xres=■
```

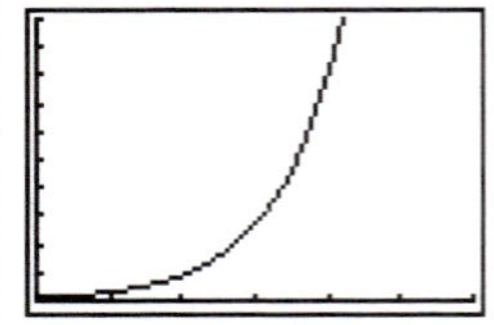

c)

```
WINDOW
 Xmin=0
 Xmax=6
 Xscl=1
 Ymin=0
 Ymax=1
 Yscl=.1
 Xres=■
```

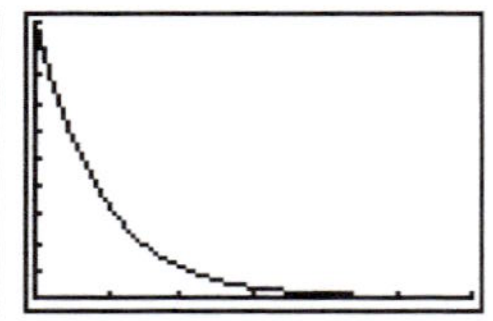

d)

```
WINDOW
 Xmin=0
 Xmax=6
 Xscl=1
 Ymin=1000
 Ymax=2000
 Yscl=100
 Xres=■
```

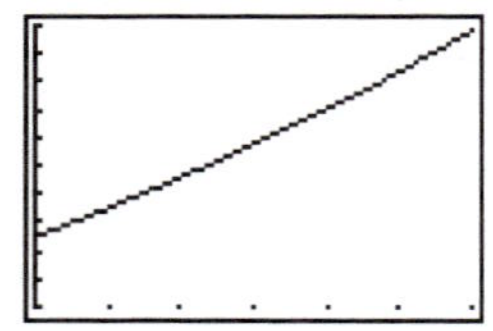

e)

```
WINDOW
 Xmin=0
 Xmax=6
 Xscl=1
 Ymin=0
 Ymax=100
 Yscl=10
 Xres=■
```

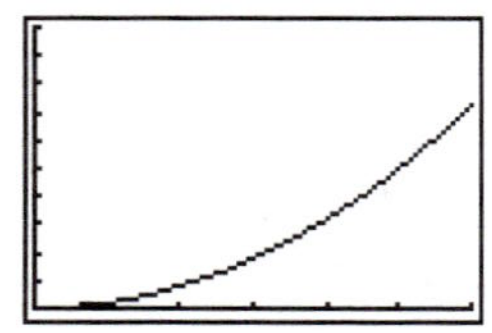

f)

```
WINDOW
 Xmin=0
 Xmax=6
 Xscl=1
 Ymin=0
 Ymax=25
 Yscl=2
 Xres=■
```

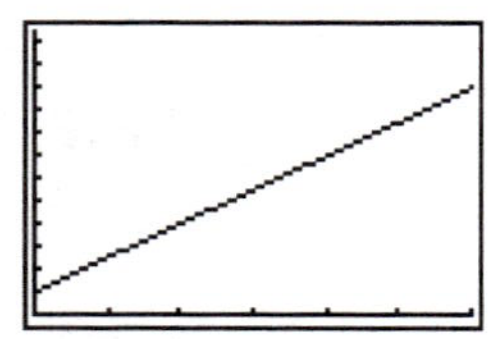

Section 6.8 Assignment

Answers

1. a) Exponential decay

   b) Exponential growth

   c) Exponential growth

   d) Quadratic

   e) Exponential decay

   f) Linear

2. a) Table a, $y = 100(0.5)^x$

   b) Table b, $y = 3^x$

   c) Table e, $y = \left(\frac{1}{3}\right)^x$

   d) Table c, $y = 1250(1.08)^x$

   e) Table d, $y = 2x^2$

   f) Table f, $y = 3x + 2$

Hour	Milligrams
0	100
1	75
2	56.25
3	42.1875
4	31.6406
5	23.7304
6	17.7979
7	13.3484

FIGURE 6.33.
Medicine concentration data.

3. Park animals may need to be given medicine. The concentration of medicine in an animal's bloodstream decreases over time. **Figure 6.33** shows the concentration of a certain medicine in milligrams for each hour after an initial dose.

   a) Do these data show a linear relationship? How do you know?

   b) Do these data show an exponential relationship? How do you know?

   c) Is this relationship an example of exponential growth or exponential decay?

   d) Use your graphing calculator to make a milligrams versus number of hours scatter plot. Record your window and sketch your graph.

   e) Write a function rule that models this situation.

   f) Graph your function rule on your scatter plot. If the rule is not a good fit use transformations to change it. Sketch your graph.

   g) How many milligrams of medication should remain in the bloodstream after 10 hours?

SECTION 6.9

Newton's Law of Cooling

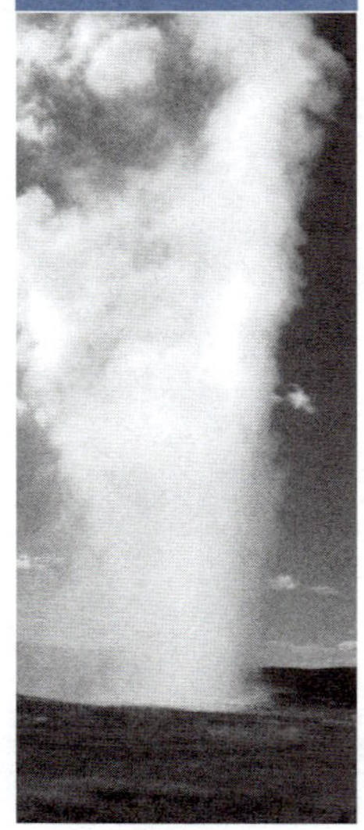

A national or state park is generally located around some natural attraction. For example, Yellowstone National Park contains geological interests such as the Old Faithful geyser. Crater Lake National Park in Oregon has a dormant volcano whose crater has filled with rainwater and become a deep lake.

Most lakes and pools fed by springs have a constant temperature. For example, Barton Springs in Austin, Texas, feeds a manmade pool in Zilker Park near downtown Austin. The water from the springs is always 68°F making the pool it feeds cool and popular during hot summer afternoons.

Since the temperature of the water is constant at 68°F, objects that are put in the pool cool to that temperature. Park managers can calculate how long it takes an object in the pool to reach 68°F using Newton's Law of Cooling.

To investigate Newton's Law of Cooling, we will perform two experiments. The first experiment will gather temperature data by putting a sensor in a cup of ice. The second experiment will gather temperature data by first putting a sensor in a cup of hot liquid then pulling it out and letting it cool to room temperature.

EXPERIMENT 1: CUP OF ICE

Before the start of the experiment, use a probe to record room temperature. This measurement is the temperature at time 0.

At the moment instructions tell you to PRESS ENTER TO START COLLECTING DATA, place the probe into a cup of ice. The calculator will collect 99 seconds of data.

1. Sketch a scatter plot of the data. Label your axes with units and record the window.

Answers (continued)

3. a) This relationship is not linear because there is not a constant rate of change.

   b) This relationship is exponential because successive quotients are constant.

Hour	Milligrams	Successive Quotients
0	100	0.75
1	75	0.75
2	56.25	0.75
3	42.1875	0.75
4	31.6406	0.75
5	23.7304	0.75
6	17.7979	0.75
7	13.3484	0.75

   c) This relationship is an example of exponential decay.

   d)

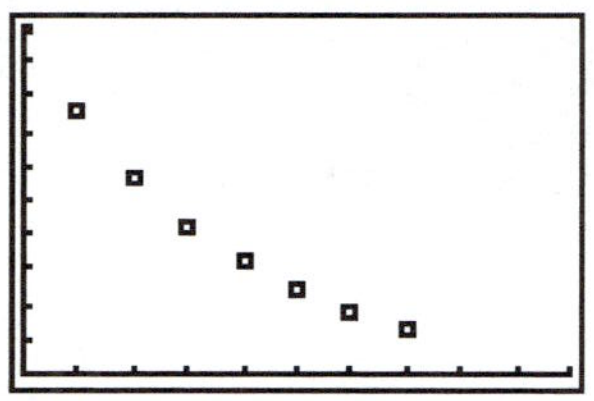

   e) $y = 100 \cdot 0.75^x$

   f)

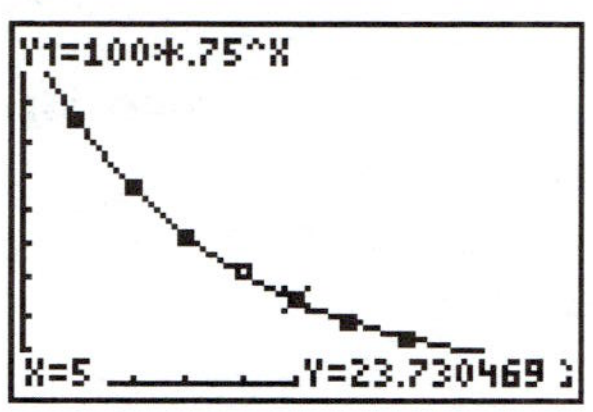

   g) 5.6314 milligrams

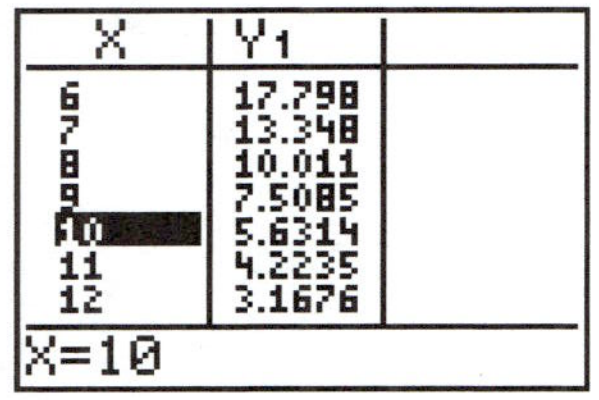

ELABORATE LESSON 2

Section 6.9: Newton's Law of Cooling

Purpose: In Sections 6.7 and 6.8, students used successive quotients to write function rules describing situations of exponential decay. In this section students will use what they have learned in previous chapters about function transformations to fit an exponential decay curve to a data set. This will enable students to apply their understanding of exponential decay to a new situation in Section 6.10.

Description: In this section students will use technology to gather temperature data to investigate Newton's Law of Cooling. According to Newton's Law of Cooling, the temperature of an object will follow an exponential decay relationship until that object cools to the temperature of the surrounding environment. Students will use their knowledge of transformations and their knowledge of exponential decay functions in order to write a function rule to describe a set of data. First, students will collect and model data describing the change in temperature over time when a thermometer is placed in a cup of ice. As a follow-up activity, students will repeat their experiment when a thermometer is placed in a cup of hot liquid (for example, water or coffee) then removed and allowed to cool to room temperature. The follow-up activity results in a vertical shift, which is a transformation of the exponential decay function.

Materials: graphing calculator, CBL2 or other calculator-based data collection device, temperature probe, data collection program

Teaching Ideas: This activity involves collecting temperature data. The data collection instructions that follow are written for use with TI-CBL2 and TI-83+ or TI-84 graphing calculators. If possible, have students work in small groups to conduct the experiment and collect the data. If there is not enough technology available, the experiment can be done as a class demonstration.

It is best to have the students use technology to collect the temperature data. However, it is a good compromise to collect the data using standard alcohol thermometers. If it is impossible to have students collect their own data, use the sample data in the answers provided to Questions 2 and 9. Students will not have the experience to connect the mathematics with, but can still perform the data analysis.

Answers to the investigation questions are based on the sample data. Actual student answers will vary, depending on the data that they collect.

The steps for using the CBL2 to collect temperature data are as follows:

1. Connect the TI-83+ to the CBL2
2. Place the temperature probe in Channel 1
3. Press the APPS button
4. Choose the CBL/CBR APP
5. Choose DATA LOGGER from the CBL/CBR menu
6. Choose the following settings for the next screen:

   PROBE : TEMP

   # SAMPLES: 99

   INTRVL(SEC): 1

UNITS: °C

PLOT: RealTme

DIRECTNS: On

7. Arrow down to GO then follow the directions.

Answers

Experiment 1: Cup of Ice

1. Sample answer:

Xmin: –5
Xmax: 100
xscl: 0
Ymin: –5
Ymax: 25
yscl: 2
xres: 1

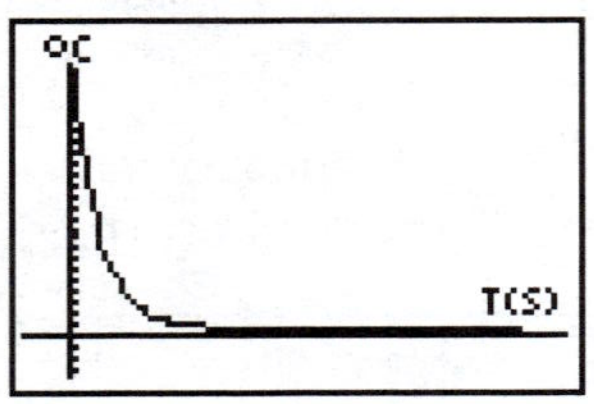

2. Copy the data for the first 7 seconds into a table like **Figure 6.34,** calculate the successive quotients and record them in the table.

Hour	Temperature (degrees Celsius)	Successive Quotients
0		
1		
2		
3		
4		
5		
6		
7		

FIGURE 6.34.
Table for data.

3. Do the data appear to be exponential? How do you know?
4. Use a graphing calculator to make a scatter plot of temperature versus seconds for the first 7 seconds. Sketch your graph.
5. If we continued this experiment, would the temperatures approach 0? Why or why not?
6. Write a function rule to model this situation.
7. Graph your function rule on the scatter plot. Sketch your graph and record your window.

EXPERIMENT 2: IT'S GETTING COLDER

For the second experiment, place the probe in a cup of hot water for 30 seconds. This time, at the moment instructions tell you to PRESS ENTER TO START COLLECTING DATA, press ENTER, take the probe out of the hot water, and hold it in the air.

8. Sketch a scatter plot of the data. Label your axes with units.
9. Copy the data for the first 7 seconds into a table like Figure 6.34. Calculate the successive quotients and record them in your table.

10. Do the data appear to be exponential? How do you know?
11. Use your graphing calculator to make a scatter plot of temperature versus seconds for the first 7 seconds of experiment 2. Sketch your graph and record your window.
12. If we continued this experiment, would the temperatures approach 0? Why or why not?
13. Transform your data from experiment 2 by subtracting room temperature from each temperature reading. Record the transformed temperatures in a table like **Figure 6.35.** Calculate the successive quotients and record them in the table.

Seconds	Temperature – Room Temperature	Successive Quotients
0		
1		
2		
3		
4		
5		
6		
7		

FIGURE 6.35.
Table for data and for successive quotients.

14. Do the data appear to be exponential? How do you know?
15. Use a graphing calculator to make a scatter plot of temperature – room temperature versus seconds. Sketch your graph and record your window.
16. Write a function rule to model the transformed data.
17. Graph your function rule on the scatter plot. Sketch your graph and record your window.
18. Use your knowledge of transformations to find a function rule for your original data.
19. Graph your function rule on the original scatter plot.

Answers (continued)

2. Sample answer:

Hour	Temperature (degrees Celsius)	Successive Quotients
0	28.5	
1	24.2093	0.85
2	21.4048	0.88
3	18.1905	0.85
4	15.122	0.83
5	12.525	0.83
6	10.4872	0.84
7	8.81081	0.84

3. Yes, the data set is exponential because successive quotients are equal.

NOTE: Point out that although all the successive quotients are not exactly the same, they are very close. Students can take the average (mean) to find a close successive quotient.

4.

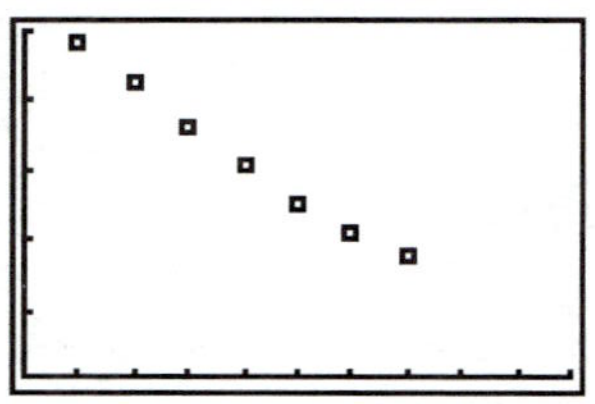

5. Yes, freezing in Celsius is 0 degrees.

6. $y = 28.5(0.84)^x$
(The temperature at time 0 was 28.5 degrees.)

7.

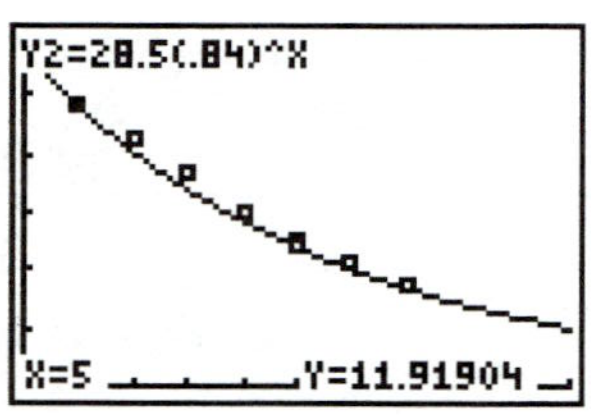

Experiment 2: It's getting colder

8.

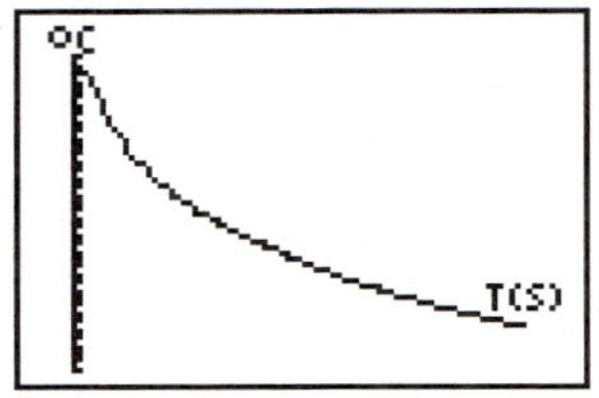

Xmin: –5
Xmax: 100
xscl: 0
Ymin: 20
Ymax: 60
yscl: 2
xres: 1

9. Sample Data:

Seconds	Temperature (degrees Celsius)	Successive Quotients
0	60.937	
1	59.9643	0.98
2	59.8214	0.998
3	58.8276	0.98
4	57.7241	0.98
5	56.3667	0.98
6	55.1667	0.98
7	54	0.98

10. Yes, the data are exponential because successive quotients are equal.

NOTE: Point out that although all the successive quotients are not exactly the same, they are very close. Students can take the average to find a close successive quotient.

11.

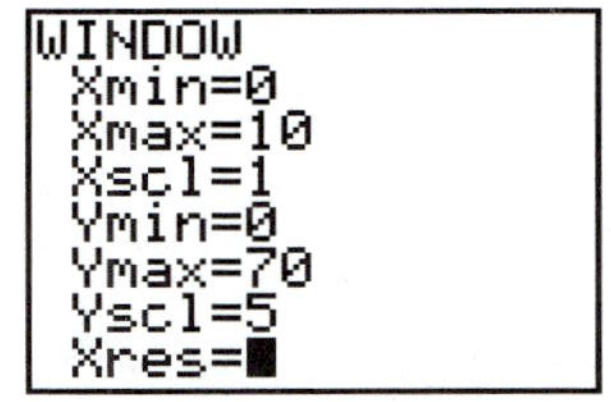

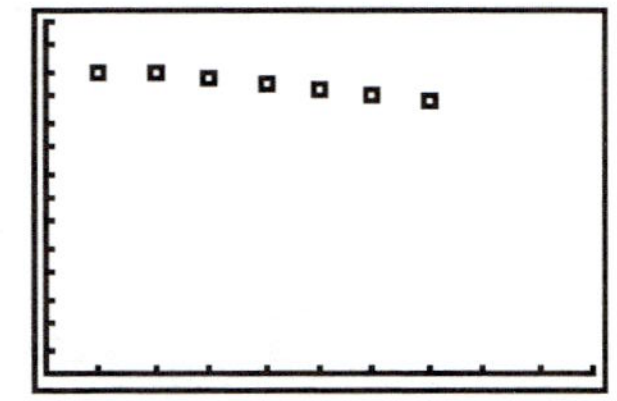

12. Although it looks like a great exponential model, this data would not approach 0 but rather would approach room temperature (the initial reading of the first experiment).

13. Sample answer assuming a room temperature of 28.5° C:

Seconds	Temperature – Room Temperature	Successive Quotients
0	32.437	
1	31.4643	0.97
2	31.3214	0.996
3	30.3276	0.97
4	29.2241	0.97
5	27.8667	0.96
6	26.6667	0.96
7	25.5	0.96

14. Yes, the data set is exponential because successive quotients are equal.

NOTE: Point out that although all the successive quotients are not exactly the same they are very close. Students can take the average to find a close successive quotient.

Answers (continued)

15.

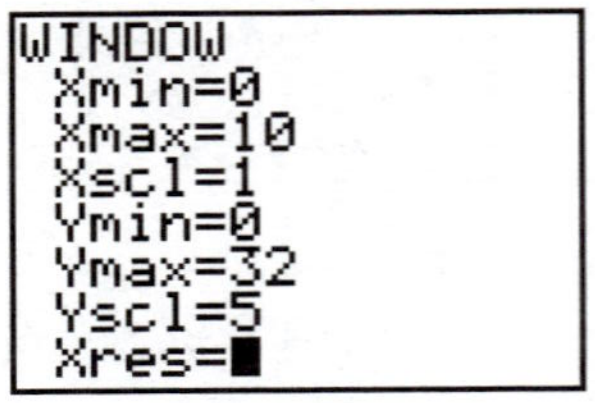

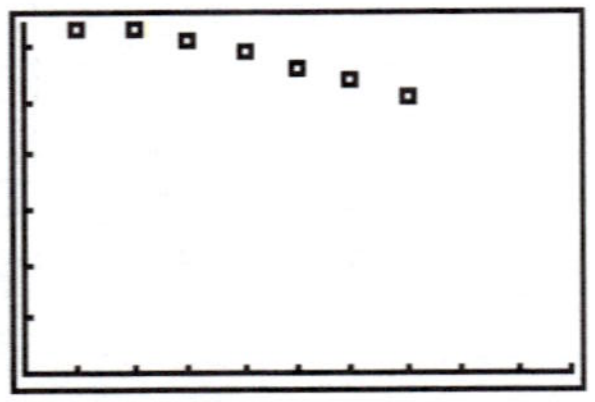

16. $y = 32.437(0.97)^x$

(As you work backward, notice the temperature at time 0 was 32.437 degrees).

17.

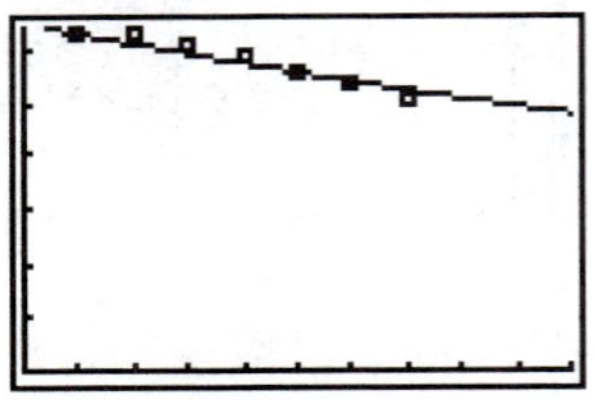

18. $y = 32.437(0.97)^x + 28.5$

19.

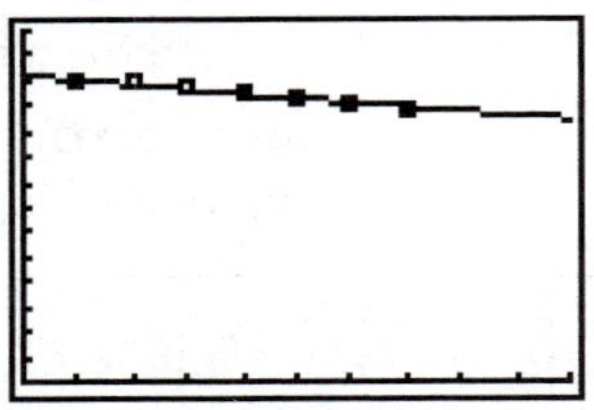

20. Graph both rules in the same window. What effect did the transformation have on the graph of your function rule from Question 16?

21. What does this transformation represent here?

SECTION 6.9 Assignment

1. Use transformations to describe the graph of each function as it relates to the function $y = (0.93)^x$.

   a) $y = (0.93)^x + 2$

   b) $y = (0.93)^x - 4$

   c) $y = 5(0.93)^x$

   d) $y = 0.3(0.93)^x$

   e) $y = 8.3(0.93)^x + 3$

   f) $y = 0.5(0.93)^x - 6$

2. In each of these, the table represents a transformation of the function $y = (2)^x$. Write the function rule that fits the transformed data.

   a)

x	y
0	6
1	7
2	9
3	13
4	21

   b)

x	y
0	3
1	6
2	12
3	24
4	48

3. Students in a math class collected the data in **Figure 6.36.** They dropped a rubber ball and measured the maximum height of each bounce.

   a) Do the data appear to be exponential? How do you know?

   b) Use your graphing calculator to make a scatter plot of height versus bounce number. Sketch your graph.

Bounce Number	Height of the Bounce (feet)
0	
1	5.231
2	4.715
3	4.436
4	4.015
5	3.788
6	3.600

FIGURE 6.36.
Rubber ball data.

Answers (continued)

20. The transformation causes a vertical shift in the graph of the function.

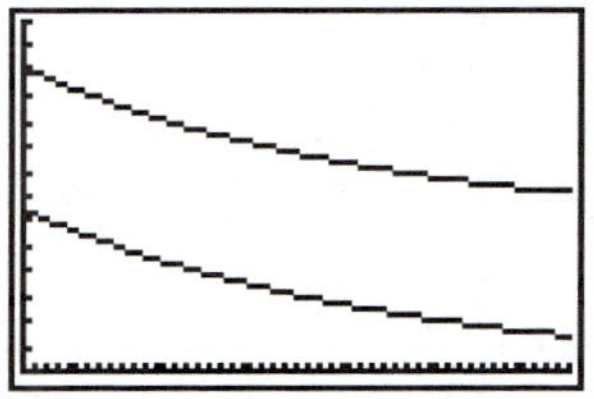

21. The transformation represents the addition of room temperature.

Section 6.9 Assignment

Answers

1. a) Vertical shift 2 units up

   b) Vertical shift 4 units down

   c) Vertical stretch

   d) Vertical compression

   e) Vertical stretch and vertical shift 3 units up

   f) Vertical compression and a vertical shift down 6 units.

2. a) $y = (2)x + 5$

   b) $y = 3(2)^x$

3. a) Yes, the data set is exponential because successive quotients are approximately equal.

NOTE: Point out that although all the successive quotients are not exactly the same, they are very close. Students should calculate the average to find a close successive quotient.

Bounce Number	Height of the Bounce (ft)	Successive Quotients
0		
1	5.231	
2	4.715	0.90
3	4.436	0.94
4	4.015	0.91
5	3.788	0.94
6	3.600	0.95

   b)

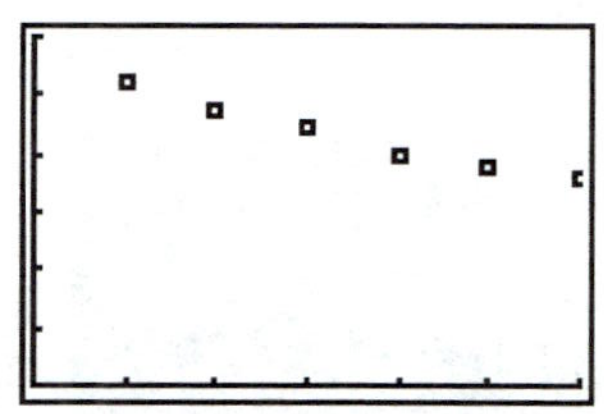

c) What is the average of the successive quotients?

d) Use the average of the successive quotients to estimate an initial value.

e) Write a function rule to fit the transformed data.

f) Graph your function rule on the scatter plot. Sketch your graph and record your window.

g) What would be the height of the 8th bounce?

h) Suppose the same ball is dropped from an initial height of 10 feet instead of 5.624 feet. What is a function rule that models this situation?

i) Predict how the graph of the function rule in h will differ from the graph of the function rule in e.

j) Graph both rules in the same window. Sketch your graph and record your window. What effect does changing the initial height to 10 feet have on the graph?

k) Suppose a different ball is dropped from an initial height of 4624 feet and that every bounce height is 1 foot less that the heights in the original data. What is a function rule that models this?

l) Predict how the graph of the function rule in k will differ from the graph of the function rule in e.

m) Graph both rules in the same window. Sketch your graph. What effect does subtracting 1 have on the graph?

SECTION 6.10 No More Foam

Cindy, Jamal, and Juanita are discussing sodas. Jamal likes Blue Cow soda best because when he pours it in a glass the foam disappears in less time than any other soda. Juanita thinks the foam in her favorite soda, Dr. Salt, disappears in less time than any other soda. Cindy thinks the same thing about her favorite soda, Diet Dr. Salt.

The three friends decide to do an experiment on the relationship between time and the amount of foam in a glass. Their data are shown in **Figure 6.37** with foam height measured in centimeters.

Time in Seconds	Blue Cow	Dr. Salt	Diet Dr. Salt
0	17.0	14.0	14.0
1	16.1	11.8	12.1
2	14.9	10.5	10.9
3	14.0	9.3	10.0
4	13.2	8.5	9.3
5	12.5	7.7	8.6
6	11.9	7.1	8.0
7	11.2	6.5	7.5
8	10.7	6.0	7.0
9	9.7	5.3	6.2

FIGURE 6.37. Foam data on three brands of sodas.

1. If each person drinks his or her favorite soda exactly 20 seconds after it is poured, who will be have the least foam?

2. Write a function rule that models each relationship.

Answers (continued)

c) The average is approximately 0.93.

d) $\frac{5.231}{0.93} \approx 5.624$

e) $y = 5.624(0.93)^x$

f)

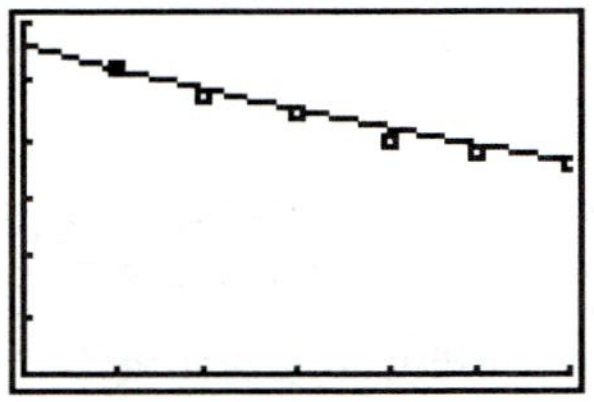

g) Approximately 3.15 feet

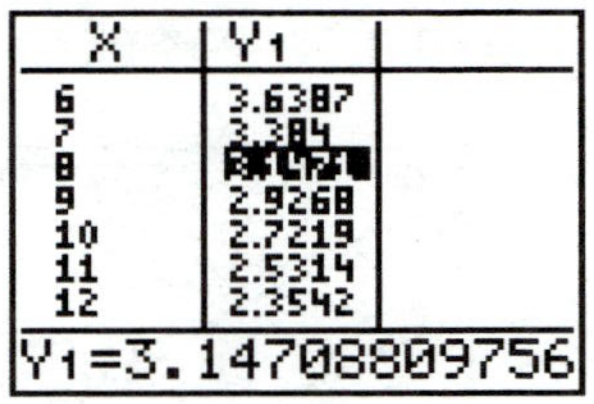

h) $y = 10(0.93)^x$

i) $y = 10(0.93)^x$ will be a vertical stretch of $y = 5.624(0.93)^x$.

j)

```
WINDOW
 Xmin=0
 Xmax=6
 Xscl=1
 Ymin=0
 Ymax=12
 Yscl=1
 Xres=■
```

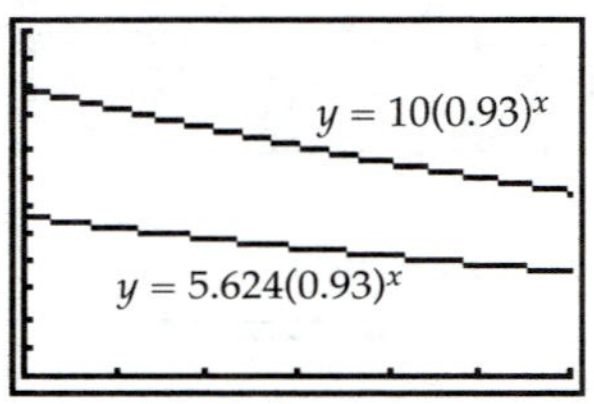

The change in initial height caused a vertical stretch of the graph (steeper appearance) and a change in the y-intercept from 5.624 to 10.

k) $y = 5.624(0.93)^x - 1$

l) $y = 5.624(0.93)^x - 1$ will be a vertical shift of $y = 5.624(0.93)^x$.

m)

```
WINDOW
 Xmin=0
 Xmax=20
 Xscl=1
 Ymin=0
 Ymax=6
 Yscl=1
 Xres=■
```

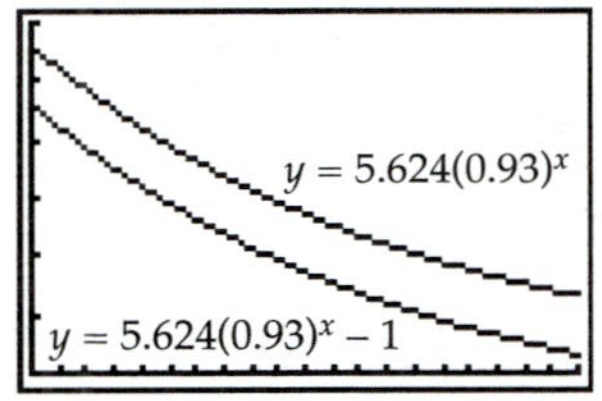

The graph was shifted vertically down one unit.

EVALUATE LESSON 3

Section 6.10: No More Foam

Purpose: In Sections 6.6 through 6.9, students developed the concept of an exponential decay function and learned to transform the parent function, $y = b^x$, to fit certain circumstances. In this section students will have the opportunity to demonstrate their knowledge by applying those concepts and skills to solve a problem in a new context.

Description: Facing a problem in a new context, students will determine and implement a method to solve the problem. This open-middle performance assessment task has only one correct answer, but there is a variety of valid methods to obtain that answer. In their justification, which is the most important part of a performance assessment task, students should be able to communicate their knowledge of mathematics and demonstrate that they have mastered the concepts presented in Sections 6.6 through 6.9.

Materials: graphing calculators

Teaching Ideas: This task may be assigned as group work or individual work. The mathematics performance assessment rubric should be used to evaluate students' work.

Answers

1. One possible solution strategy: First, find the mean of the successive quotients.

Time in Seconds	Blue Cow	Successive Quotients	Dr. Salt	Successive Quotients	Diet Dr. Salt	Successive Quotients
0	17.0		14.0		14.0	
1	16.1	0.95	11.8	0.84	12.1	0.86
2	14.9	0.93	10.5	0.89	10.9	0.90
3	14.0	0.94	9.3	0.89	10.0	0.92
4	13.2	0.94	8.5	0.92	9.3	0.93
5	12.5	0.95	7.7	0.91	8.6	0.93
6	11.9	0.95	7.1	0.92	8.0	0.93
7	11.2	0.94	6.5	0.92	7.5	0.94
8	10.7	0.96	6.0	0.92	7.0	0.93
9	9.7	0.91	5.3	0.88	6.2	0.89
Mean		0.94		0.90		0.91

2. Blue Cow: $y = 17(0.94)^x$

Dr. Salt: $y = 14(0.90)^x$

Diet Dr. Salt: $y = 14(0.91)^x$

Create a scatter plot and test the rules.

```
WINDOW
 Xmin=0
 Xmax=9
 Xscl=1
 Ymin=0
 Ymax=17
 Yscl=1
 Xres=■
```

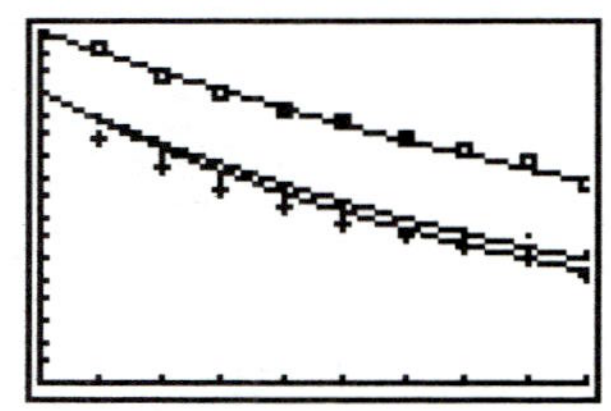

Answers (continued)

Change the window so that $x = 20$ is visible. To make it easier to read the graph, decrease the Ymax to 10.

```
WINDOW
 Xmin=0
 Xmax=22
 Xscl=1
 Ymin=0
 Ymax=10
 Yscl=1
 Xres=■
```

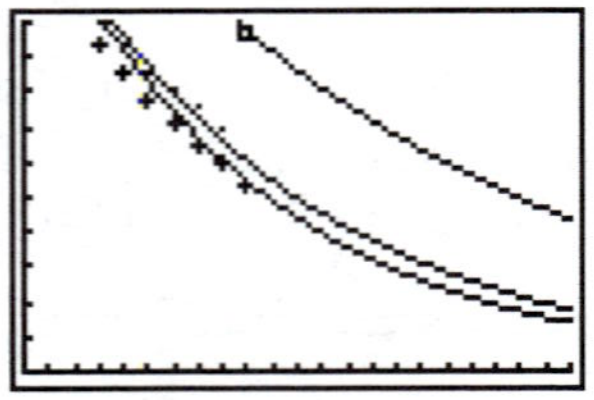

Use the trace function on the calculator.

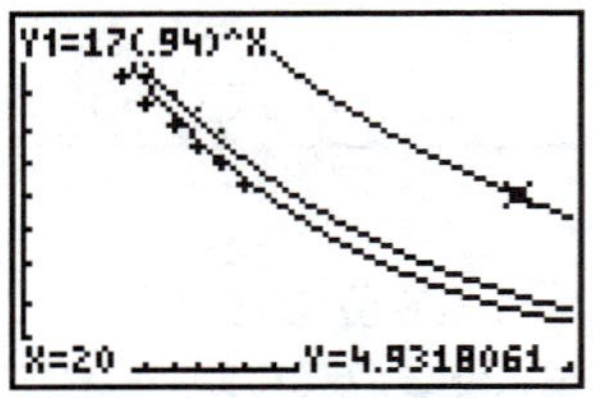

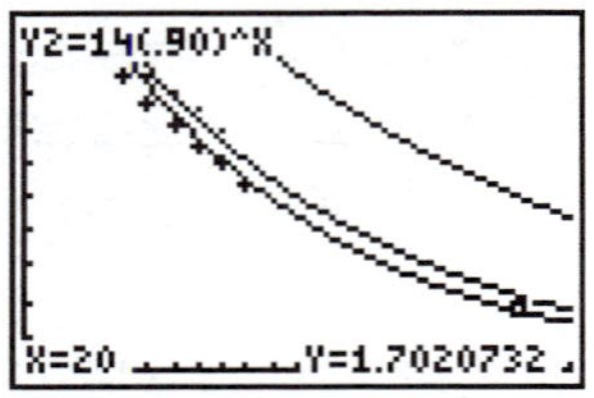

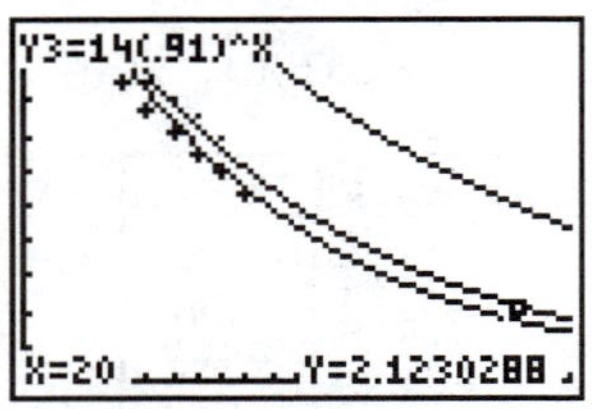

After 20 seconds, Blue Cow soda will have 4.93 centimeters of foam, Dr. Salt soda will have 1.7 centimeters of foam, and Diet Dr. Salt will have 2.12 centimeters of foam.

Juanita is correct. Her favorite soda, Dr. Salt, will have the least amount of foam after 20 seconds.

SUMMARY

The general function rule that models an exponential relationship is

$y =$ (initial value) · (common multiplier)^x

$y = a \cdot b^x$.

If $b > 1$, then the exponential function models exponential growth where successive y-values increase by a constant factor.

If $b = 1$, then the exponential function reduces to $y = a \cdot 1^x = a$, which is a constant function.

If $0 < b < 1$, then the exponential function models exponential decay where successive y-values decrease by a constant factor.

The common multiplier b can be found from a table using successive quotients as shown in **Figure 6.38.**

Number of Hours	Dissolved CO_2	Successive Quotients
0	21.87	
1	7.29	$7.29 \div 21.87 = \frac{1}{3}$
2	2.43	$2.43 \div 7.29 = \frac{1}{3}$
3	0.81	$0.81 \div 2.43 = \frac{1}{3}$
4	0.27	$0.27 \div 0.81 = \frac{1}{3}$
5	0.09	$0.09 \div 0.27 = \frac{1}{3}$

FIGURE 6.38.
Table using successive quotients.

The value of a is the y-value when $x = 0$ if there is no vertical shift.

In Figure 6.38 the value for a is 21.87 because that is the amount of dissolved carbon dioxide (CO_2) present at 0 hours. Since the successive quotients are all $\frac{1}{3}$, $b = \frac{1}{3}$; therefore this situation can be modeled using the function $y = 21.87\left(\frac{1}{3}\right)^x$.

In the context of a problem, the value for a is the initial amount of a thing that is decaying exponentially.

For example, if the initial dosage of medication is 500 milligrams and the amount of medication in the bloodstream decreases by 90% every hour, the initial value a is 500. The constant multiplier b is 0.90. The amount of medication y in the bloodstream at time t can be modeled with $y = 500(0.90)^t$.

SECTION 6.11

Switching Directions

A LOOK AHEAD

In Chapter 5 you learned that the inverse of a function is the reflection of the function over the line $y = x$. The domain of the function becomes the range of its inverse; the range of the function becomes the domain of its inverse. In the case of the quadratic function $y = x^2$, the inverse is the square root function $y = \sqrt{x}$. Likewise, the inverse operation of raising a number to a power is to take the root of that number. In other words, to un-square a number such as 9, you take the square root of the number; for example, $\sqrt{9}$.

Let's explore some ways we might *un-power* an equation such as $3^x = 27$. We will also investigate the inverse of an exponential function.

1. Make a table for $f(x) = 3^x$ like **Figure 6.39,** then sketch a graph of the function on a grid like **Figure 6.40.**

x	−2	−1	0	1	2	3	4
$f(x)$							

FIGURE 6.39.
Table for $f(x) = 3^x$.

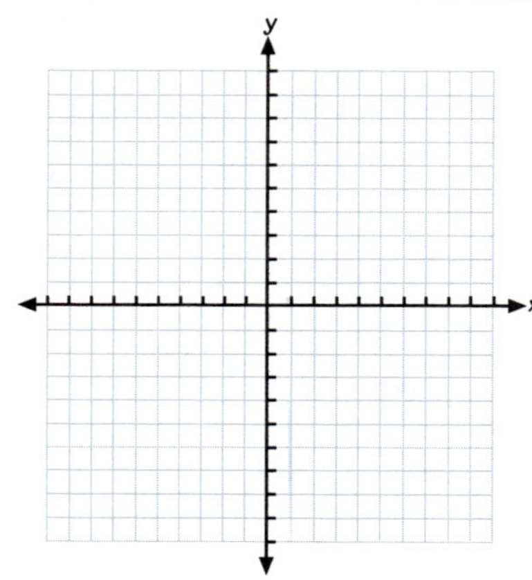

FIGURE 6.40.
Grid for graphing $f(x) = 3^x$.

2. What is the domain of this function?

3. What is the range of this function?

A LOOK AHEAD

ENGAGE LESSON 3

Section 6.11: Switching Directions

Purpose: In previous lessons students investigated and defined exponential functions using the context of growth and decay. In Section 6.11 students will find the function that is the inverse of an exponential function and define it as the logarithmic function. In Section 6.12 students will begin to explore properties of logarithms that can be used to solve exponential equations in Section 6.14.

Description: In Chapter 5 students discovered that a square root function is the inverse of a quadratic function. They also defined the inverse of a function as a reflection of the original function across the line $y = x$ and noted that the domain and range of a function become reversed when the inverse of the function is taken. In this section of the lesson, students will use their prior knowledge of inverses of functions to generate the inverse of an exponential function. They will begin with a table of values and a sketch of the graph of an exponential function. Students will use patty paper to reflect the graph across the line $y = x$, and generate the inverse of an exponential function, the logarithmic function.

Materials: patty (tracing) paper, graphing calculator (optional)

Teaching Ideas: Students will be working in small groups of 3–4 students. Each student should follow the steps of the activity, communicating with their group members as questions arise. Encourage students to share their observations and results with each other.

When discussing domain and range values, ask students about negative numbers and positive numbers. Ask about zero, and fractional values. Use a calculator to test conjectures, if necessary, substituting x-values or trying to obtain a particular y-value, such as a negative number.

Answers

1.

x	−2	−1	0	1	2	3	4
$f(x)$	$\frac{1}{9}$	$\frac{1}{3}$	1	3	9	27	81

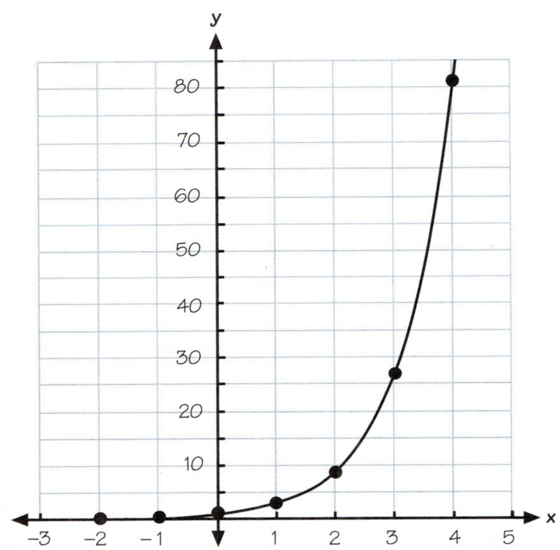

2. All real numbers

3. $y > 0$

4. Trace the x-axis, y-axis, and the graph of $f(x)$ onto a sheet of patty paper. Be sure to label the x- and y-axes. Reflect the graph of $f(x)$ across the line $y = x$ by holding the top-right and bottom-left corners of the patty paper in each hand and flipping the sheet of patty paper over (see **Figure 6.41**). Sketch what you see.

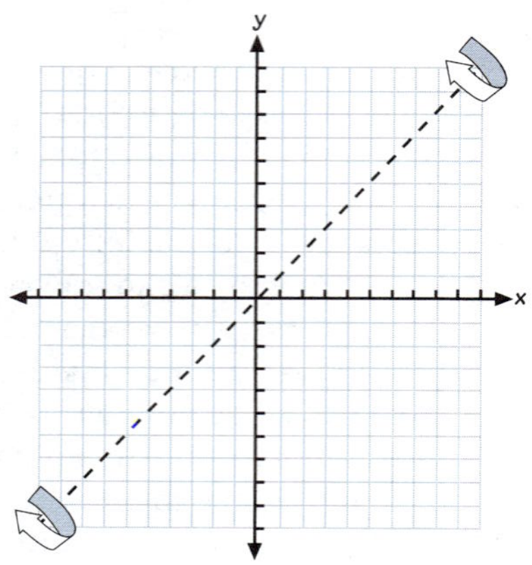

FIGURE 6.41.
Flipping over the patty paper.

CHECK THIS!

Recall that the inverse of a function is a reflection across the line $y = x$.

You can generate the inverse of a function by switching the domain and range values of the original function.

Now let's look at a function that undoes $f(x)$. Let's call that function $g(x)$. For example, when we evaluate $f(x)$ at $x = 0$, we get a result (output) of 1. In $g(x)$ when we evaluate $g(x)$ at $x = 1$, we get a result of 0.

5. Complete a table like **Figure 6.42,** and sketch the graphs of both $y = f(x)$ and $y = g(x)$.

x							
g(x)	−2	−1	0	1	2	3	4

FIGURE 6.42.
Table for $g(x)$.

6. Is the new curve that you sketched the graph of a function? How can you tell?

7. What is the domain of this function?

8. What is the range of this function?

The inverse of an exponential function has a special name; it is called a **logarithmic function**. In the next few sections you will explore and find some properties and uses for logarithmic functions.

SECTION 6.12 It All Adds Up

In the last section you developed the inverse of an exponential function, a **logarithmic function**. This function is written as **log(*x*)** where x is any positive number. In the next few sections you will explore some properties and uses for logarithmic functions.

You have studied transformations of graphs of functions. In this section you will explore transformations of graphs of logarithmic functions and make some conjectures about patterns that occur in logarithmic functions.

1. On your graphing calculator, graph $y = \log(x)$. Record your viewing window and sketch your results.

2. What is the x-intercept? What is the y-intercept? Are there any other key points? Explain how you found these points.

3. In the same viewing window, graph $y = \log(x) + 2$ with $y = \log(x)$. Sketch your graph.

4. Now graph the parent function $y = \log(x)$ and $y = \log(x) - 2$ in the same window. Sketch your graph.

5. Compare the graphs of $y = \log(x)$, $y = \log(x) + 2$, and $y = \log(x) - 2$. How are they alike? How do they differ?

6. In terms of transformations, what will the graph of $y = \log(x) + k$ look like?

7. In a new viewing window, graph the parent function $y = \log(x)$ and $y = 2\log(x)$. Sketch your graph and record your window.

8. Now graph the parent function $y = \log(x)$ and $y = \frac{1}{2}\log(x)$ in the same window. Sketch your graph.

9. Compare the graphs of $y = \log(x)$, $y = 2\log(x)$, and $y = \frac{1}{2}\log(x)$. How are they alike? How do they differ?

10. In terms of transformations, what will the graph of $y = a\log(x)$ will look like?

Answers (continued)

4. Sample answer:

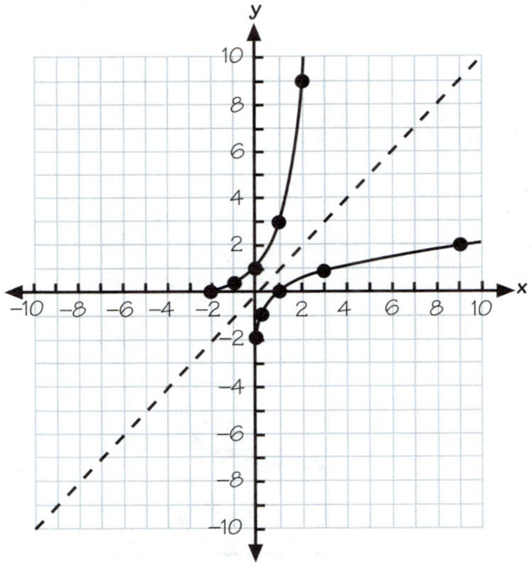

5.

x	$\frac{1}{9}$	$\frac{1}{3}$	1	3	9	27	81
$g(x)$	−2	−1	0	1	2	3	4

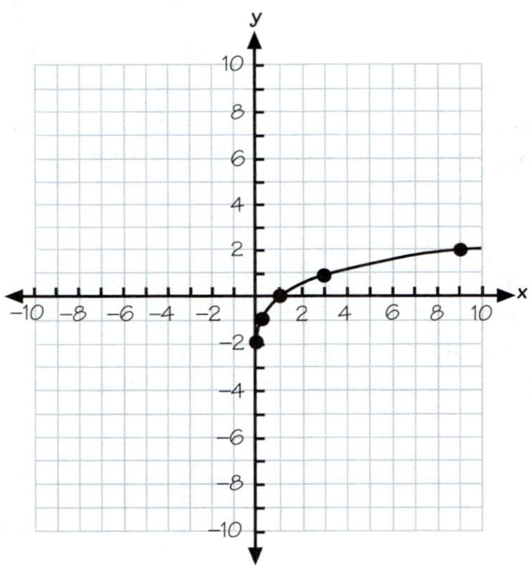

6. Yes, it is a function. Looking at a table of values or a graph reveals that for each x-value, there is only one value for $g(x)$. On the graph, this appears as passing the vertical line test.

7. $x > 0$

8. All real numbers

EXPLORE LESSON 3

Section 6.12: It All Adds Up

Purpose: In Section 6.11 students developed the logarithmic function as the inverse of an exponential function. In Section 6.12 students will explore transformations on logarithmic functions and use transformations to make conjectures about three important properties of logarithms. These properties will be formalized in Section 6.13, and will be used to solve problems in Section 6.14.

Description: In this section students will begin by applying their knowledge of transformations to the parent function of logarithmic functions, $y = \log(x)$. From these transformations, students will be asked to look for patterns to develop three properties of logarithms:

$\log_b m + \log_b n = \log_b m \cdot n$, $\log_b m - \log_b n = \log_b \frac{m}{n}$, and $\log_b m^n = n \log_b m$.

Materials: graphing calculator

Teaching Ideas: Students should work in small groups (3 to 4 students). Encourage students to communicate with each other and share their thinking and observations as they complete the lesson.

Answers

1.

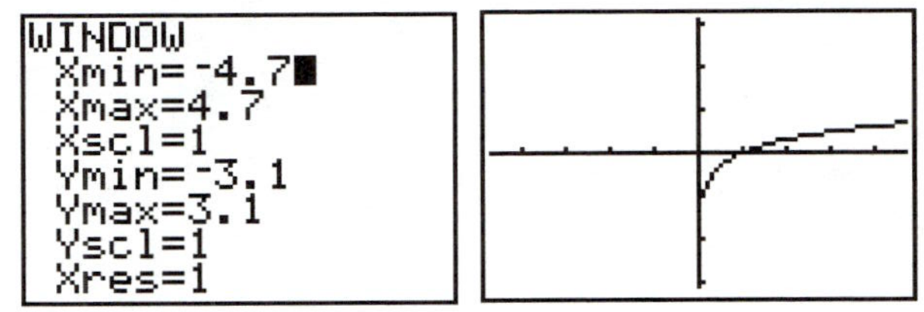

2. The x-intercept is (1, 0), which can be found on the graph using the TRACE feature.

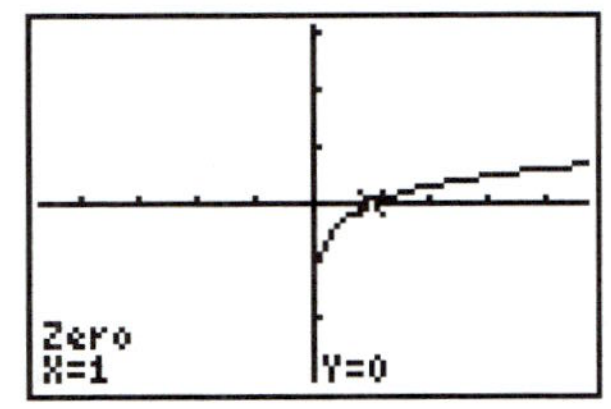

There is no y-intercept for the function $y = \log(x)$. This can be verified using a table of values, where the y-value for $x = 0$ reads "Error."

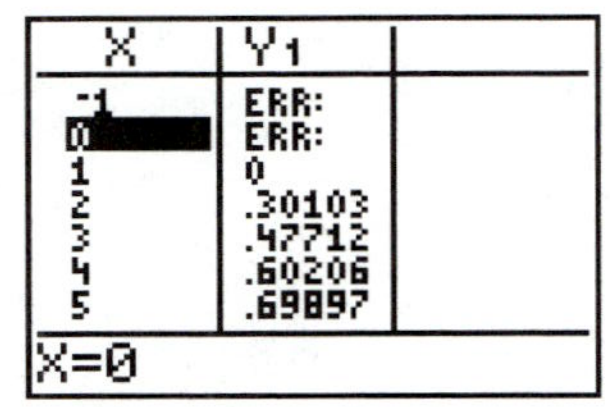

3.

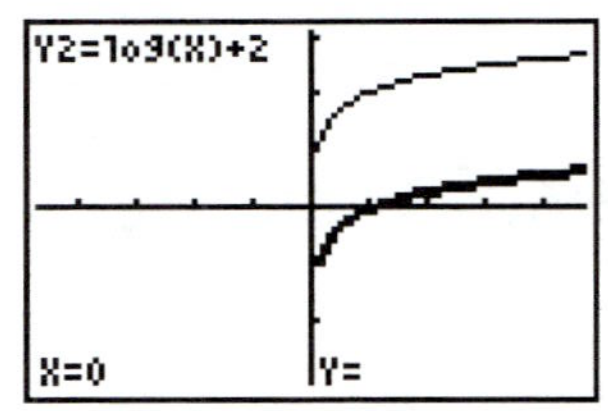

4. 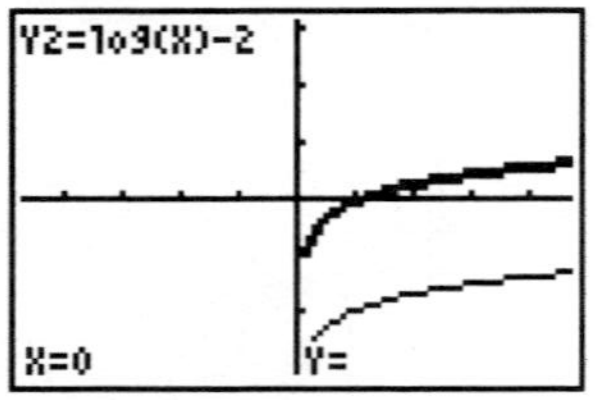

5. The three graphs have the same shape and are in the same position relative to the y-axis. However, the graph of $y = \log(x) + 2$ has been vertically shifted up 2 units, and the graph of $y = \log(x) - 2$ has been vertically shifted down 2 units.

6. The graph of $y = \log(x) + k$ will look like the parent function, $y = \log(x)$, shifted up or down k units.

7. 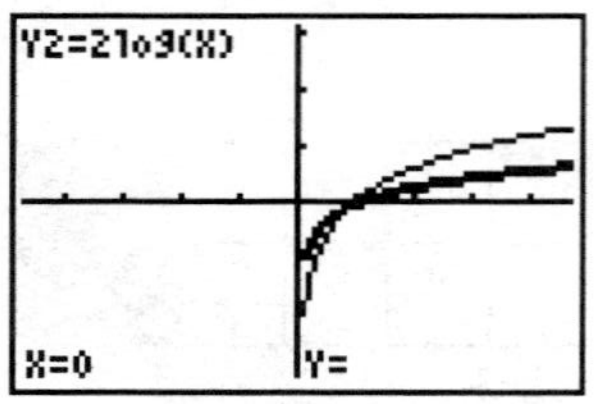

8. 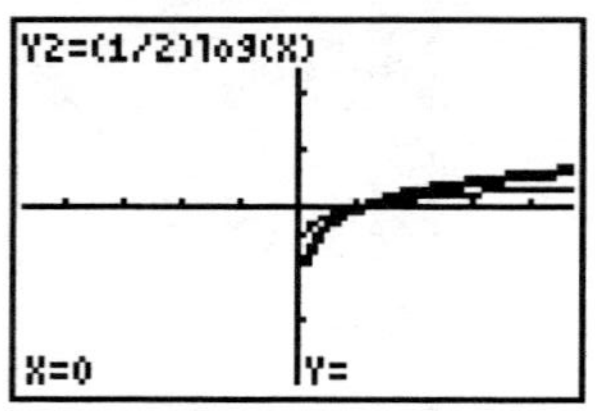

9. The three graphs have the same shape and are in the same position relative to the y-axis. However, the graph of $y = 2\log(x)$ has been vertically stretched by a factor of 2 and the graph of $y = \frac{1}{2}\log(x)$ has been vertically compressed by a factor of $\frac{1}{2}$.

If students have difficulty seeing the stretch or compression from a graph, use the table feature of the calculator to numerically verify the scale factor.

X	Y1	Y2
1	0	0
2	.30103	.60206
3	.47712	.95424
4	.60206	1.2041
5	.69897	1.3979
6	.77815	1.5563
7	.8451	1.6902

X=2

In this table, for $x = 2$, the value of log(2) in the Y1 column is 0.30103. The value of 2 log(2) in the Y2 column is 0.60206, which is twice as much as the value of log(2).

10. The graph of $y = a\log(x)$ will look like the parent function, $y = \log(x)$, vertically stretched by a factor of a units.

11. Now enter Y1 = log(3x) and Y2 = log(3) + log(x) into the function list of your graphing calculator. Change the plot style for Y2 to a tracing ball then graph the two functions. Record your window and sketch the graphs. Explain what you saw while the calculator graphed both functions.

12. Now look at a table for these two functions. What patterns between the values for Y1 and Y2 do you see? What do these patterns suggest about the relationship between Y1 = log(3x) and Y2 = log(3) + log(x)?

13. Repeat Question 11 with Y1 = log(6x) and Y2 = log(3) + log(2x). Sketch your graphs and look at a table. What patterns between the values for Y1 and Y2 do you see? What do these patterns suggest about the relationship between Y1 = log(6x) and Y2 = log(3) + log(2x)?

14. Repeat Question 11 with Y1 = log(6x) and Y2 = log(2) + log(3x). Sketch your graphs and look at a table. What patterns between the values for Y1 and Y2 do you see? What do these patterns suggest about the relationship between Y1 = log(6x) and Y2 = log(2) + log(3x)?

15. Use your answers to Questions 12, 13, and 14 to write a sentence about log(xy).

16. Now enter Y1 = log(3x) and Y2 = log(6x) – log(2) into the function list of your graphing calculator. Change the plot style for Y2 to a tracing ball then graph the two functions. Record your window and sketch the graphs. Explain what you saw while the calculator graphed both functions.

17. Now look at a table for these two functions. What patterns between the values for Y1 and Y2 do you see? What do these patterns suggest about the relationship between Y1 = log(3x) and Y2 = log(6x) – log(2)?

18. Repeat Question 16 for Y1 = log(2x) and Y2 = log(8x) – log(4). Sketch your graphs and look at a table. What patterns between the values for Y1 and Y2 do you see? What do these patterns suggest about the relationship between Y1 = log(2x) and Y2 = log(8x) – log(4)?

19. Use your answers to Questions 17 and 18 to make a generalization about $\log\left(\frac{x}{y}\right)$.

20. Now enter Y1 = log(x^3) and Y2 = 3 log(x) into the function list of your graphing calculator. Change the plot style for Y2 to a tracing ball then graph the two functions. Record your window and sketch the graphs. Explain what you saw while the calculator graphed both functions.

21. Now, look at a table for these two functions. What patterns between the values for Y1 and Y2 do you see? What do these patterns suggest about the relationship between Y1 = log(x^3) and Y2 = 3 log(x)?

22. Repeat Question 20 for Y1 = log(x^5) and Y2 = 5 log(x). Sketch your graphs and look at a table. What patterns between the values for Y1 and Y2 do you see? What do these patterns suggest about the relationship between Y1 = log(x^5) and Y2 = 5 log(x)?

23. Use your answers to Questions 21 and 22 to make a sentence about about log(x^n).

Answers (continued)

11. You may need to show students how to change the plot style for Y2. First, use the left arrow key to select the plot style to the left of Y2.

```
Plot1 Plot2 Plot3
\Y1■log(3X)
_Y2■log(3)+log(X
)
\Y3=
\Y4=
\Y5=
\Y6=
```

Next, press the ENTER key to toggle between the plot styles until the tracing ball is selected.

```
Plot1 Plot2 Plot3
\Y1■log(3X)
-0Y2■log(3)+log(X
)
\Y3=
\Y4=
\Y5=
\Y6=
```

Now, press graph. The calculator will initially graph Y1 = log(3x) normally, then the tracing ball will follow the same path. At the end, the screen will only contain "one" graph, suggesting that the two functions are equivalent. Remember to trace back and forth between the functions to verify they are the same.

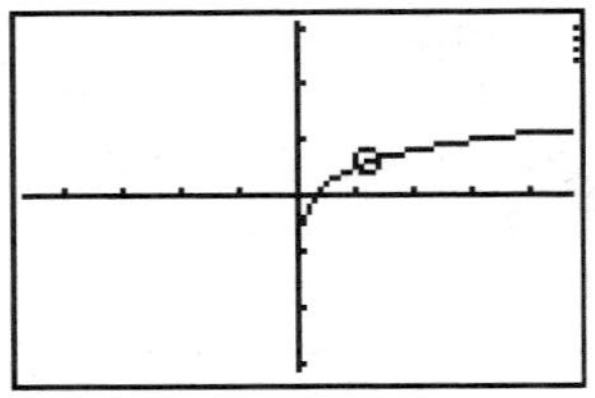

```
WINDOW
 Xmin=-4.7
 Xmax=4.7
 Xscl=1
 Ymin=-3.1
 Ymax=3.1
 Yscl=1
 Xres=1
```

12. Students should notice that the Y1- and Y2-values for every x-value are equal, indicating that the two functions are equivalent.

X	Y1	Y2
1	.47712	.47712
2	.77815	.77815
3	.95424	.95424
4	1.0792	1.0792
5	1.1761	1.1761
6	1.2553	1.2553
7	1.3222	1.3222
X=1		

13.

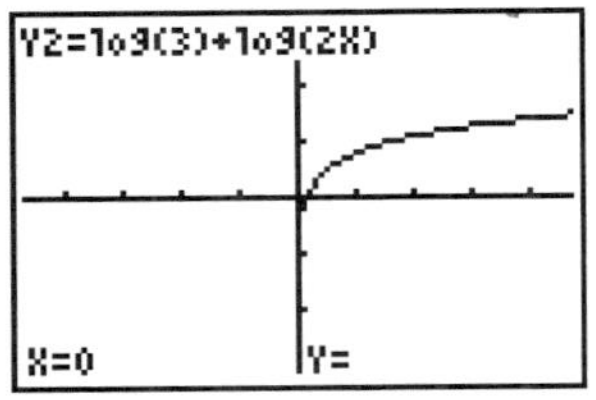

X	Y1	Y2
1	.77815	.77815
2	1.0792	1.0792
3	1.2553	1.2553
4	1.3802	1.3802
5	1.4771	1.4771
6	1.5563	1.5563
7	1.6232	1.6232
X=1		

Students should notice that the Y1- and Y2-values for every x-value are equal, indicating that the two functions are equivalent.

14.

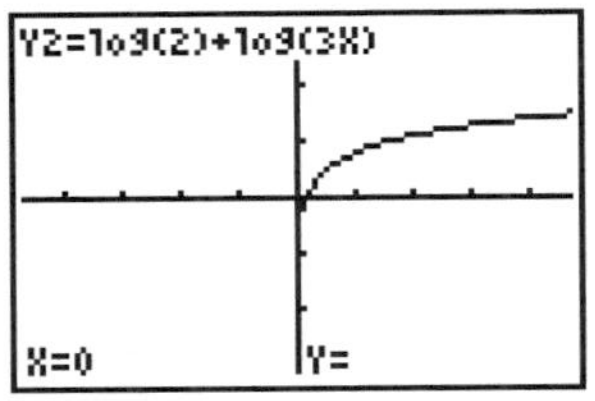

X	Y1	Y2
1	.77815	.77815
2	1.0792	1.0792
3	1.2553	1.2553
4	1.3802	1.3802
5	1.4771	1.4771
6	1.5563	1.5563
7	1.6232	1.6232
X=1		

Students should notice that the Y1- and Y2-values for every x-value are equal, indicating that the two functions are equivalent.

15. Answers may vary. However, students should observe that the logarithm of the product of x and y is equal to the sum of the logarithm of x and the logarithm of y. In other words, $\log(xy) = \log(x) + \log(y)$.

16. The calculator will initially graph Y1 = log(3x) normally, then the tracing ball will follow the same path. At the end, the screen will only contain "one" graph, suggesting that the two functions are equivalent.

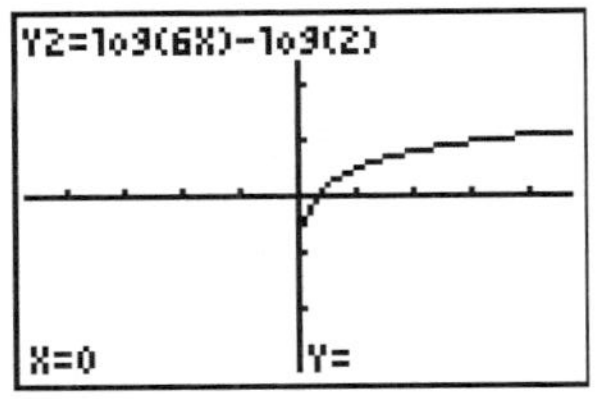

17. Students should notice that the Y1- and Y2-values for every x-value are equal, indicating that the two functions are equivalent.

X	Y1	Y2
1	.47712	.47712
2	.77815	.77815
3	.95424	.95424
4	1.0792	1.0792
5	1.1761	1.1761
6	1.2553	1.2553
7	1.3222	1.3222
X=1		

18.

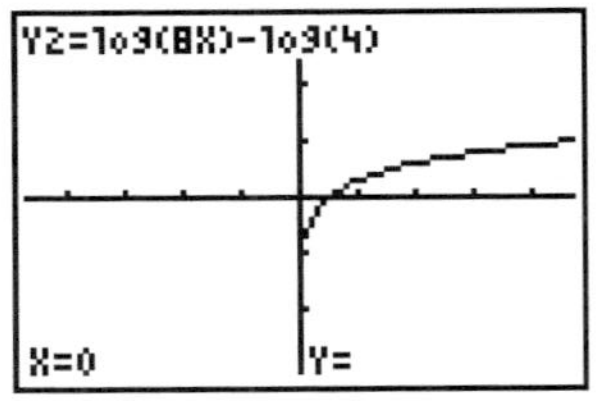

X	Y1	Y2
1	.30103	.30103
2	.60206	.60206
3	.77815	.77815
4	.90309	.90309
5	1	1
6	1.0792	1.0792
7	1.1461	1.1461
X=1		

Students should notice that the Y1- and Y2-values for every x-value are equal, indicating that the two functions are equivalent.

19. Answers may vary. However, students should observe that the logarithm of the quotient of x and y is equal to the difference of the logarithm of x and the logarithm of y. In other words, $\log\left(\frac{x}{y}\right) = \log(x) - \log(y)$.

Answers (continued)

20. 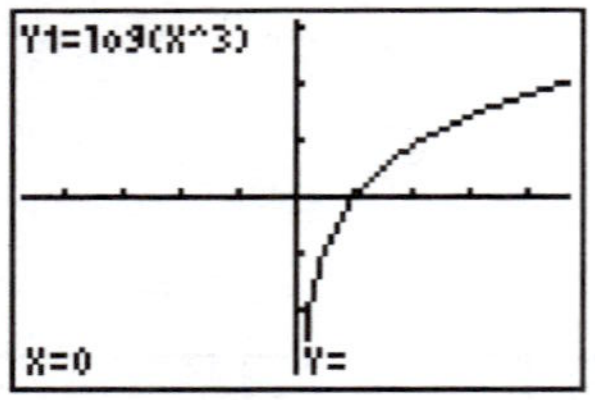

The calculator will initially graph Y1 = $\log(x^3)$ normally, then the tracing ball will follow the same path. At the end, the screen will only contain "one" graph, suggesting that the two functions are equivalent.

21.

X	Y1	Y2
1	0	0
2	.90309	.90309
3	1.4314	1.4314
4	1.8062	1.8062
5	2.0969	2.0969
6	2.3345	2.3345
7	2.5353	2.5353

X=1

Students should notice that the Y1- and Y2-values for every x-value are equal, indicating that the two functions are equivalent.

22.

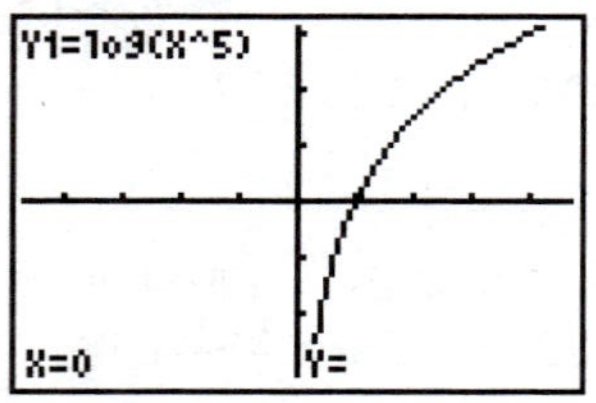

X	Y1	Y2
1	0	0
2	1.5051	1.5051
3	2.3856	2.3856
4	3.0103	3.0103
5	3.4949	3.4949
6	3.8908	3.8908
7	4.2255	4.2255

X=1

Students should notice that the Y1- and Y2-values for every x-value are equal, indicating that the two functions are equivalent.

23. Answers may vary. However, students should observe that the logarithm of x raised to the nth power is equal to the product of the logarithm of x and n. In other words, $\log(x^n) = n\log(x)$.

SECTION 6.12 Assignment

1. Sketch a graph of $f(x) = 5^x$.
2. What are the domain and range of this function?
3. Sketch the graph of $g(x) = 5^x - 3$. How is this graph of $g(x)$ related to the graph of $f(x) = 5^x$?
4. What are the domain and range of $g(x) = 5^x - 3$? How do they compare to the domain and range of $f(x) = 5^x$?
5. Sketch a graph of $f(x) = \log(x)$.
6. What are the domain and the range of this function?
7. Sketch the graph of $g(x) = \log(x - 2)$. How is this graph of $g(x)$ related to the graph of $f(x) = \log(x)$?
8. What are the domain and range of $g(x)$? How do they compare to the domain and range of $f(x) = \log(x)$?
9. Find the value of each of these:

   a) $\log(3) + \log(7) =$

   b) $\log(4) + \log(10.5) =$

   c) $\log(48) - \log(16) =$

   d) $\log(60) - \log(8) =$

SECTION 6.13 It's Not What You Say, But How You Say It

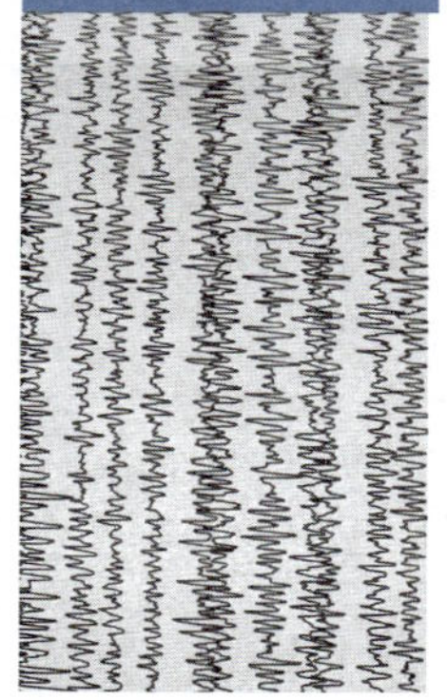

Logarithms were developed to help scientists work with large numbers in the days before calculators and computers. John Napier (1550–1617), a Scottish mathematician, developed logarithms. It has been said that science advanced quickly in the 1700s because logarithms made calculations easier.

Today, calculators and computers make calculations easy. On the other hand, logarithmic functions can still be used for modeling in science. Exponential relationships, such as energy or sound, can be expressed inversely using logarithms. For example, the Richter scale is a logarithmic scale that is used to measure the intensity of an earthquake, and the decibel scale is a logarithmic scale that is used to measure the intensity of sound.

An exponential equation has three parts as shown in **Figure 6.43.**

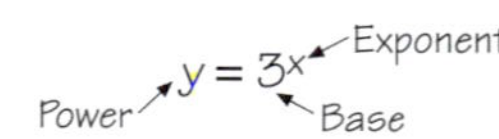

FIGURE 6.43. An exponential equation.

As you recall, a function that undoes another function is called the **inverse** of that function. One way to find an inverse of a function is to switch y and x in its equation then solve for y. A name for a function that is the inverse of $f(x) = 3^x$ is **logarithm,** base 3, of x. It is written $f(x) = \log_3 x$. The abbreviation for logarithm is log.

To write an exponential equation as a logarithm, the base of the power becomes the base of the logarithm. The power is the quantity of which you are taking the logarithm; the logarithm of that quantity is equal to the exponent (see **Figure 6.44**).

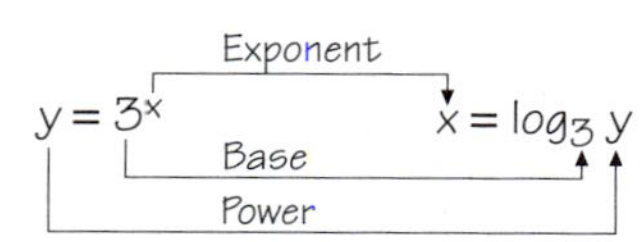

FIGURE 6.44. Writing an exponential equation as a logarithm.

Assignment for Section 6.12

Answers

1. Sample answer:

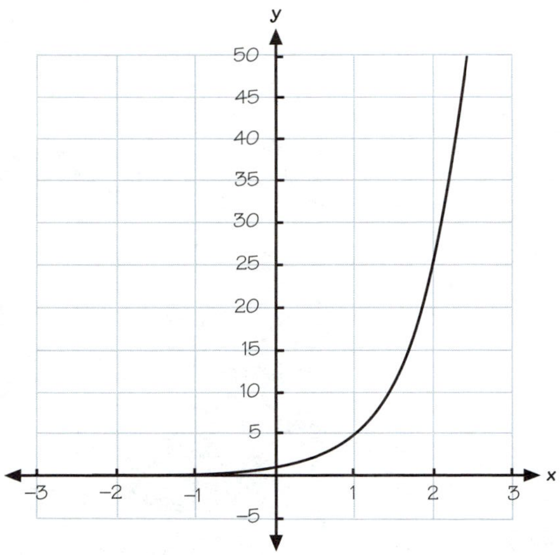

2. Domain: x is all real numbers; range: $y > 0$

3. Sample answer:

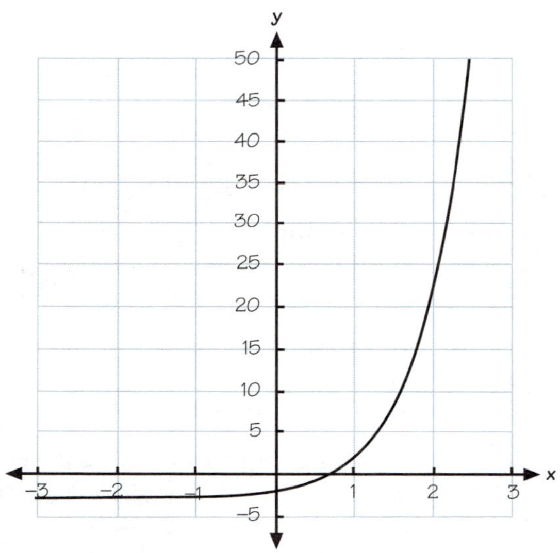

The graph of $g(x)$ is the same as the graph of $f(x)$ shifted down 3 units.

4. Domain: x is all real numbers; range: $y > -3$. The domain of $g(x)$ is the same as the domain of $f(x)$, but the range of $g(x)$ is the range of $f(x)$ shifted down 3 units.

5. Sample answer:

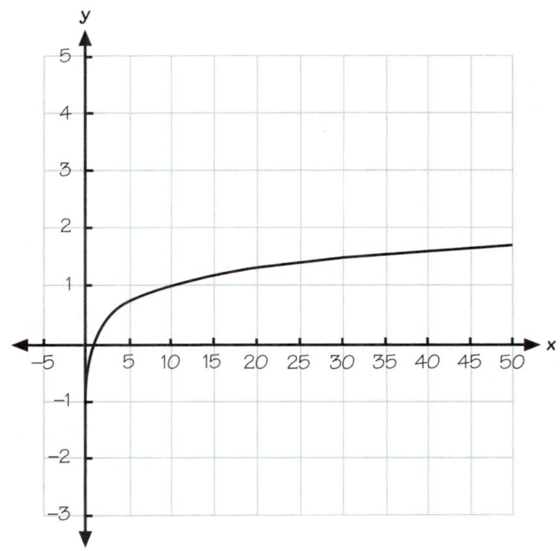

6. Domain: $x > 0$; range: y is all real numbers.

7. Sample answer:

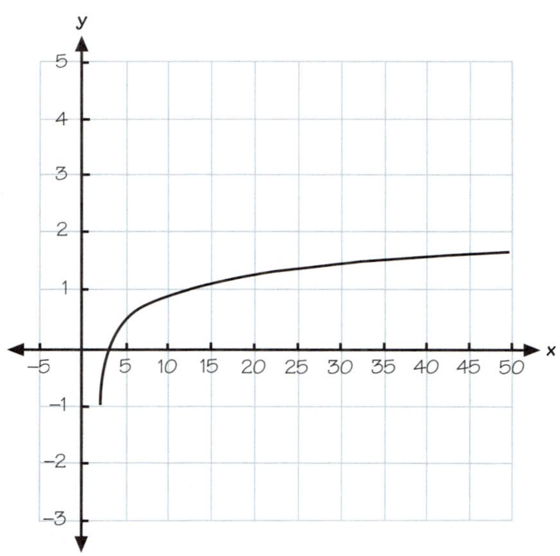

The graph of $g(x)$ is the same shape as the graph of $f(x)$, but it is shifted 2 units to the right.

8. Domain: $x > 2$; range: y is all real numbers. The range of $g(x)$ is the same as the range of $f(x)$, but the domain of $g(x)$ is the domain of $f(x)$ shifted 2 units to the right.

9. a) (21)

   b) log(42)

   c) log(3)

   d) log(7.5)

EXPLAIN LESSON 3

Section 6.13: It's Not What You Say, But How You Say It

Purpose: In Section 6.12 students numerically and graphically explored three properties of logarithms and made conjectures based on their observations. In Section 6.13 students will formalize the definition of a logarithm from an exponential equation, then formalize and algebraically prove the three properties from Section 6.12. This will prepare students to use logarithms to solve exponential equations and solve application problems in Section 6.14.

Description: In this section students will read about the uses of logarithms. They will define a logarithm from an exponential equation then use the properties of exponents to algebraically prove three properties of logarithms. This will enable students to use logarithm properties to simplify expressions and solve equations.

Materials: graphing calculators

Teaching Ideas: Debrief students after Section 6.12 by having them share their answers to the questions. As they share, be sure to fill in any missing concept, process, or skill that could have been used to answer the questions.

Work through the questions in this section with the entire class. Once again, be sure all necessary concepts, processes, or skills are addressed.

You may need to review scientific notation and the laws of exponents before or during this lesson, depending on the strengths of the students.

Every exponential equation can be rewritten in logarithmic form and vice-versa. For example, to write $y = \log_9 81$ in exponential form.

Step 1: Identify the base as 9 and the exponent as y.

Step 2: Using the base and exponent, write an exponential expression, 9^y.

Step 3: Set the expression equal to the power in the logarithmic equation, $9^y = 81$.

To evaluate a logarithm, set it equal to x and then rewrite it in exponential form. For example, evaluate $\log_7 49$.

Step 1: Set the expression equal to x: $\log_7 49 = x$.

Step 2: Rewrite the expression in exponential form: $7^x = 49$.

Step 3: Rewrite 49 as a power of 7: $7^x = 7^2$.

Step 4: Evaluate $x = 2$.

Because logarithms can be written with any base, it is important to state the base. Our number system is base-10 (that is, place value is based on powers of 10: 1, 10, 100, etc.). Base-10 logarithms are often used; thus they are called **common logarithms**. To save time, $\log_{10}$ can be written without the 10 simply as log.

CHECK THIS!

When multiplying exponents with like bases, add the powers.

$x^a \cdot x^b = x^{a+b}$

When dividing exponents with like bases, subtract the powers.

$\frac{x^a}{x^b} = x^{a-b}$

When raising an exponent to a power, multiply the powers.

$(x^a)^b = x^{ab}$

There are three key properties of logarithms that you explored in Section 6.12. First, you explored $\log(xy)$.

1. Use the log key on your calculator to find the values in **Figure 6.45.** Keep a copy of your table because you will use it later in this section.

log 2	
log 5	
log 10	
log 50	
log 100	

FIGURE 6.45. Logarithms for Question 1.

2. Use your table to evaluate and compare the quantities in **Figure 6.46.**

log 5 + log 10		log 50	
log 2 + log 5		log 10	
log 10 + log 10		log 100	

FIGURE 6.46. Quantities for Question 2.

3. What patterns do you notice? Explain these patterns. What can you conclude?

4. Do you think this might be true for all logarithms of all bases? Explain.

Second, you explored $\log\left(\frac{x}{y}\right)$.

5. Now use your table to evaluate and compare the quantities in **Figure 6.47.**

log 10 − log 5		log 2	
log 50 − log 10		log 5	
log 100 − log 10		log 10	

FIGURE 6.47. Quantities for Question 5.

6. What patterns do you notice? Explain these patterns.

7. Do you think this might be true for all logarithms of all bases? Explain.

Third, you explored $\log(x^b)$.

8. Use the log key on your calculator to evaluate the logarithms in **Figure 6.48.**

log 2	
log 5	
log 10	
log 25	
log 100	
log 125	

FIGURE 6.48. Logarithms for Question 8.

9. Now use your table to evaluate and compare the quantities in **Figure 6.49.**

2 log 5		log 25	
2 log 10		log 100	
3 log 5		log 125	

FIGURE 6.49. Quantities for Question 9.

10. What patterns do you notice? Explain these patterns.

11. Do you think this might be true for all logarithms of all bases? Explain.

Answers

1.

log 2	0.031
log 5	0.699
log 10	1
log 50	1.699
log 100	2

2.

log 5 + log 10	0.699 + 1 = 1.699	log 50	1.699
log 2 + log 5	0.301 + 1.699 = 1	log 10	1
log 10 + log 10	1 + 1 = 2	log 100	2

3. Answers may vary. Students should notice that the two quantities in each row are equal. They should also notice that 5(10) = 50, 2(5) = 10, and 10(10) = 100. This should prompt them to hypothesize that if you add logarithms of two numbers together, you should get the same answer as the logarithm of the product of the two numbers.

   For discussion, ask students to notice that log 5 + log 10 = log 50. Ask them to explain why this might be true. A sample response follows. If students do not notice this, provide this answer to set up the algebraic proof of this property of logarithms in Question 4.

   Because the logarithms are really representing exponents, we add exponents when we are multiplying numbers of the same base. This idea can be expressed symbolically.

   Let $\log_{10} 5 = a$, $\log_{10} 10 = b$, $\log_{10} 50 = c$; therefore, $10^a = 5$; $10^b = 10$; $10^c = 50$.

   We know that 5 × 10 = 50, so if we substitute

   $$5 \times 10 = 50$$
   $$10^a \times 10^b = 10^c$$
   $$10^{a+b} = 10^c$$
   $$a + b = c.$$

   Now, substitute the original logarithmic expressions for a, b, and c:

   $$a + b = c$$
   $$\log 5 + \log 10 = \log 50$$
   $$\log 5 + \log 10 = \log (5 \times 10)$$

4. Let $\log_b m = p$, $\log_b n = q$, $\log_b mn = r$; therefore, $b^p = m$; $b^q = n$; $b^r = mn$

   Since $m \times n = mn$, we can substitute our exponential expressions and simplify.

   $$m \times n = mn$$
   $$b^p \times b^q = b^r$$

   Since we are multiplying like bases, we can add the exponents to find that:

   $$b^{p+q} = b^r$$
   $$p + q = r$$

   Now, substitute the original logarithmic expressions for p, q, and r:

   $$p + q = r$$
   $$\log_b m + \log_b n = \log_b mn$$

   Thus, this property is true for all logarithms, regardless of the base.

5.

log 10 – log 5	1 – 0699 = 0.301	log 2	0.301
log 50 – log 10	1.6990 – 1 = 0.6990	log 5	0.301
log 100 – log 10	2 – 1 = 1	log 10	1

6. Answers may vary. Students should notice that the two quantities in each row are equal. They should also notice that $\frac{10}{5} = 2$, $\frac{50}{10} = 5$, and $\frac{100}{10} = 10$. This should prompt them to hypothesize that if you subtract the logarithms of two numbers, you should get the same answer as the logarithm of the quotient of the numbers.

   For discussion, ask students to notice that log 10 – log 5 = log 2. Ask them to explain why this might be true. A sample response follows. If students do not notice this, provide this answer to set up the algebraic proof of this property of logarithms in Question 7.

   Because the logarithms are really representing exponents, we subtract exponents when we are dividing numbers of the same base. This idea can be expressed symbolically.

   Let $\log_{10} 10 = a$, $\log_{10} 5 = b$, $\log_{10} 2 = c$; therefore, $10^a = 10$; $10^b = 5$; $10^c = 50$.

   We know that $\frac{10}{5} = 2$, so if we substitute the expressions for a, b, and c as powers of 10:

   $$\frac{10}{5} = 2$$
   $$\frac{10^a}{10^b} = 10^c$$
   $$10^{a-b} = 10^c$$
   $$a - b = c$$

   Now, substitute the original logarithmic expressions for a, b, and c:

   $$a - b = c$$
   $$\log 10 - \log 5 = \log 2$$
   $$\log 10 - \log 5 = \log\left(\frac{10}{5}\right)$$

7. Let $\log_b m = p$, $\log_b n = q$, $\log_b\left(\frac{m}{n}\right) = r$; therefore, $b^p = m$; $b^q = n$; $b^r = \frac{m}{n}$.

   Since any real number is equal to itself, we can write $\frac{m}{n} = \frac{m}{n}$, and substitute our exponential expressions and simplify.

   $$\frac{m}{n} = \frac{m}{n}$$
   $$\frac{b^p}{b^q} = b^r$$

   Since we are dividing like bases, we can subtract the exponents to find that:

   $$b^{p-q} = b^r$$
   $$p - q = r$$

Answers (continued)

Now, substitute the original logarithmic expressions for p, q, and r:

$$p - q = r$$
$$\log_b m - \log_b n = \log_b\left(\frac{m}{n}\right)$$

Thus, this property is true for all logarithms, regardless of the base.

8.

log 2	0.301
log 5	0.699
log 10	1
log 25	1.398
log 100	2
log 125	2.097

9.

2 log 5	2(0.699) = 1.398	log 25	1.398
2 log 10	2(1) = 2	log 100	2
3 log 5	3(0.699) = 2.097	log 125	2.097

10. Answers may vary. Students should notice that the two quantities in each row are equal. They should also notice that $5^2 = 25$, $10^2 = 100$, and $5^3 = 125$. This should prompt them to hypothesize that if you multiply the logarithms of one number by a constant factor, you should get the same answer as the logarithm of that number raised to the same constant power.

For discussion, ask students to notice that $2 \log 5 = \log 25$. Ask them to explain why this might be true. A sample response that will set up the algebraic proof of this property of logarithms in Question 11 follows.

Because the logarithms are really representing exponents, we multiply exponents when we raise a power to a power. This idea can be expressed symbolically.

Let $\log_{10} 5 = a$ and $\log_{10} 25 = b$; therefore, $10^a = 5$; $10^b = 25$.

We know that $5^2 = 25$, so if we substitute the expressions for a and b as powers of 10:

$$5^2 = 25$$
$$(10^a)^2 = 10^b$$
$$10^{2a} = 10^b$$
$$2a = b$$

Now, substitute the original logarithmic expressions for a and b:

$$2a = b$$
$$2(\log 5) = \log 25$$
$$2(\log 5) = \log(5^2)$$

11. Let $m^k = n$. Also, let $\log_b m = p$ and $\log_b n = q$; therefore $b^p = m$; $b^q = n$.

Begin with $m^k = n$, and substitute the exponential expressions and simplify.

$$m^k = n$$
$$(b^p)^k = b^q$$

Since we are raising a power to a power, we can multiply the exponents to find that:

$$(b^p)^k = b^q$$
$$pk = q$$
$$kp = q$$

Now, substitute the original logarithmic expressions for p and q:

$$kp = q$$
$$k \log_b m = \log_b(n)$$

Now, remember that $m^k = n$, so $n = m^k$ as well. Substitute m^k for n.

$$k \log_b m = \log_b(n)$$
$$k \log_b m = \log_b(m^k)$$

Thus, this property is true for all logarithms, regardless of the base.

SECTION 6.13 Assignment

1. Write each of these in logarithmic form.

   a) $x = 8^y$

   b) $x = 4^y$

   c) $x = 3^{2y}$

   d) $y = 5^x$

2. Find the value of each of these logarithms.

   a) $\log_3 27$

   b) $\log_{10} 100$

   c) $\log_2 64$

   d) $\log_8 64$

3. Simplify each of these. Leave your answers written as logarithms.

   a) $\log_3 9 + \log_3 5$

   b) $\log_2 8 + \log_2 7$

   c) $\log_2 35 - \log_2 7$

   d) $\log_7 144 - \log_7 4$

   e) $\log_{10} 125$

   f) $\log_3 64$

4. Simplify each of these as much as you can. Evaluate the logarithm when possible.

   a) $\log_2 8 + \log_2 4$

   b) $\log_2 32 - \log_2 16$

   c) $\log_2 32 - \log_2 4$

   d) $\log_3 81 - \log_3 9$

   e) $\log_5 125$

   f) $\log_2 64$

5. Can you take the logarithm of 0? Why or why not?

6. Solve each of these for x.

   a) $\log_3 x = 5$

   b) $\log_3 5 + \log_3 x = \log_3 30$

   c) $\log_5 x - \log_5 8 = 1$

   d) $\log_7 5 + x \log_7 6 = \log_7 1080$

Section 6.13 Assignment

Answers

1. a) $y = \log_8 x$

   b) $y = \log_4 x$

   c) $2y = \log_3 x$ or $y = \frac{1}{2}\log_3 x$

   d) $x = \log_5 x$

2. a) 3

   b) 2

   c) 6

   d) 2

3. a) $\log_3 45$

   b) $\log_2 56$

   c) $\log_2 5$

   d) $\log_7 36$

   e) $3 \log_{10} 5$

   f) $2 \log_3 8$

4. a) $\log_2 32 = 5$

   b) $\log_2 2 = 1$

   c) $\log_2 8 = 3$

   d) $\log_3 9 = 2$

   e) $3 \log_5 5 = 3$

   f) $2 \log_2 8 = 6$

5. No. A logarithm is an exponent that when applied to the given base will generate a number. To take the logarithm of 0 would be asking the question, "To what power can I raise this base and get 0 as my answer?" There is no real number that answers this question. Therefore, you cannot take the logarithm of 0.

6. a) 243

   b) 6

   c) 40

   d) 3

SECTION 6.14 I Feel the Earth Move

Mount St. Helens National Volcanic Monument in Washington State has an active volcano. Active volcanoes can cause earthquakes.

Park managers work closely with geologists to monitor earthquakes. Scientists who study volcanoes use earthquake data to make predictions about volcanic eruptions. Stronger or more frequent shakes can mean that magma underground is moving toward the surface and an eruption may happen soon. By measuring the magnitude or strength of an earthquake, scientists can make decisions about an impending volcanic eruption. Such decisions could include whether or not to evacuate a park or to relocate wildlife.

In Section 6.13, you learned three properties of logarithms.

- $\log_b m + \log_b n = \log_b m \cdot n$
- $\log_b m - \log_b n = \log_b \frac{m}{n}$
- $\log_b mn = n \log_b m$

These properties are useful in solving exponential equations. We will solve exponential equations by taking the base 10 logarithm of each side. (We will use base 10 since this is one of two bases that the calculator uses.)

We will use $3^x = 14$ as an example.

We need a way to isolate or *undo* the exponent. To do this we need an inverse operation for exponentiation. This inverse operation is called a **logarithm**. If we take the logarithm of both sides, the logarithm undoes the exponent.

$$3^x = 14$$
$$\log(3^x) = \log(14)$$

Now we can use the properties of logarithms to solve for x. Recall that $m^n = n \log_b m$. So,

$$\log(3^x) = \log(14)$$
$$x \log(3) = \log(14)$$

Remember that log(3) and log(14) are real numbers. Therefore, we can apply division to both sides of the equation in order to solve for x. We will divide both sides of the equation by log(3) in order to undo the multiplication.

$$x \log(3) = \log(14)$$
$$\frac{x\log(3)}{\log(3)} = \frac{\log(14)}{\log(3)}$$
$$x = \frac{\log(14)}{\log(3)}$$
$$x \approx 2.402$$

The values for logarithms are often rounded to the nearest thousandth. However, sometimes you will be asked for an exact answer. In that case leave your answer in terms of logarithms.

In Section 6.12 you saw that logarithmic functions can be graphed on a graphing calculator. Most scientific and graphing calculators have two logarithm keys: LOG, which is the base-10 logarithm, and LN, which is the **natural logarithm** having a base e. The number e is similar to *pi* that recurs throughout mathematics and science. You can graph base-10 and base-e logarithmic functions on most graphing calculators using these keys. You can also graph other logarithmic functions, but first you must change the base to either 10 or e.

Consider the function $y = 3^x$.

1. Find the inverse function of $y = 3^x$ by switching x and y. Solve for y by taking the logarithm of both sides and using the properties of logarithms.

2. How can you write the inverse of $y = 3^x$ as a logarithmic function?

3. Based on your answer to Questions 1 and 2, what is $\log_3 x$ equal to?

4. Based on your findings in Questions 1–3, what is $\log_b x$ equal to?

ELABORATE LESSON 3

Section 6.14: I Feel the Earth Move

Purpose: In Section 6.13 students formalized and used three important properties of logarithms. In Section 6.14 students use these properties to solve exponential equations, develop the change of base formula, and apply logarithms to real-world contexts. This prepares students to apply logarithmic functions in Section 6.15.

Description: In this section students accomplish three major goals. First, students use the properties of logarithms to solve exponential equations by using a logarithm as an inverse operation. Second, students algebraically develop the change of base formula in order to use graphing technology to graph a logarithmic function with any base. Third, students apply properties of logarithms and logarithmic functions to real-world settings such as the Richter scale, which is used to quantify the energy of an earthquake so that the strengths of different earthquakes can be compared, and the pH scale, which is used to compare the acidity of different solutions in chemistry.

Materials: graphing calculator

Teaching Ideas: Parts of this section are guided discussions, and parts are designed to be interactive with students. Students may work in their small groups of 3–4, discussing the important points of each part of the activity with one another. The teacher should be prepared to facilitate the problem-solving process, especially in the third part of this section when students are exploring the Richter scale.

You may need to review transformations and graphing calculator skills with your students, depending on their skill level.

Answers

1. $x = 3^y$

   $\log x = \log(3^y)$

   $\log x = y \log 3$

   $\frac{\log x}{\log 3} = \frac{y \log 3}{\log 3}$

   $\frac{\log x}{\log 3} = y$

2. $x = 3^y \rightarrow y = \log_3 x$

3. $\frac{\log x}{\log 3} = y = \log_3 x$, so $\log_3 x = \frac{\log x}{\log 3}$

4. Begin with the general exponential function $y = b^x$. Switch x and y, then solve for y.

   $x = b^y$

   $\log x = \log(b^y)$

   $\log x = y \log b$

   $\frac{\log x}{\log b} = \frac{y \log b}{\log b}$

   $\frac{\log x}{\log b} = y$

   By definition of logarithms, if $x = b^y$, then $\log_b x = y$. So, $\log_b x = \frac{\log x}{\log b}$.

The formula you found in Question 4 is called a change of base formula. It allows you to convert a logarithm of any base to a base 10. You can also use the change of base formula to convert a logarithm to any other base. In this section we are interested in changing logarithms to base 10 so that we can use the graphing calculator.

5. Use the change of base formula to graph $y = \log_5 x$ on your calculator. Record your window and sketch your results.

Dr. Charles Richter

CHECK THIS!

Dr. Charles F. Richter developed the Richter scale in 1935 at the California Institute of Technology to measure and compare the strengths of earthquakes.

Before then scientists used the Mercalli scale, which ranks earthquakes based on the damage they caused.

6. Use the Trace or Table feature to calculate the value of $\log_5 3$.

Logarithms can be used to compare large and small numbers. Dr. Charles F. Richter was the first scientist to recognize that the strength of a seismic wave could be used to describe the magnitude of the earthquake that caused it.

One of Richter's formulas relates the amount of energy released by an earthquake to its Richter magnitude:

$\log E = 11.8 + 1.5M$

In this formula, E is the seismic energy released by the earthquake and M is the Richter magnitude of the earthquake.

7. Solve Richter's formula for E in terms of M.

8. Use your graphing calculator to graph E as a function of M. Record your window and sketch your graph.

9. How useful is the graph in interpreting the relationship between energy and magnitude?

10. Use the Table feature of your graphing calculator and Richter's formula to compute the energy released for the magnitudes in **Figure 6.50.** Then find the successive ratios.

Magnitude	Energy	Successive Ratios
1		–
2		
3		
4		
5		
6		
7		

FIGURE 6.50. Energy released by earthquake magnitudes.

11. What patterns do you notice in the ratios?

12. If the magnitude of an earthquake increases by one, by what factor does the energy increase?

13. If the magnitude of an earthquake increases by two, by what factor does the energy increase?

14. If the magnitude of an earthquake increases by four, by what factor does the energy increase?

15. The Northern Sumatra earthquake on December 26, 2004, which caused a tsunami that killed hundreds of thousands of people, had a magnitude of 9.0. In 1992, an earthquake of magnitude 4.5 struck Midland, Texas. Was the Northern Sumatra earthquake twice as strong as the Midland earthquake? Explain your answer.

Answers (continued)

5.

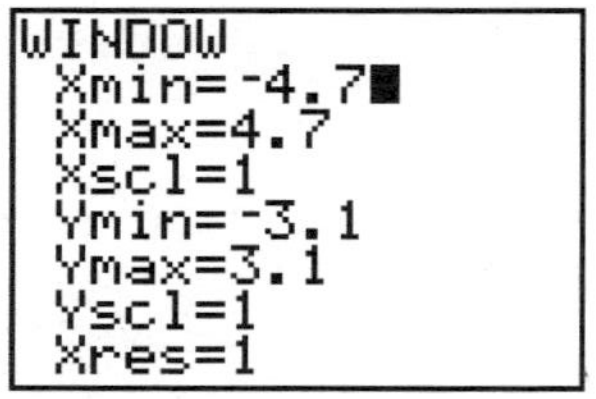

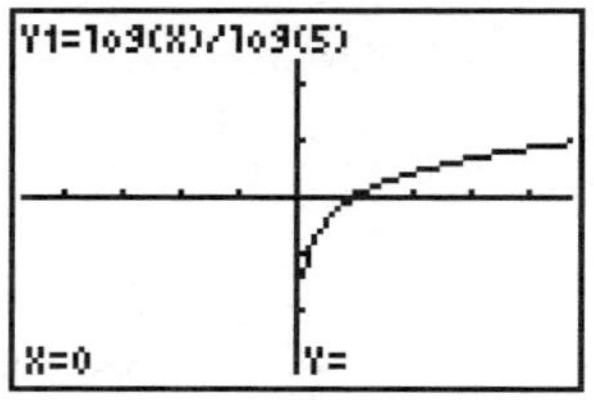

6.

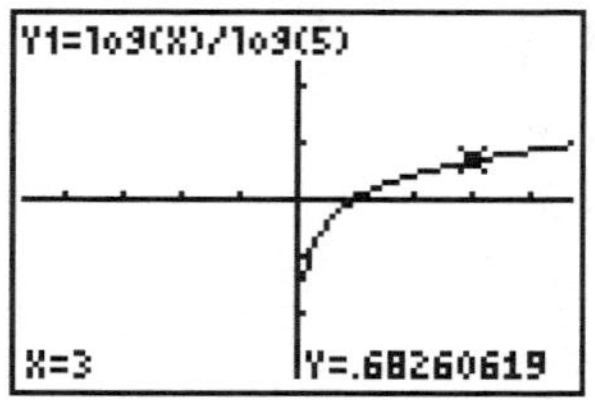

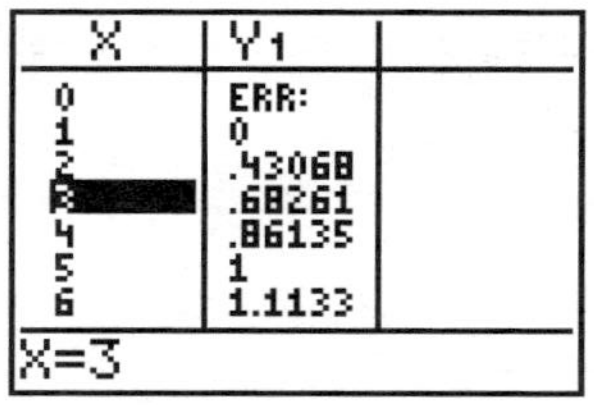

7. Using the definition of a base-10 logarithm, $10^{(11.8+1.5M)} = E$.

8. Answers may vary. If students struggle (and they likely will), suggest the following viewing window.

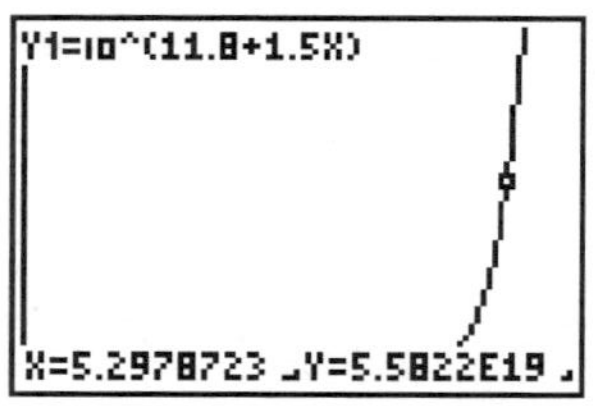

9. Answers may vary. Students should notice that they are having a hard time finding a viewing window that allows them to see the relationship clearly since the numbers are huge. A very small change in x-values generates a large change in y-values. So a graph is not very useful in interpreting this situation other than for very small changes in magnitude.

10.

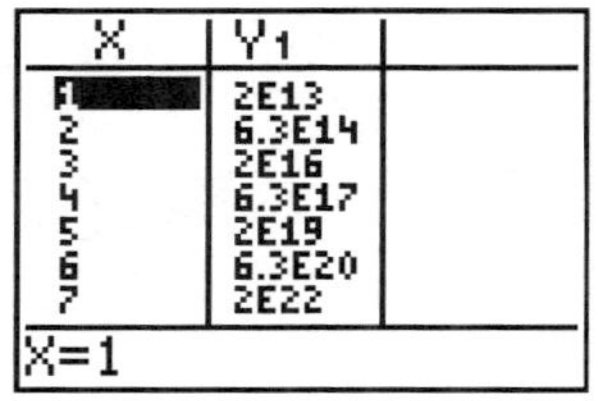

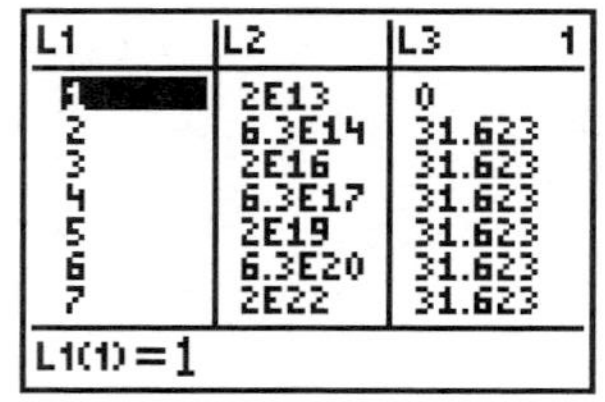

11. The successive ratios are constant at about 31.623, which is the square root of 1000.

12. About 31.623; for example, from a magnitude 1 earthquake to a magnitude 2 earthquake, the ratio of increase is $\frac{6.3 \times 10^{14}}{2 \times 10^{13}} \approx 31.623$.

13. 1000; for example, from a magnitude 1 earthquake to a magnitude 3 earthquake, the ratio of increase is $\frac{2 \times 10^{16}}{2 \times 10^{13}} = 1 \times 10^3 = 1000$.

14. 1,000,000; for example, from a magnitude 1 earthquake to a magnitude 5 earthquake, the ratio of increase is $\frac{2 \times 10^{19}}{2 \times 10^{13}} = 1 \times 10^6 = 1{,}000{,}000$.

15. No. Using Richter's formula, the 2004 Northern Sumatra earthquake released an energy of $10^{(11.8+1.5(9))} = 2 \times 10^{25}$. The 1992 Midland earthquake released an energy of $10^{(11.8+1.5(4.5))} = 3.5 \times 10^{18}$. The ratio of these two numbers is $\frac{2 \times 10^{25}}{3.5 \times 10^{18}} \approx 5.7 \times 10^6$. So, the Northern Sumatra earthquake contained over 5 million times as much energy as the Midland quake!

SECTION 6.14 Assignment

1. Solve each of these equations for x. Round to the nearest thousandth.

   a) $6^x = 8$

   b) $4^{2x} = 23$

   c) $5^x + 2 = 26$

   d) $2^x = -4$

   e) $5^x = 10$

   f) $7^x = 95$

   g) $6^x = 38$

2. Graph each of these on your graphing calculator. Record your window and sketch the graph.

   a) $y = 2 \log x + 4$

   b) $y = \log_6 x$

   c) $y = 3 \log_6 x - 2$

For Questions 3–5, use $y_1 = \log x$ and $y_2 = \log_2 x$.

3. In the same window graph $y_1 = \log x$ and $y_2 = \log_2 x$. Record your window and sketch the graph.

4. Compare the domain and range of $y_1 = \log x$ and $y_2 = \log_2 x$. How are they alike? How do they differ?

5. Use your knowledge of transformations to describe how the graphs of $y_1 = \log x$ and $y_2 = \log_2 x$ are related.

Use $10^{(11.8+1.5M)} = E$ for Questions 6 and 7.

6. On July 20, an earthquake of magnitude 2.0 was recorded on the west-southwest face of Mount St. Helens. How much energy was released by this earthquake?

7. On July 21, an earthquake of magnitude 3.4 was recorded on the same face. Compare the strength of this earthquake to the one on July 20.

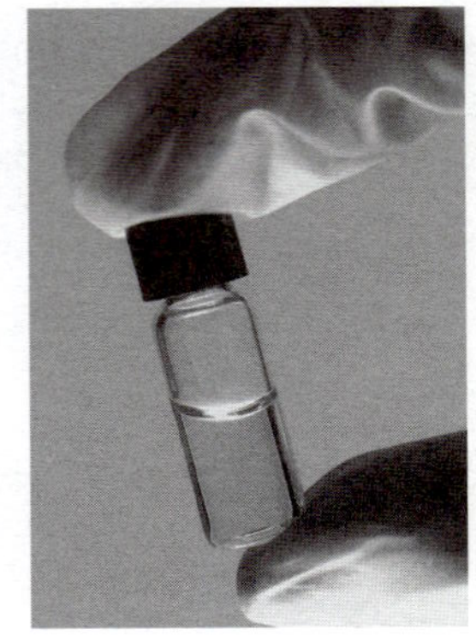

8. Chemists use the pH scale to measure acidity. The pH of a solution is measured on a scale of 1–14 and can be found using the function $y = -\log H$ where y is the pH and H is the concentration of hydrogen ions. The concentration of hydrogen ions in a solution determines how acidic the solution is; the greater the concentration, the greater the acidity.

   It is important for park managers to know the pH of lakes and streams. This information helps them decide how much air or water pollution is in the park. It also helps them assess the risk to wildlife.

   a) Use your graphing calculator to generate a pH table for the concentrations of hydrogen ions shown in **Figure 6.51.** Round your answers to the nearest tenth.

Concentration	pH
1×10^{-13}	
1×10^{-5}	
2×10^{-13}	
6.2×10^{-10}	
7.4×10^{-5}	

FIGURE 6.51. pH table.

   b) Bunsen has a solution with a hydrogen ion concentration of 0.00001. Use the properties of logarithms to find the pH of his solution. Check your answer with a calculator.

   c) The solution in beaker A has a pH of 3. The solution in beaker B has a pH of 8. How many times greater is the concentration of hydrogen ions in beaker A than in beaker B?

Section 6.14 Assignment

Answers

1. a) $\log(6^x) = \log(8)$

$$x\log 6 = \log 8$$

$$\frac{x\log 6}{\log 6} = \frac{\log 8}{\log 6}$$

$$x = \frac{\log 8}{\log 6} \approx 1.161$$

b) $x \approx 1.131$

c) $x \approx 1.975$

d) No solution

e) $x \approx 1.431$

f) $x \approx 2.340$

g) $x \approx 2.030$

2. a)

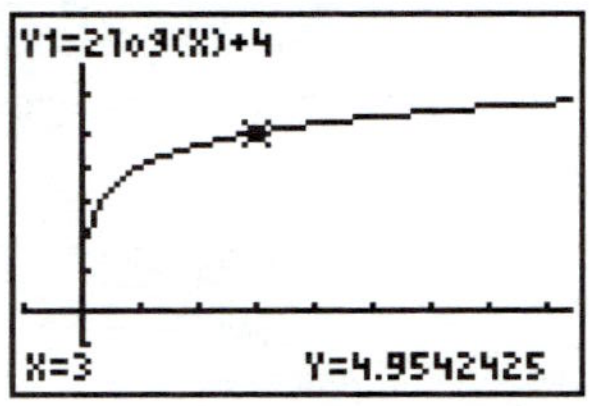

b) $y = \log_6 x$

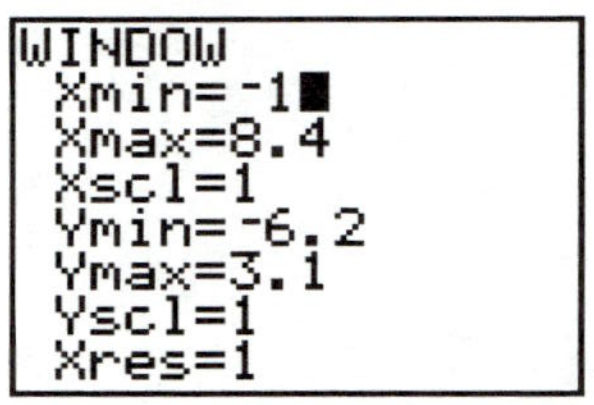

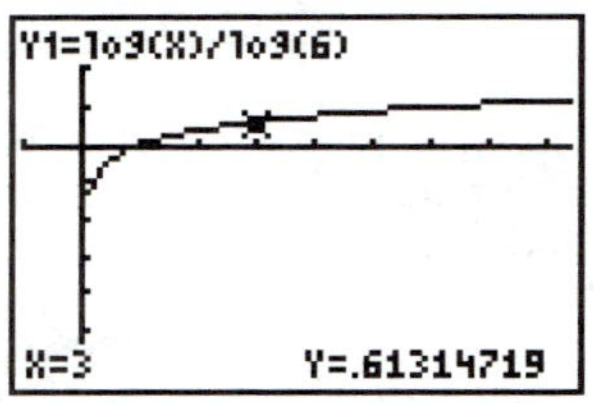

c)

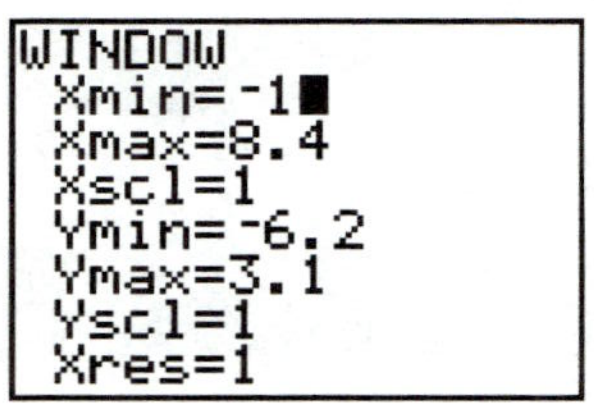

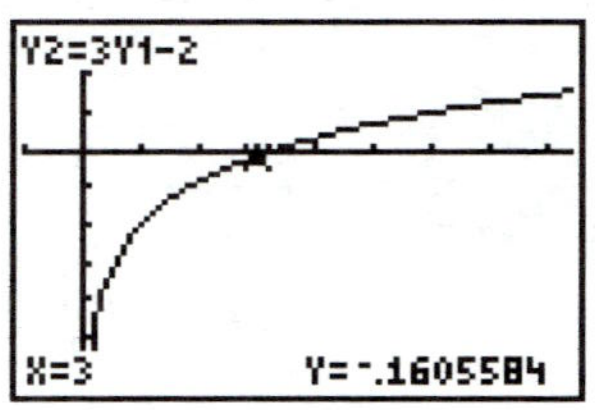

3.

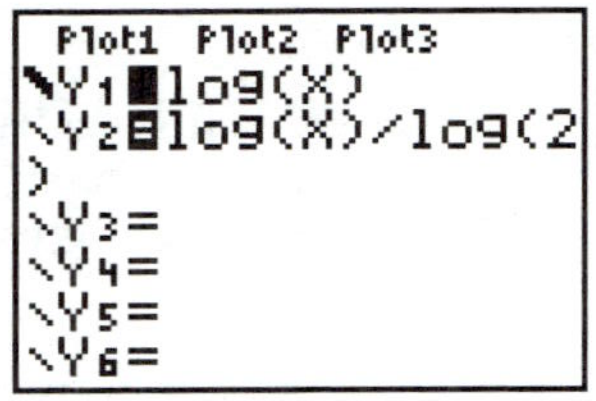

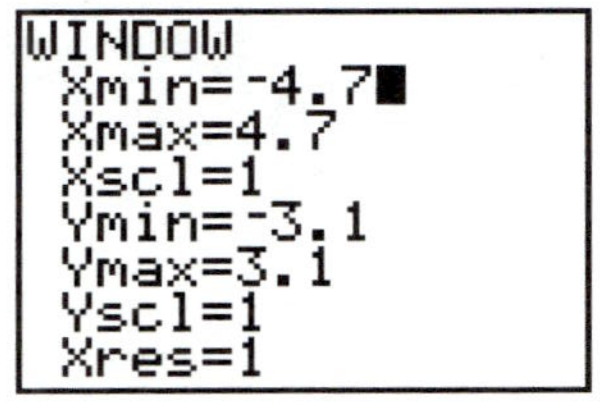

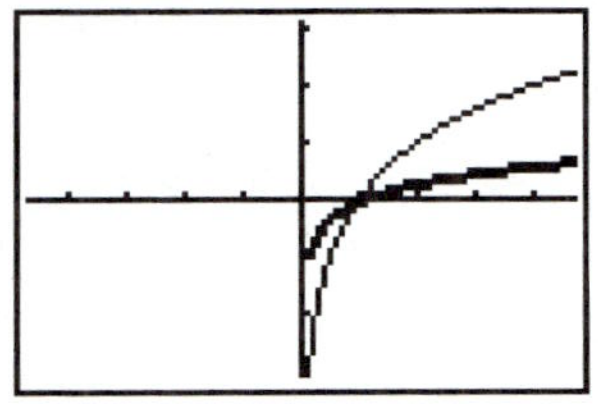

4.

X	Y1	Y2
0	ERR:	ERR:
.1	-1	-3.322
.2	-.699	-2.322
.3	-.5229	-1.737
.4	-.3979	-1.322
.5	-.301	-1
.6	-.2218	-.737

X=0

Using a table helps to identify domain and range values. The domain of both functions is $x > 0$. The range of both functions is all real numbers.

5. The graph of $y_2 = \log_2 x$ can be written using the change of base formula as $y_2 = \frac{\log x}{\log 2}$. Dividing by a real number is just like multiplying by the reciprocal: $y_2 = \left(\frac{1}{\log 2}\right)\log x$.

$\frac{1}{\log 2} \approx 3.322$, so the graph of $y_2 = \log_2 x$ is a vertical stretch of the parent function $y_1 = \log x$.

6.
$$10^{(11.8+1.5M)} = E$$
$$10^{(11.8+1.5(2.0))} = E$$
$$10^{(14.8)} = E$$
$$6.31 \times 10^{14} = E$$

7.
$$10^{(11.8+1.5(3.4))} = E$$
$$10^{(16.9)} = E$$
$$7.94 \times 10^{16} = E$$

The July 21 earthquake released about 125 times as much energy as the July 20 earthquake.

8. a)

Concentration	pH
1×10^{-13}	13.0
1×10^{-5}	5.0
2×10^{-13}	12.7
6.2×10^{-10}	9.2
7.4×10^{-5}	4.1

b)
$$y = -\log H$$
$$\text{pH} = -\log(0.00001)$$
$$\text{pH} = -\log(10^{-5})$$
$$\text{pH} = -(-5\log(10))$$
$$\text{pH} = -(-5(1))$$
$$\text{pH} = 5$$

c) 100,000 times as many

Beaker A:

$$\text{pH} = -\log H$$
$$3 = -\log H$$
$$\frac{3}{-1} = \frac{-\log H}{-1}$$
$$-3 = \log H$$
$$10^{-3} = H$$
$$\frac{10^{-3}}{10^{-8}} = 10^5 = 100{,}000$$

Beaker B:

$$\text{pH} = -\log H$$
$$8 = -\log H$$
$$\frac{8}{-1} = \frac{-\log H}{-1}$$
$$-8 = \log H$$
$$10^{-8} = H$$

SECTION 6.15

What Did You Say?

The intensity of sound is described with the **decibel** (dB) scale. Music producers and sound engineers use this scale to mix recordings of music and to achieve the right balance between layers of sound.

Park visitors often complain about noise. So park managers can use the decibel scale to set noise level limits and use sound meters to measure the intensity of noise.

CHECK THIS!

The human ear can detect a sound wave that displaces air particles by one-billionth of a centimeter.

The intensity of a sound wave that matches this displacement is called the threshold of hearing.

Figure 6.52 shows typical decibel levels for certain sounds.

Sound	Intensity (dB)
Refrigerator	50
Normal conversation	60
Headphones on full volume	100
Chainsaw	120
Jet engine taking off	150

FIGURE 6.52.
Sound intensities.

The decibel scale is defined using $\beta = 10 \log \left(\frac{I}{1.0 \times 10^{-12}}\right)$ where β is the intensity of the sound in decibels and I is the intensity of the sound wave in watts per square meter.

1. How many times more intense is the sound produced by a chainsaw than the sound produced during normal conversation? Explain your answer.

2. In the data in Figure 6.52, how many times more intense is the sound produced by a jet engine taking off than the sound produced by a refrigerator? Explain your answer.

SUMMARY

A logarithmic function is the inverse of an exponential function. Exponential functions can be written in logarithmic form. For example, $y = 3^x$ can be written as $x = \log_3 y$ as shown in **Figure 6.53.**

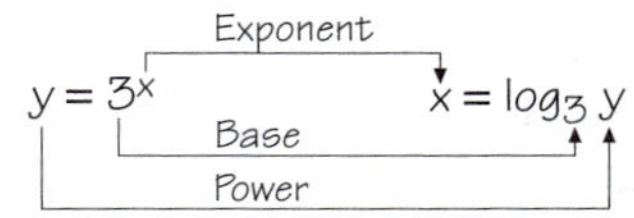

FIGURE 6.53.
Exponential expression rewritten in logarithmic form.

When expressions or equations are written in logarithmic form, three properties of logarithms can be used to manipulate them or to solve for a variable.

- $\log_b m + \log_b n = \log_b m \cdot n$
- $\log_b m - \log_b n = \log_b \frac{m}{n}$
- $\log_b mn = n \log_b m$

Logarithms can be used to solve exponential equations for an unknown exponent of the expression.

You can convert between logarithm bases using the **change of base** formula:

$$\log_b x = \frac{\log x}{\log b}$$

EVALUATE LESSON 3

Section 6.15: What Did You Say?

Purpose: In Sections 6.11 through 6.14, students defined logarithms and explored their properties. In this section students have the opportunity to demonstrate their knowledge by applying those concepts and skills to solve a problem in a new context.

Description: Facing a problem in a new context, students determine and implement a method to solve the problem. This open-middle performance assessment task has only one correct answer, but there are a variety of valid methods to obtain it. In their justification, which is the most important part of a performance assessment task, students should be able to communicate their knowledge of mathematics and demonstrate that they have mastered the concepts presented in Sections 6.11 through 6.14.

Materials: graphing calculator

Teaching Ideas: This task may be assigned as group work or individual work. The mathematics performance assessment rubric should be used to evaluate students' work.

Answers

1. Sample answer:

   Solve the decibel equation for I then find the intensity of sound produced by a chainsaw, then the intensity of sound produced during normal conversation. Find the ratio of the two.

   $\beta = 10\log\left(\frac{I}{1.0\times10^{-12}}\right)$
   $\beta = 10\log I - 10\log(1.0\times10^{-12})$
   $\beta = 10\log I - 10(-12)$
   $\beta = 10\log I + 120$
   $\beta - 120 = 10\log I + 120 - 120$
   $\beta - 120 = 10\log I$
   $\frac{\beta-120}{10} = \frac{10\log I}{10}$
   $\frac{\beta-120}{10} = \log I$
   $10^{\left(\frac{\beta-120}{10}\right)} = I$
   $10^{\left(\frac{\beta}{10}-12\right)} = I$

   Chainsaw:

   $10^{\left(\frac{120}{10}-12\right)} = I$
   $10^{(12-12)} = I$
   $10^{(0)} = I$
   $1 = I$

   Normal Conversation:

   $10^{\left(\frac{60}{10}-12\right)} = I$
   $10^{(6-12)} = I$
   $10^{(-6)} = I$
   $0.000001 = I$

   The ratio of the intensity of the sound produced by a chainsaw to the sound produced by normal conversation is:

   $\frac{\text{chainsaw}}{\text{normal conversation}} = \frac{1}{10^{-6}} = 10^6.$

   So the sound produced by a chainsaw is 10^6, or 1,000,000 times as intense as the sound produced by normal conversation.

2. Sample answer: Because the decibel scale is based on base-10 logarithms, a change in $10x$ decibels corresponds to a change in intensity by a factor of 10^x.

   The sound produced by a running refrigerator is 50 decibels, and the sound produced by a jet engine taking off is 150 decibels, a difference of 100 decibels, or 10(10) decibels.

   So the change in intensity is a factor of 10^{10}, or 10,000,000,000. A jet engine taking off produces sound 10 billion times as intense as a running refrigerator.

Modeling Project Noise Pollution

Noise pollution is a major issue in cities and towns around the country. Noise pollution is associated with certain illnesses, fatigue, and even hearing loss. Many places have passed laws limiting noise levels.

There may be sources of noise pollution at your school, home, or community that could be distracting or unhealthy. For this chapter's modeling project you will develop a report to the principal about sources of noise pollution and ways to lessen its impact on your school. You will need to back up your arguments with data and research about the cause of the noise pollution and what levels of noise are distracting or unsafe.

Look (and listen) around your school for sources of noise pollution. These could be the cafeteria, music room, or traffic on a nearby street.

Create a list of the sources of noise pollution you think are the worst.

Use a decibel meter at various distances to measure the noise of these objects. Record the data.

Use your library or the Internet to research about safe, unsafe, and distracting noise levels. Find out how far from these noise sources you have to be for them to be safe and non-distracting.

Develop a report for your principal about noise pollution at your school. Your report should contain recommendations for limiting noise pollution that are supported by data in tables and graphs. Your report should also summarize the research you did.

Practice Problems

1. Recall the migration problem in Section 6.2. In 1988, the population estimate was between 15 and 20. In 1993, the population estimate was between 25 and 30.

   a) Explain how the migration rate of two moose per year was derived from these observations.

   b) The reason for using the average migration rate is that it predicts "typical" results. Concern about overpopulation might lead a mathematical modeler to use the largest migration rate instead. Based on these observations, what would be the largest rate?

   c) Estimate the largest possible moose population in the park in the year 2000 based on the observations. (We still assume that the only change in population is due to migration.)

2. In Section 6.2 a moose model was used to predict the population in the year 2000. Use the same initial population of 27 moose in 1993, and assume the growth rate is 10% per year on average.

   a) What would be the population for the year 2000 based on this set of assumptions?

   b) What would have been the yearly percentage increase if the same population reached 75 moose in the year 2000, as shown in **Figure 6.53?**

3. Reports show that satellite-based remote sensors measured a 6% decline in the area of ice covering the earth's surface during the period from 1978 to 1994.

   a) What is the overall rate of decay over the 16 years?

Year	Population
1993	27
1994	31
1995	36
1996	42
1997	48
1998	56
1999	65
2000	75

FIGURE 6.53.
Moose population data.

Practice Problems

Answers

1. a) Using either the high or low estimates for each year, the change in population (ΔP) is 10 moose over a 5-year period. This would average to 2 moose per year.

   b) The largest change in population would occur if there were actually 15 moose in 1988, and 30 moose in 1993. Then, $\Delta P = 15$ moose (still over a 5-year period), but the migration rate is now an average of 3 moose/year.

   c) Working with the assumptions that the starting population is 15 moose in 1988, and the moose are migrating at 3 moose/year, the population in the year 2000 would be calculated as: $P = 15 + 3(12) = 51$ moose.

2. a) $P = 27(1.10)^7 \approx 53$ moose

   b) The yearly percentage increase is the same as the mean successive quotient, which is about 15.7%.

3. a) The overall growth factor is 94%, or 0.94, since it went down by 6% over the 16-year period.

b) **Figure 6.54** shows the percentage of the Earth's surface covered with ice in 1978 that was still covered with ice in the given year. What is the annual rate of decay?

Year	Percentage of Ice Coverage	Year	Percentage of Ice Coverage
1978	100%	1986	97.0
1979	99.6	1987	96.6
1980	99.2	1988	96.2
1981	98.8	1989	95.8
1982	98.5	1990	95.5
1983	98.1	1991	95.1
1984	97.7	1992	94.7
1985	97.3	1993	94.4
		1994	94.0

FIGURE 6.54.
Ice coverage data.

c) Find the annual percent of decrease in the ice coverage.

d) Use 1978 as year zero and assume the amount of ice coverage to be 100% at that time. Write an equation for the percentage of ice coverage remaining after t years.

e) Graph your equation over a large domain (several centuries). If the model holds true well into the future, what is the long-term result of this decline in the ice coverage?

4. An article reported that the buffalo population in a national park grew as shown in **Figure 6.55.**

Year	Population
2000	125
2001	160
2002	205
2003	260
2004	340

FIGURE 6.55.
Buffalo population data.

a) Find the annual growth rate for the buffalo population over that time period. Explain what it means.

b) Write a function rule to predict the population of buffalo in the park over any time period. (Use the year 2000 to represent $t = 0$.)

c) Verify your function using the data from 2000 to 2004. How well does your function predict population?

d) Use the function developed in b to predict how many buffalo will be in the park in 2010.

5. The 2000 Census determined the population used to allocate seats in Congress. As of April 1, 2000, the population of the United States was around 281,422,000. That was a 13.2% increase over the 1990 census figure. This is an average annual increase of about 1.25%.

a) Write a function rule that could be used to predict the population of the United States after the year 2000 (assume a constant growth rate).

b) At the same rate of increase, what would be the population of the United States on April 1, 2003?

c) If the population increases at a yearly rate of 1.00% instead, what effect would that have on the population estimate from b?

6. A trout farm is estimated to have around 5000 fish when it opens. Each day approximately 2% of the fish in the lake are caught. At the end of each week the lake is restocked with 1200 fish.

a) How many fish are caught by the end of the first week?

b) Will the lake run out of fish? Explain.

7. Suppose the authorities at Adirondack State Park had counted 27 moose in 1988 and 42 moose in 1993.

a) If you assume that the growth is caused by migration, what is the migration rate (average number of moose migrating each year)?

b) At the same migration rate, how many moose would be in the park in the year 2013?

Answers (continued)

b) The annual rate of decay is the same as the mean successive quotient, which is about 99.6%.

c) $1 - 0.99614 \approx 0.00386 = 0.386\%$

d) $P_t = 100(0.99614)^t$

e) The calculator display would look like this over the domain [0, 200] × [0, 100]:

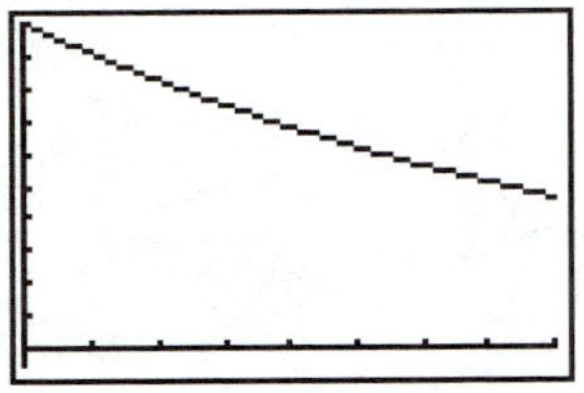

After two centuries, there would be about 40% of the ice cap remaining.

4. a) The growth rate is the same as the mean of successive quotients, which is 1.28 or 28%, meaning that the population increased by about 28% per year on average.

b) The model is: $P = 125(1.28)^t$.

c) Using $t = 4$ years, $P = 125(1.28)^4 \approx 336$ buffalo. It is supposed to predict 340 buffalo, so it is fairly accurate.

d) Substitute $t = 10$ years into the equation $P = 125(1.28)^t$.

$P = 1\ 25(1.28)^{10} \approx 1476$ buffalo

5. a) $P = 281{,}422{,}000 \cdot (1.0125)^t$, where t represents the number of years since 2000.

b) $P = 281{,}422{,}000 \cdot (1.0125)^3 \approx 292{,}108{,}000$.

c) At an annual rate of increase of 1% per year, the population estimate would be around 289,949,000. So, that slight change in the growth rate would affect the projected totals by about 2,159,000 people!

6. a) $5000 - 5000(0.98)^7 \approx 659$ fish

b) No, since the amount added to the population exceeds what was taken out.

7. a) The change in population is 15 moose over a 5-year period. So the migration rate is 3 moose/year.

b) After 20 years, the population would be $42 + 3(20) = 102$ moose.

c) If you assume that the growth is caused by reproduction, then you can calculate an annual growth factor of 1.0924. What is the annual percentage increase?

d) Let t be the number of years since 1993. Assuming a constant growth rate write an exponential function to predict the number of moose in the park.

e) Based on your answer to d predict the number of moose in the park in the year 2013.

f) How many years would it take for the population of moose to reach 84 (double the original amount)?

8. As regional director of the Fish and Game Department you estimate that there are 6000 deer in the area, and their population grows at a rate of 9% per year. To best manage this resource you would like to keep the population constant. How many deer should you allow to be harvested by hunters next year?

Glossary

KEY CONCEPTS

Base: In an exponential expression, the base is the number that is multiplied by itself. For example, in 2^3, 2 is the base because $2^3 = (2)(2)(2)$.

Common multiplier (common ratio): When multiplication by a constant is used to get successive values, the constant is called the common multiplier.

Exponent: Number that indicates how many times the base is multiplied. In 2^4 2 is multiplied four times, $2 \times 2 \times 2 \times 2 = 16$.

Exponential decay: Negative exponential growth; a growth factor in which $0 < b < 1$.

Exponential function: A function in the form $f(x) = a \cdot b^x$.

Exponential growth: A growth pattern in which multiplication by a constant larger than 1 generates successive values

Growth factor: The common multiplier in an exponential growth or decay situation.

Logarithm: A term used to describe an exponent. For example, in 2^3, 3 is an exponent and therefore a logarithm. 2^3 can be written as $\log_3(2)$.

Logarithmic function: A function of the form $y = \log(x)$. The inverse of an exponential function is a logarithmic function.

Migration: Process by which a population changes by moving from one location to another.

Natural logarithm: A logarithm with a base of e, which has a value of about 2.718.

pH: A scale used to measure the level of acidity in a substance.

Power: The value of an exponential expression. For example, in $2^3 = 8$, 8 is called the third power of 2.

Reproduction: Process by which a population (wildlife or people) increases by births.

Richter scale: A scale invented by Charles Richter used to measure and compare the strength of earthquakes.

Answers (continued)

c) 9.24%

d) $P = 42(1.0924)^t$

e) After 20 years at that same rate, the population would be $42(1.0924)^{20}$, or about 246 moose.

f) Using the closed-form for an exponential equation, $84 = 42(1.094)^t$. Dividing both sides by 42, you get: $2 = 1.0924^t$, which is solved by calculating $\log(2) \div \log(1.0924) \approx 7.84$ years, or a little more than 7 years 10 months.

8. If hunters were allowed to randomly select the deer to be killed, then 9% of 6000 = 540 deer tags should be issued. Complicating the problem is the fact that deer hunting is restricted to the male population in most areas; allowing that many males to be killed might affect the growth rate for the following year.

7 CHAPTER

Money and You: Mathematics and Finance

PREPARATION READING

Meet the Gonzales Family

Alfredo and Wilma Gonzales have been married for a long time. They both work and bring home paychecks.

Mr. and Mrs. Gonzales have three children. Their oldest, Maria, is a sophomore in college. She lives with her parents and commutes to a college in her town in order to save money. College is expensive!

Paco is a senior in high school. He plays sports, drives a car, and enjoys spending time with his friends on the weekend. He can't wait for graduation. He's planning to go to a community college for his basic coursework and then transfer to a university to major in business. He wants to go to law school after that.

Tico is a freshman in high school. He struggled in middle school last year, and his parents are grateful that he is going to a new school this year. Tico knows a lot about sports cars and can tell you how much a mint-condition 1968 Ford Mustang is worth. He doesn't know what he wants to do after high school. He is planning to spend the next four years having fun and learning what he can.

Mr. and Mrs. Gonzales have their hands full! With a daughter in college and two teenage sons, they must manage their money with care. Since they have been married for a long time, they are no strangers to careful budgeting. They make their decisions together. But sometimes, they need to consult financial planners to help them make decisions.

Financial planners spend their time studying money. They know what factors cause you to spend or lose money and what factors will help you save or earn more money. Financial planners can help a family save and plan for retirement. They can also help the family learn how to reduce expenses in order to make the most of their budget.

CHAPTER 7

Managing Your Money: Mathematics and Finance

Preparation Reading:
Meet the Gonzales Family

Description: The reading introduces the major context of the chapter, how the Gonzales family uses mathematical modeling to help make financial decisions. Students read about the family and some of the issues that they face when making decisions about their income: how they manage and invest their money, and how they use credit and financing to purchase or lease cars or homes. In the rest of the chapter, students use linear and exponential functions to model financial situations.

Materials: none

Teaching Ideas: Students can work in small groups (3 to 4 students). After the reading, ask students to describe and discuss the kinds of financial decisions that they and their families must make in a given day. Ask about budgeting: How much of their money do they think that a family must spend on housing, clothing, food, energy, etc.? Use the discussion about the Gonzales family's experience as a springboard for an inquiry into financial mathematics.

In this chapter you will act as a financial planner for the Gonzales family. You will follow the five members of the family as they encounter different situations where they earn or spend money. You will develop and use mathematical models to describe the patterns that occur in finance. Then you will use these models to help the Gonzales family make decisions.

SECTION 7.1 Keep it Simple

In this section you will begin to examine personal finance. When it comes to finances, there are many types of accounts and a variety of programs for handling your money. The Gonzales family uses Lone Star Bank, which offers several options.

Financial advisors always suggest that people keep a little money in easy reach, in case of an emergency. For interested customers, banks offer a free safe-deposit box where money can be kept under lock and key.

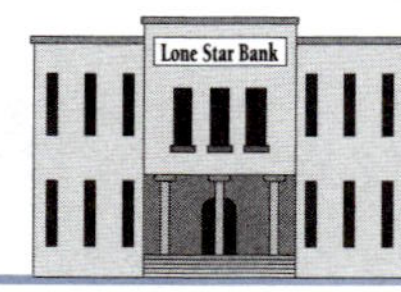

1. Lone Star Bank offers safe-deposit boxes for its long-time customers. Mr. Gonzales decided to put $200 in a safe-deposit box and leave it there. How much money would be in the safe-deposit box after one week? 10 weeks? What assumptions are you making about the situation?

2. How much does the amount of money change each week?

Putting your money in a safe-deposit box will not earn you any interest. Let's see what happens when the assumptions change slightly.

To encourage customers to deposit money in a savings account, the Lone Star Bank promotes this New Saver's account:

For every $1000 deposited for an entire year, the bank will pay the customer $50 in interest.

CHECK THIS!

The assumption of the New Saver's account is that interest is paid at a rate of 5% per year and only on the principal, or amount deposited. For each $100 you deposit, the interest you will receive is 5% of $100 or $5 each year that the money is in the account. This interest accrues, which means is paid, usually at the end of each year. When interest is calculated only on the principal, it is called simple interest.

ENGAGE LESSON 1

Section 7.1: Keep It Simple

Purpose: In previous chapters students learned the properties of linear, quadratic, and exponential functions. In this chapter students apply properties of functions in a financial context. Section 7.1 provides an opportunity to apply the concept of linear functions to an interest bearing savings account, preparing students to investigate other personal finances in Sections 7.2–7.5.

Description: Students will build a linear model to determine simple interest. Then they identify the meaning of the slope and *y*-intercept of the model in the context of the problem. Then students use the model to calculate simple interest for a savings account.

Materials: graphing calculator

Teaching Ideas: Students should work in small groups (3 to 4 students). You may need to review using the process column and first differences to determine a linear function from a table of values.

Answers

1. There would still be $200 in the box after one week or 10 weeks. It is assumed that no money is added to, or removed from, the safe-deposit box.
2. The change in the amount of money each week (ΔA) is $0.

3. Suppose Mr. Gonzales deposits $1000 in a New Saver's account. (That is, the original balance, or **principal**, is $1000.) Complete a table like **Figure 7.1** to find the **balance**, or how much money is in the account, at the end of each year for five years.

Year Number	Interest	Total Balance
0	$0	$1000
1		
2		
3		
4		
5		

FIGURE 7.1.
Account balance for five years.

4. What is a reasonable domain for this situation? Why?
5. What is a reasonable range for this situation? Why?
6. Create a scatter plot of account balance versus number of years. Record your viewing window and sketch your graph.
7. Based on your graph, what type of relationship does this appear to be?
8. How can you use your table to verify your answer to Question 7?
9. Use the process column in a table like **Figure 7.2** to write a function rule that models this situation.

Number of Years (n)	Process	Account Balance (A)
0	$1000 = 1000 + 50\,(0)$	
1	$1000 + 50 = 1000 + 50\,(1)$	
2		
3		
4		
5		
n		A

FIGURE 7.2.
Table with process column.

10. Test your function rule by graphing the rule on your scatter plot. If the rule is not a good fit, use transformations to modify it. Sketch your graph.
11. In the context of this problem, what is the meaning of the y-intercept of your function?
12. In the context of this problem, what is the meaning of the slope of your function?
13. If there are no deposits or withdrawals, how much money will be in the account at the end of 10 years?
14. If there are no deposits or withdrawals, how many years will it take for Mr. Gonzales' money to triple?
15. If there are no deposits or withdrawals, how many years will it take to have $2250.00 in the account?
16. Suppose instead of earning $50 in interest, Mr. Gonzales earns $65 in interest for every $1000 deposited for an entire year. How will this increase affect the graph of the function?

Answers (continued)

3. Answer:

Year Number	Interest	Total Balance
0	$0	$1000
1	$50	$1050
2	$100	$1100
3	$150	$1150
4	$200	$1200
5	$250	$1250

4. Since the domain is the number of years, a reasonable domain is whole numbers from 0 to 6.

5. Since the range is the account balance, a reasonable range is from 1000 to 1500.

6.

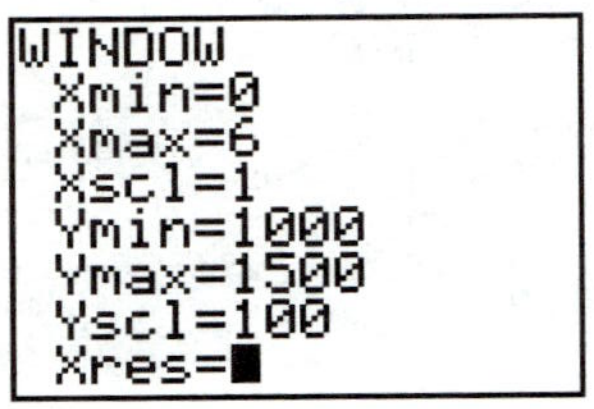

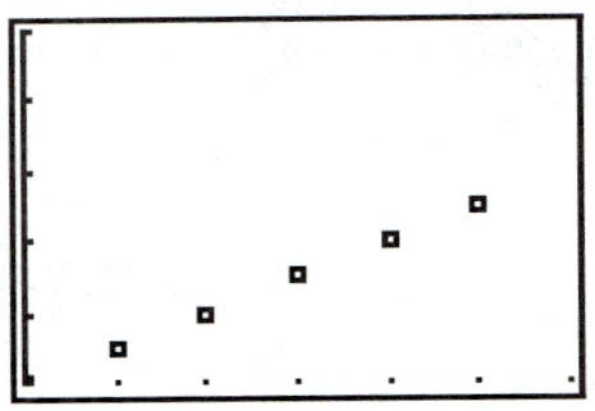

7. The graph appears to be a linear relationship.

8. Since first differences are constant, this is a linear relationship.

	Number of Years (n)	Account Balance (A)	
	0	$1000	
+1	1	$1050	+50
+1	2	$1100	+50
+1	3	$1150	+50
+1	4	$1200	+50
+1	5	$1250	+50

9. $A = 1000 + 50x$

Number of Years (n)	Process	Account Balance (A)
0	1000 = 1000 + 50 (0)	$1000
1	1000 + 50 = 1000 + 50 (1)	$1050
2	1000 + 50 + 50 = 1000 + 50 (2)	$1100
3	1000 + 50 + 50 + 50 = 1000 + 50 (3)	$1150
4	1000 + 50 + 50 + 50 + 50 = 1000 + 50 (4)	$1200
5	1000 + 50 + 50 + 50 + 50 + 50 = 1000 + 50 (5)	$1250
n	1000 + 50n	A

10.

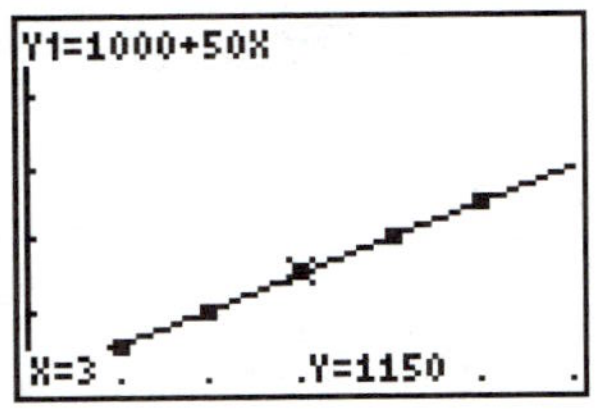

11. The y-intercept is the beginning balance of $1000.

12. The slope of the line is the $50 per year in interest.

13. In 10 years there will be $1500.00 in the account.

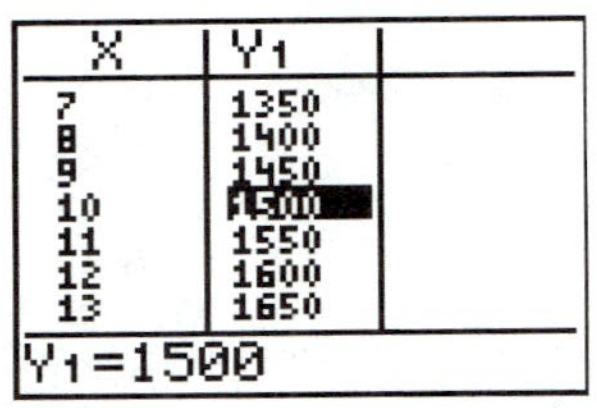

14. There will be $3000.00 in the account after 40 years.

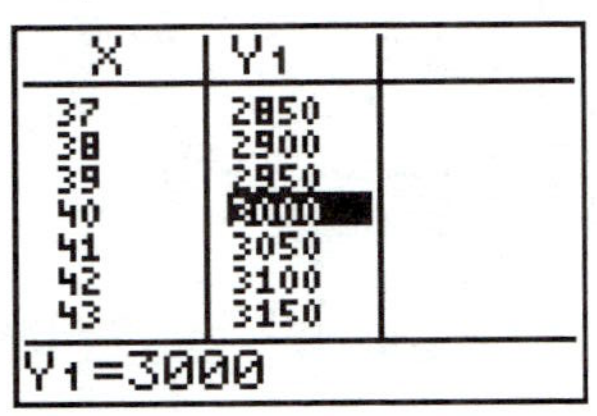

15. It will take 25 years to have $2250.00 in the account.

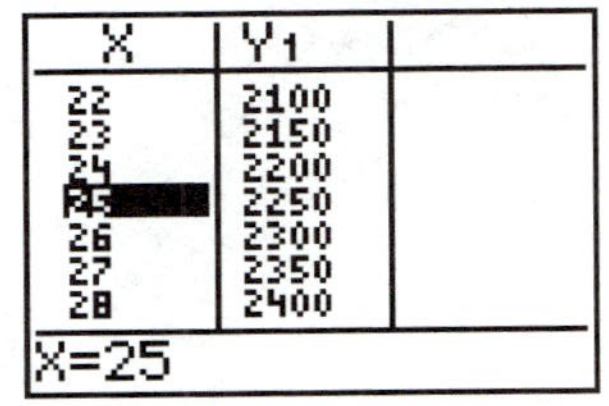

16. The increase in interest affects the slope of the line. The line will be steeper but have the same y-intercept.

SECTION 7.2

Taxing Your Earnings

Paco took a job last summer to earn money to spend in his senior year. When he got his first paycheck, he noticed it was less than his total earnings for the week. Money was deducted (subtracted) from his earnings before he got his paycheck!

In this section you will examine some of these deductions and learn how they are calculated.

Let's calculate earnings for different pay periods.

DETERMINING EARNINGS

Hourly Earnings	Weekly Earnings	Monthly Earnings	Yearly Earnings
$6.50			
$10.75			
$17.50			
	$350.00		
	$715.00		
	$1400.00		
		$1250.00	
		$1855.00	
		$2000.00	
		$3333.33	
			$15,000.00
			$23,000.00
			$32,000.00
			$54,000.00

FIGURE 7.3.
Table for calculating earnings.

1. Complete the table in **Figure 7.3.** Be sure to use these conversions:

   40 hours = 1 work week

   52 weeks = 1 year

   12 months = 1 year

2. Explain the process you used to complete the table.
3. What is the mean of the hourly wages?
4. Are the above hourly wages a good indicator of how much the average worker makes in the United States? Why or why not?

5. Listed in **Figure 7.4** are the 2004 yearly earnings for some well-known people. Complete the table to find their hourly wages.

Person	2004 Yearly Earnings	Hourly Wages
George Lucas	$200 million	
Oprah Winfrey	$150 million	
Tiger Woods	$69 million	
Adam Sandler	$47 million	
Madonna	$43 million	
Britney Spears	$39.2 million	
Derek Jeter	$18.5 million	
Sammy Sosa	$18 million	

Source: Forbes Magazine on-line www.forbes.com

FIGURE 7.4.
Yearly salaries for some well-known people.

6. What is the mean hourly wage for the people in Figure 7.4?

FEDERAL WITHHOLDING TAX

According to the Internal Revenue Service (IRS), about 37% of the income for the United States government in fiscal year 2003 came from personal income taxes. The amount of income tax that an employee pays depends on the amount of money earned in a year.

It can be hard to find the cash to pay your annual taxes all at once. The IRS requires employers to withhold money from paychecks in order to spread the tax burden over time.

The amount of withholding, or the money that an employer takes out of an employee's paycheck, can be found using a mathematical formula or a **tax withholding table.**

Many factors go into determining the amount of taxes that someone must pay. The primary factor is personal income. Another factor is marital status.

EXPLORE LESSON 1

Section 7.2: Taxing Your Earnings

Purpose: In Section 7.1 students built a linear function to investigate simple interest as a form of personal income. In this section students continue their study of personal finance by investigating income, gross pay, and deductions such as federal income taxes, social security tax, and medicare tax. In Section 7.3 students formalize methods for calculating taxes and determining net pay.

Description: In this section students investigate ways to calculate hourly, weekly, monthly, and yearly earnings. Students examine federal income tax tables and use them to determine withholding tax. They also investigate the effect of withholding allowances on the amount of tax withheld. Percents used in the calculation of social security tax and medicare tax will also be explored.

Materials: graphing calculator, tax table Handout 7.1

Teaching Ideas: Students can work in small groups (3 to 4 students). Encourage students to examine the tax tables carefully to look for patterns. Prompt them to think about their own experiences with gross pay and net pay. Those who have held part-time or summer jobs can share their experiences concerning deductions.

Answers

1. Answer:

Hourly Earnings	Weekly Earnings	Monthly Earnings	Yearly Earnings
$6.50	$260.00	$1126.67	$13,520.00
$10.75	$430.00	$1863.33	$22,360.00
$17.50	$700.00	$3033.33	$36,400.00
$8.75	$350.00	$1516.67	$18,200.00
$17.88	$715.00	$3098.33	$37,180.00
$35.00	$1400.00	$6066.67	$72,800.00
$7.21	$288.46	$1250.00	$15,000.00
$10.70	$428.08	$1855.00	$22,260.00
$11.54	$461.54	$2000.00	$24,000.00
$19.23	$769.23	$3333.33	$39,999.96
$7.21	$288.46	$1250.00	$15,000.00
$11.06	$442.31	$1916.67	$23,000.00
$15.38	$615.38	$2666.67	$32,000.00
$25.96	$1038.46	$4500.00	$54,000.00

2. Answers may vary. Students should discuss when they multiplied or divided and why? For example to find hourly earnings, divide weekly earnings by 40 hours per week.

3. $14.62

4. Answers may vary. Since the yearly salaries range from $13,520.00 to $72,800.00, $14.66 may be a good indicator of how much the average worker makes in the U.S. NOTE: According to the U.S. Department of Labor: Bureau of Labor Statistics, the average hourly wage in the U.S. for 2003 was $17.75.

5. Answer:

Person	2004 Yearly Earnings	Hourly Wages
George Lucas	$200 million	$96,153.85
Oprah Winfrey	$150 million	$72,115.38
Tiger Woods	$69 million	$33,173.08
Adam Sandler	$47 million	$22,596.15
Madonna	$43 million	$20,673.08
Britney Spears	$39.2 million	$18,846.15
Derek Jeter	$18.5 million	$8894.23
Sammy Sosa	$18 million	$8653.85

6. $35,138.22

Use the federal income tax withholding tables to answer these questions.

7. Determine the effect of marital status on federal income tax withheld. Using a withholding allowance of 1, find the amount withheld for a single person and a married person for each wage in **Figure 7.5.**

Monthly Income	Single Withholding	Married Withholding
$500		
$1000		
$1500		
$2000		
$2500		
$3000		
$3500		
$4000		
$4500		

FIGURE 7.5.
Table for Question 7.

8. What can you conclude?

9. On the same set of axes, make a scatter plot of single withholding versus monthly income and a scatter plot of married withholding versus monthly income. Record your window and sketch your graph.

10. What do you notice in the graph?

11. Who has less money withheld, a single person making $3000.00 a month or a married couple making a total of $3000.00 a month? Explain your thinking.

12. Mr. Gonzales earns $1500 per month and Mrs. Gonzales earns $3000. If they were both single, what would be their combined withholding? Since they are married, what is the withholding on their total earnings of $4500? How do these compare?

13. Make a general statement about withholding amounts for married people compared to single people. Explain your thinking.

THE EFFECT OF ALLOWANCES

Another factor that affects withholding is the number of dependents you have. The IRS allows you to reduce your withholding by a certain amount for each dependent. The IRS uses **allowances** to estimate your withholding. In general you can claim one allowance for yourself, one for your spouse, and one for each of your dependents.

CHECK THIS!
Withholding allowances are claimed on a W-4 form. They are mainly used to assist an employer in calculating the amount to withhold from an employee's paycheck.

CHECK THIS!
The more allowances you claim, the less income tax is withheld from your paycheck.

Let's explore the effect of withholding allowances. Use federal withholding tables to find the amount withheld for married people in **Figure 7.6.**

Monthly Salary	Withholding for Number of Allowances					
	0	2	4	6	8	10
$500						
$1000						
$1500						
$2000						
$2500						
$3000						
$3500						
$4000						
$4500						
$5000						

FIGURE 7.6.
Monthly salaries for finding withholding tax.

14. What can you conclude?

15. Make scatter plots of withholding for zero allowances versus monthly salary, withholding for two allowances versus monthly salary, and withholding for four allowances versus monthly salary on the same set of axes. Record your window and sketch your graph.

Answers (continued)

7. Answer:

Monthly Income	Single Withholding	Married Withholding
$500	2	0
$1000	53	9
$1500	122	57
$2000	200	109
$2500	272	174
$3000	363	252
$3500	483	324
$4000	613	402
$4500	733	474

8. Answers may vary. The withholding for single appears to be more than the withholding for married. The rate of change for single increases while the rate of change for married is fairly constant.

9.

```
WINDOW
 Xmin=0
 Xmax=5000
 Xscl=1000
 Ymin=0
 Ymax=800
 Yscl=100
 Xres=█
```

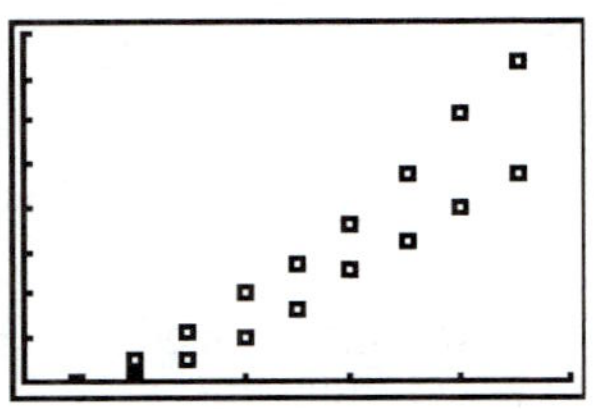

10. The graph for single withholding appears to curve while the graph for married appears to be more linear.

11. The single person has more withheld. A single person making $3000 per month has $363 withheld. A married couple making a total of $3000 per month has $252 withheld.

12. If filing single, the withholding for Mr. Gonzales is $122. The withholding for Mrs. Gonzales is $363 for a total of $485. The withholding for $4500 for married couples is $474. Being married saves them $11 a month.

13. In general, withholding will be less for married couples.

Answer for Figure 7.6.

Monthly Salary	Withholding for Number of Allowances					
	0	2	4	6	8	10
$500	0	0	0	0	0	0
$1000	35	0	0	0	0	0
$1500	83	30	0	0	0	0
$2000	142	82	29	0	0	0
$2500	214	134	77	23	0	0
$3000	292	212	132	75	22	0
$3500	364	284	204	124	70	17
$4000	442	362	282	202	122	69
$4500	514	434	354	274	194	117
$5000	592	512	432	352	272	192

14. Answers may vary. As the number of allowances increases, the amount withheld decreases.

15.

```
WINDOW
 Xmin=0
 Xmax=5000
 Xscl=1000
 Ymin=0
 Ymax=600
 Yscl=100
 Xres=█
```

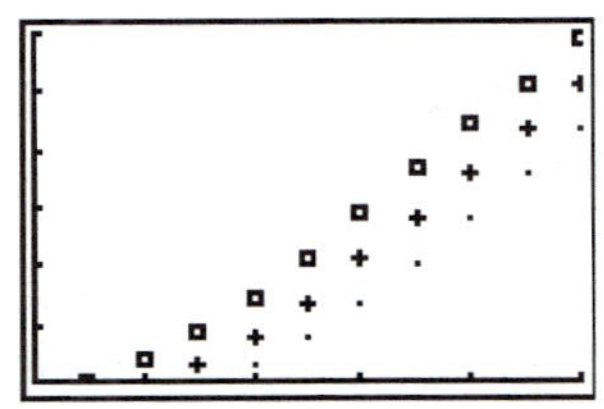

16. Make scatter plots of withholding for six allowances versus monthly salary, withholding for eight allowances versus monthly salary, and withholding for 10 allowances versus monthly salary on the same set of axes. Record your window and sketch your graph.

17. What do you notice in the graphs?

18. What is an estimate of the withholding for a married person with an income of $6000 claiming 10 allowances? Explain how you got your answer.

19. What is an estimate of the withholding for a married person earning $5000 claiming 12 allowances? Explain how you got your answer.

20. Federal withholding is not the only thing deducted from your paycheck. What are some other things?

After all deductions are made, the amount of money that remains is called your **net pay.**

Maria earned $2695.40 in February of 2005. When she received her paycheck, her net pay was much less. Along with federal withholding tax based on zero allowances, the items listed in **Figure 7.7** were deducted.

21. How much was her net pay?

22. Medicare tax and social security tax are each a certain percent of a person's income. Use Figure 7.7 and Maria's income of $2695.40 in February of 2005 to find each of these.

Deduction	Amount
Federal Withholding	?
Medicare	39.08
Health Insurance	121.23
Social Security	167.11
Life Insurance	5.65

FIGURE 7.7.
Deductions from Maria's pay.

a) What percent of her income is deducted for medicare?

b) What percent of her income is deducted for social security?

SECTION 7.2 Assignment

1. In Section 7.1 the savings account plan earned $50 in interest for every $1000 kept in the account for an entire year. The assumption was that the interest was paid at the rate of 5% per year. This kind of savings account pays **simple interest** because the interest calculation is based on the principal only.

   a) Use the 5% interest rate to find the interest earned when $1000 is deposited for one year.

   b) How much interest would the bank have paid for one year, at that same rate, if only $600 was put into the account?

   c) At that same rate, if $600 were put into the account for five years, how much interest would the bank pay?

   d) Suppose the $600 was put into an account for five years, but the simple interest rate was 4% per year. How much interest would the bank pay? At the end of the five years, what would be the balance in the account?

2. Translate the work done in Question 1 into equations that represent simple interest calculations. Let P be the principal. Let t be the number of years the money is in the account, and let r be the rate (expressed as the decimal).

   a) Write an equation to calculate the amount of interest, I, that the bank must pay.

   b) Write an equation to calculate the balance, B, in the bank at the end of that time period. (Assume interest is deposited into the account.)

3. If you deposit $1250 into a savings account that pays at a simple interest rate of 6.2%, what will be the balance after three years?

Answers (continued)

16.

```
WINDOW
 Xmin=0
 Xmax=5000
 Xscl=1000
 Ymin=0
 Ymax=600
 Yscl=100
 Xres=■
```

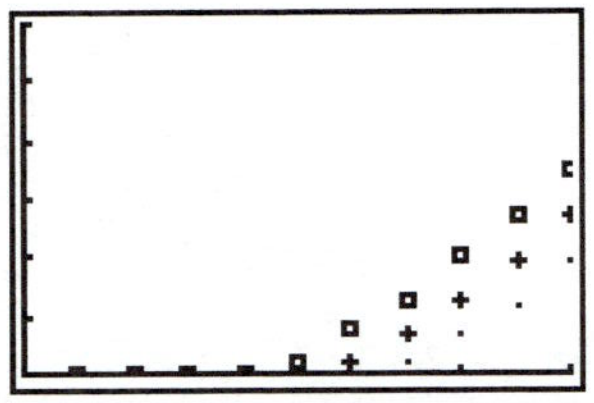

17. The graphs appear to be non-linear and confirm that as the number of allowances increases the amount withheld decreases.

18. About $350. One possible solution strategy is to follow the diagonal pattern shown in the table.

Monthly Salary	Withholding for Number of Allowances 0	2	4	6	8	10
$500	0	0	0	0	0	0
$1000	35	0	0	0	0	0
$1500	83	30	0	0	0	0
$2000	142	82	29	0	0	0
$2500	214	134	77	23	0	0
$3000	292	212	132	75	22	0
$3500	364	284	204	124	70	17
$4000	442	362	282	202	122	69
$4500	514	434	354	274	194	117
$5000	592	512	432	352	272	192
$5500						
$6000						350

19. About $115. Follow the diagonal pattern: 124, 122, 117, ….

Monthly Salary	Withholding for Number of Allowances 0	2	4	6	8	10
$500	0	0	0	0	0	0
$1000	35	0	0	0	0	0
$1500	83	30	0	0	0	0
$2000	142	82	29	0	0	0
$2500	214	134	77	23	0	0
$3000	292	212	132	75	22	0
$3500	364	284	204	124	70	17
$4000	442	362	282	202	122	69
$4500	514	434	354	274	194	117
$5000	592	512	432	352	272	192
$5500						
$6000						350

115

20. Answers may vary. From personal experience, students should realize that things like social security and medicare tax are deducted from their gross pay.

21. Maria's federal withholding tax is $348.40. The total of her other deductions is $333.07. Subtracting federal tax and other deductions from her gross pay of $2695.40 leaves $2013.93 net pay.

22. a) 1.45%

    b) 6.2%

Section 7.2 Assignment

Answers

1. a) 5% of 1000 = (0.05) * (1000) = $50

   b) 5% of 600 = (0.05) * (600) = $30

   c) 5% of 600 = $30/year * 5 yrs. = $150

   d) The interest is calculated in this way: $I = (\$600)(0.04)(5\text{ yrs}) = \$120$. The balance would be: $600 + $120 = $720.

2. a) $I = P \cdot r \cdot t$

   b) $B = P + I = P + P \cdot r \cdot t$. Alternatively, $B = P(I + r \cdot t)$.

3. Using the formula developed in Question 10: $B = 1250 + 1250 * (0.062) * (3) = \$1482.50$.

4. The Lone Star Bank features a Christmas savings account that pays a 6% simple interest rate per year, but the account can only stay open for one year. If you open the account with $500 and deposit $300 more half a year later, how much money will be in the account when it closes?

5. According to the U.S. Bureau of Labor Statistics, the average hourly wage in the United States for 2003 was $17.75. If you earn this amount find each of these:

   a) Weekly earnings

   b) Yearly earnings

   c) Monthly earnings

6. Shirley earned a total $24,750 in 2005.

   a) What was her monthly salary?

   b) If Shirley claims one allowance, what is the amount withheld each month according to the tax tables for single persons?

   c) If Shirley claims four allowances, what is the amount withheld from her pay each month?

7. The amount of income tax withheld from Ricardo's paycheck each month is $82. He claims five allowances. What is his approximate yearly income?

8. Martin earned $34,000 in 2005. He claimed no allowances and had money withheld according to the 2005 tax tables. At the end of the year when he filled out his tax return he calculated his tax to be $3840. How much money should he expect as a tax refund?

9. Huang earned $25,000 in 2005. He claimed five allowances and had money withheld according to the 2005 tax tables. At the end of the year when he filled out his tax return he calculated his tax to be $2544. How much more money does he owe in taxes?

SECTION 7.3 You Earned It!

In Section 7.2 you were given hourly, weekly, monthly, or yearly earnings. Explain how you found each of these.

1. Weekly earnings when you were given hourly earnings
2. Yearly earnings when you were given hourly earnings
3. Weekly earnings when you were given yearly earnings
4. Hourly earnings when you were given monthly earnings
5. Monthly earnings when you were given yearly earnings

Figure 7.8 lists the yearly salaries for six people in a book study group.

Name	Occupation	Yearly Salary
Duc Lee	Chemical Engineer	87,000
Samantha Rivers	Cashier	15,246
Silvia Rodriquez	Medical Doctor	150,000
Tammera Jones	Veterinarian	130,000
Jose Rubio	Teacher	45,000
Oprah Winfrey	Talk Show Host	50 million

FIGURE 7.8. Yearly salaries for people in a book study group.

6. What is the mean of the yearly salaries?
7. Is the mean salary representative of this group? Why or why not?
8. What is the median of the yearly salaries?
9. What is the mode of the yearly salaries?
10. Of mean, median, and mode, which gives the best "typical" yearly salary for this group? Why?

Answers (continued)

4. You can treat this as two separate simple interest problems. The first deposit will pay an interest of ($500)(0.06)(1) = $30. The second deposit will pay ($300)(0.06)(0.5 yrs) = $9. The account balance will be the sum of all the deposits and interest earned: $500 + $300 + $30 + $9 = $839.
5. a) $710.00
   b) $36,920.00
   c) $3076.67
6. a) $2062.50
   b) $206.00
   c) $86.00
7. Between $27,360 and $27,840
8. His monthly wages were $2833.33. From the tax tables for single persons, his withholding was $380 each month. His withholding for the year is $380(12) = $4560. $4560 – $3840 gives a refund of $720.
9. His monthly wages were $2083.33. From the tax tables for single persons, his withholding was $55 each month. His withholding for the year is $55(12) = $660. $2544 – $660 = $1884. So he owes $1884.00 to the government.

EXPLAIN LESSON 1

Section 7.3: You Earned It!

Purpose: In Section 7.2 students continued their study of personal finance by investigating income, gross pay, and deductions such as federal income taxes, social security tax, and medicare tax. In Section 7.3 students formalize methods for calculating taxes and determining net pay. In Section 7.4 they apply tax calculation methods to determine net pay.

Description: In this phase of the lesson, students formalize methods for calculating social security and medicare taxes.

Materials: graphing calculators

Teaching Ideas: Debrief students after Section 7.2 by having them share their answers to Question 9 in Section 7.2 Assignment. This review will provide an opportunity to explain any missing concept, process, or skills. Then work through the questions in this section with the entire class.

For Questions 12 and 13, students will examine a tax formula table. This table, provided by the IRS with income tax form instructions, enables people to compute their monthly income tax based on their monthly earnings. Piecewise linear functions can be built for each tax bracket. Emphasize to students how to build these functions from the table using the initial value–rate of change approach. As an extension, you can examine the piecewise functions graphically using a graphing calculator and Boolean logic operators. Instructions for graphing these functions are given in the teachers' notes with Questions 12 and 13.

Answers

1. Answers may vary. Multiply hourly earnings by 40 hours per week.
2. Answers may vary. Multiply hourly earnings by 40 hours per week then by 52 weeks in a year.
3. Answers may vary. Divide yearly earnings by 52 weeks in a year.
4. Answers may vary. Multiply monthly earnings by 12 months in a year then divide the result by 52 weeks in a year to get weekly earnings. Divide weekly earnings by 40 hours in a week to get hourly earnings.
5. Answers may vary. Divide yearly earnings by 12 months in a year.
6. $8,404,541
7. No, because the mean of $8,404,541 is much higher than everyone's salary except Oprah Winfrey. Oprah Winfrey's salary dramatically alters the mean.
8. $108,500.00
9. There is no mode.
10. Median. When people like Oprah who earn much more that the general population are included in the calculation of the mean, the mean is artificially high. The median wage is a better indicator of hourly wages since very high or very low wages will have less effect on the median.

When you calculated earnings in Section 7.2, you were calculating what is called **gross pay.**

CHECK THIS!

Gross pay is your total pay for a pay period before deductions and taxes have been subtracted.

Federal withholding tax is deducted (subtracted) from your gross pay. In Section 7.2 you used tables to find the amount withheld based on monthly earnings. Let's examine a different way to calculate withholding.

Figure 7.9 explains how to calculate federal withholding tax for single persons.

11. What patterns do you observe in the table?

MONTHLY PAYROLL PERIOD				
Single Persons				
If the Taxable Income is:		Computed Tax is:		
Over	**But Not Over**	**The total of this column PLUS>>**	**The % of taxable income**	**The amount over** (Taxable income less the amount in this column)
0	221	NO TAX WITHHELD		
221	817	0	10%	221
817	2625	59.60	15%	817
2625	5813	330.80	25%	2625
5813	12,663	1127.80	28%	5813
12,663	27,354	3045.80	33%	12,663
27,354	OVER	7893.83	35%	27,354

FIGURE 7.9. Federal withholding tax table for wages paid January 1, 2005, and after.

12. Explain in your own words how to use the table.

Let's look at an example.

A single person who earns $3285.00 in a month falls between $2625 and $5813 (see **Figure 7.10**).

Over	But Not Over	The total of this column PLUS>>	The % in this column TIMES>>	The amount over (Taxable income less the amount in this column)
0	221	NO TAX WITHHELD		
221	817	0	10%	221
817	2625	59.60	15%	817
2625	5813	330.80	25%	2625
5813	12,663	1127.80	28%	5813

FIGURE 7.10. Withholding tax for a single person earning $3285 a month.

This means the withholding tax will be $330.80 plus 25% of the amount over $2625. Calculations are shown in **Figure 7.11.**

Over	But Not Over	The total of this column PLUS>>		The % in this column TIMES>>	The amount over (Taxable income less the amount in this column)
0	221	NO TAX WITHHELD			
221	817	0		10%	221
817	2625	59.60		15%	817
2625	5813	330.80	+	25% ($3285 –	2625) =
5813	12,663	1127.80		28%	5813

Tax = initial tax + tax rate(income – limit of previous bracket)
Tax = 330.80 + 0.25(3285.00 – 2625.00)
Tax = 330.80 + 0.25(660)
Tax = 330.80 + 165
Tax = 495.80

FIGURE 7.11. Calculations for withholding tax.

13. Use the information in Figure 7.11 to write a linear function that models the relationship between a person's monthly income, x, and computed tax, T, in the 25% tax bracket.

14. What are the domain and range for this situation?

15. How does $495.80 compare to the withholding for $3285 from the tables? Does $495.80 make sense? Why?

16. Use the method in Figure 7.11 to find the withholding for a single person who claims zero allowances and earns $2500.00 per month.

17. Compare your answer to Question 16 to the tax tables. What do you notice?

18. In Section 7.2 we learned that Oprah earned $150 million in 2004. If she claims zero allowances and earns the same amount in 2005, what is her monthly federal withholding tax?

SOCIAL SECURITY TAX AND MEDICARE TAX

When you work for someone, 6.2% of your wages is withheld as social security tax. Your employer also pays a 6.2% matching amount for social security programs. In 2004 the employee tax stopped after $87,900 of wages. Also, 1.45% of your wages is withheld, and the employer pays a matching 1.45% amount to the medicare program. All wages are subject to the medicare tax; there is no ceiling.

Answers (continued)

11. Answers may vary. The percent increases as the amount of money earned increases. The last column is the same as the first column.

12. Answers may vary. First find where the taxable income will fall in the table. For example, 3000 would be over 2625 but not over 5813. Next add the value in column three, in this case 330.80, to the percent in column four (25%) times the difference in the taxable income and the amount in the last column (2625). This process will give you the amount of withholding tax.

13. $T = 330.80 + 0.25(x - 2625)$

14. The domain is real numbers (rounded to the nearest hundredth) greater than 2625 and less than 5813. The range is real numbers (rounded to the nearest hundredth) greater than or equal to 330.80 and less than 1127.80.

    As an optional extension, encourage students to graph this and they will begin to see this becomes a piecewise function using their graphing calculator and Boolean logic:

    First enter the Function editor, then type the linear function from Question 13 inside parentheses. In this example, we want to graph the function $y = 330.80 + 0.25(x - 2625)$:

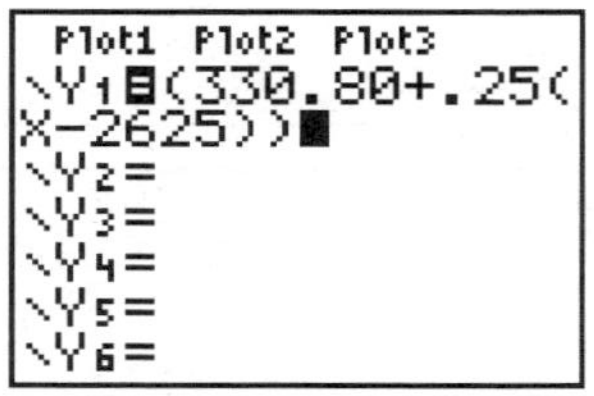

Next, after the last close parenthesis, press the division key, then inside another set of parentheses, enter the domain you want to use to generate your function. In this example, we want to graph the function over the domain $2625 < x < 5813$. So, we will enter "$2625 < x$ and $x < 5813$".

Type "/" then open the parentheses. Then type "2625":

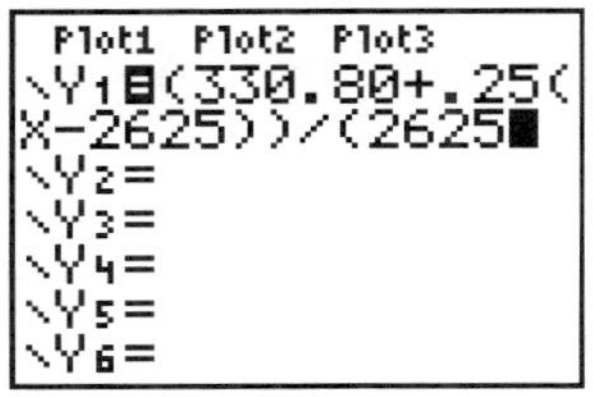

To obtain the < symbol, pull up the "TEST" menu using 2nd-MATH. Then scroll down to option 5, "<":

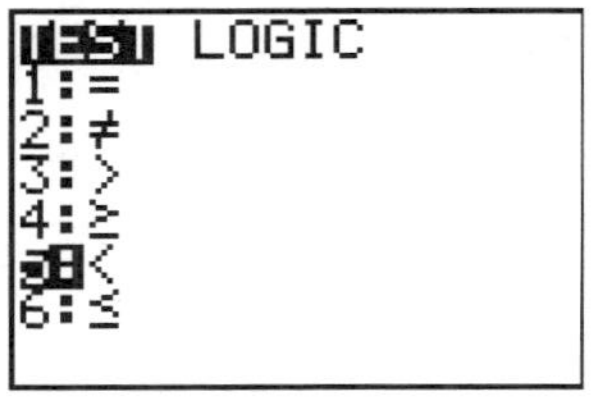

Next press ENTER, then type "X":

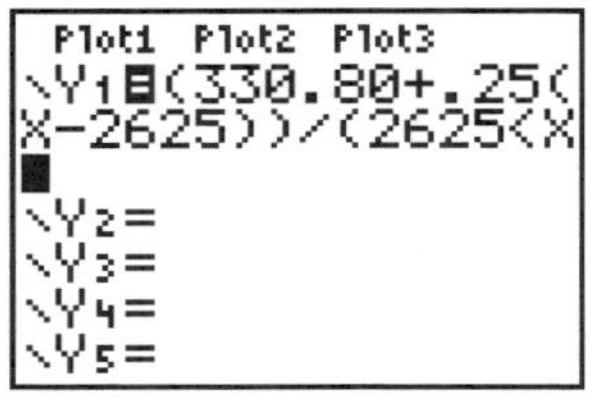

To obtain the "and" Boolean operator, pull up the "TEST" menu using 2nd-MATH, then use the right arrow key to scroll to "LOGIC". Choose option 1, "and":

Press ENTER, then enter the upper limit of the domain, "X<5813", then close the parentheses:

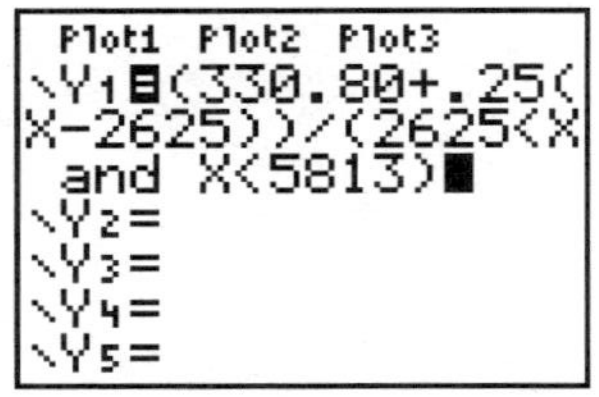

Set an appropriate window (using your domain and range from Question 14 as a guide), then press GRAPH.

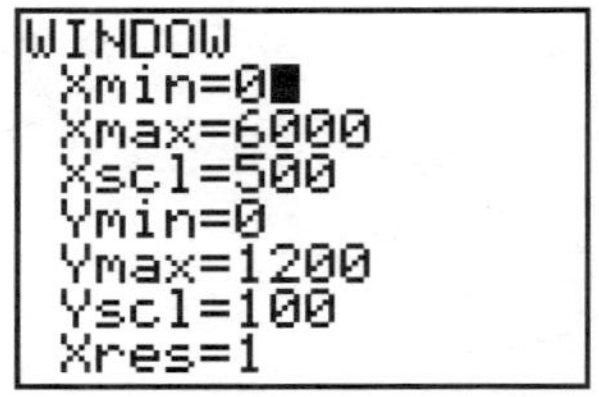

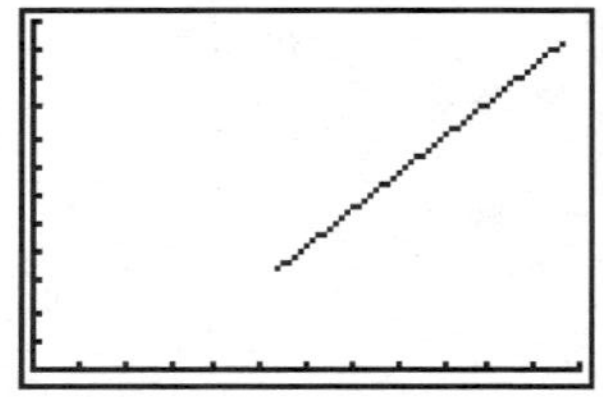

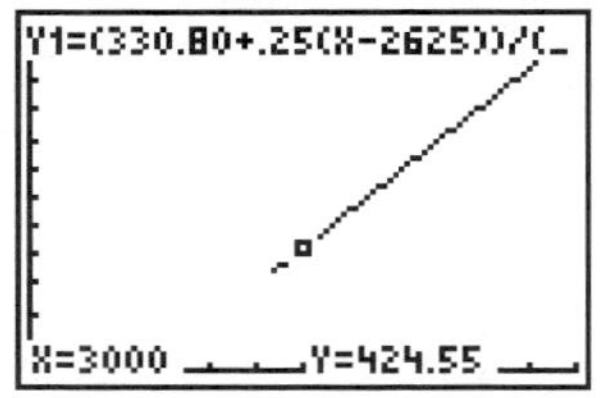

If you repeat this process for the other tax brackets, you can compare the rates of change (slopes) of the different linear functions, comparing them to the tax rates and total computed tax. The depth to which you want to take this extension investigation depends on your students' needs.

Answers (continued)

Adding the 10% and 15% tax brackets to the graph for functions Y2 and Y3 will yield the following graph:

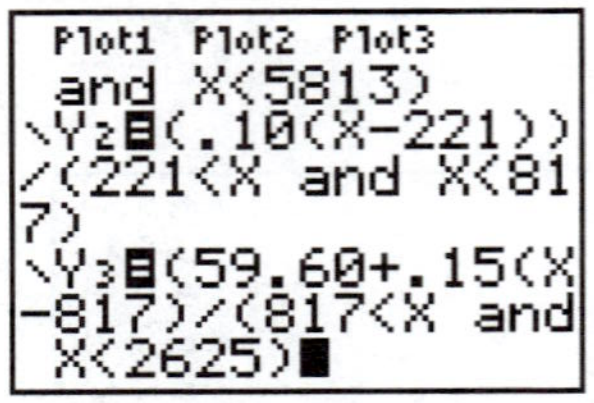

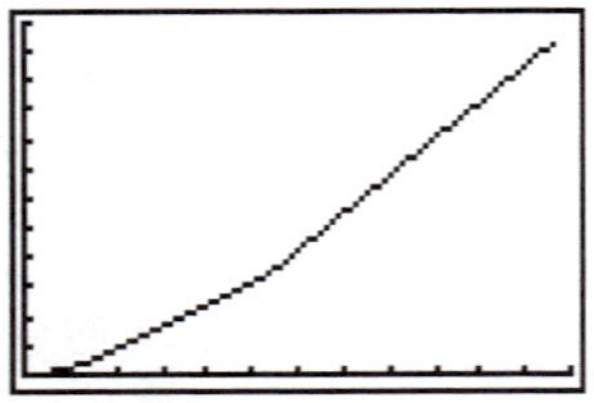

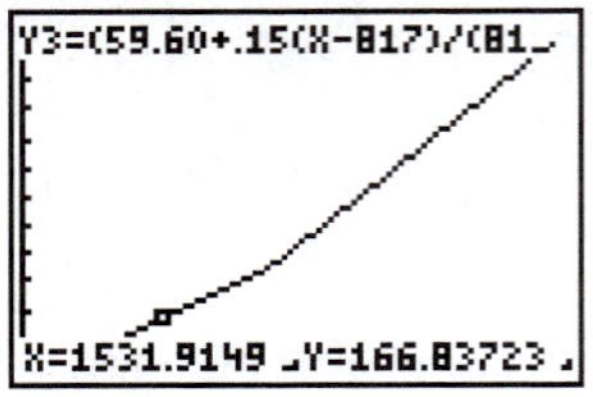

15. A single person who earns \$3285 in a month will fall between \$3280 and \$3320 in the tax table. The withholding, with 0 allowances, given in the table for this range is \$500. Since \$3285 is closer to 3280 than 3320 it makes sense that the withholding for \$3285 should be a little less than \$500.

16. tax = 59.60 + 0.15(2500.00 – 817.00)

    tax = 59.60 + 0.15(1683)

    tax = 59.60 + 252.45

    tax = \$312.05

17. The answer to Question 16 and the tax tables both give \$312. The only difference is 5 cents.

18. tax = 7893.83 + 0.35(150,000,000 – 27,354)

    tax = 7893.83 + 0.35(149,972,646)

    tax = 7893.83 + 52,490,426.10

    tax = \$52,498,319.93/12 months

    tax = \$4,374,859.94 per month

CHECK THIS!

Net pay is your take-home pay after taxes and all other deductions have been subtracted from your gross pay.

19. Assume you claim zero allowances, and your gross pay for one month is $5234.00. The cost of your health insurance is $87.00 per month and is deducted from your earnings. Federal withholding, social security, and medicare taxes are also deducted. Use your knowledge of these deductions to calculate your net pay.

20. Assume you are single and claim zero allowances; your net pay for one month is $1738. The cost of your health insurance is $26 per month and is deducted from your earnings. Federal withholding tax is $264; social security and medicare taxes are also deducted. What is your gross pay?

SECTION 7.3 Assignment

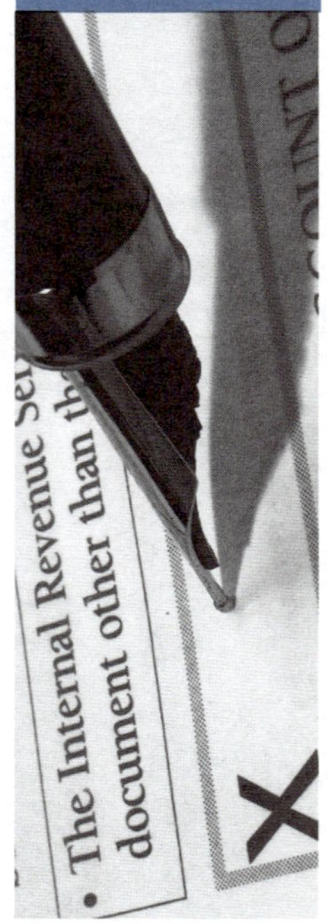

1. Calculate each of these.

   a) Weekly earnings when you make $12.75 an hour

   b) Yearly earnings when you make $525 a week

   c) Weekly earnings when you make $26,568 a year

   d) Hourly earnings when you make $2500 a month

   e) Monthly earnings when you make $75,300 a year

2. How is gross pay different from net pay?

Figure 7.12 shows how to calculate federal withholding tax for married persons.

MONTHLY PAYROLL PERIOD				
Married Persons				
If the Taxable Income is:		Computed Tax is:		
Over	**But Not Over**	**The total of this column PLUS>>**	**The % of taxable income**	**The amount over** (Taxable income less the amount in this column)
0	667	NO TAX WITHHELD		
667	1883	0	10%	667
1883	5517	121.60	15%	1883
5517	10,063	666.70	25%	5517
10,063	15,800	1803.20	28%	10,063
15,800	27,771	3409.56	33%	15,800
27,771	OVER	7359.99	35%	27,771

FIGURE 7.12. Federal withholding tax for wages paid January 1, 2005, and after.

3. What patterns do you observe in the table?

4. Explain in your own words how to use the table.

5. Use the method shown in Figure 7.12 to find the amount withheld for a married couple who claim zero allowances and make $6352 a month.

6. What is the federal withholding tax for a married couple who claim zero allowances and earn $3500 per month?

Answers (continued)

19. The total deductions are listed in the table.

Deduction	Amount
Federal Withholding	983.05
Medicare	75.89
Health Insurance	87.00
Social Security	324.51
Total	1470.45

Subtract the deductions from the gross pay: \$5234.00 – \$1470.45 = \$3763.55.

20. The deductions are:

social security tax = 0.062(gross pay)

medicare tax = 0.0145(gross pay)

withholding = 264

health insurance = 26

Substitute the known values into the equation for net pay and solve.

net pay = gross pay – deductions

1738 = gross pay – [264 + 26 + 0.062(gross pay) + 0.0145 (gross pay)]

1738 = gross pay – [290 + 0.0765(gross pay)]

1738 = 0.9235(gross pay) – 290

2028 = 0.9235(gross pay)

\$2195.99 = gross pay

Section 7.3 Assignment

Answers

1. a) \$510.00

   b) \$27,300

   c) \$510.92

   d) \$14.42

   e) \$6275

2. Gross pay is greater than net pay because net pay is the amount left after all deductions are made.

3. Answers may vary. The percent increases as the amount of money earned increases. The last column is the same as the first column.

4. Answers may vary. First find where the taxable income will fall in the table. For example, 3000 would be over 1883 but not over 5517. Next add the value in column three, in this case 121.60 to the percent in column four (15%) times the difference of the taxable income and the amount in the last column (1883). This process will give you the amount of withholding tax.

5. \$875.45

6. \$364.15

7. Mrs. Schroeder earned gross pay of $4695.40 in February 2005. When she received her paycheck, her net pay was much less. Along with federal withholding tax based on zero allowances and being married, the items listed in **Figure 7.13** were deducted.

Deduction	Amount
Federal Withholding	
Medicare	68.08
Health Insurance	201.23
Social Security	291.09
Life Insurance	15.65

FIGURE 7.13. Mrs. Schroeder's deductions.

a) How much was her net pay?

8. Assume you are married, claim zero allowances, and your gross pay for one month is $8274. The cost of your health insurance is $127 per month and is deducted from your earnings. Federal withholding, social security, and medicare taxes are also deducted. Find your net pay.

9. Assume you are single and claim zero allowances; your gross pay for one year is $25,740. The cost of your health insurance is $133 per month and is deducted from your earnings. Federal withholding, social security, and medicare taxes are also deducted. Find your net pay for one month.

10. Assume you are married and claim zero allowances; your net pay for one month is $3296.37. The cost of your health insurance is $97 per month and is deducted from your earnings. Federal withholding tax is $780.85; social security and medicare taxes are also deducted. Find your gross pay.

11. Assume you are single and claim zero allowances; your net pay for one year is $16,653.24. The cost of your health insurance is $27 per month and is deducted from your earnings. Federal withholding is $97; social security and medicare taxes are also deducted. Find your gross pay for one month.

SECTION 7.4 The Certainty of Taxes

PART A

Recall from Section 7.3 that in 2004 each person who contributed to social security and made $87,900 or less per year had 6.20% withheld for social security. They also had 1.45% withheld for medicare.

Use the above information to complete a table like **Figure 7.14.**

Base Salary (in dollars)	Social Security Taxes Withheld (in dollars)	Medicare Taxes Withheld (in dollars)	Total Taxes Withheld (in dollars)
10,000			
20,000			
30,000			
40,000			
50,000			
60,000			
70,000			
80,000			
87,900			

FIGURE 7.14. Table for Part A.

1. Use a graphing calculator to make each scatter plot using the same window. Record your window and sketch your graphs.

   a) Amount of social security taxes versus base salary

   b) Amount of medicare taxes versus base salary

   c) Total of taxes versus base salary

   d) Graph all three relationships on the same set of axes.

2. Do each of these graphs appear to describe direct variation? Why or why not?

3 a) What is a function rule that can be used to calculate the amount of social security taxes for a given salary?

Answers (continued)

7. a) Mrs. Schroeder's federal withholding tax is \$543.46. The total of her other deductions is \$576.05. Subtracting federal tax and other deductions from her gross pay of \$4695.40 leaves \$3575.89 net pay.

8. The total deductions are listed in the table.

Deduction	Amount
Federal Withholding	1355.95
Medicare	119.97
Health Insurance	127.00
Social Security	512.99
Total	2115.91

Subtract the deductions from the gross pay: \$8274.00 – \$2115.91 = \$6158.09.

9. First divide 25,740 by 12 to get monthly earnings of \$2145.00.

The total deductions are listed in the table.

Deduction	Amount
Federal Withholding	160.90
Medicare	31.10
Health Insurance	133.00
Social Security	132.99
Total	457.99

Subtract the deductions from the gross pay: \$2145.00 – \$457.99 = \$1687.01.

10. The deductions are:
social security tax = 0.062(gross pay)
medicare tax = 0.0145(gross pay)
withholding = 780.85
health insurance = 97

Substitute the known values into the equation for net pay and solve.

net pay = gross pay – deductions

3296.37 = gross pay – [780.85 + 97 + 0.062(gross pay) + 0.0145(gross pay)]

3296.37 = gross pay – [877.85 + 0.0765(gross pay)]

3296.37 = 0.9235(gross pay) – 877.85

4174.22 = 0.9235(gross pay)

\$4520.00 = gross pay

11. First divide 16,653.24 by 12 to get monthly earnings of \$1387.77.

The deductions are:
social security tax = 0.062(gross pay)
medicare tax = 0.0145(gross pay)
withholding = 97
health insurance = 27

Substitute the known values into the equation for net pay and solve.

net pay = gross pay – deductions

1387.77= gross pay – [97 + 27 + 0.062(gross pay) + 0.0145(gross pay)]

1387.77 = gross pay – [124 + 0.0765(gross pay)]

1387.77 = 0.9235(gross pay) – 124

1511.77 = 0.9235(gross pay)

\$1637.00 = gross pay

ELABORATE LESSON 1

Section 7.4: The Certainty of Taxes

Purpose: In Section 7.3 students formalized methods for calculating taxes and determining net pay. In this section students write function rules that can be used to determine social security and medicare taxes, preparing students to write and use models to make decisions in Section 7.5 about different checking account options.

Description: In this section students apply the methods for calculating taxes. They use tables and graphs to develop and test algebraic models for the calculation of taxes. Students will explore if the functions are direct variations. This will prepare them to help the Gonzales family compare different checking account options in Section 7.5.

Materials: graphing calculator

Teaching Idea: Emphasize the use of multiple representations in this section. Be sure that students understand the graphical, tabular, and algebraic representation of the functions. You may want to help students recall the properties of direct variation relationships and reconnect to the terms proportional and non-proportional relationships.

Answers

Part A

Answer to Figure 7.14:

Base Salary (in dollars)	Social Security Taxes Withheld (in dollars)	Medicare Taxes Withheld (in dollars)	Total Taxes Withheld (in dollars)
10,000	620	145	765
20,000	1240	290	1530
30,000	1860	435	2295
40,000	2480	580	3060
50,000	3100	725	3825
60,000	3720	870	4590
70,000	4340	1015	5355
80,000	4960	1160	6120
87,900	5449.80	1274.55	6724.35

1. a)

```
WINDOW
 Xmin=0
 Xmax=100000
 Xscl=10000
 Ymin=0
 Ymax=8000
 Yscl=1000
 Xres=■
```

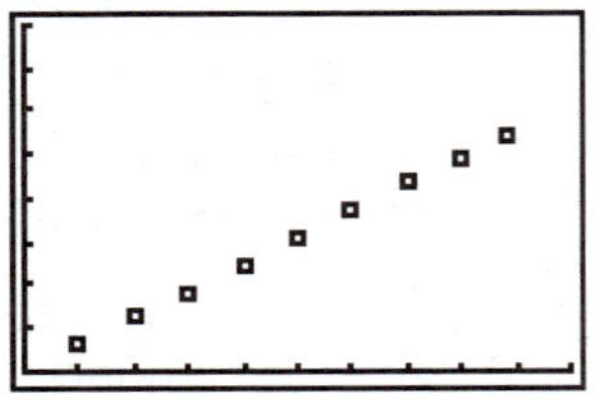

b)

```
WINDOW
 Xmin=0
 Xmax=100000
 Xscl=10000
 Ymin=0
 Ymax=8000
 Yscl=1000
 Xres=■
```

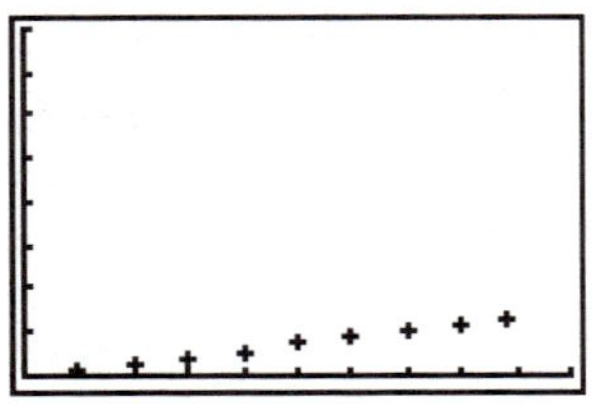

c)

```
WINDOW
 Xmin=0
 Xmax=100000
 Xscl=10000
 Ymin=0
 Ymax=8000
 Yscl=1000
 Xres=■
```

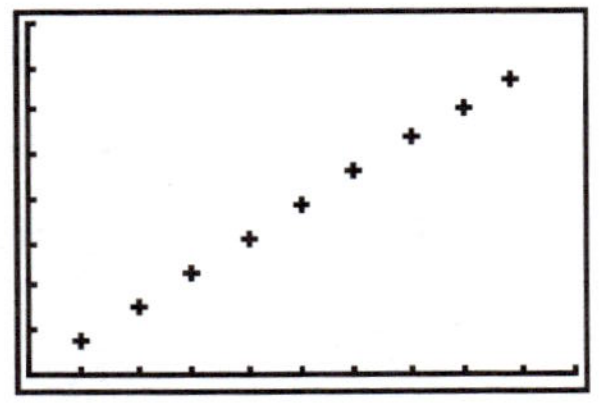

d) **NOTE:** Suggest to students that they place base salary in List 1, social security taxes withdrawn in List 2, medicare taxes withdrawn in List 3 and total taxes withdrawn in List 4.

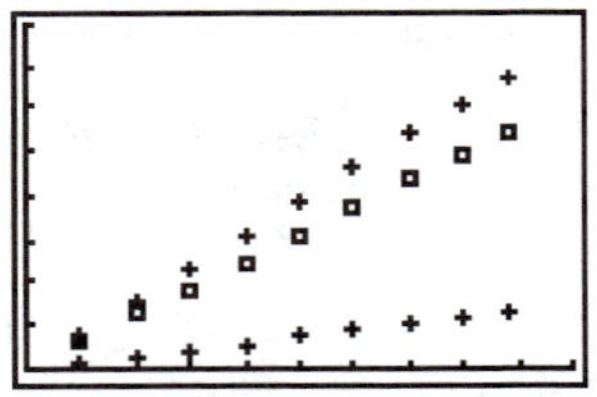

2. Each of these graphs seem to show a direct variation because the relationships are all proportional. They all are linear and pass through the origin.

3. a) $y = 0.062x$ is the equation that calculates the amount of social security taxes withheld. Remind students that this equation is of the form $y = kx$, a direct variation (proportional) equation. **NOTE:** Relate this rule to the work done in Section 7.3.

b) Graph your function rule on the scatter plot from Question 1b. Sketch your graph.

4. a) What is a function rule that can be used to calculate the amount of medicare taxes for a given salary?

   b) Graph your function rule on the scatter plot from Question 1b. Sketch your graph.

5. a) What is a function rule that can be used to calculate the total amount of taxes for a given salary?

   b) Graph your function rule on the scatter plot from Question 1b. If the function is not a good fit, use transformations to adjust it. Sketch your graph.

6. Complete a table like **Figure 7.15** to find the money remaining after taxes are deducted. Use the values from your first table to fill in the total taxes withheld column, then compute the money remaining column.

Base Salary (in dollars)	Total Taxes Withheld (in dollars)	Money Remaining (Base Salary – Total Taxes Withheld)
10,000		
20,000		
30,000		
40,000		
50,000		
60,000		
70,000		
80,000		
87,900		

FIGURE 7.15.
Table for Question 6.

7. Make a new scatter plot of money remaining versus base salary. Record your window and sketch your graph.

8. Find a function rule for the relationship between money remaining and base salary. Explain how you got it.

9. Graph your function rule on the scatter plot from Question 7. Sketch your graph.

10. Is this a direct variation function? Why or why not?

PART B

For people who make more than $87,900 per year, $5449.80 is withheld for social security, plus 1.45% of the base income is withheld for medicare. Complete a table like **Figure 7.16** based on this information.

Base Salary (in dollars)	Social Security Taxes Withheld (in dollars)	Medicare Taxes Withheld (in dollars)	Total Taxes Withheld (in dollars)
90,000			
100,000			
150,000			
200,000			
250,000			
300,000			
350,000			
400,000			
450,000			

FIGURE 7.16.
Table for Part B.

11. Use a graphing calculator to make each scatter plot using the same window. Record your window and sketch your graphs.

    a) Amount of social security taxes versus base salary

    b) Amount of medicare taxes versus base salary

    c) Total taxes versus base salary

    d) Graph all three relationships on the same set of axes.

12. a) What is a function rule that can be used to find the social security taxes for a given salary greater than $87,900?

    b) Graph your function rule on the scatter plot from Question 11d. If the function is not a good fit, use transformations to adjust it. Sketch your graph.

13. a) What is a function rule that can be used to find the medicare taxes for a given salary greater than $87,900?

    b) Graph your function rule on the scatter plot from Question 11d. If the function is not a good fit, use transformations to adjust it. Sketch your graph.

Answers (continued)

b) 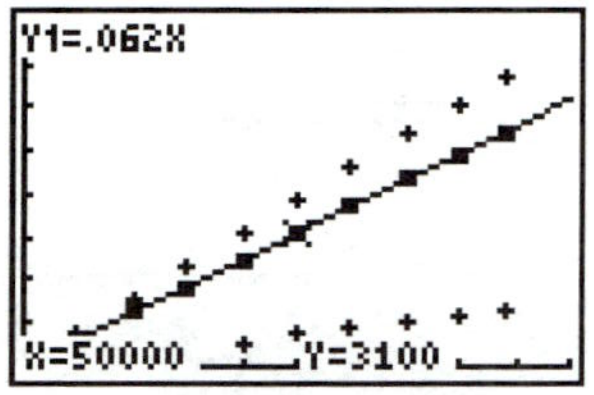

4. a) $y = 0.0145x$. Again have students understand that this is also an equation for a direct variation. NOTE: Relate this rule to the work done in Section 7.3.

b) 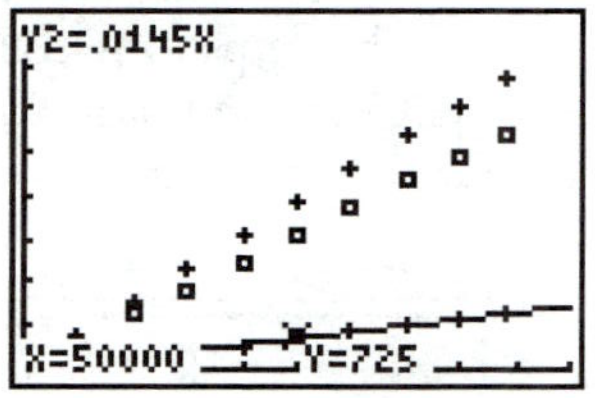

5. a) $y = 0.0765x$. This equation also represents a direct variation. The sum of two equations representing direct variations results in another equation representing a direct variation. NOTE: Relate this rule to the work done in Section 7.3.

b) 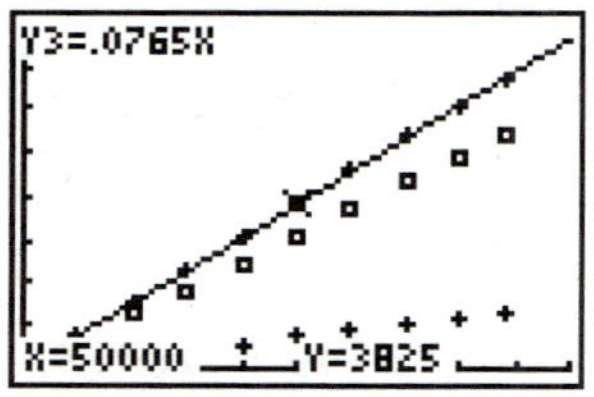

6. Answer:

Base Salary (in dollars)	Total Taxes Withheld (in dollars)	Money Remaining (Base Salary – Total Taxes Withheld)
10,000	765	9235
20,000	1530	18,470
30,000	2295	27,705
40,000	3060	36,940
50,000	3825	46,175
60,000	4590	55,410
70,000	5355	64,645
80,000	6120	73,880
87,900	6724.35	81,175.65

7. To create this scatter plot, students must create a new list that subtracts the total taxes withdrawn from the base salary, L5 = L1 – L4. The independent list is L1 and the dependent list is L5.

```
WINDOW
 Xmin=0
 Xmax=100000
 Xscl=10000
 Ymin=0
 Ymax=100000
 Yscl=10000
 Xres=■
```

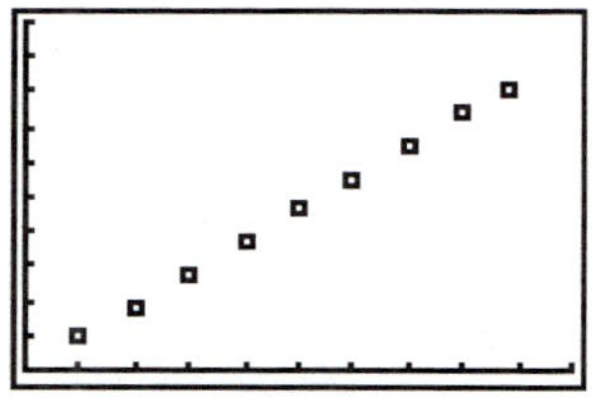

8. There are multiple ways students could arrive at this answer. They may take two points and compute the slope between the two points. They may again realize that this is a direct variation and find a y/x ratio to find the slope. Or, they may realize that the equation, based on their earlier ones, is:

$y = x - 0.0765x$

$y = 0.9235x$

9. 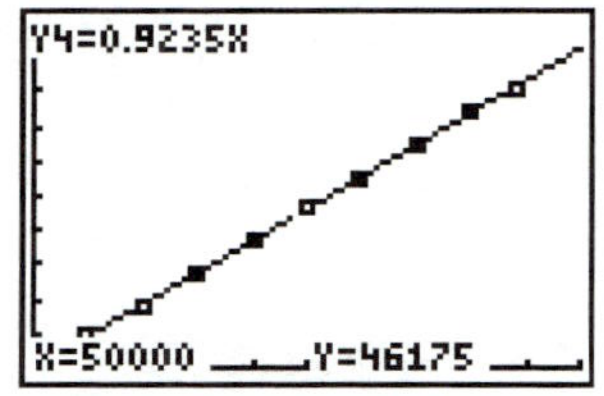

10. This line is a direct variation because it is linear and goes through the origin. The equation for the line is also in the form of a direct variation, $y = kx$.

Part B

Answer to Figure 7.16:

Base Salary (in dollars)	Social Security Taxes Withheld (in dollars)	Medicare Taxes Withheld (in dollars)	Total Taxes Withheld (in dollars)
90,000	5449.80	1350	6754.80
100,000	5449.80	1450	6899.80
150,000	5449.80	2175	7624.80
200,000	5449.80	2900	8349.80
250,000	5449.80	3625	9074.80
300,000	5449.80	4350	9799.80
350,000	5449.80	5075	10,524.80
400,000	5449.80	5800	11,249.80
450,000	5449.80	6525	11,974.80

11. a)

```
WINDOW
 Xmin=0
 Xmax=500000
 Xscl=100000
 Ymin=0
 Ymax=12000
 Yscl=1000
 Xres=■
```

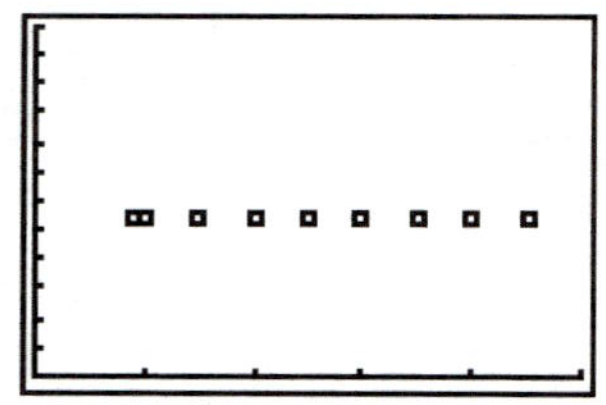

Answers (continued)

b)

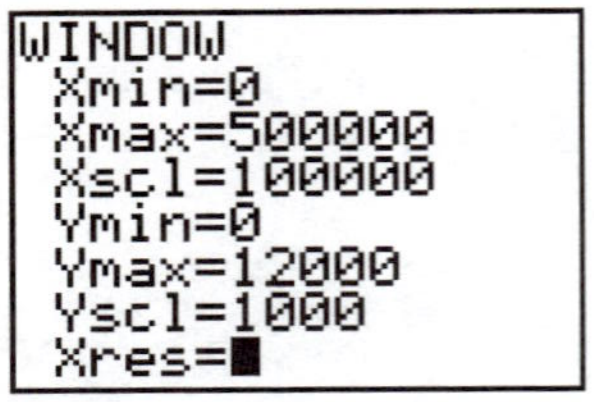

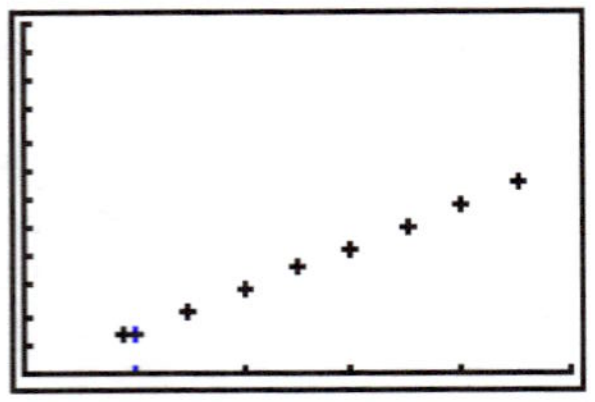

c)

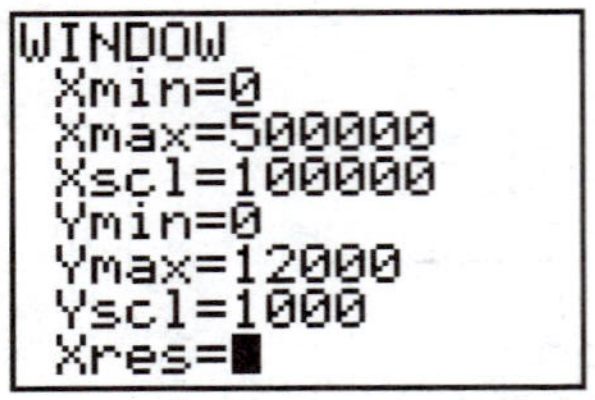

d)

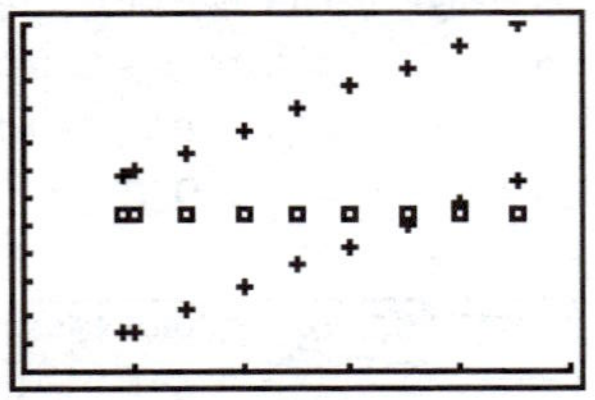

12. a) $y = 5449.80$. Students may initially be troubled about the idea of writing an equation that equals a constant, but remind them that this is the equation of a horizontal line. Help students to realize that although this equation results in a line, it is **not** a direct variation.

b)

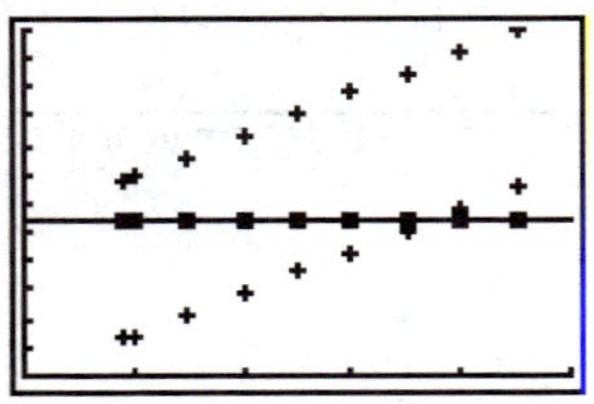

13. a) $y = 0.0145x$. This line did not change from the first part of the activity.

b)

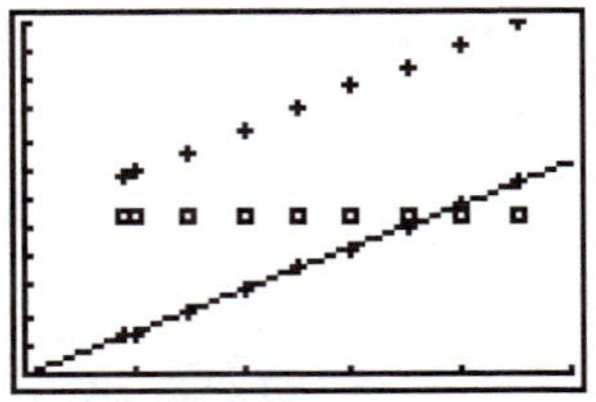

14 a) What is a function rule that can be used to find the total taxes for a given salary greater than $87,900?

b) Graph your function rule on the scatter plot from Question 11d. If the function is not a good fit, use transformations to adjust it. Sketch your graph.

15. Is this a direct variation function? Why or why not?

SECTION 7.4 Assignment

Employers often contribute a percentage of a person's salary to a retirement account. In **Figure 7.16** are graphs that show the retirement contributions of two companies as a function of salary.

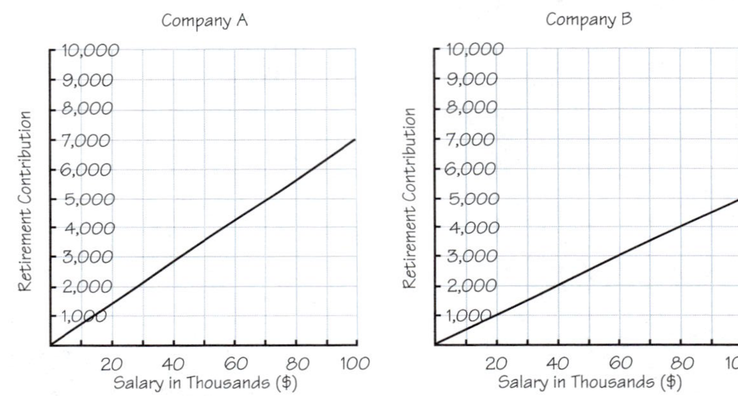

FIGURE 7.16. Retirement contributions of two companies.

1. Find a constant of proportionality for each graph. What is the real-world meaning of this number?

2. How much money would be contributed to the retirement account of a person making $48,000 under each plan?

3. How much money would need to be made to have $48,000 contributed to a retirement account under each plan?

4. Write function rules that model the contribution for each company for any salary.

5. Use your functions to find the amount each company would contribute to the retirement plan of a person making $60,000. Explain your solution.

6. Use your functions to find the amount a person would have to make to have $10,000 contributed to a retirement account under each plan. Explain your solution.

Answers (continued)

14. a) $y = 0.0145x + 5449.80$

b)

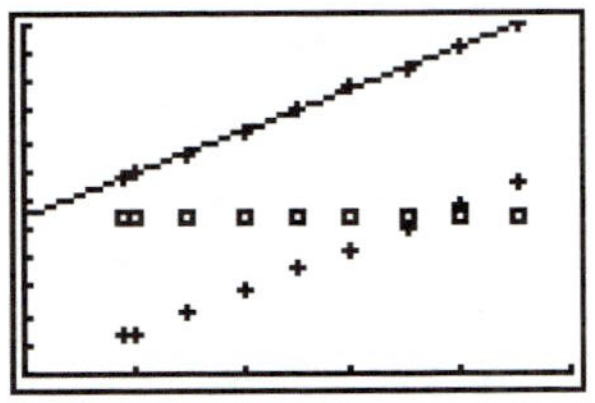

15. This line is not a direct variation since the y-intercept is not 0. The sum of a line representing a direct variation, and a line that does not represent a direct variation **will not** result in a line representing a direct variation.

Section 7.4 Assignment

Answers

1. Company A: 0.07. A constant of variation of 0.07 means that the company contributes 7% of an employee's salary to the retirement account.

   Company B: 0.05. A constant of variation of 0.05 means that the company contributes 5% of an employee's salary to the retirement account.

2. Company A: \$3360

   Company B: \$2400

3. Company A: \$685,714.29

   Company B: \$960,000

4. Company A: $y = 0.07x$

   Company B: $y = 0.05x$

5. Company A: \$4200

   Company B: \$3000

6. Company A: \$142,857.14

   Company B: \$200,000

A common example of direct variation is sales tax. The current tax rate for San Antonio, Texas has these parts:

State sales tax: $6\frac{1}{4}\%$

City tax: $1\frac{1}{8}\%$

Metropolitan Transit Authority tax: $\frac{1}{2}\%$

7. Write the equation (y_1) that gives the total sales tax for a given purchase (x). Then write an equation (y_2) that gives the final price (after taxes) for a given purchase price (x).

8. Find the sales tax and final price for each item:

   a) A DVD player priced at $149.95

   b) A pair of tennis shoes priced at $79.99

   c) A diamond bracelet priced at $5000

SECTION 7.5 Which Checking Account?

Let's apply what you have learned thus far about finances and simple interest to a new situation.

Mr. and Mrs. Gonzales want to open a checking account at Lone Star Bank. The bank offers three options: Personal Checking 1, Personal Checking 2, or Personal Checking 3. The Gonzaleses will deposit $2000 in the account and leave it there for four years. Also they plan to use the ATM machine about 10 times every month.

The details for each account are in **Figure 7.17.**

FIGURE 7.17. Account details.

Account	Monthly Interest Rate	ATM Fee	Monthly Service Fee
Personal Checking 1	0%	$1.25 each use	$12.00
Personal Checking 2	0.1%	$1.25 each use over 2	$15.00
Personal Checking 3	0.3%	No Charge	$25.00

Note: For each account, the interest is simple interest. The interest is calculated only on the initial deposit and is deposited into the account. For each account there is no charge for writing checks.

1. Complete each of these tables.

   a) Personal Checking 1

Month	Beginning Balance	Interest	ATM Fee	Service Fee	Ending Balance
1	2000				
2					
3					

   b) Personal Checking 2

Month	Beginning Balance	Interest	ATM Fee	Service Fee	Ending Balance
1	2000				
2					
3					

Answers (continued)

7. $y = 0.07875x$

   $y = 1.07875x$

8. a) $11.81; $161.76

   b) $6.30; $86.30

   c) $393.75; $5393.75

EVALUATE LESSON 1

Section 7.5: Which Checking Account?

Purpose: In Sections 7.1 through 7.4, students applied many of the concepts from previous chapters to personal finance. In this section students demonstrate their knowledge of those concepts and skills to solve a problem in a new context, preparing students to investigate other aspects of personal finance in Lesson 2.

Description: Facing a problem in a new context, students must determine and implement a method to solve the problem. In this open-middle performance assessment task there is only one correct answer, but there are a variety of valid methods to obtain it. In their justification, which is the most important part of a performance assessment task, students should be able to communicate their knowledge of mathematics and demonstrate that they have mastered the concepts presented in Sections 7.1 through 7.4.

Materials: graphing calculators

Teaching Ideas: This task may be assigned as group work or individual work. The mathematics performance assessment rubric should be used to evaluate students' work.

Answers

1. a) Answer:

Month	Beginning Balance	Interest	ATM Fee	Service Fee	Ending Balance
1	2000.00	0	12.50	12.00	1975.50
2	1975.50	0	12.50	12.00	1951.00
3	1951.00	0	12.50	12.00	1926.50

   b) Answer:

Month	Beginning Balance	Interest	ATM Fee	Service Fee	Ending Balance
1	2000.00	2.00	10.00	15.00	1977.00
2	1977.00	2.00	10.00	15.00	1954.00
3	1954.00	2.00	10.00	15.00	1931.00

c) Personal Checking 3

Month	Beginning Balance	Interest	ATM Fee	Service Fee	Ending Balance
1	2000				
2					
3					

2. Write a function rule for Personal Checking 1 where x is the month and y is the Ending Balance.

3. Write a function rule for Personal Checking 2 where x is the month and y is the Ending Balance.

4. Write a function rule for Personal Checking 3 where x is the month and y is the Ending Balance.

5. Keep in mind that Mr. and Mrs. Gonzales will keep this account for four years. Which account is the best deal for them?

SECTION 7.6

What's It Worth? A Look at the CPI

In this section you will examine the Consumer Price Index (CPI). The CPI is a measure of the change in price over time of a fixed basket of goods and services. It is based on items that cost 100 dollars in 1982. If the inflation rate is 3%, it means the CPI increased by 3% from a year earlier. You can use the CPI to find changes over time in how much you can buy with a certain amount of money. For example, something that cost about 10 cents in 1913 would cost about $1.85 in 2003.

1. Imagine that a news report stated that from 1990 to 1996 the CPI had an annual percent increase of about 3%. Based on this report, if you bought $10.00 worth of groceries in 1990, how much do you think the same groceries would cost in 1996?

To help get a better answer to Question 1, let's examine the CPI for 1990 to 1996.

2. Use the information from the news story and the actual CPI for 1990 (130.7) to complete the CPI column in a table like **Figure 7.18.**

Year	Year #	CPI
1990	0	130.7
1991	1	
1992	2	
1993	3	
1994	4	
1995	5	
1996	6	

FIGURE 7.18.
Table for CPI data.

3. Make a scatter plot of CPI versus year number. Record your window and sketch your graph.

4. Use the process column to write a function rule for this situation.

Answers (continued)

c) Answer:

Month	Beginning Balance	Interest	ATM Fee	Service Fee	Ending Balance
1	2000.00	6.00	0	25.00	1981.00
2	1981.00	6.00	0	25.00	1962.00
3	1962.00	6.00	0	25.00	1943.00

2. $y = 2000 - 12.5x - 12x$

3. $y = 2000 + 2x - 10x - 15x$

4. $y = 2000 + 6x - 25x$

5. Using a graph to compare the accounts:

```
WINDOW
 Xmin=0
 Xmax=50
 Xscl=1
 Ymin=0
 Ymax=2000
 Yscl=100
 Xres=■
```

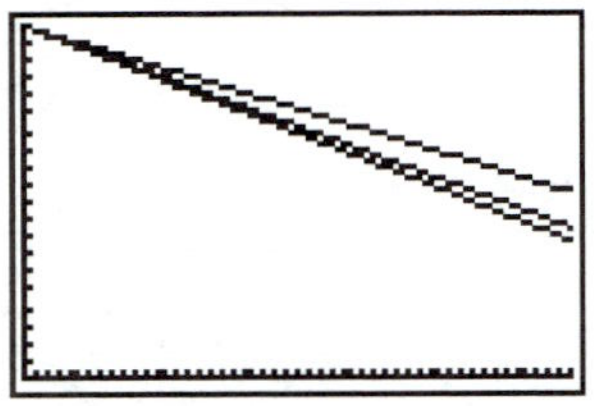

Trace to 48, Personal Checking 1 will have a balance of $824.00 after 48 months.

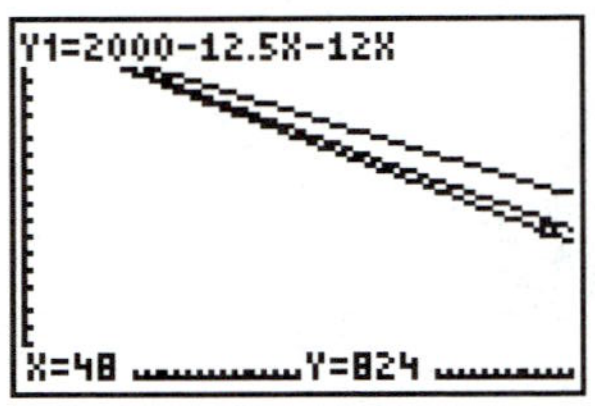

X	Y1	Y2
45	897.5	965
46	873	942
47	848.5	919
48	824	896
49	799.5	873
50	775	850
51	750.5	827

Y1=824

Trace to 48, Personal Checking 2 will have a balance of $896.00 after 48 months.

X	Y1	Y2
45	897.5	965
46	873	942
47	848.5	919
48	824	896
49	799.5	873
50	775	850
51	750.5	827

Y2=896

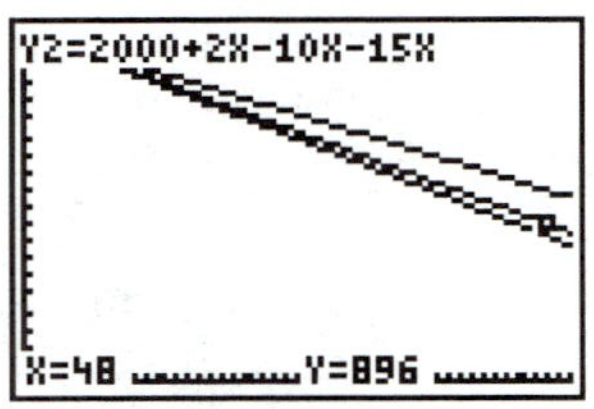

Trace to 48, Personal Checking 3 will have a balance of $1088.00 after 48 months.

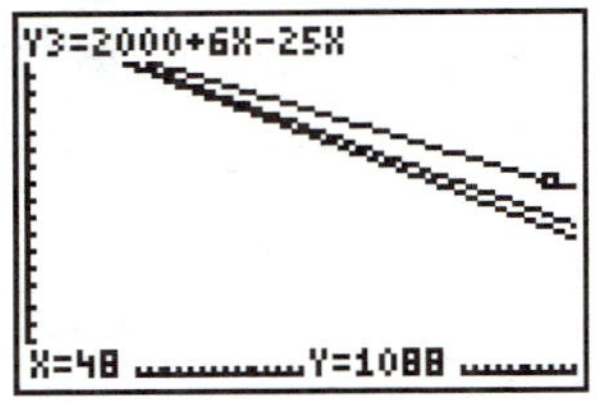

X	Y2	Y3
45	965	1145
46	942	1126
47	919	1107
48	896	1088
49	873	1069
50	850	1050
51	827	1031

Y3=1088

Although Personal Checking 3 has the greatest service charge, the Gonzales family will have more money in this type of account than any of the others after 48 months.

ENGAGE LESSON 2

Section 7.6: What's It Worth? A Look at the CPI

Purpose: In Sections 7.1 through 7.5, students helped the Gonzales family make decisions regarding personal income and budgeting. In this section students begin to investigate exponential functions in a financial context through a study of inflation. This prepares students to investigate compound interest and investment options in Sections 7.7 through 7.10.

Description: Students build an exponential growth model given an estimated Consumer Price Index rate of inflation of 3% per year. Then they compare their model to the actual CPI and compare the predicted results to the actual values, making critical judgments about a mathematical model.

Materials: graphing calculator

Teaching Ideas: Students can work in small groups (3 to 4 students). You may need to review exponential growth and the parts of an exponential function in order to write an equation to generate the first table.

Answers

1. Answers may vary. At this point students should be making an educated guess. They will use the data to evaluate their guess.

2. Answer:

Year	Year #	CPI
1990	0	130.7
1991	1	134.6
1992	2	138.7
1993	3	142.8
1994	4	147.1
1995	5	151.5
1996	6	156.1

3.

```
WINDOW
 Xmin=0
 Xmax=6
 Xscl=1
 Ymin=0
 Ymax=200
 Yscl=10
 Xres=■
```

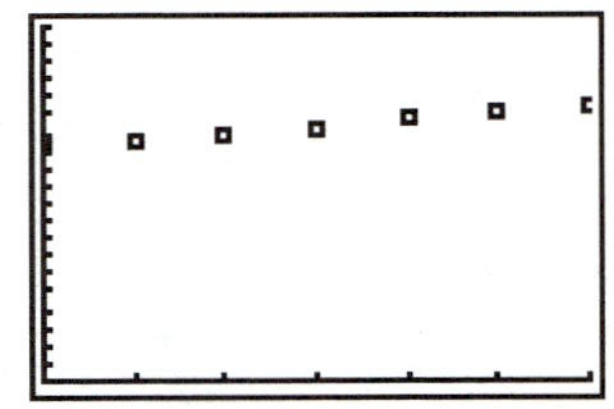

4. $f(x) = 130.7 \cdot 1.03^x$

Year	Year #	Process	CPI
1990	0		130.7
1991	1	$130.7(1.03) = 130.7 \cdot 1.03^1$	134.6
1992	2	$130.7(1.03)(1.03) = 130.7 \cdot 1.03^2$	138.7
1993	3	$130.7(1.03)(1.03)(1.03) = 130.7 \cdot 1.03^3$	142.8
1994	4	$130.7(1.03)(1.03)(1.03)(1.03) = 130.7 \cdot 1.03^4$	147.1
1995	5	$130.7(1.03)(1.03)(1.03)(1.03)(1.03) = 130.7 \cdot 1.03^5$	151.5
1996	6	$130.7(1.03)(1.03)(1.03)(1.03)(1.03)(1.03) = 130.7 \cdot 1.03^6$	156.1

5. Test your function rule by graphing it on your scatter plot. If the function rule is not a good fit, use transformations to modify it. Sketch your graph.
6. Use a function similar to the function you found in Question 4 to find out how much $10.00 worth of groceries purchased in 1990 cost in 1996.
7. Assuming this trend continues, use your function rule to predict the CPI in 2002.
8. If this trend continues, when will the CPI reach 200?

To see if the news story is accurate, let's look at real data. **Figure 7.19** shows the CPI from 1990–1996.

Year	Year #	CPI
1990	0	130.7
1991	1	136.2
1992	2	140.3
1993	3	144.5
1994	4	148.2
1995	5	152.4
1996	6	156.9

FIGURE 7.19.
CPI data from 1990–1996.

9. Make a scatter plot of CPI versus year number. Record your window and sketch your graph.
10. Use successive quotients as shown in **Figure 7.20** to write a function rule.

Year	Year #	CPI	Successive Quotients
1990	0	130.7	
1991	1	136.2	1.042
1992	2	140.3	1.030
1993	3	144.5	1.029
1994	4	148.2	1.026
1995	5	152.4	1.028
1996	6	156.9	1.030
Mean			1.031

FIGURE 7.20.
CPI data with successive quotients.

11. Test your function rule by graphing it on your scatter plot. If the function rule is not a good fit, use transformations to modify it. Sketch your graph.
12. Use a function rule similar to the function you found in Question 10 to determine how much $10.00 worth of groceries purchased in 1990 cost in 1996. How does this value compare to your answer to Question 6?
13. Use your function rule to predict the CPI in 2002. How does this value compare to your answer to Question 7?
14. If this trend continues, when will the CPI reach 200? How does this value compare to your answer to Question 8?
15. Based on your answers to Questions 12–13, is the news story accurate? Why or why not?
16. Based on the actual data, the CPI will most likely reach 200 in 2006. Were you able to conclude this using your function rule? Why or why not?
17. The CPI for 2002 was 179.9. Were you able to conclude this using your function rule? Why or why not?

Answers (continued)

5. 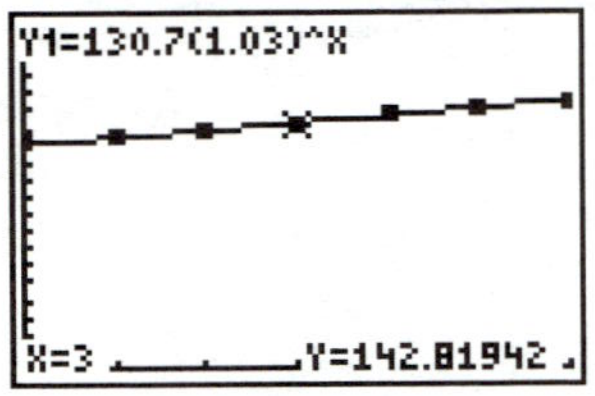

6. $f(x) = 10 \cdot 1.03^x$

$f(6) = 10 \cdot 1.03^6$

$f(6) = 10 \cdot 1.03^x$

$f(6) = 11.94$

It would take $11.94 in 1996 to equal $10.00 in 1990.

7. 2002 is year number 12 in our model. The predicted CPI is 186.35.

X	Y1
9	170.53
10	175.65
11	180.92
12	186.35
13	191.94
14	197.7
15	203.63

X=12

8. The CPI should have reached 200 about midyear in 2004.

X	Y1
9	170.53
10	175.65
11	180.92
12	186.35
13	191.94
14	197.7
15	203.63

X=14

9. 

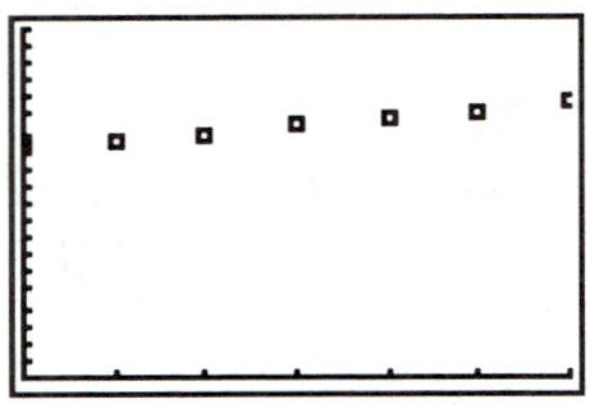

10. $f(x) = 130.7 \cdot 1.031^x$

11. 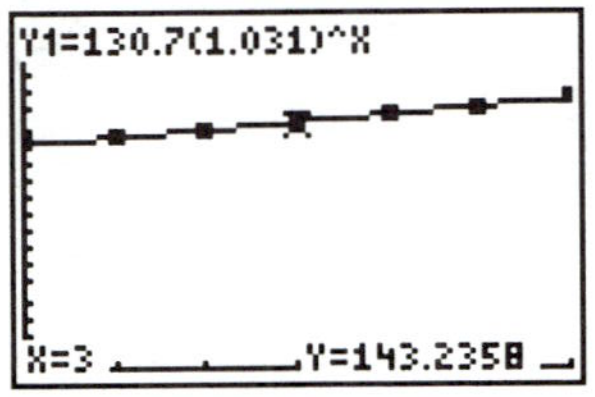

12. $f(x) = 10 \cdot 1.031^x$

$f(6) = 10 \cdot 1.031^6$

$f(6) = 10 \cdot 1.031^x$

$f(6) = 12.01$

$12.01 is $0.07 more than the answer to Question 6.

13. 2002 is year number 12 in our model. The predicted CPI is 188.53. This is 2.18 greater than the answer to Question 7.

X	Y1
9	172.03
10	177.36
11	182.86
12	188.53
13	194.37
14	200.4
15	206.61

X=12

14. The CPI should have reached 200 in 2004, which is about 6 months earlier than the answer to Question 8.

X	Y1
9	172.03
10	177.36
11	182.86
12	188.53
13	194.37
14	200.4
15	206.61

X=14

15. The news story was fairly accurate since the model the newscaster spoke of is $f(x) = 130.7 \cdot 1.03^x$ and the model that fit the actual CPI data is $f(x) = 130.7 \cdot 1.031^x$.

16. Using our function rules, the CPI reached 200 before 2005, because the growth rate remained fairly constant during the years 1990–1996; however, the growth rate decreased in later years. **NOTE:** Now would be an opportune time to reinforce the idea that we can only make reliable predictions within the range of our data.

17. Using our function rule, we estimated the CPI to be greater than 179.9 in 2002. Similar to the question above, the growth rate remained fairly constant during the years 1990–1996; however, the growth rate decreased in later years. **NOTE:** Now would be an opportune time to reinforce the idea that we can only make reliable predictions within the range of our data.

SECTION 7.7 Compound Interest

In Loan Star Bank's New Saver's account described in Section 7.1 the interest is simple interest. There is another form of interest that earns money differently. This type of interest is called **compound interest.** In this section you will explore compound interest.

CHECK THIS!

Recall that when interest is calculated only on the principal, it is called simple interest. Compound interest accounts receive interest on both the principal and any interest previously earned. The interest is compounded during a specified time period, usually daily, monthly, quarterly, or yearly. For example, if you deposit $100 at 5% interest compounded annually:

Year	Beginning Balance	Ending Balance
1	$100.00	$105.00
2	$105.00	$110.25
3	$110.25	$115.76

Looking a little closer at the New Saver's account raises an interesting question. Suppose you have $1000 in a New Saver's account. At the end of the year you should be paid $50. The bank doesn't hand over this money; they put it in your account. That means the principal should change for the next year. Let's see what happens with an account that combines the interest earned with the principal.

When interest is calculated once a year on both principal and prior interest earned we say the interest is **compounded annually.**

1. Suppose that you deposit $1000 in an account that pays 5% interest on the balance you have at the beginning of each year instead of just on the principal.

   a) How much money is in the account at the end of the first year?

   b) How much money is in the account at the end of the second year?

   c) How much money is in the account at the end of the third year?

   d) Continue this process to complete a table like **Figure 7.21.**

Year #	Account Balance
0	1000.00
1	
2	
3	
4	
5	

FIGURE 7.21. Account balance after five years.

2. What kind of function do you think best models this relationship? Use your table to support your answer.

3. Make a scatter plot of account balance versus year number. Record your window and sketch your graph.

4. Does your graph confirm your answer to Question 2? Why or why not?

5. At the end of year one, what percent of the account balance is interest? Write your answer as a percent and a decimal.

6. At the end of one year, what percent of the original balance is the new balance? Write your answer as a percent and a decimal.

7. Write a function rule using the answers to Questions 5 and 6 to relate the original balance and the balance at the end of the first year.

8. When we find the balance at the end of the first year, we are taking the starting balance plus 5% of the starting balance, which can be written as $1000 + 1000 \cdot (0.05)$. Factoring this expression gives $1000(1 + 0.05)$. Use your function rule for Question 7 to complete the process column in a table like **Figure 7.22** and write an exponential function for this situation.

Year #	Process	Account Balance
0	$1000(1 + 0.05)^0$	1000.00
1	$1000(1 + 0.05) = 1000(1 + 0.05)^1$	1050.00
2		1102.50
3		1157.63
4		1215.51
5		1276.28
x		f(x)

FIGURE 7.22. Account data with process column.

EXPLAIN LESSON 2

Section 7.7: Compound Interest

Purpose: In Section 7.6 students built an exponential function to investigate the CPI as a measure of inflation. In this section students use tables to look for patterns to investigate compound interest. In Section 7.8 students formalize a compound interest formula, and in Section 7.9 students will apply the compound interest formula to investment options and mixed-growth scenarios.

Description: In this section students expand what they learned about simple interest in Section 7.1 and build a numerical table exploring compound interest generated solely from an initial investment. Students examine their tables for number patterns that occur both recursively, from one term to the next, and explicitly, from the term number to the term value. Students look at multiple representations of compound interest formulas, including tables, graphs, and equations. Students also examine situations where compounding occurs more than once a year. From there, they will compare different banking offers in order to decide which one is a better offer for the Gonzales family.

Materials: graphing calculator

Teaching Ideas: Students can work in small groups (3 to 4 students). When working with compound interest, the numbers get large and may have many decimal places. Students may be intimidated by the use of decimal numbers, so you may need to encourage them to use the calculator to make the arithmetic easier.

This section also introduces students to the use of recursion to find patterns. Though this course does not provide an in-depth treatment of sequences and series, recursion is a helpful technique to explore number patterns. In the calculator's home screen, students are asked to use the "Ans" feature to find several consecutive terms in a sequence and to solve problems. You may need to monitor calculator skills closely and walk some students through the process.

Answers

1. a) $1050

   b) $1102.50

   c) $1157.63

   d) Answer:

Year #	Account Balance
0	1000.00
1	1050.00
2	1102.50
3	1157.63
4	1215.51
5	1276.28

2. Answers may vary. Students may determine that first differences are not constant so it is not a linear relationship, and then determine that successive quotients are constant so it is an exponential relationship.

3.

```
WINDOW
 Xmin=0
 Xmax=6
 Xscl=1
 Ymin=0
 Ymax=1500
 Yscl=100
 Xres=■
```

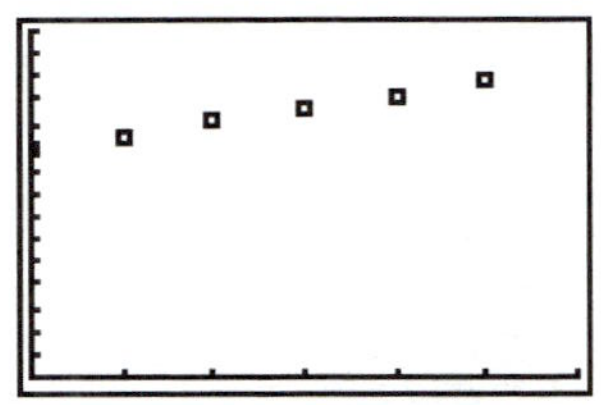

4. Answers may vary. Although this is an exponential relationship, the graph will most likely not look exponential for this range of data since the constant multiplier is close to 1.

5. 5% = 0.05

6. 105% = 1.05

7. $1000 \cdot 1.05 = 1050$

8. Answer:

Year #	Process	Account Balance
0	$1000(1 + 0.05)^0$	1000.00
1	$1000(1 + 0.05) = 1000(1 + 0.05)^1$	1050.00
2	$1000(1 + 0.05)(1 + 0.05) = 1000(1 + 0.05)^2$	1102.50
3	$1000(1 + 0.05)(1 + 0.05)(1 + 0.05) = 1000(1 + 0.05)^3$	1157.63
4	$1000(1 + 0.05)(1 + 0.05)(1 + 0.05)(1 + 0.05) = 1000(1 + 0.05)^4$	1215.51
5	$1000(1 + 0.05)(1 + 0.05)(1 + 0.05)(1 + 0.05)(1 + 0.05) = 1000(1 + 0.05)^5$	1276.28
x	$f(x) = (1 + 0.05)^x$	f(x)

9. Test your function rule by graphing it on your scatter plot. If the rule is not a good fit, use transformations to modify it. Sketch your graph.

10. If you make no deposits or withdrawals, how much money will you have in the account at the end of 10 years?

11. If you make no deposits or withdrawals, how many years will it take for your money to triple?

12. If you make no deposits or withdrawals, how many years will it take to have $2292 in the account?

CHECK THIS!

When interest is *compounded quarterly*, it is compounded every three months.

Just as you are about to open your account with Lone Star Savings, you see an ad for USA Savings Bank. They are offering a Super Savings account that pays 5% annual interest that is compounded four times a year. When interest is compounded four times a year we say the interest is **compounded quarterly.**

Let's develop a function that can be used to find an account balance when interest is compounded quarterly. Once we have a function we can use it to compare the two offers.

When interest is compounded quarterly, $\frac{1}{4}$ of the interest is added to the previous quarter's balance every three months.

13. If the interest rate is 5% per year, what percent is compounded each quarter? How did you get your answer?

14. Complete a table like **Figure 7.23** to find your balance at the end of one year.

Quarter	Beginning Balance	Process	Ending Balance
1	1000.00	1000(1 + 0.0125)	1012.50
2	1012.50		
3	1025.16		
4	1037.97		

FIGURE 7.23. Finding a bank balance after one year.

15. How does the balance at the end of one year when interest is compounded quarterly compare to the balance at the end of one year when interest is compounded annually? What is the reason for the difference?

Using the process in Question 14 is a valid way to find your ending balance for any number of years, but it takes a lot of time. A way to help speed up this process is to use a **recursive routine** on your calculator.

CHECK THIS!

A recursive routine generates a sequence of values by applying a rule to an initial value and then applying the rule to the resulting value. Continuing this process generates a sequence of values.

16. Follow the steps below to develop a recursive routine.

 a) Start with the ordered pair that represents your starting point. In this case, at time 0 you have $1000 in the account. What is the ordered pair?

 b) Enter the ordered pair into your calculator as shown in **Figure 7.24,** then press ENTER.

```
{0,1000}■
```

```
{0,1000}
         {0 1000}
■
```

FIGURE 7.24. Entering an ordered pair into a calculator.

 Each time we add 1 to the number of quarters, we need to multiply the previous balance by (1 + 0.0125). Write (1 + 0.0125) as a single decimal number without parentheses.

 c) Enter this into your calculator as shown in **Figure 7.25,** then press ENTER.

```
{0,1000}
         {0 1000}
{Ans(1)+1,Ans(2)
*1.0125}■
```

```
{0,1000}
         {0 1000}
{Ans(1)+1,Ans(2)
*1.0125}
       {1 1012.5}
■
```

FIGURE 7.25. Calculator screens from Question 16c.

CHECK THIS!

Since we want the calculator to work with two previous answers, we need to specify Ans (1) and Ans (2). The quarter number is Ans (1), and the account balance is Ans (2).

CHECK THIS!

You obtain "Ans" by pressing the 2nd key, and then the negative key (−).

 Press ENTER a second time. What ordered pair results? What is the meaning of this ordered pair in this context?

Answers (continued)

9. 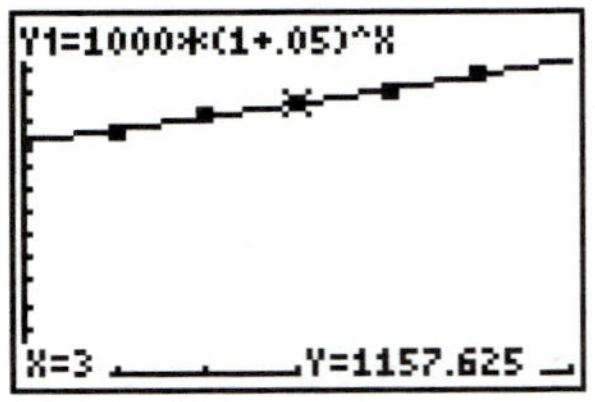

10. $1628.89

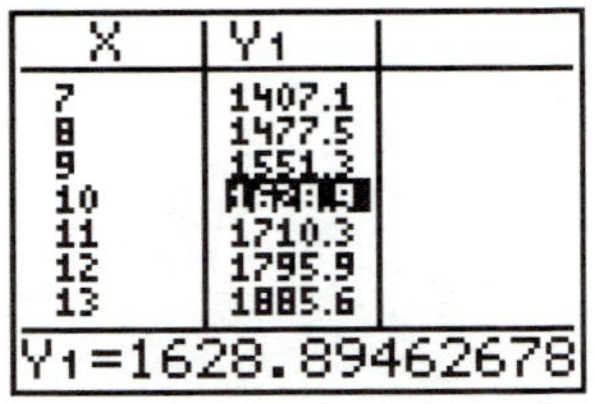

11. 23 years

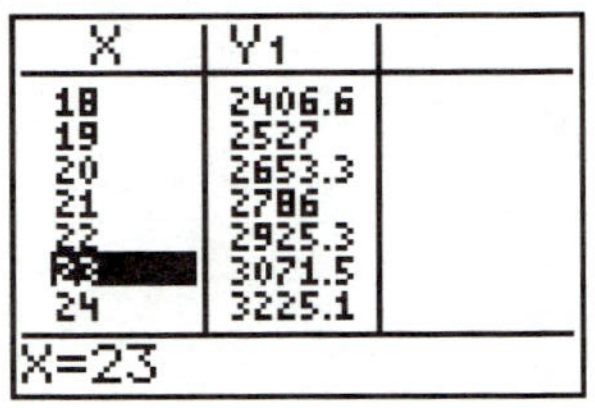

12. 17 years

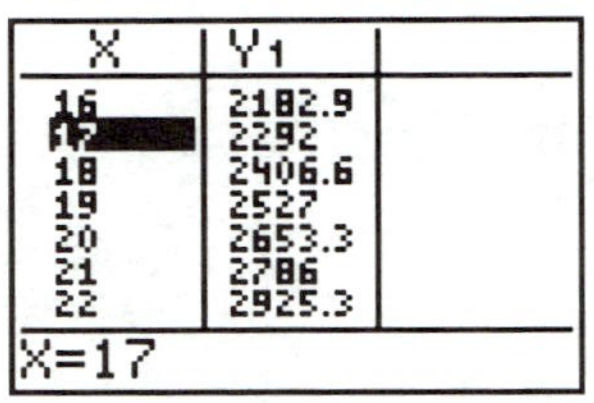

13. 1.25%. You get this by dividing 5% by 4 or $\frac{0.05}{4}$.

14. Answer:

Quarter	Beginning Balance	Process	Ending Balance
1	1000.00	1000(1 + 0.0125)	1012.50
2	1012.50	1012.50(1 + 0.0125)	1025.16
3	1025.16	1025.16(1 + 0.0125)	1037.97
4	1037.97	1037.97(1 + 0.0125)	1050.95

15. Answers may vary: The balance when 5% interest is compounded quarterly is greater than the balance when 5% interest is compounded annually. The reason for this difference is that when interest is compounded quarterly you get interest more often than when the interest is compounded annually.

16. a) (0, 1000)

b) 1.025

c) The ordered pair is (2, 1025.15625). Meaning that at the end of the second quarter there will be $1025.16 in the account.

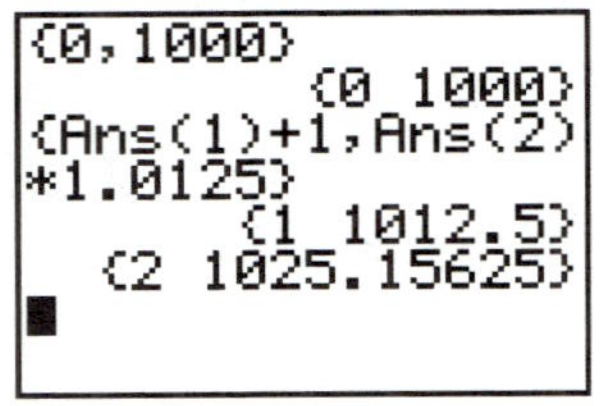

d) Continue pressing ENTER until you reach the 4th quarter. How does this compare to the account balance at the end of the 4th quarter you found in Question 14?

e) How many quarters are there in 10 years?

f) Continue using your recursive routine to find the account balance at the end of 10 years. How does your answer compare to the answer for Question 10? Why are they different?

17. Write a recursive routine to calculate the balance in an account if you start with \$500 and earn 8% interest compounded quarterly.

18. Use your recursive routine from Question 17 to find the account balance after eight years.

SECTION 7.7 Assignment

1. Suppose there is a constant 2% annual inflation for the next four years as a result of economic policy changes in 2001.

   a) How much will you have to pay in 2005 for an item that costs \$100 in 2001?

   b) What is the cost of an item in 2001 that will cost \$100 in 2005?

2. Samantha is depositing \$25,000 that her uncle left her into an account with a 12% annual interest rate that is compounded six times a year. Complete a table like **Figure 7.26** showing her balance after the first three interest periods.

Interest Period	Balance
0	\$25,000
1	
2	
3	

FIGURE 7.26. Samantha's account.

3. Use this function for calculating the balance in a savings account that earns interest compounded quarterly:

   $A(t) = 15{,}000\left(1+\frac{0.06}{4}\right)^{4t}$

   a) What is the annual interest rate?

   b) What growth factor can be used to make a sequence of yearly ending balances?

   c) Which term number corresponds to a time interval of two years, nine months?

   d) How much interest is earned during the second year?

Answers (continued)

d) Both the calculator and the table in Question 14 give the same result.

```
           {0 1000}
{Ans(1)+1,Ans(2)
*1.0125}
         {1 1012.5}
    {2 1025.15625}
   {3 1037.970703}
   {4 1050.945337}
```

e) There are 40 quarters in 10 years.

f) Answers may vary: The balance after 10 years when 5% interest is compounded quarterly is $1643.62. $1643.62 is greater than the answer to Question 10, $1628.89. The reason for this difference is that when interest is compounded quarterly you are getting interest on interest more often than when the interest is compounded annually.

```
{34 1525.566292}
  {35 1544.63587}
{36 1563.943819}
{37 1583.493116}
  {38 1603.28678}
{39 1623.327865}
{40 1643.619463}
```

17.

```
{0,500}
             {0 500}
{Ans(1)+1,Ans(2)
*1.02}
```

18. In 8 years, or 32 quarters, the balance will be $942.27.

```
{26 836.7090572}
{27 853.4432383}
{28 870.5121031}
{29 887.9223451}
{30 905.6807921}
{31 923.7944079}
{32 942.2702961}
```

Section 7.7 Assignment

Answers

1. a) $(100)(1.02)^4 \approx \$108.24$

   b) Solve the equation: $100 = A_0 \cdot (1.02)^4$. That item will be worth around $92.38.

2. Answer:

Interest Period	Balance
0	$25,000.00
1	$25,500.00
2	$26,010.00
3	$26,530.20

3. a) 6% per year

   b) 1.015

   c) Term number 11 (4 + 4 + 3)

   d) The balance after 1 year is around $15,920.45. The balance at the end of the 2nd year is around $16,897.39. The interest earned during that 2nd year is around $976.94.

4. Suppose you deposit $1800 in an account with an interest rate of 8% per year. What is the balance at the end of one year if the interest paid is:

   a) Simple interest?

   b) Compounded annually?

   c) Compounded quarterly?

   d) Compounded daily, assuming 365 days per year?

5. Each of these gives an investment in a savings account earning compound interest and the final balance after a certain time period. Find the rate that provides those earnings.

   a) Investment: $900; final balance: $1000; 2 compounds per year; $t = 1$ year

   b) Investment: $1100; final balance: $1250; 1 compound per year; $t = 3$ years

   c) Investment: $2000; final balance: $2250; 6 compounds per year; $t = 1$ year

6. Suppose $1 is deposited in an account that pays 10% interest compounded annually. The plan is to keep the money in the account for just the first year. Later, the owner decides to keep the money in the account for the second year as well.

   a) We would like to know how the individual growth factors compare with the overall one. Fill in a table like the one in **Figure 7.27** with the growth factors that correspond to the term lengths.

FIGURE 7.27. Term lengths.

1st Term Length	Growth Factor	2nd Term Length	Growth Factor	Total Term Length	Overall Growth Factor
2 years		5 years		7 years	
4 years		3 years		7 years	

   b) How can you determine the overall growth factor if you know only the individual ones? Test your idea on the two examples in Figure 7.24. Does it work?

   c) What do you notice about the overall growth factor in each case? Does that make sense in this problem?

SECTION 7.8 Very Interesting

In Section 7.6 you used a news story and the actual CPI for 1990 (130.7) to complete the CPI column in a table like **Figure 7.28.** The news story reported that from 1990 to 1996 the CPI had an annual growth rate of about 3%. We can use a recursive routine to fill a table.

1. Write the recursive routine that can complete the table.

2. Use your recursive routine to compute the values in the CPI column in Figure 7.28.

FIGURE 7.28. Table for computing CPI values.

Year	Year #	Process	CPI
1990	0		130.7
1991	1		
1992	2		
1993	3		
1994	4		
1995	5		
1996	6		

3. A function rule for the situation described in the news story in Section 7.6 is $f(x) = 130.7 \cdot 1.03^x$. How did you use this model to predict how much groceries purchased in 1990 cost in 1996?

4. In Section 7.7 you investigated Lone Star Bank's New Saver's account. For this type of account the bank pays 5% interest compounded annually. Say again that you deposit $1000 in this account. Write a recursive routine to calculate the balance at the end of any given year.

5. Use your recursive routine to verify the values in your table from Section 7.7. The table is shown in **Figure 7.29.**

Year #	Account Balance
0	1000.00
1	1050.00
2	1102.50
3	1157.63
4	1215.51
5	1276.28

FIGURE 7.29. Table from Section 7.7.

Answers (continued)

4. a) \$1944

   b) \$1944

   c) \$1948.38

   d) \$1949.90

5. a) (1000/900)^(1/2) ≈ 1.0541, so the APR should be around 10.82%.

   b) 1250/1100)^(1/3) ≈ 1.0435, so the APR should be around 4.35%.

   c) (2250/2000)^(1/6) ≈ 1.0198, so the APR should be around 11.9%.

6. a) Answer:

1st Term Length	Growth Factor	2nd Term Length	Growth Factor	Total Term Length	Overall Growth Factor
2 years	1.21	5 years	1.61051	7 years	1.9487171
4 years	1.4641	3 years	1.331	7 years	1.9487171

   b) Multiply the two numbers together. Yes, it works.

   c) They are the same answer. Since both scenarios involve an overall time period of 7 years, it makes total sense.

EXPLAIN LESSON 2

Section 7.8: Very Interesting

Purpose: In Section 7.7 students used tables and calculators to explore compound interest from a recursive perspective. In this section students develop the compound interest formula. This provides students with the concepts and skills to explore mixed-growth compound interest in Section 7.9, where students investigate the growth of money when interest is being earned while additional deposits are being made.

Description: In this phase of the lesson students will use tables and number patterns to develop the compound interest formula, $A = P\left(1+\frac{r}{n}\right)^{nt}$.

Materials: graphing calculators

Teaching Ideas: Debrief students after Section 7.7 by having them share their answers to the questions in the assignment for Section 7.7. Be sure to address any missing concept, process, or skill that could have been used to answer the questions. Then work through the questions in this section with the entire class.

Again, closely monitor calculator skills, especially as students become familiar with home-screen recursion methods.

Answers

1.
```
{0,130.7}
            {0 130.7}
{Ans(1)+1,Ans(2)
*1.03}
```

2.
```
            {0 130.7}
{Ans(1)+1,Ans(2)
*1.03}
          {1 134.621}
       {2 138.65963}
   {3 142.8194189}
   {4 147.1040015}
```

Year	Year #	Process	CPI
1990	0		130.7
1991	1		134.6
1992	2		138.7
1993	3		142.8
1994	4		147.1
1995	5		151.5
1996	6		156.1

3. Students should be able to describe the relationship between number of years and the CPI.

4.
```
{0,1000}
            {0 1000}
{Ans(1)+1,Ans(2)
*1.05}
```

5.
```
*1.05}
           {1 1050}
         {2 1102.5}
       {3 1157.625}
    {4 1215.50625}
   {5 1276.281563}
   {6 1340.095641}
```

6. Find the successive quotients for the account balances in **Figure 7.30.**

Year #	Account Balance	Successive Quotients
0	1000.00	
1	1050.00	
2	1102.50	
3	1157.63	
4	1215.51	
5	1276.28	

FIGURE 7.30.
Account balance data.

7. How can you use successive quotients and a starting balance of $1000 to write a function rule for this situation?

The function rule in Question 7 allows you to compound interest annually. We know that when interest is compounded annually there is a greater return than with simple interest.

8. What do you think will happen to the amount of interest earned over a period of years when interest is compounded quarterly?

9. Complete a table like the one in **Figure 7.31** and use the process column to write an expression that can be used to calculate the number of times interest is calculated for x years, if that interest is compounded quarterly.

Number of Years	Process	Number of Times Interest is Compounded
1	4·1	4
2		
3		
4		
5		
x		

FIGURE 7.31.
Number of times interest is calculated.

Recall that if the rate is 5% per year then $\frac{0.05}{4}$ or 1.25% is compounded each quarter. Substitute $\frac{0.05}{4}$ and the expression for the number of times interest is compounded from Question 9 into the function rule from Section 7.7, $f(x) = 1000(1 + 0.05)^x$. This

gives $f(x) = 1000\left(1+\frac{0.05}{4}\right)^{4x}$, which can be used to calculate interest compounded quarterly when the initial balance is $1000.

10. Use $f(x) = 1000\left(1+\frac{0.05}{4}\right)^{4x}$ to complete a table like **Figure 7.32.**

Year #	Account Balance: Interest Compounded Annually	Account Balance: Interest Compounded Quarterly
0	1000.00	
1	1050.00	
2	1102.50	
3	1157.63	
4	1215.51	
5	1276.28	

FIGURE 7.32.
Table for Question 10.

CHECK THIS!

When entering this function into a graphing calculator, remember to place parentheses around the exponent ($4x$).

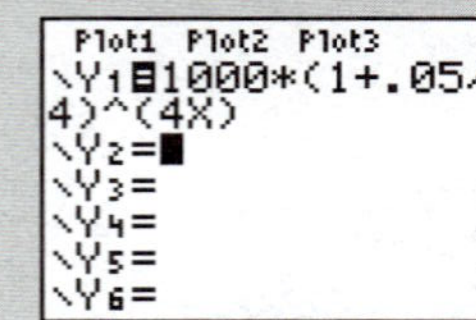

11. Based on your answer to the questions above, would it be better to open a Lone Star Bank New Saver's account or a USA Savings Bank Super Savings account? Why?

12. Use the function rule from Question 10 and substitution to write the general formula for this type of savings plan. Assume these conditions.

- You start by depositing P dollars.
- The interest rate is r.
- The number of times the interest is compounded in a year is n.
- The number of years is t.
- The ending balance is B.

13. Paco deposited $4000 in an account that earns 6% interest compounded monthly. If he makes no deposits or withdrawals, what will be his balance after eight years?

Answers (continued)

6.

Year #	Account Balance	Successive Quotients
0	1000.00	
1	1050.00	0.05
2	1102.50	0.05
3	1157.63	0.05
4	1215.51	0.05
5	1276.28	0.05

7. Using the general form of an exponential growth function, y = initial value (common multiplier), we have the function $y = 1000 \cdot 1.05^x$.

8. Answers may vary. Sample answer: the amount of interest earned should be more when interest is compounded quarterly.

9. Answer:

Number of Years	Process	Number of Times Interest is Compounded
1	4 · 1	4
2	4 · 2	8
3	4 · 3	12
4	4 · 4	16
5	4 · 5	20
x	4 · x	4x

10. Answer:

Year #	Account Balance	
	Interest Compounded Annually	Interest Compounded Quarterly
0	1000.00	1000.00
1	1050.00	1050.95
2	1102.50	1104.49
3	1157.63	1160.76
4	1215.51	1219.89
5	1276.28	1282.04

11. It would be better to open a USA Savings Bank Super Savings Account, because more interest is accumulated when interest is compounded quarterly than when it is compounded annually.

12. After substituting, the formula would be:

$$B = P\left(1+\frac{r}{n}\right)^{nt}$$

13. $B = P\left(1+\frac{r}{n}\right)^{nt}$

$$B = 4000\left(1+\frac{0.06}{12}\right)^{12 \cdot 8}$$

$$B = 4000\left(1+\frac{0.06}{12}\right)^{12 \cdot 8}$$

$$B = \$6456.57$$

SECTION 7.8 Assignment

1. Imagine you put $10,000 in a savings account for one year at 5% interest for the compounding periods in **Figure 7.33.** Determine how much money you will have at the end of the year for each compounding method.

$1000 at 5% Interest of One Year	
Compounded	Total at End of One Year
Annually	
Semiannually	
Quarterly	
Monthly	
Weekly	
Daily	
Hourly	

FIGURE 7.33.
Different compounding periods.

2. Suppose you deposit $1200 in an account with an annual rate of 4%. How much interest do you receive after one year:

   a) If the bank compounds daily using a 360-day year (called the "360 over 360" method)?

   b) If the bank compounds daily using a 365-day year (called the "365 over 365" method)?

   c) Why might a bank use the "360 over 360" method?

3. The exponential equation $A_t = A_0 \cdot b^t$ has four quantities where A_t is the amount of money in the account after t years, A_0 is the amount initially invested, and b is the yearly growth factor. Given any three of these quantities, you can solve for the missing one. In each of the following, three quantities are given. Show how to solve for the missing item, and then find its value.

   a) $A_0 = \$800$; $b = 1.08$; $t = 5.2$ years

   b) $A_t = \$1200$; $b = 1.12$; $t = 3.8$ years

   c) $A_0 = \$1500$; $A_t = \$2500$; $b = 1.05$

CHECK THIS!

A_t is an example of a subscript. The subscript is a method of showing different values of the same variable. Here it is the account balance over time.

SECTION 7.9 Annuities: Mixed Growth

Making regular deposits into an account that earns compound interest creates what is called **mixed growth**. An annuity is an example of a mixed growth account. It can be described recursively as a repeating two-step calculation of multiplying and adding (in either order, but the result is affected by which is done first). Because the basic process is two operations, it is easy to find terms of such a sequence. However, finding the total of the terms is a little more involved.

CHECK THIS!

Annuities come in two forms:

The first is an asset, things such as stocks, bonds, or cash that pay a constant amount each year to the holder until the annuity expires.

The second is a savings vehicle in which regular deposits are made for a specified period to grow an asset. Tico is creating this kind of annuity.

1. Tico deposits $500 in an annuity that earns 5% compounded annually at the beginning of each year for a six-year time period. How is this situation different from those discussed so far?

2. Fill in a table like **Figure 7.34**. For each year calculate the amount in the account that earns interest (after the deposit is made), the amount of interest earned that year, and the ending balance.

Year (n)	Deposit Made	Beginning Balance	Total Amount Earning Interest	Interest Earned During Year	Ending Balance
1	$500.00	$500.00	$500.00	$25.00	$525.00
2	$500.00	$1025.00	$1025.00		
3	$500.00	$1576.25			
4	$500.00	$2155.06			
5	$500.00	$2762.81			
6	$500.00	$3400.95			

FIGURE 7.34.
Table for account calculation.

Section 7.8 Assignment

Answers

1. Answer:

$1000 at 5% Interest of One Year	
Compounded	**Total at End of One Year**
Annually	10,500.00
Semi-Annually	10,506.25
Quarterly	10,509.45
Monthly	10,511.62
Weekly	10,512.46
Daily	10,512.67
Hourly	10,512.71

2. a) The ending balance is: $1200(1 + 0.04/360)^{360} \approx$ 1248.97, so the account earned $48.97 in interest.

   b) The ending balance is:
   $1200(1 + 0.04/365)^{365} \approx 1248.97$, so the account also earned $48.97 in interest.

   c) 360 days/year is a more convenient number with which to divide the year into months, since 360 is divisible by 12.

3. a) Evaluate the expression; $(800)(1.08)^{5.2} \approx$ $1193.70

   b) Evaluate b^t, then divide it into A_t;
   $1200 = A_0(1.12)^{3.8}$, so $1200 \approx 1.5383 \cdot A_0$.
   Finally, solve for A_0: $1200/1.5383 \approx \$780.10$.

   c) Divide the log of the overall growth factor (A_t/A_0) by the log of the yearly growth factor; $\log(2500/1500) \div \log(1.05) \approx$ $0.22185 \div 0.02119 \approx 10.47$ years.

ELABORATE LESSON 2

Section 7.9: Annuities: Mixed Growth

Purpose: In Sections 7.7 and 7.8, students explored and formalized the formula for finding the balance of an account with one initial deposit earning interest that is compounded over a given time interval. In this section students explore investment situations where interest is being earned while additional deposits are being made. This prepares students to make decisions in Section 7.10 about different investment options.

Description: In this section students apply the principles and formulas regarding compound interest to a situation where deposits are being made while interest is being earned. They use a table and home-screen recursion to calculate the balance numerically in the account at the end of different compounding periods. Then, they develop two formulas for savings situations: one where the deposits are made at the beginning of the compounding period and one where the deposits are made at the end of the compounding period. These formulas are used to help students make decisions about investment opportunities for the Gonzales family.

Materials: graphing calculator

Teaching Idea: Emphasize the use of multiple representations in this section. Be sure that students understand the graphical, tabular, and algebraic representation of the exponential function that describes compound interest.

In this section the formula for finding the sum of a specific geometric series is used to develop a savings formula. Although it is not necessary to get into the details of a geometric sequence or series, it is important that students recognize the number patterns present in the geometric series that appears in compound interest. Once they know the formula to find the sum of the series they encounter, they can build a function to predict the account balance after a certain number of compounding periods.

Answers

1. Before, a deposit was made only in the beginning. Now a new deposit is added each year.

2. Answer:

Year (n)	Deposit Made	Beginning Balance	Total Amount Earning Interest	Interest Earned During Year	Ending Balance
1	$500.00	$500.00	$500.00	$25.00	$525.00
2	$500.00	$1025.00	$1025.00	$51.25	$1076.25
3	$500.00	$1576.25	$1576.25	$78.81	$1655.06
4	$500.00	$2155.06	$2155.06	$107.75	$2262.81
5	$500.00	$2762.81	$2762.81	$138.14	$2900.95
6	$500.00	$3400.95	$3400.95	$170.05	$3571.00

3. What calculation finds the beginning balance for year two from the beginning balance in year one?

4. Does that same process work for finding the beginning balance for year three? Write a recursive routine for the beginning balances.

5. Use your recursive routine to calculate the ending balance in the account after 10 years.

6. How does the answer to Question 5 relate to the recursive routine you found in Question 4?

Let's look at Tico's annuity in a different way. This time think of each deposit as a separate account with deposits made at different times and different amounts of interest earned based on how long the money is in the account.

7. Fill in a table like **Figure 7.35**. Use the number of years each deposit is in the account to calculate the ending balance and interest that portion of the account earns.

FIGURE 7.35. Table for Question 7.

Year (n)	Deposit Made	Number of Years Deposit is in Account	Process	Ending Balance
1	\$500.00	6	$500 \cdot (1.05)^6$	≈ \$670.05
2	\$500.00	5		
3	\$500.00	4		
4	\$500.00	3		
5	\$500.00	2		
6	\$500.00	1		
Total				

8. Write the expressions from the process column that you used to calculate the ending balance for each deposit.

9. Write *one* expression that calculates the sum of the ending balances at the end of the six-year period from all the deposits made.

10. Evaluate the expression you wrote in Question 9. How does this number compare to the total of ending balances you found in the table in Question 7?

11. The terms of the expression from Question 9 have two common factors since every payment earns interest for at least one year. Identify the common factors, and then factor using the distributive property so that the common factors are in front of a set of parentheses that group what is left of your original expression.

Choosing any number as a base and raising it to exponent powers that are consecutive integers forms a **geometric sequence**. When those sequence values are added together, that is called a **geometric series**. There is a shortcut to finding the sum of a geometric series when the first term is 1 and the common ratio is b:

$$1 + b^1 + b^2 + b^3 + \ldots + b^{(n-1)} = \frac{b^n - 1}{b - 1}.$$

12. Use the formula for summing a geometric series to find the sum of the geometric series you wrote in Question 9. The part in the bracket is the geometric series.

13. Use the same formula and your work from Question 12 to write the general formula for this type of savings plan. Assume these conditions.
    - At the beginning of each time period you put d dollars into an account.
    - The account has a growth factor of b during that time period.
    - There is a total of n time periods.

14. Write an expression for the total of all the payments of d dollars made into the account n times.

15. To calculate the total interest, adjust the formula you found in Question 13 by subtracting the expression for the total of all the payments.

16. Check to see if this new formula correctly predicts the total interest earned after six years. How do you know this formula is correct?

Answers (continued)

3. Start with the beginning balance of 500, then multiply by 1.05 (the growth factor for one year).

4. Yes, it works for calculating each beginning balance from the previous year's beginning balance. The recursive routine is:

```
{1,0}
                {1 0}
{Ans(1)+1,(Ans(2
)+500)*1.05
```

5. $5788.95

```
   {4 1655.0625}
 {5 2262.815625}
 {6 2900.956406}
 {7 3571.004227}
 {8 4274.554438}
  {9 5013.28216}
{10 5788.946268}
```

6. The ordered pair you enter into the calculator first is (year, beginning balance). The 'Ans' refers to the previous value in the column. The parentheses are needed to make the calculator do the addition first, then the multiplication.

7. Answer:

Year (n)	Deposit Made	Number of Years Deposit is in Account	Process	Ending Balance
1	$500.00	6	$500 \cdot (1.05)^6$	≈ $670.05
2	$500.00	5	$500 \cdot (1.05)^5$	≈ $638.14
3	$500.00	4	$500 \cdot (1.05)^4$	≈ $607.75
4	$500.00	3	$500 \cdot (1.05)^3$	≈ $578.81
5	$500.00	2	$500 \cdot (1.05)^2$	≈ $551.25
6	$500.00	1	$500 \cdot (1.05)^1$	≈ $525.00
Total				$3571.00

8. $(500)*(1.05)^6$; $(500)*(1.05)^5$; $(500)*(1.05)^4$; $(500)*(1.05)^3$; $(500)*(1.05)^2$; $(500)*(1.05)^1$

9. Total ending balance = $(500)*(1.05)^6 + (500)*(1.05)^5 + (500)*(1.05)^4 + (500)*(1.05)^3 + (500)*(1.05)^2 + (500)*(1.05)^1$

10. The expression in Question 9 sums to $3571.00. This expression should be equal to the value in the table.

11. $(500)*(1.05)*[(1.05)^5 + (1.05)^4 + (1.05)^3 + (1.05)^2 + (1.05)^1 + 1]$

12. Substituting into the formula, we have:

$$B = (500) \cdot (1.05) \cdot \left(\frac{b^n - 1}{b - 1}\right)$$

$$B = (500) \cdot (1.05) \cdot \left(\frac{1.05^6 - 1}{1.05 - 1}\right)$$

$$B = (500) \cdot (1.05) \cdot \left(\frac{0.340096}{0.05}\right)$$

$$B = (500) \cdot (1.05) \cdot 6.8019$$

$$B = 3571.00.$$

13. After substituting, the formula for finding the ending balance would be:

$$B_n = d \cdot b \cdot \left(\frac{b^n - 1}{b - 1}\right).$$

14. Total payments = $d \cdot n$

15. The total interest earned would be found by the formula:

$$I = \text{Balance} - \text{Payments}$$

$$I_n = d \cdot b \cdot \left(\frac{b^n - 1}{b - 1}\right) - d \cdot n$$

16. When $n = 6$, this formula predicts the total interest to be $571.00. This is the same value we get when we subtract the total payments ($3000.00) from the total of the ending balances ($3571.00).

One of the powerful features of mathematics is that a formula that required a lot of work to derive can be used to solve many problems of the same type. (If you are lucky, the formula can even describe situations that aren't from the same context but *do* have the same underlying process.) The **savings formula** is one of those friendly formulas. Now that we know where it comes from, we can use it to solve a lot of problems!

Recall that b is the growth rate so $b = (1 + r)$. Substituting $(1 + r)$ for b in the *total interest earned formula from Question 15 gives us the savings formula.*

SAVINGS FORMULA

If you deposit d dollars per period (with the deposits at the beginning of the period) at an interest rate of r per period, the amount A accumulated is found by the savings formula:

$$A = d(1 + r)\left(\frac{(1+r)^n - 1}{r}\right)$$

When the deposits are made at the end of each period, the formula is even simpler.

$$A = d\left(\frac{(1+r)^n - 1}{r}\right)$$

17. Maria deposits $125 at the end of each quarter into an account that earns 5% compounded quarterly. Calculate the balance at the end of six years.

18. If Maria wants the account balance to reach $5000 in that same six-year period, how much would she have to deposit each quarter?

SECTION 7.9

Assignment

1. Find the ending balance after one year for a savings account in which $1200 principal is deposited, an additional $200 is deposited on the last day of each month, and the interest rate is 8% per year compounded monthly.

2. A couple wishes to save for the college education of their child. They want to make regular deposits at the end of each month into an account that earns 6.5% interest per year, compounded daily. Using a 360-day year, find the rate of increase for the account for one month.

3. If you invest $5000 into a savings plan that pays 6% interest compounded quarterly and wish to have $9000 after five years, how much should you deposit at the end of each quarter?

4. You decide to invest in an annuity when you are 30 years old.

   a) You will invest $25 per month for 35 years at 5% interest. How much will you have after 35 years?

   b) What if you invest $50 per month for the same time and interest?

   c) What if you invest $100 per month for the same time and interest?

Answers (continued)

17. In this situation the rate for the quarterly time period is 0.0125 ($r = 0.0125$), the deposits are \$125 ($d = 125$), and 24 payments are made ($n = 24$). So, the savings formula computes

$$A = d\left(\frac{(1+r)^n - 1}{r}\right)$$

$$A = 125\left(\frac{(1+0.0125)^{24} - 1}{0.0125}\right)$$

$$A = (125)(27.78808)$$

$$A = \$3473.51.$$

Using the calculator table:

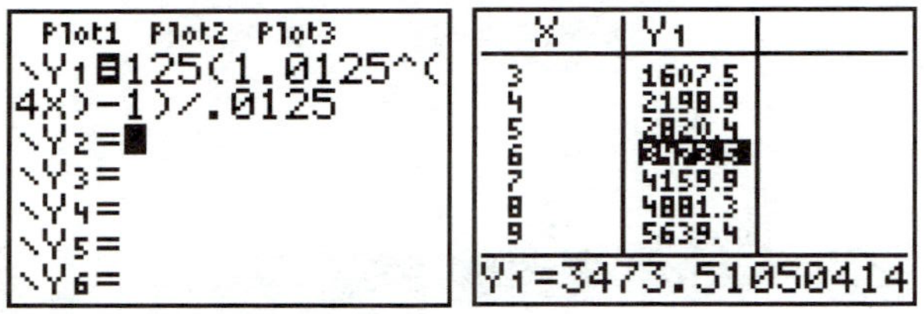

18. Now, $A = 5000$, $r = 0.0125$ and $n = 24$. Substitute I, the known information, and solve the equation for d. You would have to deposit \$179.93 each quarter.

$$A = d\left(\frac{(1+r)^n - 1}{r}\right)$$

$$5000 = d\left(\frac{(1+0.0125)^{24} - 1}{0.0125}\right)$$

$$5000 = d(27.78808)$$

$$\frac{5000}{27.78808} = \frac{d(27.78808)}{27.78808}$$

$$179.93 = d$$

Using the calculator table:

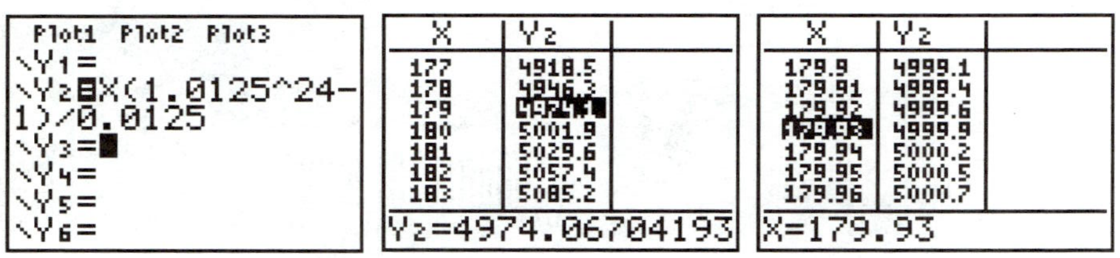

Section 7.9 Assignment

Answers

1. \$3789.58

2. The daily rate of increase is $0.065/360 \approx 0.000180556$, so the growth factor for one day is 1.000180556. The cumulative growth factor for 30 days is $(1.000180556)^{30}$, or around 1.005431. So the rate of increase would be 0.5431% per month.

3. The amount that the principal will earn in interest is $(5000)(1 + 0.06/4)^{20} - 5000$, or approximately \$1734.28. So you wish to accumulate \$2265.72. Using the savings formula and the interest rate of 1.5% per compound period, $2265.72 = d(1.015^{20} - 1)/(0.015)$. So the quarterly payments are around \$97.98.

4. a) Investing \$25 per month, you will have \$28,402.31 after 35 years.

   b) Investing \$50 per month, you will have \$56,804.62.

   c) Investing \$100 per month, you will have \$113,609.24

SECTION 7.10

Which Investment?

Let's apply what you have learned thus far to a new situation.

Investing money is a great way to make money. By making some careful decisions you can have a comfortable financial future.

Say your parents give you a graduation party and most of your guests give you money as a gift. When you count all of it you realize you have $1500.

You want to deposit the $1500 into an account at Big Bank. The account offers 6.25% compounded monthly. Your father thinks you should buy a Certificate of Deposit (CD) for $1500 that offers 6.5% compounded annually. Your aunt thinks you should deposit the $1500 in a credit union account that earns 6% compounded quarterly.

Whose idea will give you the most money at the end of 10 years? Justify your answer.

SECTION 7.11

I'm Feeling Down—Depreciation

The Gonzales family owns three cars. Mr. and Mrs. Gonzales each have a vehicle and Maria, their oldest daughter, has a car.

Maria has decided to use her old car as a trade-in on a new one. Maria and the dealer do not agree on the trade-in value of her old car, but neither feels it is worth the $30,000 it cost in 2000.

Maria learns that the value of a new car goes down as soon as it leaves the lot. Once a car has been titled in someone's name, the car's status goes from "new" to "used." Maria also learned that most cars lose value all the time, even if kept in good condition.

This loss of value is called **depreciation**, and mathematical rules determine how much value a car loses from aging.

> **CHECK THIS!**
> In accounting, depreciation is a tool used to determine the expense of replacing a capital, or expensive, item.

One method is to specify a number of years and decrease the value by the same amount of money each year until the car has no value left. The graph in **Figure 7.36** shows the change in value of Maria's car over time using this method.

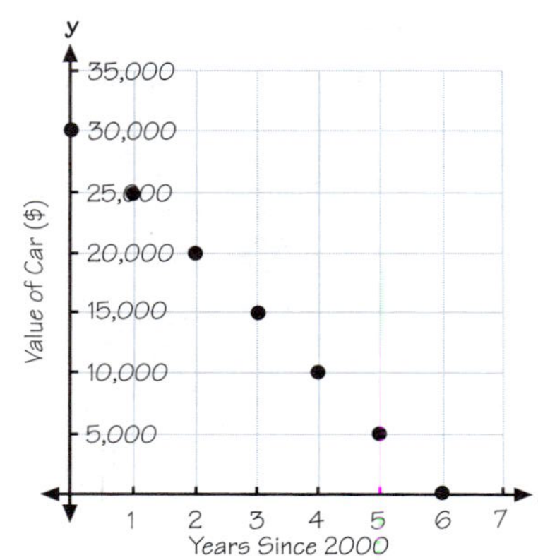

FIGURE 7.36.
Maria's car value over time.

EVALUATE LESSON 2

Section 7.10: Which Investment?

Purpose: In Sections 7.6 through 7.9, students developed the concept of compound interest as an exponential function. In this section students have the opportunity to demonstrate their knowledge by applying those concepts and skills to solve a problem in a new context. This prepares students to investigate methods of financing a car or a home in Lesson 3.

Description: Facing a problem in a new context, students will determine and implement a method to solve the problem. In this open-middle performance assessment task there is only one correct answer, but there is a variety of valid methods to obtain it. In their justification, which is the most important part of a performance assessment task, students should be able to communicate their knowledge of mathematics, and demonstrate that they have mastered the concepts presented in Sections 7.6 through 7.10.

Materials: graphing calculators

Teaching Ideas: This task may be assigned as group work or individual work. The mathematics performance assessment rubric should be used to evaluate students' work.

Answers

Sample solution:

Function Rules:

Your idea, Big Bank: $f(x) = 1500\left(1+\frac{0.0625}{12}\right)^{12x}$

Your father's idea, CD: $f(x) = 1500(1 + 0.065)^x$

Your aunt's idea, credit union : $f(x) = 1500\left(1+\frac{0.06}{4}\right)^{4x}$

Compare each idea graphically.

When all three functions are graphed on the same set of axes, they each appear to have almost the same value after 10 years.

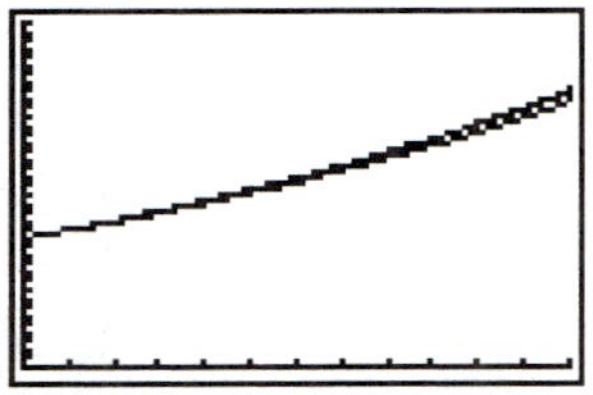

Use the trace feature to compare the values.

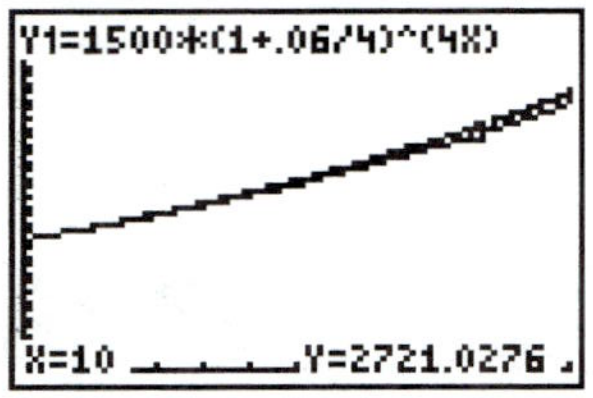

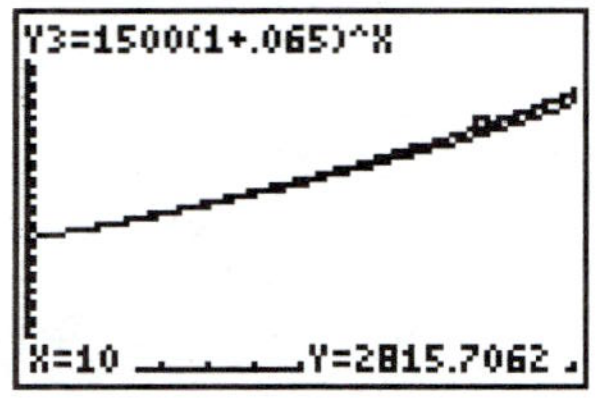

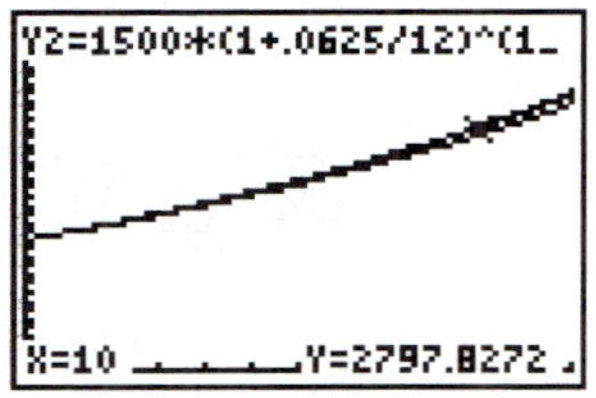

You will have $2721.02 in the credit union.

You will have 2815.71 in the CD.

You will have $2797.83 in the Big Bank account.

Your father's idea will yield the most money in 10 years.

ENGAGE LESSON 3

Section 7.11: I'm Feeling Down—Depreciation

Purpose: In Sections 7.6 through 7.10, students investigated investment options involving the growth of money through exponential functions. In Lesson 3 students investigate situations involving exponential functions and interest where they are paying interest instead of collecting it. In Section 7.11 students use tables to investigate the depreciation of a car. This will set students up to explore options for buying or leasing a car in Section 7.12.

Description: Students explore two different types of depreciation of the value of a car. First, students examine linear depreciation, where the value of a car approaches $0 rather quickly. Second, students investigate exponential depreciation, where the value of a car decreases by a certain percentage every year.

Materials: graphing calculator

Teaching Ideas: Students can work in small groups (3 to 4 students). Each student should work through the questions and make his or her own scatter plots and function rules. However, encourage students to discuss the problem together and to share their insights into solving the problem.

Encourage students to share their own experiences buying and selling cars. What other expenses are associated with owning a car? This is an excellent opportunity to discuss mathematics that is highly relevant to high school students.

1. What type of function does the graph seem to describe?

2. According to the graph, when will the value of Maria's car be $0?

This method of depreciation is called **linear depreciation**, since the car decreases in value by a constant rate each year. However, in reality, cars are rarely worth $0, even though we may joke about a worthless car.

Another method that Maria found is to have the value of the car go down by a certain percentage of its value *for that year*. Using this method, as the value of the car goes down so does the amount that the value decreases.

Maria wants to see how this method would work for a new car. She decides to use 20% as the rate of depreciation.

3. Calculate the value of the car at the end of the first year by taking away 20% of the Year 0 value. Fill in a table like **Figure 7.37** by taking away 20% of the value from the previous year. Show your calculations in the process column.

FIGURE 7.37.
Table for calculating car value.

Year	Year Number (t)	Process	Value (V) in Dollars
2005	0		30,000.00
2006	1		
2007	2		
2008	3		
2009	4		
2010	5		
2011	6		

4. Make a scatter plot of the value of the car versus time.

5. Write a function rule describing the depreciation of Maria's car.

6. Graph your function rule on your scatter plot. If necessary, use transformations to make the function rule fit the data.

7. What are the domain and range of your function rule?

8. What are the domain and range of the problem situation? How are these different from the domain and range for the function rule?

9. Using this method of depreciation, will the new car ever be worth nothing? How do you know?

Answers (continued)

1. The graph appears to be a linear function.
2. 6 years from 2000, in 2006, Maria's car will be worth \$0.
3. Answer:

Year	Year Number (t)	Process	Value (V) in Dollars
2005	0		30,000.00
2006	1	30,000(0.80) = 24,000	24,000.00
2007	2	24,000(0.80) = 19,200	19,200.00
2008	3	19,200(0.80) = 15,360	15,360.00
2009	4	15,360(0.80) = 12,288	12,288.00
2010	5	12,288(0.80) = 9830.40	9830.40
2011	6	9830.40(0.80) = 7864.32	7864.32

4. Answer:

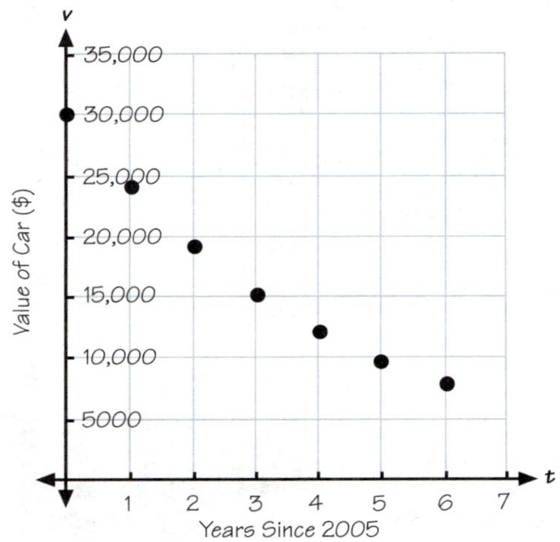

Using a graphing calculator, students should see a scatter plot similar to this one:

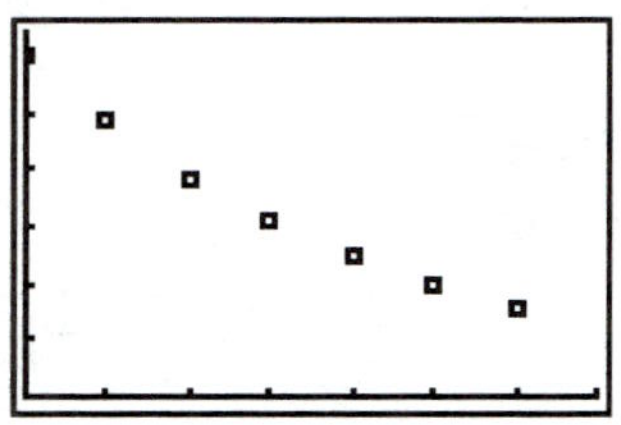

5. $V = 30{,}000(0.80)^t$
6.

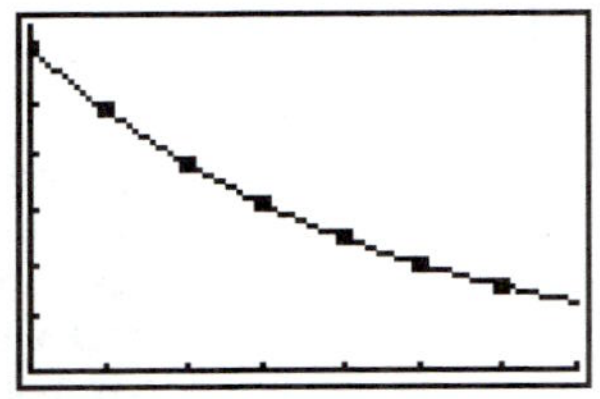

7. Domain: t is all real numbers; range: $V > 0$
8. Domain: $t \geq 0$; range: $0 \leq V \leq 30{,}000$. Because t represents the number of years since 2005, when Maria would purchase the car, it does not make sense to consider negative values for t. The range represents dollars. At $t = 0$, the year 2005, the new car would be worth \$30,000, so this is the maximum value of the vehicle. Since value is in dollars, at some point (the year 2070), the value will be rounded off to the nearest penny resulting in a value of \$0.
9. No. Each year, the car loses 20% of its value. In other words, each year, the car is worth 80% of what it was worth the previous year. Thus, there will always be some value, though it may become very small given enough time. However, due to the domain and range restrictions of this context, there will be a time (2070, to be exact) when rounding the value to the nearest cent produces an answer of 0.

SECTION 7.12 Should I Buy or Lease?

Figuring out which car to buy is an important part of owning a car. There are many factors to consider. In Section 7.11 you investigated one factor, depreciation. Other factors may include fuel economy and insurance. However, the purchase is only the first step. You also have to *pay* for it! And if you're like Maria, you don't have $30,000.

Fortunately, there are several options.

OPTION 1: PURCHASING A CAR

One option is to look into businesses that **finance** car purchases. That means that they lend you the money to buy a car and you pay them back in monthly installments. For that privilege, your car will cost more than you thought it would. In researching her car purchase, Maria found out that the **annual percentage rate**, or **APR,** plays a big role.

Assume that Maria settles on a car whose selling price is $27,000. She considers having it financed by a four-year loan for the same amount with an APR of 6.8%.

1. Assume that Maria's loan will be computed with simple interest for four years. What is a reasonable estimate of the amount of interest Maria will pay?

2. Based on your answer for Question 1, what is the total amount of money including principal and interest that Maria will pay for her new car?

That is a nice way of thinking about APR, but that is not the way that interest and monthly payments are calculated. Each month you owe some of that interest, and the payment will pay for that *plus* some of the loan. As the amount you owe goes down, you make the same payment but you owe less interest; therefore, you repay more of the loan. By the end of the **loan term** (the amount of time it takes to pay off the loan) there will be nothing left to pay.

3. Maria is going to borrow $27,000 with a four-year 6.8% APR loan. Car loans are compounded monthly. What will be Maria's monthly interest rate?

4. Suppose Maria's payments are $640 per month. If her payments are due at the end of the month, interest is compounded from the original principal. What amount of interest does Maria pay in her first monthly payment?

5. If you begin with the original principal, add the first month's interest, then subtract the monthly payment, you can calculate the principal balance at the end of the first month. What is Maria's principal balance at the end of the first month?

6. Continue this process to fill in a table like **Figure 7.38**.

Payment Number	Process	Principal Balance
0		27,000.00
1	$27{,}000\left(1+\frac{0.068}{12}\right) - 640$	26,513.00
2		
3		
4		
5		

FIGURE 7.38.
Computing the principal balance.

7. How did you calculate each monthly payment?

We could continue this process for 48 months, but that is time-consuming. Instead, we can use technology. Some Websites have monthly payment calculators. You can also use a spreadsheet. You can also use the Time Value of Money, or TVM Solver on a graphing calculator.

EXPLORE LESSON 3

Section 7.12: Should I Buy or Lease?

Purpose: In Section 7.11 students investigated depreciation in order to find out what the value of a car will be after a certain amount of time. In this section students use technology to investigate amortization so that they can evaluate whether to buy or lease a car. In Section 7.13 students develop an amortization table in order to formalize the concept of amortization and how it relates to exponential functions.

Description: In this section students investigate the mathematics behind purchasing or leasing a car. They use the TVM Solver application of a graphing calculator in order to compare different loan terms and calculate the total cost of the loan. Then, students examine the factors governing a long-term lease agreement.

Materials: graphing calculator with finance application, including TVM Solver

Teaching Ideas: Students can work in small groups (3 to 4 students). Students will be using the TVM Solver during this activity. Screen shots and instructions are based on the TI-83 or TI-84's Finance APP. Other graphing calculators have a TVM Solver feature as well. Consult the user's guide for details on how to use that feature on other graphing calculators.

In this section students use technology to investigate a situation. In the next section, the Explain phase of this lesson, students investigate amortization tables in more detail to learn why the TVM Solver works as it does. This is backwards from other Explore-Explain cycles in this book. However, the concept of amortization is much easier for the students to grasp in this order.

Students also learn the terminology and processes involved in leasing a vehicle. Many of the terms are analogous to terms involved in financing the purchase of a new vehicle. During the Explain phase of the lesson, Section 7.13, students define the terms and compare them to the terms associated with buying a car.

It is likely that this section will take several days for students to complete. Day 1 could be spent working through Option 1: Purchasing a Car, then working homework Questions 1–5. Day 2 could be spent working through Option 2: Leasing a Car, then working homework Questions 6–10. Students may need more time, depending on their calculator proficiency.

Monitor calculator use and assist as necessary.

Answers

1. $I = P \cdot r \cdot t = 27{,}000(0.068)(4) = \$7344$
2. $\$7344 + \$27{,}000 = \$34{,}344$
3. $0.068 \div 12 \approx 0.00567$
4. $27{,}000\left(\frac{0.068}{12}\right) = 153$
5. $27{,}000 + 153 - 640 = 26{,}513$
6. Answer:

Payment Number	Process	Principal Balance
0		27,000.00
1	$27{,}000\left(1+\frac{0.068}{12}\right) - 640$	26,513.00
2	$26{,}513\left(1+\frac{0.068}{12}\right) - 640$	26,023.24
3	$26{,}023.24\left(1+\frac{0.068}{12}\right) - 640$	25,530.71
4	$25{,}530.71\left(1+\frac{0.068}{12}\right) - 640$	25,035.38
5	$25{,}035.71\left(1+\frac{0.068}{12}\right) - 640$	24,537.25

7. Answers may vary. Students should observe that you take the previous month's principal balance, multiply by $\left(1+\frac{0.068}{12}\right)$, then subtract 640.

NOTE: Use the viewscreen calculator to display the TVM Solver.

Use the TVM Solver on your graphing calculator to find the exact terms of Maria's loan by using these steps.

Step 1: Press the APPS key and select the Finance application as shown in **Figure 7.39.**

FIGURE 7.39.
Finance application on a calculator.

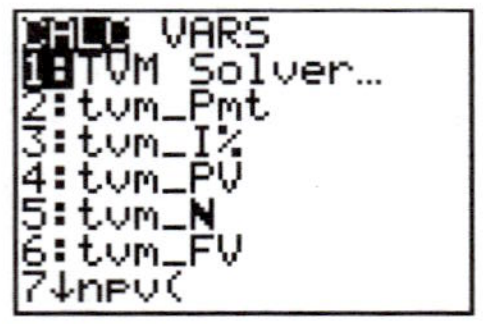

FIGURE 7.40.
TVM Solver option.

Step 2: Select the TVM Solver as shown in **Figure 7.40.**

Step 3: Enter the known information into the list. If you do not know the value, enter 0.

Let's use the TVM Solver to solve Maria's problem. If she buys a car for $27,000, finances the purchase with a loan having a 6.8% APR for four years, what will she end up paying including interest for the car? Assume that interest is compounded monthly based on the amount of the principal that Maria has not yet paid.

> Given four values, the TVM Solver solves for the fifth. Enter cash inflows as positive numbers and cash outflows as negative numbers. The variables are:
>
> - N = Total number of payment periods
> - I% = Annual interest rate (a percent, not a decimal value)
> - PV = Present value (at the beginning of the loan)
> - PMT = Payment amount
> - FV = Future value (at the end of the loan)
> - P/Y = Number of payment periods per year
> - C/Y = Number of compounding periods per year
>
> (When you store a value to P/Y, the value for C/Y automatically changes. To store a different value to C/Y, store the value to C/Y after you have stored a value to P/Y.)

8. What values would you enter into the TVM Solver for:

   a) N = Total number of payment periods?

   b) I% = Annual interest rate (a percent, not a decimal value)?

   c) PV = Present value (at the beginning of the loan)?

   d) PMT = Payment amount?

   e) FV = Future value (at the end of the loan or investment)?

   f) P/Y = Number of payment periods per year?

   g) C/Y = Number of compounding periods per year?

9. Enter these values into the TVM Solver.

10. Select the variable for which you want to solve. In Maria's case we want to find her monthly payment, so use the arrow keys to select PMT. Then press ALPHA-SOLVE. Record your results rounded to the nearest cent.

11. If Maria makes four years' worth of the monthly payments in Question 10, how much does she pay for her car including interest?

12. Based on your answer to Question 11, how much interest does Maria pay?

13. Maria decides to lower her monthly payment by financing the car over five years. Use the TVM Solver to calculate her new monthly payment.

14. If Maria finances the car over five years, how much interest does she pay?

15. How does the amount of interest Maria pays over five years compare to the amount of interest she pays if she finances the car for four years?

16. If Maria has budgeted $450 per month for a car payment, and the best interest rate she can get is 5.25% on a four-year loan, what is the most Maria can borrow?

Answers (continued)

8. a) $4 \times 12 = 48$

   b) 6.8

   c) 27,000

   d) This is the unknown quantity, so enter 0.

   e) This is the ending balance, which will be 0.

   f) 12

   g) 12

9. The graphing calculator screen should look like this:

```
N=48
I%=6.8
PV=27000
PMT=0
FV=0
P/Y=12
C/Y=12
PMT:END BEGIN
```

10.

```
N=48
I%=6.8
PV=27000
PMT=0
FV=0
P/Y=12
C/Y=12
PMT:END BEGIN
```

```
N=48
I%=6.8
PV=27000
▪PMT=644.04622…
FV=0
P/Y=12
C/Y=12
PMT:END BEGIN
```

$644.05

11. $30,914.40

12. $3914.40

13.

```
N=60
I%=6.8
PV=27000
▪PMT=532.088304
FV=0
P/Y=12
C/Y=12
PMT:END BEGIN
```

$532.09

14. $4925.40

15. If Maria finances the car for 5 years instead of 4, she will pay an additional $1011 in interest.

16.

```
N=48
I%=5.25
▪PV=19444.56576
PMT=450
FV=0
P/Y=12
C/Y=12
PMT:END BEGIN
```

$19,444.56

Maria has found that three things affect the monthly payments and therefore the total amount that the loan will cost. These things are the amount of money borrowed, the APR, and the loan term. Using technology allows Maria to compare different combinations of the three in order to find the one that best fits her budget.

OPTION 2: LEASING A CAR

17. What are some expenses that you face when owning a car?

At most dealers you can **lease** a car. Leasing is like having a long-term rental agreement. The consumer pays a fixed amount each month to the leasing company (who owns the car), and also pays for insurance and operating costs. At the end of the lease agreement, the company gets the car.

Because leasing a car is different from purchasing one, there is a new set of terms used to describe the factors that influence the cost of the lease. For now, let's work with the numbers so that we can find the costs associated with leasing. We will explain what the terms are later.

While shopping for her new car, Maria found this in an ad:

Lease a new Ford Explorer XLS with automatic transmission, 6-cylinder engine, anti-lock brakes, AM/FM radio, air conditioning, and power steering. MSRP is $25,810, and financing is available at a special APR rate of 0.9% for 48 months with a mileage allowance of 45,000 miles. $1000 down payment and $800 security deposit due on lease signing.

Maria and the dealer agreed to base the lease value of the car as $25,000. The car dealer agreed to buy Maria's old car for $5000 (this is commonly called **trading in** your old vehicle for a new one). Even though she thought it was worth more, she agreed to the offer of $5000.

Maria organized the information in the table in **Figure 7.41.**

Lease Conditions	Value
APR	0.9%
Term of Lease	48 months
Down Payment	$1000
Security Deposit	$800
Milage Allowance	45,000 miles

Leasing Figures	Amount
MSRP	25,810.00
Capitalized Cost	
Net Trade-in Allowance	
Adjusted Capitalized Cost	
Residual Value	
Money Factor	

FIGURE 7.41.
Conditions and cost of leasing an Explorer.

18. Record the selling price in a table like Figure 7.41 under the amount next to capitalized cost. Record the trade-in allowance that the dealer gave Maria under the amount next to net trade-in allowance.

19. Consider the discounts and extra fees that Maria will either receive or pay. What is the final lease price of the vehicle? Record this value in the table under the amount next to adjusted capitalized cost.

20. The dealer thinks that the car will depreciate to 48.43% of its value over the term of the lease. To the nearest dollar, what will be the value of the car at the end of the lease?

21. Begin with the APR written as a decimal. Find the interest rate for one compounding period if the interest is compounded monthly.

22. Divide your answer from Question 21 by 2. Record this value in your table under the amount next to money factor.

Maria needs to know what her monthly payments are going to be. Assume that the sales tax rate is 6.25%.

23. Calculate the monthly depreciation by subtracting the **residual value**, the value of a car at the end of a lease, from the adjusted capitalized cost from Figure 7.41, then dividing by the length of the lease in months. Record that value in a table like **Figure 7.42.**

24. Use your data from Figure 7.41 to calculate the monthly interest charge by adding the adjusted capitalized cost and residual value then multiplying that sum by the money factor. Record that value in your monthly payment table.

25. Calculate the monthly sales tax by multiplying the tax rate by the sum of the monthly depreciation and monthly interest charge. Record that value in your monthly payment table.

26. Calculate Maria's total monthly payment. Record that value in your monthly payment table.

Leasing Calculations	Amount
Monthly Depreciation	
Monthly Interest Charge	
Monthly Sales Tax	
Monthly Payment	

FIGURE 7.42.
Monthly payment table.

Answers (continued)

17. Answers may vary. Responses may include insurance, fuel, car repairs, depreciation, etc. NOTE: Ask students to answer this question as a class discussion, recording their responses on the overhead or chalkboard. This is a brainstorming session, so record all responses. You can evaluate how reasonable the responses are later in the lesson.

18. See completed table below.

19. $25,000 – $5000 – $1000 = $19,000

20. 0.4843($25,810) = $12,500

21. $\frac{0.009}{12} = 0.00075$

22. $0.00075 \div 2 = 0.000375$

Completed table after answering Questions 19–22:

Lease Conditions	Value
APR	0.9%
Term of Lease	48 months
Down Payment	$1000
Security Deposit	$800
Milage Allowance	45,000 miles

Leasing Figures	Amount
MSRP	$25,810.00
Capitalized Cost	$25,000.00
Net Trade-in Allowance	$5000.00
Adjusted Capitalized Cost	$19,000.00
Residual Value	$12,500.00
Money Factor	0.000375

23. $\frac{19,000-12,500}{48} = \$135.42$

24. (19,000 + 12,500)(0.000375) = $11.81

25. ($135.42 + $11.81) × 0.0625 ≈ $9.20

26. $135.42 + $11.81 + $9.20 = $156.43

Completed table after answering Questions 23–26:

Leasing Calculations	Amount
Monthly Depreciation	$135.42
Monthly Interest Charge	$11.81
Monthly Sales Tax	$9.20
Monthly Payment	$156.43

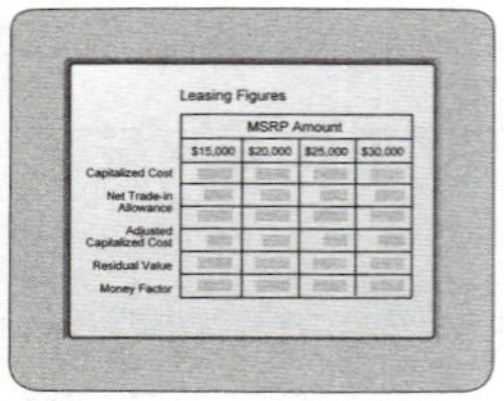

In order to compare other kinds of financing, Maria needs to know how much the leasing agreement actually cost her. Assume that she will drive the car 52,000 miles during the four years, the surcharge for extra miles is $0.12 per mile, and she will be charged an additional $1000 for excess wear and tear.

27. Calculate the total of the payments she will make.

28. Calculate the excess mileage charge.

29. Based on your answers to Questions 27 and 28, calculate the total cost of leasing the car for four years.

30. Add the trade-in allowance that Maria received to the total cost of the lease. This is the actual cost to Maria.

Maria wants to compare the leasing deal with what it would cost to buy the same car at the same selling price ($25,000), down payment ($1000), trade-in allowance ($5000), and finance rate (0.9% APR).

31. What would her monthly payments be for a four-year loan?

32. How much would it cost her to finance the car for four years?

33. Assuming that the new car will depreciate to 50% in four years, what will the new car be worth at the end of the loan?

34. Based on your answers so far, what will be the actual cost of owning the new car for the four years?

35. Based on Maria's experience, is it better to lease or buy a car? Explain your answer.

SECTION 7.12 Assignment

1. Use the TVM Solver to find the missing values in **Figure 7.43.**

	Present Value	Interest Rate	Time	Future Value
a	5000	6%	10	
b	5000	6%		10,000
c	5000		10	10,000
d		6%	10	10,000

FIGURE 7.43.
Table with missing values.

2. Compute the monthly payment, total cost of the loan, and amount of interest paid for these car loans:

   a) Joanie can get an $11,000 loan for five years at 6% APR.

   b) Mitchell can get an $11,000 loan for five years at 7% APR.

   c) Marisela can get an $11,000 loan for five years at 8% APR.

At the end of Section 7.12, Maria had compared the total cost of leasing a new car for four years with the total cost of purchasing the car and financing it for the same amount of time. Her calculations showed that for the APR rate she was offered, the two strategies would cost her just about the same amount. By leasing a car she spent a lot of money without a car to show for it. Is this a good decision? She has all the tools she needs to compare options over a longer time period and see which strategy is best.

Maria has several options available now; exploring each one will allow you to answer her question about which one is the best. In all cases assume that operating costs are the same, and focus on the cost of buying versus leasing.

3. She could lease another car in four years. Assume that the same arrangements apply. How much will it cost her to lease a new car four years in the future for another four years?

4. How much will her total leasing costs be over the eight-year period?

Answers (continued)

27. ($156.43/month)(48 months) = $7508.64

28. (52,000 – 45,000) × (0.12/mile) = $840

29. $7508.64 + $840 + $1000 = $9348.64

30. She will get the security deposit back, but the cost of ownership is the leasing costs plus the amount she gave the leasing company for the transaction or the down payment and the trade-in allowance. So it would be $6000 + $9348.64 = $15,348.64.

31. $403.15

32. ($403.15/month)(48 months) = $19,351.20

33. $12,500

34. Her total cost for the car purchase would be the $6000 (down payment and loss of equity in the car she traded in) plus the payments she made. So, the car cost her a total of $25,351.20, but she still has the value in the new car of $12,500. So, the actual cost of ownership for the four years would be $25,351.20 – $12,500 = $12,851.20.

35. Answers may vary. The actual answer is less important than the reasoning behind it. Students may observe that when you lease, you actually pay less per month and will pay less for the use of a car for four years. However, students may also observe that when you lease a car, at the end of the lease agreement, you will either have to purchase the car or return it to the leasing company. If you purchase a car, then when the loan is paid off, you have equity in the car and thus have something of value.

Section 7.12 Assignment

Answers

1.

	Present Value	Interest Rate	Time	Future Value
a	5000	6%	10	8954.24
b	5000	6%	11.9	10,000.00
c	5000	7.2%	10	10,000.00
d	5583.95	6%	10	10,000.00

Below are the screen shots used to solve each of the questions above:

Question 1a

```
N=10
I%=6
PV=-5000
PMT=0
FV=8954.238483
P/Y=1
C/Y=1
PMT:END BEGIN
```

Question 1b

```
▪N=11.89566105
I%=6
PV=-5000
PMT=0
FV=10000
P/Y=1
C/Y=1
PMT:END BEGIN
```

Question 1c

```
N=10
▪I%=7.177346254
PV=-5000
PMT=0
FV=10000
P/Y=1
C/Y=1
PMT:END BEGIN
```

Question 1d

```
N=10
I%=6
▪PV=-5583.947769
PMT=0
FV=10000
P/Y=1
C/Y=1
PMT:END BEGIN
```

2. a) On a 6% loan, the monthly payment is $212.66, the total cost of the loan is $12,759.60, and the interest paid is $1759.60.

   b) On a 7% loan, the monthly payment is $217.81, the total cost of the loan is $13,068.60, and the interest paid is $2068.60.

   c) On an 8% loan, the monthly payment is $223.04, the total cost of the loan is $13,382.40, and the interest paid is $2382.40

3. Assuming that she had the same back end charges as before, it will cost her $15,348.64. Without the penalties, it will cost $13,508.64.

4. The higher estimate would be $30,697.28 while the lower estimate would be $28,857.28.

5. At the end of the lease, Maria could buy the purchase option price and finance it over the next four years at the same APR (0.9%). Recall that the purchase option price was $11,500. How much would Maria's monthly payments be?

6. Assume that after eight years the resale value of the car will be only $5000. How much will it cost Maria to keep the car another four years?

7. How much did the lease and buy option cost her over the entire eight-year period?

8. Had Maria just bought the car, financed it over four years, and kept it for a total of eight years, how much would it cost her? (Hint: the value of the car is $5000 at the end of the eight years.)

SECTION 7.13 Amortization of a Car Loan

In Section 7.12 you explored two ways that Maria could get a new car. One was to purchase the car and the other was to lease it. You also looked at ways Maria could borrow money to pay for her new car.

When you finance a purchase such as a new car, you borrow money and pay interest. Interest is often compounded monthly based on a yearly rate. As you pay off the loan, you pay less interest because the amount you owe is less. This process of computing the amount of interest and the amount of principal that you pay each month is called **amortization**.

Suppose you are buying a used car for $8000. You put $1600 down and finance the remainder at 9.5% interest for 36 months. The monthly payment is $205.01. Create an amortization table for the loan.

Handout 7.2 contains a copy of the table you just built. Use your table to answer these questions.

1. How is interest computed on a loan?

2. How much total interest do you pay?

3. How much do you end up paying for the car?

4. How much money do you save if you pay off the loan at 24 months?

5. How much money do you save if you pay off the loan at 12 months?

6. Compare your answers to Questions 3 and 4. Explain what you find.

7. Is it to your advantage to pay off the loan early? Why or why not?

Answers (continued)

5. $244.01

```
N=48
I%=.9
PV=11500
▪PMT=▮244.01153...
FV=0
P/Y=12
C/Y=12
PMT:END BEGIN
```

6. The total of her payments would be ($244.01 / month)(48 months) = $11,712.48, but she would still have $5000 in equity in the car. Therefore, it would cost her $6712.48.
7. The lease arrangement cost her $15,348.64. Buying the car and financing it for an additional 4 years cost another $6712.48. Altogether it cost $22,061.12.
8. Assuming that there were no major repair bills, the new car cost Maria $25,351.20; however, after 8 years, she still has $5000 in equity in the car. So, in this scenario, it cost her a total of $20,351.20 over the 8 years.

EXPLAIN LESSON 3

Section 7.13: Amortization of a Car Loan

Purpose: In Section 7.12 students investigated amortization using technology and compared buying a car with leasing a car in order to make a decision about which was a better method to get a new vehicle. In Section 7.13 students use a spreadsheet to formalize the mathematics behind amortization. This prepares students to investigate home loans and apply what they have learned about car loans in Section 7.14.

Description: In this phase of the lesson students use a spreadsheet to build an amortization table. By inputting the formulas into the spreadsheet cell, students get a better feel for the recursive nature of amortization. They also look at the amount of interest that a person pays when they obtain a loan for a new car and how that interest payment changes over time.

Materials: computer with projection device or computer lab, spreadsheet program, copies of Handout 7.2 for each student

Teaching Ideas: Debrief students after Section 7.12 by having them share their answers to the questions in the assignment for 7.12. Be sure to address any missing concept, process, or skill that could have been used to answer the questions. Then work through the questions in this section with the entire class.

This activity is written using technology. It can be adapted in several ways. If you have access to a computer lab or several workstations, you can have the students use a spreadsheet to explore exactly what the numbers in an amortization table mean. Alternatively, you can use a computer projector to demonstrate the use of the spreadsheet as a whole-class discussion. A student volunteer can enter the data and formulas into the spreadsheet cells while you facilitate the discussion.

Using a computer projection device, work with students to create a spreadsheet amortization table for the car loan. A sample spreadsheet follows. The second spreadsheet shows the formulas used to create the first spreadsheet. After row three is filled in, a "fill down" command can be performed to create the rest of the table.

	A	B	C	D
1	Payment Number	Interest Portion	Principal Portion	Balance
2				6400
3	1	50.67	154.34	6245.66
4	2	49.44	155.57	6090.09
37	35	3.21	201.80	203.44
38	36	1.61	203.40	0.04
39		980.40	6399.96	

	A	B	C	D
1	Payment Number	Interest Portion	Principal Portion	Balance
2				6400
3	1	=D2*(0.095/12)	=205.01 – B3	=D2 – C3
4	=A3 + 1	=D3*(0.095/12)	=205.01 – B4	=D2 – C4
37	=A36 + 1	=D36*(0.095/12)	205.01 – B37	=D36 – C37
38	=A37 + 1	=D37*(0.095/12)	205.01 – B38	=D37 – C38
39		=SUM(B3:B38)	=SUM(C3:C38)	

Answers

1. The interest is compounded monthly based on the amount of the principal that is still owed, or the balance of the loan.
2. $980.40
3. $1600 down payment + $980.40 interest + $6400.00 balance = $8980.40
4. $122.06 is the sum of the interest amounts from payment 25 through payment 36.
5. 455.21 is the sum of the interest amounts from payment 13 through payment 36.
6. Your savings almost quadruples if you pay off after 1 year as opposed to 2 years.
7. Answers may vary, but should be yes. Sample reason: Since interest is calculated based on the balance of the principal due, the faster you can pay the principal down, the less interest you will pay. Students might raise the issue of investing the money, rather than paying off the loan. In most cases, however, your return on an investment would be lower than 9.5%.

Consider a leasing option.

Maria negotiates a $25,000 price for the car, and she chose $2000 in options. The **capitalized cost** is the combination of all expenses related to buying the car:

$25,000 base sticker price + $2000 options = $27,000.

Maria is prepared to give a **down payment** of $1000. She is also being given a **net trade-in allowance** of $3000 for her old car. To find the **adjusted capitalized cost**, subtract these credits from the selling price, or capitalized cost.

8. After applying the down payment and the net trade-in allowance, how much is Maria's adjusted capitalized cost?

Assume that the **lease term** (the loan term when purchasing) is 36 months. For that interval it is estimated that the residual value, or value of the car at the end of the lease term, is $13,955. Dealers calculate the residual value as a percentage of the MSRP (manufacturer's suggested retail price). This value is often between 50% and 60% of the MSRP, depending on the length of the lease.

Maria organized the information as a formula:

(capitalized cost) – (down payment) – (net trade-in allowance) = (adjusted capitalized cost)

There are three factors that determine the monthly payment for a lease:

- monthly depreciation
- finance charge or lease charge
- sales tax

The first step is to find the **monthly depreciation**, which measures the monthly decrease in the value of the car:

$$\text{monthly depreciation} = \frac{\text{adjusted capitalized cost} - \text{residual value}}{\text{lease term}}$$

9. In this case, what is the monthly depreciation?

The second step is to find the **monthly interest charge**, which is also called the rent charge or the lease charge. (This is like the finance charge in purchasing a car.) To find the monthly interest charge, you add the residual value to the adjusted capitalized cost and multiply that sum by something called the **money factor**. A money factor is used to find the finance charge, which is found as follows:

$$\text{finance charge} = \frac{(\text{adjusted capitalized cost} + \text{residual value})}{2} \times \frac{\text{interest rate}}{12}$$

$$\text{finance charge} = (\text{adjusted capitalized cost} + \text{residual value}) \times \left(\frac{\text{interest rate}}{2 \times 12}\right)$$

$$\text{finance charge} = (\text{adjusted capitalized cost} + \text{residual value}) \times \text{money factor}$$

The money factor includes dividing the interest rate by 12 (since there are 12 months in a year and interest is compounded monthly), then dividing by 2 (to get a mean value by averaging the adjusted capitalized cost and residual value). While this is a common method, the money factor is not calculated the same way by all lease companies.

10. Suppose, for example, that Maria can get an interest rate of 8.04%. What is the money factor?

11. In this case what is Maria's monthly interest charge?

The third part is sales tax. You pay sales tax on both the monthly depreciation and the monthly interest charge.

12. Assume that the tax rate is 6.25%. How much sales tax does Maria pay per month?

13. Based on these three parts, what is Maria's monthly payment?

14. Over the term of the lease, how much does Maria pay to the lease company?

15. List some reasons why leasing is popular.

16. List some reasons why leasing might not be a good idea.

17. Write a formula that describes how to calculate monthly depreciation.

Answers (continued)

8. $27,000 – $1000 – $3000 = $23,000
9. ($23,000 – $13,955) = $9045/36 months = $251.25
10. $\frac{\text{interest rate}}{2 \times 12} = \frac{0.0804}{24} = 0.00335$
11. ($23,000 + $13,955) × 0.00335 ≈ $123.80
12. ($251.25 + $123.80) × 0.0625 ≈ $23.44
13. ($251.25 + $123.80 + $23.44) = $398.49
14. ($398.49/month) × (36 months) = $14,345.64
15. Answers will vary. Monthly loan payments are high (because the cost of a vehicle is high) compared to lease payments. You get to drive a fairly new car. You don't have to worry about age-related repairs.
16. Answers will vary. You always have monthly payments. You won't own the car at the end of the term. You might pay more in the long run.
17. (adjusted capitalized cost – residual value)/(term of lease) = monthly depreciation

18. Write a formula that describes how to calculate the monthly interest charge.

19. Write a formula that describes how to calculate the monthly sales tax.

20. Write a formula that describes how to calculate the monthly payment.

21. Summarize the vocabulary in this section by completing **Figure 7.44.**

Definition	Finance Term	Leasing Term
Price that the buyer and seller of a car agree upon after negotiation		Capitalized cost
Amount of money that the buyer pays toward the purchase or lease before the loan or lease begins	Down payment	
Amount of money that the dealer will pay for the buyer's old car		Trade-in allowance
Amount of money that the buyer will either finance or lease, after any credits and charges are applied to the negotiated price	Principal balance	
Length of time required to make payments		Lease term
Rate used to calculate the interest portion of the monthly payments	Annual percentage rate	
Loss in value of the car over time		Depreciation
Value of the car after the loan has matured	Equity	

FIGURE 7.44. Vocabulary table.

SECTION 7.13 Assignment

Jennifer is considering how to finance a Sport Utility Vehicle (SUV). She isn't sure if the Special Edition, or SE option package, is worth the extra money so she considers both. With sales tax the list price for a standard SUV (XE) is $25,095.19 and for a fully loaded SE the price is $26,786.69.

1. Use the rounded figure of $25,000 for the selling price of the XE.

   a) If the dealer offers financing at 7.4% APR over four years, what does the XE cost Jennifer?

   b) If the loan is financed over five years instead of four years, how much does the payment go down?

   c) How much more does the car cost Jennifer if the loan is for five years rather than four?

2. Use the rounded figure of $27,000 for the selling price of the fully loaded SE.

   a) A local bank offers five-year loans at 7.0% APR. How much does the car cost Jennifer with this financing?

   b) A local lending company offers five-year loans at 6.6% APR. How much does Jennifer save by borrowing the money from this company instead of the bank?

3. Jennifer *really* wants the SE. She has been saving money for the day when she can get one and has $1000 for a down payment. She also has a car that the dealer will take as a $3000 trade-in allowance. By shopping around she found the lowest financing of 6.4% APR for a four-year term. Assuming the rounded $27,000 price for the SE, if she applies both the down payment and the trade-in and finances the rest, how much does the car cost her?

Answers (continued)

18. (adjusted capitalized cost + residual value)*(money factor) = monthly interest charge

19. (monthly depreciation + monthly interest charge)*sales tax rate = monthly sales tax

20. (monthly depreciation + monthly interest charge + monthly sales tax) = monthly payment

21. Answers:

Definition	Finance Term	Leasing Term
Price that the buyer and seller of a car agree upon after negotiation	Selling price	Capitalized cost
Amount of money that the buyer pays toward the purchase or lease before the loan or lease begins	Down payment	Down payment
Amount of money that the dealer will pay for the buyer's old car	Trade-in allowance	Trade-in allowance
Amount of money that the buyer will either finance or lease, after any credits and charges are applied to the negotiated price	Principal balance	Adjusted capitalized cost
Length of time required to make payments	Loan term	Lease term
Rate used to calculate the interest portion of the monthly payments	Annual percentage rate	Money factor
Loss in value of the car over time	Depreciation	Depreciation
Value of the car after the loan has matured	Equity	Residual value

Section 7.13 Assignment

Answers

1. a) Financing $25,000 at 7.4% APR for 4 years, she will have monthly payments of $603.31. The total of the payments will be ($603.31 / month)(48 months) = $28,958.88.

   b) For the same amount and APR, the monthly payment for a 5-year loan term is $499.76. So, the monthly payment went down by $103.55.

   c) Making monthly payments of $499.76 for 60 months, she will pay a total cost of $29,985.60 for a 5-year loan term. When compared to the 4-year loan amount, she pays an extra $1026.72.

2. a) Financing $27,000 over 5 years at 7.0% APR, she will have monthly payments of $534.63. The total cost is ($534.63 / month)(60 months) = $32,077.80.

   b) Financing $27,000 over 5 years at 6.6% APR, she will have monthly payments of $529.55. The total cost is ($529.55 / month)(60 months) = $31,773.00. Jennifer would save $304.80 by financing at the lower APR at the lending company.

3. The amount to be financed is (27,000 – 1000 – 3000) = $23,000. Borrowing that amount at 6.4% for 4 years, she will have monthly payments of $544.38. The total of the payments is ($544.38 / month)(48 months) = $26,130.24. So the car costs Jennifer a total of $30,130.24, including the cash from the down payment and the cash value from the car trade-in.

4. A tip that consumer groups suggest for reducing costs is to purchase options like car stereos *after* you buy the car. Assume Maria buys the basic-model car; with sales tax included, the selling price is around $30,000. She will use her old car's trade-in value of $5000 and finance the rest with a five-year loan at 7.0% APR. She has three options for getting the sound system.

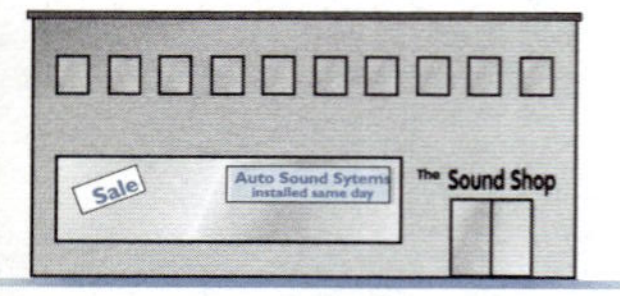

a) One option is to drive her new car to a store that sells car stereos and buy a system with 6-CD changer for $1000 (installation included). How much will that option cost her above the cost of the car?

b) A second option is to have the dealer install the system and 6-CD changer as options. The selling price of the car will go up by $1000. How much extra will she pay as a result of having to borrow more money to buy the car?

c) A third option is to have the dealer install the sound system as an option, but to match the additional cost by making a $1000 down payment. How much will that option cost her above the cost of the car?

d) What assumptions is the consumer group making when they recommend that you buy the stereo *after* you buy the car?

5. Jennifer also found dealers who would lease an SE. One dealer offered her these terms: MSRP of $25,211; 36-month lease at 6.4% APR; 10,000 miles per year; excess mileage surcharge $0.18/mile; residual value of $14,500.

a) Under those terms, how much is Jennifer's monthly lease payment?

b) Excluding operating costs, how much does the car cost Jennifer to drive for the three years?

6. Jennifer really liked the car and negotiated a selling price of $24,500. She extended the lease term to 48 months. (The residual value was adjusted to be $12,200.) She agreed to make a $1000 down payment and applied the $3000 trade-in value of her old car.

a) Under these terms, how much is Jennifer's monthly lease payment?

b) Assume Jennifer put 53,000 miles on the car during the lease term. Assume she also paid $2100 for excess wear on the car. Excluding operating costs, how much does the car cost Jennifer to drive for four years?

c) If she had *bought* the car, and financed what she needed at the same APR, how much would the car have cost Jennifer during those four years? (Use the residual value to estimate the worth of the car after the lease term expired.)

d) Do you think leasing the car is a good option? Explain.

7. Jennifer decided that shopping around for the best lease made sense. She checked with two car dealers using the same assumptions: capitalized cost of $24,500, down payment of $2000 with financing of 6.9% APR. Her information is in **Figure 7.45.**

	Leasing Company A			Leasing Company B		
Lease Term	36 mo	48 mo	60 mo	36 mo	48 mo	60 mo
Monthly Payment	$402	$344	$316	$474	$412	$373
Money Factor	0.00308	0.00312	0.00319	0.00436	0.00431	0.00427
Mileage Penalty	$0.17/mi	$0.17/mi	$0.17/mi	$0.15/mi	$0.15/mi	$0.15/mi
Purchase Option	$12,849	$12,102	$10,857	$11,747	$10,486	$9,225
Residual Value	$12,699	$11,952	$10,707	$11,597	$10,337	$9,076

FIGURE 7.45.
Two leasing options.

a) Which company makes more money from monthly interest charges? Explain.

b) Which company makes more money from penalty charges? Explain.

c) Which company makes more money from re-selling leased vehicles as used cars? Explain.

d) Which company do you think Jennifer should select? Why?

e) For the company you picked in d, which lease term costs the least? Why?

Answers (continued)

4. a) That will cost her $1000 (the cost of the stereo).

   b) At 7% for 5 years, the extra $1000 financed will increase the monthly payment by $19.80 ($514.83 – $495.03). Over 60 months, that will cost her an extra $1188.00.

   c) Since the amount financed remains constant, it won't cost her anything extra. It is as if she bought the stereo from the car dealership instead of from the other company.

   d) Answers may vary. Since dealers know the stereo is being financed with the car, the price of the stereo may seem low to the person buying the car (only $20 a month), but may not represent a good value when compared to buying the stereo somewhere else.

5. a) The money factor would be $6.4/2400 \approx 0.0026667$.

   The monthly depreciation would be $(25211 - 14500)/36 \approx \$297.53$.

   The monthly interest charge would be $(25211 + 14500)*0.0026667 \approx \$105.90$.

   The sales tax would be $(297.53 + 105.90)*0.0625 \approx \$25.21$, making the monthly lease payments equal to $(297.53 + 105.90 + 25.21) = \$428.64$.

   b) ($428.64/month)(36 months) = $15,431.04

6. a) The money factor would be $6.4/2400 \approx 0.0026667$.

   The adjusted capitalized cost would be $(24500 - 4000) = \$20,500$.

   The monthly depreciation would be $(20500 - 12200)/36 \approx \$230.56$.

   The monthly interest charge would be $(20500 + 12200)*0.0026667 \approx \$87.20$.

   The sales tax would be $(230.56 + 87.20)*0.0625 = \$19.86$, making the monthly lease payments equal to $(230.56 + 87.20 + 19.86) = \$337.62$.

   b) She would have put $4000 into the car from the down payment and the trade-in value of her old car.

   The payments would cost ($337.62/month)*(48 months) = $16,205.76.

   Her back end charges would be (13,000 miles)($0.18/mile) = $2340.00 + $2100.00 = $4440.00.

   Altogether, the car would have cost (4000 + 16205.76 + 4440) = $24,645.76.

   c) Financing $20,500 at 6.4% APR over 4 years would have monthly payments of $485.21. So, the car would have cost $4000 (down payment + trade-in value) + (48)(485.21) – 12,200 = $15,090.08.

   d) Sample answer: No. Leasing the car would have cost her an extra $9555.68, which is accounted for by the end-of-lease charges that she accrued, and the fact that she gained no equity in the car by leasing it.

7. a) Company B is using larger money factors in comparing each lease term and is also coming in with lower estimates for the residual value.

   b) Company A has a higher mileage penalty surcharge.

   c) If you assume that the companies use the purchase option as the sticker price for the used car, both companies are making around $150 on each car. However, the purchase option is only the price they give the person holding the lease. Company B can charge more than that and still be competitive with Company A. But they will have a higher profit margin as a result.

   d) Assuming she drives the car within the mileage and normal wear-and-tear guidelines, she should choose Company A. The monthly payments will reflect a lower depreciation cost since Company A uses the higher residual value and a lower interest charge and the lower money factor. However, students can make different assumptions about Jennifer's driving habits and draw a different conclusion, provided, they defend their answer.

   e) Just like with new-car financing, Jennifer should go with the shortest lease term if she can make the payments. For 36 months, it will cost a total of $14,472, as opposed to $16,512 for 48 months or $18,960 for 60 months.

SECTION 7.14

There's No Place Like Home

Buying a home is like buying a car. Because it is a high-cost item, most people have to borrow money in order to buy a home. There are many types of home loans available for consumers.

All home loans calculate interest monthly based on the remaining principal, which is just like a car loan. Some home loans use a fixed rate for the entire life of the loan. These loans are called **conventional fixed-rate loans.**

Lenders offer fixed-rate loans of different term lengths. Common term lengths are 15 years and 30 years.

The Gonzales family is considering a conventional fixed-rate loan to buy a new home. After a down payment, they need to finance $150,000. They consider two options:

Option A: 30-year home loan for $150,000 at 7.25% interest

Option B: 15-year home loan for $150,000 at 7.25% interest

1. What is the monthly payment for each option?
2. How much money do they save per month with the 30-year loan?
3. How much money does the house cost, including principal and interest, with the 30-year loan?
4. How much money does the house cost, including principal and interest, with the 15-year loan?
5. Since the time doubles, does the cost of the house double, too? If not, then how much did the cost of the house change?
6. How long does it take to pay off half of the 30-year loan?
7. How long does it take to pay off half of the 15-year loan?

8. When would someone want to use a 15-year conventional loan?
9. When would someone want to use a 30-year conventional loan?

Mrs. Gonzales found this in a magazine:

WHAT A DEAL!!!
Move into your $150,000 home today for only $750 a month! Adjustable Rate Mortgages are available today for only 4.39% APR.
CALL NOW!

Mrs. Gonzales called the bank offering this deal. They told her that the ad was for a 30-year loan term. The APR for the adjustable rate loan is based on the six-month US Treasury Bill index, which is currently 3.11%. The bank calculates the APR by adding 1.27% to the index. Right now that means an APR of 4.39%.

10. If the Gonzales family chooses this type of loan, what is the principal balance in one year?
11. Suppose that after one year the index rises to 4.5%. What is the Gonzales family's new interest rate?
12. What is the new monthly payment? How does this compare to their original monthly payment?
13. What is the principal balance of their loan after three years?
14. After the third year of their mortgage, the index rises to 6.1%. What is their new interest rate?
15. What is the new monthly payment? How does this compare to their original monthly payment?
16. What is the percent change in the Gonzales' monthly payment from when they first got the loan?
17. When would someone want to use an Adjustable Rate Mortgage? When would someone NOT want to use one?

ELABORATE LESSON 3

Section 7.14: There's No Place Like Home

Purpose: In Sections 7.12 and 7.13, students used amortization to make decisions about car loans and to compare purchasing and leasing a car. In this section students apply that knowledge to compare the purchase or lease of a home. This enables students to apply their understanding of credit and financing to a new situation in Section 7.15 dealing with using credit cards to borrow for retail purchases.

Description: In this section students use the TVM Solver to make calculations and decisions about purchasing a home with different types of mortgages. Then, they use newspapers and other periodicals or the Internet to look for and evaluate advertisements to rent a home.

Materials: graphing calculator, newspaper or local advertisements for sales and rental property, Internet access (optional)

Teaching Ideas: This activity involves collecting data from print or Internet sources. Students should, in their small groups, look through local publications or browse the Internet for information to compare the purchasing or leasing of homes in comparable neighborhoods in their communities. The purpose of this activity is to make students aware of the financial implications of both purchasing a home and renting one.

Answers

1. Option A: $1023.26

```
N=360
I%=7.25
PV=150000
▪PMT=■1023.26442
FV=0
P/Y=12
C/Y=12
PMT:END BEGIN
```

Option B: $1369.29

```
N=180
I%=7.25
PV=150000
▪PMT=■1369.2943…
FV=0
P/Y=12
C/Y=12
PMT:END BEGIN
```

2. $346.03
3. $368,373.60
4. $246,472.20
5. No. But it does increase significantly, by about 50%.
6. 262 months (21 years 10 months)

   One way to find this is to use the balance calculation feature for a given payment number:

```
        70880.47247
bal(246)
        84136.80651
bal(262)
        75513.80893
bal(263)
         74946.7782
■
```

7. 114 months (9 years 6 months)

```
        68742.63706
bal(100)
        86663.50074
bal(110)
        77972.85265
bal(114)
        74347.34535
■
```

8. Answers may vary. The monthly payments are higher, so the principal balance is paid off more quickly. So if someone wants to pay the loan off more quickly and quickly build up equity in the home, then a 15-year loan would be good.
9. Answers may vary. The monthly payments are lower with a longer loan term. So if someone wants low monthly payments, a 30-year loan is a better option.
10. $147,531.43
11. 4.5% + 1.27% = 5.77%
12.

```
N=348
I%=5.77
PV=147531.43
▪PMT=■874.03091…
FV=0
P/Y=12
C/Y=12
PMT:END BEGIN
```

   $874.03, which is $124.03 higher than the original monthly payment.

13. $143,353.40
14. 6.1% + 1.27% = 7.37%
15.

```
N=324
I%=7.37
PV=143353.4
▪PMT=■1020.8386…
FV=0
P/Y=12
C/Y=12
PMT:END BEGIN
```

   $1020.84, which is $270.84 more than their original monthly payment of $750!

16. 36% increase in their monthly payments.
17. Answers may vary. If the index that the interest rate is attached to is low and will remain low, then an adjustable rate mortgage is a good idea. If the interest rate is low and the person plans to keep the loan for only a few years, then an adjustable rate mortgage is a good idea. However, if the market index is volatile and changes quickly, or if the person plans to keep the loan for a long time, then they would not want to use an adjustable rate mortgage.

RENTING A HOUSE OR AN APARTMENT

As with cars, many people choose to rent a home. There are several reasons that someone might do this.

18. What costs are involved with owning a home?

19. Of these costs, which ones are fixed amounts each month? Which ones vary from month to month?

Use newspapers, magazines, or the Internet to find two houses in a similar neighborhood. One house should be for sale and one for rent.

20. Calculate the monthly payment for a 30-year loan at 6.25%.

21. How does this monthly payment compare to the monthly rent for a similar house?

CHECK THIS!

A monthly mortgage payment includes only principal and interest payments. Other expenses that a homeowner must pay include property taxes and insurance. Depending on where they live, some homeowners must also pay for flood or other special types of insurance.

Renters are generally not required to pay for insurance but can buy renter's insurance to cover their personal items.

22. When is it better to own a home? When is it better to rent a home or apartment?

SECTION 7.14 Assignment

1. Find the monthly payment, total amount of interest paid, and total cost of the loan for a 30-year home loan for $175,000 at 7.25%.

2. Find the monthly payment, total amount of interest paid, and total cost of the loan for a 15-year home loan for $175,000 at 7.25%.

3. The Washington family is buying a new home. They plan to get a 30-year adjustable rate mortgage beginning at 4.75% for $125,000.

   a) What is their initial monthly payment?

   b) At the end of three years, what is their principal balance?

   c) After three years the interest rate increases to 6.25%. What is the monthly payment now?

   d) At the end of the fourth year, what is the principal balance?

   e) At the end of the fourth year, the interest rate decreases to 5.75%. What is the monthly payment now?

   f) When the Washingtons bought their home, they could have gotten a 30-year conventional fixed-rate loan for 5.625%. What would their monthly payment have been?

   g) After four years of payments, what is the principal balance for this loan?

   h) Which option do you think is better for the Washingtons?

Answers (continued)

18. Answers may vary. Students may observe that if you own a home, you have to pay for repairs, appliances that break, property taxes, homeowner's insurance, etc. They may also include bills such as cable television and utilities. If students do not mention them, be sure to add property taxes and homeowner's insurance to the list of costs.

19. Answers may vary. Insurance and taxes are fixed within a calendar year. Repairs and utilities vary from month to month based on use.

NOTE: If you do not have access to print or electronic resources, provide students with the following information:

Why rent? Home for Sale!

112 Elm Street—$145,000 for a beautiful 3 bedroom, 2 bath home. Great yard! Excellent schools. No money down. Call today.

Why own? Rent this House!

116 Elm Street—$995 a month for a beautiful 3 bedroom, 2 bath home. Beautiful backyard! Excellent schools. Call today.

20. Answers will vary depending on students' findings. Using the sample data, the monthly payment for principal and interest alone is $892.79.

21. Answers will vary depending on students' findings. Using the sample data, the monthly payment is about $100 less than the rent.

22. Answers may vary. When you own a home, you are building equity in the property (i.e., if you sell it, you can make money from the sale). When you rent a home or apartment, you can avoid costly repairs and possibly save on property taxes and insurance costs.

Section 7.14 Assignment

Answers

1. The payment on a 30-year loan will be $1193.81 per month. The total cost of the loan will be $429,771.60 and the interest paid will be $254,771.60.

```
N=360
I%=7.25
PV=-175000
▪PMT=1193.80849
FV=0
P/Y=12
C/Y=12
PMT:END BEGIN
```

```
1193.81*360
          429771.6
Ans-175000
          254771.6
```

2. The payment on a 15-year loan will be $1597.51. The total cost of the loan is $287,551.80 and the interest paid is $112,551.80.

```
N=180
I%=7.25
PV=-175000
▪PMT=1597.510042
FV=0
P/Y=12
C/Y=12
PMT:END BEGIN
```

```
1597.51*180
          287551.8
Ans-175000
          112551.8
```

3. a)

```
N=360
I%=4.75
PV=125000
▪PMT=652.05917…
FV=0
P/Y=12
C/Y=12
PMT:END BEGIN
```

$652.06

b)

```
N=360
I%=4.75
PV=125000
▪PMT=652.05917…
FV=0
P/Y=12
C/Y=12
PMT:END BEGIN
```

```
bal(36)
       118928.0007
```

$118,928.00

c)

```
N=324
I%=6.25
PV=118928
▪PMT=760.76092…
FV=0
P/Y=12
C/Y=12
PMT:END BEGIN
```

$760.76

d)

```
N=324
I%=6.25
PV=118928
▪PMT=760.76092…
FV=0
P/Y=12
C/Y=12
PMT:END BEGIN
```

```
bal(12)
       117182.4283
```

$117,182.43

e)

```
N=312
I%=5.75
PV=117182.43
▪PMT=724.56241…
FV=0
P/Y=12
C/Y=12
PMT:END BEGIN
```

$724.56

f)

```
N=360
I%=5.625
PV=125000
▪PMT=719.57050…
FV=0
P/Y=12
C/Y=12
PMT:END BEGIN
```

$719.57

g)

```
N=360
I%=5.625
PV=125000
▪PMT=719.57050…
FV=0
P/Y=12
C/Y=12
PMT:END BEGIN
```

```
bal(48)
       117825.453
```

$117,825.45

h) Answers may vary. At the end of the fourth year with the ARM, the Washingtons have a principal balance of $117,182.43. At the end of the fourth year with the fixed-rate mortgage, the Washingtons have a principal balance of $117,825.45. The principal balance is less with the ARM, so the ARM is a better choice for the Washingtons.

4. Bryan is looking for a townhouse in the city where he has a new job. He does some research and finds these offers:

MUST BUY!	Now Leasing!
2108 Elmcroft Drive— a bargain at \$85,500. 2 bedrooms, 1-1/2 bathrooms. No money down. CALL TODAY!	2115 Maplecroft Drive— a great neighborhood! Lease for only \$850 per month. 2-bedroom specials. Going Fast! Call now.

a) Bryan can get a 15-year fixed-rate mortgage at 6.25%. Which offer gives him a lower monthly payment?

b) Insurance will cost Bryan \$1100 per year. Property taxes are \$3200 per year. If Bryan buys the townhouse on Elmcroft Drive and pays both insurance and taxes monthly, what are his total monthly costs?

c) Bryan spoke to the leasing agent for the Maplecroft townhouse. The agent said that renter's insurance costs \$450 per year, and there is a residents' fee of \$25 per month. If Bryan leases the townhouse on Maplecroft Drive, what are his total monthly costs?

SECTION 7.15 Gotta Have My Tunes!

Paco, Tico, and Maria have convinced their parents to buy a \$2400 home theater system, but they only have \$100 per month budgeted for the purchases. The Gonzales family has two options:

Option 1: Place \$100 a month into a savings account that pays 5% annual interest compounded monthly and purchase the system when they have saved \$2400.

Option 2: Use a credit card to purchase the system now. The credit card charges 11.5% annual interest compounded monthly. They plan to pay \$100 a month until the stereo is paid off.

Which will happen first: Will they save enough money to buy the stereo with money in their savings account or will they be able to pay the credit card balance entirely? Explain your answer.

SUMMARY

Vehicles: Purchase or Lease?

There are two ways to acquire a vehicle.

First, you can buy one. If you need to borrow money you can finance the purchase. You will pay a certain amount per month depending on several factors: the principal, or amount of money you borrow; the annual percentage rate (APR), which is compounded monthly; and the length of the loan.

Second, you can lease a car. When you do so, you agree to pay for the depreciation in the car's value over the length of the lease. To calculate the monthly payment the following formulas are used:

- monthly depreciation $= \frac{\text{adjusted capitalized cost} - \text{residual value}}{\text{lease term}}$
- monthly finance charge = (adjusted capitalized cost + residual value) × money factor

Answers (continued)

4. a)

```
N=180
I%=6.25
PV=85500
▪PMT=█733.09655…
FV=0
P/Y=12
C/Y=12
PMT:END BEGIN
```

The $733.10 monthly payment is less than the $850 per month lease payment. Assuming no other expenses, Bryan will have a lower monthly payment if he buys the townhouse.

b) $1100 ÷ 12 = $91.67 for insurance and $3200 ÷ 12 = $266.67 for taxes

$733.10 + $91.67 + $266.67 = $1091.44

c) $450 ÷ 12 = $37.50 per month for renter's insurance

$850 + $37.50 + $25 = $912.50

EVALUATE LESSON 3

Section 7.15: Gotta Have My Tunes!

Purpose: In Sections 7.11 through 7.14, students applied what they learned about exponential functions in a financial context with compound interest to the concepts of financing and credit. In this section students apply those concepts and skills to solve a problem in a new context.

Description: Facing a problem in a new context, students determine and implement a method to solve the problem. This is an open-middle performance assessment task; consequently there is only one correct answer, but there are a variety of valid methods to obtain it. In their justification, which is the most important part of a performance assessment task, students should be able to communicate their knowledge of mathematics and demonstrate that they have mastered the concepts presented in Sections 7.11 through 7.14.

Materials: graphing calculators

Teaching Ideas: This task may be assigned as group work or individual work. The mathematics performance assessment rubric should be used to evaluate students' work.

Answer

Sample Answer:

First, find the length of time it will take to save the money in a savings account by using recursive iteration in the home screen of the graphing calculator.

Set up a table to calculate the first few months' balance in the account.

Month	Process	Balance
1	$100\left(1+\frac{0.05}{12}\right)=100.42$	$100.42
2	$(100.42+100)\left(1+\frac{0.05}{12}\right)=201.25$	$201.25
3	$(201.25+100)\left(1+\frac{0.05}{12}\right)=302.51$	$302.51
4	$(302.51+100)\left(1+\frac{0.05}{12}\right)=404.18$	$404.18

In the process column, to find the new balance, take the previous month's balance, add one hundred, then multiply by the interest multiplier. To take advantage of the graphing calculator's recursion capabilities, we can write this as (Ans + 100)(1+0.05/12).

```
(Ans+100)*(1+.05
/12)█
```

To set the initial term, type 0, then press ENTER. Then, enter the recursion formula into the calculator. When you press ENTER, you will get the balance for the end of month 1. Pressing ENTER again yields the balance for month 2.

```
0
                 0
(Ans+100)*(1+.05
/12)█
```

```
0
                 0
(Ans+100)*(1+.05
/12)
         100.4166667
█
```

```
0
                 0
(Ans+100)*(1+.05
/12)
         100.4166667
         201.2517361
█
```

Repeat this process until the balance reaches $2400. Be sure to count the number of times you press ENTER.

After the 23rd month, the balance will exceed $2400.

```
         1765.189063
         1872.960684
         1981.181353
         2089.852942
          2198.97733
         2308.556402
         2418.592053
█
```

NOTE: You can use home screen recursion to keep track of the terms for you, but the formula is a little more complex. The calculator can keep track of two answers simultaneously. The table representation would be:

Month Ans (1)	Balance Ans (2)
0	0
1	100.42
2	201.25
3	302.51
$n+1$	$(100+n)\left(1+\frac{0.05}{12}\right)$

Answer (continued)

On the calculator screen, first set the Answer lists to {0, 0}, which corresponds to the 0th month having a balance of $0.

```
{0,0}
                {0 0}
```

Next, enter your recursion formulas. We want the month to increase by 1, so we will add 1 to the previous month, or Ans(1) +1.

```
{0,0}
                {0 0}
{Ans(1)+1,
```

We want to take the previous balance, Ans(2), and add 100 to it, then multiply that sum by our interest rate factor, $\left(1+\frac{0.05}{12}\right)$. That part of our recursion formula is (100 + Ans(2))(1 + 0.05/12).

```
{0,0}
                {0 0}
{Ans(1)+1,(100+A
ns(2))(1+.05/12)
}
```

If you press ENTER now, it will calculate the first month's balance.

```
{0,0}
                {0 0}
{Ans(1)+1,(100+A
ns(2))(1+.05/12)
}
 {1 100.4166667}
```

Pressing ENTER repeatedly will generate a list, in brackets, of the month number and the balance.

```
{Ans(1)+1,(100+A
ns(2))(1+.05/12)
}
 {1 100.4166667}
 {2 201.2517361}
 {3 302.5069517}
  {4 404.184064}
```

Continue pressing ENTER until the balance exceeds $2400, then record the month number.

```
{17 1765.189063}
{18 1872.960684}
{19 1981.181353}
{20 2089.852942}
 {21 2198.97733}
{22 2308.556402}
{23 2418.592053}
```

Second, find the length of time it will take to pay the credit card balance. Then compare the two.

Use the TVM Solver to find the number of months it will take to pay off $2400 at 11.5% interest if the interest is compounded monthly.

The Solver window should look like this:

```
N=0
I%=11.5
PV=2400
PMT=-100
FV=0
P/Y=12
C/Y=12
PMT:END BEGIN
```

Now, solve for the number of payments (tvm_N):

```
tvm_N
          27.4033196
```

After 28 months of payments (27 is not enough), the balance will be paid.

Therefore, saving money in a savings account will pay for the stereo more quickly.

- monthly sales tax = (monthly depreciation + finance charge) × tax rate
- monthly payment = monthly depreciation + monthly interest charge + monthly sales tax

Home: Purchase or Lease?

There are two options to pay for housing costs.

First, you can buy a home. Most people need to borrow money, or obtain a **mortgage,** to do so. Mortgages are long-term loans, often 15 or 30 years, and can have fixed interest rates as in a **conventional fixed-rate loan** or variable rates as in **adjusted rate mortgages** (**ARMs**).

Second, you can rent or lease a home. As with cars, leasing a home has advantages. Since you do not own the home, you are not responsible for major repairs or property taxes.

Modeling Project Money for Life

One problem that faces all people is managing income and planning for retirement. Do you buy a house or rent, and at what point? Do you keep your money secure and easily accessed, or do you put it in an account that pays higher interest, but with restrictions on its use?

The world of finances is complex, with interest rates that change almost daily, speculative investments that might lose all value, and a multitude of costs—some changing with the seasons, others constant. Even the decision to raise a family has implications for your finances. As good mathematical modelers, it is better to keep the problem reasonable and work on making it more realistic in the next version of the model.

For that reason, this Modeling Project includes only these assumptions:

- You just finished school, have no outstanding debts, and have $10,000 in a savings account.
- You just took a job at a firm where you will remain for the rest of your career. The starting salary is $32,000 and you receive pay raises of $2500 every two years.
- You are 25 years old and can work for up to 40 years. After that you have to retire.
- The selling price for a typical house is $180,000 and is going up by 2% per year (the value of the house goes up at the same rate).
- You need at least 20% of the price of a house as a down payment, and the closing costs are 5% of the price. The down payment must come from assets but the closing costs can be financed with the house, which is financed at 6.5% per year compounded monthly over 30 years.
- You will be in the 15% tax bracket until your salary goes above $44,000; then it will be 20%.
- You need $1400 a month in living expenses including the monthly rent on your apartment, which is $900 but goes up $100 every two years. Other expenses go up by 3% per year.
- Your savings account pays 5% interest compounded monthly.

- Certificates of Deposit (CDs) pay 7% interest compounded quarterly but you have to keep the money in the bank for two years.
- You may put up to $2000 in an Individual Retirement Account (IRA) every year, which pays 10% per year. Any contribution to an IRA is not taxed until the money is taken out of the account. However, you may not take out any money until you retire.

Prepare a written plan for your financial future. You know the mathematics. You can explore this problem with spreadsheets as another option. What investments will you make? At what point in time will you make them? When you retire what will your worth be? Try and make as much money as you can—but remember, *"Money doesn't buy happiness!"*

Practice Problems

1. Eduardo wants to buy a computer that costs $1200. The store offers a one-year financing plan that allows him to pay $100 as a down payment and then $100 per month for 12 months.

   a) What interest rate is he paying?

   b) Is he paying simple interest?

   c) He can get a loan for the same computer from his college at a rate of 5% but he can only finance 85% of it. Is this loan a better deal? Why?

2. Lucy buys a used car for $15,000. The dealer tells her she will make monthly payments of $360.94 for 48 months.

   a) What interest rate is she paying?

   b) Lucy can only afford $300 per month. If she extends the financing to 60 months, can she get her payment under $300?

   c) Lucy's parents lend her $1500 so that she can make a down payment. The new amount financed is $13,500. Also the dealer tells her that the down payment gets her a lower rate. Her payments are now $317.05 for 48 months. What is the new interest rate?

3. A principal amount of $3500 is deposited in an account that earns 4.5% interest. Calculate the ending balance after five years if the interest is:

   a) Simple interest

   b) Compounded annually

   c) Compounded monthly

   d) Compounded daily (using 360 days)

Practice Problems

Answers

1. a) 10%

   b) Yes, there is only one compounding period (annually).

   c) Answers may vary. Students should realize that even though his down payment will be higher ($180 versus $100) he will save $11.75 per payment ($89.25 versus $100). This saves him a total $49 ($129 in payment savings over the year less the $80 extra down payment). Except for needing the extra $80 down payment, the university loan is better.

2. a) 7.25% APR

   b) Yes. Her new payment will be $298.79.

   c) 6.0% APR

3. a) $(\$3500)(0.045)(5) + (\$3500) = \$4287.50$

   b) $(\$3500)(1 + 0.045)^5 \approx \$4361.64$

   c) $(\$3500)(1 + 0.045/12)^{60} \approx \$4381.29$

   d) $(\$3500)(1 + 0.045/360)^{(360 \cdot 5)} =$
   $(\$3500)(1 + 0.045/360)^{1800} \approx \$4383.07$

4. Find the missing values in **Figure 7.46.**

FIGURE 7.46. Table for Question 4.

	APR	Compounding Condition	Time Interval	Growth Factor	APY
a)	5%	annually	3 years		
b)	5%	bi-monthly	6 months		
c)	5%	monthly	2.5 years		
d)	5%	continuously	1/4 year		
e)		bi-monthly	1 year	1.10	
f)		quarterly	1.5 years	1.25	
g)		monthly	5 years	2.00	
h)		semiannually	4 years		8%
i)		bi-monthly	3 years		10%

5. A credit card company offers this plan to new customers: no annual fee, an initial rate of 6% for the first year, and then 9% after the first year.

   a) Robert decides to take the deal and transfers a balance of $2000. He also charges $1000 more on the card. Assuming Robert pays $200 per month and does not charge anything more, how long will it take him to pay the bill in full?

   b) If after the first month he charges $25 per month, how long will it take him to pay the bill in full?

6. If you invest $5000 into a savings plan that pays 6% interest compounded quarterly, and you wish to have $9000 after five years, how much should you make in payments at the end of each quarter? (Hint: How much will the principal earn? The rest needs to come from the savings plan accumulation.)

7. You wish to buy a house for $220,000. Closing costs are $5600. You pay 20% of the total cost of the house as a down payment, and you want to finance the rest over 30 years with a loan that charges 7.6% interest per year compounded monthly.

   a) What is the total cost for the house?

   b) How much do you need for the down payment?

   c) What will your monthly payments be?

8. a) Elena opened a savings account with a deposit of $50, and she deposits $25 per month from her earnings. The bank pays interest at the rate of 0.25% per month. That is, each month the bank adds 0.0025 × (current balance) to the account and then Elena adds $25 of her own money. Write a function rule describing her balance under this process.

   b) Use a calculator or a spreadsheet program to simulate the growth of her account using this saving scheme. Round to the nearest cent.

9 a) Tia plans to retire in 20 years and expects inflation to average about 3% per year between now and then. If she retired now, she would want $1500 per month in income. How much income will she need in 20 years in order to maintain the same standard of living?

   b) Tia can secure an investment that pays 1% per month. She wants to use the interest from this investment as her sole income during retirement. (She will remove each monthly interest payment as soon as it is put into the account; the principal will never be touched.) How much does she need in the account when she retires to meet the goal in a?

Answers (continued)

4. Answer

	APR	Compounding Condition	Time Interval	Growth Factor	APY
a)	5%	annually	3 years	1.157625	≈ 5.25%
b)	5%	bi-monthly	6 months	≈ 1.0252089	≈ 5.04%
c)	5%	monthly	2.5 years	≈ 1.1328542	≈ 5.31%
d)	5%	continuously	1/4 year	≈ 1.0125785	≈ 5.03%
e)	≈ 9.60712%	bi-monthly	1 year	1.10	10%
f)	≈ 15.15633%	quarterly	1.5 years	1.25	≈ 16.67%
g)	≈ 13.94333%	monthly	5 years	2.00	20.00%
h)	≈ 7.06264%	semiannually	4 years	1.32	8%
i)	≈ 8.80952%	bi-monthly	3 years	1.30	10%

5. a) Approximately 16 months

   b) Approximately 22 months

6. $96.53

7. a) 220,000 + 5600 = $225,600

   b) (0.20)(225,600) = $45,120

   c) The principal on the loan is 225,600 – 45,120 = $180,480, and the rate for each compound period is 7.6/12 ≈ 0.63333%. Using the amortization formula, $180{,}480(1.00633333)^{360} = d*(1.00633333^{360} - 1)/(0.00633333)$. Therefore, 1,751,852.6 ≈ d*1374.73, and $d \approx \$1274.32$

8. a) $b_{new} = b_{current} + (0.0025) \times b_{current} + 25$

   b) Here are the first few: $50, $75.13, $100.31, $125.56, $150.88, $176.26

9 a) 1500 (1.03)20 = 2709.17, or about $2700 per month

   b) approximately $270,000

Glossary

KEY CONCEPTS

Accumulation: Combination of deposits made on a regular basis and the interest that those deposits make.

Adjusted Rate Mortgages (ARMs): A loan where the interest rate changes over the term of the loan.

Allowances: A means of determining your federal income tax withholding. Increasing the number of allowances decreases the amount of withholding taken from gross pay.

Amortization: Process in which a debt is repaid in regular installments.

Amortization formula: Formula for situations involving installment loans:

$P(1 + r)^n = d\,\frac{(1+r)^n+1}{r}$, where P is the principal, r is the effective interest rate per compounding period, d is the payment amount made at the end of each period, and n is the number of payments required to pay off the loan.

Annual Percentage Rate (APR): The cost of a loan or the amount of interest received on an investment expressed as a yearly percentage rate.

Annuities: Assets that make periodic payments to the holder. Also can be an investment method.

Balance: Amount remaining in a savings account, or amount remaining to be repaid on a loan.

Capitalized cost: The purchase price of a capital item that forms the basis for leasing transactions. Net capitalized cost is the capital cost less any deductions, such as a trade-in.

Compound interest: Method of paying interest on both the principal amount and the interest previously earned. The compound interest formula is: $A = P\left(1+\frac{r}{n}\right)^{nt}$.

Compounding period: Time interval between interest calculations.

Consumer Price Index (CPI): A measure of the change in price over time of a fixed "basket " of goods and services. CPI is a measure of inflation. In general, things get more expensive over time.

Continuous compounding: Payment of interest based on using the limiting rate from compounding as frequently as possible.

Depreciation: The loss of value of an item over time.

Down payment: An initial payment that lowers the capitalized cost of a car, house, or other capital item.

Equity: The value of an individual's ownership of a capital item.

Fixed rate loan: A loan where the interest does not change over the term of the loan. Also known as conventional fixed rate loans.

Geometric sequence: Sequence of numbers that can be described recursively by a multiplication operation, or that is characterized by a repetitive multiplicative process.

Geometric series: A sum of consecutive terms (either some or all of them) of a geometric sequence.

Gross pay: Amount earned before deductions for taxes, benefits, and other reductions.

Inflation: Rate at which the cost of goods increases.

Installment: Amount paid on a regular basis to repay a loan.

Internal Revenue Service (IRS): The federal agency responsible for collecting federal taxes from individuals and businesses.

Lease: A rental agreement.

Lease term: The length of time (usually in months) that an item will be leased.

Medicare: A government health care program that serves people primarily 65 years of age and over. A medicare tax is deducted (and matched by the employer) from an individual's gross pay.

Mixed growth: Sequence of numbers that can be described recursively by a combination of one multiplication and one addition operation.

Net pay: Amount earned after taxes, benefits, and other reductions. Also known as take-home pay.

Present value: The value of money today that is to be received in the future.

Principal: Initial deposit or balance on a savings account or loan.

Quarterly: Four times per year.

Recursive: Description of change during a time interval, based upon the previous value.

Residual value: The value of a capital item after deducting depreciation.

Salary: The amount of money you receive for work.

Savings formula: Formula for the amount in an account in which regular deposits are made: $A = d\,\frac{(1+r)^n-1}{r}$, where A is the amount accumulated, d is the amount deposited at the end of each compounding period, r is the interest rate over the compounding period, and n is the number of compounding periods (and payments).

Savings plan: Savings account with compound interest in which deposits are made regularly for the purpose of accumulating money.

Semiannual: Term used to describe two compounding periods per year.

Sequence: A list of variables.

Simple interest: Method of paying interest on only the initial balance in an account and not on any interest that is earned.

Social security: A government retirement and disability program. A social security tax is deducted (and matched by the employer) from an individual's gross pay.

Subscript: Method of showing different values for the same variable, such as an account balance (the variable) over time (the subscript). In x_t the subscript is t.

Tax: A payment to a federal, state, or local government that is based on earned income, unearned income, the value of a purchase, and other transactions.

Trade-in allowance: The dollar value realized from the sale of a vehicle to an automobile dealer for the purpose of lowering the net purchase price of another vehicle.

TVM Solver: A graphing calculator utility for solving financial problems.

US Treasury bill index: An index that measures the cost of buying the debt of the United States government. This index is often linked to variable interest loans such as ARM.

Withholding (tax): An amount deducted from your gross pay to cover anticipated taxes due. Withholding taxes are based on allowances.

CHAPTER 8

Cycles: Trigonometry and Sinusoidal Functions

PREPARATION READING

It is Coming Around Again!

Certain things in nature follow a regular repeating pattern known as a cycle. Since these cycles occur again and again, they are predictable. One cycle you have seen is that of the moon, which is about a thirty-day cycle that begins and ends with a new moon. This process is constantly repeating. The National Weather Service can predict the dates of each phase of the moon for years to come. There are many cycles in nature, including tides, the position of planets in our solar system, and the populations of certain animals. In music, frequencies of a sound wave, or cycle, determine the sound, or note.

Cycles are not only in the natural world. For practical and social reasons cycles occur in many manmade things. For example, electricity use during the day, the purchase of items like heating oil over the course of a year, and, some people argue, public opinion on issues over time can be described as cycles. Sometimes the reasons for these cycles are easy to explain but sometimes they are a mystery.

The ability to make predictions about cycles is important. If local officials know that a storm will hit at high tide, they are more likely to worry about coastal flooding. If a retail store buys 10,000 bathing suits in October, they may not be able to sell them.

Mathematicians have created tools for describing patterns. You may find it surprising that these tools are based on similar triangles, like the ones you explored in Chapter 4. In this chapter we will expand these tools to explore cycles and understand the relationships that create them.

CHAPTER 8

Cycles: Trigonometry and Sinusoidal Functions

Preparation Reading:
It is Coming Around Again!

Description: The reading introduces the major context of the chapter: natural and manmade phenomena cycles. This initial definition of a cycle is expanded to understanding a common sinusoidal function, the periodic function.

Purpose: Students learn the definition of a cycle and develop a list of manmade and natural cycles.

Materials: none

Teaching Ideas: Students can work in small groups (3 to 4 students). After the reading ask students to describe and discuss either natural or manmade events and phenomena that might be cycles (we will develop the term periodic later in the chapter). Then come together as a class and have each group report some examples of real-world phenomena that may be cycles. Have the class debate whether they are actually cycles or just appear to be. Keep a list of the cycles your class develops for use in the modeling project.

SECTION 8.1

Right Triangle Relationships: Pythagorean Theorem and Similar Triangles

In Chapter 4 you saw pictures like **Figure 8.1**.

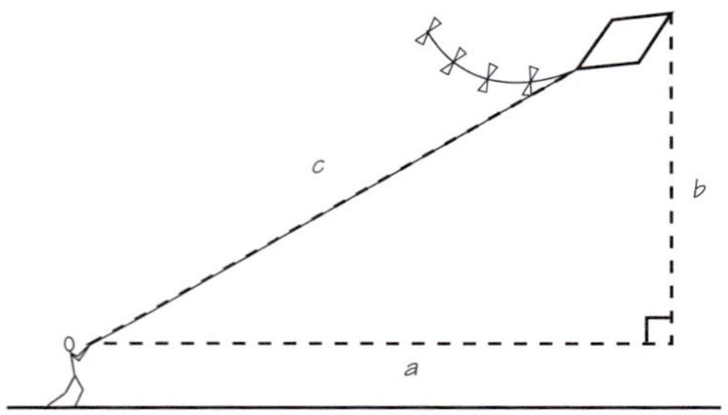

FIGURE 8.1. Picture similar to those in Chapter 4.

1. What type of problem does Figure 8.1 suggest?
2. What do you think the problem would ask you to find?
3. What information would you expect to be given?

Stand up and face the clock or some other object that is located near the top of your classroom wall. Look straight ahead and extend both arms out in front of your body. Your line of vision and your arms should be parallel to the floor. Pretend the clock on the wall is the kite; rotate your head and one arm upwards until the clock is directly in your line of vision. The angle you create with your arms is called the angle of elevation (**Figure 8.2**).

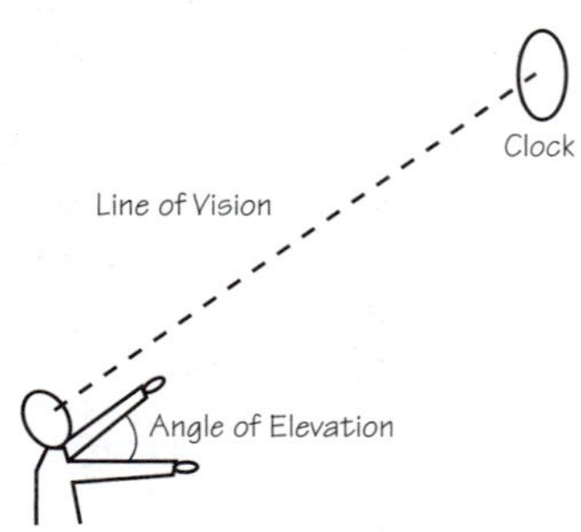

FIGURE 8.2. Angle of elevation.

Once again think about the kite. Use this information.

The kite string is 100 yards long.

The kite's angle of elevation from the ground is 30° (**Figure 8.3**).

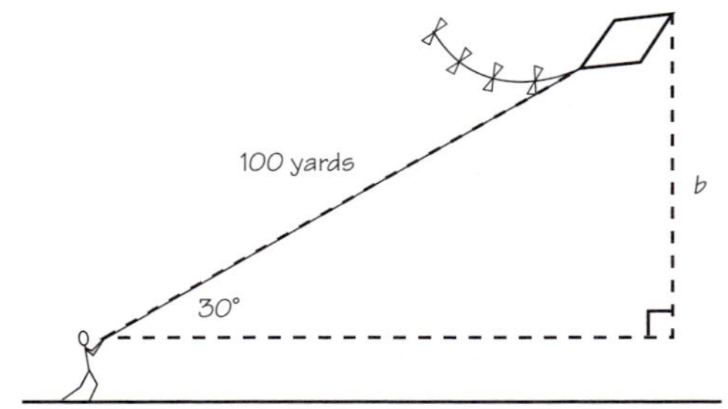

FIGURE 8.3. The kite problem.

4. Use your knowledge of similar triangles from Chapter 4 and **Figure 8.4** to find the height of the kite.

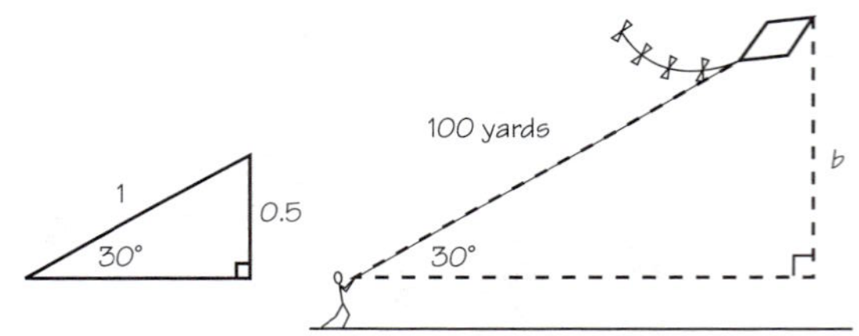

FIGURE 8.4. Similar triangles.

CHECK THIS!

Recall from Chapter 4 that two triangles are similar if their corresponding angles are congruent. Corresponding sides of similar triangles are proportional.

5. 50 yards is the length of the missing leg of the right triangle. Is it actually the height of the kite? Why or why not?
6. Use two methods to find the length of the third side of each triangle in **Figure 8.5**.

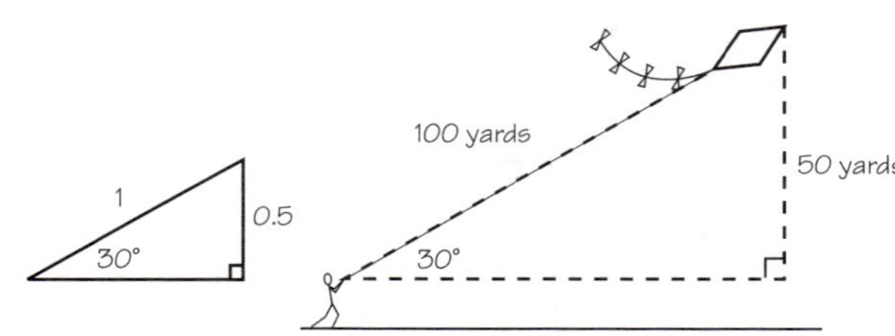

FIGURE 8.5. Finding the length of the third side.

ENGAGE LESSON 1

Section 8.1: Right Triangle Relationships: Pythagorean Theorem and Similar Triangles

Purpose: In previous chapters students used algebraic modeling to make predictions and decisions. In this chapter students use geometric and trigonometric modeling. In this section students investigate ways to solve right triangle problems where the Pythagorean theorem does not work, motivating students to investigate similar triangles and trigonometric ratios in Sections 8.2 through 8.5.

Description: This section begins with a problem that suggests using the Pythagorean theorem, with which most high school students are comfortable. Then students are confronted with a problem that looks like a Pythagorean theorem problem but cannot be solved using the Pythagorean theorem. Instead, they need to consider relationships between angles and sides of the triangle. Students develop the idea of an angle of elevation, which they use in later sections to apply trigonometric ratios.

Materials: classroom wall clock, flag, or other object for large-group discussion

Teaching Ideas: Students should work in small groups (3 to 4 students). You may need to review the Pythagorean theorem and the parts of a right triangle during the discussion of the activity.

During the first part of Section 8.1, especially Questions 1–3, student answers are not as important as their reasons for making their statements. Most students will recognize the setup as a Pythagorean theorem application, but some students may have had different experiences that cause them to make conjectures that other students do not.

Answers

1. Answers may vary, but most students will recognize this type of problem as a classic Pythagorean theorem application.
2. Answers may vary, but students should observe that they will be asked to find the length of one side of the triangle.
3. Students will need to know the lengths of two of the three sides of the right triangle in order to use the Pythagorean theorem to solve the problem.
4. Set up the proportion: $\frac{b}{0.5} = \frac{100}{1}$. Solve for b: $b = 50$.
5. No. The ground would actually be another 3 to 5 feet away.
6. First triangle: Using the Pythagorean theorem: $1^2 - (0.5)^2 = x^2$, so $x = 0.866$.

   Second triangle: Using the property of similarity: $\frac{x}{100} = \frac{0.866}{1}$, so $x = 86.6$ yards.

7. Use **Figure 8.6** to write a word problem with the answer of 86.6 yards.

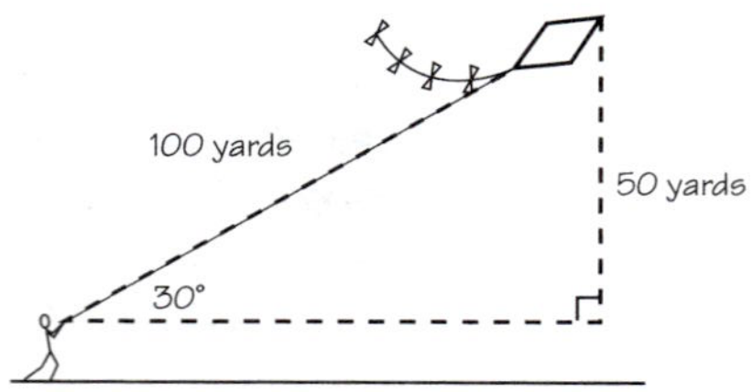

FIGURE 8.6.
Write a word problem.

SECTION 8.2 Right Triangle Relationships: Ratios

SAY WHAT YOU MEAN, MEAN WHAT YOU SAY

CHECK THIS!

Recall that the hypotenuse is the side of a triangle opposite the right angle.

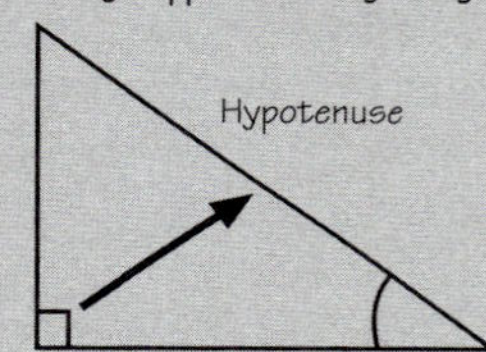

Identifying a side of a right triangle can be awkward. If you say **hypotenuse**, your meaning is clear. On the other hand, when you say **leg**, there are two sides possible. To avoid confusion, there must be names for the sides of the triangle. The legs are named according to their relationship to a **reference angle**.

The **adjacent leg** is the leg that touches or is next to the reference angle (**Figure 8.7**).

FIGURE 8.7.
Adjacent leg.

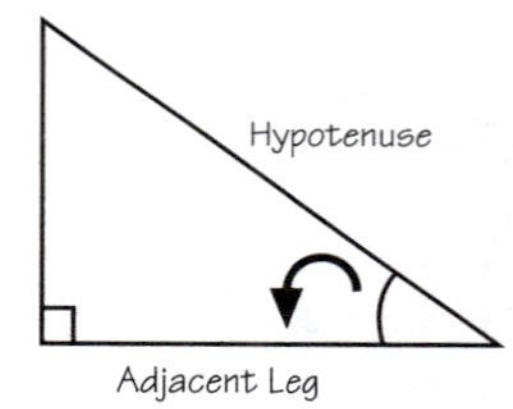

The **opposite leg** is the side that is opposite the reference angle (**Figure 8.8**).

FIGURE 8.8.
Opposite leg.

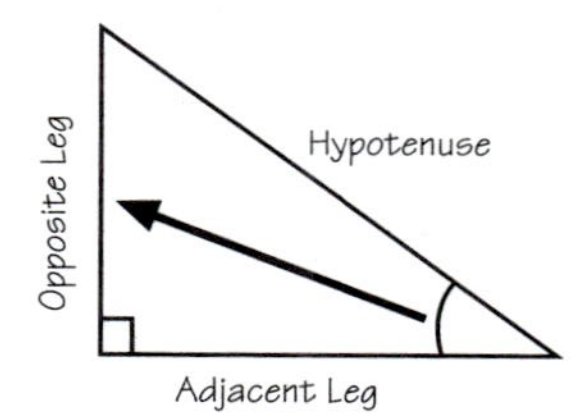

Answers (continued)

7. One possibility: Billy Bob's kite is flying directly over a water tower. How far is Billy Bob from the tower?

EXPLORE LESSON 1

Section 8.2: Right Triangle Relationships: Ratios

Purpose: In Section 8.1 students investigated situations when the Pythagorean theorem would or would not help them solve a problem. In this section students use similar triangles to develop three particular ratios that can be used to solve problems. In Section 8.3 students formalize the sine, cosine, and tangent ratios and use them to solve problems.

Description: In this section students construct similar triangles then use geometry to verify that the triangles are similar. Using these similar triangles, students measure the lengths of corresponding sides then examine three important ratios within one right triangle: the length of the opposite leg to the length of the hypotenuse, the length of the adjacent leg to the length of the hypotenuse, and the length of the opposite leg to the length of the adjacent leg. Students should notice that for every right triangle with a certain angle, these ratios are unchanged. Hence, these ratios can be used to solve problems when a triangle has that particular angle measure.

Materials: graphing calculator, graph paper, ruler, protractor

Teaching Ideas: Students should work in small groups (3 to 4 students). Students may be intimidated by the use of decimal numbers, so you may need to encourage them to use the calculator to make the arithmetic easier.

1. Sketch two triangles like those in **Figure 8.9**. Label the hypotenuse of each, then label the side adjacent to the marked angle and the side opposite the marked angle.

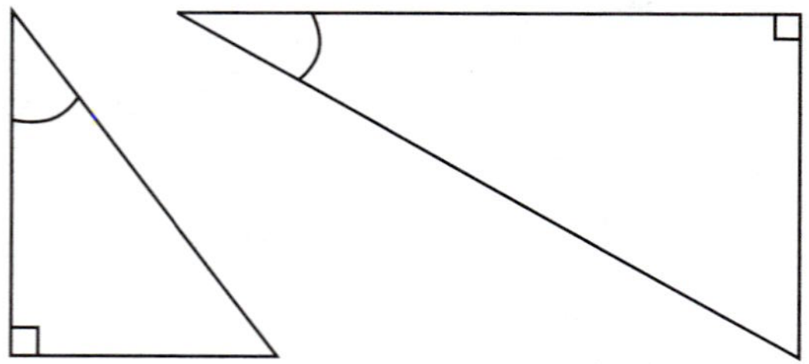

FIGURE 8.9.
Two triangles.

LET'S INVESTIGATE!

2. Use a protractor to draw a 55° angle on a piece of paper. Label it ∠A and extend the rays to the edges of the paper (**Figure 8.10**).

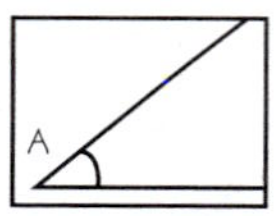

FIGURE 8.10.
Angle A.

3. In your drawing, add a small right triangle, ΔABC, where ∠C is the right angle (**Figure 8.11**). Use a centimeter ruler to measure the lengths of the sides of the triangle to the nearest tenth of a centimeter. Record the measurements in a table like **Figure 8.12**.

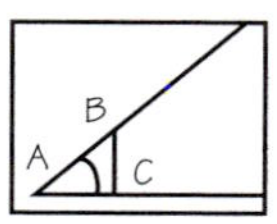

FIGURE 8.11.
Triangle ABC.

Triangle	Hypotenuse	Opposite Side	Adjacent Side
ΔABC	AB =	BC =	AC =
ΔADE			
ΔAFG			

FIGURE 8.12. Measurements.

4. Add right triangles ADE (**Figure 8.13**) and AFG (not shown) in a way similar to how you drew ΔABC in Question 3. Measure the lengths of the sides of both triangles and record the measurements in your table.

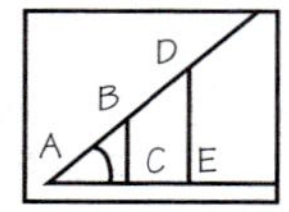

FIGURE 8.13.
Triangle ADE.

5. Are all three triangles, ΔABC, ΔADE, and ΔAFG, similar to each other? Explain.

6. Use the measurements from your table in Figure 8.12 to find and compare these ratios. Record your results in a table like **Figure 8.14**.

Ratio	ΔABC	ΔADE	ΔAFG
$\frac{\text{opposite}}{\text{hypotenuse}}$			
$\frac{\text{adjacent}}{\text{hypotenuse}}$			
$\frac{\text{opposite}}{\text{adjacent}}$			

FIGURE 8.14.
Recorded results.

CHECK THIS!

Mathematicians tire of writing the word *leg* when they work with right triangles. So they agree to use just *adjacent* when they mean *adjacent leg* and *opposite* when they mean *opposite leg*.

7. What patterns do you observe in the table?

8. What can you conclude?

9. Create another right ΔAHI with a hypotenuse of 10 cm. Use the ratios above to predict the length of the opposite leg.

$$\frac{\text{opposite}}{\text{hypotenuse}} = \_\_\_\_ \rightarrow \frac{\text{opposite}}{10} = \_\_\_\_ \rightarrow \text{opposite} = \_\_\_\_$$

Answers

1. Answer:

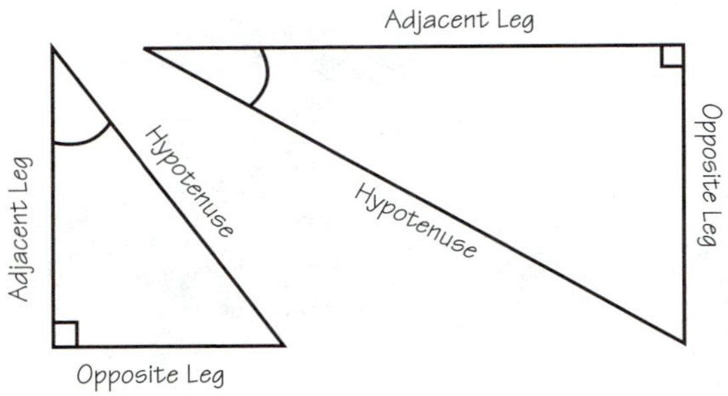

3. Answers may vary depending on where students choose to draw segment BC. Sample answers are given to illustrate pattern recognition in future questions.

Triangle	Hypotenuse	Opposite Side	Adjacent Side
ΔABC	AB = 7.5 cm	BC = 6.1 cm	AC = 4.3 cm
ΔADE			
ΔAFG			

4. Answers may vary depending on where students choose to draw segment BC. Sample answers are given to illustrate pattern recognition in future questions.

Triangle	Hypotenuse	Opposite Side	Adjacent Side
ΔABC	AB = 7.5 cm	BC = 6.1 cm	AC = 4.3 cm
ΔADE	AD = 13.1 cm	DE = 10.7 cm	AE = 7.5 cm
ΔAFG	AF = 20.7 cm	FG = 17.0 cm	AG = 12.0 cm

5. Yes. For triangles to be similar, it is sufficient to know that two pairs of corresponding angles are congruent.

All three right triangles share a 55° angle and have a 90° angle by definition. Since the measures of the angles of a triangle total 180, all three pairs of corresponding angles must be congruent.

Some students might decide to compare ratios of corresponding sides.

Compare ΔABC and ΔADE. Corresponding sides are $\frac{AB}{AD}$, $\frac{BC}{DE}$, and $\frac{AC}{AE}$. Compute the ratios of the lengths of the corresponding sides:

$\frac{AB}{AD} = \frac{7.5}{13.1} \approx 0.57$, $\frac{BC}{DE} = \frac{6.1}{10.7} \approx 0.57$, and

$\frac{AC}{AE} = \frac{4.3}{7.5} \approx 0.57$.

Repeat the process for ΔABC and ΔAFG.

$\frac{AB}{AF} = \frac{7.5}{20.7} \approx 0.36$, $\frac{BC}{FG} = \frac{6.1}{17.0} \approx 0.36$, and

$\frac{AC}{AG} = \frac{4.3}{12.0} \approx 0.36$.

Since the ratios are all equal, the lengths of the corresponding sides are proportional.

6. Sample answers:

Ratio	ΔABC	ΔADE	ΔAFG
$\frac{\text{opposite}}{\text{hypotenuse}}$	$\frac{BC}{AB} = \frac{6.1}{7.5} \approx 0.81$	$\frac{DE}{AD} = \frac{10.7}{13.1} \approx 0.82$	$\frac{FG}{AF} = \frac{17.0}{20.7} \approx 0.82$
$\frac{\text{adjacent}}{\text{hypotenuse}}$	$\frac{AC}{AB} = \frac{4.3}{7.5} \approx 0.57$	$\frac{AE}{AD} = \frac{7.5}{13.1} \approx 0.57$	$\frac{AG}{AF} = \frac{12.0}{20.7} \approx 0.57$
$\frac{\text{opposite}}{\text{adjacent}}$	$\frac{BC}{AC} = \frac{6.1}{4.3} \approx 1.42$	$\frac{DE}{AE} = \frac{10.7}{7.5} \approx 1.43$	$\frac{FG}{AG} = \frac{17.0}{12.0} \approx 1.42$

7. Answers may vary. Students should notice that each ratio is about the same for each triangle.

8. Answers may vary. Students should conclude that based on the data in the table, the ratio of any corresponding pair of sides for similar triangles should be the same; in other words, for any set of similar triangles, the ratio of the length of the opposite leg to the length of the hypotenuse should be the same. The same applies for the other two ratios.

9. Answer:

$\frac{\text{opposite}}{\text{hypotenuse}} = 0.82$

$\frac{\text{opposite}}{10} = 0.82$

$10\left(\frac{\text{opposite}}{10}\right) = 10(0.82)$

opposite = 8.2

SECTION 8.2 Assignment

1. In **Figure 8.15**, ΔABC is similar to ΔEFG. Write three proportions based on the similarity.

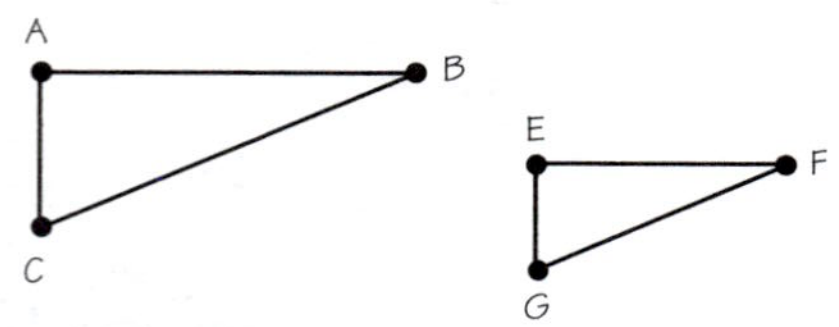

FIGURE 8.15. Triangles ABC and EFG.

2. In relationship to the marked angles in **Figure 8.16**, label the hypotenuse, opposite leg, and adjacent leg.

FIGURE 8.16. Three triangles.

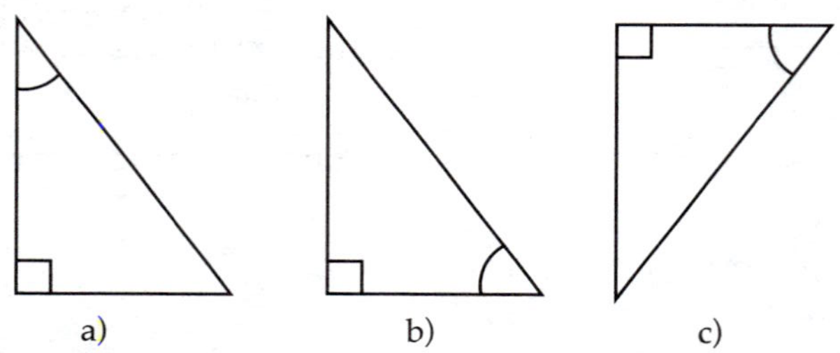

3. Set up proportions and find the length of the hypotenuse and adjacent side of the larger triangle in **Figure 8.17**.

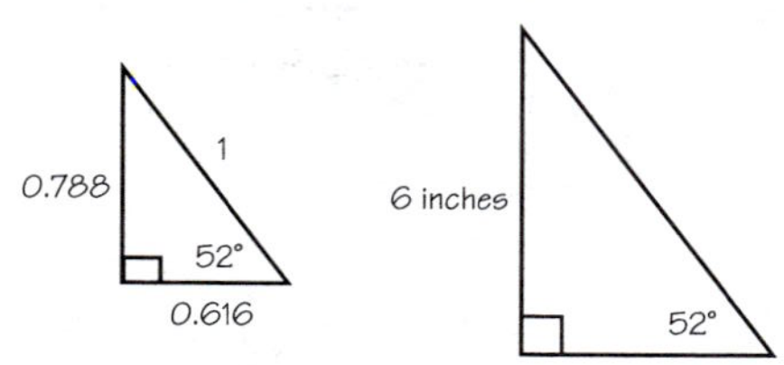

FIGURE 8.17. Two triangles.

4. Use **Figure 8.18** to answer these questions.

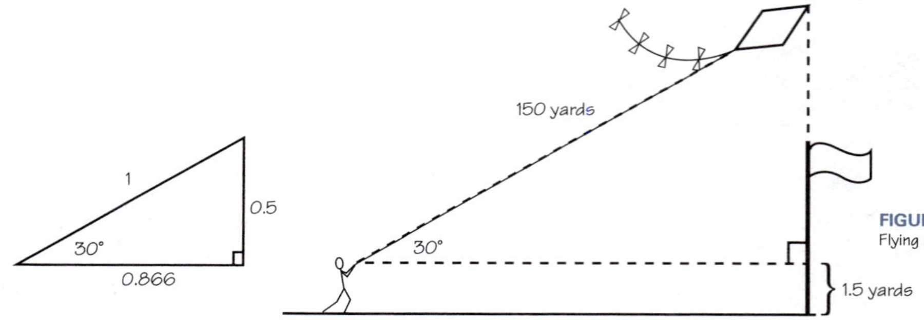

FIGURE 8.18. Flying a kite.

a) About how high is the kite?

b) The kite is flying directly over a flagpole. How far is the kite flyer from the pole?

5. In **Figure 8.19**, find the width of the stream.

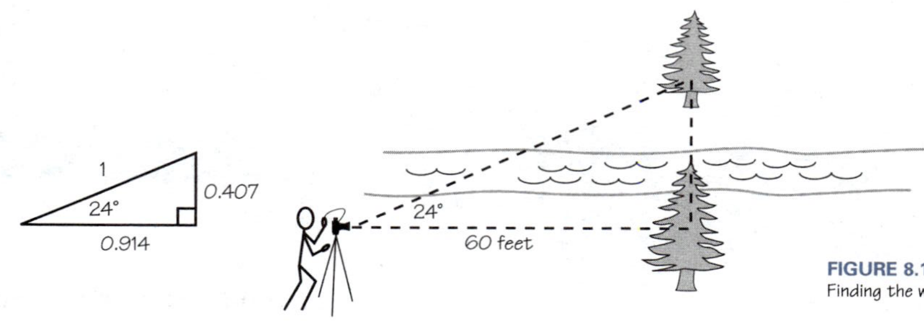

FIGURE 8.19. Finding the width of a stream.

Section 8.2 Assignment

Answers

1. Possible answers:
$\frac{AC}{EG} = \frac{AB}{EF}$, $\frac{AC}{EG} = \frac{BC}{FG}$, or $\frac{AB}{BC} = \frac{EF}{FG}$.

2. Answers:

a)

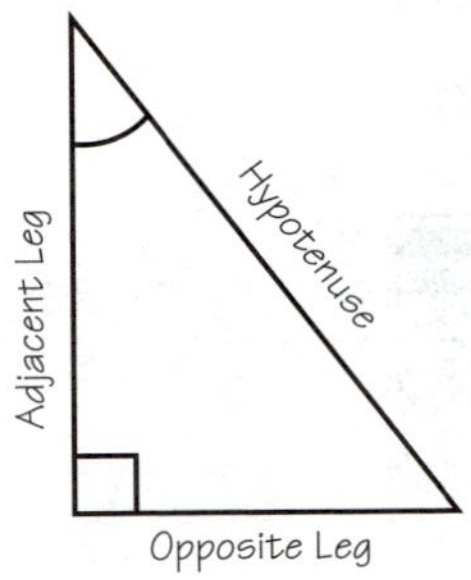

b)

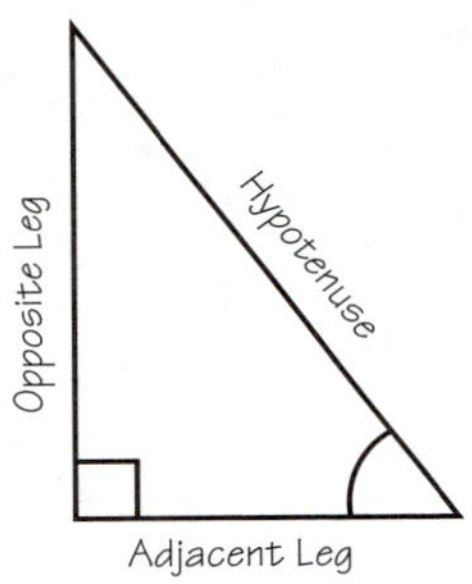

c)

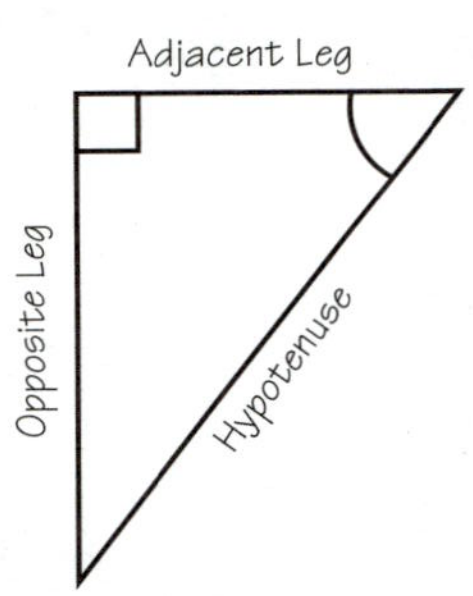

3. Possible answers: $\frac{\text{adjacent}}{6} = \frac{0.616}{0.788}$ and $\frac{\text{hypotenuse}}{6} = \frac{1}{0.788}$.

4. a) 76.5 yards (remember to add the height of the string from the ground).

   b) 129.9 yards

5. $\frac{x}{60} = \frac{0.407}{0.914}$, then $x = 26.7$ feet.

SECTION 8.3

Trigonometric Ratios

1. In Section 8.1 you saw problems that cannot be solved with the Pythagorean theorem. Why was it not possible to solve these problems?

2. Many of the problems that cannot be solved with the Pythagorean theorem can be solved with similar triangles. How did you use similar triangles to solve problems?

3. In Section 8.2 you explored three ratios that are the same for similar right triangles. What were these ratios?

These right triangle ratios are called **Trigonometric Ratios.** Each of the trigonometric ratios you found has a special name:

The sine ratio = $\frac{\text{leg opposite the reference angle}}{\text{hypotenuse}}$ or sin = $\frac{\text{opposite}}{\text{hypotenuse}}$

The cosine ratio = $\frac{\text{leg adjacent to the reference angle}}{\text{hypotenuse}}$ or cos = $\frac{\text{adjacent}}{\text{hypotenuse}}$

The tangent ratio = $\frac{\text{leg opposite the reference angle}}{\text{leg adjacent to the reference angle}}$ or tan = $\frac{\text{opposite}}{\text{adjacent}}$

Use a graphing calculator to investigate triangle ratios. Since there are several units used to measure angles, first check your calculator's MODE to be sure that it is in the DEGREE mode. If DEGREE is not selected, use the arrow keys to highlight DEGREE. Press ENTER to select (**Figure 8.20**).

FIGURE 8.20.
Checking the MODE.

For example, to find the sine ratio for a 30° angle, press SIN, 30, ENTER. You should get 0.5 (**Figure 8.21**):

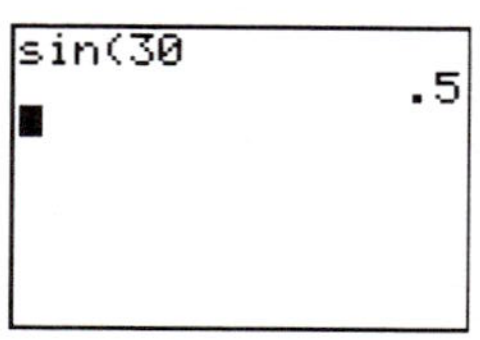

FIGURE 8.21.
Finding the sine.

4. What does sin 30° = 0.5 tell you about the relationship between the length of the leg opposite a 30° angle and the length of the hypotenuse?

CHECK THIS!

The values of trigonometric ratios are often rounded to the nearest ten-thousandth.

However, to get a more precise answer when solving problems, use the ratio as written instead of a decimal approximation. For example, use "sin(41)" instead of 0.6561.

5. Find sin 32°.

6. Find cos 32°.

7. Find tan 32°.

8. Compare the ratios you found in Section 8.2 for a 55° reference angle with the values that your calculator gives you. Record your data in a table like **Figure 8.22**.

$m\angle A = 55^\circ$	Calculated Ratio	Calculator Value
sin 55°		
cos 55°		
tan 55°		

FIGURE 8.22.
Ratios and values.

9. How do your calculated ratios compare to the calculator values for the sine, cosine, and tangent of ∠A?

There are many stories to help remember the trigonometric ratios. One story follows:

Prince Rahib escorted his girlfriend, Princess Rahana, as she rode her camel. He impressed her with his new running shoes and his athletic prowess. The sun beat down on them as they made their way through the burning sands of the desert. Suddenly Rahana's camel lay down and refused to move. All of their efforts to get the camel back on its feet were to no avail. The sun was hot, and they were going to have to walk back to the palace. Princess Rahana was barefoot, so Prince Rahib insisted that she wear his shoes to protect her feet from the hot sands. They hurried, but the Prince's feet were soon blistered. When they rushed onto the palace grounds, Rahana cried, "My hero!" and went to give Rahib a kiss. "Get back!" he cried, "I have no time for you. I just want to SOH-CAH-TOA!"

EXPLAIN LESSON 1

Section 8.3: Trigonometric Ratios

Purpose: In Section 8.3 students investigated three special ratios in a right triangle then used patterns to discover that these ratios are the same in similar right triangles. In this section students formalize these special ratios as sine, cosine, and tangent trigonometric ratios then use them to solve right triangle problems. These applications prepare students to work backwards in the Elaborate phase, Section 8.4, when they are given side lengths and need to determine the angle using inverse trigonometric ratios.

Description: In this phase of the lesson students define the sine, cosine, and tangent ratios. They use the mnemonic device "SOH CAH TOA" to help remember the ratios. Students begin using trigonometric ratios to solve simple problems from a diagram then progress to analyzing real-world situations and applying trigonometric ratios to solve angle of elevation and angle of depression problems.

Materials: graphing calculators, Transparency 8.1

Teaching Ideas: Debrief students on Section 8.2 by having them share their answers to the questions. As they share, be sure to address any missing concept, process, or skill that could have been used to answer the questions. Then work through the questions in this section with the entire class. Once again, be sure all necessary concepts, processes, or skills are addressed.

Calculator skills should be closely monitored, especially if students are unfamiliar with using the SIN, COS, and TAN features of a graphing calculator. Emphasize the use of the 2nd, SHIFT, or INV feature of a calculator as a means of undoing the parent function, just like the squaring or square root keys. You may need to assist students as they learn this calculator feature.

Answers

1. Answers may vary. Students should observe that the Pythagorean theorem can be used only to solve problems involving missing side lengths if you know the other two sides. Also, the Pythagorean theorem can only be used for right triangles.

2. If you know angle measures, you can use proportions to find missing side lengths when you know the lengths of the corresponding sides.

3. The ratio of the length of the opposite leg to the length of the hypotenuse

   The ratio of the length of the adjacent leg to the length of the hypotenuse

   The ratio of the length of the opposite leg to the length of the adjacent leg

4. In a right triangle with one acute angle measuring 30°, the ratio of the length of the side opposite the 30° angle to the length of the hypotenuse is 1:2. In other words, the hypotenuse is twice as long as the side opposite the 30° angle.

5. 0.5299

6. 0.8480

7. 0.6249

8. Sample answers:

$m\angle A = 55°$	Calculated Ratio	Calculated Value
sin 55°	0.82	0.8191
cos 55°	0.57	0.5736
tan 55°	1.42	1.4281

9. The values are almost the same if you round to the nearest hundredth.

NOTE: After the Prince Rahib story, show Transparency 8.1.

Trigonometric ratios can be used to solve problems involving angle measures and side lengths. Using trigonometric ratios allows you to take advantage of the power of similar triangles without using a second triangle.

For example, consider this problem.

In **Figure 8.23**, a flagpole's shadow is 23.5 feet long. How tall is the flagpole?

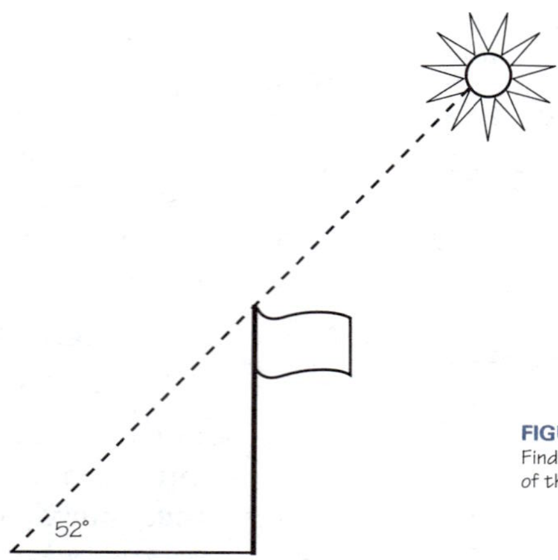

FIGURE 8.23. Finding the height of the flagpole.

When solving problems with trigonometric ratios, it is helpful to write all known information on the figure and label the hypotenuse, the adjacent leg, and the opposite leg (**Figure 8.24**).

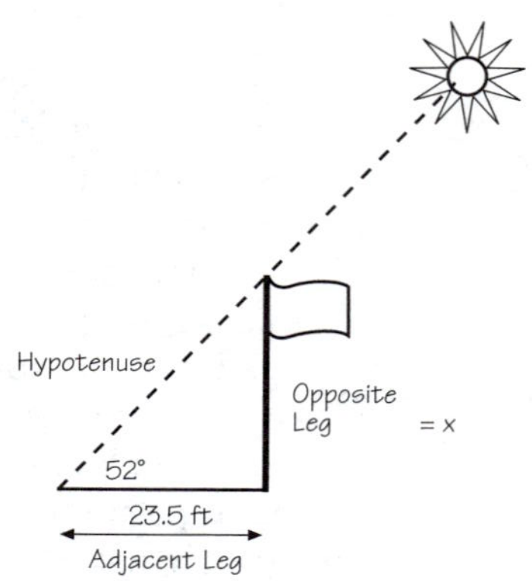

FIGURE 8.24. Labeling all information.

Next, decide which ratio can be used to solve the problem. In this case, you know the length of the adjacent leg and want to know the length of the opposite leg. Therefore, use the ratio relating the adjacent leg to the opposite leg, which is the **tangent** ratio.

$\tan = \frac{\text{opposite}}{\text{adjacent}}$

Now you can substitute the known information into the ratio equation and then solve for x.

$\tan 52° = \frac{\text{opposite}}{\text{adjacent}}$

$\tan 52° = \frac{x}{23.5}$

$23.5(\tan 52°) = 23.5\left(\frac{x}{23.5}\right)$

$23.5 \tan 52° = x$

$30 \approx x$

So, the height of the flagpole is about 30 feet.

Next, consider this problem:

A right triangle has a 15° angle and a short leg of 18 inches. How long is the hypotenuse?

10. First, draw and label a picture. Be sure to label the reference angle, all known information, the unknown length, and the hypotenuse, adjacent leg, and opposite leg.

11. Which side length is given? Which side length is unknown? Which ratio relates these two?

12. Substitute the known values into the ratio equation and then solve for the unknown.

Answers (continued)

10. Sample answer:

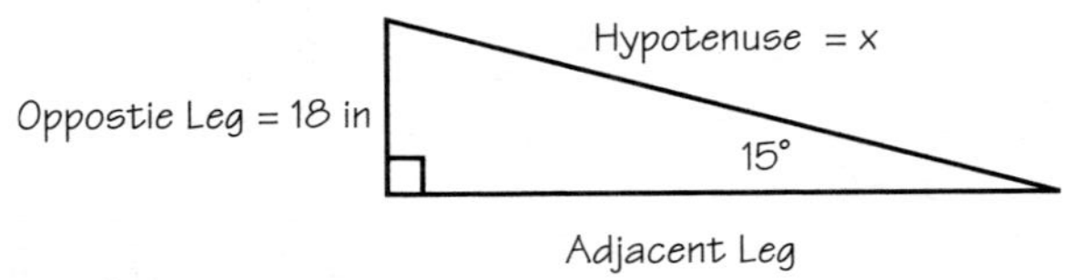

11. The opposite leg is given, and the hypotenuse is unknown; the sine ratio is required.

12. $\sin 15° = \frac{\text{opposite}}{\text{hypotenuse}}$

$\sin 15° = \frac{18}{x}$

$\frac{\sin 15°}{1} = \frac{18}{x}$

$x(\sin 15°) = 1(18)$

$\frac{x(\sin 15°)}{\sin 15°} = \frac{18}{\sin 15°}$

$x \approx 69.5$ inches

SECTION 8.3 Assignment

1. Use your calculator to find each of these:

   a) sin 76° = ______

   b) cos 31° = ______

   c) tan 16° = ______

   d) sin 60° = ______

   e) cos 60° = ______

   f) tan 45° = ______

2. Write each ratio in **Figures 8.25 and 8.26** in both words and numbers.

   a)

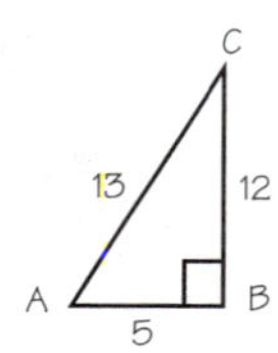

FIGURE 8.25.
Figure for Question 2a.

   sin ∠A = ___ = ___

   cos ∠A = ___ = ___

   tan ∠A = ___ = ___

   b)

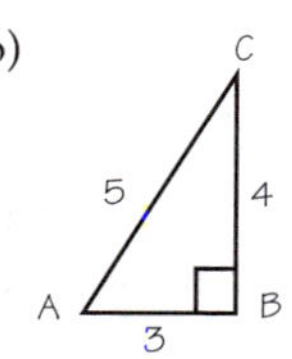

FIGURE 8.26.
Figure for Question 2b.

   sin ∠C = ___ = ___

   cos ∠C = ___ = ___

   tan ∠C = ___ = ___

3. An isosceles right triangle has a hypotenuse that is 10 units long. About how long are its legs?

4. The stairs of the Kukulcan Pyramid (**Figure 8.27**) rise at a 48° angle. When you get to the top, you are 100 feet above the ground (about ten stories). How long is the stairway?

FIGURE 8.27.
Kukulcan Pyramid, Mexico.

In Questions 5 and 6 draw a picture and solve the problem.

5. A right triangle has a 22° angle and a hypotenuse of 10 units. Find the length of each leg.

6. Rigo is holding his kite spool about 3 feet off the ground. The kite is flying at a 55° angle with the horizontal. The kite is directly over his friend Milo's head. If Milo is about 50 feet from Rigo, how high is the kite?

Section 8.3 Assignment

Answers

1. a) $\sin 76° = 0.9703$

   b) $\cos 31° = 0.8572$

   c) $\tan 16° = 0.2867$

   d) $\sin 60° = 0.8660$

   e) $\cos 60° = 0.5$

   f) $\tan 45° = 1$

2. a) $\sin \angle A = \frac{\text{opposite}}{\text{hypotenuse}} = \frac{12}{13}$

   $\cos \angle A = \frac{\text{adjacent}}{\text{hypotenuse}} = \frac{5}{13}$

   $\tan \angle A = \frac{\text{opposite}}{\text{adjacent}} = \frac{12}{5}$

   b) $\sin \angle C = \frac{\text{opposite}}{\text{hypotenuse}} = \frac{3}{5}$

   $\cos \angle C = \frac{\text{adjacent}}{\text{hypotenuse}} = \frac{4}{5}$

   $\tan \angle C = \frac{\text{opposite}}{\text{adjacent}} = \frac{3}{4}$

3. An isosceles triangle implies the base angles are congruent, thus 45° each. Either sine or cosine is appropriate.

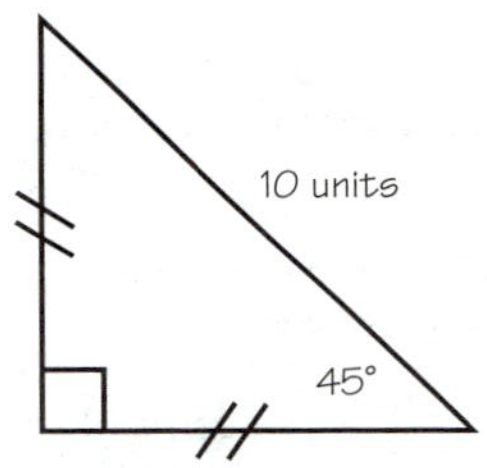

$\cos 45° = \frac{\text{adjacent}}{\text{hypotenuse}}$

$\cos 45° = \frac{x}{10}$

$10 \cos 45° = 10\left(\frac{x}{10}\right)$

$10 \cos 45° = x$

$x = 7.071$ or around 7

This problem could also be solved using the Pythagorean theorem: $x^2 + x^2 = 10^2$, then $2x^2 = 100$, $x^2 = 50$, and x is about 7 units long.

4. The length of the opposite side is given, and the length of the hypotenuse is unknown, so use the sine ratio to relate the two lengths.

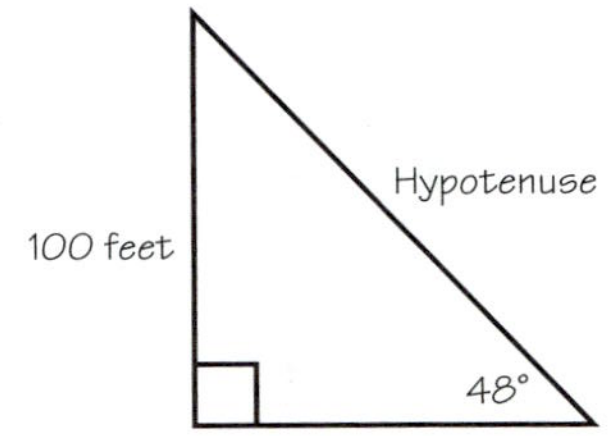

$\sin 48° = \frac{\text{opposite}}{\text{hypotenuse}}$

$\sin 48° = \frac{10}{x}$

$\frac{\sin 48°}{1} = \frac{100}{x}$

$x \sin 48° = 1(100)$

$\frac{x \sin 48°}{\sin 48°} = \frac{1(100)}{\sin 48°}$

$135 \text{ feet} \approx x$

5. Draw a right triangle and label the known information. Label the hypotenuse, opposite leg, and adjacent leg.

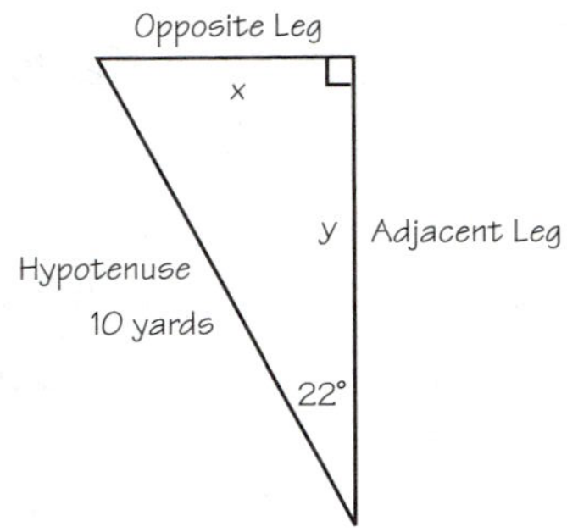

First find x. The length of the hypotenuse is given, and x represents the length of the opposite leg, so use the sine ratio to relate the two lengths.

$\sin 40° = \frac{\text{opposite}}{\text{hypotenuse}}$

$\sin 22° = \frac{x}{10}$

$10 \sin 22° = 10\left(\frac{x}{250}\right)$

$10 \sin 22° = x$

$3.75 \approx x$

$\cos 22° = \frac{\text{adjacent}}{\text{hypotenuse}}$

$\cos 22° = \frac{x}{10}$

$10 \cos 22° = 10\left(\frac{x}{10}\right)$

$10 \cos 22° = x$

$9.27 \approx x$

Next find y. The length of the hypotenuse is given, and y represents the length of the adjacent leg, so use the cosine ratio to relate the two lengths.

The opposite side is 3.75 units; the adjacent side is 9.27 units.

6. Draw a picture. Label all known information. Label the hypotenuse, opposite leg, and adjacent leg.

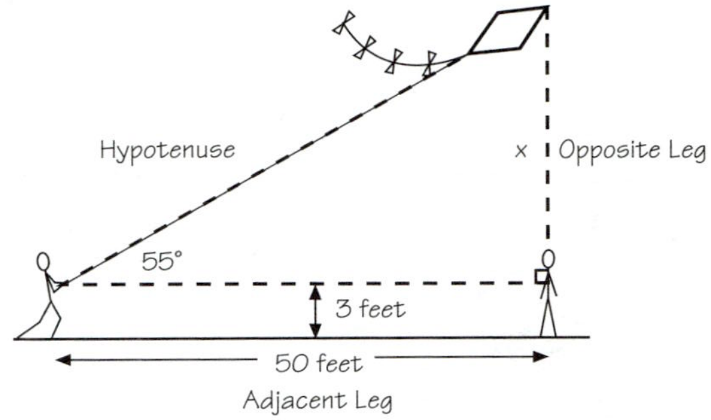

The length of the adjacent leg is given, and the length of the opposite leg is unknown, so use the tangent ratio to relate the two lengths.

$\tan 55° = \frac{\text{opposite}}{\text{adjacent}}$

$\tan 55° = \frac{x}{50}$

$50 \tan 55° = 50\left(\frac{x}{50}\right)$

$50 \tan 55° = x$

$71 \text{ feet} \approx x$

The kite is about 71 feet above Rigo's line of sight. Since Rigo is holding the kite about 3 feet off the ground, add 3 feet to this height. The kite is about 74 feet above the ground.

SECTION 8.4

A New Angle on This ...

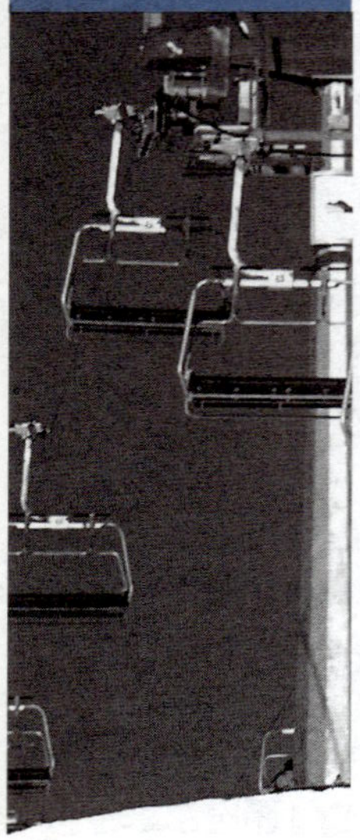

In Section 8.3 you used sine, cosine, and tangent to find side lengths of triangles.

Consider this problem:

A ski lift rides a 250-meter track up the side of a mountain to the top, which is 160 meters tall. What is the ski lift's angle of elevation?

1. Draw a picture of this situation. Label the known information.
2. Relative to the angle of elevation, label the hypotenuse, adjacent leg, and opposite leg.
3. What does the problem require you to find?
4. How does this problem differ from those you saw in Section 8.3?
5. Which ratio is formed by the lengths of the two known sides?
6. What is the sine ratio for the angle that you want to find?
7. How can you find an angle measure if you know the sine ratio for that angle?

Sometimes mathematicians use tables of trigonometric values to find sine, cosine, and tangent ratios. These tables contain the ratios for certain angles rounded to a certain decimal place. A portion of a trigonometric table is in **Figure 8.28**.

Angle	Sine	Cosine	Tangent
38°	0.616	0.788	0.781
39°	0.629	0.777	0.810
40°	0.643	0.766	0.839
41°	0.656	0.755	0.869
42°	0.669	0.743	0.900

FIGURE 8.28.
Trigonometric table.

8. In the table in Figure 8.28, what angle measure has a sine ratio closest to 0.64? How did you find this value?

CHECK THIS!

Inverse sine is notated as $\sin^{-1}$.

Inverse cosine is notated as $\cos^{-1}$.

Inverse tangent is notated as $\tan^{-1}$.

When you see $\sin^{-1}(0.5)$, think of it as "the angle whose sine is 0.5."

This process of working backwards to find an angle measure whose sine ratio is known is called finding the **inverse sine**. Likewise, you can find the **inverse cosine** or **inverse tangent** if you know an angle's cosine ratio or tangent ratio.

9. The sine ratio of ∠A is 0.64. How would you write this as an equation?

Graphing calculators have built-in features to allow you to compute an inverse trigonometric ratio. You need to press the 2nd or SHIFT key before pressing the SIN, COS, or TAN key to obtain the inverse sine, cosine, or tangent, respectively.

10. Use your graphing calculator to find the angle whose sine ratio is 0.64.

An advantage to using a graphing calculator to find angles is that you do not need to find the decimal value for a ratio before finding the angle measure.

11. Find the measure of ∠M in **Figure 8.29**.

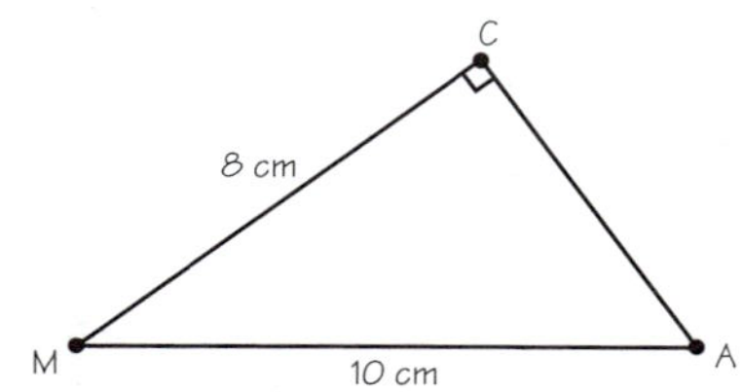

FIGURE 8.29.
Angle M.

ELABORATE LESSON 1

Section 8.4: A New Angle on This…

Purpose: In Sections 8.2 and 8.3 students explored and formalized the sine, cosine, and tangent ratios, then used them to find the missing side lengths of right triangles. In this section students use trigonometric ratios to find missing angle measurements using inverse sine, inverse cosine, and inverse tangent ratios, preparing students to solve a different kind of right triangle problem in Section 8.5.

Description: In this section students apply their knowledge of inverse functions to solve trigonometric equations for angle values. Again, students first solve problems involving right triangle diagrams then progress in levels to solving real-world application problems.

Materials: graphing calculator

Teaching Ideas: Emphasize the use of inverse functions in this section. Just as students "un-multiply," "add the opposite," or "take the square root" when solving other types of equations, students will use the inverse sine, inverse cosine, and inverse tangent to work backwards.

Answers

1. Sample answer:

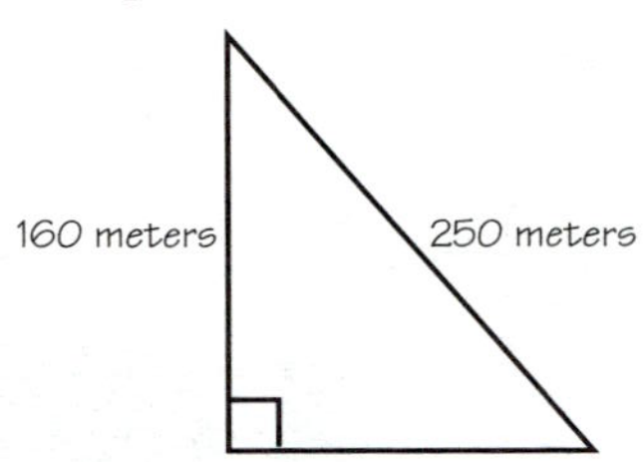

2. Sample answer:

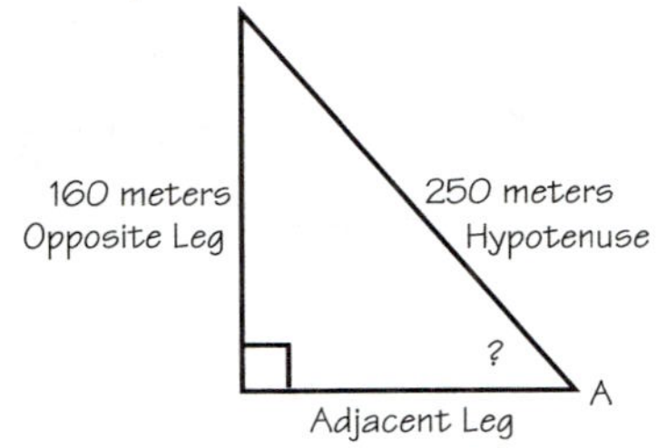

3. The problem requires you to find the angle of elevation. In the sample drawing this angle is the measure of angle A.
4. We know the lengths of two sides of the triangle instead of one side and an angle.
5. sine ratio
6. $\sin A° = \frac{\text{opposite}}{\text{hypotenuse}} = \frac{160}{250} = 0.64.$
7. Answers may vary. Students might suggest doing some kind of inverse sine. In fact, that is exactly what we will do.
8. A 40° angle has a sine ratio of 0.643, which is the closest to 0.64 in the table. This value can be found by reading down the sine column until you find a number close to the target sine ratio of 0.64. Then read over to the angle column to find a value of 40.
9. $\sin A = 0.64$
10. Answer:

```
sin⁻¹(0.64)
          39.7918195
```

11. The hypotenuse of triangle MAC is $\overline{MA}$, which has a length of 10 cm. The adjacent leg to ∠M is $\overline{MC}$, which has a length of 8 cm. Use the cosine ratio to relate the two lengths.

```
cos⁻¹(8/10)
          36.86989765
```

$$\cos M = \frac{\text{adjacent}}{\text{hypotenuse}}$$

$$\cos M = \frac{8}{10}$$

$$M = \cos^{-1}\left(\frac{8}{10}\right)$$

$$M \approx 37°$$

SECTION 8.4 Assignment

1. What are the angle measures of a 3-4-5 right triangle?

2. A staircase, which takes up 12.5 feet of horizontal floor space, connects two floors that are 10 feet apart (**Figure 8.30**). What is the angle of elevation of the staircase?

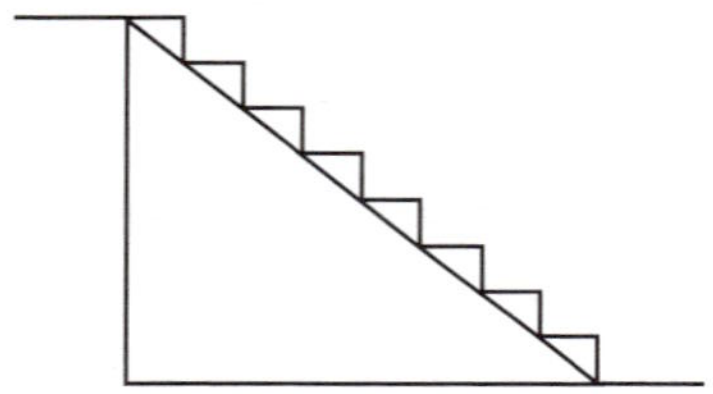

FIGURE 8.30. Staircase.

3. Go fly a kite! (Draw pictures with your work.)

   a) Shelia is standing 225 feet from a long line of trees 70 feet high. If she is holding the reel of string 4 feet off the ground, how far is it from the reel to the top of the trees?

   b) If Shelia lets out 300 feet of string so that the kite is now directly over a small building 195 feet away, what angle does the string make with the horizontal?

   c) Suppose Shelia moves away from the building so that her kite string makes a 40° angle with the horizontal. If Shelia still has 300 feet of string out, what is the ground distance between Shelia and the building?

4. A right triangle has a hypotenuse of 10 centimeters and one angle that measures 28°. What is the area of the triangle?

SECTION 8.5 The Two Towers

A distance of 2000 feet separates two towers in a state forest. One of the towers is directly west of the other. An observer in the West Tower notices a fire at an angle of 40° from the line of sight between the two towers. An observer in the East Tower notices the same fire due north from her tower (see **Figure 8.31**). Due to the terrain around the towers and through the forest, a fire truck from the West Tower travels at an average speed of 20 miles per hour. A fire truck from the East Tower travels at an average speed of 15 miles per hour. If both leave their towers at the same time, which one will get to the fire first? Justify your answer.

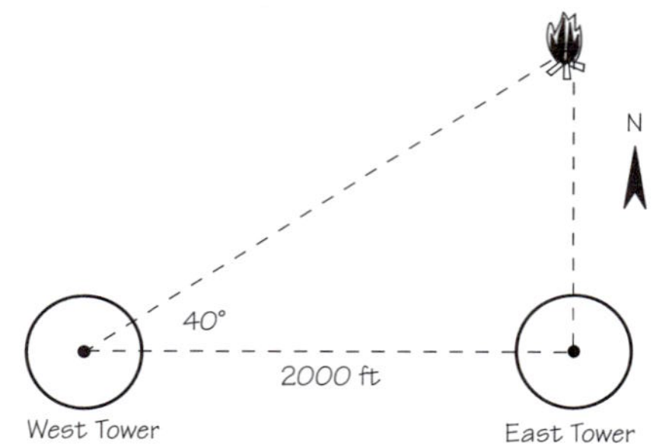

FIGURE 8.31. Two towers.

Section 8.4 Assignment

Answers

1. Draw a 3-4-5 right triangle. For reference label the vertices A, B, and C.

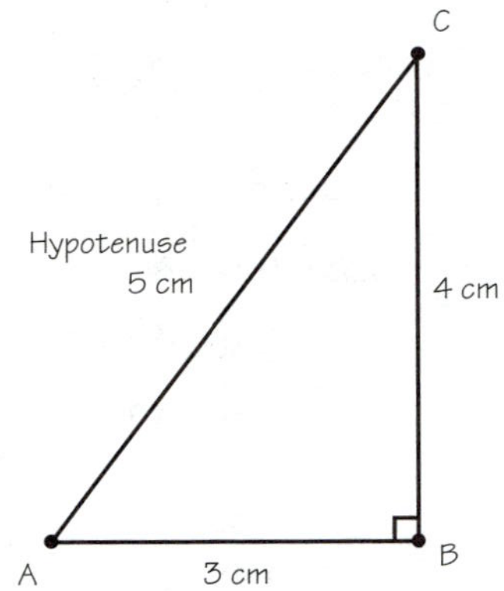

First find the measure of ∠A. Using ∠A as the reference angle, label the adjacent and opposite legs.

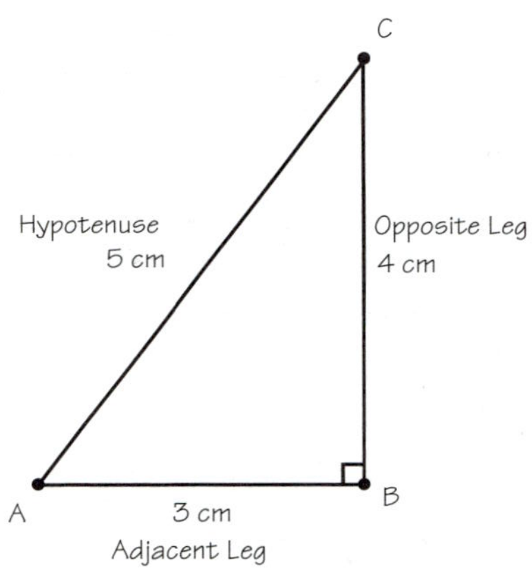

Second find the measure of ∠C. Using ∠C as the reference angle, label the adjacent and opposite legs.

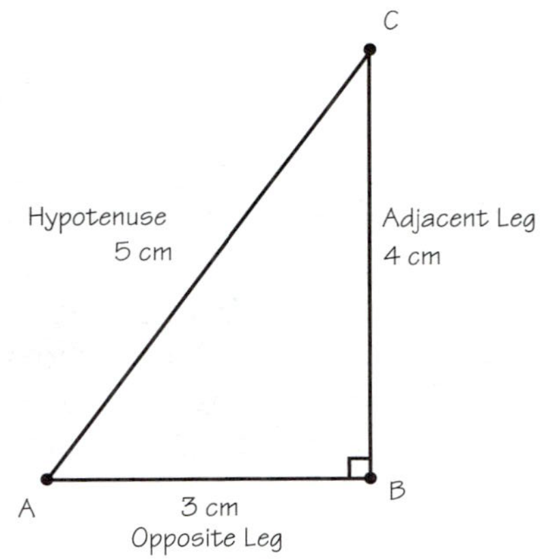

The lengths of both the opposite leg and the adjacent leg are known, so use the tangent ratio to relate the two lengths.

$\tan A = \frac{\text{opposite}}{\text{adjacent}}$

$\tan A = \frac{4}{3}$

$A = \tan^{-1}\left(\frac{4}{3}\right)$

$A \approx 53°$

The lengths of both the opposite leg and the hypotenuse are known, so use the sine ratio to relate the two lengths.

$\sin C = \frac{\text{opposite}}{\text{hypotenuse}}$

$\sin C = \frac{3}{5}$

$C = \sin^{-1}\left(\frac{3}{5}\right)$

$C \approx 37°$

2. Answer: about 39°

   Sample solution:

   Draw and label a picture.

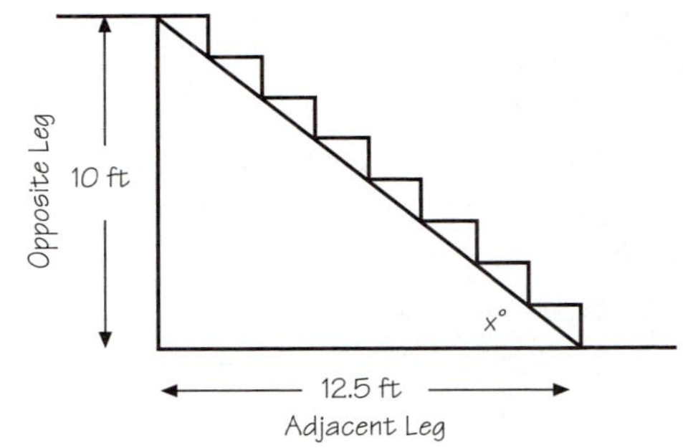

Since the lengths of the opposite leg and adjacent leg are both given, use the tangent ratio to relate the two lengths.

$\tan x = \frac{\text{opposite}}{\text{adjacent}}$

$\tan x = \frac{10}{12.5}$

$x = \tan^{-1}\left(\frac{10}{12.5}\right)$

$x \approx 39°$

3. a) Strategy: Use the Pythagorean theorem:

   $225^2 + 66^2 = c^2$

   $c = 234.5$ ft

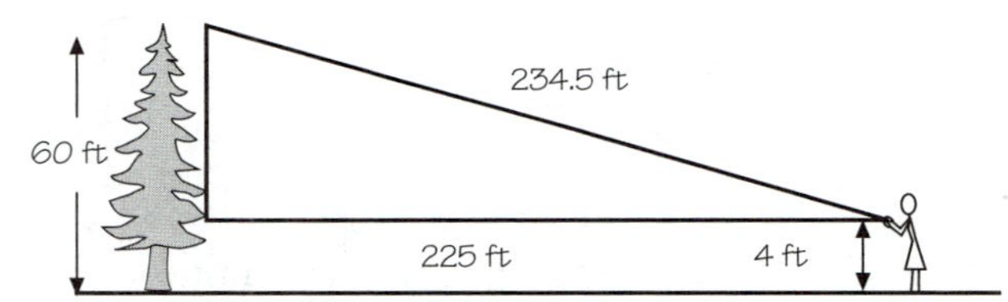

Answers (continued)

b) Answer: about 50°

Sample solution:

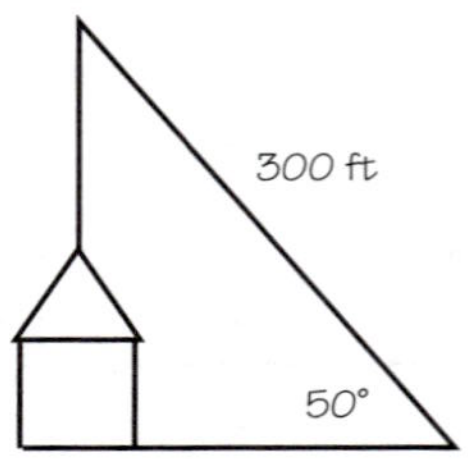

The hypotenuse is the string so the length of the hypotenuse is the same as the length of the string, which is 300 feet.

The adjacent leg to the angle between the string and the horizontal is the distance between Shelia and the building, which is 195 feet.

Since the length of the hypotenuse and the length of the adjacent leg are given, use the cosine ratio to relate the two lengths.

$$\cos x = \frac{\text{adjacent}}{\text{hypotenuse}}$$

$$\cos x = \frac{195}{300}$$

$$x = \cos^{-1}\left(\frac{195}{300}\right)$$

$$x \approx 50°$$

c) Answer: about 230 feet

Sample solution:

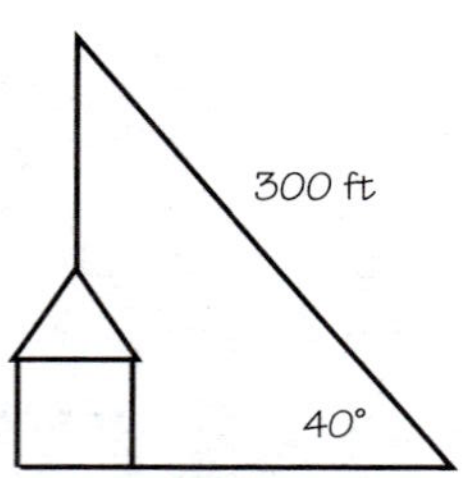

The hypotenuse is the string so the length of the hypotenuse is the same as the length of the string, which is 300 feet.

The adjacent leg to the angle between the string and the horizontal is the distance between Shelia and the building, which is the unknown.

Since the length of the hypotenuse is given and the length of the adjacent leg is the unknown, use the cosine ratio to relate the two lengths.

$$\cos 40° = \frac{\text{adjacent}}{\text{hypotenuse}}$$

$$\cos 40° = \frac{x}{300}$$

$$300 \cos 40° = 300\left(\frac{x}{300}\right)$$

$$230 \approx x$$

4. Answer: 20.7 square centimeters

Sample solution:

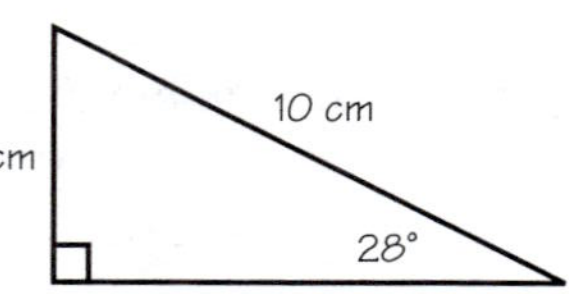

In order to find the area of the triangle, you need to know the base length and the height. In a right triangle, the two legs that are perpendicular are the base length and the height.

First find the length of the opposite leg, which is the height of the triangle:

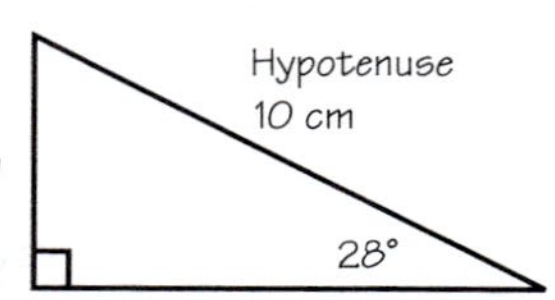

$$\sin 28° = \frac{\text{opposite}}{\text{hypotenuse}}$$

$$\sin 28° = \frac{h}{10}$$

$$10 \sin 28° = 10\left(\frac{h}{10}\right)$$

$$4.69 \approx h$$

Second find the length of the adjacent leg,

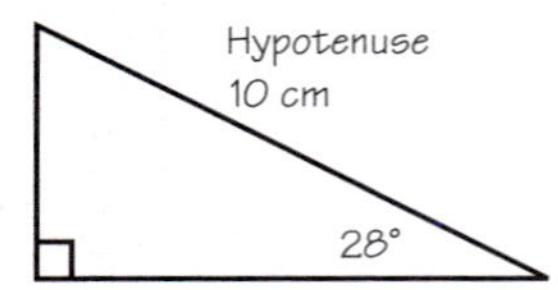

which is the base length of the triangle:

$$\cos 28° = \frac{\text{adjacent}}{\text{hypotenuse}}$$

$$\cos 28° = \frac{b}{10}$$

$$10 \cos 28° = 10\left(\frac{b}{10}\right)$$

$$8.83 \approx b$$

To find the area of a triangle, multiply the base length by the height then divide by 2:

$$A = \frac{bh}{2}$$

$$A \approx \frac{(8.83)(4.69)}{2}$$

$$A \approx 20.7$$

EVALUATE LESSON 1

Section 8.5: The Two Towers

Purpose: In Sections 8.1 through 8.4, students developed the concept of trigonometric ratios and formalized skills and procedures necessary to solve problems using right triangles and trigonometric ratios. In this section students have the opportunity to demonstrate their knowledge by applying those concepts and skills to solve a problem in a new context. In Sections 8.6 through 8.10, students expand on trigonometric ratios in order to explore and describe periodic motion.

Description: Facing a problem in a new context, students will determine and implement a method to solve the problem. In this open-middle performance assessment task, there is only one correct answer, but there are a variety of valid methods to obtain it. In their justification, which is the most important part of a performance assessment task, students should be able to communicate their knowledge of mathematics and demonstrate that they have mastered the concepts presented in Sections 8.1 through 8.4.

Materials: graphing calculators

Teaching Ideas: This task may be assigned as group work or individual work. The mathematics performance assessment rubric should be used to evaluate students' work.

Answers

Sample Solution:

First identify the key information. Draw and label a picture to help organize the information and form a strategy to solve the problem. Since the fire is located due north of the East Tower and the towers are on an east/west line, the angle between the line of sight to the West Tower and the line of sight to the East Tower is a right angle, and the triangle is a right triangle. Label the hypotenuse. Using the angle at the West Tower as a reference angle, label the line of sight from the East Tower to the fire as the opposite leg and the line of sight from the East Tower to the West Tower as the adjacent leg.

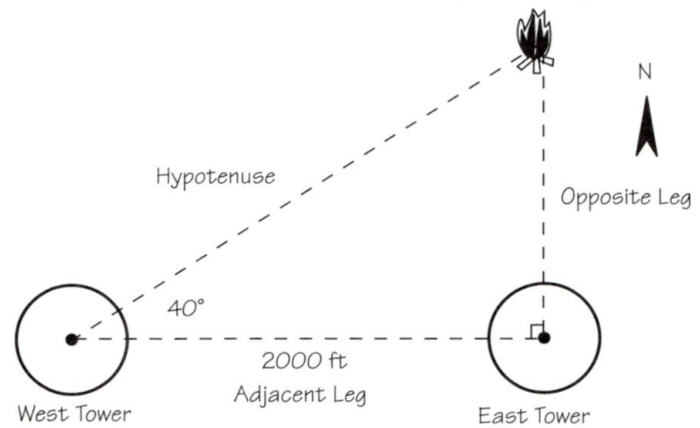

Next formulate a strategy. We need to know the time it takes for each fire truck to travel from the tower to the fire. Given the information in the figure, we can find the distances between each tower and the fire using trigonometric ratios. Once we know the distance, we can use the formula $d = rt$ to find the time it takes for each fire truck to travel its required distance.

First find the length of the opposite leg, x. We know the length of the adjacent leg, so use the tangent ratio to relate the two.

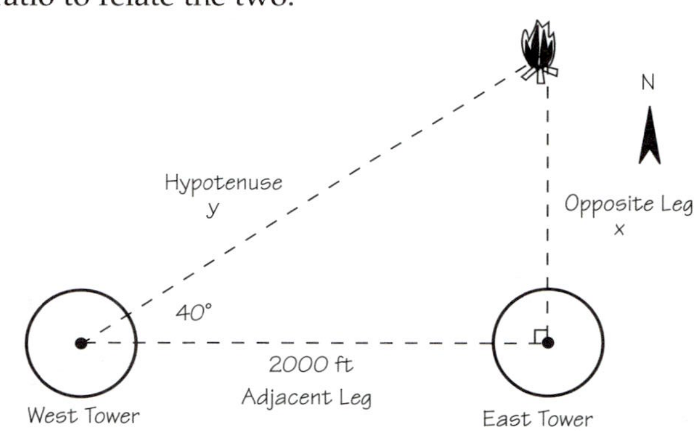

$$\tan 40° = \frac{\text{opposite}}{\text{adjacent}}$$
$$\tan 40° = \frac{x}{2000}$$
$$2000 \tan 40° = 2000\left(\frac{x}{2000}\right)$$
$$1678 \approx x$$

So the distance between the East Tower and the fire is about 1678 feet.

Next find the length of the hypotenuse. We know the length of the adjacent leg so use the cosine ratio to relate the two.

$$\cos 40° = \frac{\text{adjacent}}{\text{hypotenuse}} \qquad y \cos 40° = 2000$$
$$\cos 40° = \frac{2000}{y} \qquad \frac{y \cos 40°}{\cos 40°} = \frac{2000}{\cos 40°}$$
$$\frac{\cos 40°}{1} = \frac{2000}{y} \qquad y \approx 2611$$

So the distance between the West Tower and the fire is about 2611 feet.

Now, we are given the speed of the West Tower's fire truck as 20 miles per hour and the speed of the East Tower's fire truck as 15 miles per hour. However, our distances are in feet. We must first convert the "miles per hour" to "feet per second."

West Tower truck:

$$\frac{20 \cancel{\text{miles}}}{\cancel{\text{hour}}} \times \frac{5280 \text{ feet}}{\cancel{\text{mile}}} \times \frac{1 \cancel{\text{hour}}}{60 \cancel{\text{minutes}}} \times \frac{1 \cancel{\text{minute}}}{60 \cancel{\text{seconds}}} \approx 29.33 \frac{\text{ft}}{\text{sec}}$$

East Tower truck:

$$\frac{15 \cancel{\text{miles}}}{\cancel{\text{hour}}} \times \frac{5280 \text{ feet}}{\cancel{\text{mile}}} \times \frac{1 \cancel{\text{hour}}}{60 \cancel{\text{minutes}}} \times \frac{1 \cancel{\text{minute}}}{60 \cancel{\text{seconds}}} \approx 22 \frac{\text{ft}}{\text{sec}}$$

Answers (continued)

Use the $d = rt$ (distance equals rate times time) formula to find the time it takes for each fire truck to travel to the fire:

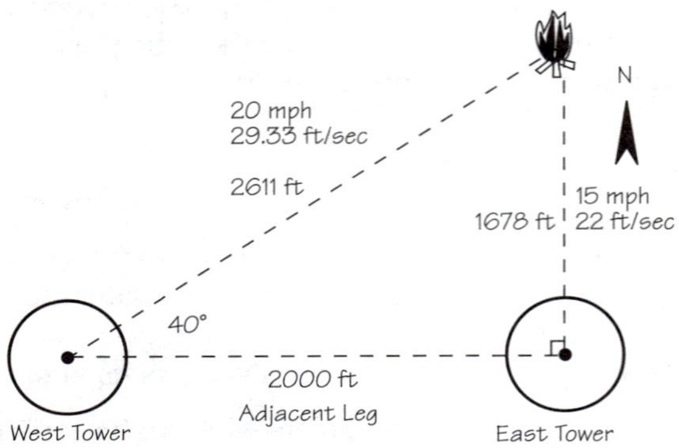

West Tower Fire Truck:

$$d = rt$$

$$2611 \text{ ft} = \left(29.33 \frac{\text{ft}}{\text{sec}}\right) t$$

$$\frac{2611 \text{ ft}}{29.33 \frac{\text{ft}}{\text{sec}}} = \frac{\left(29.33 \frac{\text{ft}}{\text{sec}}\right) t}{29.33 \frac{\text{ft}}{\text{sec}}}$$

$$89 \text{ seconds} \approx t$$

East Tower Fire Truck:

$$d = rt$$

$$1678 \text{ ft} = \left(22 \frac{\text{ft}}{\text{sec}}\right) t$$

$$\frac{1678 \text{ ft}}{22 \frac{\text{ft}}{\text{sec}}} = \frac{\left(22 \frac{\text{ft}}{\text{sec}}\right) t}{22 \frac{\text{ft}}{\text{sec}}}$$

$$76 \text{ seconds} \approx t$$

It takes the West Tower fire truck 89 seconds to make it to the fire, and it takes the East Tower fire truck 76 seconds to make it. Therefore the East Tower fire truck will make it to the fire first.

SECTION 8.6

Making the Sine Function Connection

You just explored three important ratios in right triangles. Next, you will use a circle to investigate the sine, cosine and tangent ratios for angles larger than 90°.

MEASURING ANGLES IN THE UNIT CIRCLE

Draw x- and y-axes on a sheet of chart paper with grid lines. Place the origin near the paper's center. Tie one end of a piece of string around a pushpin (**be careful**!). Measure off 1 foot of the string then wrap the end of the string around a marker so that the marker is exactly 1 foot from the pushpin. Place the pushpin at the origin of your coordinate plane and draw a large circle with a radius of 1 foot (see **Figure 8.32**).

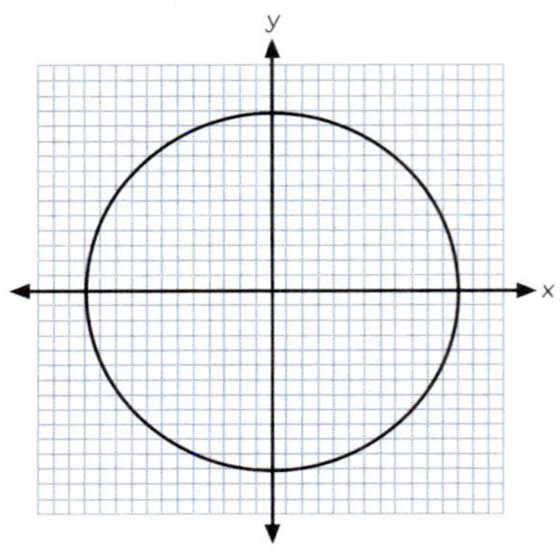

FIGURE.8.32.
Circle with radius of 1 foot.

Label the positive end of the x-axis 0°. Use a protractor to measure a 30° angle from the positive x-axis and draw this radius. Label the point where the radius touches the circle. Now, use a protractor to measure a 60° angle from the positive x-axis and draw this radius. Label the point where the radius touches the circle. Continue this process constructing angles in 30° steps until you reach the other side of the circle (see **Figure 8.33**).

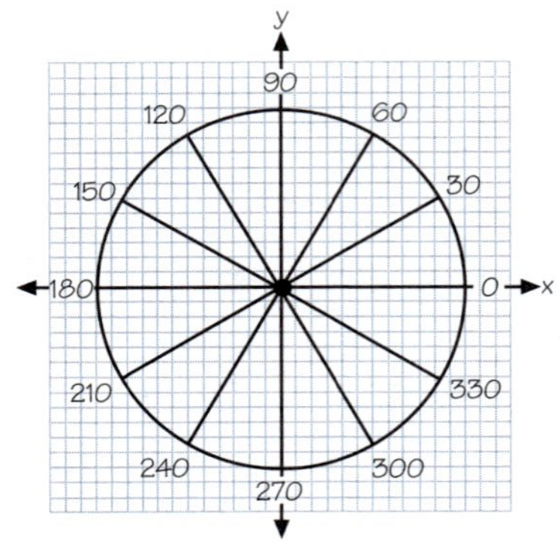

FIG.8.33.
Circle with 30° intervals

1. If you extend the 0° line to the other side of the circle, what is the measure of that angle?

2. Did you measure in a clockwise or counterclockwise direction from the positive x-axis?

3. Create a table like **Figure 8.34**. For each angle you drew, measure the vertical distance in **feet** between the x-axis and the endpoint of the radius that lies on the circle as shown in **Figure 8.35**. If the point is above the x-axis, record that distance as *positive*. If the point is below the x-axis, record that distance as *negative*.

CHECK THIS!
Remember that to convert inches to feet, divide by 12.

Measure of Angle	Vertical Distance
0°	
30°	
60°	
90°	
120°	
150°	
180°	
210°	
240°	
270°	
300°	
330°	
360°	

FIGURE 8.34.
Data table.

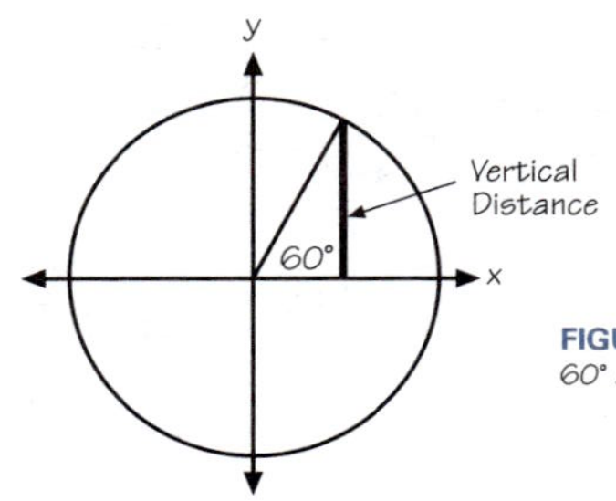

FIGURE 8.35.
60° angle.

ENGAGE LESSON 2

Section 8.6: Making the Sine Function Connection

Purpose: In Sections 8.1 through 8.5, students investigated and applied the basic trigonometric ratios to calculate indirect measurements. In this lesson students use trigonometric functions to model periodic motion. In this section students develop periodic behavior from a circle, motivating them to look for a function that describes periodic phenomena.

Description: This section begins by asking students to use a string as a compass to generate a large circle. Then students use a protractor and meter stick to mark off the circle in 30° sectors. They measure the distance between each radius' endpoint and the x-axis in order to generate a scatter plot of the parent sine function.

Materials: chart paper with 1-inch grid, markers, protractor, meter stick, pushpin, and about 2 feet of string for every group

Teaching Ideas: Students should work in small groups (3 to 4 students). Instead of using a protractor, students can fold the circle into 12 equal sectors by folding diameters and thirds. However, using a protractor provides an opportunity for students to practice their measurement skills.

Answers

1. 180°
2. counterclockwise
3. Answers:

Measure of Angle	Vertical Distance
0°	0
30°	0.5
60°	0.87
90°	1
120°	0.87
150°	0.5
180°	0
210°	−0.5
240°	−0.87
270°	−1
300°	−0.87
330°	−0.5
360°	0

4. Use your table to make a scatter plot of vertical distance versus angle measure. If you use a graphing calculator, be sure to state your window.

5. Describe the shape of the scatter plot. What would happen if you continued measuring vertical distances for angles beyond 360°, such as 390° or 420°?

Consider the vertical distance that you measured and that is shown in **Figure 8.36**. Look at the right triangle made by the vertical distance, the x-axis, and the radius. For the 60° reference angle, label the hypotenuse, adjacent leg, and opposite leg.

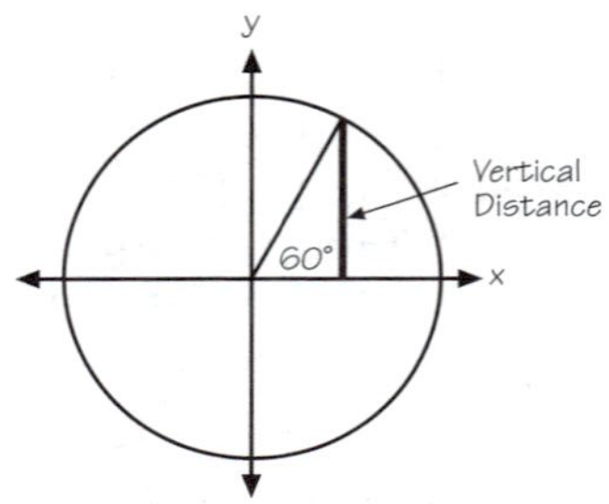

FIGURE 8.36.
A measure of vertical distance in a circle.

6. Which trigonometric ratio involves the opposite side and the hypotenuse?

7. What is the length of the hypotenuse?

> **CHECK THIS!**
> The terminal **side** of the angle is the hypotenuse of the right triangle.

8. How does the vertical distance compare to the sine ratio of a 60° angle?

9. Make a general statement about the relationship between the sine ratio and the vertical distance in a circle with a radius of 1 unit, or **unit circle**.

SECTION 8.7

Spinning Your Wheels

In 1893, George W. Ferris built the world's first Ferris wheel for the World's Fair in Chicago. In 1985, the Texas Star opened at the Fair Park in Dallas. Today, the Texas Star is the largest Ferris wheel in the United States. From its maximum height of 212 feet, a rider gets a great view of the Dallas skyline.

One Texas Star rider wondered how he could find his height above ground at any given time.

To help solve this problem, first consider a simpler one. You will need something smaller than a Ferris wheel to model the motion of a rider. Instead of an actual Ferris wheel, you will collect data from a hula-hoop or some other round object.

COLLECTING THE DATA

1. What is the diameter of your wheel? What is its circumference?

   Set up the experiment as follows:

   a) Extend the tape measure and tape it to the floor with masking tape. (Later, you will roll your wheel alongside the tape measure.)

   b) Place another piece of tape on the floor perpendicular to the tape measure at its zero end. Place the wheel directly on top of this marker. The wheel should touch the masking tape; its center should be directly above the tape.

   c) Place a dot on the wheel (a sticker or piece of tape will do) at the point where the wheel sits on the tape.

Answers (continued)

4. Answer:

```
WINDOW
 Xmin=0
 Xmax=360
 Xscl=30
 Ymin=-1.5
 Ymax=1.5
 Yscl=.5
 Xres=1
```

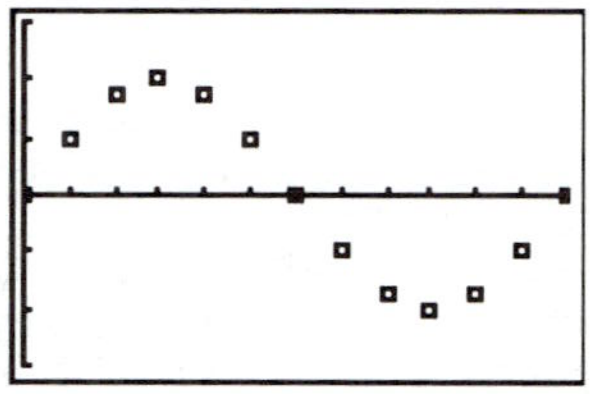

5. Students should observe that the graph has a wave-like appearance. If they continued measuring angles, the wave would repeat the shape in the scatter plot.

6. $\sin A = \frac{\text{length of the opposite side}}{\text{length of the hypotenuse}}$

7. The hypotenuse is 1 foot long.

8. $\sin 60° \approx 0.88$. The vertical distance is the same as the sine ratio of the angle.

$$\sin 60° = \frac{\text{opposite}}{\text{hypotenuse}} = \frac{\text{vertical distance}}{\text{radius}} = \frac{\text{vertical distance}}{1} = \text{vertical distance}$$

9. The sine ratio in a unit circle is equal to the vertical distance.

EXPLORE LESSON 2

Section 8.7: Spinning Your Wheels

Purpose: In Section 8.6 students used a unit circle to create a scatter plot of the sine function. In this section students use a circular object such as a wheel or hula-hoop in order to obtain actual vertical distance data and identify periodic patterns. In Section 8.8 students formalize terminology and properties of periodic functions, especially the sine parent function.

Description: In this section students roll a hula-hoop or other wheel-like object. They record the height of a marked spot as a function of the distance that the wheel rolls. Using tables and graphs, students interpret the data collection situation in terms of properties that are important to periodic motion.

Materials: graphing calculator, meter stick, hula-hoop (or some other round object such as a bicycle tire), tape measure, masking tape (or duct tape), Handout 8.1

Teaching Ideas: Students should work in small groups (3 to 4 students). As students collect and analyze data, emphasize the cyclical nature of periodic motion. Encourage students to notice patterns in their data.

Encourage students to use chart paper to generate a large scatter plot of their data. This scatter plot will help them see the critical attributes of the resulting function more easily.

Answers

1. Sample answer: The diameter is 26 inches; the circumference is $26\pi \approx 81.7$ inches (all other answers are based on this sample data).

   Sample data are based on a wheel with a 26-inch diameter. The data are recorded to the nearest eighth-inch.

Units of measurement: inches

Distance Wheel Moves	0	12	24	36	48	60	72	84	96
Height Above the Floor of Dot	0	5.25	16.5	25.125	24.125	14.25	3.375	0.25	7.125

Distance Wheel Moves	108	120	132	144	156	168	180	192	204
Height Above the Floor of Dot	18.75	25.75	22.75	11.875	2.0	0.75	9.25	20.75	26.0

Distance Wheel Moves	216	228	240	252					
Height Above the Floor of Dot	21.0	9.75	1.0	1.75					

Roll the wheel forward alongside the tape measure. Stop the wheel every 12 inches. When you stop, measure and record two things: the height of the dot above the floor and the total distance the wheel has traveled. (See **Figure 8.37**.) Record your data in Handout 8.1. (See Handout 8.1 for more instructions.) Be sure to save Handout 8.1 because you will need the data in Section 8.8.

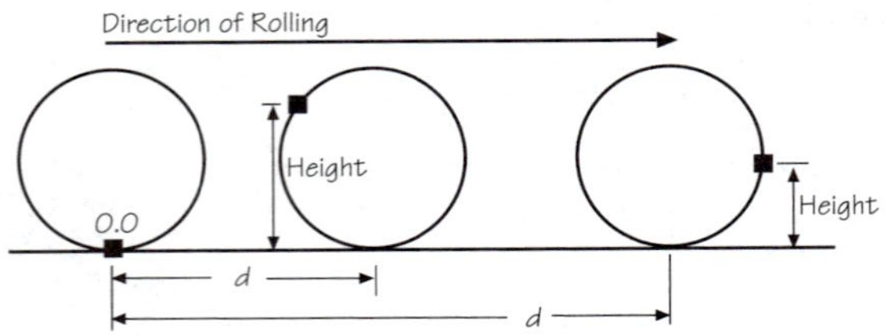

FIGURE 8.37. Measuring the distance the wheel rolls and height of the dot.

GRAPHING THE MOTION

CHECK THIS!

A function that has a graph that repeats itself on intervals of equal length is called periodic.

2. Make a scatter plot of height versus distance. If you use a graphing calculator, be sure to state your window.
3. Sketch your scatter plot on paper. On your plot, sketch a smooth curve (no sharp corners) that you think best represents the relationship between height and distance.
4. Do you think that the relationship between the height of a marked spot on the wheel and the distance rolled is periodic? Explain.
5. Does your scatter plot appear to be periodic? Explain.
6. What is the height of the center of your wheel above the ground? How do you know?
7. Based on the height of the center of your wheel above the ground and the diameter of the wheel, what would you expect the maximum and minimum heights of your marked spot to be? Explain.
8. From your data, what is the difference between the maximum and minimum heights? How does this compare to what you expected? Explain any differences.
9. How much distance did your wheel cover before your mark on the heights began repeating?

10. What is the horizontal length of one repetition of your graph?
11. How do your answers in Questions 9 and 10 compare?
12. For what distances is the marked spot on the wheel at the top of the wheel? How can you get this answer from your graph?
13. For what distances is the marked spot horizontally level with the center of the wheel? How can you get this answer from your graph?

Answers (continued)

2. Answer:

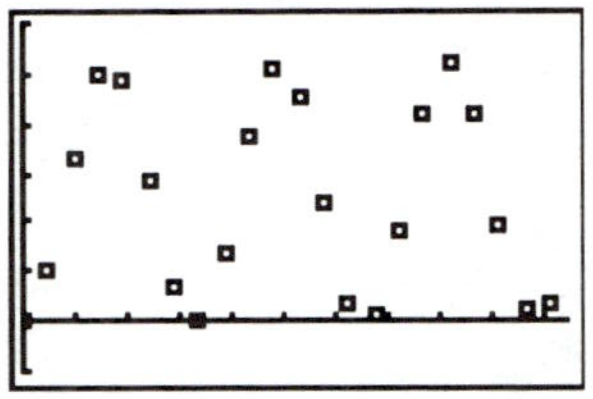

3. Answer:

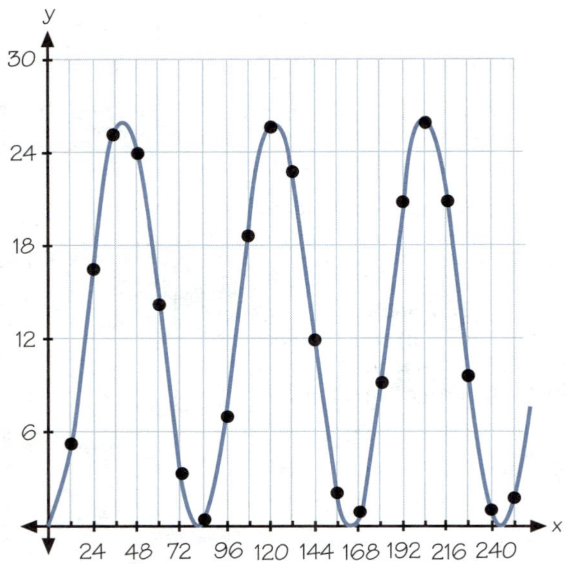

4. Yes. A wheel can roll an infinite number of times. The height of the marked spot rises and falls as the wheel rolls in the same pattern for every turn of the wheel.

5. Yes. There are three "waves" that look to have the same width and height.

6. Answers may vary depending on the diameter of the wheel. The center should be the same height above the ground as the radius of the wheel (half the diameter). In the sample situation the center of the wheel is 13 inches above the ground.

7. The minimum height of the marked spot is the ground, which is a height of 0 inches. The maximum height should be equal to the diameter of the wheel, which in the sample data is 26 inches.

8. Answers may vary. According to the sample data the maximum height is 26 inches and the minimum height is 0 inches, which matches expectations.

9. Answers may vary depending on the wheel. The wheel will cover a distance on the ground that is equal to its circumference before the pattern of heights begins repeating. In the sample data this distance is equal to 26π, or about 82 inches.

10. The horizontal length can be estimated on the graph, but the table reveals a more precise answer. In the sample data, at a distance of 36 inches the spot is 25.125 inches above the ground. At a distance of 120 inches, the spot is at 25.75 inches above the ground. Both of these heights are approximately the same. Since the horizontal length represents a difference in distance, subtract 120 – 36, so the horizontal length of one repetition in the graph is about 84 inches.

    If students use a graphing calculator, the TRACE feature can be used to determine an approximate horizontal length from the graph.

11. The answers should be fairly close.

12. From the sample data the distances are approximately 40, 120, 204. You get these answers from your graph by determining the x-values that correspond to the peaks in the graph.

13. From the sample data the distances are approximately 20, 62, 102, 143, 183, and 225. The marked spot is horizontal to the center when it is 13 inches high (half of the wheel's diameter). Draw the line $y = 13$ and then find the x-values that correspond to the points of intersection between the graph and the line.

SECTION 8.7 Assignment

Time (sec)	Height of Right Pedal (in)
0	
0.25	
0.5	
0.75	
1	
1.25	
1.5	
1.75	
2	
2.25	
2.5	
2.75	
3	

FIGURE 8.38. Height-time data for a bicycle pedal.

Angle Measure	Sine Value

FIGURE 8.39. Table of sine values.

Suppose you pedal a bicycle so the pedals revolve once a second. Each pedal is 5 inches from the ground at its lowest and 18 inches from the ground at its highest. You start with the right pedal in its lowest position.

1. Copy the table in **Figure 8.38** and complete the height entries.
2. Make a scatter plot of height of the right pedal versus time. Be sure to label the axes.
3. Use your scatter plot to sketch a graph of a function that you think best represents the relationship between height and time.

The graph that you drew for Question 3 oscillates between 5 inches and 18 inches. The up-down pattern repeats itself each second. Any function that repeats itself on intervals of equal length is called **periodic**.

4. Create a table of sine values (rounded to two decimal places) of angles from 0° to 360° in 15 degree steps. (You can copy and fill the table in **Figure 8.39**.)
5. Make a scatter plot of sine versus angle measure.
6. What kind of graph is this?
7. What would you expect to happen if you continued the graph to 720°?

SECTION 8.8 Periodic Motion and Sinusoidal Regression

Carousel horses like those in **Figure 8.40** move up and down as the carousel spins. The graph in **Figure 8.41** represents the height above the floor of a carousel horse's back hooves versus time.

FIGURE 8.40. Carousel horses.

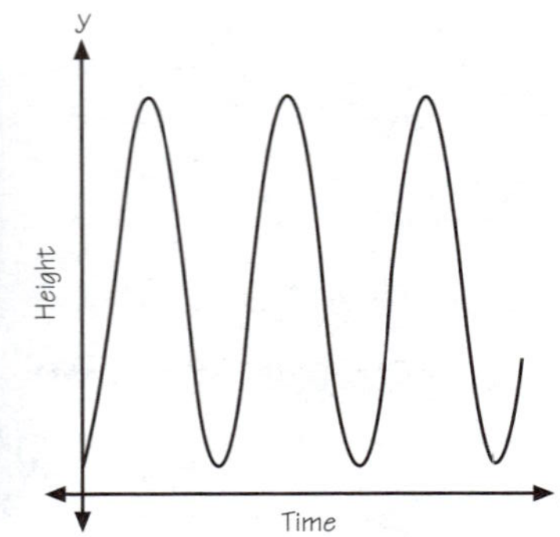

FIGURE 8.41. Height vs. time graph.

CHECK THIS!

An oscillating pattern is one that swings back and forth or varies between two alternate extremes.

1. According to this graph, what happens to the height of the hooves over time?
2. What do the maximum values (the peaks) in the graph represent? What do the minimum values (the troughs) in the graph represent?
3. Where in the graph are the hooves in the middle? Explain.

The centerline of the **oscillation** is a line halfway between the graph's peaks and troughs. The centerline, also called the **axis of oscillation**, is shown in **Figure 8.42** with a dotted line.

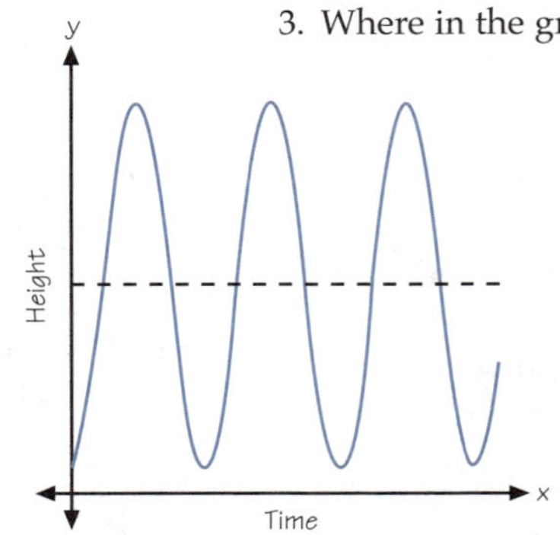

FIGURE 8.42. Axis of oscillation.

Section 8.7 Assignment

Answers

1. Answers:

Time (sec)	Height of Right Pedal (in)
0	5
0.25	11.5
0.5	18
0.75	11.5
1	5
1.25	11.5
1.5	18
1.75	11.5
2	5
2.25	11.5
2.5	18
2.75	11.5
3	5

2.

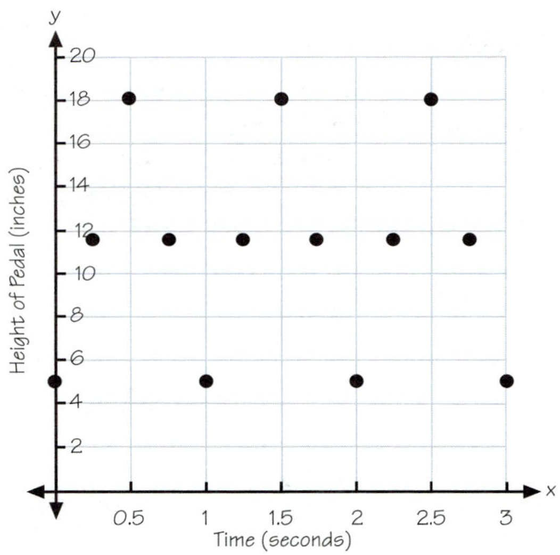

3.

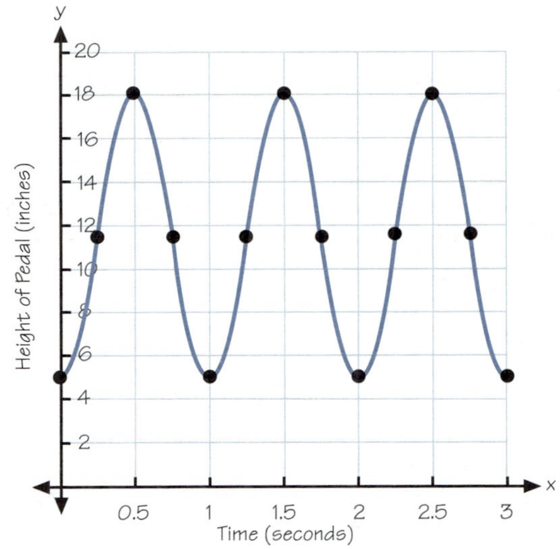

4. Answers:

Angle Measure	Sine Value	Angle Measure	Sine Value	Angle Measure	Sine Value
0°	0	135°	0.71	270°	–1
15°	0.26	150°	0.5	285°	–0.97
30°	0.5	165°	0.26	300°	–0.87
45°	0.71	180°	0	315°	–0.71
60°	0.87	195°	–0.26	330°	–0.5
75°	0.97	210°	–0.5	345°	–0.26
90°	1	225°	–0.71	360°	0
105°	0.97	240°	–0.87		
120°	0.87	255°	–0.97		

5.

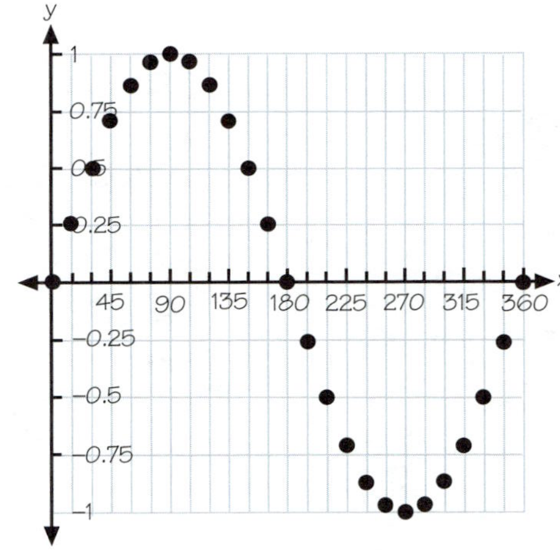

6. Periodic

7. The graph would repeat the same pattern.

EXPLAIN LESSON 2

Section 8.8: Periodic Motion and Sinusoidal Regression

Purpose: In Sections 8.6 and 8.7 students investigated the sine function and periodic motion described by the sine function. In this section students formalize the attributes of the sine function and sinusoidal regression. This foundation prepares students to model another form of periodic motion, the distance a pendulum swings, in Section 8.9.

Description: In this phase of the lesson students define attributes of the sine function including vertical shift, horizontal shift, amplitude, and period. They also use sinusoidal regression on the graphing calculator to generate a sine curve from the parent function that models a set of data.

Materials: graphing calculators, Handout 8.1

Teaching Ideas: Debrief students on Section 8.7 by having them share their answers to the questions. As they share be sure to address any missing concept, process, or skill that could have been used to answer the questions. Then work through the questions in this section with the entire class. Once again be sure all necessary concepts, processes, or skills are addressed.

Answers

1. The height oscillates between a high value and a low value in a smooth repeated pattern.
2. The maximum values represent the height of the hooves at their highest point above the floor of the carousel. The minimum values represent the height of the hooves at their lowest point above the floor of the carousel.
3. The hooves are halfway between the highest and lowest points.

In Section 8.7 you rolled a wheel along the floor and measured the height of a marked spot as the wheel rolled.

4. What happened to the height of the center of your wheel as the wheel rolled along the ground?

5. What were the minimum and maximum heights that the marked spot reached?

The height of the marked spot above the floor oscillated between the minimum and maximum values. This oscillation can be described with a **periodic function**. A periodic function is a function in which the value of the dependent variable, in this case the height of a marked spot, oscillates between a minimum value and a maximum value.

Think back to Section 8.6 where you used a large unit circle with a radius of 1 foot to explore the relationship between vertical distance and angle measure (**Figure 8.43**). You noticed that the vertical distance is equal to the sine ratio for any given angle:

$$\sin(x°) = \frac{\text{opposite}}{\text{hypotenuse}}$$

$$\sin(x°) = \frac{\text{vertical distance}}{\text{radius}}$$

$$\sin(x°) = \frac{\text{vertical distance}}{1}$$

$$\sin(x°) = \text{vertical distance}$$

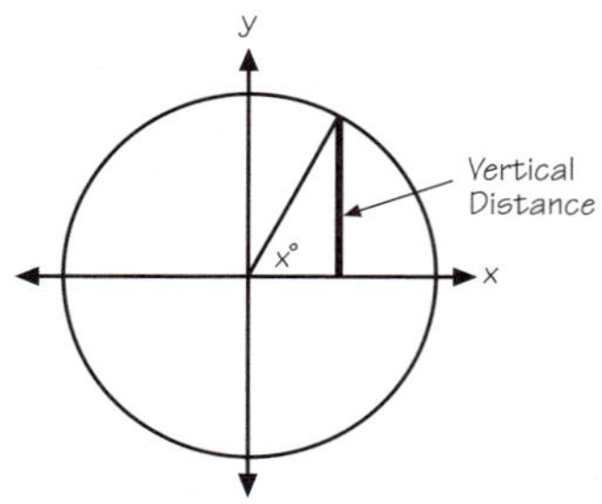

FIGURE 8.43. x° angle.

In Section 8.6 you made a scatter plot of the sine function, $y = \sin(x)$, when you plotted the vertical distance (length of the opposite leg of the right triangle whose hypotenuse is 1 unit) versus angle measure (**Figure 8.44**).

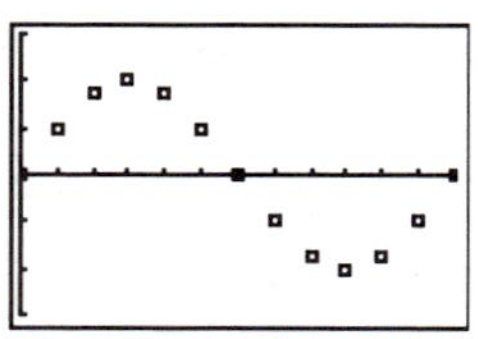

FIGURE 8.44. Scatter plot.

6. Refer back to your data in Handout 8.1. Also, look at the scatter plot that you made of the data. How does this scatter plot compare to the scatter plot of the function $y = \sin(x)$?

One way to model periodic relationships is to use the **sine function** as a parent. You can use your graphing calculator's regression features to find a function in this family. This type of regression is called **sinusoidal regression**.

7. Use your graphing calculator to find a sine function to fit your wheel data in Handout 8.1.

8. Write down your function rule. Round to the nearest ten-thousandth.

CHECK THIS!

When the independent variable is measured in degrees, the calculator's degree mode is used for sinusoidal regression and for related calculations and graphing. Otherwise, the radian mode is used.

Once you have a function rule that models a periodic relationship, you can use the TRACE feature of your calculator to answer questions. Use your function rule to answer these questions.

9. Find the minimum and maximum values of your function. How do they compare to the minimum and maximum values of your data?

10. Subtract the maximum and minimum values. What does this difference represent?

11. Take half of the difference from Question 10. Add this number to your minimum value. How does this value compare to your axis of oscillation?

The **amplitude** is half of the vertical length of the basic repeating shape or half of the difference between the maximum and minimum y-coordinates of the graph.

12. Use the TRACE feature to find the x-values that correspond to consecutive maximum values. Record these x-values.

13. Subtract the two x-values from Question 12. In terms of the cycle of repetition, what does this difference represent? What does this difference represent in terms of the situation?

Answers (continued)

4. The height of the center of the wheel did not change over time.

5. The minimum height was 0, and the maximum height was the diameter of the wheel.

6. Answers may vary. Students should notice that their scatter plots from the rolling wheel data have a similar shape, though shifted upward, as the scatter plot of $y = \sin(x)$.

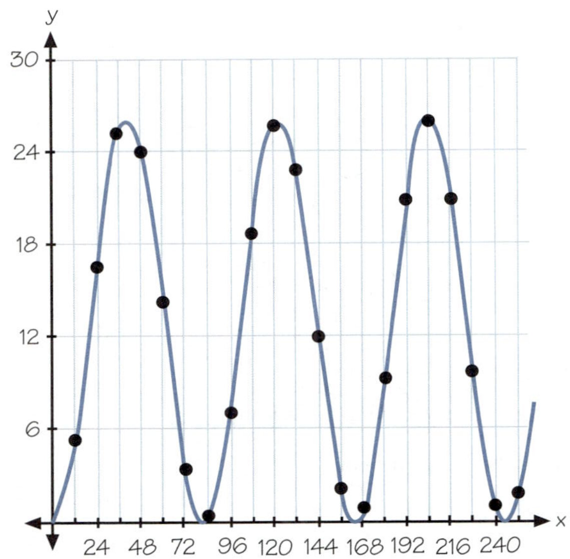

7. NOTE: Actual instructions may vary depending on the graphing calculator that you use. Consult your User's Manual for more detailed instructions.

a) Enter your data into List 1 and List 2.

b) Generate a scatter plot of height versus distance. Be sure to identify your viewing window.

```
WINDOW
 Xmin=0
 Xmax=260
 Xscl=25
 Ymin=-5
 Ymax=30
 Yscl=5
 Xres=1
```

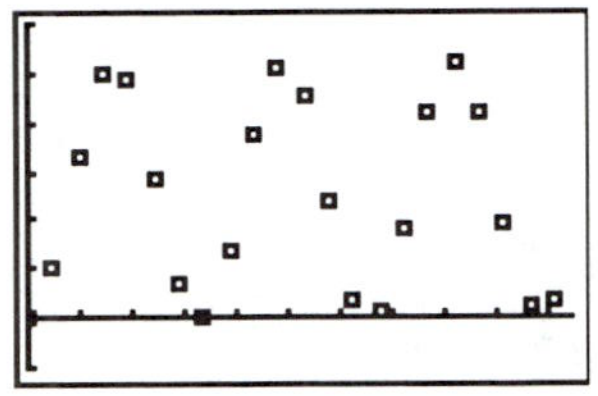

c) Be sure that the calculator is in RADIAN mode. From the STAT menu, use the right arrow to select CALC. Then scroll down to SinReg.

```
NORMAL SCI ENG
FLOAT 0123456789
RADIAN DEGREE
FUNC PAR POL SEQ
CONNECTED DOT
SEQUENTIAL SIMUL
REAL a+bi re^θi
FULL HORIZ G-T
01/01/30 2:09AM
```

```
EDIT CALC TESTS
7↑QuartReg
8:LinReg(a+bx)
9:LnReg
0:ExpReg
A:PwrReg
B:Logistic
C:SinReg
```

d) Press ENTER to return to the home screen. Enter the lists containing your data and the function where you want the calculator to store the regression equation. Press ENTER.

```
SinReg
```

```
SinReg (L1,L2,Y1
)
```

```
SinReg
 y=a*sin(bx+c)+d
 a=13.0175859
 b=.0768972003
 c=-1.567016203
 d=13.0039485
```

e) Press GRAPH to view the scatter plot overlaid with the sinusoidal regression equation.

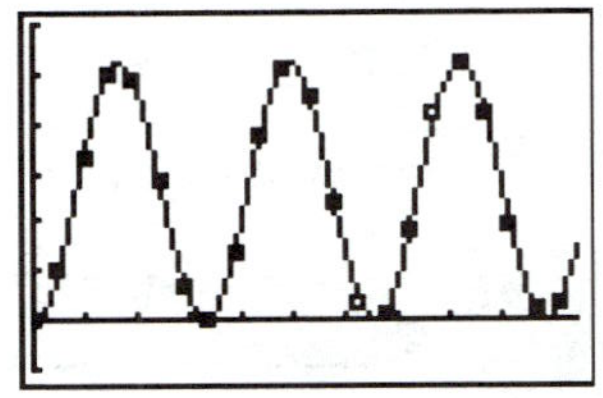

8. Answer based on the sample data: $y = 13 \sin(0.0769x - 1.5670) + 13$.

9. Minimum Value: Maximum Value:

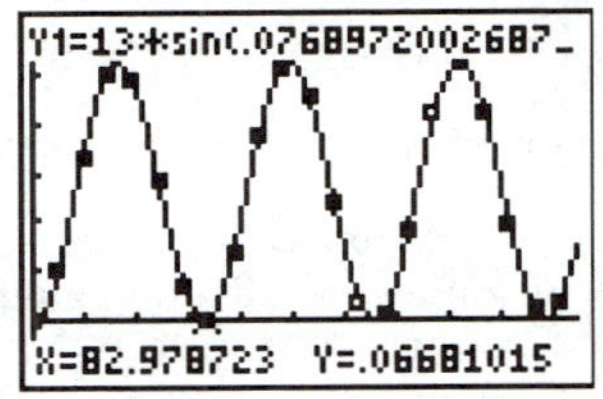

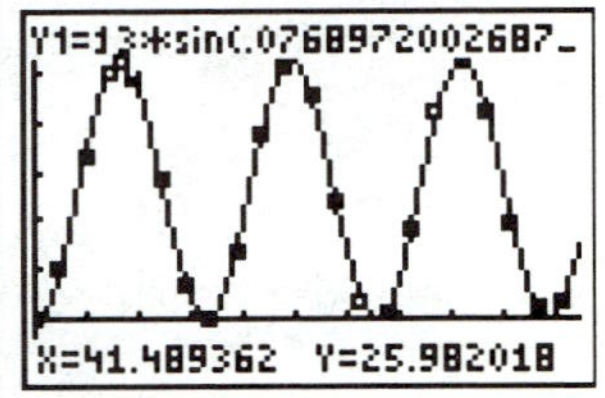

The traced minimum and maximum values should be very close to the data.

10. From the sample data, $26 - 0 = 26$. This value represents the size of the interval over which the heights range.

11. $26/2 = 13$. $13 + 0 = 13$. $y = 13$ is the axis of oscillation.

12. Answer:

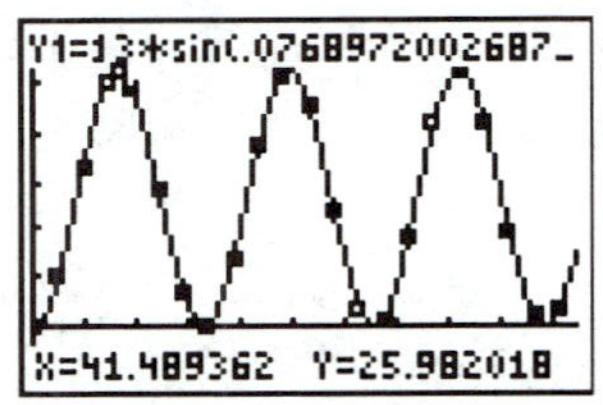

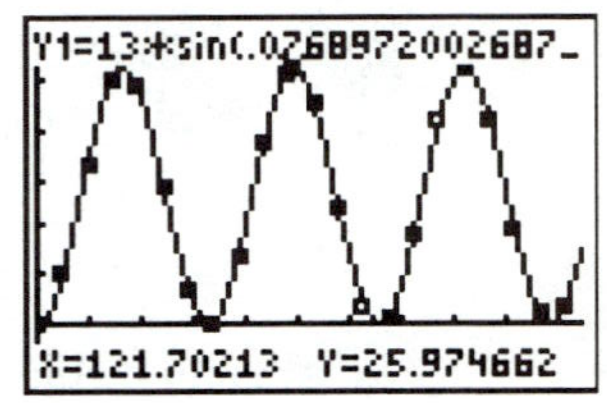

13. $121.7 - 41.5 = 80.2$; this value represents the width of one cycle of the periodic graph. This value also represents the horizontal distance that the wheel traveled before the marked spot returned to a maximum height.

The shortest horizontal length of a basic repeating shape is called the **period**. On a graph, you can find the period by tracing to get the x-coordinates of consecutive corresponding points, as you did in Question 12.

As with many other functions, the sine family has a general form:

$y = A \sin(B(x - C)) + D$

You can use a graphing calculator to see the effect of each control number A, B, C, and D.

14. To understand the impact of A, graph the parent function $y = \sin(x)$ and several others with different values of A. For example, graph $y = \sin(x)$, $y = 2 \sin(x)$, and $y = 3 \sin(x)$. Use your calculator's radian mode. What can you conclude?
15. Conduct a similar investigation into the role of B.
16. Conduct a similar investigation into the role of C.
17. Conduct a similar investigation into the role of D.

SECTION 8.8

Assignment

Use this information for Questions 1–3.

Suppose you are pushing your sister on a swing. If you sketch a graph of her distance from you over time it might look like **Figure 8.45**.

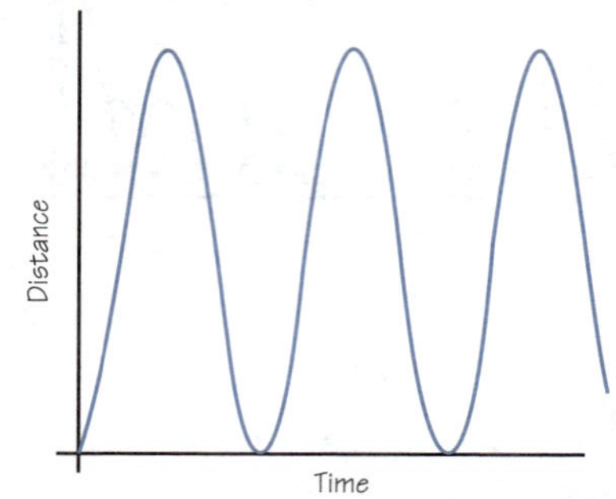

FIGURE 8.45.
Graph of swing distance.

1. What happens to the distance between you and your sister over time?
2. What do the maximum values in the graph represent? What do the minimum values in the graph represent?
3. Where in the graph are the swing's chains perpendicular to the ground? Explain.

Use **Figure 8.46** for Questions 4 and 5.

Figure 8.46 shows graphs of three periodic functions. Each function's graph repeats some basic shape.

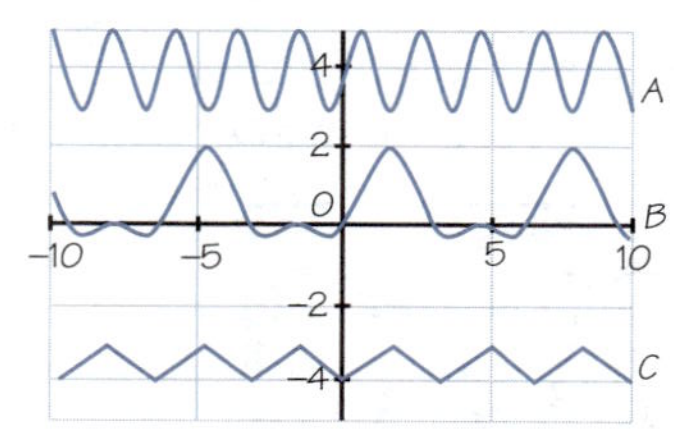

FIGURE 8.46.
Graphs of three periodic functions.

Answers (continued)

14. Sample answer:

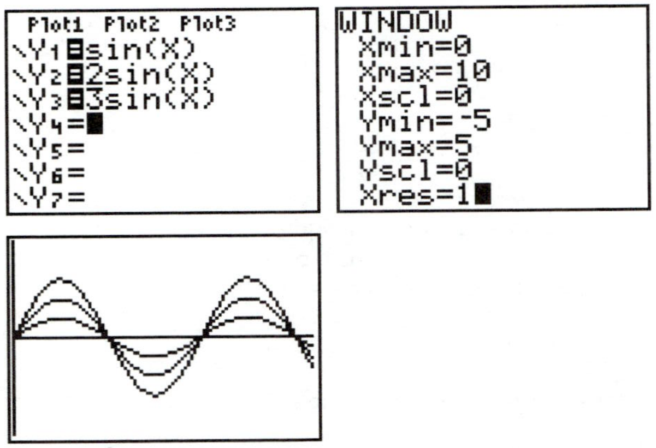

A controls the amplitude of the graph. The larger the value of A, the larger the amplitude.

15. Sample answer:

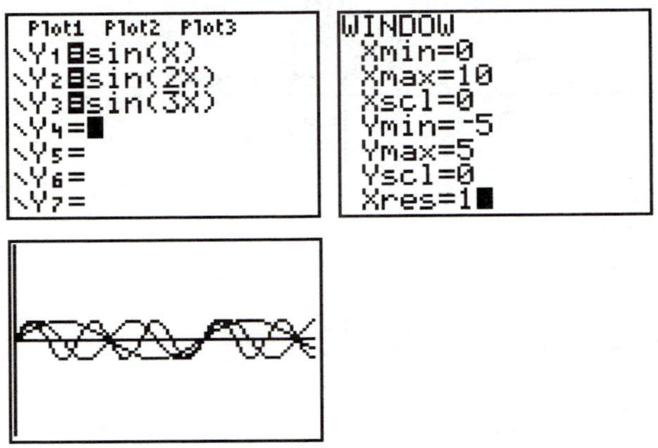

B controls the graph's period.

16. Sample answer:

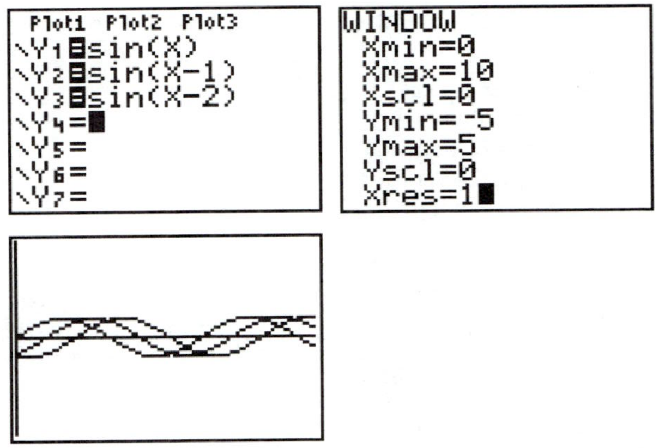

C shifts the graph horizontally.

17. Sample answer:

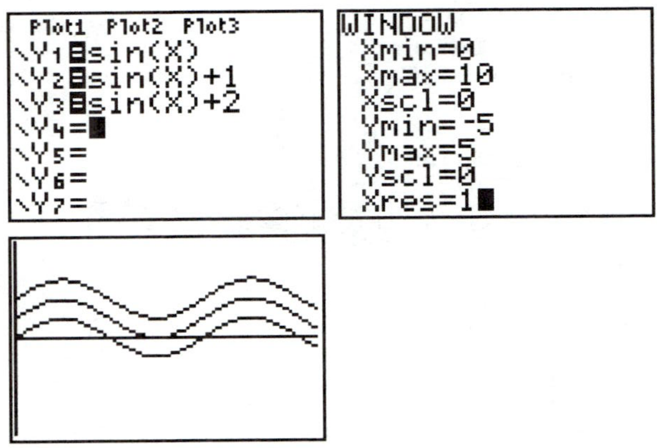

D shifts the graph vertically.

Section 8.8 Assignment

Answers

1. The distance oscillates between two values, a maximum distance away from you and a minimum distance right next to you.
2. The maximum value is the distance from your sister when she is farthest away from you, and the minimum value is the distance to your sister when she is closest to you.
3. The swing is perpendicular to the ground exactly halfway between the maximum and minimum values. This corresponds to the axis of oscillation.

4. Find an approximate period for each graph.

5. What is the amplitude for each graph?

Use **Figure 8.47** for Questions 6–8.

The height of a carousel horse's stirrup changes periodically over time as the carousel spins. The graph of this motion is shown in Figure 8.47.

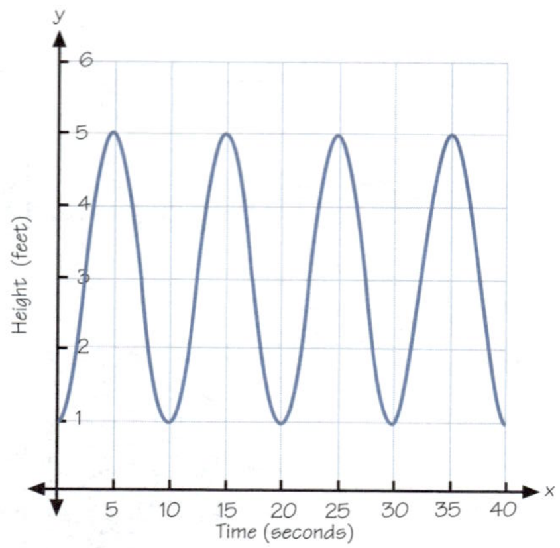

FIGURE 8.47. The height of the stirrup of a carousel horse versus time.

6. What is the period?

7. What is the maximum height of the stirrup? What is the minimum height?

8. What is the amplitude?

Use this information for Questions 9–13.

In Section 8.7 you drew periodic graphs by plotting the heights of dots on a rolling wheel. Although wheels come in only one shape—a circle—they come in many sizes. If different groups used different sized wheels, then different height-versus-distance graphs would have been drawn.

9. For Section 8.7 Jerry's group used a wheel with a 26-inch diameter. Their height-versus-distance graph is shown in **Figure 8.48**.

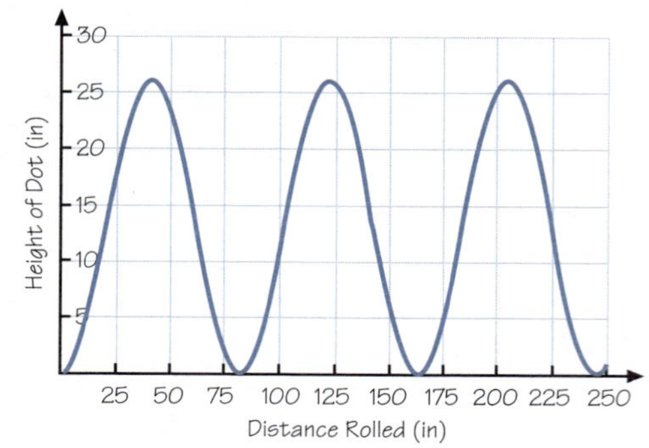

FIGURE 8.48. Height-versus-distance graph from Jerry's group.

10. Based on the wheel diameter, what are the period and amplitude? Explain how you got your answer. Check that your answer is consistent with the graph in Figure 8.48

11. What is the axis of oscillation?

Imagine doing Section 8.7 with a wheel that has a one-foot radius (or two-foot diameter).

12. How far would you have to roll the wheel for the dot to make one complete turn? Give both an exact answer (involving π) and a decimal approximation.

13. How far would you have to roll the wheel for the dot to move from the bottom of the wheel to the top of the wheel for the first time? And then the second time? Give both exact and approximate answers.

Use **Figure 8.49** for Questions 14–16.

Assume that the Ferris wheel in Figure 8.49 does one revolution every 30 seconds. The ride lasts about 5 minutes.

14. Thia is the last person to board the Ferris wheel. How far has she traveled after the wheel makes one complete revolution? (To find the circumference of the wheel, use $c = 2\pi r$, where r is the radius and c is its circumference.)

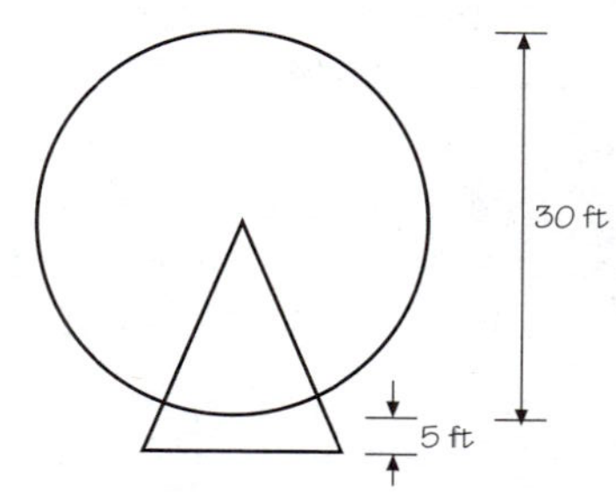

FIGURE 8.49. Ferris wheel rotating at a constant speed.

Answers (continued)

4. A: period is ≈ 2; B: period is ≈ 6.3; C: period is ≈ 3.

5. A: amplitude is $0.5(5 - 3) = 1$; B: amplitude is $0.5(2 + 0.4) = 1.2$; C: amplitude is $0.5(-3 + 4) = 0.5$.

6. about 10 seconds

7. The maximum height is 5 feet and the minimum height is 1 foot.

8. 2 feet

10. The period is equal to one turn of the wheel and hence is 26π inches ≈ 81.7 inches; the amplitude is the same as the wheel's radius and hence is 13 inches. These measurements are consistent with the graph in Figure 8.48. One complete up-down cycle occurs over a distance of a little more than 80 inches.

11. The axis of oscillation is $h = 13$.

12. 2π feet ≈ 6.28 feet

13. $\pi \approx 3.14$ feet; $3\pi \approx 9.42$ feet

14. 30π feet ≈ 94.2 feet

15. Plot Thia's height above the ground every 7.5 seconds for the first two minutes of the ride. Then use these data points to draw a graph of a function that you think best represents the relationship between height and time.

16. The graph that you drew for Question 15 is periodic. What is the period? What is the amplitude? What is the axis of oscillation?

SECTION 8.9

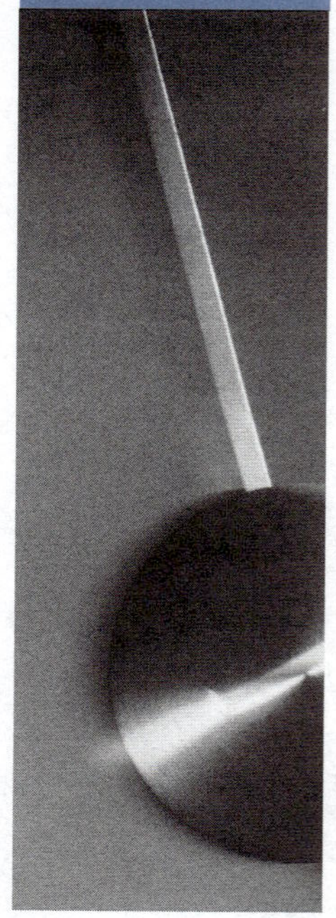

It's Just the Swing of the Old Pendulum

In the last three sections you modeled relationships using the sine function as a parent function. These situations involve **periodic motion**, in which motion oscillates, or cycles, back and forth between two values. Most of these situations involve circular motion, where an object is moving along a circle and you measure something, such as height, that changes over time or distance.

Other kinds of motion can be modeled with the sine function. One of these involves the distance between a pendulum and a stationary object over time.

THE PENDULUM STATIONS

There are three pendulum stations. Your teacher will assign student pairs to each station. Each station has a CBR and a pendulum with a length that differs from the other stations.

Place a CBR about half a meter from the maximum swing of the pendulum and at the same height as the pendulum at rest. You might need to use books to adjust the height of the pendulum or the CBR. (See **Figure 8.50**.)

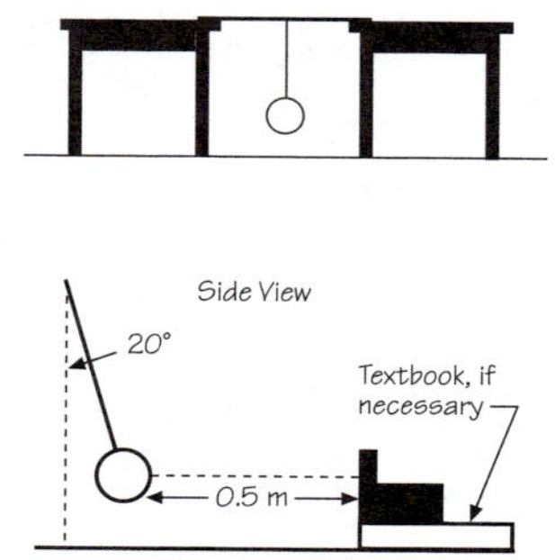

FIGURE 8.50. Pendulum stations.

Answers (continued)

15. Answer:

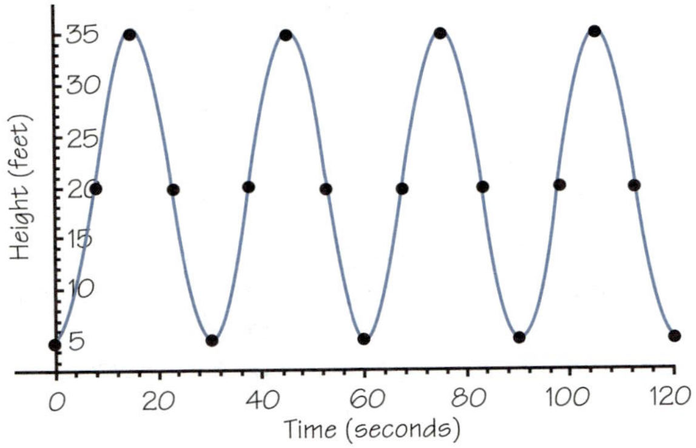

Note that some students may have connected the points with line segments.

16. The period is 30 seconds; the amplitude is 15 feet. The axis of oscillation is $y = 20$. The axis of oscillation is the horizontal line midway between the maximum height of 35 feet and the minimum height of 5 feet.

ELABORATE LESSON 2

Section 8.9: It's Just the Swing of the Old Pendulum

Purpose: In Sections 8.6 through 8.8 students investigated periodic motion resulting from situations that can be modeled using a circular function. In this section students extend their ideas about periodic motion to other situations that can be modeled using trigonometric functions, which prepares them to solve problems and make critical judgments in Section 8.10.

Description: In this section students use a CBR to collect distance-versus-time data that can be described with a sine function. Students collect pendulum motion data from swings beginning at different angles. They measure and compare period and amplitude for each swing.

Materials: 3 CBRs, 3 meter sticks, 3 1-meter lengths of string, 6 flat desks or small tables, 3 pendulums (such as a 3-inch diameter fishing float ["cork"] with washer for weight), graphing calculator, Transparency 8.2

Teaching Ideas: In this section students collect data at one of three pendulum stations. They debrief the activity by meeting with different groups of students. The following suggestions will help facilitate grouping processes.

Role of the Students:

Students, working in pairs, make two pendulum data collections using the CBR at their assigned station. One data collection is with the pendulum in a fuller swing than the second data collection. The collections are made on different calculators so each student has data to work with. One-third of the pairs collect data from each of the stations.

Once the data is collected, students determine and compare the periods and amplitudes of the collection for which they are responsible. They record the period and amplitude.

Students Meet by Collection Station to Compare Similar Collections:

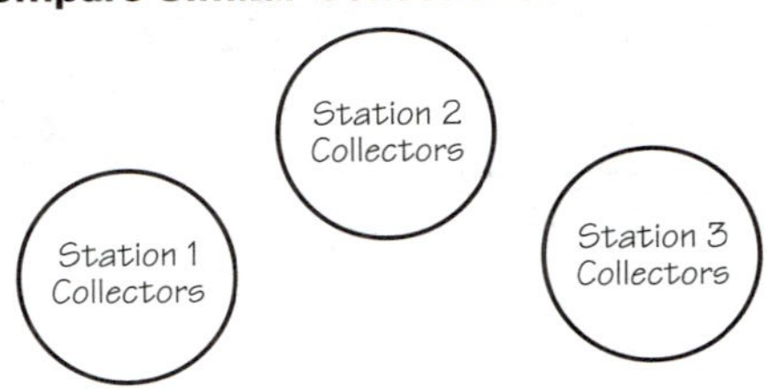

Gather students by collection station. All Station 1 collectors compare findings for period and amplitude. Can you tell from the findings who pulled the pendulum back the farthest? The least? How do the periods compare? Station 2 and 3 collectors do the same.

The larger station groups are looking for information they can gather from similar data [periods are the same; amplitudes are determined by angle of pendulum at start]. Students should speculate about what they will find when they evaluate the data for the different stations.

Students Meet in Groups with Representatives from Each Station to Compare Different Collections:

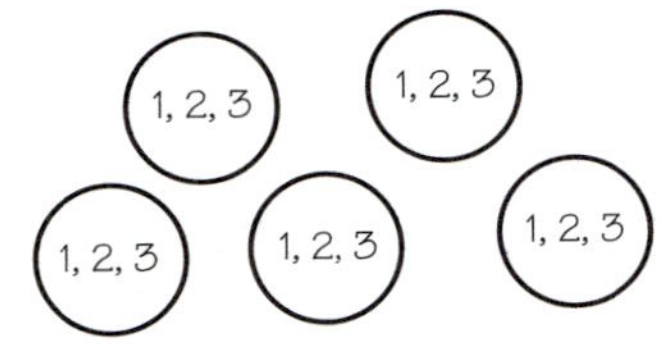

Gather students in groups of 3, each one has data from a different collection station. The small mixed groups are looking for information they can gather from different data. [Periods are dependent on the length of the pendulum. The longer the string is, the longer the period.]

Whole-Class Debrief:

Come back together for a large group discussion. Reveal the completed Transparency 8.2. Can you determine from the list which data applies to each station? How are you making that judgment? If we were to repeat the experiment using a 0.3-m pendulum, how would the period and amplitude be different from the data we have collected? Using a 0.7-m pendulum? Test your predictions by doing the experiment.

Pendulum Setup:

- The pendulum you use will impact the quality of the data collected by the CBR. The most reliable shape is a sphere. Large fishing floats (about 3 inches in diameter) weighted down with $\frac{3}{8}$-inch washers work well. You do not need to weigh down the pendulum for the data collection, but the pendulum does not bounce around as much when it is weighted a little bit.

Answers (continued)

- Construct the pendulum by tying a piece of string about 1 meter long to the fishing float and wrapping the other end around a meter stick until there is about 0.5 meters of string between the meter stick and the float.
- Rest the ends of the meter stick on two flat-topped desks or tables.

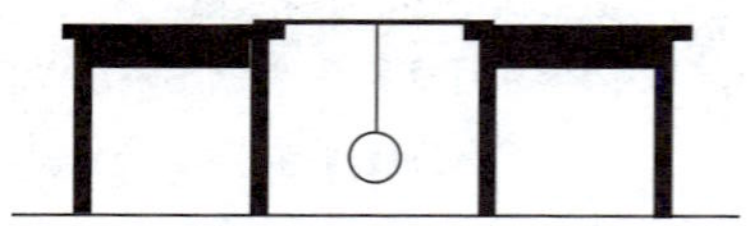

- You have constructed your first pendulum station. Construct two more stations using different lengths of string. The first station has a 0.5-m string. Depending on the height of your tables or desks, make a station with a 0.6-m pendulum and one with a 0.4-m pendulum. Number the stations 1, 2, and 3 and assign student pairs to each station.

CBR Setup:

- Place a CBR about 2 feet from the maximum swing of the pendulum and at the same height as the pendulum at rest. If necessary, use books to adjust the height of either the pendulum or the CBR as needed.
- Aim the CBR at the float. Pull the pendulum back about 20° and release it so that it swings back and forth in the motion detector's path. Get out of the way and start the CBR.

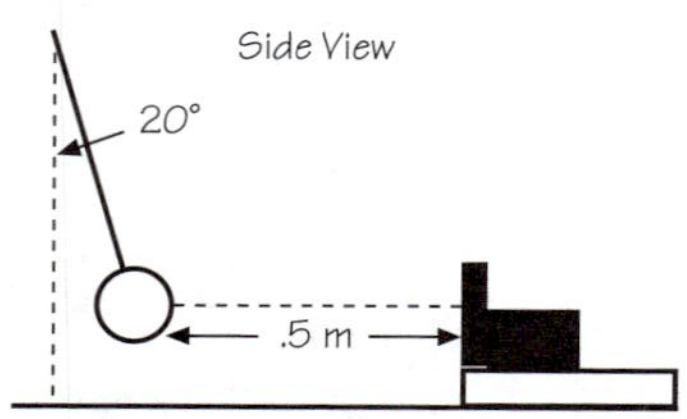

CBR SET UP

1. Connect the CBR to the calculator using the link cable.
2. Run the **RANGER** program on your calculator:

   a) Press **APPS** (**PROGRAM** on some calculators).

   b) Arrow down to **CBL/CBR**. Press **ENTER**.

   c) Press any key.

   d) Arrow down to **RANGER**. Press **ENTER**.

   e) Arrow to **SET UP SAMPLE**. Press **ENTER**.

   f) The Main Menu window should be set up as shown. If it is not, arrow down and press **ENTER** until the correct label is showing.

MAIN MENU	START NOW
REAL TIME:	YES
TIME(S):	15
DISPLAY:	DIST
BEGIN ON:	[ENTER]
SMOOTHING:	NONE
UNITS:	METERS

   g) Arrow up to **START NOW**. When you are ready to collect your data after you release the pendulum, press **ENTER**.

1. Record the length of your pendulum from the meter stick to the bottom of the pendulum.

For the first data collection, pull the pendulum back about 30° from vertical and release it so that it swings back and forth in the motion detector's path. Practice pulling back the pendulum a few times.

COLLECTING THE DATA

You and a partner will make **two** data collections using the CBR at your assigned station:

- One collection is with the pendulum in about a 20° swing.
- The second collection is with the pendulum in about a 40° swing.
- The collections should be made on different calculators.
- Each person in the pair is responsible for one of the data collections.

2. Record the angle of the swing for which you are responsible.
3. Before collecting your first set of data, sketch what you think a distance-versus-time graph will look like. Explain why you drew the graph as you did.

HINTS FOR COLLECTING DATA WITH THE CBR

If you do not get good data, there are two likely reasons: The most likely is that the CBR is placed too close to the pendulum. Less likely, is that the pendulum swings a path that is not in line with the CBR, so that data collection is skewed.

Set up your CBR using the directions on page 530. The CBR should be **perpendicular to the floor** and aimed directly at the pendulum. When you are ready to begin collecting data, release the pendulum, get out of the way, and start the CBR.

Collect both sets of data, then answer Questions 4–9.

4. Make a scatter plot of your data. State the window. Sketch the scatter plot.
5. Does the scatter plot match your sketch in Question 3? Why or why not?
6. Mark one period on the sketch. Use the TRACE key to find the length of that period. Be sure to record the units of measure.
7. Mark the amplitude on your sketch. Use the TRACE key to find the minimum and maximum distances of the pendulum. Use these values to compute the amplitude. Be sure to record the units of measure.
8. How do your period and amplitude compare with your partner's?
9. How can you tell who used the larger angle from the graph?

CHECK THIS!

A pendulum completes a *cycle* as it moves from one extreme position to another and back again. The time required for the pendulum to complete a cycle is called the *period*.

CHECK THIS!

The *amplitude* is the distance the pendulum moves from at rest to the farthest position away. The amplitude can be found by dividing the total distance the pendulum moves by two.

For Questions 10–14, compare your findings with other students who collected at the same station.

10. What is similar about the data? What do you think causes the similarities?
11. What is different? What do you think causes the differences?
12. Can you tell from the graphs who pulled the pendulum back the farthest? The least?
13. How do the periods compare?
14. You are about to compare data with students who collected data from stations with pendulum lengths different from yours. How do you think their data will compare to yours?

Answers

1. Students should record 0.4 meters, 0.5 meters, or 0.6 meters depending on the pendulum to which they are assigned.
2. Students should record 20° or 40° depending on which one they chose to perform.
3. Answers may vary. Students should draw a sinusoidal graph since the distance between the pendulum and the CBR oscillates over time between the maximum and minimum values. Some students may show decreasing amplitudes over time if they expect the pendulum to slow down. In the real world, due to friction between the string and the meter stick, energy is lost from the pendulum and it really does slow down.
4. Sample answer for 20° pendulum swing:

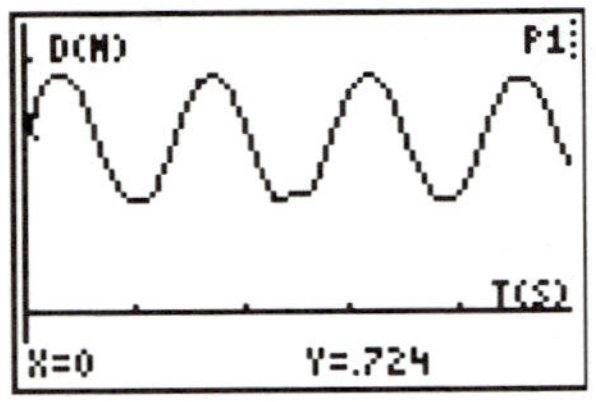


```
WINDOW
 Xmin=0
 Xmax=4.99968
 Xscl=1
 Ymin=-.2337796
 Ymax=1.128898
 Yscl=1
 Xres=1
```

NOTE: the RANGER program automatically sizes the window. An estimate from reading the axes is sufficient for this question. In the sample image the time values (domain) are between 0 and 5 seconds and the distance values (range) are between 0 and 1 meter.

5. Answers and explanations may vary. The accuracy of the prediction is not as important as students' interpretations and understanding of the situation at this juncture in the lesson.
6. Sample answer for 20° pendulum swing:

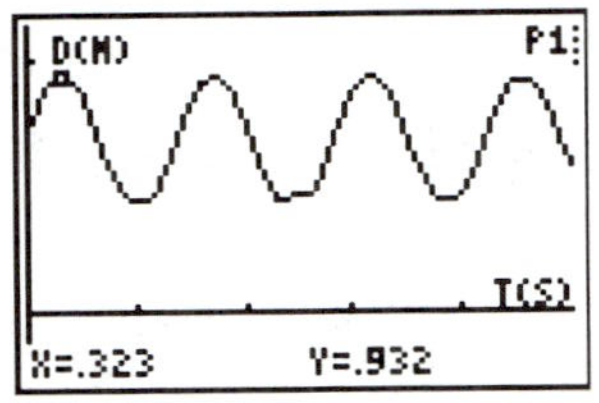

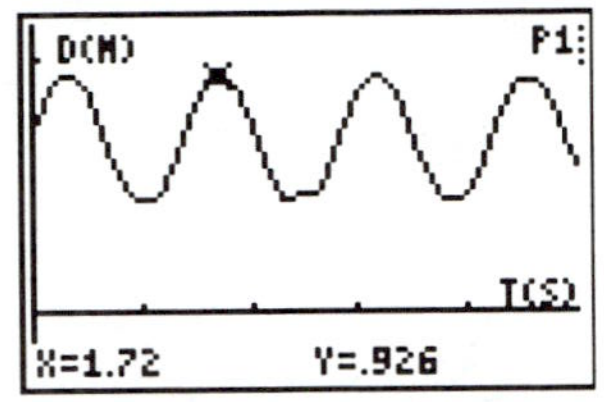

1.72 – 0.323 = 1.397 seconds

7. Sample answer for 20° pendulum swing:

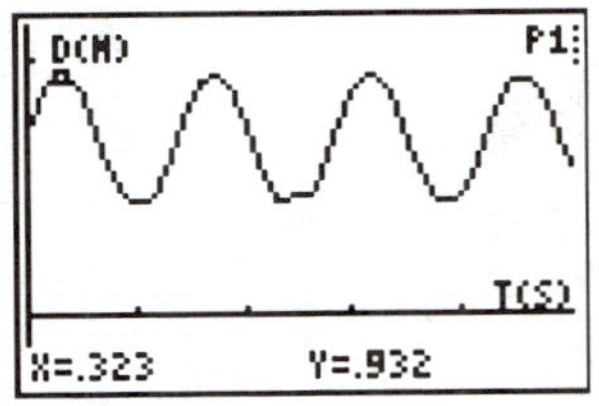

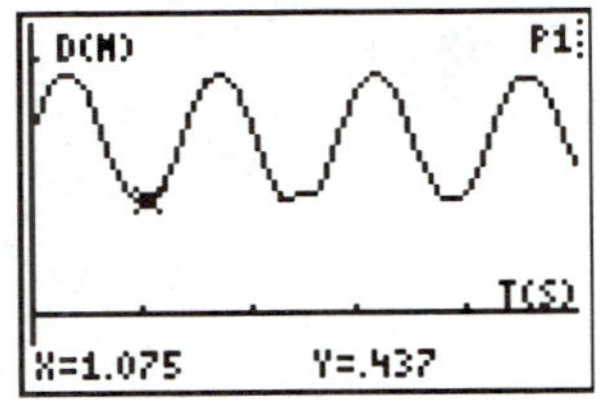

$\frac{0.932 - 0.437}{2} = 0.2475$ meters

8. Answers may vary. The periods should be the same (within a reasonable range due to measurement and rounding errors), and the amplitude of the larger swing should be larger than the amplitude of the smaller swing.
9. The larger amplitude indicates the larger angle.
10. The periods are all the same because the length of the pendulum was the same. The period of a pendulum depends on the length of the pendulum.
11. The amplitudes are different and appear to be determined by the angle between the pendulum and vertical at release. The larger the angle, the greater the distance the pendulum swings. The greater the distance the pendulum swings, the larger the amplitude.
12. The graph(s) with the largest amplitude(s) indicate the students who pulled the pendulum back the farthest.
13. The periods should be approximately the same. There may be minor variations due to precision of measurement and rounding errors.
14. Speculations may vary. Students may hypothesize that the longer the pendulum the greater the amplitude. Or they may hypothesize that the greater the angle of the swing the longer the period. This question gives students an opportunity to speculate about the comparisons between data from different stations before actually investigating.

For Questions 15–17, compare your findings with those of students who collected from different stations.

15. Did their data meet your expectations? Explain.

16. What is similar about your data? What do you think causes the similarities?

17. What is different? What do you think causes the differences?

18. You can adjust a pendulum clock by changing the length of the pendulum. If a clock runs slow, how should you adjust the pendulum? Should it be longer or shorter? Why?

SECTION 8.9 Assignment

1. Over a year, the length of daylight (the number of hours from sunrise to sunset) changes. **Figure 8.51** shows the length of daylight every 30 days from 12/31/97 to 3/26/99 for Boston, Massachusetts.

FIGURE 8.51. Data on length of day.

Date	Day Number	Length (hours)	Date	Day Number	Length (hours)
12/31	0	9.1	8/28	240	13.3
1/30	30	9.9	9/27	270	11.9
2/1	60	11.2	10/27	300	10.6
3/31	90	12.7	11/26	330	9.5
4/30	120	14.0	12/26	360	9.1
5/30	150	15.0	1/25	390	9.7
6/29	180	15.3	2/24	420	11.0
7/29	210	14.6	3/26	450	12.4

a) Draw a set of axes similar to **Figure 8.52**. Then plot the data from Figure 8.51.

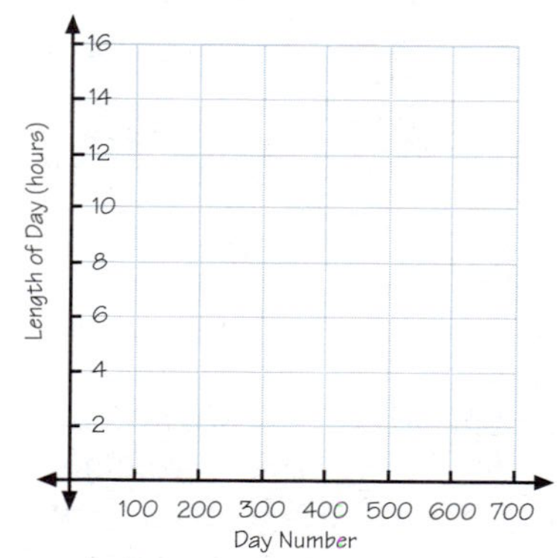

FIGURE 8.52. Axes for length of day versus day number.

b) Draw a smooth curve through the points on your graph. Extend your graph to show how you think the graph should look over two years. Explain why you think the graph should look this way.

Answers (continued)

15. Answers and explanations may vary. The accuracy of the prediction is not as important as students' interpretations and understanding of the situation at this juncture in the lesson.

16. In general, the greater the angle of the swing is, the larger the amplitude.

17. Periods are dependent on the length of the pendulum. The longer the string is, the larger the period.

18. The length of a pendulum determines its period. A clock having a period of one second means that the pendulum will swing back and forth in one second. If a clock is running slow then it takes longer than one second for the pendulum to swing back and forth, meaning that the period is too long. Shortening the length of the pendulum will shorten the period, making the clock run faster.

Section 8.9 Assignment

Answers

1. a) Answer:

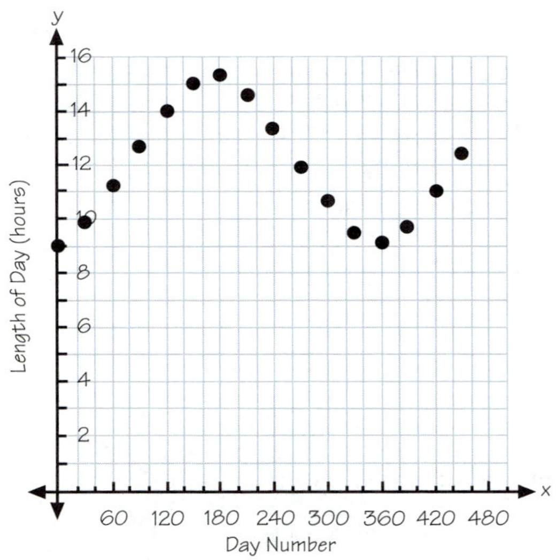

b) The length of daylight over time is periodic because the Earth orbits the sun once a year. The longest day of the year is on June 21st, and the shortest day is on December 21st. The length of daylight oscillates between these two extremes. Below is a periodic graph. The length of the period is one year, approximately 365 days.

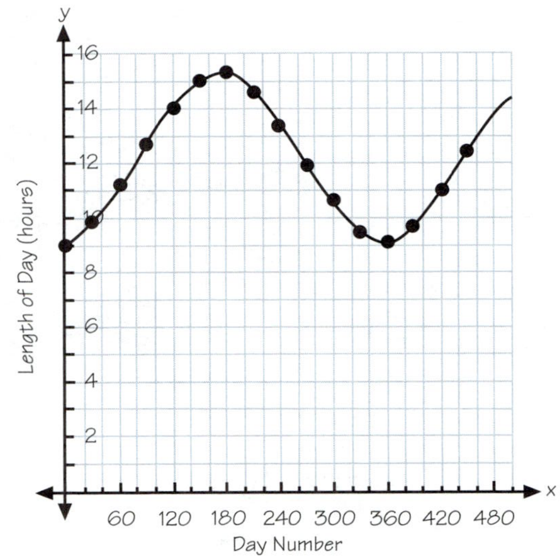

c) What is the period of your graph?

d) What is the graph's amplitude?

e) What is the axis of oscillation?

2. Figure 8.51 gives data on length of daylight for Boston, Massachusetts. A plot of the data indicates that the relationship between length of daylight and day number is periodic.

   a) Write a sinusoidal function that models this relationship. Explain how you got your function.

3. Ariana and Terry are turning a long jump rope for Tracy. Ariana and Terry turn the rope once per second. The maximum height is seven feet. At its lowest the rope just touches the ground. Assume the rope is on the ground moving up and away from Tracy at starting time.

   a) What is the amplitude for the relationship between the height of the rope and time?

   b) What is the period?

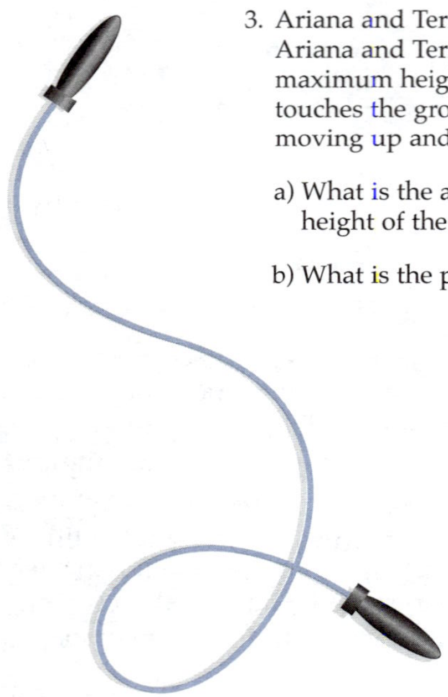

SECTION 8.10 The Sun is Shinning Down

Brockport, New York, is a small town on the Erie Canal in western New York state. Winters there can be very cold. Average high temperatures in January are 31°F. Summers, on the other hand, can be quite warm, with average high temperatures in July rising to about 82°F.

One factor that affects temperature is the amount of sunlight received during the day. Scientists can measure and record the amount of sunlight, or solar radiation. They often use the units of calories per square centimeter. **Figure 8.53** contains data from Brockport, New York. The values for each month are the average daily solar radiation for the 15th day of that month.

Month	Solar Radiation
January	132
February	190
March	274
April	365
May	446
June	500

Month	Solar Radiation
July	495
August	418
September	310
October	210
November	117
December	92

FIGURE 8.53. Solar radiation data.

Based on these data, predict the approximate solar radiation received at the beginning of March. Justify your prediction.

Answers (continued)

c) approximately 365 days

d) The maximum length of daylight in the data is 15.3 hours and the minimum is 9.1 hours. Using these values, amplitude = (15.3 – 9.1)/2 = 3.05 hours.

e) The average of the maximum and minimum daylight hours is 12.2. The axis of oscillation is $y = 12.2$.

2. a) Students probably used sinusoidal regression. The student's calculator will show an answer of $y = 3.02\sin(0.0172x - 1.396) + 12.2$.

3. a) Amplitude = 3.5 feet

   b) 1 second

EVALUATE LESSON 2

Section 8.10: The Sun is Shining Down

Purpose: In Sections 8.6 through 8.9, students developed the concept of periodic motion and formalized skills and procedures necessary to solve problems involving periodic motion. In this section students have the opportunity to demonstrate their knowledge by applying those concepts and skills to solve a problem in a new context.

Description: Facing a problem in a new context, students determine and implement a method to solve the problem. In this open-middle performance assessment task there is only one correct answer, but there is a variety of valid methods to obtain it. In their justification, which is the most important part of a performance assessment task, students should be able to communicate their knowledge of mathematics and demonstrate that they have mastered the concepts presented in Sections 8.6 through 8.9.

Materials: graphing calculators

Teaching Ideas: This task may be assigned as group work or individual work. The mathematics performance assessment rubric should be used to evaluate students' work.

Students may obtain climate data for the city of their choice via the World Meteorological Organization's Website at www.worldweather.org or the National Climate Data Center at www.ncdc.noaa.gov.

Answers

Sample answer:

Use the data to construct a scatter plot of solar radiation versus month:

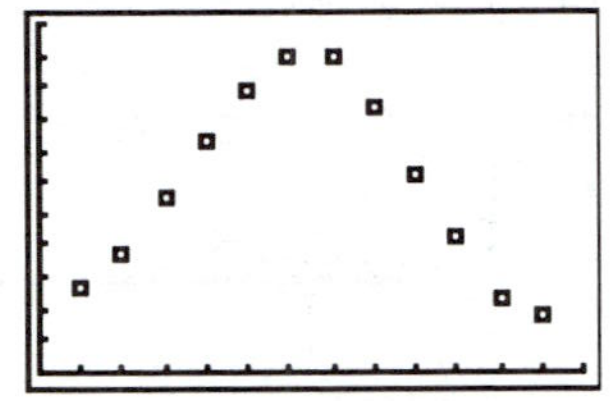

```
WINDOW
 Xmin=0
 Xmax=13
 Xscl=1
 Ymin=0
 Ymax=550
 Yscl=50
 Xres=1
```

Use the calculator to find a sinusoidal regression model:

NOTE: Be sure the calculator is in radians instead of degrees in the MODE screen.

```
SinReg L1,L2,Y1
```

```
SinReg
 y=a*sin(bx+c)+d
 a=199.0795153
 b=.5330533324
 c=-1.728130258
 d=299.2716013
```

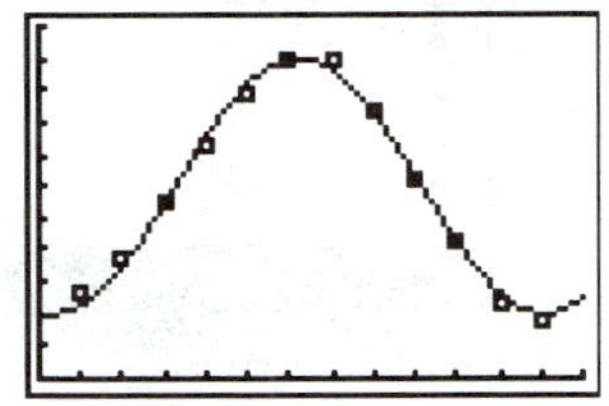

Trace between the February and March data points and estimate the value for solar radiation:

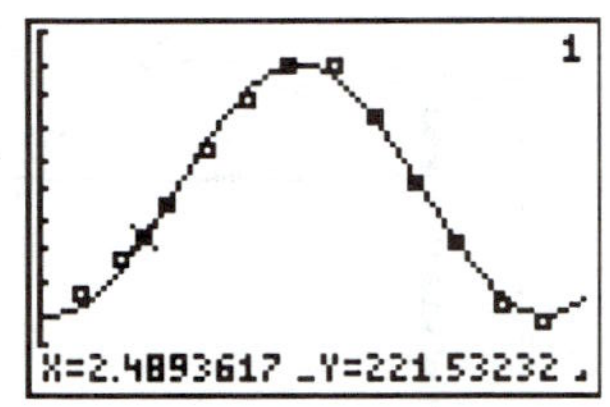

There will be about 221 calories per square centimeter of solar radiation in Brockport, New York, at the beginning of March.

SECTION 8.11

Mathematics and the Sound of Music

Music is played in many forms all over the world. Instruments vary from the didgeridoo of the Australian aborigines (a hollow tree branch that plays only two tones) to large concert organs. People everywhere appreciate good music.

When you pluck a violin or guitar string, it vibrates back and forth. To see this effect, try plucking a stretched rubber band once. The blur you see is the rubber band vibrating back and forth.

The vibrations of a violin string push on nearby air particles. These air particles push on other nearby air particles. Pressure waves are created. Sound waves are pressure waves. The energy from a vibration causes changes in air pressure. The pressure variations create a wave that travels through the air to reach the eardrum, which then vibrates and causes us to hear sound.

Sound waves are longitudinal waves. You can create a longitudinal wave with a Slinky.

- Attach a Slinky to a stationary object on a table.
- Tie a string to one of the coils of the Slinky.
- Stretch the non-stationary end of the Slinky out about 2 meters.
- Push and pull back once on the non-stationary end of the Slinky.

1. What happens as the non-stationary end of the Slinky is pushed?
2. How is the movement of the Slinky similar to the movement of sound waves? (See **Figure 8.54**.)

Music is, of course, not the only place where sound waves are used. Sound waves are used in ultrasound, CDs, and cell phones. Not only does sound travel through air, it also travels through liquids, solids, and other gases. You may have tried to make a phone with two cans and string when you were a child. When one person speaks into a can sound travels through the string and the other person can hear the sound in the other can.

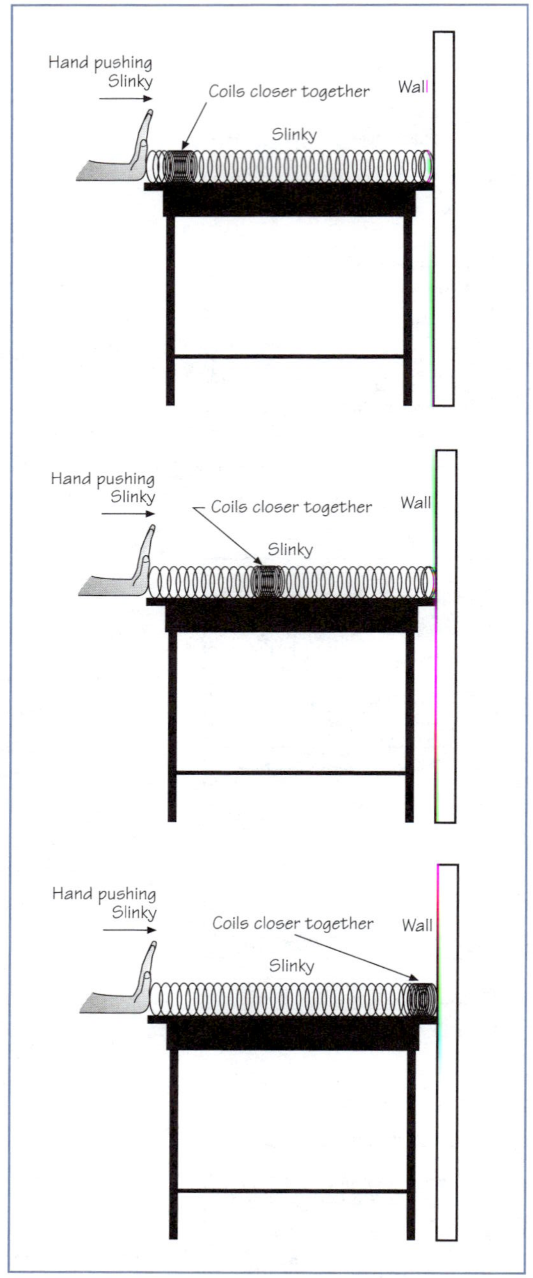

FIGURE 8.54.
The Slinky sound wave experiment.

ENGAGE LESSON 3

Section 8.11: Mathematics and the Sound of Music

Purpose: The purpose of Section 8.11 is to engage students in the study of sound. Students are then ready to investigate the period of the graph of a sound wave in Section 8.12.

Description: Students read the material in Section 8.11 of their textbook. The teacher demonstrates in a concrete physical situation what a sound wave looks like by stretching a Slinky and having students observe as a longitudinal wave travels through the coils.

Materials: Slinky, table, something to which the Slinky can be attached

Teaching Ideas: The main ideas that students should grasp from Section 8.11 are that varying pressure produces sound, and that sound waves are longitudinal. Musical notes can be produced by these variations. Have students read the material in small groups and discuss what they read. The Slinky might be attached to a ring stand or the wall. You could even have a student hold one end of the Slinky stationary. Instead of actually doing this presentation you can ask students to visualize it; however, actually seeing the Slinky move does help students understand sound waves.

Answers

1. A longitudinal type wave travels through the Slinky. (A longitudinal wave is a wave in which the vibration moves in the same direction as the wave is traveling.)
2. Sound travels as a longitudinal wave in much the same way as energy travels through the Slinky.

Sound waves produced by music have a regular pattern. Sound waves produced by noise do not have a regular pattern. You may have heard of singers who can break a glass with their voice. By singing a perfect note the singer's voice causes the glass to vibrate so much that it shatters from the pressure.

The harp is the oldest stringed instrument in the world. Each string vibrates to produce a different note. Some string instruments use a few strings to produce many notes. For example, violins and guitars produce different notes by changing the length of the string. Wind instruments, such as trombones, clarinets, and oboes play different notes by changing the length of the column of air that vibrates inside the instrument. Percussion instruments, such as xylophones and steel drums, create vibrations that ring at a certain pitch when struck.

SECTION 8.12

Do You See the Music?

Have you ever blown across the top of a soda bottle? What happened?

You may have noticed that a sound is produced when you blow across the top of the bottle.

You can use a microphone collection device to measure the pressure variations that cause us to hear sound.

1. Sketch what you think a sound pressure versus time graph might look like.

2. Your teacher will demonstrate how to collect sound pressure data. After the demonstration, record the scatter plot.

3. Label one cycle on the scatter plot.

CHECK THIS!

A cycle is one complete pattern.

4. On the scatter plot, mark and label two points. One point should be at the beginning of a cycle and the other at the end. Record these values in a table like **Figure 8.55** and compute the period. Show your work.

CHECK THIS!

The period is the number of seconds to complete one cycle. That is,

$$\text{period} = \frac{\text{number of seconds}}{\text{number of cycles}}$$

Point	x	y
A		
B		

FIGURE 8.55.
Table for Question 4.

5. Find the frequency of the graph. Explain the meaning of this value.

CHECK THIS!

The frequency is the number of cycles completed in one second. That is,

$$\text{frequency} = \frac{\text{number of cycles}}{\text{number of seconds}}$$

EXPLORE LESSON 3

Section 8.12: Do You See the Music?

Purpose: In Section 8.12 students explore the concept of collected sound data as a periodic function. Students discover the relationship between period and frequency. They calculate both from points on a graph. They explore the effect of changing volume on a graph. They will be prepared to explain formally these concepts in Section 8.13. In Section 8.14 they will use this knowledge to tune a bottle to a given frequency and participate in a bottle band.

Description: Set up the data collection unit with a microphone probe on an overhead graphing calculator to demonstrate the setup and to collect sound data for group discussion. Students should predict what they think a sound-versus-time graph looks like and make a sketch. Have a volunteer make a musical note by blowing across an empty glass bottle. Use the graph to promote understanding of the lesson vocabulary. Then place students in groups to collect data with various heights of water in the bottles.

Materials: overhead graphing calculator, graphing calculator with DataMate application, CBL2 with microphone probe, unit-to-unit link cable, glass pop or water bottles of equal size, pitcher of water, funnel, straws (optional), tuning fork (optional)

Teaching Ideas: Sanitation is an important part of this experiment. Have alcohol wipes available for students to use as necessary. Some students may find it difficult to blow across the top of a bottle; blowing through a straw may help.

Answers

1. Answers will vary. Some students might think that the graph would be linear since they hear one constant note. Ask students why they predicted what they did.

2. Have a student volunteer blow across an empty soda bottle. Collect the sound data onto the overhead graphing calculator for the entire class to observe. (You may want to use a tuning fork instead of the glass bottle.) Use the graph to promote understanding of the lesson vocabulary.

   Sample data:

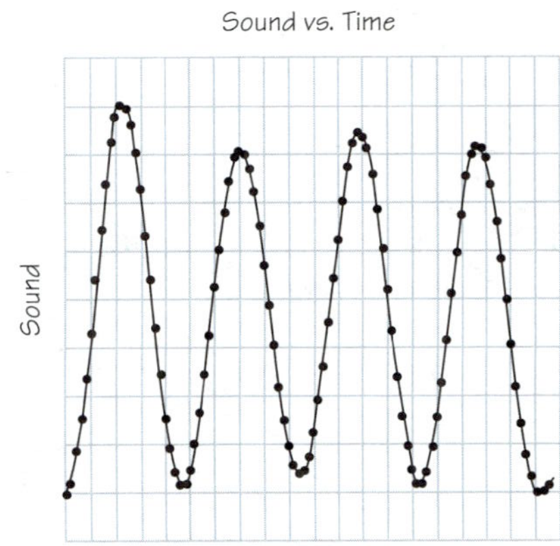

   Discuss any surprises between student predictions and the actual graph. Ask for student input on characteristics of the graph, including what quadrants the graph is located in and why. In this real-life data collection, time (the independent variable) is always positive and increasing. The sound (pressure from the sound wave) is both increasing and decreasing.

   Ask students how this graph is similar to and different from other functions with which they are familiar. If students do not generate the characteristic of a repeating pattern, ask them if they would feel confident about predicting what the graph would look like beyond the edge of the view screen, and elicit the concept of periodic function from the discussion. Remind students that periodic functions have repeating patterns.

3. Have students mark a cycle on their paper and discuss different possible ways to mark a cycle (from maximum point to adjacent maximum point or from minimum point to adjacent minimum point).

   Ask students if they think it is significant where the cycle is marked. It is not significant as long as you can exactly identify the beginning and the end of the cycle.

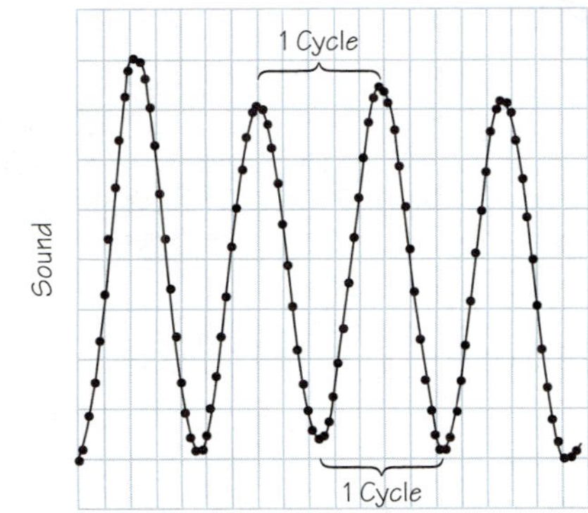

4. Trace and compute the period. Talk about how much time that really is.

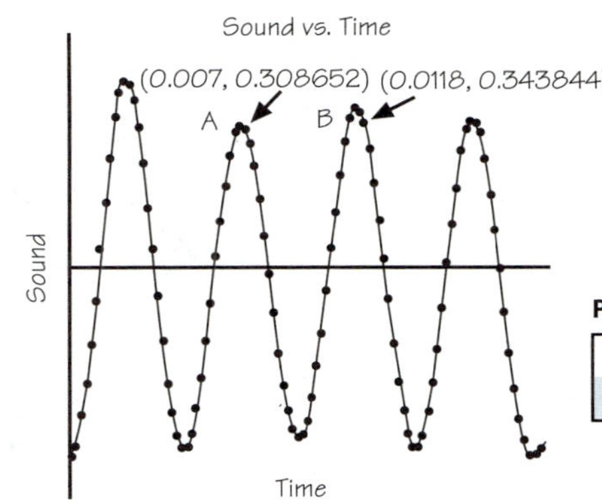

Point	x	y
A	0.007	0.309
B	0.0018	0.334

   From the sample data above, the period is found by subtracting the time values (x-values) of adjacent peak points.

Answers (continued)

0.0118 sec – 0.007 sec = 0.0048 sec per cycle (time to complete one cycle)

0.0048 is 48 ten-thousandths of a second per cycle.

5. Since frequency is cycles per second, simply find the reciprocal value of the period:

$$\text{period} = \frac{0.0048 \text{ sec}}{1 \text{ cycle}}$$

$$\text{frequency} = \frac{1 \text{ cycle}}{0.0048 \text{ sec}} = 208.33 \, \frac{\text{cycles}}{\text{second}}$$

The unit of cycles per second is also the unit of Hertz (Hz). The frequency of the sample data is 208.33 Hz. The pattern in the graph will repeat approximately 208 times every second.

A standard unit of sound measure is the Hertz. Hertz means "cycles per second." FM radio waves are given in megahertz. AM radio waves are given in kilohertz. For example, 97.9 FM is 97.9 megahertz. The frequency of a wave that a 97.9 FM station sends out is 97,900,000 cycles per second. Frequencies of radio waves are outside the range of human hearing. Only after a radio receives the waves can we hear the music. The Hertz is named after Heinrich Hertz, the first person to successfully experiment with the transmission and reception of radio waves.

CHECK THIS!

Pitch refers to the highness or lowness of a tone. For example, a low note at the left end of the piano keyboard has a low pitch. A high note at the other end of the keyboard has a high pitch.

6. Repeat the experiment in your group with an assigned water height. Before adding water, blow across the top of the bottle and listen to the sound. Now add the assigned height of water. How does adding water change the sound?
7. Collect new data with water in the bottle. Make a scatter plot.
8. Use the TRACE feature of the calculator to mark and label two points on your scatter plot. Find the period. Show your work.
9. How does the new period compare to the period of the original data?
10. Find the frequency of your new data. Label your answer with appropriate units.
11. How does the frequency of the new data compare to the frequency of the original data?
12. What is the relationship between frequency and pitch?

Complete the following:

13. If the period of a sound vs. time graph decreases, the frequency ____.
14. If the period of a sound vs. time graph increases, the frequency ____.
15. To decrease the frequency of a bottle, you would ____.
16. Consider this: Sound wave A has a shorter period than sound wave B. What can you say about their relative frequencies? Justify your answer.

SECTION 8.12 Assignment

1. Use the data collected by groups in your class to fill in a table like **Figure 8.56**.

Height of Water (inches)	Period (sec/cycle)	Frequency (cycles/sec)

FIGURE 8.56.
Table for Question 1.

2. Make a scatter plot of the relationship between period and height of the water.
3. Which quantity did you use as the independent variable?
4. Which quantity did you use as the dependent variable?
5. What do you notice about a period versus height of water scatter plot?
6. If the height of the water is 3.75 inches, what do you predict the period to be?
7. Make a scatter plot of frequency versus height of water.
8. What do you notice about the frequency versus height of water scatter plot?
9. If you want the frequency of a bottle's note to be 400 cycles per second, about what should the height of the water be?
10. Make a scatter plot of frequency versus period.
11. What do you notice about the relationship between frequency and period? Explain your answer.

Answers (continued)

6. Assign each group a different height to fill the bottle. For example, you could assign heights in intervals of 0.5 inches (0.5 in, 1 in, 1.5 in, 2 in, 2.5 in, 3 in, 3.5 in etc.). The pitch of the sound is higher with water in the bottle.

7. Sample data:

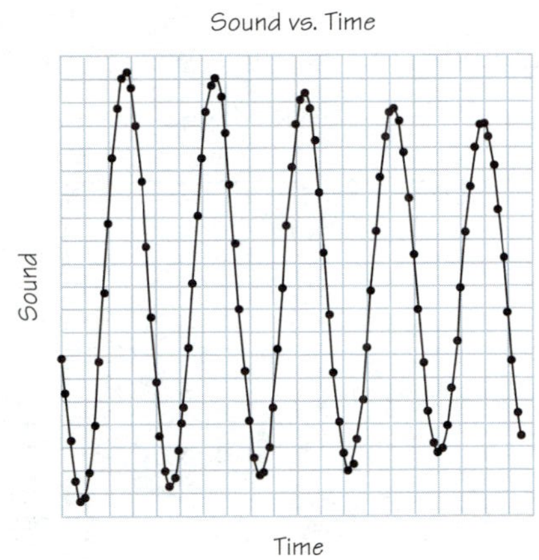

8. Sample data:

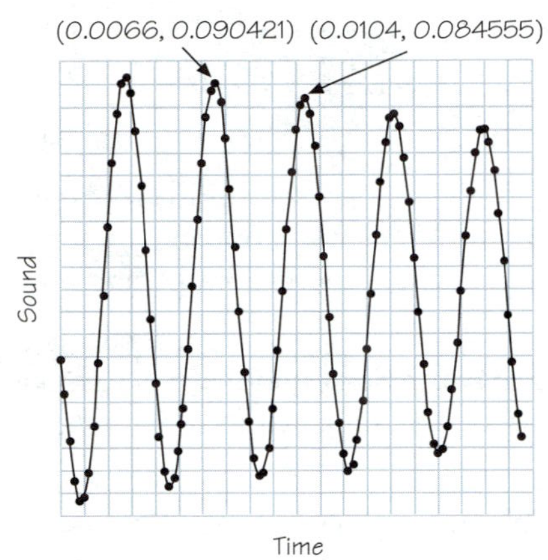

$$\text{period} = 0.0104 - 0.0066 = 0.0038\ \frac{\text{seconds}}{\text{cycle}}$$

9. The new period should be smaller than the original. In the sample data provided, the original period was 0.0048 sec/cycle, and the new period is 0.0038 sec/cycle.

10. The frequency is the reciprocal of the period: $\text{frequency} = \frac{1 \text{ cycle}}{0.0038 \text{ seconds}} = 263.158\ \frac{\text{cycles}}{\text{second}} =$ 263.158 Hz.

11. The frequency of the new data is larger than the frequency of the original data. This difference makes sense for two reasons. First, the period is a smaller value; therefore, the frequency, the reciprocal of the period, should be larger. Second, the sound the bottle made when air was blown across it had a higher pitch because of the higher frequency.

12. As the frequency increases, the pitch gets higher. As the frequency decreases, the pitch gets lower.

13. increases

14. decreases

15. remove water from the bottle

16. The shorter the period, the higher the frequency (the higher the note). Since sound wave A has a shorter period than sound wave B, sound wave A will have a higher frequency and, in fact, is a higher note.

Section 8.12 Assignment

Answers

1. NOTE: Have one student from each group fill in their data values on a transparency so that all groups have the opportunity to record all of the data points. You may give students the sample data in the table instead of classroom-collected data.

   Sample data:

Height of Water (inches)	Period (sec/cycle)	Frequency (cycles/sec)
0.0	0.004575	218.6
0.5	0.004375	228.6
1.0	0.00435	229.9
1.5	0.004	250
2.0	0.003775	264.9
2.5	0.0034	294.1
3.0	0.00325	307.7
3.5	0.002975	336.1
4.0	0.0027	370.4
4.5	0.00235	425.5
5.0	0.001933	517.2

2. Answer:

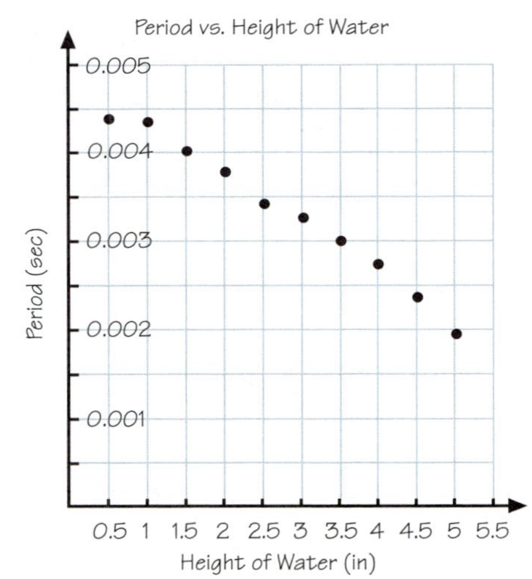

Answers (continued)

3. The variable that you control is the height of the water, so that is the best choice for the independent variable.
4. The period depends on the height of the water, so period is the best choice for the dependent variable.
5. As the height of the water increases, the period decreases, demonstrating a negative relationship.
6. The period would be between 0.0027 and 0.002975 seconds. A good guess might be 0.0028 seconds.
7. Answer:

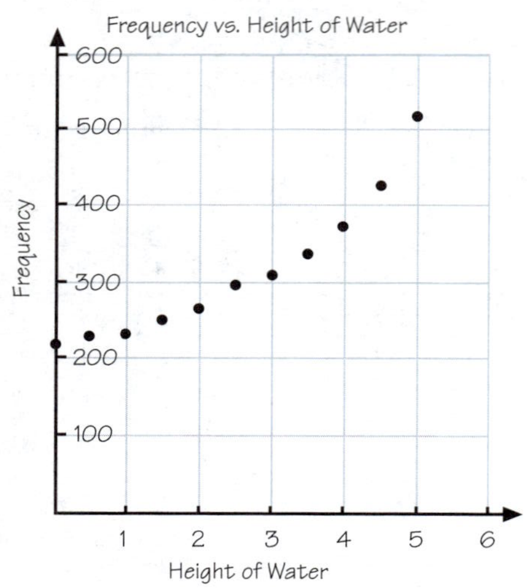

8. Answers may vary. As the height of the water increases, the frequency increases. This set of data does not appear to be linear.
9. Since 400 cycles per second is approximately halfway between 370.4 and 425.5, the height of the water would be approximately 4.25 inches for the sample data.
10. Answer:

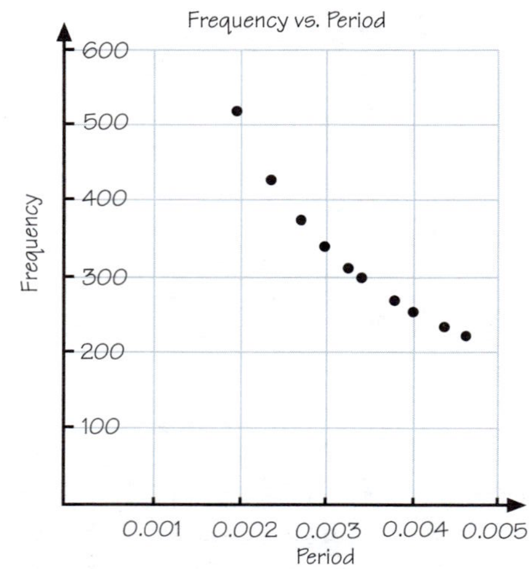

11. As the period gets larger, the frequency gets smaller. The graph appears to be an inverse variation. Since we found the frequency by finding the reciprocal of the period, the function should be inverse variation.

SECTION 8.13

Name That Note

Blowing across the top of a bottle produces sound pressure waves. These waves look like a scatter plot of the data you generate when you roll a wheel or when you swing a pendulum.

Sound waves have properties that determine the note that your ears hear. In Section 8.12 you investigated these properties of sound waves.

The *period* is the number of seconds per cycle. The period can be measured by finding the time it takes to get from one crest of the graph to the next crest (**Figure 8.57**).

$$P = \frac{\text{number of seconds}}{\text{number of cycles}}$$

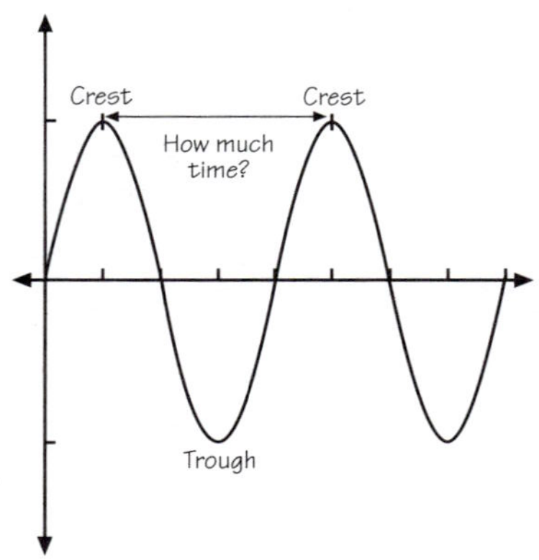

FIGURE 8.57.
The period of a cycle.

The *frequency* is number of cycles per second (**Figure 8.58**).

$$f = \frac{\text{number of cycles}}{\text{number of seconds}}$$

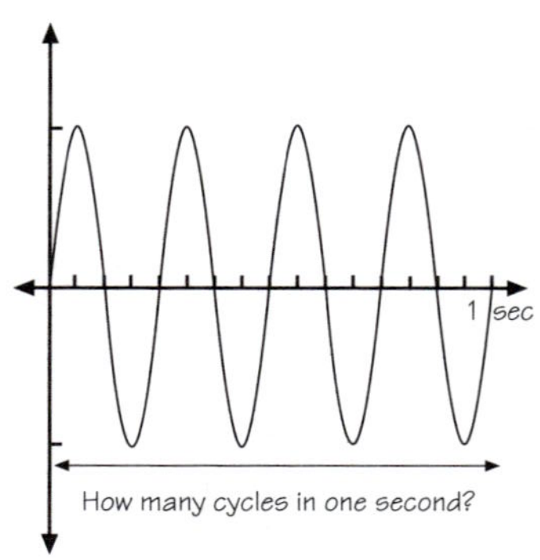

FIGURE 8.58.
The frequency of a cycle.

1. Suppose the period is 0.0048 seconds. Find the frequency.
2. Suppose the frequency is 153.85 cycles per second. Find the period.

Period and frequency have a reciprocal relationship.

Another property of sound is loudness. The larger the amplitude of a sound wave, the more energy the sound wave has and the louder the note (**Figure 8.59**).

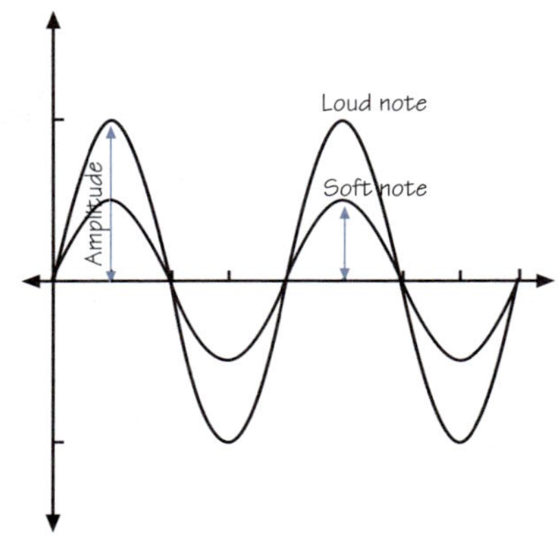

FIGURE 8.59.
The loudness of a sound wave.

In Section 8.8 you investigated properties of periodic functions such as amplitude and period.

In a sinusoidal function for a musical note, $f(t) = A \sin(B(t - C)) + D$, t represents time, and A represents the amplitude or loudness of the note. B is equal to the frequency multiplied by 2π. Assuming that the function includes the point (0, 0), we can use the simpler $f(t) = A \sin Bt$.

CHECK THIS!

For a pure note, with data that includes (0, 0), a sinusoidal function that models the note has the form $f(t) = A \cdot \sin(\text{frequency} \cdot 2\pi \cdot t)$.

EXPLAIN LESSON 3

Section 8.13: Name That Note

Purpose: In Sections 8.11 and 8.12, students investigated sound waves produced by blowing across the top of a soda bottle. They found the period and frequency of the sound waves. In this section students formalize the concepts of period, frequency, and amplitude. They develop a simple sinusoidal model for musical notes. They will be ready to "tune" a bottle with water in order to play in the bottle band in Section 8.14.

Description: With given data, students find a model for the sound wave of a musical note by finding the period, frequency, and amplitude with data from either a table or graph.

Materials: graphing calculator

Teaching Ideas: In order to avoid horizontal shifts to the sine function, data are given with a starting point of (0, 0). Emphasize the concept that the period and the frequency are inverses of each other.

Answers

1. $P = \frac{0.0048 \text{ seconds}}{1 \text{ cycle}} = 0.0048$ seconds/cycle

   $f = \frac{1 \text{ cycle}}{0.0048 \text{ seconds}} = 208.33$ cycles/second

2. $f = \frac{153.85 \text{ cycles}}{1 \text{ second}} = 153.85$ cycles/second

   $P = \frac{1 \text{ second}}{153.85 \text{ cycles}} = 0.0065$ seconds/cycle

3. The data in **Figure 8.60** were collected when three tuning forks were struck.

Tuning Fork 1		Tuning Fork 2		Tuning Fork 3	
Time, t (seconds)	Sound Pressure, $f(t)$	Time, t (seconds)	Sound Pressure, $f(t)$	Time, t (seconds)	Sound Pressure, $f(t)$
0.00000	0.00000	0.00000	0.00000	0.00000	0.00000
0.00029	0.00285	0.00022	0.00116	0.00057	0.00285
0.00057	0.00400	0.00057	0.00200	0.00114	0.00400
0.00096	0.00190	0.00093	0.00110	0.00191	0.00190
0.00114	0.00000	0.00114	0.00000	0.00227	0.00000
0.00134	−0.00214	0.00137	−0.00121	0.00287	−0.00295
0.00171	−0.00400	0.00171	−0.00200	0.00341	−0.00400
0.00201	−0.00265	0.00204	−0.00119	0.00402	−0.00265
0.00227	0.00000	0.00227	0.00000	0.00455	0.00000
0.00259	0.00304	0.00252	0.00127	0.00488	0.00180

FIGURE 8.60.
Tuning fork data.

Make a scatter plot of all three data sets. Put all three on the same grid, and use a different color pencil for each set.

4. For each data set find the amplitude, period, frequency, and a function rule. Fill in a table like **Figure 8.61**.

	Tuning Fork 1	Tuning Fork 2	Tuning Fork 3
Amplitude			
Period (seconds/cycle)			
Frequency (cycles/second)			
Function model using sinusoidal regression on graphing calculator	$y = 0.004 \sin(2766x - 0.005)$	$y = 0.002 \sin(2766x - 0.003)$	$y = 0.004 \sin(1382x - 0.004)$

FIGURE 8.61.
Table for Question 4.

5. Sketch graphs of the sine function that models each tuning fork note on your scatter plot.

6. Which note(s) was (were) played the loudest?

7. Which note(s) has (have) the lowest frequency?

8. Which note(s) has (have) the highest pitch?

Answers (continued)

3. An example of a student's final scatter plot with functions that model all three notes.

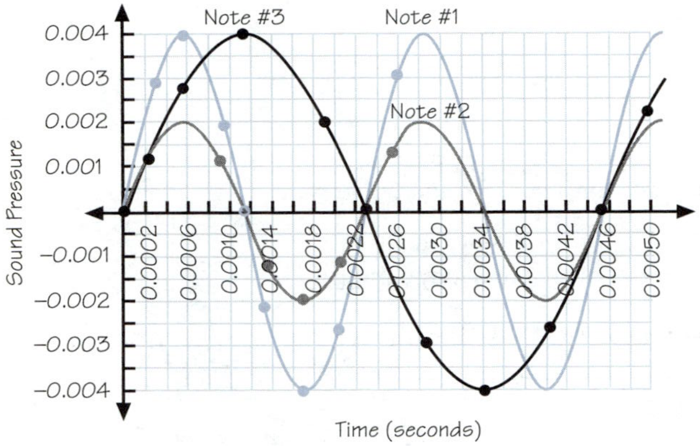

4. Answers:

	Tuning Fork 1	Tuning Fork 2	Tuning Fork 3
Amplitude	0.004	0.002	0.004
Period (seconds/cycle)	0.00227	0.00227	0.00455
Frequency (cycles/second)	440	440	220
Function model using sinusoidal regression on graphing calculator	$y = 0.004 \sin(2766x - 0.005)$	$y = 0.002 \sin(2766x - 0.003)$	$y = 0.004 \sin(1382x - 0.004)$

5. NOTE: You might want to demonstrate for students the similarities of the graphs of the two function models (with and without the graphing calculator) students found by using the "motion ball" for one of the functions. For example, using Note #1 as the calculator graphs $Y2$, the ball will move along the graph of the second function.

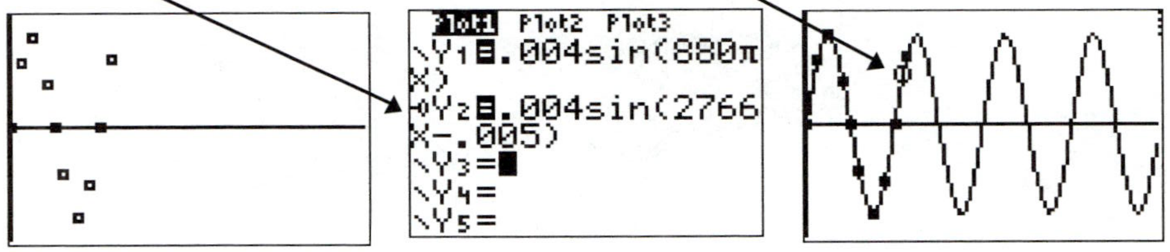

6. Notes #1 and #3 have the same amplitude, so they are equally loud.

7. Note #3 has the lowest frequency, 220 cycles per second.

8. Even though Note #1 is louder than Note #2, Notes #1 and #2 have the same frequency so they are the same note. They have a higher frequency and are, therefore, higher pitched than Note #3.

SECTION 8.13 Assignment

1. Instead of identifying notes by their frequencies, musicians have given them names. **Figure 8.62** displays the frequencies for the notes in the middle octave of a tuned piano. Make a scatter plot of frequency versus note number.

Name of the Note	Note Number	Frequency
C_4	0	261.63
$C_4^{\#}$	1	277.18
D_4	2	293.66
$D_4^{\#}$	3	311.13
E_4	4	329.63
F_4	5	349.23
$F_4^{\#}$	6	369.99
G_4	7	392.00
$G_4^{\#}$	8	415.30
A_4	9	440.00
$A_4^{\#}$	10	466.16
B_4	11	493.88
C_5	12	523.25

FIGURE 8.62.
Table for Question 1.

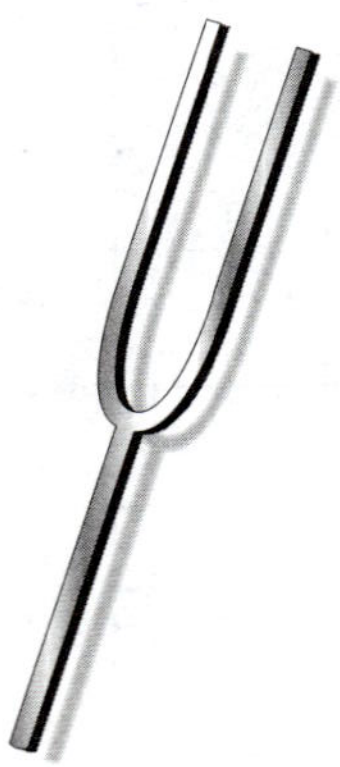

2. Do these data represent a function?

3. Find a function rule to model the relationship between frequency and note number.

4. What note do you think was played by tuning fork in Section 8.13?

Match each term with its description.

5. high amplitude a) a sound wave with a short period

6. low amplitude b) the number of cycles of a sound wave in one second

7. period c) a loudly played note

8. frequency d) a sound wave with a lower frequency

9. low note e) the length of time for one complete cycle of a periodic function

10. high note f) a softly played note

11. Joseph collected the data in the scatter plot in **Figure 8.63** from a tuning fork using a CBL. Find the period, the frequency, the amplitude, and a function rule.

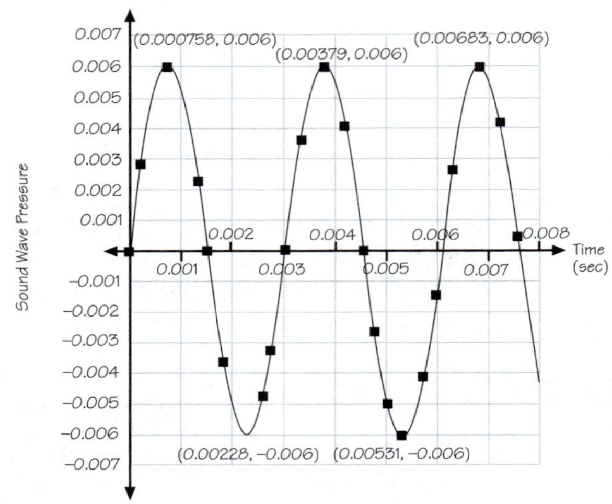

FIGURE 8.63.
Scatter plot for Question 11.

Section 8.13 Assignment

Answers

1. Sample answer:

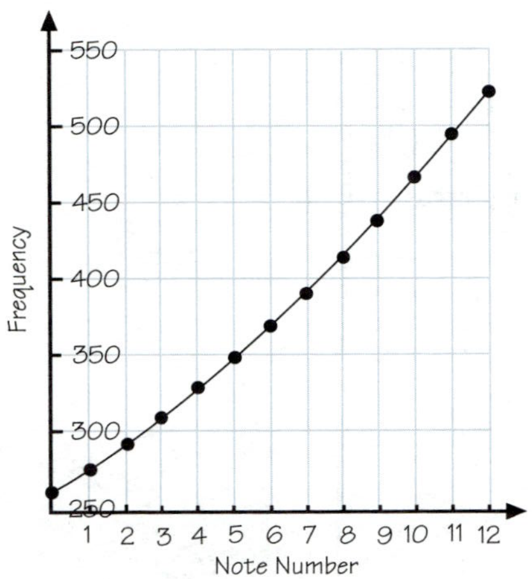

2. Yes, for each value in the domain there is one value in the range of the function.

3. By finding successive quotients, students should decide that the relationship is exponential. The exponential regression function that models the relationship between frequency and note number is $f(x) = 261.63(1.0595)^x$. The value of 1.0595 is not arbitrary; it is the approximate value of $2^{\frac{1}{12}}$. Musicians intentionally have planned for the frequency of C_5 to be twice as large as the frequency of C_4. Twelve steps occur between the two notes. In order for the final value in the octave to be twice as large as the initial value in an exponential function, there are twelve repeated multiplications. So $\left(2^{\frac{1}{12}}\right)^{12} = 2$.

4. Since the frequency was 440 hertz, the note is A_4.

5. c

6. f

7. e

8. b

9. d

10. a

11. The period of the note is approximately 0.00303 seconds per cycle. The frequency of the note is approximately 330 cycles per second. The amplitude of the note is 0.006.

    A sinusoidal model without using the graphing calculator is

    $f(t) = 0.006 \sin(330 * 2\pi t)$ or $f(t) = 0.006 \sin(660\pi t)$.

SECTION 8.14 The Big Bottle Band

In Section 8.12 you filled a soda bottle with a certain height of water and found the frequency of the note you played. Your teacher will give you a card with a frequency, and you will tune your bottle to this frequency. Use what you learned in Section 8.12 and 8.13 to help you tune your note. When all of the groups tune their bottles, the class will play a song. As in a bell choir, each person is responsible for playing a certain frequency.

1. My assigned frequency is ________.
2. Record your first graph and find the frequency. Explain how you found the frequency.
3. Do you need to add or remove water from the bottle? Explain your reasoning.
4. Repeat the process until your bottle is tuned. Record the trial number and frequency for each trial in a table like **Figure 8.64**.

Trial Number	Frequency (note)

FIGURE 8.64. Table for Question 4.

5. Measure the final amount of water you used to tune your bottle.
6. How many trials did it take to tune your bottle?
7. What have you learned about bottle tuning that would help you the next time you need to tune a bottle?

SECTION 8.14 Assignment

1. Group 1 collected data for the bottle band activity. They used the TRACE feature of the calculator to find the two sets of coordinates shown in **Figure 8.65**. The note that Group 1 was supposed to tune to has a frequency of 300. Should they add or remove water to get closer to this frequency?

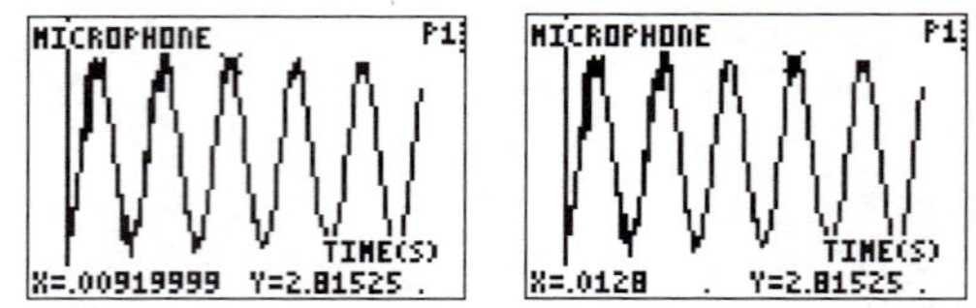

FIGURE 8.65. Coordinates for Question 1.

2. Group 2 collected data for the bottle band activity. They used the TRACE feature of the calculator to find the two sets of coordinates shown in **Figure 8.66**. The note that Group 2 was supposed to tune to has a frequency of 400. Should they add or remove water to get closer to this frequency?

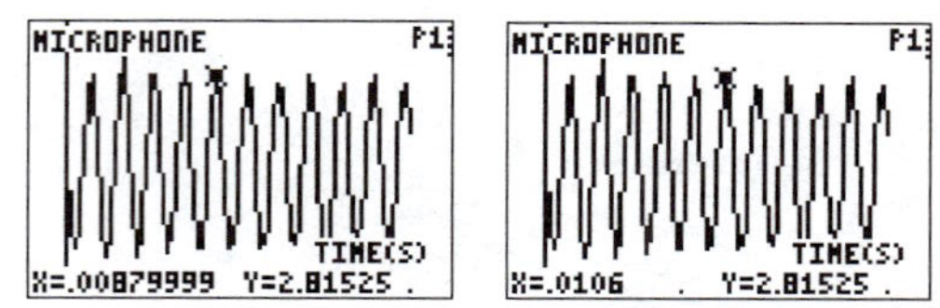

FIGURE 8.66. Coordinates for Question 2.

ELABORATE LESSON 3

Section 8.14: The Big Bottle Band

Purpose: In Section 8.13 students continued their study of periodic functions. In this section students demonstrate their understanding of the connection between frequency and period and use this connection to produce notes of the correct frequency to play a simple, familiar song.

Description: Each student is assigned a frequency at which to tune his or her bottle. (Although each student needs a graphing calculator with the DataMate application, small groups of students can share the CBL2s and microphone setups.) Students apply their knowledge of period and function to the task of tuning the bottle. When the class has finished tuning, have students perform the "Bottle Band Music" (on Transparency 8.3). As you point to each frequency, students tuned to that frequency play their bottles.

Materials: TI-83+ graphing calculator, CBL2 unit, microphone probe, unit-to-unit link cable, TI application DataMate, glass pop or water bottles of equal size, pitcher of water, frequency cards, Transparency 8.3, titration pipettes, graduated cylinders, alcohol wipes, straws (optional), the Bottle Band Cards located on the Teacher Resource CD-ROM

Teaching Ideas: It may take a couple of tries before students get a smooth sine wave. When done correctly; however, the graph should be extremely smooth, so encourage students to keep trying. It works best to start the tone a split second before the Enter key is pressed and to keep the tone steady for as long as possible. Songs are written using the frequency of the notes on the "Bottle Band Music" page. Using the frequencies of the notes instead of the names of the notes allows students who do not read music to participate fully.

Answers

2. Answers will vary, but students should find the period by using the trace feature of the graphing calculator. Then they should be able to calculate the frequency.
3. Some students may remember the table from Question 1 in the assignment for Section 8.12 and use the information to help themselves get started with an approximate height. In general if the frequency needs to be higher, the students should add more water. If the frequency needs to be lower, the students should take out water. Using titration pipettes allows students to either remove or add small amounts of water.
5. Answers will vary. As an extension activity, you could have students plot the heights of their water on chart paper along with the frequency and height data they obtained in Section 8.12.
6. Answers will vary.
7. Answers will vary. For example, some groups may decide that if they find the period by tracing over several periods they may get a more accurate value. For example, find the period over five cycles and divide by 5 to get an average.

Section 8.14 Assignment

Answers

1. Group 1 should add a small amount of water. The frequency they currently have is approximately 278, and the frequency they want to be able to play is higher, 300.
2. Group 2 should remove some water. The frequency they currently have is approximately 556, and the frequency they want to be able to play is lower, 400.

SECTION 8.15 Sound Assessment

1. Describe the graph of a periodic function.
2. What is meant by the period of such a graph?
3. How can you find the period of a graph? Why might someone choose to find the average of several values instead of using just one?
4. What is the frequency of a periodic graph? How is it related to the period?
5. Find the period and frequency of the sound wave from the graph and data in **Figure 8.67**.

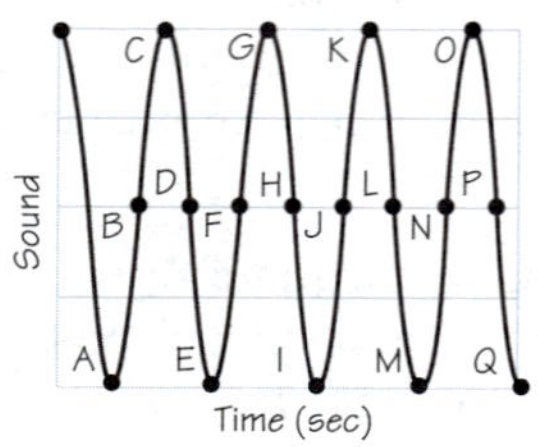

Point	x	y
A	0.0028	–1.917
B	0.0042	0
C	0.0056	1.917
D	0.0069	0
E	0.0081	–1.917
F	0.0095	0
G	0.0109	1.917
H	0.012	0
I	0.013	–1.917
J	0.0148	0
K	0.0159	1.917
L	0.0174	0
M	0.0187	–1.917
N	0.02	0
O	0.0212	1.917
P	0.0227	0
Q	0.0239	–1.917

FIGURE 8.67. Figure and table for Question 5.

6. The graphs in **Figure 8.68** are of different tones. Which one represents the tone with the lower pitch? Justify your answer.

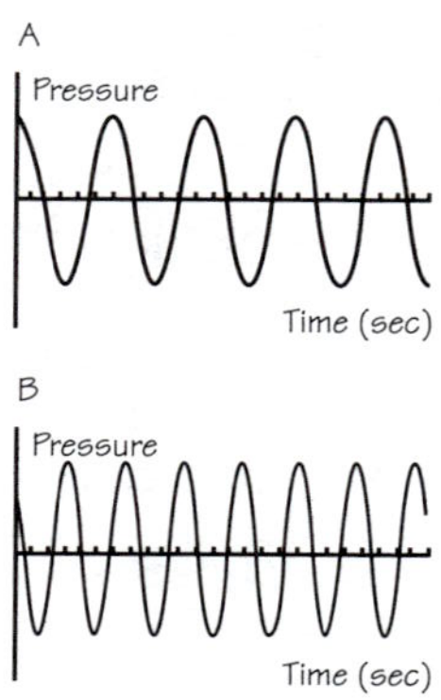

FIGURE 8.68. Graphs for Question 6.

Modeling Project How Does Your Cycle Go?

You have learned that some things can be modeled with periodic functions. Because periodic functions have consistent cycles, they can be used to answer questions such as:

- How high will the tide be next September 15th?
- What will be the demand for air conditioners next April?
- How much coffee should a coffee shop have ready for each hour of the day?

However, in these examples there are things that could interrupt the cycle, such as a change in customers' habits or unusual weather.

For this modeling project you will choose a relationship that you believe is periodic. You can gather information from the Internet, library research, or by interviewing people in your community. You will create a presentation for your class that:

- Describes how your choice is cyclical, both mathematically (using terms like period and amplitude), and in words;
- Discusses the strengths and weaknesses of a model for describing your choice and for making predictions based on your model;
- Discusses the usefulness of your model to people (i.e. how it helps them run a business, protect people from harm, or some other purpose).

EVALUATE LESSON 3

Section 8.15: Sound Assessment

Purpose: In Sections 8.11 through 8.14, students developed the concept of periodic functions as they relate to sound waves. In this section students have the opportunity to demonstrate their knowledge.

Description: Students should be able to explain what they have learned and justify their answers to the questions below. The most important part of this task is that students communicate their knowledge of mathematics and demonstrate that they have mastered the concepts presented in Sections 8.11 through 8.14.

Materials: graphing calculators

Teaching Ideas: These questions may be assigned as group work or individual work.

Answers

1. The graph of a periodic function has a repeating pattern.
2. The period is the time it takes to complete one cycle or pattern of the graph.
3. Subtract the maximum x-values of adjacent peaks. Averaging several peaks "smoothes out" the inexactness of the data.
4. The frequency is the number of cycles in one second, the reciprocal of the period. Frequency values are assigned to notes in the musical scale.
5. Period $= 0.0081 - 0.0028 = 0.0053\,\frac{\text{seconds}}{\text{cycle}}$

   Frequency $= \frac{1}{0.0053} = 188.679\,\frac{\text{seconds}}{\text{cycle}} = 188.679$ Hz
6. Graph A represents the graph of a lower tone. Lower tones have longer cycles and generate fewer cycles per second than higher tones. Graph A has fewer cycles per second than Graph B.

Practice Problems

1. In triangle ABC, $\angle C$ is a right angle, c measures 13 cm, and a measures 5 cm.

   a) Make a sketch of the triangle. Label the hypotenuse and the legs that are opposite and adjacent $\angle A$.

   b) Find the length of b.

   c) List the sine, cosine, and tangent ratios for $\angle A$.

   d) To the nearest degree, what is the measure of $\angle A$?

2. Find x and y for the right triangle in **Figure 8.69.**

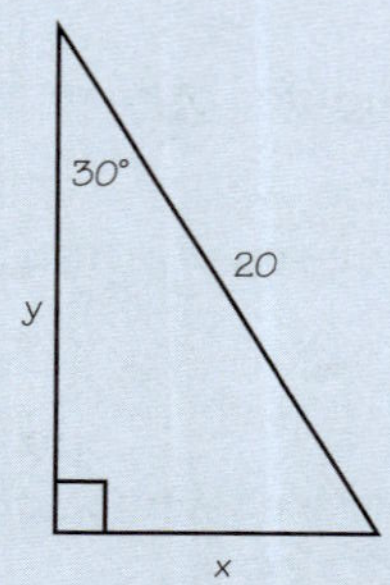

FIGURE 8.69.
Figure for Question 2.

3. Two holes are drilled in a metal plate as shown in **Figure 8.70.**

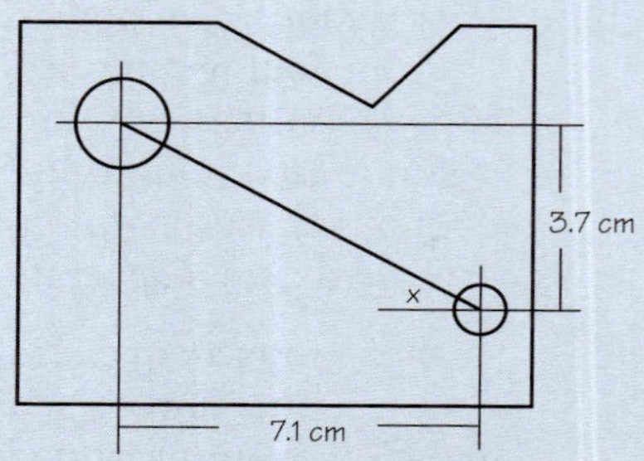

FIGURE 8.70.
Figure for Question 3.

   a) Find the distance between the center of the large hole and the center of the small hole.

   b) Find $\angle x$.

4. A wheelchair ramp makes an angle of 5° with the ground. It rests on the ground at one end and is 4 feet off the ground at the other end.

   a) How long is the ramp?

   b) Federal disability guidelines say that ramps should rise no more than 1 foot for every 12 feet of horizontal distance. Does this ramp meet the guidelines? Explain.

5. Metals and other materials expand and contract with temperature. As a result, railroad tracks must be laid in sections with small gaps between them to avoid buckling due to expansion. Suppose a track 1 mile long is laid in one continuous piece and that its length increases 0.1% in warm weather. If the track buckles straight up at its center, how far off the ground is the middle of the track? Show your work and give your answer to the nearest foot.

6. **Figure 8.71** is a graph of a sinusoidal function.

   a) What is the period?

   b) What is the amplitude?

   c) Find the equation of the axis of oscillation.

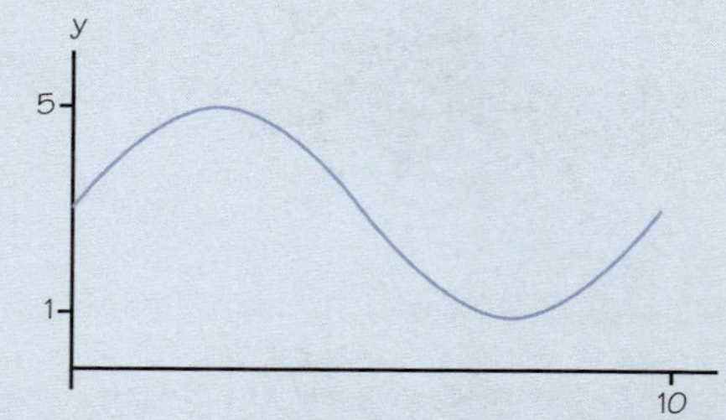

FIGURE 8.71.
Figure for Question 6.

7. A population of gray wolves in Yellowstone Park varies periodically over a year (365 days). A graph of a function that models the relationship between wolf population and time is shown in **Figure 8.72.**

   a) State the graph's period and discuss what it means in this situation.

   b) Give the graph's axis of oscillation and discuss what it means in this situation.

   c) Give the graph's amplitude and discuss what it means in this situation.

   d) When is the wolf population largest, and how many wolves are there at that time?

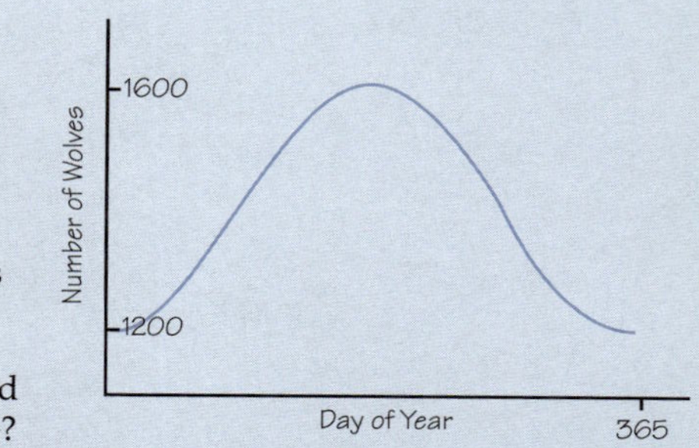

FIGURE 8.72.
Graph for Question 7.

Practice Problems

Answers

1. a) Sample sketch:

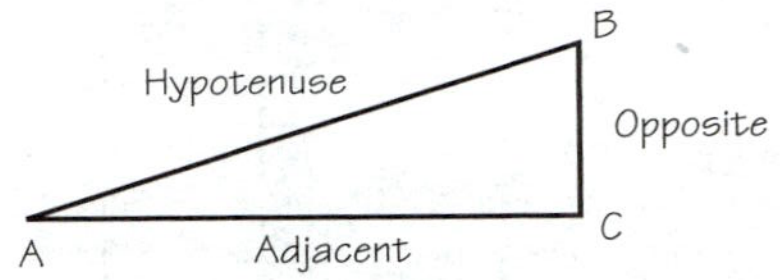

b) $b = \sqrt{13^2 - 5^2} = 12$

c) sin A = 5/13; cos A = 12/13; tan A = 5/12

d) Sample answer: The measure of ∠A is $\sin^{-1}(5/13)$ or approximately 23°.

2. $x = 10$; $y \approx 17.3$

3. a) $d^2 = 3.7^2 + 7.1^2$; $d = 8.0$

b) $\text{Tan}(x) = \frac{3.7}{7.1}$; $\angle x = \tan^{-1}\left(\frac{3.7}{7.1}\right) \approx 27.5°$

4. a) Sample answer:

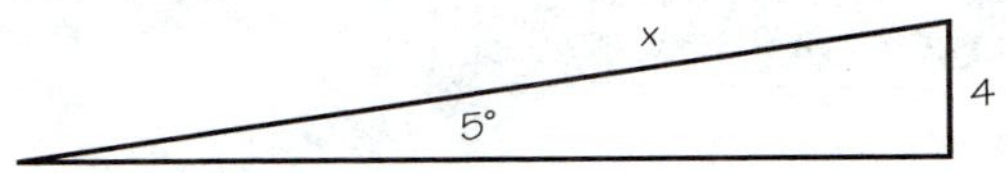

$\sin 5° = \frac{4}{x}$, so $x = \frac{4}{\sin 5°} \approx 46$ feet

b) No. Sample explanation: $\tan^{-1}\left(\frac{1}{12}\right) \approx 4.8°$. The ramp's angle of 5° is slightly more than the guidelines allow.

5. Answer:

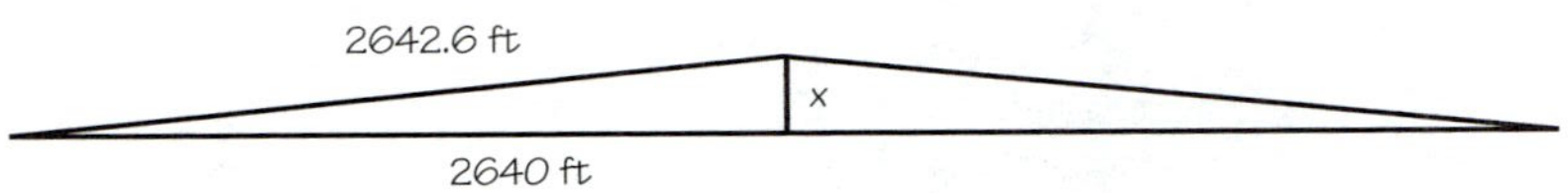

$x = \sqrt{2642.6^2 - 2640^2} \approx 117$ feet!

6. a) 10

b) 2

c) $y = 3$

7. a) The period is 365 days, the number of days in a year, which is the time it takes the population to return to a low of 1200.

b) The axis of oscillation has the equation $y = 1400$. 1400 is the number of wolves in the park around the end of March and the end of September.

c) The amplitude is 200, which is the amount the population rises or falls from its middle value of 1400.

d) There are 1600 wolves around July 1.

8. A group of students do a rolling wheel experiment. They mark a spot on a wheel, roll the wheel along a tape measure, and collect data on the height of the spot. They model the relationship between height of the spot and distance rolled with a sinusoidal function whose graph is shown in **Figure 8.73.**

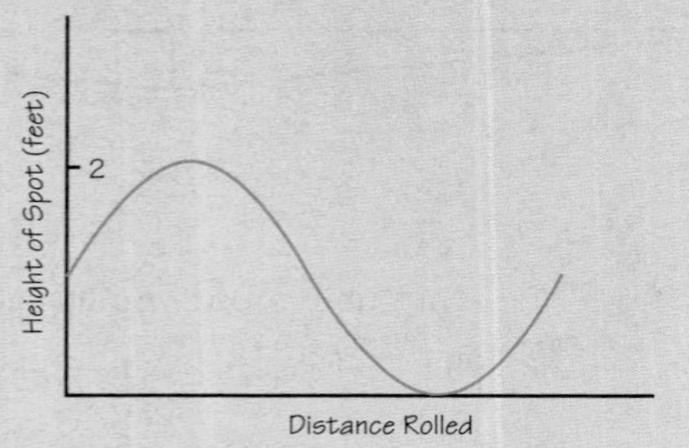

FIGURE 8.73.
Graph for Question 8.

a) What is the diameter of their wheel?

b) What is the wheel's circumference?

9. Ryan is watching his sister Hannah ride a carousel that turns once every 30 seconds. Hannah's distance from Ryan varies between 5 and 25 feet.

a) Sketch a graph of Hannah's distance from Ryan vs. time.

b) Identify your graph's period, amplitude, and axis of oscillation.

10. Tides are examples of periodic events in nature. The table in **Figure 8.74** has data of the depth of water at an ocean beach.

Time (hours)	Depth (inches)
0	7
8	4
16	1
24	7
32	4
40	1
48	7
56	4
64	1
72	7

FIGURE 8.74.
Data table for Question 10.

a) Create a scatter plot for these data.

b) Find a sinusoidal regression model for the relationship between depth and time. Graph the model on your scatter plot.

c) Use your model to predict the water's depth 11 hours after data collection began.

11. A group of students are trying to tune a bottle to a note that has a frequency of 311 cycles per second. They put water in the bottle and blow across it while collecting sound pressure data with a CBL. They estimate the period of the scatter plot to be 0.003. Should they add or remove water from their bottle?

12. **Figure 8.75** shows graphs of sound pressure waves collected from two sources.

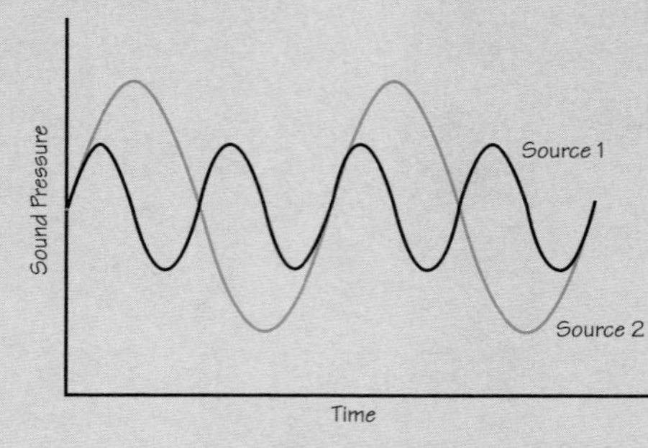

FIGURE 8.75.
Sound pressure graphs.

a) Which source is louder? How do you know?

b) Which source has the highest pitch? How do you know?

Answers (continued)

8. a) 1 foot

   b) $2\pi \approx 6.28$ feet

9. a) Sample sketch:

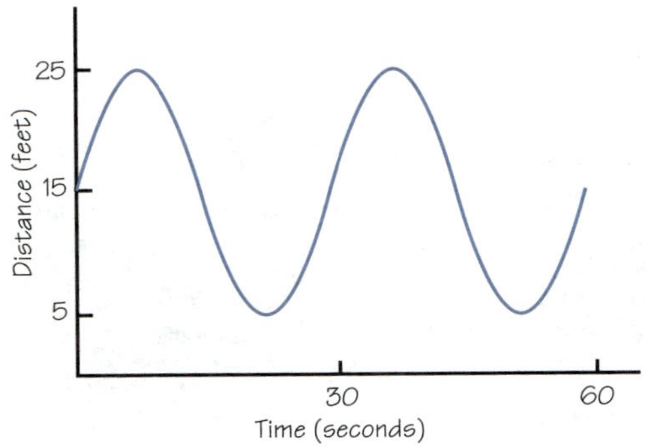

   b) The period is 30 seconds. The amplitude is 10 feet. The axis of oscillation has equation $y = 15$.

10. a) Sample plot:

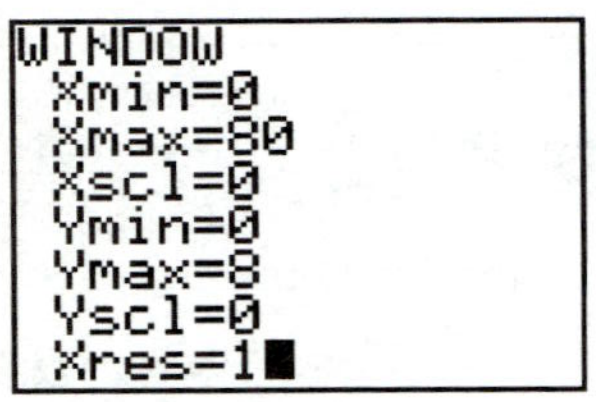

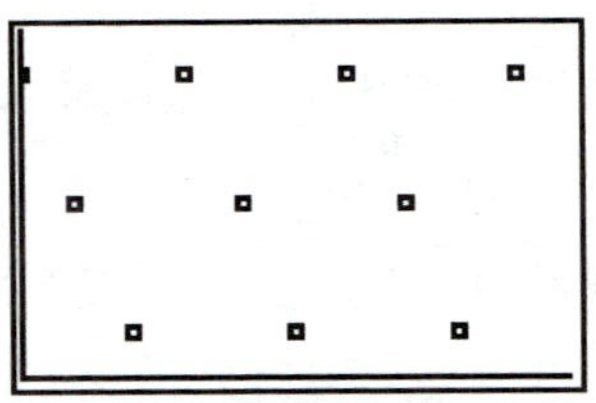

   b)

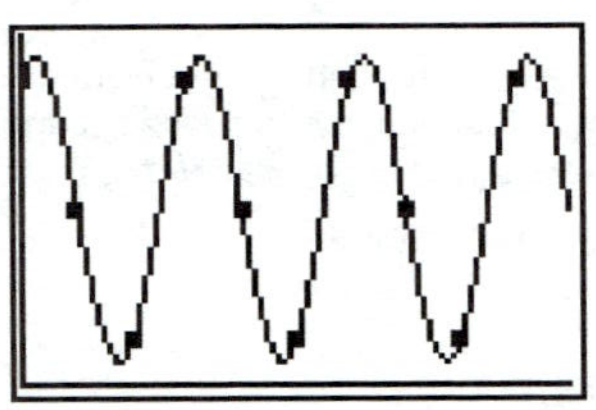

   c)

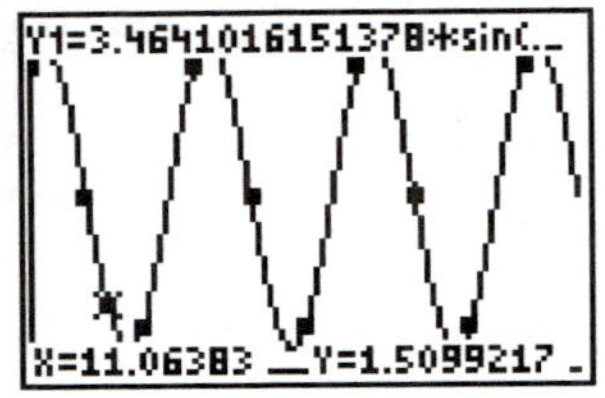

   Approximately 1.5 inches

11. The frequency of their bottle's tone is currently $\frac{1}{0.003} \approx 333$ cycles per second, which is too high. Therefore, they should remove water.

12. a) Source 2 is louder because its graph has the larger amplitude.

    b) Source 1 has the higher pitch because it has the smaller period.

Glossary

KEY CONCEPTS

Amplitude (of a periodic function): Half the vertical height of the basic repeating shape; that is, half of the vertical distance between maximum and minimum values of the periodic graph.

Axis of oscillation (of a periodic function): The center line between a periodic graph's peaks and troughs.

Central angle: An angle whose vertex is the center of a circle.

Cosine (cos) : In a right triangle, the ratio of the length of the side adjacent to an acute angle to the length of the hypotenuse; or $\cos = \frac{\text{adjacent}}{\text{hypotenuse}}$.

Cycle: One complete pattern.

Frequency (sound): The number of cycles a sound wave completes in one second.

Inverse sine: The process of working backwards to find an angle measure whose sine ratio is already known. It is written as $\sin^{-1}$ When you see $\sin^{-1}$ (0.5), it is the angle whose sine is 0.5.

Oscillation or oscillating pattern: Where data points consistently swing back and forth between two alternate extremes.

Period (of a periodic function): The shortest horizontal length of a basic repeating shape.

Periodic function: A function that repeats itself on intervals of a fixed length (equal to the period).

Periodic motion: Motion that follows a predictable back and forth pattern between two extremes. A swinging pendulum is an example of this type of motion.

Pitch: The highness or lowness of a tone.

Sine (sin): In a right triangle, the ratio of the length of the side opposite an acute angle and the length of the hypotenuse; or $\sin = \frac{\text{opposite}}{\text{hypotenuse}}$.

Sine function: A function that describes the relationship between the y-coordinate of a point on the unit circle and a central angle.

Sinusoidal function: Any function that can be expressed in the form $y = A\sin(B(x - C)) + D$, where neither A nor B is zero.

Sinsoidal regression: A method for finding a sinusoidal function for a periodic data set.

Tangent (tan): In a right triangle, the ratio of the length of the side opposite an acute angle and the length of the side adjacent to the acute angle; or $\tan = \frac{\text{opposite}}{\text{adjacent}}$.

Unit circle: Circle with radius one unit and centered at the origin of a rectangular coordinate system.

SPECIAL SECTION

- **PART 1** THE PROBABILITY OF A DEFECT
- **PART 2** HOW MANY FAULTY PUMPS?
- **PART 3** PROBABLE CAUSE

Probability: What Are the Chances?

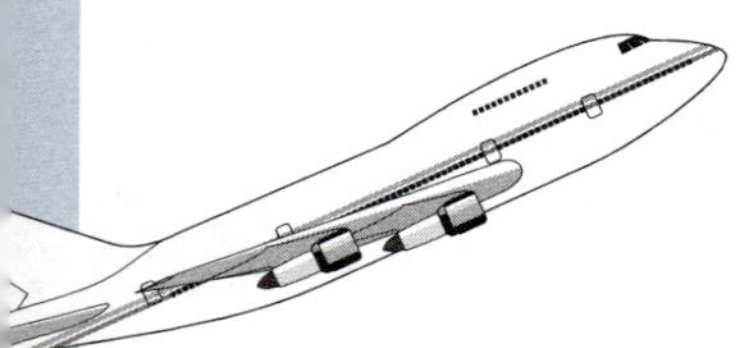

In 2002 Boeing Corporation found that fuel pumps on some airplanes were faulty. A company in California made the pumps for Boeing. The pumps were installed in large commercial jets during January and April. No serious incidents resulted from the faulty pumps, but all the planes that had pumps installed during those months had to be inspected.

The problem was that wires in some pumps were too close to a rotor. Friction with the rotor might create a spark that could cause a fuel explosion. Only a small percentage of the pumps were actually faulty. So most planes that had these pumps were not at risk. But how much risk was there for any particular plane? There are several important questions that could be asked:

- What is the chance that a particular pump had the misplaced wiring?
- If a pump had misplaced wiring, what was the chance of an explosion?
- Some large planes had several fuel tanks and pumps. What was the chance that one of the pumps on such a plane was faulty?

Answers to questions like these are based on **probability**. In this section you will learn to

- find the probability for a single event by doing an experiment.
- compute the probability for a combination of events.
- determine the probable cause of a failure event.

PREPARATION READING PART 1 Determining the Probability

Any company that sells a product or service must be concerned about quality. If a product doesn't work well or has flaws, people will not buy it. Many companies have *quality control* departments. Items produced on assembly lines are inspected on a regular schedule by quality control inspectors or machines.

A manufacturing operation cannot produce perfect items 100% of the time. The inspection process determines what fraction of the total are defective and what fraction are not defective (or nondefective). These fractions also describe the probabilities that a single item is defective or not defective.

You may already know that a probability can sometimes be easily predicted. For example, the probability of spinning the number 1 (or any other number) on the spinner in **Figure 1** is $\frac{1}{4}$. This means that, on the average, we expect one-fourth of all spins to result in the number 1. The probability can also be expressed as a decimal, 0.25, and as a percent, 25%.

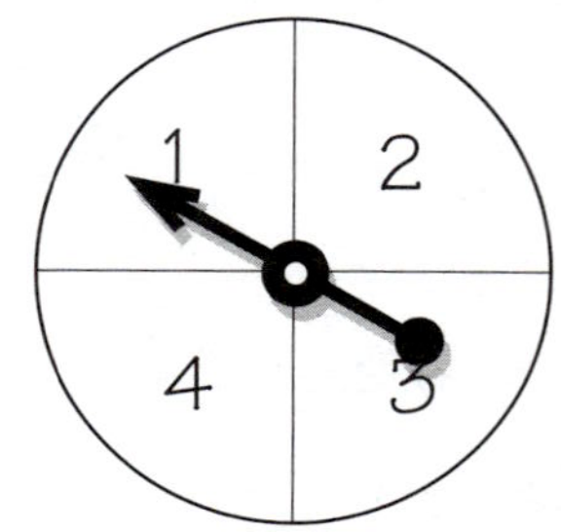

FIGURE 1. Spinner.

THEORETICAL PROBABILITY

The probability that a particular event will occur when an experiment is performed is found by counting all the different possible outcomes of the experiment:

$$\text{Probability of an event} = \frac{\text{Total number of ways the event can occur}}{\text{Total number of equally likely outcomes}}.$$

But sometimes theory and logic are unable to provide an answer. In such cases, only the collection of data can provide the information necessary to determine probabilities.

SPECIAL SECTION

Probability: What Are the Chances?

Mathematics Objectives:

Find experimental probabilities from relative frequencies.

Compute conditional probabilities.

Understand the difference between theoretical probability and experimental probability.

Find the probability of the complement of an event.

Compute joint probabilities.

Use area models to visualize probabilities.

Construct tree diagrams.

Compute posterior probabilities.

Contextual Objectives:

Understand the probability basis of quality control.

Relate product reliability to probability.

Use tree diagrams to find probable cause of failure.

Prerequisites:

Basic probability concepts

Areas of rectangles

Percent

Linegraphs

Technology Note: This module requires the use of a graphing calculator.

Background Reading: *Reliability Evaluation of Engineering Systems: Concepts and Techniques*, 2nd ed., by Billinton and Allan (New York: Plenum Press, 1992) gives a complete introduction to reliability and failure calculations. Some reliability information on simple tasks was taken from *Reliability Maintainability and Risk*, 6th ed., by David J. Smith (Boston: Butterworth Heineman, 2001).

A complete high school level treatment of conditional probability and probability trees in the context of testing for performance-enhancing drugs can be found in Chapter 7 of *Mathematics: Modeling Our World*, Course 1, by COMAP (New York: W.H.Freeman & Co., 1998).

Section Opener: Probability is introduced in the context of defective fuel pumps that were discovered in 2002 on Boeing aircraft. The pumps, made by Hydro-Aire Inc. of California, initially necessitated testing of 1440 jets, but ultimately 30,000 planes had to be inspected. The context is revisited in Part 2.

Preparation Reading Part 1:
Determining the Probability

This sets up the quality control context of the part, and introduces the concept of a defective item. Nondefective is a term used for an item that is not defective, and therefore is the complement of defective. (These days the term non-conforming is often used instead of nondefective.) Throughout this section, the shorthand OK is frequently used in place of nondefective for simplicity, especially in diagrams.

The type of quality control discussed here is referred to in the industry as inspection for attributes, which are categorical properties that can be classified as defective or nondefective. Inspection for variables, on the other hand, involves measurement of some value, such as strength of a metal, time to failure of a fuse, etc.

Students are reminded of some basic probability concepts. It is assumed that they have had some exposure to probability fundamentals in the context of spinners, coin flipping, number cubes, etc. Terms such as experiment, event, outcome, etc., are used in this part without being formally defined. You may wish to discuss their definitions with students if necessary.

PART 1

The Probability of a Defect

Materials: two different brands of jumbo paper clips, graph paper

1. Hold a paper clip, and open it up (see **Figure 2**). Did it break?

FIGURE 2.
An opened paper clip.

Paper clips are not intended to be opened all the way. But suppose that was their purpose. If one breaks, we could say it is defective. A quality control inspection would involve testing several clips to see if they break when opened.

2. A common sample size for inspection is five. Test five clips, and record the number of defective ones.
3. Now repeat your test on many more samples of size five. Record your results in a table similar to **Figure 3.** The third column should contain the total number of defective clips found in all samples up to that point. The fourth column should contain the percent of defective clips up to that point.

FIGURE 3.
Table for Question 3.

Sample Number	Number of Defects in Sample	Cumulative Number of Defects	Cumulative Percent of Defects
1			
2			
3			
etc.			

4. Construct a line graph of the percent of defects versus the total number of clips tested.

As the number of paper clips tested increases, your graph should begin to level off. The height of the graph as it levels off provides an approximate value for the probability that a given clip is defective. This type of probability is called an **experimental probability**.

EXPERIMENTAL PROBABILITY

The experimental probability that a particular event will occur is found by collecting data. The probability that a randomly selected member of a group has a certain property is equal to the fraction of the group that has that property.

$$\text{Experimental probability} = \frac{\text{Number of members with the property}}{\text{Total number of members in the group}}.$$

5. Estimate the probability of a defective paper clip. Express your answer as a percent, a decimal, and a fraction.
6. Now repeat Questions 2–5 for a different brand of paper clip.

PROBABILITY NOTATION

A shorthand notation is often used to denote probabilities. The symbol P(A) stands for "the probability of event A" or "the probability that event A occurs." For example, the probability that a paper clip is defective can be written P(paper clip is defective).

You may have found that the probability of a defective paper clip was different for the two brands you tested. If so, then it is not meaningful to try to identify a single value for P(paper clip is defective). The probability in this case depends on the brand. When the probability of an event depends on whether another event occurs, it is called a **conditional probability**.

CONDITIONAL PROBABILITY

A conditional probability measures the likelihood of an event based on whether another event occurs.

The probability of event A, given that another event B occurs, is symbolized by $P(A \mid B)$.

EXAMPLE 1

Two machines produce optical lenses for cameras (like the one in **Figure 4**). Of 200 lenses made by Machine 1, six are defective. Five of the 250 lenses produced by Machine 2 are defective.

FIGURE 4.
A camera lens.

a) Explain what is meant by P(defective lens | Machine 1), and find its value.

b) Explain what is meant by P(defective lens | Machine 2), and find its value.

Part 1: The Probability of a Defect

Materials: two brands of jumbo paper clips, pennies, pasta, thumbtacks

A somewhat contrived, though inexpensive and easy, experiment to test for defectives is constructed from bending paper clips. This is clearly not normal use for a paper clip, which is what quality control inspections usually test. Small (#1) paper clips can be straightened and bent many times without failure, which is why jumbo clips are suggested here. There is a surprising difference in brittleness among different brands of jumbo paper clips: some will break almost every time, while some have a low breakage rate.

Constructing a graph of relative frequency for accumulating samples should give students an appreciation for sampling error, the natural variability of proportions in a random sample. But they should see the percent of defects level off to a fairly stable value if enough samples are taken. If desired, you may want to have several groups pool their data to identify the defect rate more precisely as an experimental probability.

As an alternative to the paper clip experiment, students could perform either Question 8 (flipped versus spun pennies) or 9 (regular versus wet pasta) as a substitute for Part 1.

Probability Notation

The P(A) notation for the probability of an event, and the P(A | B) notation for conditional probability are introduced. Most students will probably find these convenient shorthands, but some may understand the module better if verbal descriptions are always written out in their entirety.

Independent events are informally defined. But you may wish to discuss with some students, either here or later in the module, that independence can be demonstrated by showing that unconditional and conditional probabilities of an event are equal: If $P(A \mid B) = P(A)$ and $P(B \mid A) = P(B)$, then A and B are independent events.

The Complement of an Event

The notion of complement is introduced here, before any mention of mutually exclusive events, because of its intuitiveness. Note also that the concept of "the sum of the probabilities of all possible outcomes of an experiment is 1" is not explicitly mentioned, although it is implied here with regard to complementary events. Formality is kept to a minimum where possible.

Answers

1. Sample answer: No
2. Sample answer: 1
3. Sample answer:

Sample Number	Number of Defects in Sample	Cumulative Number of Defects	Cumulative Percent of Defects
1	1	1	20%
2	0	1	10%
3	1	2	13%
4	1	3	15%
5	0	3	12%
6	2	5	17%
7	0	5	14%
8	0	5	13%
9	1	6	13%
10	0	6	12%

4. Sample answer:

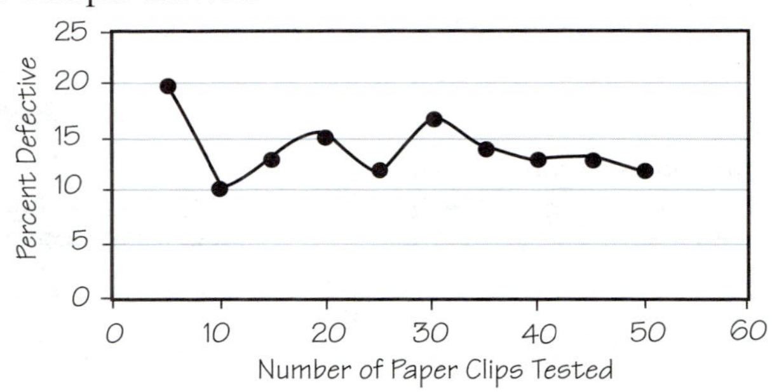

5. Sample answer: 13%, 0.13, 13/100
6. Sample answer:

Sample Number	Number of Defects in Sample	Cumulative Number of Defects	Cumulative Percent of Defects
1	2	2	40%
2	4	6	60%
3	2	8	53%
4	3	11	55%
5	1	12	48%
6	3	15	50%
7	2	17	49%
8	3	20	50%
9	4	24	53%
10	3	27	54%

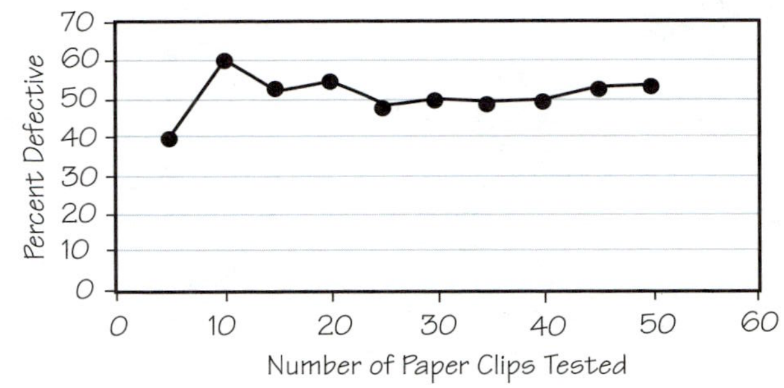

Probability of a defective paper clip is about 54%, or 0.54, or $\frac{27}{50}$.

SOLUTION:

a) P(defective lens | Machine 1) means the probability that a lens is defective if it was made by Machine 1. Its value is $\frac{6}{200} = 0.03$, or 3%.

b) P(defective lens | Machine 2) means the probability that a lens is defective if it was made by Machine 2. Its value is $\frac{5}{250} = 0.02$, or 2%.

Two events are called **independent** if the probability of one does not depend on the occurrence of the other. For example, we expect that when a coin is tossed there is a 50% chance that it will land heads up. The type of coin used should not affect the outcome. That is, P(heads | coin is a penny) = P(heads | coin is a dime) = P(heads | coin is a quarter) = 50%, or $\frac{1}{2}$. Whether a coin lands heads or tails is independent of the type of coin used.

THE COMPLEMENT OF AN EVENT

An **event** is a particular outcome of an experiment or process. The **complement** of the event is all *other* possible outcomes. Here are a few examples:

- The complement of heads is tails for the tossing of a coin.
- The complement of winning a game is tying or losing.
- The complement of the door is open is the door is closed.
- The complement of red is green or yellow for a traffic light.
- The complement of even is odd for the rolling of a six-sided die.
- The complement of one or two is three, four, five, or six for the rolling of a six-sided die.

CHECK THIS!

The sum of the probabilities of an event and its complement is 1, or 100%

EXAMPLE 2

A manufacturing process makes circuit boards for computers (like the one in **Figure 5**). The probability that a circuit board made by this process is defective is 0.001. What is the probability that a circuit board is *not* defective?

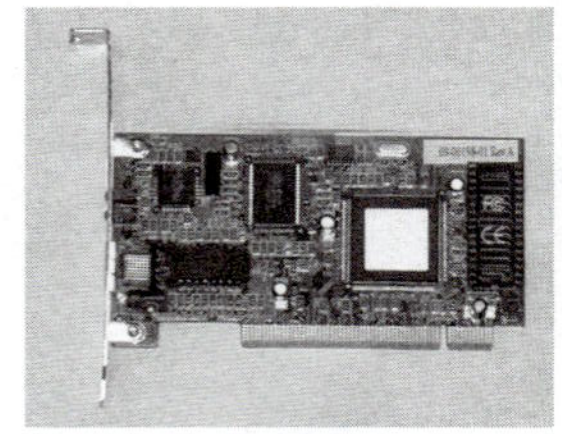

FIGURE 5. A circuit board.

SOLUTION:

A circuit board is either defective or it is not defective. So nondefective is the complement of defective.

P(nondefective) = 1 – P(defective) = 1 – 0.001 = 0.999.

PART 1

Assignment

1. A parts cabinet in an electronics laboratory has a drawer that contains resistors of various sizes, all mixed together. The drawer's contents include 40 one hundred ohm resistors, 20 two hundred ohm resistors, and 30 five hundred ohm resistors. If a student reaches into the drawer and randomly picks one, what is the probability that it will be a one hundred ohm resistor?

2. The median of a numerical data set is the middle number of the set when it is arranged in numerical order. In a large data set, what is the probability of randomly choosing a value that is below the median?

3. Veteran's Day is one of the few non-religious holidays that is celebrated on its original date (November 11), rather than always being on a Monday. In any arbitrary year, what is the probability that Veteran's Day will be on a weekend?

4. If you randomly guess at the answer to a multiple choice test question for which there are five choices, what is the probability that you will guess correctly?

5. According to a NASA space scientist, a city could be destroyed once every 30,000 years by an asteroid hitting the Earth. If he is correct, what is the probability of a city being destroyed by an asteroid in any year?

6. When the euro coin was issued, a German newspaper reported that a coin-flipping experiment with the Belgian one-euro coin showed that heads came up 140 times out of 250. Because of this, a ban on using euros at the start of soccer matches was considered. According to the experiment, what is the probability that a flipped euro results in heads?

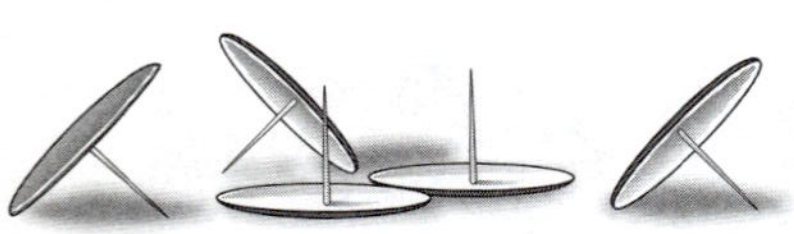

7. Design and conduct an experiment to determine the experimental probability that an ordinary thumbtack will land with its point facing up when dropped on a horizontal surface.

8. When a penny is tossed, we expect that the probabilities of heads and tails are equal, at 50% each. And if the coin is actually tossed many times, the observed fraction of heads (or tails) should approach 0.5. Will the same results hold true if the penny is spun instead of tossed? Is there any reason to think that the fraction of heads might not be 50% in this case? You can experiment to find out.

   a) Use one finger to hold a penny vertically so its edge touches a table. "Flick" the penny with a finger of your other hand so that it spins. Notice which side lands face up. Repeat for a total of at least 100 spins and record your results. How many times did the penny land with heads face up?

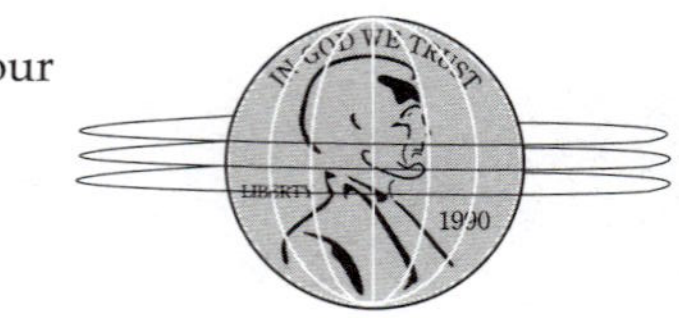

   b) Based on your experiment, what is the conditional probability that a penny will land heads up if it is spun?

   c) What is the conditional probability that a penny will land heads up if it is flipped? See if you can verify your answer experimentally.

9. Some materials must be flexible in order to perform correctly under normal use. For example, the steel in a bridge must be flexible enough to bend, but not break, when subjected to stresses. The battery compartment to a calculator has a flexible plastic clasp that can be pressed to open it.

   a) You can perform a flexibility test on a piece of pasta. Press the pasta firmly onto a table top so that 4 inches hang over the edge of the table. Press the free end of the pasta down until it is 3 inches below the edge of the table. Did it break?

   b) Repeat the experiment for many pieces of pasta. How many pieces of pasta can be bent downward 3 inches without breaking?

   c) Now repeat the experiment for similar pasta that has gotten wet but has been allowed to dry out.

   d) What is the conditional probability that a piece of pasta can be bent downward 3 inches without breaking if it has gotten wet?

Part 1 Assignment

Teaching Ideas: Questions 1–6 provide practice with computing elementary probabilities, both experimental and theoretical.

Question 7 is an activity to find the experimental probability of a dropped thumbtack landing with its point up. It is helpful for students to construct a graph of relative frequency for increasing numbers of 10- or 20-tack samples (dropping 10 or 20 tacks at a time), as in Part 1. If you'd prefer not to have students dropping loose tacks in the classroom, you can place a group of tacks in a sealed clear plastic box, and have students shake the box to conduct their sampling.

Question 8 is similar to Part 1, as it compares probabilities of a coin landing heads for two different conditions. Although all students know the theoretical probability of a flipped coin landing heads, it may be instructive to have them find the experimental probability as well, and compare it with that of the spun penny.

Question 9 is another activity, and is perhaps a better simulation of a conditional probability experiment, albeit a bit trickier to carry out. You'll need to experiment with some pasta yourself to see what test is appropriate for your situation. The bending test works well with fettucini, which tends to be more brittle and to break more cleanly than, say, spaghetti. To prepare the second batch of pasta, simply submerge it in water for less than 5 minutes and then let it dry. (It will tend to curl unless you hold down the ends.) Pasta that has been exposed to water and then dried is much more brittle and less flexible than pasta out of the box. You may wish to conduct a test of ability to support a fixed load, instead of flexibility.

Questions 10–17 involve practice with conditional probability and complements. Question 10 includes questions about independence.

Questions 18–21 introduce the concept of product reliability, which can be thought of as quality control with a time component. Reliability is defined as the probability that an item will perform a required function, under stated conditions, for a stated period of time. One speaks of the "100-hour reliability," or the "one-million-hour reliability," or the "one-year reliability," of a product. Alternatively, a set of tires might be described in terms of "50,000-mile reliability." Or, especially in military usage, the term "mission reliability" describes the probability that a piece of equipment will perform correctly throughout the duration of a mission.

Reliability is quantified in terms of failure rate. For simple items and tasks, with failure rate per unit time expressed by λ, the reliability for time t is given by the formula $R = e^{-\lambda t}$.

Answers

1. $40/90 = 4/9 \approx 0.444 = 44.4\%$
2. $1/2 = 0.5 = 50\%$, since half the numbers are below the median.
3. $2/7 \approx 0.222 = 22.2\%$
4. $1/5 = 0.2 = 20\%$
5. $\frac{1}{30,000} \approx 0.00003$
6. 0.56 or 56%
7. Sample answer: 100 thumbtacks are dropped, and 62 landed point up.
   P(tack lands point up) = 0.62 or 62%
8. a) Sample answer: 39

   b) Sample answer: P(heads | spun) = 0.39

   c) Theoretical probability is P(heads | flipped) = 0.50. Experiment should result in a fraction very close to 0.50 if the penny is flipped enough times.
9. a) Sample answer: No

   b) Sample answer: 46 out of 50

   c) Sample answer: 9 out of 50 bend without breaking.

   d) Sample answer: P(bends | wet) = 9/50 = 0.18 or 18%

e) What is the conditional probability that a piece of pasta can be bent downward 3 inches without breaking if it has *not* gotten wet?

10. Air travelers occasionally find that their luggage is missing when they arrive at their destination.

a) In May 2004, there were 199,788 passengers on major airlines in the United States whose checked baggage was missing. The total number of passengers on these airlines was 48,368,609. What was the probability that any one passenger's luggage was lost on a flight?

b) Do you think that a traveler's chance of losing his or her luggage is independent of the choice of airline? What do you need to know in order to decide?

c) **Figure 6** shows May 2004 data on lost luggage for major airlines.

Airline	Number of Passengers	Number of Reports of Lost Luggage
JetBlue Airways	971,841	2115
Hawaiian Airlines	460,080	1304
Airtran Airways	1,136,947	3265
America West Airlines	1,741,982	5122
Southwest Airlines	7,321,279	22,473
Alaska Airlines	1,217,055	3801
Delta Air Lines	7,407,135	23,597
ATA Airlines	839,193	2708
Continental Airlines	2,786,245	9497
United Airlines	5,408,451	19,302
US Airways	3,202,200	12,883
Northwest Airlines	3,393,330	13,852
American Airlines	6,399,856	27,199
Expressjet Airlines	1,067,563	5815
Skywest Airlines	1,069,487	8487
American Eagle Airlines	1,267,581	10,618
Comair	1,110,230	9874
Atlantic Southeast Airlines	877,237	9601
Atlantic Coast Airlines	690,917	8275

FIGURE 6. Data are from the July 2004 "Air Travel Consumer Report" of the Office of Aviation Enforcement and Procedings, Aviation Consumer Protection Division.

Find the conditional probabilities of losing luggage for travel on JetBlue Airways and for travel on Atlantic Coast Airlines. Round your answers to four decimal places.

d) Do you think whether an airline passenger lost luggage in May 2004 was independent of whether he or she flew on JetBlue or Atlantic Coast? Explain.

e) Do you think whether an airline passenger lost luggage in May 2004 was independent of whether he or she flew on Delta or ATA? Explain.

11. In 2000, for every 1000 licensed 16-year-old girls in the U.S., 175 got in car accidents. That same year, 210 of every 1000 licensed 16-year-old boys got in accidents.

a) Find the conditional probability that a licensed 16-year-old got into a car accident in 2000 if the driver was a girl.

b) Find the conditional probability that a licensed 16-year-old got into a car accident in 2000 if the driver was a boy.

12. A study of poor African-American children who grew up in Ypsilanti, Michigan in the 1960s looked at the effects of attending preschool. The study found that 71% of the children who went to preschool earned a high school diploma or GED. Of those who didn't go to preschool, 54% earned a high school diploma or GED. Use conditional probability notation to express these results.

13. The following probability statements involve late flights from U.S. airports during 2002. Write a sentence to explain each one.

a) $P(\text{late departure} \mid \text{Detroit}) = 0.20$

b) $P(\text{late departure} \mid \text{Boston}) = 0.14$

14. United Airlines had the best on-time arrival record of the major U.S. airlines in 2002. The probability of a United flight being on time was 0.84. What was the probability that a United flight was *not* on time in 2002?

15. An injection molder is used to make plastic bicycle reflectors. If the probability of a defective reflector is 0.13, what is the probability that a reflector is not defective?

Answers (continued)

e) Sample answer: P(bends | not wet) = 46/50 = 0.92 or 92%

10. a) about 0.004 or 0.4%

    b) Sample answer: Yes, but you would need data on lost luggage for different airlines.

    c) P(luggage lost | traveled on JetBlue Airways) = 0.0022

    P(luggage lost | traveled on Atlantic Coast Airlines) = 0.0120

    d) Sample answer: No, because passengers on Atlantic Coast were about 6 times as likely to lose their luggage as passengers on JetBlue.

    e) Sample answer: Yes, because the chance of losing luggage was almost exactly the same on both airlines, about 0.0032 or 0.32%.

11. a) 0.175 or 17.5%

    b) 0.210 or 21.0%

12. Sample answer: P(graduated high school | attended preschool) = 0.71; P(graduated high school | didn't attend preschool) = 0.54

13. a) Sample answer: The probability that a plane departed late if it took off from Detroit was 0.20.

    b) Sample answer: The probability that a plane departed late if it took off from Boston was 0.14.

14. 0.16 or 16%

15. 0.87 or 87%

16. Many companies use automated telephone voice systems to provide their customers with account information. A study was made of such systems at Massachusetts utility companies in 2003. For one gas company, the probability that the automated system provided a caller with the desired information was 0.932. What was the probability that the system failed to provide the desired information?

17. Write your results from Part 1 using conditional probability notation.

Questions 18–21 involve the concept of product *reliability*. The reliability of a product is defined as the probability that it will perform its intended function correctly for a specified time interval, under normal conditions of use.

18. A control valve for a water feed line has a one-year reliability of 0.92. What is the probability that the valve will fail sometime during the year?

19. The probability that a hair-dryer's blower motor will fail during its first year of use is 21%. What is the one-year reliability for the blower motor?

20. The probability that a typical thermocouple will fail during 1000 hours of use is 1%. What is the 1000-hour reliability of such a thermocouple?

21. The 100-hour reliability of a steam turbine is 0.996. What is the probability that the steam turbine will fail within 100 hours?

PREPARATION READING PART 2

Airline Pump Failures

The airplane fuel pump problem mentioned at the beginning of this section required inspection of 30,000 pumps. The initial study suggested that only 3% of the pumps may have had faulty wiring. So there was a 3%, or 0.03, probability that any one pump was faulty.

But many planes had more than one pump. Large planes like the Boeing 747 have fuel tanks in the fuselage and also in the wings, and sometimes in the tail section. And each tank may have several pumps. In fact, one plane can have as many as 17 or 18 fuel pumps.

If the probability of one pump being faulty is known, how can we find the probability that more than one pump in a plane might be faulty? If a plane has several pumps, what are the chances that all are faulty, or that none are faulty? To explore these questions, we will first consider a situation in which a plane has two pumps. And to make it easier to visualize, we will assume a high defect rate.

Answers (continued)

16. 0.068 or 6.8%
17. Sample answer:
    $P(\text{defective paper clip} \mid \text{Brand A}) = 0.13$;
    $P(\text{defective paper clip} \mid \text{Brand B}) = 0.52$
18. 0.08 or 8%
19. 0.79 or 79%
20. 0.99 or 99%
21. 0.004 or 0.4%

Preparation Reading Part 2:
Airline Pump Failures

The Boeing fuel pump problem is revisited in the context of a plane with multiple fuel pumps. This sets up consideration of joint probabilities.

PART 2

How Many Faulty Pumps?

Materials: graph paper, ruler

Imagine that a pump manufacturer has such poor quality control that one-fourth of its pumps are defective. Also imagine that a certain type of airplane is fitted with two fuel pumps. Such a plane might have no defective pumps. But it's possible that just one or even both pumps might be defective. You can use an area model to see the combined probabilities.

1. Draw a square to represent a total probability of 1, or 100%. Probabilities for all possible pump combinations will be contained in the square. Draw a vertical line that divides the square into two rectangles. One rectangle should contain $\frac{1}{4}$ of the total area. The other should contain $\frac{3}{4}$ of the total area. These rectangles represent the probabilities that Pump #1 is either defective or OK. Write appropriate labels along the top edge of the square. (Include the probability values.)

2. Now draw a horizontal line that divides the left edge of the square in a similar way. Write labels along the left edge to show the probabilities that Pump #2 is either defective or OK.

3. The square is now divided into four regions. The smallest one represents the event that both pumps are defective. Label each region according to the combination of pump conditions it represents.

4. Compute the areas of the regions and complete **Figure 7.**

State of the System	Probability
Both pumps defective	
Pump #1 defective, Pump #2 OK	
Pump #1 OK, Pump #2 defective	
Both pumps OK	

FIGURE 7.
Table for Question 4.

5. The probabilities in the table are called *joint probabilities*. Each of them is a combination of the probabilities of two events. In each case, the state of the system depends on two separate events *both* occurring. Look at the way you computed each joint probability. Then write a rule for finding the probability that two events both occur.

6. Suppose an airline flies 80 aircraft of this type. Use your answer to Question 4 to help you complete **Figure 8.** Fill in the number of airplanes in each category.

		Fuel Pump #1	
		OK	Defective
Fuel Pump #2	OK		
	Defective		

FIGURE 8.
Table for Question 6.

7. How many of the 80 planes have only one defective fuel pump?

8. What is the probability that a plane has exactly one defective pump?

9. Explain how you could use the information in your answer to Question 4 to find the probability that a plane has exactly one defective pump.

INDEPENDENT AND DEPENDENT EVENTS

In Part 2 you found that the joint probability that both fuel pumps fail could be found by multiplying the failure probabilities for the individual pumps. In fact, even for a plane that contains more than two fuel pumps, the probability that all of them fail is equal to the product of the individual failure probabilities.

But this assumes that fuel pump failures are independent events. That is, the probability that one of the pumps fails does not depend on whether the other pump fails. For example, if failure of Pump #1 made Pump #2 more (or less) likely to fail, the events would be *dependent*. In such a case, a conditional probability would have to be used for the failure probability of Pump #2. And the probability that both pumps fail would be found from

P(both pumps fail) = P(Pump #1 fails) × P(Pump #2 fails | Pump #1 fails).

Part 2: How Many Faulty Pumps?

Materials: graph paper, ruler

Joint probabilities for two independent events are examined using an area model, in which probabilities are proportional to rectangular areas. Many students are helped by visualizing joint probabilities this way. Those who are should be encouraged to use area models in the questions when appropriate. (Of course, this type of model is not feasible for joint probabilities of more than two events.)

NOTE: The defect rates used here are unrealistically high in order to make the area models legible. The actual Boeing pump data are addressed in Question 22 of the assignment.

Students are led to discover the multiplication rule for the probability that two independent events both occur. They then interpret the probabilities in concrete terms using a hypothetical sample.

The section ends with a brief look at the addition rule for mutually exclusive events, which is then formalized somewhat afterwards.

NOTE: Students should be continually made aware of the different ways of expressing probabilities: as fractions, as decimals, and as percents.

Independent and Dependent Events

The concept of independence is extended here. Example 3 illustrates, using the same context, the use of the multiplication rule for independent (a and b) and dependent (c) events. Note, however, that in c we must assume that the father does not have allergies. If both parents have allergies, the probability that a child has allergies rises to about 70%, according to current research. You could use this fact to extend the example.

Adding Probabilities

Joint probabilities for any one of a number of mutually exclusive events are discussed, and illustrated with an example. Joint probabilities for non-mutually exclusive events are not explicitly mentioned. However, area models and tree diagrams (see Question 23) can be used to analyze some non-mutually exclusive event situations.

Answers

1. Sample answer:

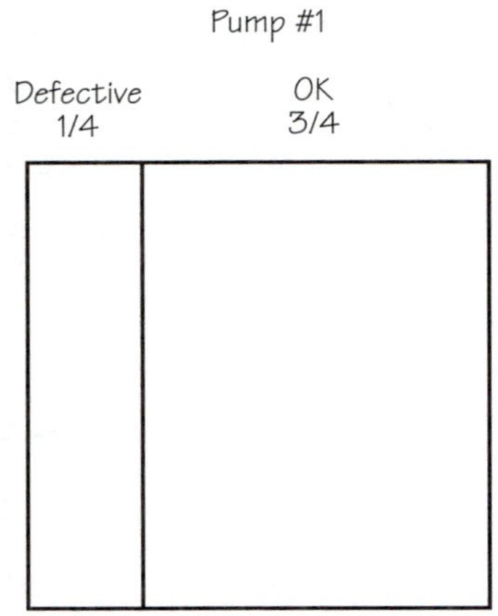

2. Sample answer:

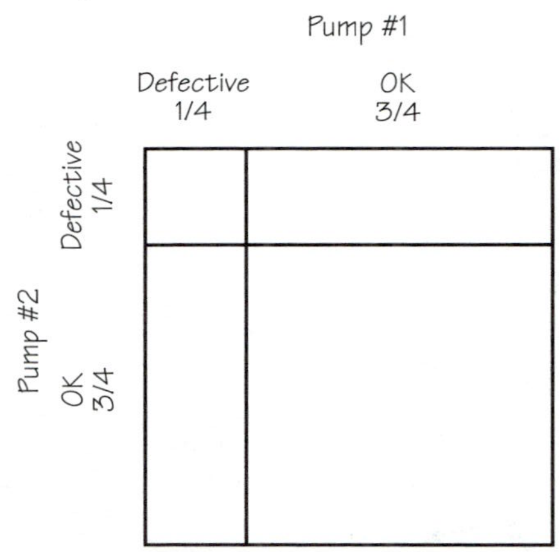

3. Sample answer:

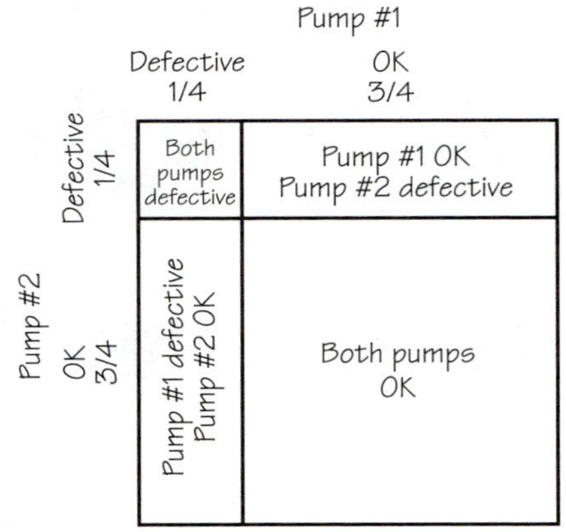

4.

State of the System	Probability
Both pumps defective	1/16
Pump #1 defective, Pump #2 OK	3/16
Pump #1 OK, Pump #2 defective	3/16
Both pumps OK	9/16

5. Sample answer: To find the probability that two events both occur, multiply the probabilities of the events.

6.

		Fuel Pump #1	
		OK	Defective
Fuel Pump #2	OK	5	15
	Defective	15	45

7. 30

8. $30/80 = 3/8$

9. Sample answer: Add the probabilities for the two situations that have one defective pump and one OK pump.

EXAMPLE 3

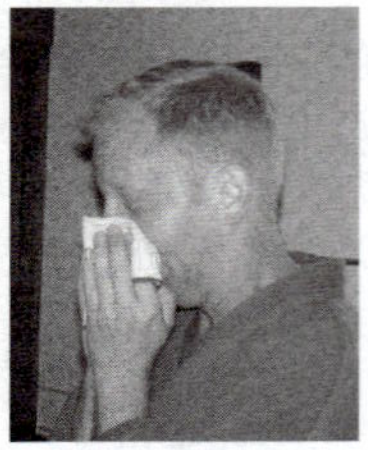

According to the Asthma and Allergy Foundation of America, one-fifth of all Americans have allergies of some kind. But if a person has one parent with allergies, that person has a one-third chance of being allergic.

a) What is the probability that two randomly selected Americans have allergies?

b) What is the probability that four randomly selected Americans all have allergies?

c) What is the probability that a mother and child both have allergies?

SOLUTION:

a) If the two people are randomly selected, then we can assume that whether each has allergies or not are independent events. That is, the probability that one person is allergic does not depend on whether the other is allergic. So the joint probability that they both have allergies is

P(first person has allergies) × P(second person has allergies) = $\left(\frac{1}{5}\right)\left(\frac{1}{5}\right) = \frac{1}{25}$ or 4%.

b) Assume independent events:

P(four people have allergies) = $\left(\frac{1}{5}\right)\left(\frac{1}{5}\right)\left(\frac{1}{5}\right)\left(\frac{1}{5}\right) = \frac{1}{625}$ or 0.16%.

c) These are dependent events. A child is more likely to have allergies if his or her mother has allergies.

P(mother and child have allergies)

= P(mother has allergies) × P(child has allergies if mother has allergies)

$= \left(\frac{1}{5}\right)\left(\frac{1}{3}\right)$

$= \frac{1}{15}$ or about 6.7%.

EXAMPLE 4

Studies have been made of human error rates for simple tasks. For example, a computer technician will insert a printed circuit board incorrectly about 0.4% of the time. If a technician has to replace 20 circuit boards, what is the probability that at least one of the 20 will be incorrectly inserted?

SOLUTION:

The events "at least one of 20 incorrectly inserted" and "all 20 correctly inserted" are complementary. And if there is a 0.4% chance of incorrect insertion, there is a 99.6% chance of correct insertion. So

P(at least one of 20 incorrectly inserted) = 1 – P(all 20 correctly inserted)

= 1 – P(one correctly inserted)^{20}

$= 1 - (0.996)^{20}$

$\approx 1 - 0.923$

= 0.077 or 7.7%.

In this solution it is assumed that correct insertions of the 20 boards are all independent.

ADDING PROBABILITIES

In Question 9 of Part 2, you found that the chance of only one pump being defective is equal to the sum of two probabilities. In many cases, the joint probability that one or the other of two events occurs is equal to the sum of their individual probabilities. But this is only true if the events *cannot both occur*. Such events are called **mutually exclusive** events.

In general, the probability that any one of many events occurs is equal to the sum of their individual probabilities, provided they are all mutually exclusive.

Number of Days	Probability
5	0.02
6	0.04
7	0.11
8	0.22
9	0.25
10	0.19
11	0.13
12	0.04

FIGURE 9.
Slate roof installation.

EXAMPLE 5

A project scheduler for a roofing company has data on the number of days it takes to install a slate roof on a certain type of building. (See **Figure 9.**)

What is the probability that such a roof will take more than 9 days to install?

SOLUTION:

Only one length of time is possible for a particular installation. For example, it is impossible that a particular roof installation will take both 10 days *and* 11 days. All of the possibilities in the table are mutually exclusive.

P(more than 9 days) = P(10 days) + P(11 days) + P(12 days)

= 0.19 + 0.13 + 0.04

= 0.36

CHECK THIS!

Rules for Combining Probabilities

P(A and B) = P(A) × P(B | A) But if events A and B are independent, P(A and B) = P(A) × P(B). In general, to find the joint probability that any number of independent events all occur, multiply their individual probabilities.

P(A or B) = P(A) + P(B) only if events A and B are mutually exclusive. In general, to find the probability that any one of a number of mutually exclusive events occurs, add their individual probabilities.

PART 2 Assignment

1. **Figure 10** shows data on the performance of two lathes in a machine shop.

	Total Number of Pieces Produced	Number of Defective Pieces Produced
Lathe #1	500	40
Lathe #2	300	30

FIGURE 10.
Lathe data.

a) Find the probability that a randomly selected piece was produced on Lathe #1.

b) Find the probability that a piece produced on Lathe #1 was defective.

c) Find the probability that a randomly selected piece was produced on Lathe #1 and was also defective.

d) Repeat a–c for pieces produced on Lathe #2.

e) Are the choice of lathe and the occurrence of a defective piece independent or dependent?

2. **Figure 11** shows data on items made on two assembly lines in a manufacturing plant.

	Total Number of Pieces Produced	Number of Defective Pieces Produced
Line #1	7520	188
Line #2	10,680	267

FIGURE 11.
Assembly line data.

a) Find the probability that a randomly selected item was produced on Line #1.

b) Find the probability that an item produced on Line #1 was defective.

c) Find the probability that a randomly selected item was produced on Line #1 and was also defective.

d) Repeat a–c for items produced on Line #2.

Part 2 Assignment

Teaching Ideas: Questions 1–10 provide practice with the multiplication rule. Questions 1 and 2 stress the distinction between independent events and dependent events. (Question 10 also includes dependent events.)

Questions 11–14 involve the addition rule. Questions 11 and 12 stress the concept of mutually exclusive events.

Questions 15–22 are somewhat more challenging. Many involve the use of complements as well as the multiplication rule. Several also provide practice with some of the language often used in the context of probability. The interpretation of phrases like "at least" can be difficult for students. Suggest that students write out the actual numbers of events involved: in Question 17, "at least one" means "one or two" in this context. And in several cases, "at least one" must be recognized as the complement of "none."

In Question 21a, the actual answer is approximately 1.3×10^{-26}. Students may not be familiar with numbers like this, and especially with the relevant calculator notation, so the answer is given as "almost zero."

Questions 23 and 24 introduce the use of tree diagrams purely for identification and enumeration of outcomes. Trees will be used for a different purpose in Part 3.

Answers

1. a) P(Lathe #1) = 500/800 = 0.625

   b) P(defective | Lathe #1) = 40/500 = 0.08

   c) P(defective and Lathe #1) = P(Lathe #1) · P(defective | Lathe #1) = (0.625)(0.08) = 0.05

   d) P(Lathe #2) = 300/800 = 0.375

   P(defective | Lathe #2) = 30/300 = 0.10

   P(defective and Lathe #2) = P(Lathe #2) · P(defective | Lathe #2) = (0.375)(0.10) = 0.0375

   e) Dependent

2. a) P(Line #1) = 7520/18,200 ≈ 0.413

   b) P(defective | Line #1) = 188/7520 = 0.025

   c) P(defective and Line #1) = P(Line #1) · P(defective | Line #1) = (0.413)(0.025) ≈ 0.010

   d) P(Line #2) = 10,680/18,200 ≈ 0.587

   P(defective | Line #2) = 267/10,680 = 0.025

   P(defective and Line #2) = P(Line #2) · P(defective | Line #2) = (0.587)(0.025) ≈ 0.0147

e) Are the choice of assembly line and the occurrence of a defective item independent or dependent?

3. On January 28, 1986, the space shuttle *Challenger* exploded over Florida. Many authorities believe this tragedy could have been prevented if closer attention had been paid to the laws of probability. The rocket that carried the shuttle aloft was separated into sections that were sealed by large rubber O-rings. Experts believe that at least one of the O-rings leaked burning gases that caused the rocket to explode. Studies after the tragedy found that the probability a single O-ring would work was 0.977. If the six O-rings were truly independent of one another, what is the probability that all six would function properly during the mission?

4. A particular structural beam in a building has a 0.02 probability of failure when subjected to a load of a certain critical size. If the probability of such a load occurring during the design life of the building is one in 100, what is the probability of beam failure during the building's design life?

For Questions 5–9, assume that events are independent.

5. In 2002, 85% of Massachusetts high school students passed the MCAS mathematics test on their first attempt. Of four randomly selected test-takers, what is the probability that all passed the test on their first attempt?

6. One type of computer modem has a one-year reliability of 98%. What is the probability that five such modems will all work correctly for one year?

7. A radio receiver consists of five assemblies. The 100-hour reliability of each is given in **Figure 12.**

   Each component assembly must function correctly for the receiver to work. (This kind of system is called a "series" system.) What is the 100-hour reliability of the receiver?

8. A simple electronic circuit consists of 4 transistors, 3 capacitors, 5 resistors, and 1 switch. The reliabilities for each type of component are listed in **Figure 13.**

   All components must operate for the circuit to work. What is the 1000-hour reliability for the circuit?

Assembly	Reliability
Power supply	0.892
rf stage	0.943
Mixer	0.961
if stage	0.907
Power output	0.939

FIGURE 12.
Data for Question 7.

Component	1000-hr Reliability
Transistor	0.9990
Capacitor	0.9998
Resistor	0.9950
Switch	0.9980

FIGURE 13.
Data for Question 8.

9. Four electronic components are wired in series. Their 100-hour reliabilities are 0.98, 0.995, 0.96, and 0.975. What is the 100-hour reliability of the system; that is, what is the probability that all components will work for 100 hours?

10. In Part 1 you compared the quality of two brands of paper clips. For that type of test, suppose 20% of the Brand A paper clips and 45% of Brand B are defective. Also suppose 100 Brand A clips and 150 Brand B clips are mixed together. A randomly selected paper clip is tested.

    a) What is the probability that the selected paper clip is Brand A and it breaks?

    b) What is the probability that the selected paper clip is Brand B and it breaks?

    c) What is the total probability that the selected paper clip breaks?

11. **Figure 14** shows the U.S. car production shares of major companies for the first half of 2004.

    a) What is the probability that a randomly selected car produced in the U.S. in the first half of 2004 was made by Toyota?

    b) What is the probability that a randomly selected car produced in the U.S. in the first half of 2004 was made either by Ford or General Motors?

    c) Are the events "made by Ford" and "made by General Motors" mutually exclusive? Explain.

12. **Figure 15** shows data on U.S. traffic fatalities for 2002.

    a) What is the probability that a randomly selected traffic fatality was either under 16 or over 54 years old?

    b) Are the events "person is under 16" and "person is over 54" mutually exclusive? Explain.

    c) Are the events "person is under 25 years old" and "person is from 16 to 24 years old" mutually exclusive? Explain.

Company	Percent of U.S. Cars
Daimler Chrysler	8.7
Ford	18.6
General Motors	28.7
Honda	13.0
Nissan	7.9
Toyota	10.6
Other	12.5

FIGURE 14.
Data are from the Automotive News Website.

Age Group	Percent of Fatalities
Under 16	6
16–24	25
25–54	45
Over 54	24

FIGURE 15.
Data are from the National Highway Transportation Safety Administration.

Answers (continued)

e) Independent

3. $(0.977)^6 \approx 0.870$

4. Probability of critical load and beam failure = P(critical load) · P(beam failure | critical load) = (0.01)(0.02) = 0.0002

5. $(0.85)^4 \approx 0.522$

6. $(0.98)^5 \approx 0.90$ or 90%

7. $(0.892)(0.943)(0.961)(0.907)(0.939) \approx 0.688$, or about 69%

8. $(0.9990)^4(0.9998)^3(0.9950)^5(0.9980) \approx 0.9688$

9. $(0.98)(0.995)(0.96)(0.975) \approx 0.913$

10. a) P(Brand A and breaks) = P(Brand A) · P(Breaks if Brand A) = (0.4)(0.2) = 0.08

    b) P(Brand B and breaks) = P(Brand B) · P(Breaks if Brand B) = (0.6)(0.45) = 0.27

    c) P(breaks) = P(Brand A and breaks) + P(Brand B and breaks) = 0.08 + 0.27 = 0.35

11. a) 10.6%

    b) 18.6% + 28.7% = 47.3%

    c) Yes, a car can have only one manufacturer.

12. a) 6% + 24% = 30%

    b) Yes, no one can be both under 16 and over 54.

    c) No, it is possible for someone to be in both categories.

13. A quality control inspector checks electrical transformers for defects. **Figure 16** shows the probabilities of different numbers of defects.

Number of Defects Observed	Probability
0	0.03
1	0.07
2	0.18
3	0.23
4	0.16
5	0.14
6	0.09
7	0.06
8	0.04

FIGURE 16.
Data for Question 13.

a) What is the probability that a transformer has fewer than three defects?

b) What is the probability that a transformer has at least five defects?

14. The probabilities of a randomly selected blood sample in the New York Blood Center being type O, type A, type B, or type AB are 0.45, 0.40, 0.10, and 0.05, respectively. What is the probability that a sample is either type A or type B?

15. A student takes a ten-question multiple choice quiz. Each question has five choices.

a) If the student guesses each answer, what is the probability that the student will get all the questions wrong?

b) What is the probability that the student will get all the questions correct?

16. A company that makes refrigerator ice-making units has a 4% defect rate. What is the probability that a quality control inspector might examine a sample of five randomly selected units and find no defectives in the sample?

17. Two jeeps are used on a military mission that is expected to take 150 hours. The motor in each jeep has a 150-hour reliability of 0.94. The mission can succeed provided at least one of the jeeps completes the mission. What is the probability that the motor in at least one of the two jeeps lasts for the entire mission?

18. Hinges are stamped from a punch press in a process that has a 10% defect rate. What is the probability that a random sample of five hinges contains no defective hinges?

19. Work studies suggest that there is a probability of 0.002 that a worker will incorrectly calibrate a dial. Assume that a certain technician has to calibrate a dial on a meter at the start of each workday. In a month that has 21 workdays, what is the probability that the dial will be incorrectly calibrated at least once?

20. An air traffic control system has three identical central processing units (CPUs). Each unit has a 5% probability of failure in a year. The system will function normally unless all three CPUs fail. What is the one-year reliability of the system?

21. A McDonnell Douglas MD-11 airplane has 17 fuel pumps. Assume each pump has a 3% probability of being defective.

a) What is the probability that all of the fuel pumps on one MD-11 are defective?

b) What is the probability that at least one of the pumps is defective?

22. The fuel pumps used by Boeing were thought to have a 3% chance of having faulty wiring. Consider a plane with two such pumps.

a) What is the probability that both pumps are faulty?

b) What is the probability that neither pump is faulty?

c) What is the probability that exactly one of the pumps is faulty?

23. Consider an airplane with three fuel pumps. Assume each has a 3% chance of having faulty wiring.

a) What is the probability that all three pumps are faulty?

Answers (continued)

13. a) 0.28

    b) 0.33

14. Blood types are mutually exclusive. Probability of type A or type B = (probability of type A) + (probability of type B) = 0.40 + 0.10 = 0.50

15. a) Probability of none correct = (probability of 1 incorrect)^{10} = $(4/5)^{10} \approx 0.107$

    b) Probability of all correct = (probability of 1 correct)^{10} = $(1/5)^{10} = 0.0000001024$

16. Probability of no defects in five units = (probability of a nondefective unit)^5 = (1 – probability of a defective unit)^5 = $(1 - 0.04)^5 = (0.96)^5 \approx 0.815$

17. P(at least one motor lasts) = 1 – P(both motors fail) = $1 - (0.06)^2 = 0.9964$

18. P(no defectives in a sample of 5) = (P(one nondefective hinge))^5 = $(0.90)^5 \approx 0.59$

19. P(at least one incorrect calibration) = 1 – P(no incorrect calibrations) = $1 - (0.998)^{21} \approx$ 0.04 or 4%

20. P(system functions normally) = 1 – P(system fails) = $1 - (0.05)^3 \approx 0.9999$

21. a) Almost 0

    b) $1 - (0.97)^{17} \approx 0.404$

22. a) $(0.03)^2 = 0.0009$

    b) $(0.97)^2 = 0.9409$

    c) $1 - (0.0009 + 0.9409) = 0.0582$

23. a) $(0.03)^3 = 0.000027$

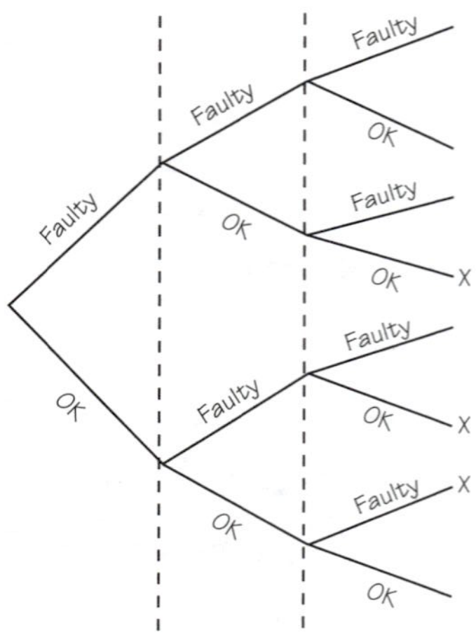

FIGURE 17.
Tree diagram.

b) What is the probability that none of the pumps are faulty?

c) Finding the probability that only one of the pumps is faulty is a little more involved than in the case of two pumps. The various combinations that include one faulty pump and two OK pumps must be counted. A tree diagram can be of help in the counting. **Figure 17** shows a tree in which each stage of branches represents one of the pumps.

Each path through the tree that is marked with an X represents a case where one pump is faulty, with probability 3%, and two pumps are OK, with probability 97%. Find the total probability that exactly one pump is faulty.

d) Use the tree diagram in Figure 17 to help you find the probability that exactly two of the three pumps are faulty.

24. When a basketball player is fouled, he or she shoots free throws. Over the years, there have been several different formats for awarding free throws. Use your knowledge of probability rules to determine the probability that an 80% free throw shooter will make two free throws in each of these formats. (Assume that free throws are independent.)

a) A player fouled in the act of shooting is awarded two free throws.

b) A player is allowed one free throw, and then can try another only if the first one is made. (This is called a one-and-one situation, and is currently not used above the college level.) It will be helpful if you first list all of the possible outcomes for this case.

c) A player is given three chances to make two shots. If the first two are made, there is no third shot. (This format was used in the NBA for a few years and then discarded.) A tree diagram may be helpful.

PREPARATION READING PART 3

What is the Cause?

Charles De Gaulle International Airport is near Paris, France. A section of a new terminal at the airport collapsed in May 2004. Several people were killed, and the terminal was closed.

It is often very difficult to find the cause of such a major disaster. But a knowledge of probability can sometimes be helpful.

New Cracks Stop Search At Terminal After Collapse
(*New York Times*, May 25, 2004)

PARIS, May 24 – Cracking noises and new fissures at an airport terminal at Charles De Gaulle International Airport here Monday interrupted the search for survivors in a section of the building that collapsed early Sunday.

With more cracks discovered at the terminal on Monday, airport officials suggested that much of the $900 million structure may have to come down, forcing the airport to shift as many as 25,000 passengers a day to other terminals for months, if not years, to come.

While it is too early to say whether a design flaw, engineering error or construction mistake caused the collapse, architects and engineers asked whether the fashion for increasingly innovative buildings has strained the limits of what a group of specialized construction companies can safely build together.

Critics warn that the increasingly extreme designs rely on complex engineering that creates many opportunities for critical errors in implementation. About 400 contractors were involved in the construction of Terminal 2E.

The collapsed terminal's design was hailed as groundbreaking when it opened last year, using technology developed for concrete-lined tunnels adapted to be free-standing. The graceful elliptical shell was built in prefabricated 12-foot sections by the French firm GTM Construction and installed on reinforced concrete pylons by the company Hervé.

Each 60-ton section rested on four pylons, and at least seven of the sections collapsed.

One engineer who worked on the project told French television that the pylons had shown signs of weakening before the accident, supporting a hypothesis that the pylons were not sufficient to bear the load placed upon them, or that shifting soil may have played a part in the collapse. Guy Nordenson, a structural engineer in New York who helped evaluate the causes of the World Trade Center collapse, said that in most building collapses there is a "cascade of problems" that could come from design or construction errors or omissions.

"Often what appears to be at the beginning the cause that triggered the problem is, in the end, not the real one," said Roberto Leon, a professor of structural engineering at the Georgia Institute of Technology. In most cases, he said, the underlying cause of a building collapse comes down to improper construction or quality control rather than the initial design.

Answers (continued)

b) $(0.97)^3 \approx 0.913$

c) $(0.03)(0.97)(0.97) + (0.97)(0.03)(0.97) + (0.97)(0.97)(0.03) = 3(0.03)(0.97)^2 \approx 0.0847$

d) $(0.03)(0.03)(0.97) + (0.03)(0.97)(0.03) + (0.97)(0.03)(0.03) = 3(0.03)^2(0.97) \approx 0.0027$

24. a) Probability of making both free throws = $(0.80)(0.80) = 0.64$ or 64%

b) There are three possible outcomes, which are not equally likely: Miss the first shot, no second shot; probability 0.20. Make the first shot *and* miss the second shot; probability $(0.80)(0.20) = 0.16$. Make the first shot *and* make the second shot; $(0.80)(0.80) = 0.64$ or 64%.

c)

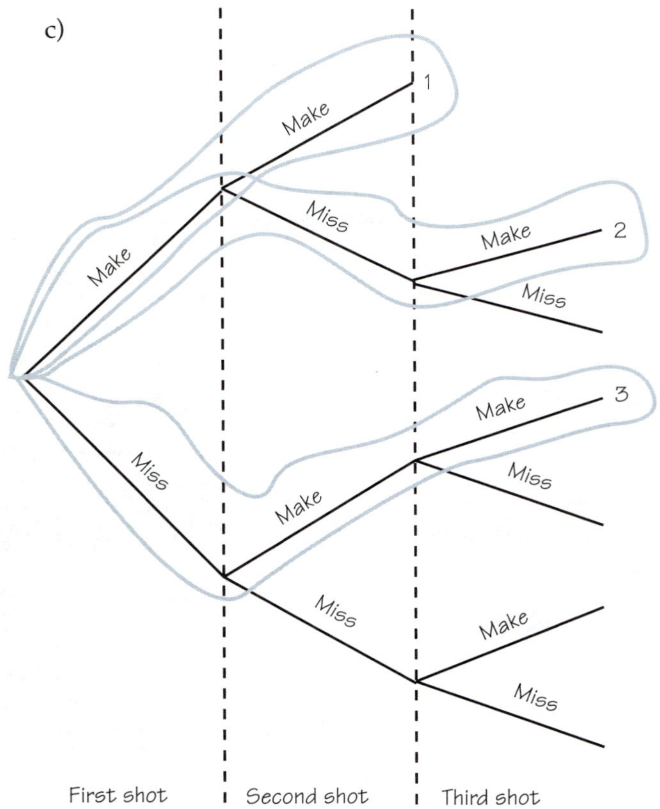

There are three paths through the tree diagram that result in two made free throws. Since the paths are mutually exclusive (only one of them can happen), the probability of making two free throws equals the sum of the probabilities along the three paths, or $(0.80)(0.80) + (0.80)(0.20)(0.80) + (0.20)(0.80)(0.80) = 0.896$ or about 90%.

Preparation Reading Part 3:
What is the Cause?

The collapse of part of a new terminal with a radical design at De Gaulle Airport provides the contextual introduction for this part. As discussed in the news clip, probable cause analysis for such a catastrophe is a vastly more complicated process than the simple examples explored in this part. But the basic mathematical concepts, albeit in more extensive form, can inform the failure analysis process.

PART 3

Probable Cause

Materials: none

If a piece of equipment is misused, it could wear out sooner than it should. Suppose you lend a one-year-old electric drill to a friend, and it is returned with a broken motor. Maybe the drill was old and the motor was ready to go. But your friend might have overloaded the motor by trying to drill into something that was too hard for the drill to penetrate. How can you know what caused the failure?

In order to analyze the drill's failure, you would need information about the drill and the way people use drills. Suppose that research on a sample of 200 drills shows that:

- Under normal use, only 5% of drill motors wear out during the first year.
- Of drills that are overloaded, 30% of their motors wear out during the first year.
- 10% of users overload their drills.

1. A tree diagram can be used to picture this situation. The box at the left contains the sample size. The upper branch in **Figure 18** represents overloaded drills. The lower branch represents drills that weren't overloaded.

   Fill in the boxes with the number of drills in each category.

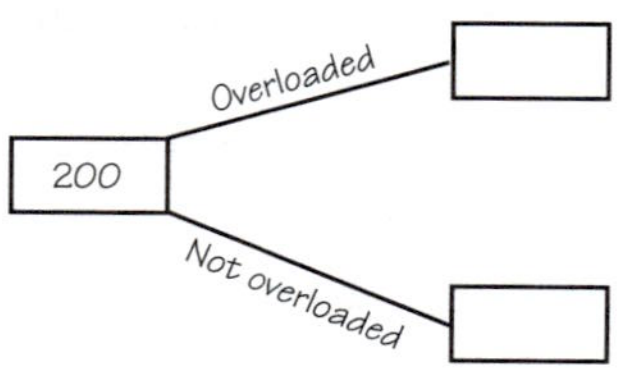

FIGURE 18.
Diagram for Question 1.

2. Consider only the overloaded drills. **Figure 19** shows a second stage of tree branches. In this stage, the branching factor is whether or not a drill fails within a year. The upper branch represents the overloaded drills that fail. The lower

branch represents the overloaded drills that don't fail. Use the fact that 30% of overloaded drills fail to fill in the boxes at the end of the branches.

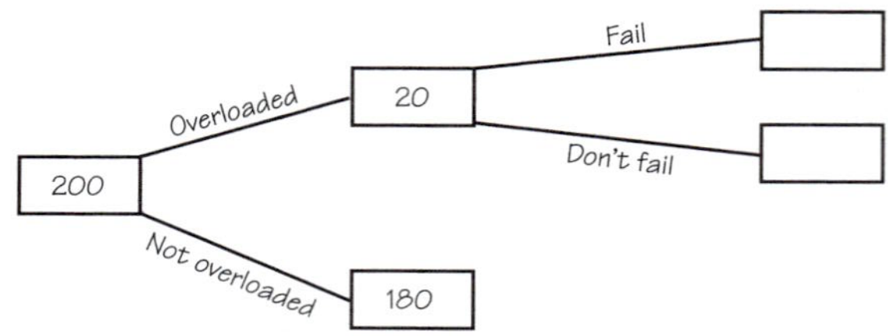

FIGURE 19.
Diagram for Question 2.

3. **Figure 20** completes the second stage of branches. Fill in the boxes at the ends of these branches. They represent the results for drills that are *not* overloaded.

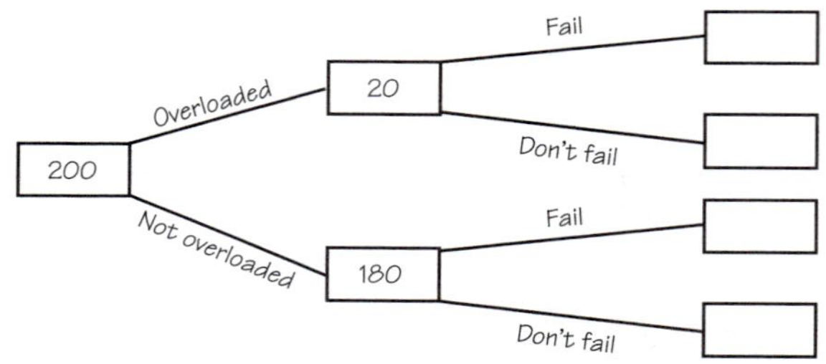

FIGURE 20.
Diagram for Question 3.

4. Altogether, how many drills fail?

5. How many of the drills that fail were overloaded?

6. If a certain drill has failed, what is the probability that it was overloaded?

7. Your tree can be redrawn as a probability tree. Each branch is labeled with the probability of a drill being on that branch. For example, the upper branch in Stage 1 is labeled P(O), for the probability that a drill is overloaded. The top branch in Stage 2 is P(F | O), for the conditional probability that a drill fails if it was overloaded (see **Figure 21**). Label the remaining branches, along with the probability values.

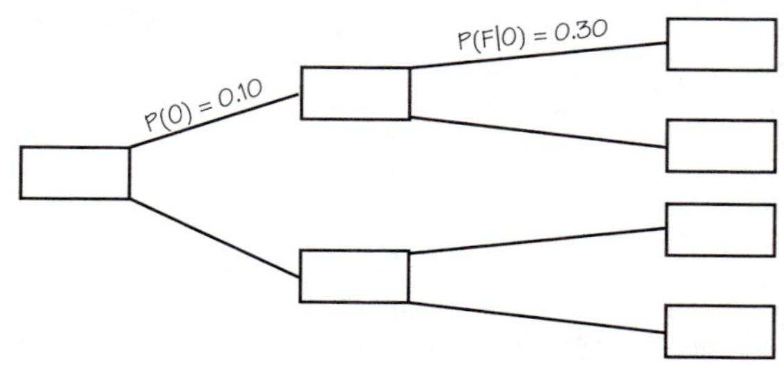

FIGURE 21.
Diagram for Question 7.

Part 3: Probable Cause

Materials: none

Simple probability trees are introduced for the assessment of posterior probabilities: using observed results to obtain information about the (unknown) prior state of a system.

In this part, as well as the subsequent examples and exercises, data and probabilities are hypothetical as such information is usually proprietary. (However, the probabilities for the testing situations in the exercises are based on real data; although even there, sample sizes are hypothetical.)

Constructing Tree Diagrams

Note that these guidelines implicitly assume that the sum of probabilities of all outcomes of an experiment is 1. You may wish to make this fact more explicit.

Tree diagrams are visual models that many students find helpful as part of a strategy for analyzing problems that involve conditional probability. In particular, using concrete numerical information at the ends of the branches tends to be much more useful to students than a tree on which only probabilities are written. Some students may even prefer to compute the joint probabilities directly from the numbers in the boxes, as in Part 3, rather than explicitly using products of probabilities.

Note that there can be more than two stages in a probability tree, although no examples are presented here.

Answers

1. Sample answer:

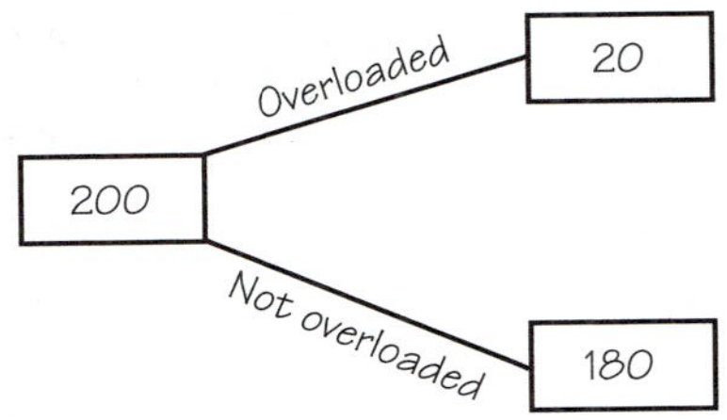

2. Sample answer:

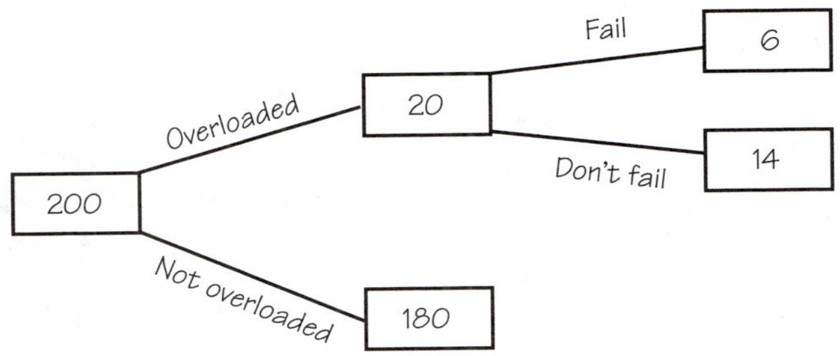

3. Sample answer:

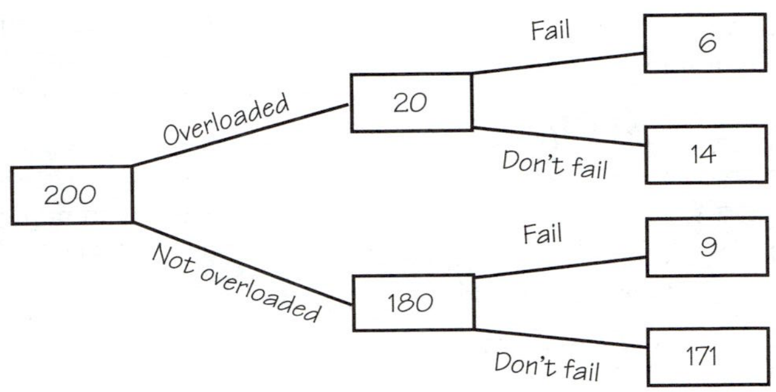

4. 15

5. 6

6. $6/15 = 0.4$ or 40%

7. Sample answer:

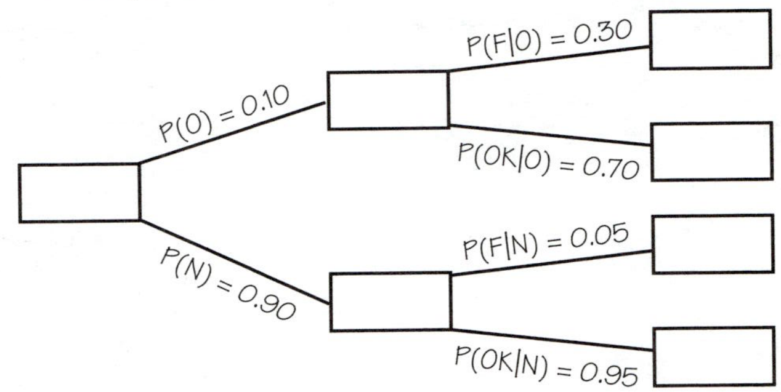

8. The tree contains four different paths. Each one begins at the box on the left and ends on one of the boxes on the right. The probability that a drill follows the top path is a joint probability. It is the probability that a drill is overloaded *and* it fails. This probability is found from the product P(O) × P(O | F). What is this probability?

9. Find the joint probabilities for each of the four paths. Write each value next to the appropriate box.

10. You can use your answer to Question 9 to calculate the probability that a failure was caused by an overload. First, find the total probability of failure by adding the probabilities at the ends of paths that include failure.

11. If your drill was returned broken, then you know it followed one of these two paths. To find the probability that it was overloaded, divide the probability of following the overloaded and failed path by your answer to Question 10.

Notice that there are two ways to use a tree diagram that give the same result. The probability that the drill failure was due to an overload was found using only probabilities and also by using a fixed initial sample size. In fact, *any* sample size could have been placed in the leftmost box. The result would have been the same.

CONSTRUCTING TREE DIAGRAMS

When tree diagrams are used to investigate the probable cause of an accident or other failure, a few guidelines should be followed:

- The tree branches out from left to right;
- The branching factor for Stage 1 is based on the conditions preceding the failure;
- The branching factor for Stage 2 is whether or not failure occurred;
- The sum of the probabilities for all branches that leave any branch point is 1;
- The sum of the probabilities for all complete paths is 1.

EXAMPLE 6

A production line in a parts factory operates around the clock. There are three shifts of workers: day, evening, and night. The day shift makes 40% of the parts. Each of the other two shifts

makes 30% of the parts. Only 1% of the parts made on the day shift are defective. The evening shift has a 2% defect rate, and the night shift produces 5% defects.

Suppose the parts from all three shifts are mixed together.

a) If a randomly selected part is defective, what is the probability that it is from the night shift?

b) If a randomly selected part is not defective, what is the probability that it is from the night shift?

SOLUTION:

a) The tree diagram in **Figure 22** represents this situation.

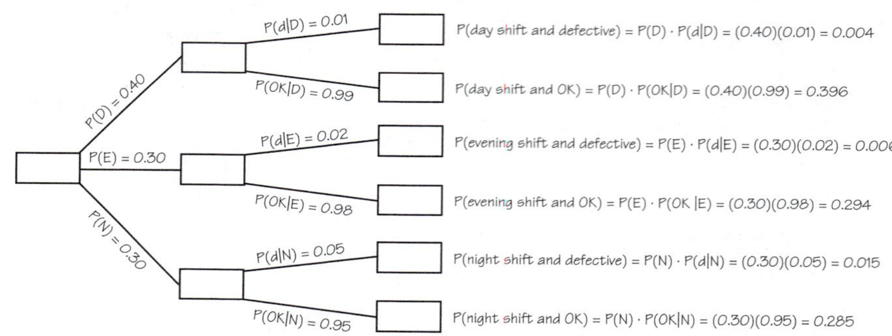

FIGURE 22.
Tree diagram.

In the figure, P(D), P(E), and P(N) represent the probabilities that a part is made by the day shift, evening shift, or night shift, respectively. P(d | D) represents the conditional probability that a part is defective if it is made by the day shift, etc.

If a part is known to be defective, then it must have followed one of the three paths that are circled in **Figure 23.**

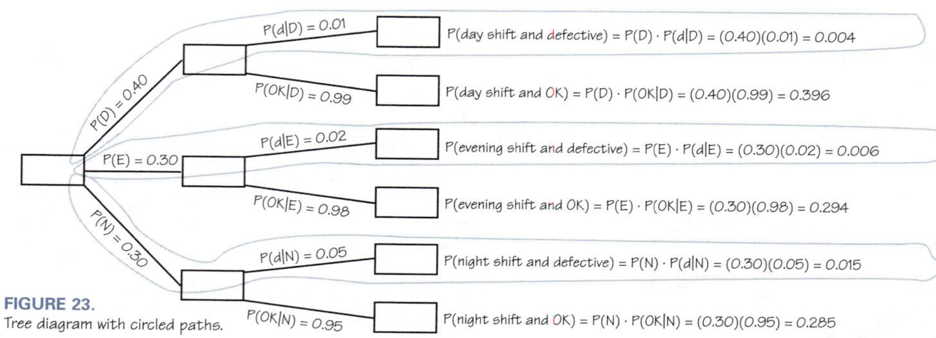

FIGURE 23.
Tree diagram with circled paths.

Answers (continued)

8. $(0.10)(0.30) = 0.03$

9. Sample answer:

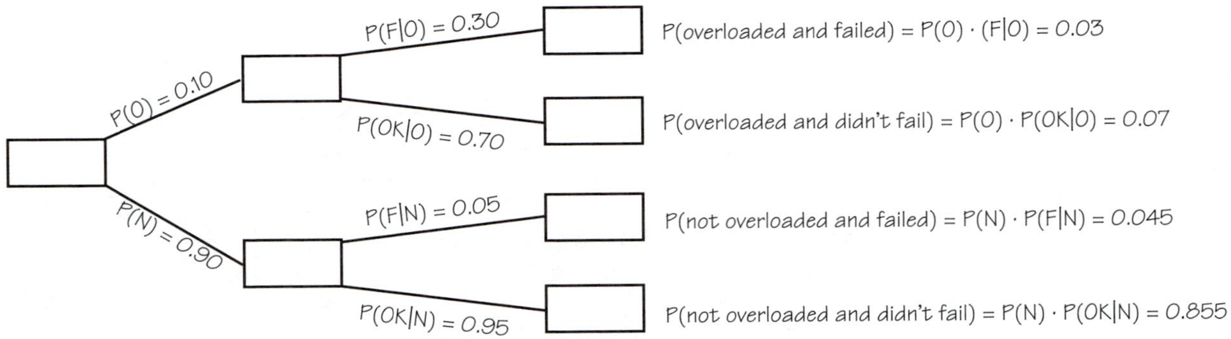

10. $0.03 + 0.045 = 0.075$

11. $\frac{0.03}{0.075} = 0.40$

The total probability that a part is defective is 0.004 + 0.006 + 0.015 = 0.025. The amount contributed to this total by the night shift is 0.015. So the fraction of defective parts that were made on the night shift is

$$\frac{\text{P(night shift and defective)}}{\text{P(defective)}} = \frac{0.015}{0.025} = 0.60, \text{ or } 60\%.$$

The probability that a particular defective part was made on the night shift is 60%.

The same result could be found by assuming a sample size of, say, 2000 parts (see **Figure 24**).

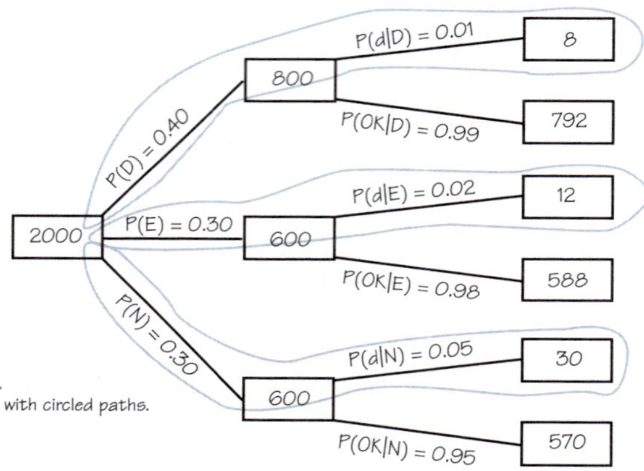

FIGURE 24.
Tree diagram with circled paths.

The probability that a particular defective part was made on the night shift is

$$\frac{\text{Number of defectives produced on the night shift}}{\text{Total number of defectives produced}} = \frac{30}{8+12+30} = \frac{3}{5} \text{ or } 0.60.$$

b) The other three paths represent nondefective parts. The probability that a nondefective part came from the night shift is

$$\frac{\text{P(night shift and OK)}}{\text{P(OK)}} = \frac{0.285}{0.396+0.294+0.285} \approx 0.29, \text{ or } 29\%.$$

PART 3 Assignment

1. A particular brand of electric hedge clippers is intended for use on branches no larger than $\frac{1}{4}$-inch diameter. They carry a one-year warranty. If used as intended, their one-year reliability is 92%. If used on branches larger than $\frac{1}{4}$-inch, there is a 60% probability of failure. Only 20% of owners use the clippers on branches larger than $\frac{1}{4}$-inch diameter.

   a) Draw a tree diagram for this situation. Use a sample size of 400. Label probabilities on the branches, and fill in all boxes.

   b) Of 400 sets of clippers, what is the total number of failures expected in a year?

   c) What is the probability that a particular set of failed hedge clippers was used on branches larger than $\frac{1}{4}$-inch?

2. A parts bin contains 500 switches. One-fourth of the switches are from Vendor A, and the rest are from Vendor B. The switches are mixed randomly. The defect rate for Vendor A is 16%, and the defect rate for Vendor B is 5%.

   a) Draw a tree diagram for this situation.

   b) If a particular switch from the bin is defective, what is the probability that it came from Vendor A?

3. Suppose that, on the average, you do your math homework four out of every five nights, or 80% of the time. Also suppose your math teacher gives you a short quiz daily. If you do your homework the night before a quiz, the probability that you get a good score on the quiz is 5/6. If you don't do your homework, the probability that you get a good score on the quiz is only 1/3.

   a) Draw a tree diagram for this situation for a school year of 180 days.

   b) Your math teacher grades one of your quizzes, and finds that you got a good score. What is the probability that you did your homework last night?

Part 3 Assignment

Teaching Ideas: In all questions where a sample size is not given, students can be encouraged to use an arbitrary sample size in order to make the problem more concrete.

Questions 1–5 provide practice with the type of probable cause analysis presented in this part. Exercise 5 involves a tree that has three branches (three conditions) in Stage 1.

Questions 6–8 involve testing situations. The existence of false positives and false negatives is a problem inherent in virtually all tests. Students may be surprised at the high probability that some test results are wrong.

Question 6 concerns the PSA (Prostate-Specific Antigen) test for prostate cancer, one test for which considerable information on false positives and false negatives is available. However, recent research suggests that the false negative rate may be much higher than the 20% previously accepted and used here.

Question 7 deals with tests for arsenic in water. There has been recent controversy over how high an arsenic level is acceptable.

Question 8 looks at vehicle emissions tests. The inaccurate equipment cited in this question has since been improved.

Question 9 is thought-provoking and asks students to apply the language of testing to the U.S. judicial system. It can help provide insight into the meaning of false positives and false negatives.

Answers

1. a) Sample answer:

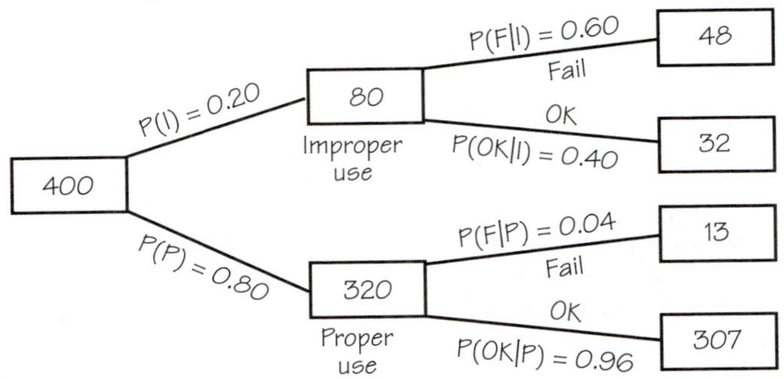

b) 48 + 13 = 61

c) 48/61 ≈ 0.79 or 79%

2. a) Sample answer:

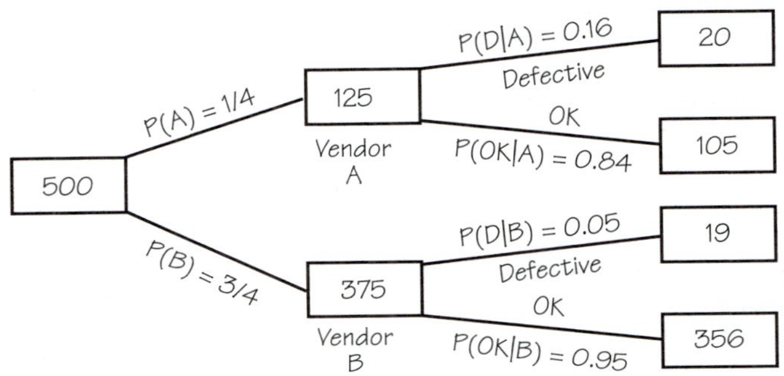

b) 20/39 ≈ 51%

3. a) Sample answer:

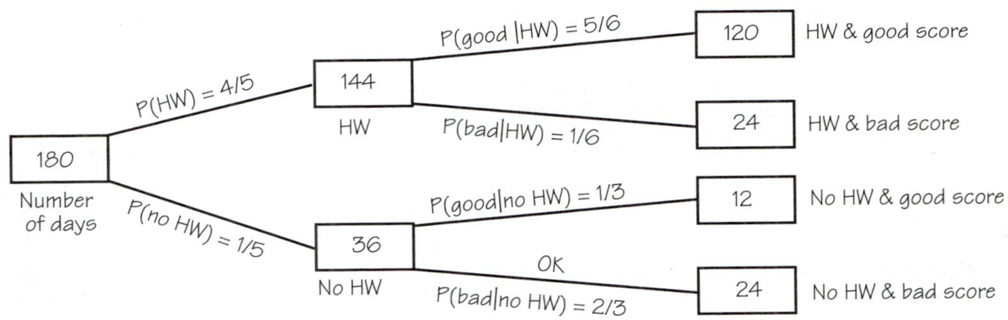

b) 120/132 = 10/11 ≈ 0.91 or 91%

4. A welding machine is used for welding the parts of chairs. If a weld is properly made, it will fail under stress with a probability of 0.002. But 3% of the welds are improper, and as a result are weak. A weak weld will fail when stressed with probability 0.30.

   a) Draw a tree diagram for this situation.

   b) If a weld fails under stress, what is the probability that it was improperly made?

5. The mortar (a cement and water mixture) that is used to bond bricks in a wall, staircase, or chimney slowly wears away. When this happens, the bricks are repointed, with fresh mortar applied after loose mortar is removed. Assume that:

   - Mortar that is properly mixed and cured fails within one year with a probability of 0.03.
   - Mortar that is mixed with the wrong proportions fails within one year with a probability of 0.75.
   - Mortar that is improperly cured fails within one year with a probability of 0.50.
   - Of all home owners that attempt brick repointing themselves,
     - 50% do it correctly;
     - 20% don't mix the mortar with the right proportions;
     - 30% don't properly cure the mortar.

   a) Draw a tree diagram for this situation.

   b) If the mortar fails within a year, what is the probability that it was mixed in the wrong proportions?

   c) If the mortar fails within a year, what is the probability that it was improperly cured?

6. Tests for medical conditions and drug usage are usually not 100% accurate. And even if a test has a high accuracy, the results can be misleading.

One test is often used to test for a type of cancer that occurs only in men. It is approximately 80% accurate in identifying the presence of cancer. In other words, if a man has cancer, there is a 0.80 probability that he will test positive. This is called a true positive test. On the other hand, if he doesn't have cancer, the test will be negative (and correct) about 97% of the time.

   a) If a test on a healthy person turns out positive for a disease, the result is called a false positive. What is the probability that a healthy man will receive a false positive for this test?

   b) Of all the men who take the test, 4% have cancer. What is the probability that a man who takes the test has cancer and has a true positive test?

   c) Draw a tree diagram for this situation for a sample of 1000 men. The branching factor for Stage 1 should be whether or not a man has cancer. The branching factor for Stage 2 should be whether the test is positive or negative.

   d) What is the probability that a man who takes the test doesn't have cancer and tests positive (a false positive)?

   e) Out of every 1000 men who take the test, how many will have a true positive?

   f) Out of every 1000 men who take the test, how many will have a false positive?

   g) What is the probability that a man who tests positive for cancer on this test is healthy?

7. Some areas of the United States have levels of arsenic in the drinking water that pose health risks. Suppose that a test for arsenic in the water supply correctly reports the presence of arsenic with probability 0.95, and correctly reports the absence of arsenic with probability 0.82. In reality, only about 4% of the drinking water in the United States is thought to contain arsenic.

   If a water sample tests positive, what is the probability that it actually contains arsenic?

Answers (continued)

4. a) Sample answer:

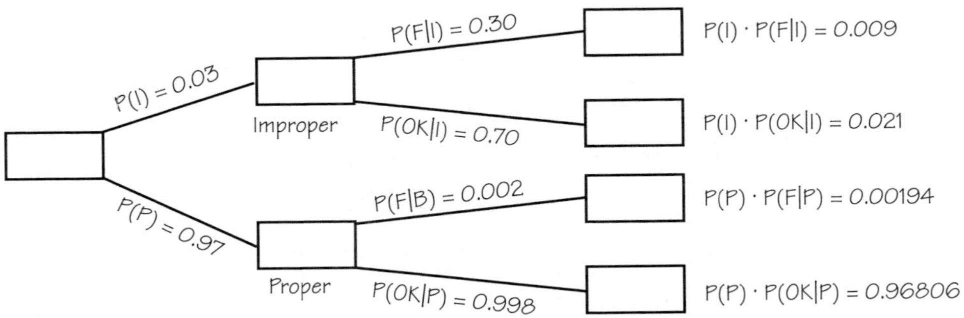

b) $\frac{0.009}{0.009+0.00194} \approx 0.82$

5. a) Sample answer:

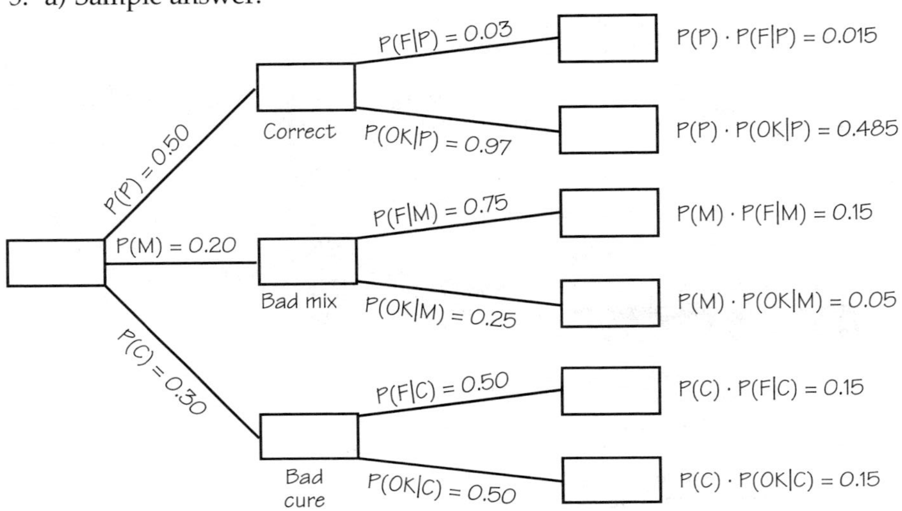

b) $\frac{0.15}{0.015+0.15+0.15} \approx 0.48$

c) $\frac{0.15}{0.015+0.15+0.15} \approx 0.48$

6. a) Since 97% of tests on healthy men are negative, 3% or 0.03 is the probability of a false positive test.

b) (Probability of having cancer)(Probability of a positive test if he has cancer) = (0.04)(0.80) = 0.032

c) Sample answer:

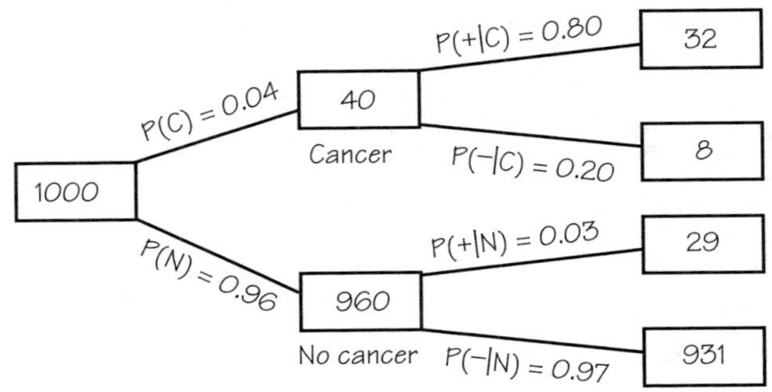

d) (Probability of not having cancer)(Probability of a positive test if he doesn't have cancer) = (0.96)(0.03) = 0.029

e) (0.032)(1000) = 32 men

f) (0.029)(1000) = 29 men

g) 29 men out of a total of (32 + 29) = 61 men with positive tests are actually healthy, for a probability of $29/61 \approx 48\%$.

a) Complete a tree diagram for this situation, if a random sample of 5000 water supplies is tested.

b) Estimate the fraction of all tests that are positive.

c) Is the percentage of positive test results a good estimate for the percentage of water supplies in the United States that contain arsenic?

d) What is the probability that a water sample that tests positive contains arsenic?

8. A 2001 federal study of emissions testing equipment in Arizona found that many results were not correct. Test results for all cars in the study were compared with actual emissions levels as determined by more thorough measurements. **Figure 25** shows results similar to those found in the study.

FIGURE 25. Emission testing data.

		Actual Emissions Condition	
		Met standards	Didn't meet standards
Test	Passed	627	24
	Failed	384	18

a) Draw a tree diagram for this situation.

b) If a car passed the test, what was the probability that it met emissions standards?

c) If a car failed the test, what was the probability that it did not meet emissions standards?

9. In criminal trials in the United States, a person is innocent until proven guilty. Assume that a guilty verdict is like a positive test result. Also assume an acquittal (not guilty verdict) is like a negative test result.

a) For a criminal trial, what is a false positive?

b) What is a false negative?

c) For a criminal trial, which do you think is worse: a false positive or a false negative?

Practice Problems

These practice problems are taken from state tests. Many of these tests decide whether or not a student graduates from high school. Your experience in this module should help you do these problems.

1. Graduate students are selected for grants from a pool of applicants. Half of the students in the pool are selected and half are rejected. Because some students decline grants, the students who are rejected have a second chance. The probabilities of being selected and rejected are shown in the diagram below.

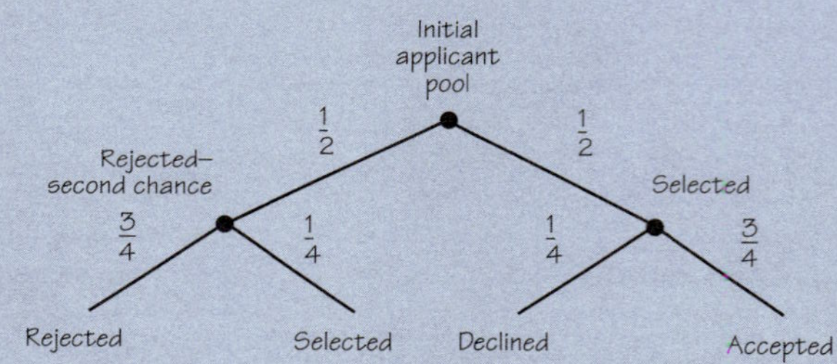

What is the probability that a student will be rejected in the first round and selected in the second round?

A. 1/8

B. 1/4

C. 3/8

D. 7/8

2. The two spinners at right are used for a math game. Each arrow has an equal chance of landing on any one number on its spinner. If both arrows land on even numbers, one point is given. Which of these fractions is the probability of getting two even numbers?

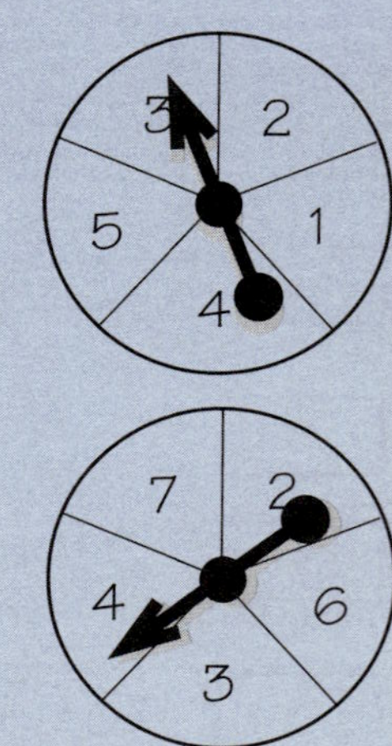

A. 5/25

B. 6/25

C. 10/25

D. 19/25

Answers (continued)

7. a)

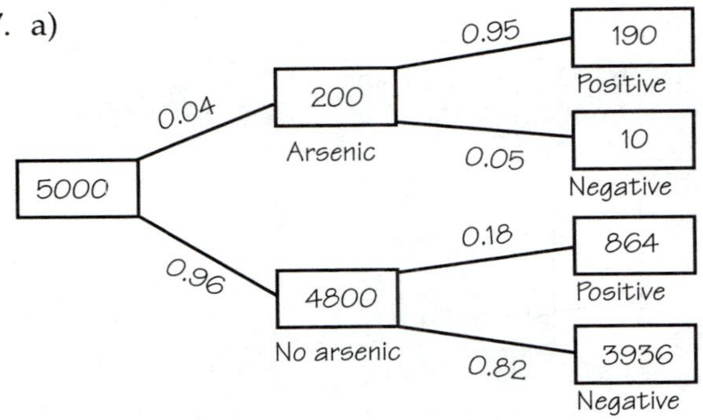

b) $\frac{190+864}{5000} \approx 0.21$

c) No. The percentage of positive results is a poor estimate of the percentage of contaminated water supplies. It is more than five times as large.

d) $\frac{190}{190+864} \approx 0.18$

8. a) Sample answer:

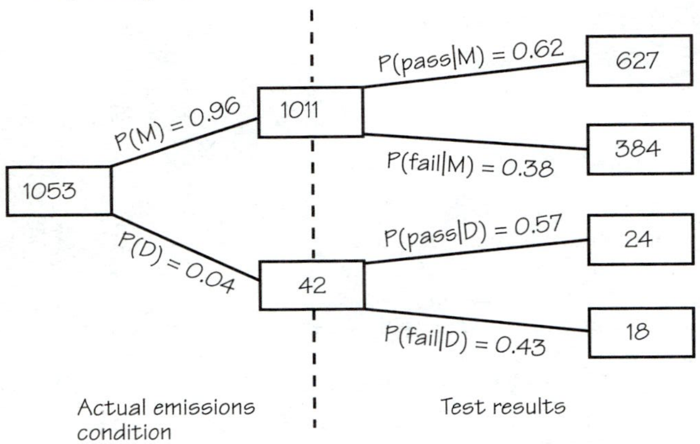

b) $\frac{627}{627+24} \approx 0.96$

c) $\frac{18}{18+384} \approx 0.045$

9. a) Sample answer: A false positive occurs when an innocent person is found guilty.

b) Sample answer: A false negative occurs when a person who committed a crime is acquitted.

c) Sample answer: A false positive is worse, since it means an innocent person could go to jail or be executed.

Practice Problems

Answers

1. A
2. B

3. Cindy tossed a coin onto the playing board shown below.

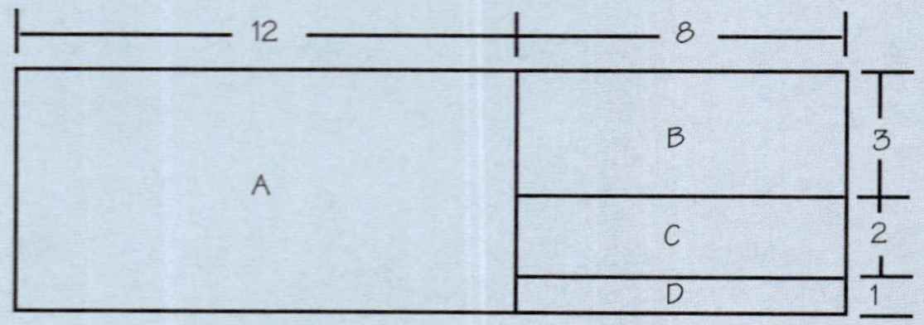

What is the probability the coin will land in region C?

4. A quality control engineer for a toy company tested 800 video games and found 3 defective. The company plans to make 500,000 video games this year. Based on the findings, how many can be expected to be defective?

A. 16

B. 20

C. 1875

D. 2000

5. Tara plays a game using 2 bags of game pieces. One bag has 6 blue game pieces and 6 red game pieces. The other bag has ten game pieces numbered 1 through 10. On her turn, Tara must draw one game piece from each bag. What is the probability that she draws a red piece and an even-numbered piece?

6. To get home from work, Curtis must get on one of the three highways that leave the city. He then has a choice of four roads that lead to his house. In the diagram below, each letter represents a highway, and each number represents a road.

		Highway		
		A	B	C
Road	1	A1	B1	C1
	2	A2	B2	C2
	3	A3	B3	C3
	4	A4	B4	C4

If Curtis randomly chooses a route to travel home, what is the probability that he will travel Highway B and Road 4?

A. 1/16

B. 1/12

C. 1/4

D. 1/3

7. What is the probability that the spinner will NOT stop on red if it is spun one time?

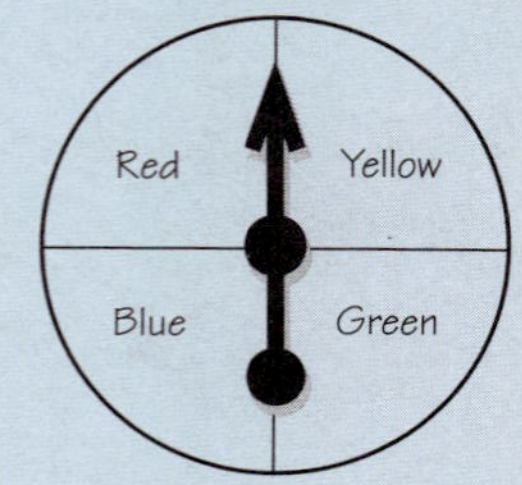

A. 1/4

B. 1/3

C. 3/4

D. 4/3

8. Mrs. Penner has 7 keys for her school on her key ring. All of the keys are different but look alike. One of these keys is for the bathroom and two are for her classroom. If she picks one key at random, what is the probability that the key is *not* for the bathroom or her classroom?

9. Jared has a white cube and a red cube. The surfaces of each cube are numbered with a unique number from 1 to 6. If Jared tosses the cubes, what is the probability he will get a 4 on the white cube and an even number on the red cube?

A. 1/12

B. 1/3

C. 1/2

D. 2/3

10. Pedro designed a board game with a spinner that is divided into four equal sections: 1, 2, 3, and "Lose a Turn." He wanted to determine how many times a player would spin "Lose a Turn" in 10 spins. He conducted a simulation where one trial consisted of 10 spins. The table at right shows his results for 50 trials.

Based on these results, what is the probability that a player will spin "Lose a Turn" 3 or more times in 10 spins?

A. 7/50

B. 15/50

C. 22/50

D. 43/50

Simulation Results

Number of "Lose a Turn" Spins	Frequency
0	5
1	12
2	11
3	15
4	4
5	2
6	0
7	1
8	0
9	0
10	0

Answers (continued)

3. $16/120 = 2/15 \approx 0.133$
4. C
5. 1/4
6. B
7. C
8. 4/7
9. A
10. C

11. Lani had a box that contained
    - 1 blue marble;
    - 1 green marble;
    - 1 purple marble;
    - 1 yellow marble; and
    - 2 red marbles.

    Lani removed one marble without looking, and she recorded the result. She placed the marble back in the box and repeated the procedure one more time. What is the probability that Lani removed a red marble followed by a blue marble?

    A. 1/36

    B. 1/18

    C. 1/3

    D. 1/2

12. Weatherpersons predict tomorrow's weather based on what has happened in the past. During the past 50 years, there have been 380 days that have been just like today, and of those, 200 have been followed by a clear day. Which of these is the approximate probability of a clear day tomorrow?

    A. 13%

    B. 34%

    C. 53%

    D. 66%

13. Anita and Ajay play a game of chess. The probability that Anita will win is 48%, and the probability that Ajay will win is 42%. What is the probability that this game will end in a tie?

    A. 6%

    B. 10%

    C. 52%

    D. 58%

14. Jeremiah is doing an experiment in his math class. He flips four pennies in the air. What is most likely to happen?

    A. Two of the pennies will be heads and two will be tails.

    B. Three of the pennies will be heads and one will be tails.

    C. All four pennies will be heads.

    D. None of the pennies will be heads.

15. When Joe bowls, he gets a strike (knocks down all the pins) 60% of the time. How many times more likely is it for Joe to bowl *at least* three strikes out of four tries as it is for him to bowl zero strikes out of four tries? Round your answer to the *nearest whole number*.

16. The rectangular garden shown at right has a rectangular, brick walkway.

    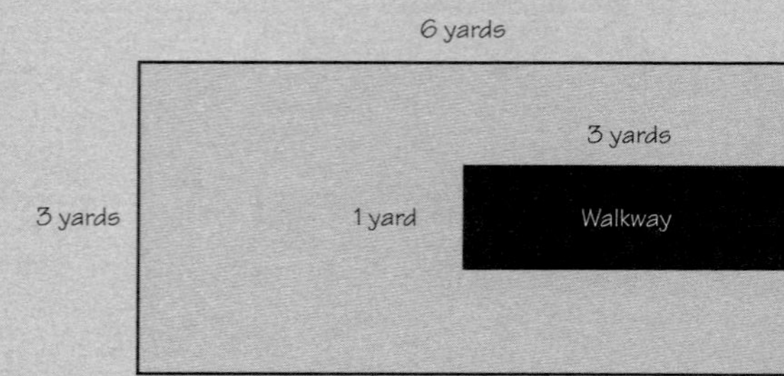

    What is the probability that a seed tossed randomly into the garden will land on the walkway?

    A. 1/6

    B. 1/5

    C. 1/4

    D. 1/3

17. If a number is selected at random from the set {–3, –2, –1, 0, 1, 2, 3}, what is the probability that it is odd?

18. A computer simulated tossing 3 coins 400 times. The results are shown at right.

HHH	41
HHT	54
HTH	48
THH	57
TTH	50
THT	53
HTT	45
TTT	52

    a) Calculate the experimental probability of tossing 2 heads and a tail as shown by this simulation.

    b) Find the theoretical probability of tossing 2 heads and a tail.

    c) Compare the two probabilities and explain any differences.

19. What is the probability of a team losing a game if the probability of the team winning a game is 0.735?

Answers (continued)

11. B
12. C
13. B
14. A
15. 19
16. A
17. 4/7
18. a) 159/400 = 0.3975

    b) 3/8 = 0.375

    c) Sample answer: The experimental probability will only approach the theoretical probability after many trials.
19. 0.265

20. A bag contained four green balls, three red balls, and two purple balls. Jason removed one purple ball from the bag and did not put the ball back in the bag. He then randomly removed another ball from the bag. What is the probability that the second ball Jason removed was purple?

A. 1/36

B. 1/9

C. 1/8

D. 2/9

21. A spinner has spaces labeled X, Y, and Z. The spinner was spun 20 times. The results are shown in the table below.

Letter	Number of Times Spinner Landed on Letter
X	4
Y	3
Z	13

According to these data, what is the experimental probability that the spinner will land on Y?

A. 3/20

B. 4/20

C. 13/20

D. 17/20

22. The table below shows information about the students in Ms. Murphy's algebra class.

	Male	Female
Rides the bus	6	8
Does not ride the bus	9	5

What is the probability that a randomly selected male student does <u>not</u> ride the bus to school?

A. 9/28

B. 9/14

C. 15/28

D. 9/15

23. The probability that Caitlin gets an A on a math test is 0.4. Find the probability that she earns an A on exactly two of three math tests.

24. Janet is playing a game using the two spinners shown below. She will spin the arrow on each spinner once and will move a specified number of steps forward or backward according to the results.

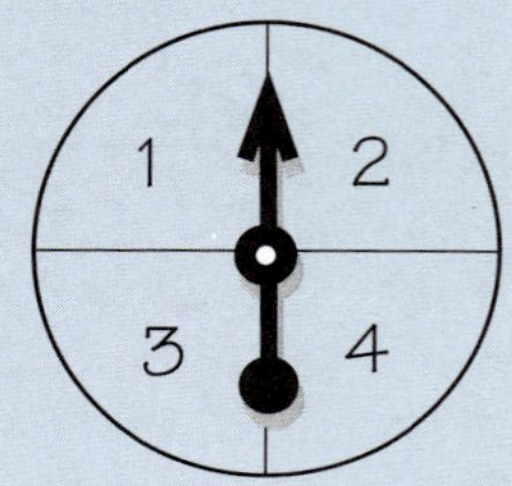

Spinner 1

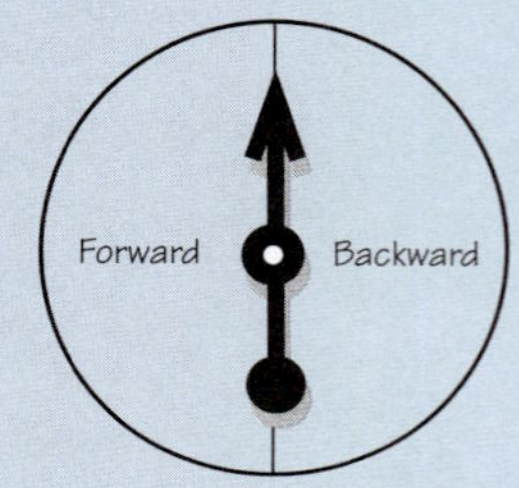

Spinner 2

What is the probability that Janet will have to move backward **less than** 4 steps?

A. 1/8

B. 3/8

C. 1/2

D. 3/4

25. For quality control, a light bulb company conducted a random sampling of their light bulbs. The results are shown below.

Number of Defective Bulbs	Number of Nondefective Bulbs
4	96

The light bulb company makes 6000 light bulbs in a day. Based on this sample, how may defective light bulbs can the company expect to make in a day?

A. 240

B. 250

C. 1500

D. 2400

Answers (continued)

20. C
21. A
22. D
23. 0.288
24. B
25. A

26. Captain Crunch contains one of 6 different prizes in each box of cereal. The prizes are distributed evenly and randomly. If two boxes of the cereal are purchased, what is the probability that the prizes are different?

    A. 2/3

    B. 5/6

    C. 6/5

    D. 1/6

27. Mark wants to compare theoretical and experimental probabilities. He plans to roll two cubes with faces numbered 1 through 6 and add the numbers showing on the top face of each cube.

    a) What is the **theoretical** probability of rolling a sum that is less than or equal to 4?

    b) Mark performed 100 trials of this experiment and recorded the results in the table below. Based on Mark's data, what is the **experimental** probability of rolling a sum that is less than or equal to 4?

Sum	2	3	4	5	6	7	8	9	10	11	12
Number of Occurrences	2	4	6	10	14	17	13	14	11	5	4

    c) The experimental probability can be used as an estimate of the theoretical probability. How could Mark alter his experiment so that the experimental probability is likely to be closer to the theoretical probability?

Answers (continued)

26. B

27. a) $1/12 \approx 0.083$

    b) $6/100 = 0.06$

    c) Perform more trials.

INDEX

Symbols and Numerics

ACKNOWLEDGMENTS

Project Leadership

Jo Ann Wheeler
Region 4 ESC, Houston, TX

David Eschberger
Region 4 ESC, Houston, TX

Solomon Garfunkel
COMAP Inc., Lexington, MA

Roland Cheyney
COMAP Inc., Lexington, MA

Lead Authors

Gary Cosenza
Region 4 ESC, Houston, TX

Paul Gray
Region 4 ESC, Houston, TX

Julie Horn
Region 4 ESC, Houston, TX

Authors

Sharon Benson
Region 4 ESC, Houston, TX

Roland Cheyney
COMAP, Lexington, MA

David Eschberger
Region 4 ESC, Houston, TX

Jo Ann Wheeler
Region 4 ESC, Houston, TX

Publisher
Craig Bleyer

Acquisitions Editor
Cyndi WeissGoldner

Marketing Manager
Mike Saltzman

Supplements and New Media Editor
Victoria Anderson

Associate Editor
Brendan Cady

COMAP Staff
Laurie Aragón, Rafael Aragón, Roland Cheyney, Kevin Darcy, Gary Feldman, Gary Froelich, Solomon Garfunkel, Daiva Kiliulis, Tim McLean, Sheila Sconiers, Anne Sterling, John Tomicek, George Ward, Pauline Wright

Index Editor
Seth Maislin
Focus Publishing Services, Arlington, MA

Illustrations
Lianne Dunn, George Ward

Photo Research
COMAP Production

PHOTO/ILLUSTRATION CREDITS

Rafael Aragón: 572

David Barber: 112, 120, 147, 411, 564

Jackson Barber: 195

Bridgeman Art Library International LTD: 216, 217, 239, 244

Vin Catania Studios: 45, 220, 221, 234, 237

Corbis Images: 62, 70, 82, 94, 96, 100, 114, 117, 119, 125, 127, 130, 337, 455, 538

Dover Publications Inc.: 76, 147, 352

Lianne Dunn: 41, 71, 74, 88, 93, 112, 122, 123, 417, 441, 470, 477, 478

EyeWire: 109, 439, 461, 464, 529

Mark Finkensteadt: 131

Gary Froelich: 200, 201

Hewlett-Packard: 258

Image Bank: 233, 243

Daiva Kiliulis: 123, 124, 468, 480

NASA: 282

National Anthropological Archives: 89

Novica.com: 197

Purdue University: 71

Shutterstock: 2, 6, 8, 14, 16, 18, 20, 22, 23, 30, 35, 42, 47, 57, 59, 79, 98, 132, 137, 142, 143, 145, 149, 154, 163, 164, 165, 166, 171, 172, 176, 177, 182, 184, 186, 187, 190, 191, 194, 206, 248, 256, 257, 261, 266, 280, 285, 294, 295, 302, 314, 323, 332, 340, 344, 345, 347, 351, 356, 359, 361, 362, 376, 381, 395, 400, 405, 406, 414, 415, 426, 431, 444, 459, 487, 494, 495, 510, 513, 517, 521, 525, 536, 542

Syi Tong: 215

Susan Van Etten: 106, 107, 266, 418, 460, 473, 489

George Ward: 215, 534, 546, 548, 565, 582

F. G. Wilson: 559